Sourcebook of Control Systems Engineering

L.C. Westphal

Department of Electrical and Computer Engineering
University of Queensland
Australia

CHAPMAN & HALL

London · Glasgow · Weinheim · New York · Tokyo · Melbourne · Madras

Published by Chapman & Hall, 2-6 Boundary Row, London SE1 8HN, UK

Chapman & Hall, 2-6 Boundary Row, London SE1 8HN, UK

Blackie Academic & Professional, Wester Cleddens Road, Bishopbriggs, Glasgow G64 2NZ, UK

Chapman & Hall GmbH, Pappelallee 3, 69469 Weinheim, Germany

Chapman & Hall USA, One Penn Plaza, 41st Floor, New York, NY10119, USA

Chapman & Hall Japan, ITP-Japan, Kyowa Building, 3F, 2-2-1 Hirakawacho, Chiyoda-ku, Tokyo 102, Japan

Chapman & Hall Australia, Thomas Nelson Australia, 102 Dodds Street, South Melbourne, Victoria 3205, Australia

Chapman & Hall India, R. Seshadri, 32 Second Main Road, CIT East, Madras 600 035, India

First edition 1995

© 1995 Chapman & Hall

Commissioned by Technical Communications (Publishing) Ltd

Printed in Great Britain at the University Press, Cambridge

ISBN 0 412 48460 9

∞ Printed on acid-free text paper, manufactured in accordance with ANSI/NISO Z39.48-1992 (Permanence of Paper).

Contents

xviii *Contents*

Preface

This book joins the multitude of Control Systems books now available, but is neither a textbook nor a monograph. Rather it may be described as a resource book or survey of the elements/essentials of feedback control systems. The material included is a result of my development, over a period of several years, of summaries written to supplement a number of standard textbooks for undergraduate and early post-graduate courses. Those notes, plus more work than I care right now to contemplate, are intended to be helpful both to students and to professional engineers.

Too often, standard textbooks seem to overlook some of the engineering realities of (roughly) how much things cost or how big computer programs for simple algorithms are, of hardware for sensing and actuation, of special systems such as PLCs and PID controllers, of the engineering of real systems from coverage of SISO theories, and of the special characteristics of computers, their programming, and their potential interactions into systems. In particular, students with specializations other than control systems are not being exposed to the breadth of the considerations needed in control systems engineering, perhaps because it is assumed that they are always to be part of a multicourse sequence taken by specialists. The lectures given to introduce at least some of these aspects were more effective when supported by written material: hence, the need for my notes which preceded this book.

By design and because of its background, this book is different and unusual. A detailed outline is given in Chapter 1, but here note that:

- the coverage of topics is exceptionally broad for a book which does not claim to be a handbook, although the theory is mostly restricted to linear time-invariant dynamic systems;
- the multiplicity of chapters almost constitutes modularization, a property which makes the book a useful elementary reference and which leads to some overlap of chapters and repetition of basics;
- the level of the book is mostly undergraduate and elementary, with references to more advanced and complete presentations added for those wishing to progress further – examples are very simple ones, intended to show how the theory translates into usable algorithms; and

- the modularization is mostly on the basis of usefulness, as I am convinced that so-called unifying theories are appropriate, even for non-specialists and students just beginning to learn about the subject.

Thus the book will be helpful to several classes of readers:

- students, especially undergraduates and early postgraduates – to supplement their textbooks and as a handy overview;
- engineers who are not control specialists – to summarize in one place the nature of the field so they can interact with specialists or, with this as a starting place, learn enough to help with a particular job; and
- control systems specialists – as a refresher concerning topics outside their speciality.

It is appropriate to acknowledge a number of influences here. Although students who made it clear that standard textbooks were not entirely satisfactory were the original motivation, the Electrical Engineering Department provided helpful support, and my industrial experience undoubtedly influenced my attitudes. Several colleagues, notably Drs Gerry Ledwich, Gerry Shannon and Pra Murthy, contributed by discussing the concepts and conveying their experiences in trying various aspects of teaching.

LCW

1

Introduction and overview

The feedback control systems specialist has a multifaceted job involving several different types of tasks; an alternative point of view is that control systems encompass several different sub-specialties from which an engineer might choose to emphasize only one or two. The aspects of the systems to be understood include:

1. the **process** being controlled, whether it be a beer brewing plant, a refinery, an aircraft, or an artificially paced human heart;
2. the **hardware**, including instruments for sensing, wiring carrying data, computers processing data, and motors implementing commands; and
3. the **algorithms** used and the computer coding which implements them.

People tend to specialize in one or another aspect, and personal inclinations tend to lead to at least one traditional demarcation, that between the 'control engineer' and the 'control theorist'. The former is nominally concerned with hardware, instrumentation, and 'practical' engineering; the latter is devoted to mathematics and computer utilization, with applications, if any, being those with advanced performance demands such as in aerospace systems.

This hardware vs. theory split is common to many fields, but with control systems the availability of inexpensive computer power is bringing these together again, and in particular is making possible many algorithms formerly thought too sophisticated for applications such as machine and process control. For example, observer and filter algorithms now allow inferences to be made about the internal workings of processes without actually measuring internal variables, and those algorithms are well understood and easily computed with commonly available computers.

At the other extreme, the desire of engineers to improve control of some long established systems, such as steamships, plus the need to succeed with new systems such as robots and nuclear power plants, are making demands on theorists for more applicable results.

The meetings of known theory with present applications and the demands for better performance make it necessary for all control systems engineers to have an overview of the field. This book is an introduction to the elements of the field for those who will have future contact with it, either as specialists in control or in other fields for which control is a means to an end.

In this first chapter we present a brief overview of computer control systems engineering. In section 1.1 we look at feedback control systems, in order to distinguish our usage of the word 'control' from those who work in what we might call situation supervision. Then section 1.2 gives an outline of the sources of influences which have contributed to the control systems field, with section 1.3 giving three standard problems which are discussed in the light of small and large systems. Section 1.4 gives a breakdown of the field, similar to that followed later in this book and section 1.5 provides a list of 'tools of the trade', i.e. a selection of the many learned society journals and magazines in the field plus useful and available computer programs for studies on personal computers. Finally, section 1.6 considers the outline structure and philosophy of the later chapters of this book.

1.1   WHY HAVE FEEDBACK CONTROL SYSTEMS?

Control systems are needed for simple tasks (e.g. turning motors on and off, starting furnaces) and for complicated tasks (such as robot motion co-ordination and aircraft autopiloting). Feedback control systems are needed because, for various reasons (such as imprecision of components and disturbances by external events), the outputs of a system cannot be relied upon to be suitably precise – so they are measured and the inputs adjusted to allow for the inaccuracies found. It is precisely this measurement of output and using the information to influence the input, i.e. this feedback, that makes control systems fascinating and difficult.

Primitive examples of the need for simple on–off commands augmented by a bit of feedback appear with automatic washing machines. A portion of the wash cycle has the drum being filled with water and agitation started. The sequence of

a)   water valves opened
b)   water valves closed
c)   gears set to cause agitation (rather than drum spin as in the spin

dry portion of the cycle)
d)   motor started

could in principle be a timed one, requiring only a set of cogs on a clock drum. However, variations in inlet water pressure mean that the time for the drum to fill will vary, so the solution is to sense water level and use this before step b) rather than pure timer activity to cause the next step. Thus feedback is needed even at this level.

More elaborate examples occur with aircraft autopilots, which must keep the craft on course and stable in the presence of head winds, gusts, varying fuel and passenger loads, etc. Here, course and vehicle attitude and altitude must be measured and the craft flown by the autopilot utilizing control algorithms to keep these quantities as desired.

Many such problems can be represented by the block diagram form of Fig. 1.1.

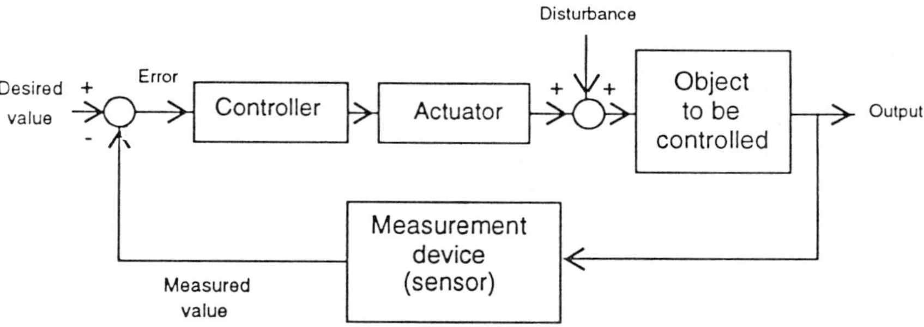

Figure 1.1 A typical control system block diagram. The boxes represent objects and the lines represent communication paths. Desired value, etc. represent signals.

1.2   INFLUENCES ON DEVELOPMENT OF THE FIELD

Perspective on a field is difficult to obtain without some knowledge of its history. History may be considered as a chronological sequence, but we choose to present it as a meeting of many influences, and observe that therefore there are a number of tools and approaches which do not necessarily sit well together.

One influence has been that of the **inventors**. Devices which demonstrate a feedback control property date from antiquity, and include water clocks and geared windmill pointers. The control systems field is usually taken to date from James Watt's invention of the steam engine governor in 1769. This clever device, still observable on vintage steam engines from less than one hundred years ago, is shown in Fig. 1.2. The principle is simply one of using the centripetal forces associated with the rotation of the motor shaft coupled rotating balls to adjust the steam valve admitting steam energy to the cylinders. Among the important sensor inventions was Sperry's gyroscope for aircraft in 1912–14.

Figure 1.2 A Watt's governor, used for more than a century to regulate the speed of engines, particularly steam engines. This one is on a turn-of-the-century traction engine. The engine shaft is geared to a shaft which rotates the two heavy balls. These latter are on swinging arms whose radius of rotation increases with speed, and these arms are coupled using levers to the engine throttle.

A second influence has been that of the **process control engineers**, particularly chemical engineers. Many control tasks in batch and continuous flow processing have used pneumatic or hydraulic controllers. This has limited the range of control algorithms to those possible with such elements, and in particular to proportional, integral, and derivative control (PID). This technique dates back to the mid-19th century and is highly developed and widely understood.

The **electrical engineers** have contributed their understanding of frequency response. This derives from their history in electrical power generation, which is usually alternating current (AC), and later their work in communications and the associated amplifiers. Key advances were those made by Black, whose feedback amplifiers were patented in the late 1920s and reported in (Black, 1934), and by Nyquist (1932) in a seminal paper on stability and Bode whose studies in the 1930s of gain-phase relationships made extensive use of the graphical representation associated with his name and were widely disseminated in (Bode, 1940, 1945).

Mathematicians have been a part of the field since Maxwell in 1868 applied linearization and differential equation theory to demonstrate why some of the Watt governed engines demonstrated an oscillatory speed profile called 'hunting'. Not long afterward Routh related stability to the differential equation coefficients, and Lyapunov formed a stability analysis approach which was not rediscovered until the late 1940s. Mathematicians also applied some of their concepts to creating new approaches to design and analysis. Among these were the filter theories developed by Wiener during the second World War and later publicly reported in (Wiener, 1949), the state estimation filter of Kalman (1960) and Kalman and Bucy (1961), the optimal control theories of Bellman (see e.g. Bellman, 1957) and Pontryagin and his colleagues (an English transaltion is Pontryagin *et al.*, 1962) in the 1950s, and the sampled data theory originating with the physical problem posed by radar in World War II and developed by Ragazzini, Zadeh, Jury, Franklin, and others (see e.g. Ragazzini and Zadeh, 1952, Ragazzini and Franklin, 1958, Jury, 1958). These also lead to the branch of control systems studies known as **control theory**.

Although electrical engineers influenced control systems in many ways, their development of the **digital computer** has transformed the field. Computers were proposed for real-time process control in 1950, were demonstrated for flight and weapons control systems by the mid-1950s, were used in supervisory roles (i.e. in calculating and commanding set points for analog control loops) in industry in the late

1950s, and replaced the analog elements to become inline direct digital controllers shortly thereafter; by the mid-1960s minicomputers such as the DEC PDP-8, were coming into common use. The now universal programmable logic controllers (PLCs) were first used in 1969, and since their appearance in 1974 microprocessors have gradually come to make the advantages of digital control available at almost all levels.

Mechanical engineers and instrumentation specialists have designed many of the devices used in control systems, from refined valves to sensitive accelerometers, and their machine tools and the robots themselves need control engineering and theory to work well.

User needs have been spurs in particular directions ever since the industrial revolution's demand for mechanical power at constant speed motivated Watt. Wartime may emphasize performance regardless of cost; industrial users may look for improved profitability. Many users look to feedback control to make their tasks possible, just as the feedback amplifier made long distance telephony practical. Present needs for manufacturing flexibility are driving the development of **flexible manufacturing systems** (FMS), with both devices and theory in demand.

Finally, **university academics** have contributed to the shaping of the field by alternately subdividing the field into special interests according to application (chemical, electrical, mechanical engineering; mathematics; computer applications) and combining it into a single department called 'Control Systems', 'Systems Science', or something with a similar name. While subdivision is perhaps natural, the combining seems to date from the 1950s and the Massachusetts Institute of Technology.

A legacy of the above influences is that the field is not unified. There are many techniques which are widely used in pockets of industry, for example. Thus PID controllers are the mainstay of process control, whereas Kalman filters and optimal control are important tools for the aerospace businesses. Large companies, research institutes and universities have the luxury of specialist theoreticians for advanced studies, and sometimes a concomitant reputation for impracticality. In other situations, process controllers know their processes and devices very well, but may either be unfamiliar with advanced theories and computer controllers or consider them uneconomic to implement. Furthermore, instrumentation specialists are important to both fields and to advances into new areas such as biomedicine, while computer specialists may prove crucial to all applications. The control systems field encompasses all of these, partly because of history and partly because of their common core problem: to control an object or set of objects.

1.3   TYPES OF CONTROL – REGULATION, TRACKING, TRAJECTORY DESIGN

The typical control system problem is presented as one of feedback control of some object or group of objects, with the aim of the controller to make the output match the input. The presence of disturbances, if any, is also modelled as an input to the object, and it is usually unspoken that the object may not have precisely the properties for which the design was made. A typical block diagram from an undergraduate text is as in Fig. 1.1. Also unspoken is the source of the input and how it came about.

Although with some effort many systems can appear to have this characteristic form, or 'structure', it is not always helpful to think of them this way. For one thing, real systems may have many (100s or 1000s) such loops in them with inputs and outputs coupled. More importantly, the design techniques will vary with the overall goals of the designer.

Only a few useful distinctions are made with the terminology. Most study, and the easiest, is concerned with single-input–single-output (SISO) systems; the alternative is multi(ple)-input–multi(ple)-output (MIMO) systems, sometimes called multivariable systems. Another important distinction we shall make is between regulators, servomechanisms, and trajectory designers.

1.3.1 Regulation

When the error between input and measured output is operated upon by the controller to generate the commands to the plant (controlled system), there are two common possibilities. The first occurs when the input is a constant which is occasionally changed. The value of the input is called the **set point**, and the designer's goal with the controller is to generate commands so that the plant output rapidly and smoothly, without drama in the form of wild oscillations, takes on the value of the set point and then maintains that value in the presence of exogenous disturbances.

A simple example of a regulator is presented in Fig. 1.3. Here a water tank is to have its level maintained at a specified depth while water is being drawn off (both purposely and through evaporation). A motor drives a pump which adds water from some reservoir. The problems for the engineer include how to instrument the tank so that water depth is known, whether to specify a constant speed motor used in an on–off fashion or a variable speed motor, and what algorithms to use in deciding when, and in the variable speed case at what speed, to run the pump motor. All of this is for the relatively simple task of maintaining water depth at a fixed level, the set point level, in the presence of water usage, leakage, evaporation, and other disturbances.

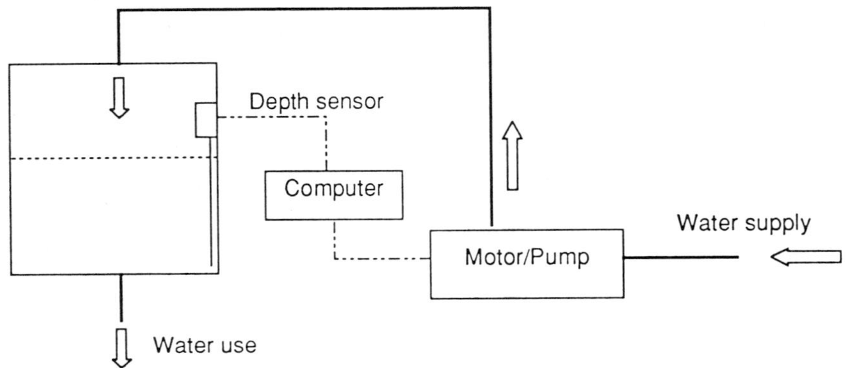

Figure 1.3 A very rudimentary control system for regulating water depth in a tank.

Another typical example of a regulator is one of the earliest control systems: the Watt's governor was a clever arrangement of mechanical devices to sense and control the speed of stationary steam engines. As such, it was a set point system with the desired speed being the input and the actual speed the output. Measurement was performed by means of flyballs rotated by geared output of the engine shaft; radius of rotation of the flyballs indicated speed, and this radius was coupled via levers to steam valves on the engine. The set point became explicit and changeable when a spring system for varying the desired speed setting was added.

In current versions of the Watt problem, the set point may well be a number commanded by a supervisory technician through a console dial or a number in a buffer loaded by a supervisory computer; the sensor might be a tachometer, and the actuator a solenoid moving the engine's throttle.

The regulation problem is so common that a range of methods is applicable: off-the-shelf controllers; classical control theory using both steady-state and transient performance indicators; modern theories of the linear (dynamics) quadratic (performance index) regulator (LQR) type. Disturbance rejection is presented differently with these approaches, however.

1.3.2 Tracking and servomechanism problems

An alternative to the above is to have the input vary in a somewhat unstructured manner and require the output to track this input. A paradigm problem here (and the one which produced the terminology) is the naval gun turret pointing problem of a century ago. A gunnery officer rotates a dial to command a gun pointing direction. This dial reading is then input to a set of motors which rotate the gun turret until the guns point in the required direction. As distinct from the set point problem, the input could in principle change frequently and unpredictably and the output be required to match it. Such tracking problems are called servomechanism problems or **servo problems**, and the designer's task is to have the output track the input rapidly for at least a specified class of inputs (such as all those which have a rate of change less than a specified value).

To see the ubiquity of servo-type problems, we notice that in section 1.3.1 in the mention of the updated Watt steam engine control a modern implementation might require a small motor or solenoid to operate the steam valve. We can well envisage that this actuator might be required to move the valve frequently and accurately, thus giving a servo problem in an overall regulator milieu. A modern example is provided by robots, whose manipulators are required to move accurately along a designed path.

Servo motors are among the most common actuators in control use, and much effort goes into developing the control electronics and electrical devices so that they work well. Servo motors drive the moving arms of computer disc drives and the pens of plotters in applications that are familiar to most of us; an alternative for small tasks is the **stepper motor**.

The theories applied here have been classical (of the frequency domain or root locus type) and modern (using linear quadratic tracking methods).

1.3.3 Trajectory design/targeting problems

There are a number of problems in which the task is neither to maintain a level nor to track a signal. One such problem is a navigation problem: the automatic navigation from city LA to city NY is unlikely to be solved by taking into consideration only the miles to go. Rather the system must know where it is, where it is to go, and establish some path for going from 'here' to 'there' efficiently. Other problems of this type abound: a robot arm may need to move from one position to another, carrying a heavy load and avoiding obstacles in its work area, in a rapid manner which does not overly stress the components; a processing plant may need to be put through a sequence of operations to configure from one product line to another; a spacecraft may be required to change orbits with minimal use of fuel.

The problems in these cases are often approached using more advanced ideas of control theory, such as state estimators and optimal control; we shall call them trajectory or target problems.

1.3.4 Systems

An important failing of the above classification is that many real problems have aspects of all three in their operation. We have already seen a simple regulator problem which has a servo motor subproblem (i.e. the valve actuator of a constant speed motor). A robot, with its problem of suitable arm trajectories, also has servo problems in the actuation of particular joints. Process control systems have several levels of regulator-type (i.e. constant required output) problems with servo actuators.

A space launch booster rocket needs trajectory planning in real-time to lay out the most efficient path from sensed position to desired orbit, converts that path into a flight plan track to be followed, and implements its plans by commands to subsystems which are themselves control systems having command following, i.e. servo type, specifications.

1.4   THIS BOOK: ELEMENTS OF CONTROL SYSTEMS ENGINEERING

Many textbooks concentrate on one or two aspects of control engineering, such as theory or computer applications, to the exclusion of others; some even specialize as to the level, such as optimal control theory for linear systems with quadratic performance indicators, algorithms for system identification, or basics of computer hardware. A control engineer, however, is necessarily systems oriented, and some of the trades to be understood are given below.

* Systems engineering
* Hardware
    Instrumentation for sensing
    Hardware for actuation
    Computers – hardware and software
    Communications between elements of the system
* System modelling
    Mathematical and simulation models
    System behaviour – desired and actual
* Control theory and algorithm design
    Classical and modern methods, both time domain and frequency domain approaches
    Implications of the theory and appropriate applications

This book can be viewed as having three major subdivisions, incorporating the above:

* Engineering
* Modelling
* Theory

1.4.1 Engineering

In Chapters 2–8, we consider systems engineering in overview and then the practical application material concerning sensors, actuators, computers and their programming, and finally a bit of 'theoryless' control tuning.

Systems engineering is the rather ambiguous name given to a number of levels of studies of engineering problems. We will take it to mean three levels:

1.   the top level which is the management oriented study of system requirements necessary to make basic decisions of strategic type;
2.   the second level which is one the engineer will face frequently – this entails system planning (but we do not consider the problem of schedules and plan implementation); and
3.   the third level which is the component level, and concerns choices of hardware, software, and algorithms.

In most general respects, systems engineering is no different for control systems engineers from what it is for other engineers, except for the particular problems addressed and for the need to emphasize 'systems'. Partly for that reason, we address it only briefly and at the two lower levels (in Chapter 2).

The components for implementing computer control systems

The control task requires measurement, decision, and actuation. Hence, the hardware needed consists of *instrumentation* for sensing the process's state, the *computer* and its interfaces for deciding what commands to send, and *actuators* for implementing the commands. The structure of the control may require *communications* between components; the control algorithms are implemented in the computer using *software.*

To control a quantity, first that quantity must be directly measurable or inferable from measurements. This requires sensors and, in many cases, transducers (i.e. devices to translate the sensor output into a form which is useful for processing – in our case, into electrical signals for input to computers). Measuring devices abound: it is possible to measure temperatures, radiation of various kinds, displacements, speeds, accelerations, forces, voltages and currents, etc. Often quantities are to be measured and recorded, i.e. data logs are to be kept, even though these quantities are not themselves directly controlled. This area is the subject of Chapter 3.

If quantities are to be actively controlled, then there must be a way to influence their values and there must be devices, called actuators, to operate the influencers, called control elements. Thus a temperature can be maintained only if heat can be added and/or removed from the process; turning the heater or cooler on or off, or allowing the process to come under the influence of a hot or cold substance, is done using some sort of actuator. Simple actuators include on/off switches for motors; a complex actuator might use a valve, opened and closed by a motor, to allow steam through a pipe to heat a liquid for which the temperature is to be maintained in the presence of heat losses. Complex control elements/actuators such as the latter are sometimes themselves feedback control systems of the regulator or servo type. Chapter 4 briefly considers actuators and control elements.

In computer control systems, computers are the key and defining element. A control systems engineer can have little hope of being an expert in this field in which specialties are proliferating. Knowledge of how the computer communicates with devices connected to it is essential, however. Furthermore, the control task is not the usual data processing or number crunching task of elementary computer courses: it must operate in real time, and for large systems it is inherently in a multitasking environment. These features mean that the control systems engineer needs some appreciation of interrupts, of real-time operating systems, and of the programming language implementation being used. In many cases, these tasks are taken care of with off-the-shelf systems, particularly those allowing only logical control, (or programmable logic controllers, PLCs) or simple PID (proportion–integral–derivative) controls handling only one loop, and those allowing structured supervision and monitoring of such straightforward algorithms. Chapters 5–6 are concerned with computer hardware and software respectively.

In a large plant, there may be hundreds, if not thousands, of control loops operating simultaneously and interacting. The associated computer power may be centralized in one location, may be in interconnected cells of activity, or may be in isolated locations with no direct automatic interaction. In all cases involving computers, however, sensors must send signals to the computers, and actuators must receive commands from the computers. If the computers interact, they must also have communications channels arranged, perhaps at several levels: sensor to computer, computer to computer, and computer to actuator. These can all be attacked in several ways, as we see in Chapter 7.

Finally, the engineer must make all this work together. This may be done using experience, *ad hoc* techniques, and rules of thumb, or by theory-based methods involving studies of mathematical models. The former constitutes much of control engineering and relies upon theoryless controllers such as PLCs and rule of thumb tuning of available control parameters, as in Chapter 8.

1.4.2 Process characteristics and modelling

There is no substitute for an understanding of the process to be controlled – not even an elaborate mathematical model – because real processes are not equivalent to a set of integro-differential-difference equations (the most common types of models). The designer of an aircraft autopilot need not necessarily be an aerodynamics expert, but he must understand how airplanes fly and manoeuvre. A consultant advising on a controller for a process such as beer-making need not be an expert on the product, but he should understand what is essential in the process. The designer of a resuscitator unit need not be a medical doctor, but he should know some physiology. On the other hand, the experts on the application should also appreciate the essentials of control systems engineering if a useful collaboration is to result.

Particular process characteristics which are likely to be important regardless of the application are numerous. These include time response characteristics of controlled quantities to inputs (e.g. delays in response, speed of response once it begins, stability characteristics, and steady-state errors), measures of accuracy of response (e.g. mean square error) and responses to particular classes of inputs (particularly sinusoidal inputs, yielding frequency response).

In some cases, it may prove necessary to construct a model of the system; this could be an analog, scale, pilot plant, or mathematical model. The last is the most common in the literature. Here a set of equations, often differential, is constructed for which the dependent variables represent quantities within the actual system. A simple such model relates altitude and down-range distance of an exo-atmospheric rocket to the motor thrust and steering law: it is called simple because the construction is based upon well-known physical laws, not because a useful model is likely to involve only a couple of equations. Similarly, a process controller requiring mass balance and temperature control may be straightforward to derive. On the other hand, the attempt to construct models for some chemical processes may be difficult, and appropriate models in biology, such as a quantitative model of human intermediate metabolism suitable for use in designing an artificial pancreas control law, have proven elusive.

Models and process understanding are application specific. Fortunately, most processes have existing models suitable for use in the field involved. Models and their representations are the subjects of Chapters 9–12. Performance characterization is in Chapter 13.

1.4.3 Control theory and algorithms

Control theory as a discipline separate from applications in, for example aeronautical engineering, chemical engineering, electrical engineering and mechanical engineering, probably arose in the 1950s, partly out of the recognition that the separate branches were overlapping and complementary in their approaches, and partly because the knowledge base was becoming too large to handle in only a course or two. The needs of the aerospace industry for high performance pushed theoreticians into new applications.

The result is debatable, but clearly, from scanning the offered textbooks in several fields and several countries, certain knowledge is considered fundamental in the area of theory. These fundamentals are concerned with techniques for linear time-invariant systems: root locus methods, frequency response methods, and state–space methods for the analysis and design of controllers for sets of equations which purportedly model real devices. In the elementary courses, the systems are of low order, are linear, and are known.

Chapters 13–34 outline some of the basic notions of both classical and modern control theory at the undergraduate and early post-graduate level before introducing intermediate post-graduate topics and finishing with a brief look at recent research level developments. In this, we use *classical* to refer to methods based upon Laplace, z- and Fourier transforms, and *modern* to refer to state–space methods; we remark that the classical methods date to before the turn of the century and the modern ones became popular over 30 years ago. The 'theory' chapters (13–34) cover the following ground.

1. Introduce performance criteria (in Chapter 13).
2. Present various viewpoints, especially differential/difference equation coefficient tests, Nyquist methods, and Lyapunov's direct method, of stability analysis (Chapters 14–17). These are seen as the basis for much of linear system theory, with representation of classical transform domain pole checking and steady-state error prediction, the core notion of frequency domain analysis, and the alternative of generalized examination of differential/difference equation trajectories.
3. Present root locus and Bode methods of analysis and design (Chapters 18–20). These are, when done for continuous time systems, the standard methods taught in first undergraduate courses. The former studies the system pole locations (choice of poles is given in Chapter 19 and a very special case is treated in Chapter 21); the latter studies frequency response.

4. Discuss controllability and observability and the associated notions of pole placement controllers and state observers (Chapters 22–25). These are standard modern undergraduate course topics. The technical terms of controllability and observability are explained and tests described. Applications include the pole placement design of control laws and the first meeting with state estimator filters, i.e. observers.

5. Introduce optimal control, with emphasis on Lagrange multiplier and the related Pontryagin methods for discrete time systems and examples relating to the standard minimum time and linear quadratic (LQ) problems for linear systems (Chapters 26–27). The results are standard, but we also lead into numerical methods.

6. Introduce the topic of noisy measurements state estimation. The Kalman filter, its engineering extensions to non-linear systems, and the observer for explicitly noisy systems are met (Chapter 28). The issue of the interactions of filters and control laws is addressed (Chapter 29).

7. System identification is looked at in several different ways, with algorithms shown to illustrate some of the concepts (Chapter 30).

8. Adaptive control, particularly parameter adaptive control, is introduced in a couple of variations (Chapter 31). The important notions associated with self-tuning regulators, which are commercially available, are among those met.

9. Some of the notions and goals of learning control systems and robust control systems are presented (Chapters 32–33). Both of these are advanced topics of intense research interest, and the development here is superficial, intended to present only the points of view applied in the research.

10. Some theory associated with control system structuring, such as model approximation and control separation, is introduced (Chapter 34). Some of this is dated, but is still included, partly to show that not all systems use SISO ideas or direct extensions of those.

It must be repeated that most of the material is presented only for linear systems which do not vary with time. Rarely are non-linear systems more than commented upon (because there are so few general results). This book attempts to indicate the current situation for each topic, to serve as a source book for those wishing to progress further in their own studies of a particular topic and as a review for those seeking a reminder of things once known.

1.5   TOOLS OF THE TRADE

Any engineer's principal tool is a brain and the knowledge and judgement stored in it; engineering remains partly a science and partly an art form. The engineer will know the process being controlled – its physics and/or chemistry and/or biology – and have a library of appropriate techniques to handle the problem. These latter tools will at a minimum include the following.

1. Knowledge of the standard hardware used in a particular range of applications.
2. Classical design and analysis methods for linear systems. Stability testing using algebraic methods such as Routh or, for discrete time systems, Schur–Cohn or Jury testing. Compensation of the poles using root–locus methods. Compensation in the frequency methods, with lag, lead, and lead–lag networks designed using ideas based upon Nyquist–Bode–Nichols charts.
3. Modern design and analysis methods, mainly for linear systems and operating principally with state–space models. Methods include optimization using Pontryagin's maximum principle and Bellman's principle of optimality, with particular emphasis on the linear–quadratic–gaussian problem (i.e. a problem with linear dynamics, quadratic optimization criterion, and gaussian noise), and the filtering ideas of Kalman and Luenberger. Structural ideas of observability and controllability and the stability ideas of Lyapunov are also important.

Beyond this, a library of textbooks, monographs, and perhaps parts specifications and reports will be available; computational tools such as computer aided design (CAD, sometimes CACSD for computer aided control system design) and simulation programs will probably be available.

To maintain knowledge, the engineer will take two types of magazines – technical journals and advertising-containing magazines. The technical journals most commonly used are listed here.

• *IEEE Transactions on Automatic Control*, published by the Control Systems Society of the Institute of Electrical and Electronics Engineers, US. *Control Systems Magazine* is a more survey- and application-oriented publication by the same group.

- *IEEE Proceedings*, from the Institute of Electrical and Electronics Engineers, occasionally has survey special issues summarizing theoretical advances in control systems.
- *IEE Proceedings-D on Control Theory and Applications*, published by the Institution of Electrical Engineers, UK.
- *ASME Journal on Dynamic Systems, Measurement, and Control*, published by the American Society of Mechanical Engineers, US.
- *Automatica*, the journal of IFAC, the International Federation on Automatic Control, published by Pergamon Press.

Others include:

- *International Journal on Control*, published by Taylor and Francis Ltd., UK, and
- *SIAM Journal on Control and Optimization*, published by the Society for Industrial and Applied Mathematics, US.

The above tend to be theory and research oriented. A great many more specialized publications have articles which may be important to engineers working on particular applications, e.g.

- *Journal of Process Control*, a relatively new journal published by Butterworth–Heinemann Ltd., London, UK; and
- *IEEE Transactions on Robotics and Automation* and *IEEE Transactions on Aerospace and Electronic Systems*, both published by the Institute of Electrical and Electronics Engineers, US.

For the process engineer attempting to keep up with developments in hardware and system design trends, there are several magazines containing system advertising and short, direct applications articles.

- *Control Engineering*, published monthly by Cahners Publishing Company, a division of Reed Publishing, US.
- *Automation and Control*, published monthly in New Zealand by Associated Group Media Ltd.
- *I&CS* (Instrumentation and Control Systems), published by Chilton Co., US.
- *Control and Instrumentation*, published by Morgan–Grampian, UK.

Beyond the above, the engineer will have study and evaluation tools. Laboratory instrumentation may or may not be part of this, but, just as 30 years ago engineers used slide rules and 15 years ago they used calculators, they now use personal computers and workstations for many design and study calculations. The software for these, once available only on main-frame computers, allows detailed calculations and simulations. Provided mathematical models of systems are available, one may use readily available packages, such as:

- MATLAB®, from The MathWorks, Inc., US, for various design calculations such as root locus plots, frequency response plots in various forms (Nyquist, Bode, Nichols), conversions between continuous-time and sampled-data models, and time response computations (essentially simulations). Additional packages are available for robust control, system identification, and digital signal processing studies. A student version is available.
- CTRL-C®, from Systems Control Technology, Inc., Palo Alto, California, which is similar to MATLAB®.

These are mainly suited for the computations used in design and analysis. Simulations are also very important, and computer packages which make simulations easy to design and operate include the following.

- SIMNON®, from Lund Institute of Technology, Sweden, and available from ESC (Engineering Software Concepts, Inc.), US, and SSPA Systems, Sweden, is a system simulation program allowing continuous time, sampled-data, and mixed systems to be simulated.
- PSI® is a simulation tool from Delft University of Technology which is particularly easy for undergraduates to use.
- EASY5® is a powerful program available from Boeing, Seattle, Washington, US, which exploits workstation power to create, analyse, and operate elaborate simulations.
- SIMULINK® is a product from The MathWorks, Inc., US integrated with their MATLAB® package.

In larger companies, it is not unusual that they have created their own, more specialized programs. More and more textbooks are also including at least some programs for analysis or simulation of special methods.

Most engineers will find it useful to have at least some appreciation of standard computer languages. A knowledge of one language of a certain level is usually transferable to others at that level. Among the likely possibilities are the following.

- **Pascal** is a common teaching language with many extensions to allow it to be useful. Its structuring is its obvious feature.
- **Fortran** is the language of choice for many number crunching applications. A close and common relative is **BASIC**.
- **Modula-2** and **ADA** are the newest candidates for a number of reasons, including their structuring and (in the case of ADA) their sponsorship.
- **C**, an intermediate level language intended at first for systems programmers, has a good capability for direct dealing with the hardware, and is in some respects like a cross between assembly language and an ordinary compiler language.
- Assembly languages and associated programs exist for most common processors.

The above are all for digital computation. An additional simulation tool, becoming less common because of the increasing power of mainframe digital computers, is the analog computer. This is a set of electronic operational amplifiers, arranged so that they may be interconnected using simple patch cords to yield solutions of ordinary differential equations. Voltages within the computer are taken as analogs of physical quantities, and the time histories of these voltages are viewed on oscilloscopes or recorded on x-y plotters. Outputs are displayed on 'real' meters, etc., and hence the analog computers are often used in training operators (and pilots) under simulated conditions.

In this book, many of the references are to *IEEE Transactions on Automatic Control*, *IEEE Proceedings*, *Automatica*, and *Automation and Control*. Books frequently referred to are Åström and Wittenmark (1990) and standard first texts such as Phillips and Harbor (1991), Dorf (1989), and Franklin *et al.* (1990). Many of the calculations used PC-Matlab.

1.6 THIS BOOK – SCOPE AND GOALS

This book portrays two elements of the control systems engineer's work. First is the engineering aspect, including systems decisions, hardware and software, communications, and elementary process controller tuning. After a transition incorporating a chapter on the sources of models and three chapters on model representations, we proceed to theory, working through performance indicators, stability and its testing, and pole location design methods and goals. We continue into essentials of modern theory with information extraction (filtering), optimal control, system identification, and adaptive control, and conclude with introductions to specialized and advanced topics. All of the theory is primarily about methods for systems which are modelled by linear constant coefficient differential or difference equations, except when extensions are easily done.

In all cases, the book's goal is to introduce issues and basic approaches. Leads into more thorough or advanced treatments are given for each topic, and for each we attempt a brief summary of the essential ideas.

1.7 FOR MORE INFORMATION

History can be found in fragments in various texts, depending upon the detail sought. The article by MacFarlane (1979) is very instructive. Åström and Wittenmark (1990) give an interesting perspective on the influence of digital computers. Jury's reminiscences (1987) are also helpful. Bennett (1979) has one of the few books devoted to a history of the field; he covers the period 1800–1930.

This book is not deep, intentionally, since full and detailed books can be found concerning almost any single topic mentioned, and the individual chapters here offer specific references. The reader might consider this book as a somewhat condensed version of the following textbooks.

- *Sensors and actuators*
  - Hunter (1987)
- *Computer control*
  - Bollinger and Duffie (1988)
  - Bennett (1988)
- *Analog control theory*
  - Dorf (1989)
- *Digital control theory*
  - Åström and Wittenmark (1990)
  - Kuo (1980)

-     Franklin, Powell and Workman (1990)
■ *Advanced topics*
-     Bryson and Yu-Chi Ho (1969)
-     Anderson and Moore (1989)
-     Ljung and Söderström (1983)
-     Maciejowski (1989)
-     Åström and Wittenmark (1989)

This list is by no means a complete bibliography, nor is any mentioned book necessarily the best in the field for all topics considered. Instead, we mention them here to give the reader some notion of what books represent a class of books relevant to the topics listed.

2

Elements of systems engineering of digital control

Control systems are the subsystems of plants which generate and send the commands to the plants' 'working' components. Hence, they are the elements which turn motors on and off, regulate inputs, record (or **log**) data on the processes, send messages to operators, etc. The level of sophistication is decided at the systems engineering stage, with the goal of using control components and techniques appropriate to the task – neither using a supercomputer to turn a heater on and off nor a commercial personal microcomputer for a sophisticated satellite antenna pointing system. The various decisions involved are aspects of systems engineering, and require decisions at a number of levels. This chapter explores only two levels: system structuring and component selection.

2.1 SYNOPSIS

The objective of systems engineering is to provide orderly overall management of the development and operation of systems. It is this management aspect that the engineer must keep in mind, often at several levels simultaneously. Among the intermediate level problems are the selection of the layout, or structure, of the control system. Here it is decided whether control is centralized or distributed, and whether in multiple computer cases the computers interact at all, and if so, whether hierarchically.

At the lowest level are the problems of component and algorithm selection. Objective evaluation is possible in several different formats: qualitative checklists, qualitative/quantitative ranking and weighting, and quantitative scoring. A knowledge of how project costing is evaluated is essential to contributing to business decisions.

2.2   MANAGEMENT-LEVEL CONSIDERATIONS

The top level of systems engineering is the program planning level. It is likely to deal heavily in general rather than specific systems elements, with specifics derived as needed from lower level studies. The study may be of whether a new or refurbished plant or assembly line or a new ship control system should have an overall digital computer control system and even whether a flexible manufacturing system (FMS) is to be installed; the impetus may have come from one or more of several sources, such as salesmen talking to management extolling the virtues of specific systems or a researcher finding that further automation of the assembly line might be expected to produce better quality control.

At this level are investment decisions and operations management, both of which are beyond the scope of this book.

2.3   SYSTEMS CONFIGURATION ENGINEERING – THE INTERMEDIATE LEVEL

The intermediate level in computer control systems is the configuration design. Choices must be made as to quantities to be measured and logged, quantities to be controlled, how automatic operations are to be structured, how communications are to be carried out, etc. All of these choices are specific to the task at hand, but do not necessarily involve particular component selection. Configuration design can be among the most difficult parts of design: the answers must allow for a great many, possibly only partly known, factors.

2.3.1 Closing the loops

For many control problems, the system to be controlled is a relatively uncomplicated one: a motor to be speed controlled or an amplifier to be frequency-response compensated. For these, the theoretical classical control methods, which are largely single-input–single-output (SISO) and which require well structured problems, are directly and obviously applicable.

Larger problems will typically require many more decisions, with choices which may or may not be obvious. Among the choices are the following:

1.   measured variables;
2.   controlled variables; and
3.   control technique, both algorithms and structure.

Tying these together are the system requirements, and the final choices are often arrived at through an iterative process. This iteration may proceed from the top down or the bottom up, i.e. from the specifications through increasingly lower level decisions as to detail or from step-by-step build-up from component control up to a major system.

The choice of controlled variables is sometimes obvious, but even simple systems can show subtleties. Hence, fluid flow can be controlled by controlling a pump's speed, by using a constant speed pump plus a valve, or by maintaining a suitable head in the liquid supply tank. Another example is that a boat's course can be affected by rudder action or by the differential speed of dual propellers.

Similarly the choice of measurements may be obvious, subtle, or simply expedient. For the fluid flow problem mentioned above, one might measure the actual flow or the supply tank depth, for example. Additionally, the need for instrumental or other compensation may dictate that temperature be measured. Furthermore, it may be necessary to measure disturbances and compensate for them.

The control approach must be selected. One common method is to use pair-wise matching of measurements and control variables; this tends to yield thousands of control loops in large process controllers. In other applications such as aerospace navigation and control, the inputs and outputs are treated in groups using multiple-input–multiple-output (MIMO) control algorithms.

Associated with algorithm choice, but different from it, is the choice of control structures. MIMO controllers may use a modern state estimator cascaded with an optimal controller, for instance (see Chapter 29). In basically SISO approaches feedback control will almost certainly be used. Beyond feedback are other possibilities, including cascade (nested) and feedforward controllers (see Chapter 34).

For a large plant, many such choices will be made. The result of all of these choices, when combined, is a system. Even a portion of the system will look complicated, as can be seen in the following example.

Example

A detailed and instructive example is given by Singh *et al.* (1980, pp. 185+) based on a simplification of regulators used by Foxboro Company on ships. The ships are turbine powered and require superheated steam at constant pressure in the presence of a need for variable quantities of steam. The main control variable is the fuel flow rate to the burners of the boiler, and the approach is to act on the enthalpy of the saturated steam. As presented by the reference, the design progresses in a bottom up manner as follows.

The basic situation is shown in Fig. 2.1, and the progression of loop installations is given in Fig. 2.2(a)–(d).

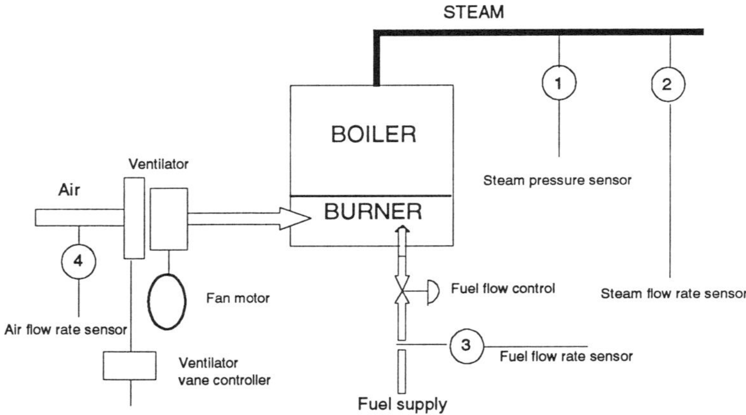

Figure 2.1 A steam boiler tank, with its various sensors and actuators. The object is to supply a regulated amount of steam energy to an engine.

The reasoning behind the development is straightforward.

1. The first loop to install is the principal one: a regulator with a measurement of steam pressure as input and a command to the fuel flow rate valve as output (Fig. 2.2(a)).
2. Evaluation shows that, since the above command is to the valve, the flow rate can still vary because of pressure variations in the fuel supply. Although these pressure variations would eventually affect the steam pressure and cause valve commands to change, improved regulation is achieved by measuring the actual fuel pressure and using this to generate a secondary command to the valve (Fig. 2.2(b)).

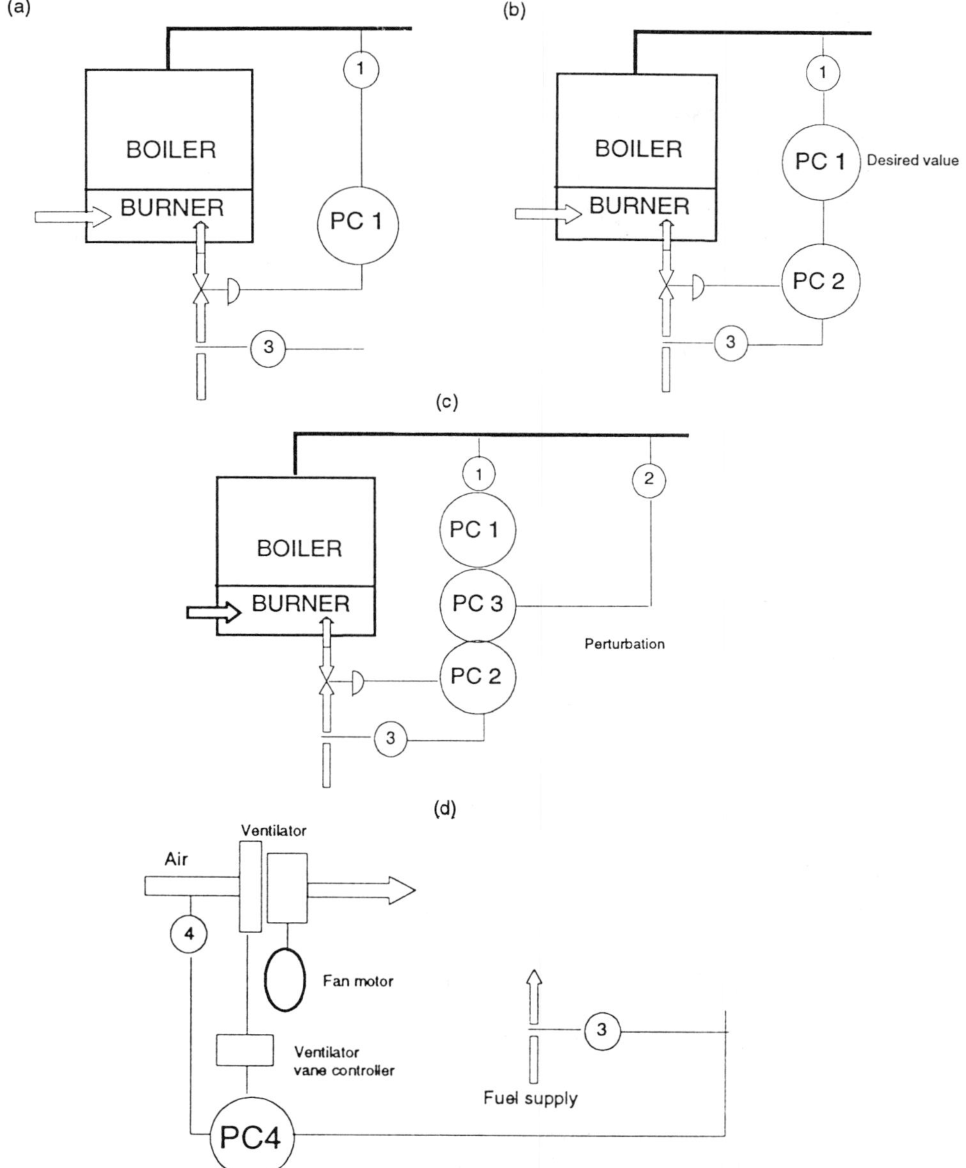

Figure 2.2 Stages in development of feedback control for the boiler using process control (PC) elements: (a) steam pressure feedback to fuel valve; (b) addition of fuel flow feedforward; (c) addition of steam pressure feedforward; and (d) control of air ventilator based upon air flow rate and fuel flow rate demands.

3.   If the motor demands more steam suddenly, this may be seen as a disturbance to the steam pressure maintenance system. Because of the size of the boiler, however, there may be a time constant in sensing this disturbance using pressure. To achieve a rapid response to this situation, a flow rate sensor is installed to provide a further (feedforward) input to the fuel valve controller (Fig. 2.2(c)).

4.   Combustion also requires air to burn with the fuel. Maintenance of the proper air/fuel ratio is therefore also necessary and requires control loops. The amount of air needed is provided by a constant speed fan directed at variable-opening vanes (a classic case of using constant supply plus valving rather than variable supply). Hence the primary control of the air supply is provided by a measurement of fuel flow rate input to a controller of the vane angles. In addition, air flow reaching the vanes may vary, e.g. due to variations in fan speed or to excess or deficit pressure due to the burner. To compensate for such effects, airflow is measured and used as an auxiliary adjustment to the vane angles. These two effects both influence the vane setting (Fig. 2.2(d)).

In addition, engineering refinements are added, including a limiter to process both fuel flow rate and valve opening command and base the command on the more demanding value (to overcome time delays and avoid smothering the flame in transients due to lack of air) and an auto/manual option for vane control. The final situation is represented in Fig. 2.3. The figures are based upon Singh *et al.* (1980).

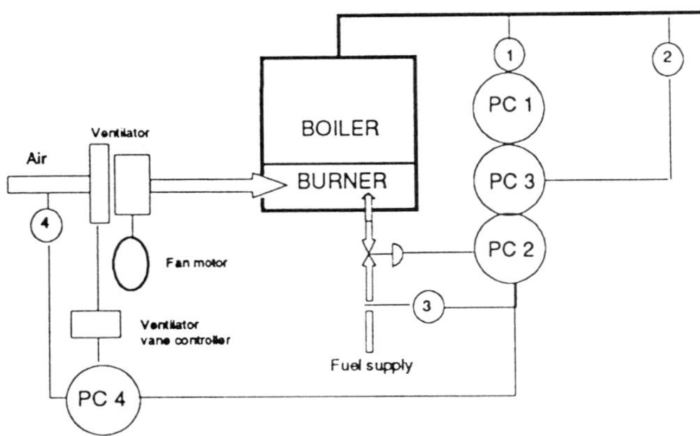

Figure 2.3 Completed multiloop multivariable control system.

We remark that an alternative is to note that steam pressure, steam flow rate, fuel flow rate, and air flow are measured variables, while fuel flow valve opening and air flow vane opening are the command variables. In principle the variables can be taken as interacting and a MIMO controller can be designed.

A MIMO controller analysis can show that SISO controllers are sufficient, depending upon the situation. For example, although in an aircraft autopilot the roll, yaw, and pitch axes are coupled in an important way, a space launch missile will ordinarily be flown in such a manner that yaw and roll manoeuvres are small after lift-off. When this is so, guidance may well be designed assuming the angles are uncoupled; this significantly reduces the navigation problem, for pitch commands and engine commands are then the only ones of major importance in the orbital injection accuracy.

2.3.2 Computer control configurations – central, distributed, and hierarchical control

The extremes of the possible computer configurations are shown in Fig. 2.4: one has a single central computer and control room with all instruments sending data to it and all control commands issuing from it; and the other has each loop having its own controller, probably based upon a microprocessor, and no communication between loops. The two extremes might be called completely centralized control and completely distributed control, respectively.

The arguments for and against the two extremes help to define the rationales for the intermediate systems which are more common.

- **Centralized computer** This system may benefit from the economy of scale in having one large computer rather than many small ones. It might be expected to provide 'better control' because the commands for all the components of the system can, in principle, be co-ordinated ones. However, a back-up computer might be required to gain the necessary system reliability, the cabling to and from the central control room will be expensive, and the system may be expensive to trouble shoot and maintain if the software is complex.
- **Completely distributed control** Each control loop is easy to understand and tune because the loops are wired independently. Wiring costs are minimized. Trouble shooting is straightforward (at least as far as breakdowns are concerned), and changes to the

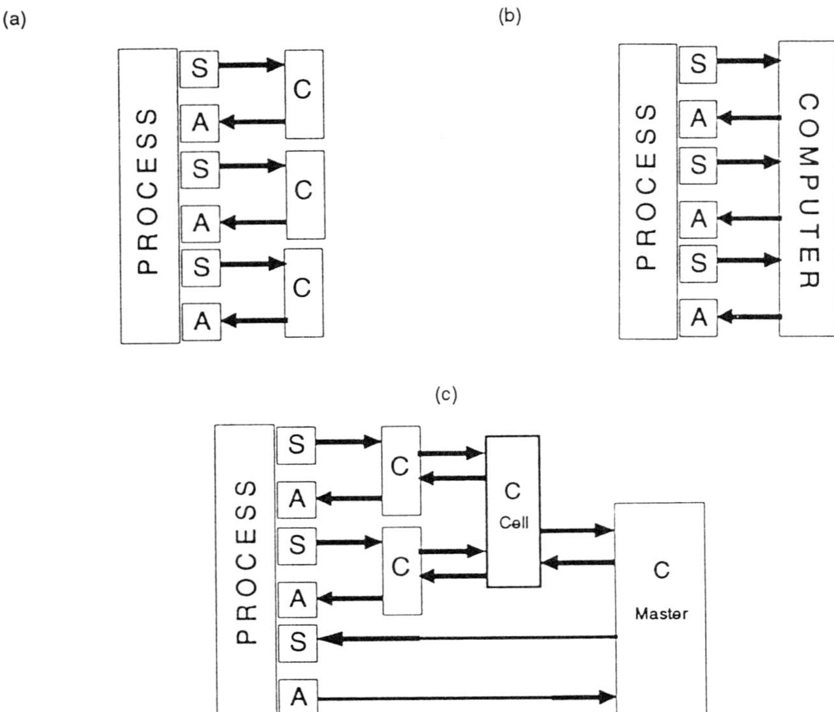

Figure 2.4 Possible computer control configurations: (a) distributed; (b) centralized; and (c) heirarchical. S are sensors, A are actuators, C are computers.

plant are relatively easy to make because co-ordination with other parts of the plant is not done at the automatic control level. However, lack of co-ordination may lead to strange interactions of control loops and significant inefficiencies in machine use. Tuning of hundreds of scattered control loops is unlikely to be done frequently. At least some cabling to a central location will probably be needed so that the operation of the system can be monitored.

- **Hierarchical control** In this scheme, the computers form a hierarchy, with some handling direct measurement–controller–actuator loops, supervisory computers co-ordinating their actions, planning computers generating parameters for the supervisors, etc. It has characteristics between those of the two methods above and is of increasing interest because of the potential flexibility available.

The choice of structure will be influenced by the plant configuration and projections for future changes. The options are growing partly because of the rapid lowering of the price of computer hardware.

Hierarchical control structures are worth more discussion at this point. A terminology, still only crudely defined, is developing which attempts to give a basis for discussing the systems and hence for giving a rough notion of the computer power necessary at each level; the notions are those of the work cell, the area, and the plant controllers.

A **work cell** consists of a few machines or devices which are 'obviously' best controlled in a co-ordinated manner. Thus the elevators, ailerons, and rudder of an aircraft should be moved in a co-ordinated way during turning manoeuvres. In a plant in which a robot is feeding raw materials to a stamping press, the two should be synchronized so that the material is properly placed and the robot manipulator is out of the way when the stamping is performed. A cell might be expected to require a modest computer controller, and it might or might not be operated independently of other cells. Its computer might be programmed to communicate only modest amounts of data to a central control room. It would also co-ordinate lower level controllers such as the individual motor controllers of machines in its cell.

The next level in control is the **area** control. It might be thought of as co-ordinating a number of cells, possibly only in a supervisory manner. The overall control of an assembly line, with several stages of welding robots, painting robots, assemblers, etc., might be vested in an area controller, with the painters in one cell, welders in others, etc. Office building air conditioning may have room controllers, floor cell controllers, and building area control levels.

In principle, above this level is **plant** control, in which the operations of the entire factory are co-ordinated.

Other factors can also affect the configuration choice. The decision to have a flexible manufacturing system (FMS), consisting as it does of several computer numerically controlled (CNC) machines plus connecting parts movers such as conveyors and robots, will require at least some central co-ordination and an ability to transmit instructions or instruction sets from the central computer to the machines; coupling such systems with the computer aided design (CAD) office will place even more burdens on the computer system, but will lead to a goal of computer integrated manufacturing (CIM). The system is necessarily hierarchical or centralized. Similarly, the addition of sensors and 'intelligence' to a robot cell may determine the computer requirements. The point is that other factors may force certain parts of the configuration decision.

Beyond the aspects of philosophy, convenience, and history which affect configurations are other considerations.

1. In any configuration except completely distributed controllers, the control elements must communicate with other elements of the control system. Here both the distances and philosophy involved become important. The extreme choices here are to make each communication channel an *ad hoc* affair so that some signals are over instrumentation type cables (heavy cables carrying 4–20 mA, say, with shielding), others are computer pairs in master/slave configurations, others are computers and smart instruments in a computer local area network (LAN). The other extreme is a complete LAN for everything. Such communication is discussed in Chapter 7.

2. Almost certainly some decision making will be left to the human operators, whether plant technicians or pilots or other operators. A centralized controller may be able to handle most set-up and operation of the plant, with monitoring and emergency interventions done by the operators. This is the basic scheme of electric power plants, for instance. On the other hand, a distributed system will require operators for each subsystem for at least tuning, adjustment to be compatible with other subsystems, etc. The burdens, responsibilities, and duties of the operators in the configurations are different, obviously, and so are their information needs and training.

2.3.3 The top-down alternative

If the above approach seems reasonable, it also hints at being inefficient. In applications demanding high performance, it may be necessary to account for the coupling of loops explicitly. For example, in the flight control of high performance aircraft the air speed, vehicle attitude, and altitude may all be both controlled variables and sensed variables. They are certainly coupled and proper aircraft control requires consideration of this coupling. Furthermore, these are really only intermediate variables in a situation in which the ultimate task is to go from point A to point B or situation C to situation D. Simple uncoupled loops may well be inadequate for such applications, and the engineer will need all of modern control theory to design the control system.

The alternative approach is called top-down design, and it starts with overall problem definition and breaks down the problem into sub-problems successively until the level of solved problems is reached. In principle, this would appear to lead directly to modern control theory, multiple-input–multiple-output (MIMO) systems, centralized controllers, etc. In fact, it partly does so in the case of flight controllers and similar systems, where the models of the system dynamics are well known and performance criteria are easily established. In the case of process control, however, the breakdown may come to quite low levels for the simple reason that low-level problems have existing solutions but often do not have the structure needed for modern MIMO methods.

The first choices at any level are an interlocking net of choices on what is to be controlled, what is to be measured, and what the control variable is to be. The designer will be aware of several possibilities for each choice.

2.4 CHOOSING SYSTEM ELEMENTS

The time finally comes to buy hardware or choose control algorithms and start closing the loops. When closing the loops, components must be selected, as must control algorithms, communications and its cabling, etc. Each step requires a conscious choice among real alternatives: no longer is the choice between central computer and co-ordinated cell controllers, but between brand A and brand B of cell controllers, between two methods of temperature sensing and components which do them, etc. Making these choices can be made at least somewhat orderly by considering technical factors, price, vendor support, and even personal prejudices in a straightforward way. The factors can be combined qualitatively, quantitatively, and by combinations of the two. In this section, these approaches are demonstrated.

2.4.1 Assumption: configuration known

The assumptions made at this level of systems engineering are that a set of specifications exists and that a configuration has been chosen. Thus, an actual choice of methods or components is to be the result. The choice could be between two control laws, with criterion to be the quality of control as measured by variance of the controlled variable, and with the control computer (and hence available computer power) 'given'. It could be between five programmable logic controllers (PLCs), with several factors included in the choice, as in the minicomputer choice below. It could be between several temperature-sensing methods, with criteria centering on reliability and maintainability. The point is that the trade-offs are now precise because the situation is well defined.

For computer selection, for example, the configuration definition would include the computer size and speed, its I/O elements, and the types of computations to be performed. These definitions help to define the class of candidates. Thus it would be known at evaluation time whether a microcomputer, minicomputer, mainframe, or supercomputer is required, and how much storage is required. Knowing the I/O elements, particularly the number of digital and analog inputs and outputs in the control systems case, may narrow down the architectural choices. The type of computations, e.g. whether the utilization is predominantly number-crunching, logical operations, or input–output in nature, also defines the system requirements. This knowledge is typically derived from knowledge of similar systems, analysis of the task, and from simulation.

2.4.2 Selection factors – technical, support, financial

The factors influencing selection of a system fall into three categories: technical performance, vendor support, and financial aspects.

Technical performance must be established for the device sought. For example, in the case of computer systems, there are several technical performance factors to be considered. These include both hardware and software elements of the system.

Hardware considerations start with the raw architecture of the machine. The number of bits in words of various types (addresses, data, bus size) will influence many aspects of performance. The user will be interested in speed of operation and throughput, particularly for certain defined operations. The throughput and capability of I/O devices and peripherals will also be a consideration.

Software factors will include the number of assemblers, compilers, and special programs available. A critical point may be whether an appropriate operating system is available. Off-the-shelf systems may include a software package; this should be examined for general capability, and for ease of programming. De-bugging aids should be available.

Related to both of the above, but perhaps more of a hardware consideration, is the ability to expand the system under consideration by adding more memory, more interface elements, and communications, etc. *Modularity* is often prized for this reason, as well as for maintenance ease. *Upward compatibility* with more powerful systems may be a factor if increased demands are anticipated.

Performance can be difficult to predict for actual installed systems, even though instruction speeds are well known. This is for a number of reasons, but mainly because applications have a mix of instructions and because compilers may vary in their usage of powerful instructions and code optimizations. For this reason it is not uncommon to demand benchmark tests of candidate computer systems. Such tests usually involve prediction of execution times, or actual running of benchmark cases, for particular problems.

Support factors in the decision to select a piece of equipment are a variety of influences which are real but perhaps difficult to quantify precisely. One of the first among these is reliability, usually given in terms of the mean time between failures (MTBF). This can be very difficult to discover, and may require interviewing other users of the equipment in question. Somewhat related is the mean time to repair (MTTR), i.e. the time to repair a non-functioning device. Also associated with the issue of reliability and maintenance is the availability of parts and access to trained service personnel.

Any purchaser should attempt to acquire proper documentation along with purchased equipment: user instructions, repair manuals, and, for electronic equipment, wiring diagrams. The latter two are particularly important for buyers who intend to perform their own maintenance.

Related to hardware maintenance is software maintenance, in that similar problems arise. The buyer would certainly like to have good documentation. Even more, perhaps, (since software always seems to have bugs and improvements are frequent) the buyer would like assurance that software will be delivered on time and will be kept up to date by the manufacturer.

Other factors will include warranties, if any, and the general co-operation of the seller. Computer warranties are typically 60–90 days, for example. Seller co-operation in installation, rectification of problems, etc. should be assured, but sometimes the sellers are sales personnel working for a distributor rather than manufacturers. Judgments in this area are intangible, but can be crucial.

The obvious **financial factor** is price, but this must be considered along with possible discounts, contract terms, and eventually life-cycle cost. Even overlooking the high-level factors in systems financing such as profitability and pay-back periods, the predicted costs associated with a particular piece of hardware may diminish substantially if the equipment is reliable (implying low repair costs and small loss of production due to down-time) and good contract terms are available.

Another direct financial factor is the cost of installation, with it being important whether the buyer or the vendor does the installation, and in the latter case whether the cost is already included in the price. Similar comments apply to maintenance: who does it and what is it to cost?

Intangibles among the financial factors are the ability of the vendor to fill orders on time and at quoted price, the marketing support which the vendor has (is there supplier back-up?), and the vendor's future competitive position (is there risk of going bankrupt?). The engineer is not necessarily expected to be a financier concerning these matters, but should certainly be aware that they should influence the decision-making process.

2.4.3 Combining the factors – qualitative, rankings, quantitative

The above section presents many factors to be considered in choosing a system. Since it is rare that a single candidate is clearly best on all counts, a rationale is needed for combining the considerations. This should be done as an aid to decision making, rather than a commitment to choosing the 'best' by some arbitrary weighting of factors. To this end, the factors may be jointly considered in several different ways.

1. **Qualitative evaluation** In qualitative evaluation, each factor of interest is considered for each candidate and a subjective rating assigned. The subjective ratings may take on three, up to five, or seven different levels, depending upon the situation. A three-category rating, for example, is 'good', 'adequate', and 'deficient'. Two possible additional categories are 'excellent' and

'inadequate'. Table 2.1 shows such a ranking for four candidate systems using the factors discussed above.

Table 2.1 Evaluation summary using qualitative ratings

Criteria	*Device*			
	A	*B*	*C*	*D*
Technical				
General design	+	+	√	-
Ease of use	+	√	√	√
Modularity/growth potential	√	+	√	√
Support				
Expected reliability	√	+	+	√
Documentation	-	√	+	√
Warranty	√	√	√	√
Vendor operation	+	√	-	-
Financial				
Ability to fill order	√	+	√	√
Prices	+	+	-	√
Vendor stability	-	√	+	√

+ = Good/excellent
√ = Adequate/satisfactory
- = Poor/deficient

With the guidance of this table (which might be a compilation of results from having several experts consider factors in their areas of expertise – a Delphi method) the decision maker hopes to be closer to a decision. In this particular case, device D is clearly a poor candidate. Although at first glance, device A appears superior to device C, one should note that device C is satisfactory for most factors. Is there an engineering advantage in having device A exceed the levels required? One is inclined to extend this question to device B, and finally choose device B not because of all the 'goods' but because it shows no deficiencies on the factors considered.

2. **Ranking and weighting** If the above seems unsatisfactory because there is no final winner, and more so because it appears to make 'ease of use' equal in importance to 'prices', a more

quantitative approach may be taken. In this system, each factor is assigned a weighting to indicate its relative importance and, for each factor, each candidate is placed on a numerical scale. Thus relative weightings might be 0–10 for 'unimportant' to 'very important', and factors might be 0–4 for 'bad' to 'very good'. Table 2.2 shows example weightings and the bases for the scoring of factors in technical and vendor evaluations.

Table 2.2 Criteria for ranking/weighting type evaluation

Criteria	*Weighting*
Technical	
General design	10
Ease of use	8
Modularity/growth potential	4
Support	
Expected reliability	8
Documentation	5
Warranty	4
Vendor operation	3
Financial	
Ability to fill order	8
Prices	7
Vendor stability	5

An application of these criteria to the four devices is shown in Table 2.3. For each factor, scoring was on the basis of 0–4, with 0 for inadequate and 4 for superior.

3. **Quantitative evaluation** Strict quantitative evaluation is a further continuation of the scheme of the previous section. Its paradigm is the price/performance ratio, e.g. dollars per digital I/O channel of a programmable logic controller (PLC), but it can have many forms. The point is that each system is judged according to some function of its characteristics. The problem is: what function is appropriate?

Table 2.3 Performance score

Criteria	A	B	C	D
Technical				
General design	40	30	20	10
Ease of use	32	32	16	8
Modularity/growth potential	8	16	8	8
Support				
Expected reliability	16	24	32	16
Documentation	0	10	20	10
Warranty	4	4	4	4
Vendor operation	12	6	3	3
Financial				
Ability to fill order	16	24	16	16
Prices	28	21	7	7
Vendor stability	0	6	12	6
	156	173	138	88

(Column header above A B C D is *Device*.)

A possible function for price/performance of computer hardware is

$$P_h = \frac{\text{cost}}{f}$$

where f is some function of internal storage, bits in address, number of registers, memory cycle time, arithmetic instruction set power, logic instruction set power, and I/O capacity.

A similar function for software is

$$P_s = \frac{\text{cost}}{g}$$

where g is some function of availability of diagnostic routines, of debugging routines, of loader routines, number of assemblers, number of compilers, and operating system power.

Such a function was used by Butler (1970) with

$$f = 0.1*(\text{memory size in bits}) * \left\{\frac{1}{2} + \frac{\text{no. of address bits}}{2*\text{word length in bits}}\right\}$$

$$+ \frac{20}{T} \{f_1 \text{ (arithmetic, logic, I/O instruction types)}\}$$

$$+ 100*(\text{no. of extra features})$$

$$+ 50*(\text{no. of general purpose registers})$$

where T = computer read/write cycle time, and f_1 = valuation from 0 to 300, depending on hardware instruction capabilities, and with

$$g = 500*B(\text{debuggers}) + B(\text{diagnostic routines}) + B(\text{loaders})$$

$$+ 1000*(\text{no. of assemblers}) + 2000*(\text{no. of compilers})$$

$$+ 50*g_1 \text{ (operating system properties)}$$

where $B(x) = 1$ if at least one x routine is present, and 0 otherwise and g_1 = valuation from 0 to 100.

This particular example is arguably little more objective than those of the qualitative and the ranking and weighting methods; although many of the factors are demonstrably objective, others and the combining are not necessarily entirely objective.

The point of the above is not to find the 'winners' in these studies, but to indicate how scoring, prices, etc., are guides to evaluation. Decisions are judgements made after considering the data sensibly and from several viewpoints.

2.5   COSTING

It is worthwhile for the engineer to have some knowledge of costing and financial evaluation of projects. We touch on those by indicating the cost of some components and systems, and by presenting one such method (internal rate of return).

2.5.1 Examples of costs

It is necessary for the student to appreciate the amounts of money at stake in computer control and computer co-ordinated data logging (a less demanding task in that the outputs are allowed occasional errors because they are not real-time effective). To this end, Table 2.4 is presented. Other prices will appear later in the book.

Table 2.4 Rough prices of alternatives in computer control

System or component	Cost ($)	Comments/examples
Microprocessor chip	10–100	Chip only
Microcomputer PC	500–10 000	Includes a few I/O channels. Not industrially hardened
PLC	1000 up	Cost around $100 per digital I/O pair
PID controller	600–2000	Single channel of input and output
CAD workstation	20 000 up	Not an on-line device. Included to give perspective
Controller system	15 000 up	Basic system with several digital I/O channels and some analog capacity. Expandable at around $100 per digital I/O and $1000 per several PID channels
Minicomputer	50 000 up	DEC VAX system
Main-frame computer	500 000 up	Moderate IBM system
Supercomputer	30 million	
Common instrument/ sensor	500–1000	A temperature sensor
Special instrument/ sensor	50 000	
Actuator/control element	500–700	Small valve with motor

The reader will appreciate that these are 'order of magnitude' numbers only. Computer system peripherals, for example, can add a factor of two to the costs of the computer systems. Systems costs are very particular, and different systems may have different costs.

Some system costs are $300 000 for a control system refit of a small power station, $150 000 for an industrial robot with a vision system, and millions of dollars for advanced aircraft electronics (of which the control systems would be a fraction). A small simple university pilot plant may incorporate $25 000 in control computers, sensors, and actuators.

2.5.2 Internal rate of return

Costing principles should be considered by engineers at all stages, even though their primary inputs may be technical. In making a capital decision, a common parameter is the internal rate of return, i.e. the rate of interest the expenditure would have to generate to provide equivalent after-tax profits. Equivalent to this is the **pay-back period**, the time by which the profit increase will repay the capital expenditure. Thus an initial (capital) investment of $C on an object with n-year lifetime, if it leads to an annual cash savings of $S, will show an internal rate of return I given by the solution of

$$\$C = \$S \sum_{k=1}^{n} \frac{1}{(1 + I)^k}$$

and a pay back period of $1/I$.

Example

Examples of the calculations are given by Hall and Hall (1985) for a robot with vision in a materials handling application (Tables 2.5–6). The initial investment has two components: hardware (capital expenditure) and start-up costs.

The system is assumed to have a five-year life with no residual value (actually, robots are being found to have working lives in the range of 8–10 years), with straight-line depreciation of the hardware costs, yielding ($124 800/5 =) $24 960 per year depreciation.

The cash flow effects are due to taxation, assumed to be at a 50% rate, and to productivity increase of 10 parts per shift at $50 per part; manpower is unaffected because present workers are either retrained to operate the robot or transferred to other tasks due to the increased workload attributable to the robot. Two-shift operation is envisaged, and 200 working days are assumed.

Cash flow in the first year then has the components shown in Table 2.6.

Table 2.5 Costing for a robot installation[*]

Capital investment	
Robot	$66 800
Manipulator	4 000
Safety equipment	4 000
Vision system	40 000
Conveyors	5 000
Other	5 000
Total	$124 800
Installation costs	
Feasibility study (400 hr @$20/hr)	$8 000
Engineering (200 hr @ $20/hr)	4 000
Site preparation (80 hr @ $15/hr)	1 200
Installation (80 hr @ $20/hr)	1 600
Total	$14 800
Total initial costs	$139 600

[*] Derived from Hall and Hall (1985).

Table 2.6 Cash flow for robot evaluation[*]

Productivity (20 parts/day × 200 days/yr × $50/part)	$200 000
Less: Energy (electricity) usage	
(30 kW × $0.04/kW-hr × 3200 hr/yr)	−3 840
Maintenance (3.75% of equipment cost)	−4 680
Insurance (5% of equipment cost)	−6 240
Plus: Depreciation	24 960
Total before taxation	$210 200
Taxation effects:	
Tax savings due to start-up expenses (first year)	$7 400
Tax credit for investment (10% first year)	12 480
Tax on cash flow	−105 100
Total tax paid	$85 220
NET CASH FLOW	$124 980

[*] Derived from Hall and Hall (1985).

The numbers remain essentially the same for succeeding years, except that all but depreciation are increased by 5% due to (assumed) inflation. Hence, we have net cash flow for the second to fifth years as follows

Second year	$109 731
Third year	114 594
Fourth year	119 699
Fifth year	125 060

From this, the internal rate of return is approximately $I = 0.80$, a very good figure. The pay-back period is about 1.25 years. (Less than a 2-year pay-back is often sought, although it is claimed that the investment-minded Japanese businessman may settle for five years.)

2.6   SUMMARY

These last few pages are included primarily to give an introduction to the nature of the problem faced by the systems engineer in the control systems field and to indicate that it is possible to be systematic in evaluating alternatives. It should be stressed that there is no algorithm for evaluation at any level of the systems engineering process, i.e. there is no checklist which, faithfully followed, will yield a best solution. Rather, there is a systematic way of thinking about the problems which may be helpful in design and procurement stages.

2.7   FURTHER READING

This chapter has considered only a minor subset of systems engineering – intermediate design exemplified by the closure of loops and structuring of the computer system, and alternatives in semi-objective component selection. A minor addition, useful both for an example of costing and the handling of finance, was also presented. Further reading is in a variety of areas:

1.   Systems engineering approaches can be pursued in Sage (1977) and in *IEEE Transactions on Systems, Man, and Cybernetics.* There have been attempts to give a general perspective of systems engineering; Sage is one starting point for this. Systems engineering has also had a more particular meaning for some

writers, and really means *computer* systems engineering to them. The computer science literature contains several texts of the specialized computer systems engineering variety.

2. Control systems engineering for process control systems is covered at many different levels in magazines such as *Control Engineering, Control and Instrumentation,* and *I&CS.*

3. Flexible manufacturing systems (FMS), work cells, and area control are popular topics of the above-mentioned magazines. The recent text by Groover (1987) is very helpful concerning manufacturing processes. Hierarchical control is the topic of at least one monograph: Singh (1980).

4. Example trade-off studies are difficult to come by, but helpful information is usually to be found in magazines such as *Control Engineering* and in manufacturers' literature. The magazines periodically present data on topics such as PLCs.

5. An important topic which we have not covered here, but which influences system layout, is **human factors.** System operation can be strongly affected by information displays to human operators, for example. A text in this area is McCormick and Sanders (1983). Operator reactions to new process controllers have been reported by the magazines. A brief mention is given later in Chapter 5.

More general approaches into management are in the operations and production management texts such as Heizer and Render (1988). Management for engineers is given in Samson (1989).

3

Sensors and instrumentation

In this book we choose to work from the process back to the controller on a system such as is indicated abstractly in Fig. 3.1 and more specifically in Fig. 3.2. The immediate connections to the process are the sensors which measure the state of the process and the actuators which influence the control elements to adjust the process.

For this example, to know the temperature, the computer must be connected to a transducer such as a thermocouple through devices which convert the sensor's output to a binary number represented in bits of 0 or 5 volts, and this number must be made available to the computer in an orderly manner. The computer output is also one or more binary numbers; for these to be useful, they must at least be displayed and in control loops they must lead to an action. This action is done by actuators such as motors and solenoids which manipulate control elements such as valves to implement the computer's commands.

The devices which gather the data and which interface them to the control computer are the subject of this chapter; the devices which implement the commands are the subject of Chapter 4.

3.1   SYNOPSIS

Of itself, a computer is incapable of dealing directly with other devices – it must be attached to them with interface equipment. The idea is shown in Figs 3.1–2.

There are many types of transducers for many physical quantities, and these directly or indirectly give electrical outputs which can be sampled for use by the computer. When chosen they typically have a number of characteristics which are specified, including accuracy, size, typical application, and price. After looking at these, this chapter briefly outlines a few of the common sensor types with emphasis on transduction, that is, how they operate to give electrical outputs

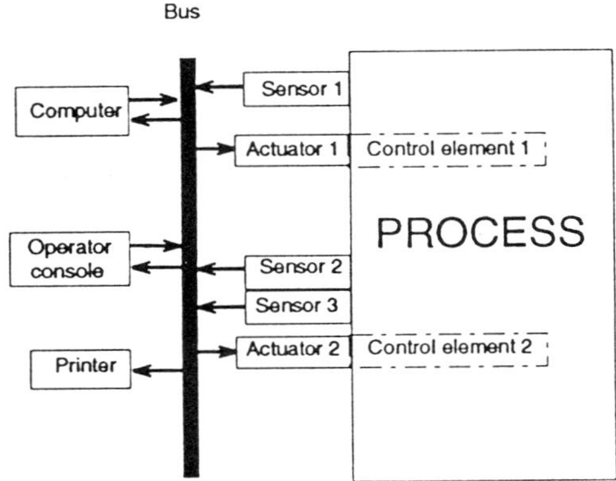

Figure 3.1 Block diagram showing computer connected via a bus to its own peripherals and to the sensors and actuators associated with the controlled process.

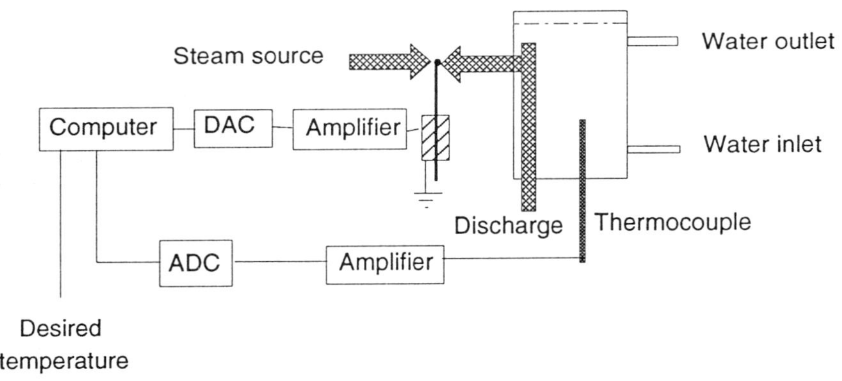

Figure 3.2 A more specific version of Fig. 3.1. The process is the supply of heated water, the controlled variable is water temperature, and the controller variable is the valve admitting steam to the heat exchanger.

corresponding to the physical quantity being measured. We notice that a typical process control application may have a great many sensors accurate to better than 1% at a cost of several hundred dollars each.

Since the computer requirements are ultimately for low voltage binary signals, signal conditioning and possibly conversions from analog to digital and back are often necessary. We briefly consider such conditioning and conversions.

3.2   SENSORS AND INSTRUMENTATION

The computer senses the external world, and in particular the plant to be controlled, by means of transducers which convert physical phenomena to electrical signals. These signals are then converted as necessary to a form which the computer can process, i.e. to binary words of some number of bits, each bit having voltage of 0 or 5 V and appropriate impedance matching. The data are carried to the computer using pairs of wires, coaxial cables, or increasingly by optical fibres, depending upon the environment. Sensors are readily available to measure light, temperature, position, velocity, chemical composition, force, acceleration, motion, etc. Clever use of raw transducers and electronics allows many other variables to be sensed.

3.2.1 Transduction principles

It is the ability of the control system to sense what is happening that gives it the knowledge to behave intelligently in its role. The basis of sensing is the use of a transducer to transform a physical phenomenon into an electrical signal, which may then be processed into a form suitable for the computer. The sequence is shown in Fig. 3.3, but it must be observed that many of the steps may be optional. For example, a microswitch may need only a buffer to contain its result for the computer.

Sensors may be classified in many ways, but it is most straightforward to classify by the quantity sensed, as this is arguably one of the most useful approaches. Starting with this, we describe the transducers by the following:

1.   measurand;
2.   transduction principle;
3.   (optional) special features, special provisions, sensing element; and
4.   range and units.

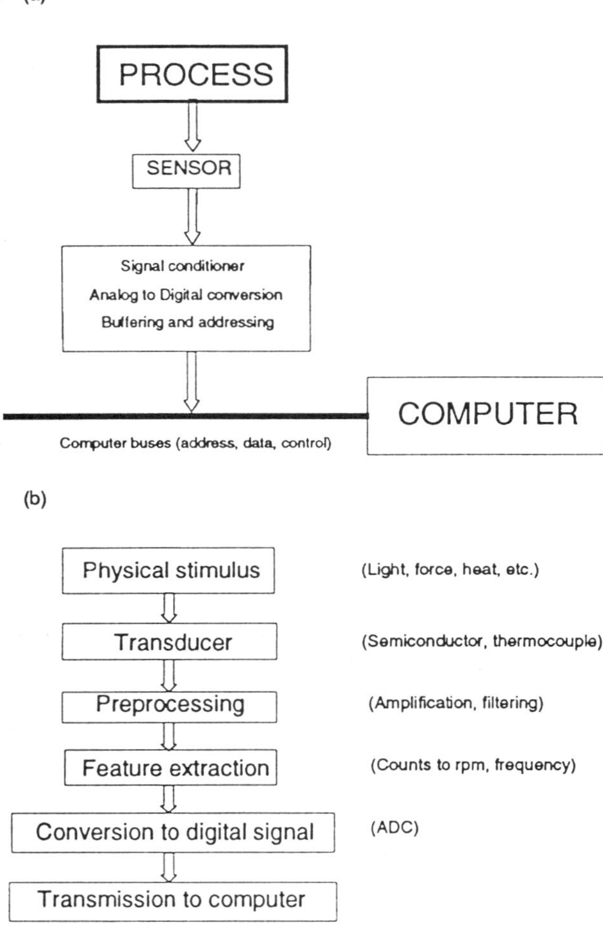

Figure 3.3 Two points of view of sensing procedure: (a) the computer-based view; and (b) the functional view.

The measurand, of course, is the quantity such as flow rate, velocity, or pressure which is being measured. The transduction principle is the scheme by which the measurand or one of its physical effects is converted to an electrical signal by exploiting special properties of an object such as a wire or a semiconductor.

Transduction principles in common use often measure the change in some electrical circuit property and from this infer the change in the measurand as follows.

1. **Resistive**   The resistance of a metal changes with its temperature, cross-sectional area, and length. One or more of these is employed in developing a great many transducers: for temperature using the thermal response, for displacement using length as in potentiometry, for force or pressure using strain gauges in which the wires are stretched, etc.
2. **Capacitive**   Capacitance may be changed by moving a capacitor's plates relative to each other or by changing the dielectric. One application uses a liquid dielectric, so that changes in liquid level yield changes in capacitance.
3. **Electromagnetic**   Relative motion between a magnet and a sensing or pick-up coil will produce an output voltage. One use is in tachometry.
4. **Inductive**   Similar to capacitive in philosophy, the inductive transducer uses the displacement of a coil's core relative to a coil to change its measured inductance.
5. **Photosensors**   The measurand of photoconductors is converted to a change in the resistance of a semiconductor material by a change in the amount of light incident upon that material. The common CdS cell used in photography is an example. In photovoltaic transducers, the impinging light produces a voltage proportional to the light intensity.
6. **Piezoelectric**   Certain crystals will, when mechanically stressed, produce a change in electrostatic charge which can be measured. This is useful for force and pressure measurement.
7. **Thermoelectric**   A voltage is produced (Seebeck effect) when a joining of dissimilar metals is heated relative to another such joining (the reference node). This is the property exploited in thermocouples.

3.2.2 General properties

For any sensor, both the static and dynamic properties will be important. The static properties include accuracy and precision, reproducibility, and the functional relationship between measurand and input. The character of the time response, including how long it takes for the device to settle to its new output value after a change in the measurand, constitutes the dynamic properties. We examine the properties by considering what might be in typical device specifications, as in Table 3.1.

Table 3.1 Typical specification block

| Model No. | TC-10 Temperature Sensor |
Type	Thermocouple K
Input signal	Temperature
Range of operation	−200°C to 1200°C
Output signal	4–20 mA or 0–10 V
Accuracy	±0.2% of span
Repeatability	±0.1%
Speed of response	Fast
Reliability	5 years w drift compensator
Size	Probe lengths: 10 cm–1 m
	Body: 12 cm dia. × 7 cm
Mounting	On tank wall
Environment	−40°C to 100°C
Power requirements	12–35 VDC
Special properties	Low cost
Guarantee	90 days
Design date	1987
Typical application	Chemical vat temperature

Shows typical (there are no standards) block as might be seen in manufacturer's brochure.

Let us consider the information given.

Model Number is the company's part identification for the figures quoted. **Type** is the nature of the sensor: thermocouple, strain gauge, accelerometer, etc.

The device's basic characteristices are defined in terms of its **input signal** (e.g. temperature, force, flow rate), **range of operation** (e.g. 0–500 °C, 0–5 kg, 0–50 l/min), and **output signal**. Common, near standard, options for the latter include analog signals such as 4–20 mA, 0–5 VDC, 0–10 VDC, and digital signals with connections satisfying RS–232C or IEEE 488 (HPIB) (Chapter 8).

Various operational accuracy and reliability characteristics are sometimes given. Often specified are **accuracy**, usually meaning percentage deviation from the true value and given as either percentage of measured value or of full range, with typical numbers the order of 0.1%. Particularly when the output is non-linearly related to the actual value, its **repeatability** may also be specified in a similar manner; thereby the possibility of calibration is presented.

The **precision**, often called **sensitivity**, defined as the minimum input change needed to register a change in output, and indications of **reliability** or of **lifetime** may also be given.

Physical characteristics of the sensor system such as **size** and **mounting** are of interest for most applications. Necessary **power supplies** should be indicated, including both character (e.g. pneumatic, electric) and amount (e.g. 20 psi, 10 watts at 24 VDC). **Environment** characteristics of interest are at least allowable temperature range and often include humidity, shock, vibration, and other properties.

The summary will usually make a brief claim about other characteristics of the device or its builder. **Special properties** may include the meeting of certain standards (e.g. MIL-SPEC in the US, other standards association rules), availability of support personnel, claims about cost effectiveness (actual costing is probably subject to quotation). **Guarantees** may be stated, and the **design age** may indicate something about the technology used. Finally, it is common to indicate **typical applications**, so that the purchaser may be able to gauge whether the sensor is suitable for the proposed use.

We elaborate on certain aspects of the instrument's performance which may be only alluded to in specifications. First, and obviously, the sensor output will be some function of the measurand (Fig. 3.4). It is usually most convenient if the relationship is linear over the values of interest, and closeness to linearity may be included in the specifications.

A non-linear relationship, or one which varies due to some disturbance (e.g. many devices are temperature sensitive), can usually be compensated by calculations in the computer, provided the relationships are reproducible, that is, that they are known, repeatable, and one-to-one to within some limits which may be specified, often as a percentage. Significant hysteresis or non-linearity of the device will lead to poor reproducibility. The above are static response properties. The other important part of the response characteristics is the dynamic, or time, response. The question here is: How rapidly does the sensor output take on a new value when the measurand changes? The typical model of the device response is that the output $y(t)$ to an input quantity $x(t)$ satisfies a differential equation of the form

$$\tau \frac{dy(t)}{dt} = -y(t) + bx(t) \tag{3.1}$$

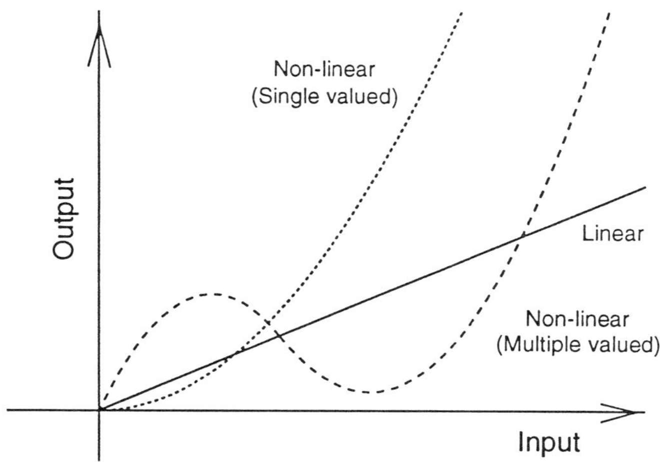

Figure 3.4 Input/output relationships possible with a sensor. Linear is preferred, but known non-linear relationships can be compensated in a computer. Multi-valued sensors are ambiguous unless special action is taken or input variable constraints are imposed.

Eventually, if $x(t) = X$, a constant, then $y(t) \rightarrow Y = bX$ and the output is proportional to the input. However, if the time constant τ is large, it may take a long time for this to happen, as the solution of the above is

$$y(t) = (y(0) - Y)\, e^{-(t/\tau)} + Y \tag{3.2}$$

and hence $y(t)$ will show an error of 37% of $(y(0) - Y)$ after a time τ. More complex dynamic response is possible, but will not be pursued here. We mention, however, that accelerometers are often mass–spring–damper mechanical systems and exhibit a response associated with a second order differential equation.

The importance of time lags such as that above depends, as does almost everything in systems, on the actual system under study. Five seconds can be a long delay in some systems, whereas 20 minutes is short in others. An inferred value, computed from a relationship of two measurements containing lags, can be seriously wrong even if both measurements are 'almost' correct, as can be seen from Fig. 3.5.

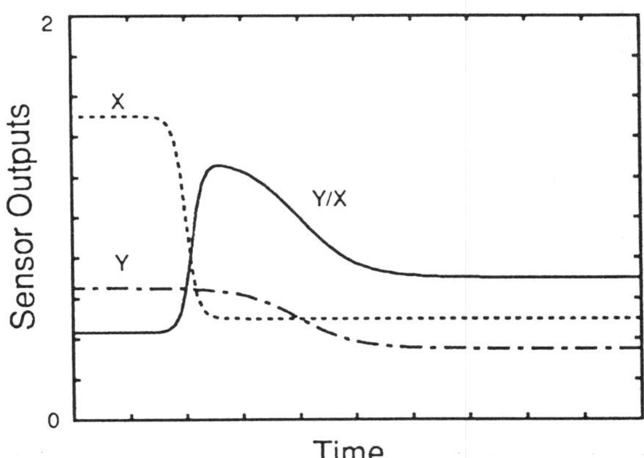

Figure 3.5 It is not infrequent that a ratio of sensed values is used in control. If the sensors have differing time responses, the ratio can be temporarily misleading.

3.3   COMMERCIAL SENSOR PACKAGES

Commercial sensors usually come packaged, including some signal conditioning, robust packaging and mounting, and perhaps with digital output. It is becoming more common for the sensor package to contain a microprocessor, and some manufacturers have programmed this to give an intelligent (or smart) sensor by incorporating unit conversion, linearization, communication capability, and sometimes even simple control laws.

3.3.1 Sensor examples

We present here a few examples of sensors, grouped essentially by measurand or application. The list is nowhere near exhaustive, and the information concerning the types is not uniform.

Example sensors

- On-off
- Temperature

- Thermocouples
  - Resistance thermometers
  - Quartz thermometers
  - Radiation pyrometers
- Displacement
  - Contacting sensors
  - Non-contacting sensors
  - Direct digital readout sensors
  - Synchro transformers
- Ranging devices (radar, etc.)
- Time
- Pressure and force
- Liquid level sensing
  - Float valves
  - Immersed sensors
  - Ultrasonic sensors
  - Differential pressure sensors
- Flow measurement devices
  - Orifice type devices
  - Turbine meters
  - Electromagnetic flowmeters
  - Positive displacement meters
- Tachometers – velocity and speed
  - DC tachogenerators
  - Counters – digital tachometry
- Accelerometers
- Attitude sensors – gyros
- Light sensors – basic methods and devices
- Others

Of these, the most common sensors in process control are probably those that measure temperature and flow. In aerospace applications, the most important are those that measure acceleration and attitude.

3.3.2 Temperature sensors

Temperature is the most commonly measured quantity in process control, and arguably is the most commonly measured of all variables, although it would seem that size or distance might be a candidate for the latter prize. The most common temperature sensors are thermocouples, resistance thermometers, quartz thermometers, and radiation pyrometers. Of these, the first three are contact-type sensors, in which the sensor is in contact with the substance whose temperature is being measured, and the fourth is non-contact.

Thermocouples are quite stable. They operate over wide temperature ranges, having electrical signals as output. Their advantages include their small size, ease and flexibility of mounting, and low cost for materials. Disadvantages include the need for a reference junction or simulation thereof, low DC output requiring amplification, and drift in calibration as the alloys gradually change composition. Typical packaging places the sensor in a long slender rod to be inserted into the substance measured and includes a protected head containing amplifiers and other signal conditioning electronics. Range and other characteristics vary with the choice of metals, as in Table 3.2.

Table 3.2 Thermocouples

Type	*Materials*	*Range* (°C)	*Sensitivity* µV/°C	*Accuracy*[1] (°C)
K	Chromel/Alumel	−200 – 1200	38.8	±0.7
J	Iron/constantan	0 – 760	52.6	"
T	Copper/constantan	−200 – 370	40.5	±0.5
R	Platinum/Pt	0 – 1450	12.0	

[1] Using polynomial over part of range

Figure 3.6 shows thermocouple connections, which are simply bondings of two different wires (actually two such are needed, the second as a reference called the cold junction), and a transducer mounted on a probe containing the sensor.

Other methods are also used.

1. Resistance thermometers depend upon the fact that the electrical resistance of materials varies (usually increases with metals) in a specific and reproducible way as temperature increases.
2. For the quartz thermometer, the conversion principle is that the device will change (slightly) in size with temperature changes; this is sensed indirectly, as a change in resonant frequency.
3. The radiation pyrometer is basically a simple optical system with a means, such as a thermocouple, for measuring the heat focused by its lenses. Since the temperature of the surfaces is inferred

(a)

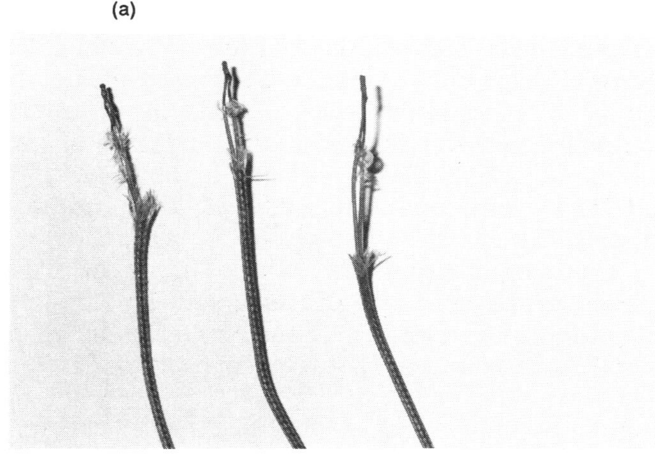

(b)

Figure 3.6 Temperature sensing: (a) thermocouple tips are only bonded wires; and (b) when mounted in probes and inserted in pipes, they look like many other sensors.

from the heat radiation they emit, the measurement is contactless and the device is not subject to wear through heat, corrosive action of the substance being measured, or frictional heat and damage from measuring moving surfaces (such as hot steel in a rolling mill). It is required, however, that the surface have an emissivity high enough to give reasonably accurate measurement.

Other possibilities include mercury and other thermometers and bi-metallic strips.

Depending upon the technology chosen, temperature can be measured with accuracy/reproducibility in the order of $\pm 0.05\,°C$ to $\pm 1\,°C$, and environmental ranges as subsets of $-200\,°C$ to $1500\,°C$. Time lag for an infra-red sensor may be of the order of $50\,ms$, while thermocouples may be of the order of seconds or, with cladding, minutes. Prices are of the order of $500 for thermocouples and resistance thermometers.

3.3.3 Displacement measurement

Displacement is measured either for its own sake or as part of the sequence of inferring other quantities such as pressure and acceleration. There are many variations on displacement sensing, although one typically measures either linear displacement or angular displacement. Even this split is not clear-cut, for it is quite possible to convert linear displacement to angular or vice-versa (e.g. using crankshafts and rods); a common instance of the former is the measurement of distance in an automobile by, in effect, counting wheel revolutions.

Most displacement transducers have a sensing shaft which is mechanically connected to the point or object whose position is to be sensed. This mechanical connection is then also attached to a fixed sensor mounting, and the position of the shaft relative to the mount is transduced. Simple transducers may have the shaft physically move capacitor plates or dielectrics, wiper arms on either potentiometers or rheostats, a core in a coil (inductor), etc. Slightly more elaborate are schemes in which the shaft moves a shutter between a light source and a photoconductive or photovoltaic cell or cells.

The basic principle of contacting displacement sensors – that a change in position will lead to a change in resistance (or capacitance or inductance or transformer characteristic) and that the resulting electrical characteristic can be measured and from it the displacement inferred – is straightforward. The variations of materials, sizes, and gearing (to change operating range) for different requirements of accuracy and environmental conditions seem endless. Resistance potentiometers, for example, can exhibit ranges in angular measurement from a fraction of a turn to 10 turns, and similar sized versions (say 5 cm in diameter and 3 cm long plus shaft) may have accuracies of 1% or 0.01%, with corresponding costs ranging from a few dollars to many tens of dollars. Linear variable differential transformers (LVDTs) may be miniatures accurate to $\pm 0.001\,\mu$m or large versions with a range of 15 cm.

Although it is common to have analog readouts followed by analog to digital converters (ADCs) for displacement measurement, the above methods can usually be adapted to give a direct digital readout of the displacement, either incrementally or absolutely. A common example is the absolute shaft-angle encoder, in which light sensors record the presence or absence of light through a slotted shaft-mounted disc, and the array of the on–off signals, converted either photovoltaically or photoconductively, together with knowledge of the position of the holes in the disc gives the shaft angle to within $360/2^n$ deg, where n is the number of photoreceptors.

A linear contact displacement transducer using LVDT technology might be of the order of 10 mm diameter by 50 mm long, with accuracy around ± 0.4 mm. Environmental constraints would limit frequency to less than 200 Hz and temperatures to 0–50 °C.

3.3.4 Ranging devices

These are devices such as radar and sonar. The principle here is to radiate pulses at ultrasonic, radio, or other frequency and measure the time delay until a reflection of that pulse returns. An ultrasonic ranger can be accurate to better than 0.01% full range, work in –5 °C to 65 °C, and cost the order of $2000.

3.3.5 Time measurement

Time is critical in many control applications. It is often not so much absolute time (day of the week, time of day) which is of importance, but relative time (from one valve opening to another, from one output command to the next input sample). Time can either be sensed from the external world or counted internally.

External sensing is usually based upon using a synchronous AC motor running from the electricity supply as provided by the local electrical power utility (at 120 V or 240 V and 50 Hz or 60 Hz) plus appropriate gearing.

Internal clocks are based upon the counting of oscillations of a resonating electronic circuit. A common oscillator is a simple quartz crystal plus amplifiers and voltage supplies. Simple circuitry can be used to count the oscillations and output pulses periodically for a variety of uses. Digital watches illustrate how small and inexpensive such devices can be.

3.3.6 Pressure and force sensors

Pressure sensing is done in two stages: conversion of pressure to displacement or force, and conversion of the displacement or force to an electrical signal. In a typical system, as installed in an industrial environment with temperature extremes, shock, vibration, occasional overload, corrosive agents, etc., accuracy may be expected to be of the order of 1%, although higher accuracies such as 0.1% are obtainable and 0.01% might be possible in principle.

A typical pressure or force sensor is a device accurate to ±0.25% of full range, 2 ms response time, based upon strain gauge technology, and costing several hundred dollars; part of the cost, as in much industrial equipment, is for ruggedness sufficient to cope with expected temperature extremes, shock, vibration, occasional overload, corrosive agents, etc.

A commercial differential pressure transducer is shown in Fig. 3.7. A pair of strain gauges, mounted so that the difference of their outputs represents the twist of a shaft, can be used for torque measurement and are shown in Fig. 3.8. Both operate by conversion to an electrical signal of the displacement of sensors mounted on a surface of known displacement vs. force characteristics.

3.3.7 Level sensors

Level sensing, for applications such as measuring the depth of liquid in a tank, is generally done by the mechanical conversion of level to a displacement, using for example a float in a liquid, followed by the measurement of the displacement by a displacement transducer mechanically connected to the float. When this is not possible, there are a variety of other methods available, including the following.

Figure 3.7 Differential pressure sensors measure the difference of two input pressures and have many applications. In this case, one of the inputs is at atmospheric pressure and the other is liquid at the bottom of a tank. This allows the inference of liquid depth in the tank.

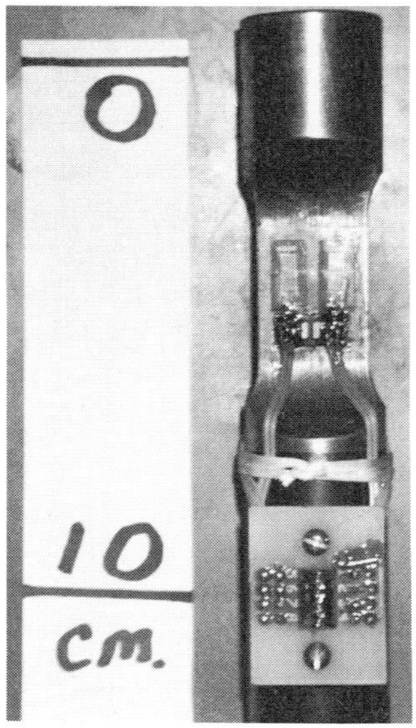

Figure 3.8 These strain gauges are mounted on a shaft to allow the measurement of shaft twist and the inference of torque.

- **Electrode methods** The principle is that the liquid is allowed to rise and fall between two electrodes. Depending upon the liquid involved, the accuracy needed, and range of measurement, the resistance, conductance, or inductance between the electrodes is measured and used to infer the level of the liquid (see Fig. 3.9).

Figure 3.9 The outside of a capacitive liquid depth sensor shows nothing of its action. A probe is inserted into the liquid, which is the capacitor's dielectric, and the capacitance is measured by simple circuits in the head shown. The resulting inferred depth is, in this case, transmitted to a PLC.

- **Ultrasonic sensor** This in essence measures the distance from a fixed mounting to the surface of the contents of a container of known depth.
- **Hydrostatic head methods** The pressure of the liquid at some point (say, the bottom) of the containing tank is measured using pressure measuring instruments. The depth of liquid above the mounting point is inferred from this pressure.
- **Thermal methods** The fact that heat transfer from a wire to a

liquid is different from the transfer to the vapour above the liquid is exploited.

3.3.8 Flow measurement devices

Flow measurement devices are widely used and appear in a variety of applications. Measurands include gases, liquids, and slurries. Flow measurement is sometimes claimed to be the most important parameter in a variety of industries dependent upon the transference of bulk liquids.

Flow may be measured either as a rate or a quantity (the latter being equal to the integral of the former). The various devices used are ultimately either inferential (i.e. the output implies but does not measure the flow directly) or direct (i.e. the quantity is the flow, as in piston-type meters below).

- **Orifice type devices** Here Bernoulli's principle is exploited to obtain a pressure drop through an orifice. This drop is measured using pressure sensors and the flow rate is inferred.
- **Turbine meters** These are based upon the principle that an impeller placed in a fluid flow will rotate at a rate proportional to the flow velocity of the fluid. Flow rate is then inferred from a measurement associated with the impeller rotation rate.
- **Electromagnetic flowmeters** These are particularly suitable for measuring flow of sludges, slurries, and electrically conducting liquids. In such flowmeters, a magnetic field is created perpendicular to the flow of the liquid. Since the liquid is then a conductor moving in a magnetic field, an elementary generator is created whose output voltage is proportional to the speed of motion of the fluid.
- **Doppler flowmeters** These provide a non-invasive flow measuring technique. They use an ultrasonic transmitter, an ultrasonic receiver, and a Doppler frequency conditioning unit, all mounted outside the pipe. The transmitter emits a continuous tone which is reflected back to the receiver by particles and discontinuities in the liquid stream.
- **Positive displacement meters** These actually measure all of the liquid flowing, instead of inferring it as in the above methods. The input fluid is trapped briefly in a container of fixed size which is then emptied into an output pipe or channel; all the fluid must pass through the container, so the system simply records the number of containers-full to find the flow.

Two small flow transducers are shown in Fig. 3.10.

(a)

(b)

Figure 3.10 Two industrial flowmeters: (a) an electromagnetic type; and (b) a rotating impeller type. Both are on 2.5 cm pipes.

One Coriolis effect mass flowmeter is accurate to around $\pm 0.2\%$ full range, with that range being selected from 0–1 to 0–9000 kg/min. For liquids, a volumetric flowmeter with range chosen from 0–100 to 0–17000 l/min at an accuracy of $\pm 1\%$ full scale is available.

3.3.9 Tachometry – speed measurement

Speed control, to wit, the Watts governor, constituted one of the earliest problems in machine control. Mechanical devices for governing or indicating speed are still common, but they are ultimately of limited accuracy – a few per cent at best. The measurement of speed of rotation of a revolving shaft is common today both for direct speed indication, e.g. of automobiles, and for sensing of quantities such as flow in turbine flowmeters. Two types of transducers are common: generators whose output voltage is proportional to speed of rotation of the shaft, and counters which respond as a point on the shaft passes a sensor.

- **DC tachogenerators** use the fact that the output voltage of a DC generator varies linearly with the rotation rate of the shaft.
- **Digital tachometers** use one of several possible methods to generate pulses as the shaft rotates and then count the pulses. The number of pulses per time interval divided by the number of pulses per revolution yields the rotation rate in revolutions per time interval.

3.3.10 Accelerometers

The basic sensing element of acceleration transducers is the 'seismic mass', a mass restrained by a spring and having a damper on its motion. When the transducer is accelerated along the allowed axis, the mass moves relative to the case. The displacement of the mass is measured by any one of several methods and from this the acceleration is inferred.

A somewhat different conversion approach is evidenced by strain-gauge and piezo-electric accelerometers, with the measurement being more directly of force rather than of displacement. The basic idea of the piezo-electric design is that the seismic mass compresses a piezo-electric crystal, while in strain-gauge transduction, the seismic mass is essentially suspended from the strain gauge.

A commercial accelerometer utilizing the piezo-electric effect can show accuracy of ±0.1% full scale for a scale chosen from ranges 0–2G to 0–200G. Such a device may be small (1 × 2 × 0.3 cm) and work in a temperature range −40°C to 85°C at a cost of a couple of hundred dollars. A more elaborate and sensitive device for seismic event sensing might trigger on as little as 1 mG, be around 300 × 300 × 100 mm, and cost around $1000; it would need special mounting to avoid triggering on, e.g. passing truck traffic.

3.3.11 Attitude sensing – gyros

The sensing of the angle an object makes with some reference falls in the realm of attitude sensing. For example, we use a compass to determine the direction of magnetic north and thus which direction we are facing. In fact, there are a number of references possible and hence a number of possible instrument types. Few of these are important in process control, but one is crucial in aerospace applications and this last, the gyro, is briefly reviewed here.

Inertial reference sensing is based upon the fact that a rotating body will continue turning about a fixed axis without change in rotation speed unless it is acted upon by an external torque. Thus a gimbaled rotating wheel will maintain its axis of rotation; by reading the gimbal angles, we may find the attitude of the gimbal frame relative to the axis of the wheel. This is the principle of the gyroscope, perfected by Sperry in about 1912–14.

The free (or two-degree-of-freedom) gyro is commonly used in attitude sensing. It supplies two angles whose interpretation depends upon knowledge of the gyro spin axis orientation. For example, if the spin axis is vertical, then the two axes supply pitch and roll attitude information; if the spin axis is horizontal, the angles supply yaw information and either roll or pitch or a combination, depending upon initial orientation.

3.3.12 Light measurement

Very useful non-contacting measurement devices are often based upon sensing of light radiation (including infra-red). There are a large variety of devices based upon each of several potential methods: photoemissive methods, photoconductive, photovoltaic, and phototransistor. Each of these has its advantages and disadvantages.

3.3.13 Others

Many other quantities can be sensed, but have not been described above. Material compositions can be determined or inferred; one common example is the measurement of pH in liquids using membrane and electrochemistry methods. Magnetic field strength measurement is essential in nuclear magnetic resonance (NMR) devices and has many uses. Mechanical properties such as hardness and density are often instrumented. Humidity is measured using several types of hygrometers. New in the industrial context, if not in medical laboratories, are sensors for biological variables. Some very interesting sensor studies are associated with robotics, where vision and tactile systems are receiving intense research (Fig. 3.11).

Figure 3.11 Advanced sensing: a CCD camera mounted on a robot arm and aimed at the manipulator.

3.4   COMPUTER TO SYSTEM INTERFACES – ADCs AND SIGNAL CONDITIONERS

The digital computer is restricted to use numbers represented in binary with two voltage levels, typically 0 and 5 V, to represent 0 and 1 (off and on) for each binary digit (bit) of the number. Very few sensors and actuators work directly in this binary system with these levels. Such sensors include shaft encoders; a common actuator is the stepper motor. For most sensors, the transduced output to the computer is a voltage or current (proportional to the variable sensed) lying in a continuous range over the instrument's full scale; similarly, actuators such as servomotors expect a voltage or current input, possibly a fairly powerful one. Under these circumstances, before computers can process the data, sensor outputs must be converted to binary using analog to digital converters (ADCs) while computer outputs must be converted back to continuous levels using digital to analog converters (DACs) before the data can be transferred to another device. Furthermore, it may be necessary to 'condition' the signals by changing their levels, converting from current signals to voltage signals (or vice-versa for the outputs) and possibly by filtering to remove noises or interferences from AC power lines and devices. In the following subsections we consider ADCs and some of the associated circuitry.

3.4.1 Analog to digital converters (ADCs)

There are several types of converters which take a single signal in the range 0–5 V (usually) and output several pulses in parallel, each either 0 or 5 V and each representing one bit in a numerical binary representation of the input level. Depending upon the approach used and details of its implementation, the representation usually has 8–16 bits, with 10 or 12 bits commonly found, and the conversion takes a few tens of nanoseconds up to tens of milliseconds.

The popular types of converters include successive approximation converters, flash converters, ramp and dual slope converters, and voltage-to-time and voltage-to-frequency converters. For our purposes, the last two are intermediate stages for the conversion to a binary word, although they may be directly used in some instrumentation applications.

3.4.2 Associated components

Although the actual converters are the dominant elements in ADC and DAC operations, practical systems usually have several other elements associated with them. We briefly review them here.

Sample and hold

Although some ADCs are sufficiently faster than the signals they must convert, so that the signal is essentially constant during the interval during which conversion occurs, often it is better to freeze the signal during conversion. This freeze not only prevents strange errors (such as could occur if the signal moved from $v_{ref}/2 + \varepsilon$ to $v_{ref}/2 - \varepsilon$ during successive approximation conversion), but allows precise definition of the time instant at which the sample applies. The operation of freezing the value, called 'sample-and-hold' and often implemented as 'track-and-hold', is implemented by straightforward electronic circuitry with the generic characteristics of tracking error, settling time, and holding time.

In operation, the units are typically integrated circuits (ICs) with solid state switching. The control signals come from the computer or from some other source, typically a clock.

Along with their virtue of freezing the signal for conversion, sample-and-hold units are also useful for ensuring simultaneity of sampling of several signals.

Multiplexers

An ADC may only be needed for conversion of a particular signal for a few milliseconds every 1–20 s. A way to share the converter and the addressing, buffering, etc., circuitry between it and the computer, is to use a multiplexer. This device is essentially a multi-input–single-output switch – the output line is time-shared by the incoming signals in a manner called time division multiplexing in communications systems. The multiplexer is usually controlled by the computer, which decides which input is fed through to the output at any given time; another possibility is a simple periodic time sequencing of the allocation.

It is quite possible to multiplex at other levels; it is not uncommon to multiplex raw thermocouple outputs prior to conditioning, amplification, and conversion.

Signal conditioners

Signal conditioners are used to provide amplification, filtering, linearity adjustment, and perhaps type conversion for the arriving signals. Thermocouples, for example, must be amplified from their level of millivolts to the 0–5 V range typically required by the ADC.

Current signals (usually 4–20 mA) must be converted to voltage signals (basically by passing them through a resistor). Linearity adjustment may be done either by the conditioner or the computer. Offsets such as the 4 mA in current signals are usually removed.

One important element of the signal conditioning is frequently the low-pass analog filter. Often called a **guard filter**, its role is to make certain the signal to the ADCs contains no frequencies higher than $(1/2T)$ Hz, where T secs is sampling period. An effect called **aliasing** could arise because it is impossible from the samples to distinguish a signal such as $\sin(\omega t)$ from $\sin(\omega t + 2\pi kt/T)$, where k = integer > 1. This problem may be avoided only by insuring that the signal received by the ADC contains no components at the higher frequencies.

3.4.3 Commercial signal conditioning systems

A user can always build up a signal conditioner/sample-and-hold/multiplexer/ADC and buffer/DAC systems, with the required addressing from the computer, use of control lines, etc. Cost for elements other than ADCs are typically in the $2–20 range per IC package. ADCs cost $5 up to $100s depending upon speed and number of bits; 8 bits at 40 μs conversion are among the low priced units.

The alternative is usually attractive: purchase a commercial data acquisition unit. These are modules, often immediately compatible with popular computers and 'hardened' besides, which perform all of the above tasks. They vary in size, capabilities, and cost, but with 8 analog I/O and 4 digital I/O a cost in the range $500–1500 for an IBM PC compatible unit could be expected. Related to these are the modules readily available for programmable logic controllers (PLCs), discussed in Chapters 5 and 8; a cost of $1000 for 8 channels of input or output, analog or digital, could be expected.

3.5 INTELLIGENT INSTRUMENTS

Sensors/transducers convert a physical change into an electrical signal; except for some modest conditioning, interpretation is left to the computer systems, and calibration may be left to manual means. Smart sensors do this, plus convert the signal into a directly useful form or even a decision. Such sensors are quite new and are undergoing rapid development. The key notion is that such devices combine a sensor/transducer with signal conditioners and a microprocessor in a single package.

The capabilities added by the microprocessor include the following:

1. linearization and conversion to engineering units;
2. compensation for environmental factors (usually temperature and sometimes pressure) through the use of auxiliary built-in sensors;
3. communication with other control or data logging system elements in a standard format;
4. sensor self-checking and diagnosis with appropriate indicating outputs;
5. decision making and (perhaps) control actuation; and
6. remote reprogrammability or parameter adjustment.

One typical application is in temperature sensing using a non-contact infra-red (IR) sensor. The reprogrammability can be used to change the parameters when the product is changed, thus allowing for the differing emissivity of the new components.

Correction for sensor non-linearity is obviously needed before display of the inferred variable, in this case temperature, and correction for ambient temperature of the sensor and electronics may be necessary. Conversion to standard communication format removes a burden from the supervisory computer, as does self-checking. Decision making, such as determining and sending commands (for heating or cooling, say), may not only remove a burden on the supervisor but lead to tighter control loops, to improved stand-alone capability for part of the system, and to a decrease in factory wiring requirements.

Smart sensors are available which determine presence or absence of objects through shape or colour recognition, object positioning through switches and proximity sensors, measurements such as thickness, and several other quantities. Important applications include inspection.

3.6 FURTHER READING

Discussion of instrumentation, actuators, and their interfaces appears in many textbooks; much of the above general overview is based upon Mansfield (1973). Extensive treatment of measurement, including devices and signal processing, is presented by Doebelin (1990).

A recent textbook with much practical engineering information is Derenzo (1990), and a more general process control text is Hunter, (1987). One book devoted to intelligent instruments is Barney (1988).

Recent texts which are helpful include Borer (1985). Those wishing to explore the electronics aspects might consider Jacob (1988) or Webb and Greshock (1990).

Definitive information on particular devices must come from the manufacturers and to a lesser extent from the trade magazines such as *Instrumentation and Control*, *Control Engineering* and *Automation and Control*. The latter are particularly helpful in keeping up with the rapidly changing aspects of applications, such as intelligent instrumentation.

4
Control elements, actuators, and displays

The reverse operation of computer data gathering is information output, particularly data display and control commands. We now look at the interface aspects of these – the transduction of the computer words to signals appropriate for operating valves, moving dials, running motors, etc. In this context, a control element is a device such as a valve, an actuator is a motor or solenoid which opens and closes the valve, and a display is an indicator on the operator's console.

4.1 SYNOPSIS

Because a computer works with binary words consisting of $0\,V$ and $5\,V$ bits, its outputs are rarely of direct use. A plant needs motors turned on and off, valves opened and closed, heaters adjusted, operator information displayed, etc. Hence the computer outputs must be processed.

The computer words must be transduced to other forms and are usually first conditioned. Thus the computer words are first amplified, perhaps converted to another electrical signal using modulation, and then used to control electromagnetic fields, heating of wires, lighting, and small motions, using various physical effects. These effects are either the final process control elements themselves, or are used to operate actuators which affect elements such as valves which adjust the plant variables.

The key ideas then are those associated with transduction of electrical signals to other physical signals and with the specification of control elements; the latter is the primary feature of this chapter.

An alternative to control output is the provision of information to operators using dials, flashing lights, and audio signals. The transduction here and the human factors characteristic of the displays are both of interest.

4.2   ACTUATORS AND TRANSDUCTION

There are few direct control actuators, i.e. devices which take the low voltage electrical signal and convert it to fluid flow rate, temperature, acceleration, etc. Rather, the signals often command valve openings, switch openings and closings, amplifier outputs, etc. The valves, switches, airplane flaps, etc., are the actual control elements, the devices which affect the process; many of them have local feedback to maintain a commanded opening, and thus the elements are themselves servomechanisms. When packaged as such, they are smart controllers and are analogous to smart sensors.

Among the few direct transductions of electricity to other physical quantities are:

1.   force or torque via EMF
2.   heat
3.   light
4.   displacement

These appear to be the principal quantities to which electricity may be transduced, but usually the low-power output of the digital computer is inadequate for them. Amplifiers and signal conversions, such as to frequency or analog values via digital to analog converters (DACs), are therefore required, and such signal conditioning becomes part of the computer output process.

The transduction principles from electric signals to motion, heat, etc., are applied either directly to control a process or themselves are used in controlled subsystems to command the control elements (as in use of a motor to open/close a valve). Five simple principles of transduction are given below.

4.2.1 Linear-acting devices

A very common device which converts an electric current to linear motion (and hence to sound) is the **audio loudspeaker**. The relay is similar in operation, but it is used for switching of electrical signals and is in effect a sort of binary amplifier or signal conditioner. An electromagnetic device, used for translation of an electrical command into a mechanical straight-line motion, ordinarily has a moving core which is connected to the moved system, but fixed core moving electromagnet devices are not uncommon.

4.2.2 Rotating devices

When the magnets and fields are arranged so that the force is applied about an axis so that a torque is obtained, we enter the class of devices leading to motors, where great variety is obtainable and the applications seem endless. Motors may be applied in such a way that the important output is the torque, speed, or angular displacement. With eccentric cams they can be used as counters, and with crankshafts a conversion to linear force or motion is possible.

4.2.3 Transduction via piezo-electricity

When the surface of a piezo-electric material is displaced, a small EMF is generated; alternatively, when a small voltage is applied to such a material, the material will expand. This property is difficult to use effectively, but it does have applications as a displacement transducer for precision alignment.

4.2.4 Heating

Electrical heating is the result of passing currents through conductors. Often an unwanted side-effect of the use of electrical components such as amplifiers (where the high temperatures can affect accuracy of operation and reliability), the effect is sometimes sought for heating of small amounts of liquids and gases. In such cases the conductor is chosen for its electricity to heat conversion properties. The heating is proportional to the applied electrical energy, so variable control is possible, but it often seems that on–off control such as in domestic water heaters is the most common form.

4.2.5 Light

As with heaters, special conductors will, when electricity is applied, yield an output energy of a different form, in this case visible light. An alternative to using incandescence is to use the gases, such as fluoron and neon, which will emit visible light when electricity is passed through them. Also, special materials struck by an electron beam will either emit light (as in cathode-ray oscilloscope devices) or change their light reflecting properties. In all of these, the light tends to be used in signalling and communications rather than as a direct control quantity.

4.3   CONTROL ELEMENTS AND ACTUATORS

High power requirements in many applications make direct use of electrically-driven actuators uneconomic if not nearly impossible. Large amounts of heat are best produced by combustion of biochemical fuels such as gases, petroleum, coal, or wood products (which are typically the source of the electricity in any case). Moving large valves, for many reasons, is often done hydraulically or pneumatically. In such cases the computer control system commands small electrical transducers which then become the inputs to large hydraulic, pneumatic, mechanical, etc., controllers. For example, a small stepper motor may position the valve of a hydraulic actuator for a very large door on a gravel bin feeding a crusher. With a balance of some kind in the valve, it itself becomes a control system.

In this section we look briefly at some common actuators or controller devices. The first of these are necessarily the amplification devices, since few devices other than display outputs are driven directly by the computer power supply.

4.3.1 Amplifiers

One of the simplest devices supplying a sort of binary amplification effect is the mechanical relay: a small voltage input may move the contacts engaging a high power signal. The relay is a binary amplifier – on or off – but is very common and inexpensive. Solenoids may be latched or in pairs, used basically for on–off applications such as valve opening–closing, or may be opposed by a spring so that the displacement is proportional to the applied electrical signal. In the latter instance, typical applications are again valve opening or closing, with the relaxed position of the electromagnet being typically 'valve closed' (for safety reasons). A small solenoid activated valve is shown in Fig. 4.1.

Ordinarily the signal to a solenoid, loudspeaker, relay, etc., must at least be in analog form and often it is amplified. Frequently it is even an AC (alternating current) signal.

Solenoids can be simple and inexpensive, being a few cubic centimetres in size and costing tens of dollars, but costs can vary

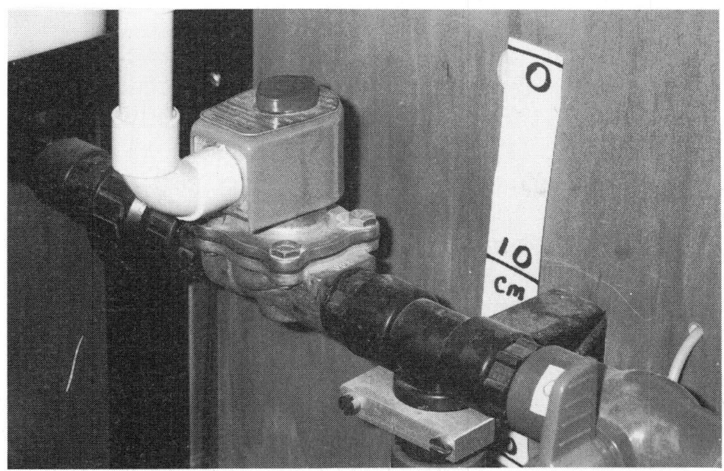

Figure 4.1 A simple solenoid on/off valve.

upwards with higher prices charged for environmental sealing (often hermetic), heavier contacts, and more robust devices. Many small on–off valves for liquids are solenoid actuated and, including the valve mechanism, cost the order of a hundred dollars. Actuation signals can be mains AC or various levels of DC: common input voltages are 24 or 40 VDC, but 12 VDC is becoming more common as computer control becomes more common, because 12 V is often readily available within the computer power supply.

Solid-state relays, essentially semiconductor switches, are also possible and are frequently used.

Amplifiers which take the DC analog outputs of digital to analog converters (DACs; see below) and give higher power DC outputs are also straightforward at lower power levels. Single chips are sufficient for the amplification necessary to drive loudspeakers, for example. The problem is that in all amplifiers, the input is essentially 'opening a gate' to let through an externally supplied higher power capability. If this must be DC, the supply of such a power source can itself be expensive. This is because the supply will ordinarily be from an AC source, which must be rectified and controlled to give the necessary direct current. For this reason, other means are used at higher power levels.

4.3.2 Motors

Motors are either AC or DC, with advantages to each; AC machines dominate when larger power requirements are needed, whereas DC motors are often used for positioning when lower power is demanded. A special case of the DC motors is the class of stepper motors, which have many poles and can be accurately moved one pole position at a time, in steps.

Motors have a great many applications and hence there are a great many sizes and styles of motors. We distinguish two types of tasks here: power suppliers and position controllers. This is rather arbitrary and is done mostly to give the flavour of the tasks involved.

In the class of **power suppliers** are motors applied for powering pumps, conveyor belts, etc. Also included are the large and important traction motors, used for example in railway locomotives. In applications, many of the former class are simply on–off devices for which it is only necessary to supply a steady power source when operation is desired. Although they may need special start-up procedures, these can often be applied locally with manufacturer supplied hardware. The control task is one of sequencing the start-up events.

Table 4.1 Actuator motor specification – typical

Type	*Electric, part-turn actuator motor*
Torque	300 nm
Travel	120°
Speed	30 s full arc
Operating period	Intermittent, 5% time, to 1200 c/h
Size	50 × 80 × 70 cm; 50 kg
Power requirement	220 V 50 Hz AC
Environment	−20°C to 60 °C
Typical application	Valve operation
Extras available	Shaft position feedback transducers
Special features	
Price	$2000
Warranty	

The second class typically requires speed control in addition to start-up and similar sequencing. The precise application of power to the traction motors of a locomotive is critical to smooth and efficient operation of railways. The use of variable speed pumps, as opposed to constant speed pumps with throttled outputs, is an energy saver in the process control industry. Electric pumps of small size are shown in Fig. 4.2.

The **position control** motors are used in a number of applications in which rotary motion, or a geared derivation of it, is to be precisely controlled. These include small applications such as pens on x–y plotters, larger ones such as joints on robots, and still larger ones for valve opening and closing. The motors in these applications are somewhat specialized and fall into two general classes: servomotors and stepper motors. Each is a design problem to create, and a control problem to drive properly. A servomotor with most of its control electronics is shown in Fig. 4.3.

At its heart, a **stepper motor** is a many-pole DC motor, often with a permanent magnet in the small sizes. Careful selection of the windings to be energized means that a more or less precise alignment of the shaft may be obtained; switching to another set then leads to motion to a new configuration. With appropriate interfacing, the device can be used in a motion control system. They have several advantages:

1. they give the appearance of being digital in nature and are therefore easily interfaced with digital controllers;
2. since positioning is accurately controllable, sequences of positions can be carefully timed to lead to precise control of velocity and acceleration profiles without the need of encoders, tachometers, etc.; and
3. they are fundamentally simple and rugged, so reliability and long life come naturally to them.

Stepper motors are common in computer disc drives and are often used in small robot applications; larger robots use electric servomechanisms for joint movements, while the largest may have pneumatic or hydraulic elements.

Problems with the devices lie with their basically underdamped stepping characteristic, but manifest themselves at the possibility of getting out of step with the commands and hence no longer having the shaft angle the controller assumes they have.

(a)

(b)

Figure 4.2 Electric powered pumps: (a) a small centrifugal pump; and (b) a modest motor driving a positive displacement pump. The latter is needed because the pumped quantity is a slurry, which is too thick for a simpler pump.

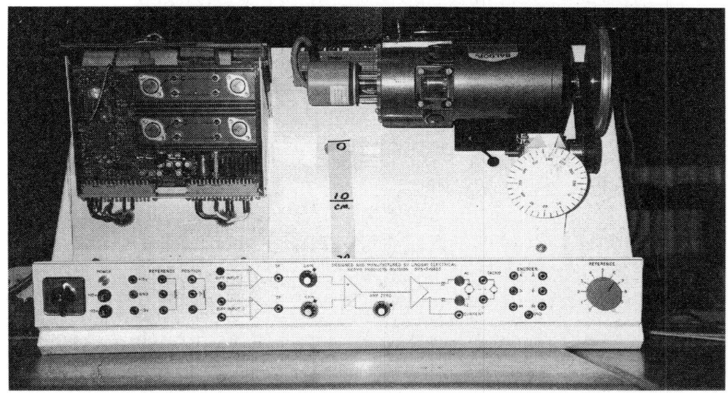

Figure 4.3 A servomotor. Mounted on the motor is an electric tachogenerator for speed sensing, while the belt drive is attached to a potentiometer for shaft angle sensing. The necessary electronics for the control of the servo are mounted next to the motor.

Servomotors are intrinsically AC or DC electric motors specially designed and controlled for applications in which shaft position or velocity and the attainment of specified values of such are of importance. Thus it is really the system, rather than the isolated motor, which is important to the control system designer (unless the task is to design a servomotor system).

The advantages of the servomotor systems over stepper motors are typically those of degree. Servomotor systems are available, using either AC or DC motors (and hence requiring DC or AC power supplies), capable of the order of up to 8 kW of power and 50 Nm of torque in the range of a few thousand rpm rotation speed. Some can accelerate quite rapidly if necessary, e.g. at up to 3000 rad/s^2. Special circuitry can even make them appear as stepper motors with tens of thousands of counts per revolution. For small precisely controlled positioning applications, manufacturers are making these a viable alternative to stepper motors.

Electric motors vary widely in price, depending upon power and type; large ones require auxiliary circuitry for starting, which adds to the price. The servomotor in the figure was configured for educational use at a cost of a few thousand dollars complete, whereas quite small servos may cost a few tens of dollars.

4.3.3 Heater systems

Heating can be controlled in two ways, depending upon the heat source. Electric heaters are directly controlled, whereas heat exchangers are controlled using valves, motors, etc.

The simplest heaters are probably the small resistive coils operating directly from the AC power supply and controlled by an on–off switch. These are useful for small applications (such as home hot-water heaters) in powers up to several kW.

Large scale heating in industry seems to use steam or hot water, piped around the plant as necessary. The heat supply to a particular location is then controlled by a valve. We remark that the production of the steam is itself a control system problem, with a boiler to be regulated and water and fuel to be supplied to it. In fact, boiler control is one of the major control system problems in electric power stations.

4.3.4 Coupled transducers and control elements

Some final control elements are electrically controlled in a way that closely couples the electrical transduction to the element. This is particularly true in force-balance arrangements, in which typically a small solenoid is leveraged to balance and, through mechanical linkages, control a power source. We elaborate somewhat upon this.

Many actuators, particularly in process control industries for valve control applications, are pneumatic or hydraulic ones. This is partly 'traditional' now, in that electric actuation was once more expensive, less safe (because of sparking, for example), or less capable than now is the case, and partly because such devices can still be quicker and more powerful in certain applications because of the storage of energy involved in having high pressure air or liquid on-line.

The use of an electrical signal to control a pneumatic (or hydraulic) signal can be done from a low-power electrical actuator using either a force-balance or motion-balance technique, in a manner which essentially reverses the transduction problem of pressure sensing. Thus an external (i.e. computer commanded) force can be generated as the external set point using a motor or a solenoid. This force must be balanced by the measured force, which is related to the actuation force as a constricted value or leveraged value of the on-line force. Such a scheme is utilized in the pneumatically powered, electrically controlled valve in Fig. 4.4(b), and might be compared to the directly motor controlled valve of Fig. 4.4(a).

(a)

(b)

Figure 4.4 Electrically operated valves: (a) a small electric motor is geared directly to a valve shaft; and (b) an electric motor is leveraged against the motion of a pneumatically powered valve in a force-balance configuration.

In such cases, the control actuator is itself a servomechanism. Its design is mechanical in nature and is beyond the scope of this book, even though the goals (such as rapid and accurate response) are the same as those for any control system.

4.4   DISPLAYS

Some of the computer gathered data will be recorded for archival purposes, and some will have an important indirect control role in informing operators of system status. This latter in particular means that the computer outputs will be displayed on dials (using perhaps a electromagnet to deflect a needle), with command display lights and alarms, etc. These, while in principle straightforward, are always worth some thought because of the human factors involved: the ability of operators to use the information conveniently and properly.

Underlying all thinking about the problem are the following 'rules':

1. the goal is to present essential information clearly and unambiguously, and in emergencies to prompt clear-cut sequences of actions; and
2. process operators are likely to be neither engineers nor computer programmers, and they should not be expected to respond as such.

Visual displays of information may be presented with numbers, dials, coloured shapes, flashing lights, etc. This may be done either with discrete elements (such as individual lights or dials) or on computer terminals with special displays created using graphics techniques. We comment on some of the factors.

First, there is the matter of quantitative vs. qualitative displays. Here the choice is basically between displaying numerical values and displaying attention getters such as flashing lights which indicate that, e.g. a preset value has been exceeded. In fact, a single display unit may have both, and in addition use an analog presentation of the quantitative data in which the scale is vertical and an indicating pointer is toward the top of the scale for 'high' values, etc. (Fig. 4.5).

For quantitative displays, there is also the choice of scale shapes, pointer styles, moving scale vs. moving pointer, full scale vs. windowed scale, and others. There is now also the choice of a purely digital readout: the choice between an analog clock with hands and a digital clock with numbers is one paradigm; analog moving needle speedometers vs digital readouts is another. Here a number of studies have been performed, and some general rules of thumb are summarized by McCormick and Sanders (1983):

1. for qualitative readings, use fixed scale/moving pointer displays, as operators use the pointer to derive both trend and rate information; and

Figure 4.5 For output display, meters directly on the hardware, in this case a PID controller, are common.

2. warning is best done with flashing lights, shutters, or other clearly visible signals.

Audible signals seem like a good idea, but one must remember that they are not always reliable in stress situations: many airplanes have landed gear up with warning horns blaring. In spite of this caveat, it is known that audible signals can be useful. Two things to remember are:

1. use only a few different sounds – this means <5 frequency levels, <4 intensity levels, and <3 durations of beeps, buzzes, etc., and
2. choose the sound direction, with sound coming from a source near instruments to be read or equipment to be operated.

It may be remarked that the computer can use VDUs to present very elaborate displays, with colour block diagrams of the processes, flashing lights, and large blocks of text; one of the simpler examples is shown in Fig. 4.6. It is not clear that this approach is always helpful in a difficult situation, as operator information overload has been blamed for poor system response in failure situations.

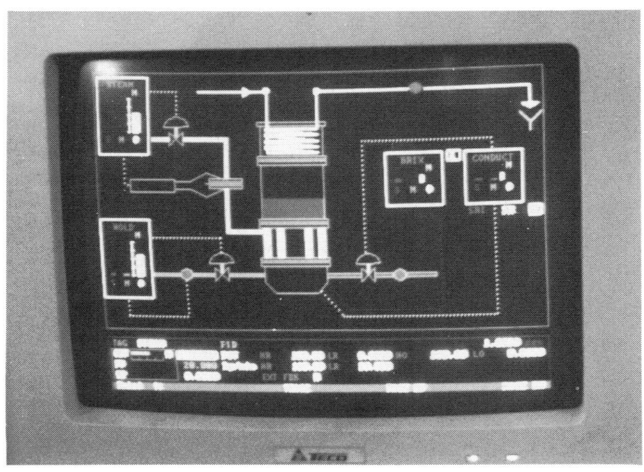

Figure 4.6 Very elaborate displays, of which a bank of the form of Fig. 4.5 are only a simple special case, are possible using computer VDU screens.

4.5   DIGITAL TO ANALOG CONVERTERS AND SIGNAL CONDITIONERS

The basic concept of the semiconductor digital to analog converter (DAC) is that switches are used in parallel to gate a reference voltage through precision resistors with values $R, 2R, ..., 2^{n-1}R$ to yield currents $I, I/2, ..., I/2^{n-1}$ which are summed. The summation is done at the virtual ground point of a feedback amplifier to provide an analog voltage output proportional to the binary word represented by the gated currents.

The switches are actually transistor switches in an integrated circuit (IC) and they are controlled (opened and closed) by parallel inputs from the computer or a computer buffer (see below). The range of resistances needed may be reduced using special methods, such as binary ladder networks. Voltage output DACs with 8–12 bit inputs have settling times to reach constant voltage output of 1–20 microseconds. Current output DACs can be faster.

4.5.1 Output buffers

The nature of the DAC is to convert the signal represented by the switch settings at a given instant. Many of the converters do not latch these settings, and hence without special interfacing – i.e. a latch on the computer output – the DAC output may drift or be undefined except at the instant of transmitting the value from the CPU. The usual way around this is to latch that value until a new one arrives; since the latch buffer is then piecewise constant in contents, so is the output of the DAC to which it is connected. In mathematical models of the computer and its outputs, the transmission signal or strobe may be interpreted as an impulse and the buffer/DAC as a sample-and-hold, or zero order hold (ZOH). (See Chapter 12 for use of these models.)

4.5.2 Signal conditioning

Output signal conditioning may also prove necessary. Typical requirements are to convert the low-power 0–5 V DAC output to a more powerful signal, perhaps even an AC command signal to a servomotor, or to a current 4–20 mA signal. Some of this has already been considered.

4.6   EXAMPLES OF COSTS

We give only a few examples here of small equipment. A small water pump and motor costs a few hundred dollars and is around 20 cm diameter × 30 cm long.

A small valve with electrical motor for servo control might cost several hundred dollars for a unit with 25 mm pipe and 2 min response time (full open to full closed). A modest servomotor with associated electronics and sensors can cost a few thousand dollars.

We saw digital–analog interface subsystems in section 3.4.3. A DAC itself (8 bits, slow) may cost only a few dollars, and associated amplifiers a few dollars more for the IC circuits.

Gauges for displays depend partly upon the ruggedness desired, but needles on dials can be had for tens of dollars. This may not be an issue itself – many sensors and some process controllers will have their own displays – but the human factor aspects may well influence choice of the latter.

The modern alternative of computer VDU displays in place of individual dials and gauges will cost a few thousand dollars for the hardware, but the software can be expensive. One is likely to puchase these in a system rather than as individual elements, except for large or highly specialized operations.

4.7   FURTHER READING

Discussion of instrumentation, actuators, and their interfaces appears in many textbooks. Definitive information on particular devices must come from the manufacturers and to a lesser extent from the trade magazines such as *Instrumentation and Control* and *Control Engineering.* The process control texts are often helpful within their own field of expertise. One such is Hunter (1987).

Electro-hydraulic, electro-pneumatic, and electro-mechanical devices are, except for the simplest types, not commonly described in elementary textbooks, although such devices are often themselves servo-mechanisms needing some control-theoretic study. Of interest sometimes are all-pneumatic or -hydraulic systems such as those described by McCloy and Martin (1980).

Human factors are interesting and important. One text is McCormick and Sanders (1983).

We have not touched on the final control element here, but should observe that one of the very important classes of such elements is that of valves for liquid and gas flow control. Valves and their sizing may be found in books such as Borer (1985).

5

Computer systems hardware

The connecting hardware element between sensors and actuators is the computer system, with connections being performed using various communications strategies (Chapter 7). In this chapter some of the essential aspects of computers in a real-time environment are introduced.

5.1 SYNOPSIS

A very broad overview has the control computer communicating with the plant, with its peripheral devices such as memory and operators, and possibly with other computers as in Fig. 5.1.

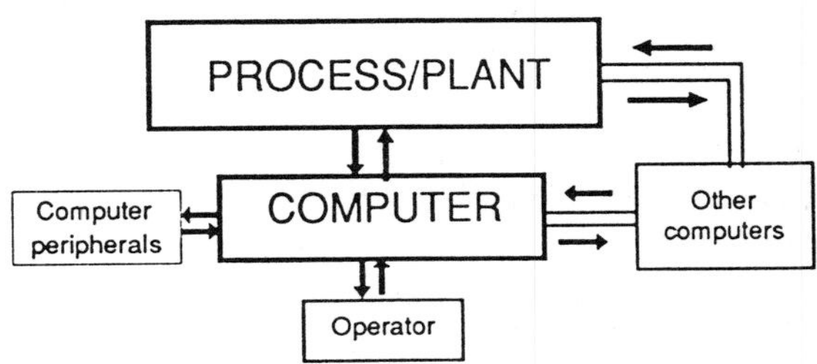

Figure 5.1 Conceptual block diagram of the connection of a computer to the process, its operator displays and keyboard, its other peripherals, and to other computers.

The computer system consists of hardware such as

- central processing unit (CPU)
- input devices: keyboards, mouse
- output devices: printers, visual display unit (VDU)

- memory: read-write memory, read-only memory (ROM)
- mass storage: magnetic discs, magnetic tape
- interfaces: communications devices (to local area networks (LANs), modems to use telephone lines, special data carriers)

plus software (Chapter 6).

A closer look at the computer system shows the 'brains' of the system, the central processing unit (CPU), using sets of wires called buses to deal with all other aspects of the system and with devices external to the system. These buses are grouped into a data bus, an address bus, and control lines as shown in Fig. 5.2.

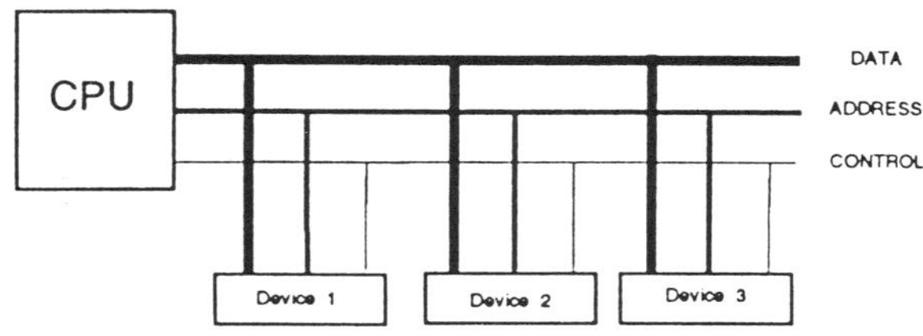

Figure 5.2 The computer is bus orientated, with the CPU connected to other computer devices using (typically) 8–16 parallel address lines, 8–16 data lines, and several control (signalling) lines.

The CPU is the central device and in the case of microprocessors is a single semiconductor chip. It typically has a functional breakup of its circuits as in Fig. 5.3.

The suitability of a computer for control tends to depend upon its ability to handle I/O to various devices such as ADCs, DACs, and operator displays, to react quickly to external unscheduled events, and to meet speed and arithmetic word length requirements. These typically require an adequate structure for addressing external devices and exchanging data with them, a special hardware interrupt capability, and at least 8-bit words and instruction cycles of the order of no more than several microseconds.

Commercial computer-based systems for control range from programmable logic controllers (PLCs) and process controllers (PID controllers) costing several hundred to a couple of thousand dollars to systems with base prices of several thousand dollars. Major installations may use mini-computers valued at many tens of thousands of dollars each.

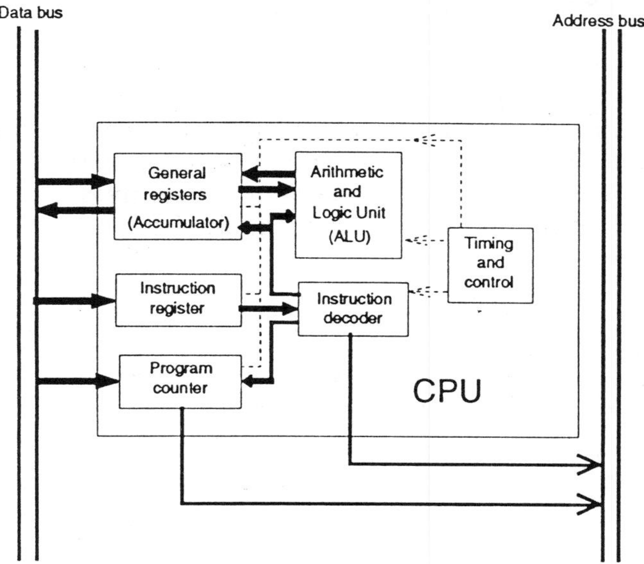

Figure 5.3 The functional composition of the CPU shows several blocks of registers and logical units.

5.2   THE GENERIC PROCESSOR ELEMENT

Discussion of the computer system starts with a single central processing unit (CPU) and appropriate additional elements. Figure 5.3 shows the essential elements of a generic CPU.

The CPU is the core element of the computer, characterized by its speed, its technology (e.g. CMOS), the extent and power of its instruction set, the amount of memory it can easily address (the number of bits in the address bus) and the number of bits in data words (width of data bus). For control systems applications, features such as number of I/O ports and number and hardware processing of interrupts may also be important.

Some CPUs come in single chip configurations, whereas others may require auxiliary hardware. A common example of the latter is the arithmetic co-processor to perform multiplication and other arithmetic operations in hardware; the same functions can usually be performed in software by the CPU chip alone, but at a considerably slower rate.

The CPU hardware operation is basically a simple repetitive cycle, comprising seven steps as follows.

1. Using the contents of the program counter (PC) as an address, **fetch** an instruction from memory, i.e. 'place the address on the address lines and, using the control lines for synchronization, obtain on the data lines the contents of the memory at the addressed location'.
2. **Route** the fetched data into the CPU instruction register (IR), i.e. 'use logic settings within the CPU to allow the IR to become the same as the data lines in step 1'.
3. **Increment** the PC by 1 location (so that the next instruction is fetched from the next location after the current instruction).
4. Interpret (or **decode**) the contents of the IR in the instruction decoder, i.e. 'set logic gates to accept various portions of the IR contents and hence enable other logical elements (e.g. adders)'.
5. **Execute** the decoded instruction (which may involve addition, fetching data, modifying one of the registers, etc., depending upon the instruction) i.e. 'execute the logic paths set up in the interpretation of the IR'.
6. If a special control line (**interrupt**) is set, change the PC to an address preset in the hardware, i.e. 'automatically and in the hardware, override the usual sequencing to start a special sequence if a particular control input line has a signal on it'.
7. Go to step 1.

The CPU **cycle time** (the time for one such sequence of seven steps) is perhaps a fraction of a microsecond or so, with each separate step taking some number of nanoseconds. Step 5 (executing the decoded instruction) is typically the longest and most variable (depending upon the exact instruction) portion of the cycle. Note that instruction execution is sequential; although instructions may be partially overlapped, multiple instructions are not executed simultaneously (in parallel) in most control applications.

It is worth emphasizing the communication of an external device with the CPU. When an external device requires attention, it signals using a control line. If checking this control line is an ordinary CPU instruction, and hence done only when the CPU is ready, then the line is called a **sense line**. If the line is automatically checked by the CPU hardware *every* cycle, and leads to a special reaction by the CPU, then the line is an **interrupt line** to which the interrupt response is in essence a forced subroutine call to a location fixed by the hardware. Because the interrupt response may utilize various registers of the CPU (the PC must at least be used to call the interrupt routine), the relevant registers must be saved upon entering the routine and restored to their pre-interrupt values before returning to the main program. The sense line approach allows deferment of recognition of the signal and hence allows delays in reacting to it, while use of interrupt requires at least minimal action immediately.   Hence interrupts are preferable for rapid response, but because of their unpredictability in time they must be carefully used.

5.3   COMPUTER SYSTEM

The CPU is placed with other components to form a system. A rather minimal computer (the CPU alone is just a processor) will have memory and some input and output (I/O) capability as indicated in Fig. 5.4.

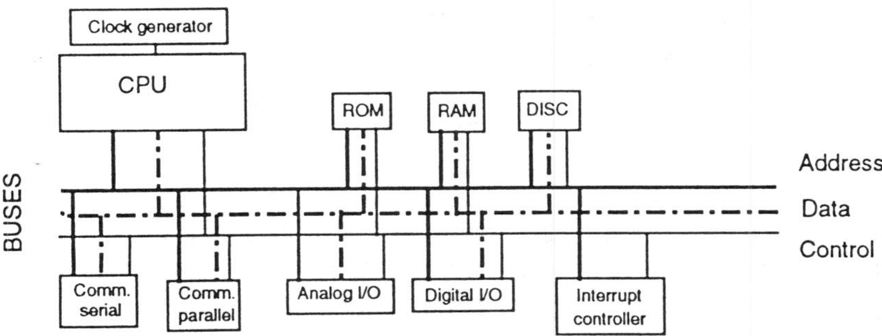

Figure 5.4 A CPU as in Fig. 5.3 is connected to buses and thereby to the interfacing units, as well as to the computer's memory, which are attached to the buses.

Some common elements of a computer for control applications include:

1. random access read/write memory (RAM) for data storage;
2. read only memory (ROM) for the program, including control algorithms;
3. digital I/O interface for reading or commanding on/off type settings;
4. analog I/O interface for reading sensors or commanding, e.g. valve settings or motor speeds;
5. interrupt controller for handling the receipt of interrupts for emergencies, plant errors, system real-time clock, etc.;
6. serial communication interface for communications with printers for data logging, some types of instruments and actuators (those using RS–232C, see Chapter 7); and
7. clock generator, for the computer's internal timing of instructions (but not the same as the real-time clock).

General purpose microprocessor chips include the 8-bit Intel 8080, the 16-bit Intel 80x86 series, and the 32-bit Motorola 16000 series. For special applications, the RAM or ROM or other elements are sometimes placed on the same integrated circuit (IC) chip as the CPU, creating a microcomputer. This can be particularly useful in control applications because it lowers parts counts and can be hoped to improve reliability. One such microcomputer is the Intel 8096 (a special relative of the ubiquitous 8088/8086 family used in many personal computers).

5.3.1 Components of the system – specialized for control

The computer control system components are in one sense little different from standard ones: RAM, ROM, bulk memory (discs and tapes), operator interfaces (keyboards and VDUs), printers. However, it becomes immediately apparent that there are differences.

1. **Hardening** The system may be mounted in metal rather than plastic cases and have special cooling fans. Power supplies may also be able to stand occasional fluctuations. If the computer environment is to be dirty, vibrate, or have other problems, the system mounting must protect it.
2. **Standard sizing** The equipment will fit standard industrial equipment racks, typically 19" (0.5 m) wide.
3. **Special input/output** Special I/O cards deal with instruments

and actuators of standard levels. Thus input of 4–20 mA, thermocouples of various types, and outputs of 0–24 VDC will be readily available, with power supplies, rack mounting, etc.

4. **Test hardware**, such as LCD terminals for readouts of programs, will be available.
5. **Instrument buses** of standard types will be catered for. This will probably include IEEE 488 (GPIB or HPIB; see Chapter 7).

Special CPU characteristics which may be important include the following.

1. **Interrupts** Number and handling of interrupts may be important for higher level controllers, but nearly irrelevant for smart instruments.
2. **Special instructions or I/O capabilities** This is becoming less important as CPUs become more powerful in general. I/O ports (essentially special control lines for I/O) can be important for efficient memory use, but control applications will seldom have extensive memory requirements; the effects and convenience of ports are sometimes provided in other ways using memory-mapped interfaces.
3. **Number and nature of special registers** Having a large number of registers for intermediate data handling can be both a speed and convenience advantage.

5.3.2 How the CPU maintains supervisory control

As shown in Figs 5.2 and 5.4, there are three sets of wires to which all of the devices – CPU, memory, clock, etc. – are attached. To avoid a Tower of Babel, the CPU (or occasionally a designated other device) supervises the traffic on these lines.

1. **Control bus** This is used for signalling. One of these lines may, for example, carry a short pulse (called a **strobe** signal) to indicate that the device addressed by the address lines should accept the data on the data lines 'now'.
2. **Address bus** Each device has a unique address, expressed as a binary number. Signals on the address lines are set to a device address to show that the information on the data lines is intended for that device.
3. **Data bus** This carries the actual messages between devices. The simplest configuration has all messages routed to and from

the CPU.   The information carried may be numbers from memory (such as algorithm parameters) or from a sensor (such as a flow meter) or to an actuator (such as a valve).   They may also be characters to the console display, commands to devices (such as a reset command to the clock), and many others.

The control lines are the primary means of signalling from other devices to the CPU, and may operate in those situations in one of two different modes: as interrupts and as sense lines.   The distinction is in the CPU's response mechanism, as in both cases the external signaller places a signal on an appropriate control wire.   Sense lines are checked by the CPU when there is a software instruction to do so; hence, in writing the code for the algorithm one could include a wait loop to check for the clock signal.   If there is no such check, the clock's information will be unknown to the CPU in this mode.

The alternative is to use the CPU's hardware interrupt capability. This is somewhat like the sense lines but with a crucial distinction: the line is checked automatically by the hardware after each instruction executed.   If the interrupt line has been set – in our case if the clock device has set the line – the hardware will automatically cause program execution to transfer to a designated location in the program. This location, or a pointer to it, is built into the hardware; from that location, further aspects of the interrupt response are specified by user-written software.   Other aspects of the response, such as saving of registers, may be done either automatically by the hardware or optionally by user-specified software, depending upon the CPU architecture.

Although interrupt handling is built into the hardware, it has very important ramifications for the system and the software used, and we return to some of those in Chapter 6.   The underlying problem is that a interrupt can cause a program transfer at any time (provided interrupt inhibit commands have not been given), and hence it can lead to the possibility of performing a particular instruction sequence which has not been tested (since not all such sequences will have been tested).   Interrupts are used when fast response to an external event is required or when very rare events must be handled.

Interrupts come in many varieties: priority interrupts, software interrupts, and internally generated interrupts are among the possibilities.   Responses also vary in detail: for example, sometimes the hardware will automatically save all registers and inhibit further interrupts prior to actually performing the interrupt-demanded program transfer.

5.3.3 Device interfaces – the core ideas

There are two possibilities for message I/O: direct ports and memory-mapped I/O. In the former, data placed on the data lines are addressed using the address lines and have special control line signals (such as IN and OUT) with their own special CPU instructions. Advantages include speed and precision of addressing. It is possible, if enough IN and OUT commands are available, to avoid address decoding and so simplify the interface. Ports may have special CPU instructions associated with them.

The alternative situation is called **memory-mapped I/O** and has the I/O device addressed just as memory is. This can be wasteful of allowable locations. Also, there is a burden on the interfacing so that the I/O connection indeed looks like memory to the CPU. On the other hand, the number of I/O devices is virtually unrestricted (up to the limit of the number of memory cells addressable by the CPU). Sometimes memory-mapped I/O can be used for special interfaces to give the latter the appearance of ports; the cost is the usage of memory addresses for things other than data and program memory, but the gain is that the CPU and its instruction set are simpler.

A typical output connection configuration is shown in Fig. 5.5.

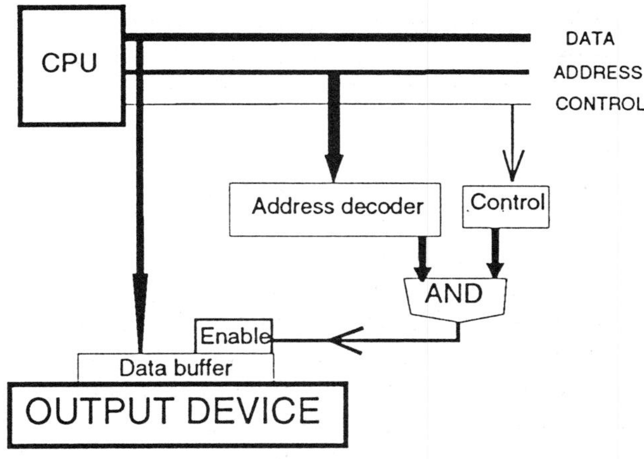

Figure 5.5 The CPU outputs a data word to a device by placing the device's address on the address lines, the data on the data lines, and a WRITE command on the control lines. The device gates the data to its buffer (i.e. enables the buffer to accept the data) if the address is correct and the proper control line settings are present.

If the address desired equals the device address and if the control line carries the proper signal (such as WRITE DATA LINES command), then the AND gate will admit the data line information to the data buffer, where it will be available for conditioning, conversion to analog levels, etc., as required.

Data input is similar. Here the input device must request attention using control lines (interrupts or sense lines) or by replying affirmatively when polled, must make its identity known (using its address), and on command gate its data from the buffer onto the data lines.

In both cases, the CPU must have all of the lines in a proper state for the transfer: addresses on the address lines, proper control signals (such as timing strobes), and data lines carrying data or ready to receive data.

Typical electronic devices needed for these and other interfaces are given below.

1. **Address decoder** This chip 'knows' its own address and puts out an indication to attached devices whenever the address lines from the CPU contain this address.
2. **Buffer** This will, on command from its associated address decoder and the CPU control lines, interact with the data lines to the CPU. It will read the data lines or hold data for reading as appropriate to the commands and to the function of its attached peripheral device.
3. **Serial/parallel converters** Data on the computer system bus is bit-parallel (all of the bits of a byte, or word, are carried simultaneously on separate wires). Electronic devices consisting of a register plus a clock do the conversion from one to the other.
4. **Controllable interfaces** These ICs are of a general type for signalling but have many varieties called variously PIAs (peripheral interface adapters), UARTs (universal asynchronous receiver–transmitters) and ACIAs (asynchronous communication interface adapters), etc. The latter two incorporate serial/parallel conversion.

5.4 COMMERCIAL SYSTEMS

It is possible to buy digital-computer-based controllers 'off-the-shelf', with only tuning or other job specific parameter setting to be done by the user. It is instructive to look at some of these, particularly the two most common types of elementary systems: the PID (proportional-integral-derivative) process controller and the PLC. Following this we consider systems which are more obviously computer systems; the above are digital implementations of analog and relay systems, respectively.

5.4.1 Microcontroller

The microcontroller is a special computer system for special applications. It is special because virtually all of its requirements for CPU, RAM, and ROM are on one or two ICs. Power must be supplied, and interfacing to the system is necessary but may be simplified. Basically a specialized microprocessor, and costing about the same (a few dollars to a few tens of dollars) but requiring special manufacturing to program the ROM, the microcontroller is particularly attractive when large quantities of goods with identical small computing loads are to be constructed. This occurs with toys, with consumer goods such as washing machines and microwave ovens, and with motor controllers which are supplied with the motors. Large systems will need more computer power and flexibility than the microcontroller is able to supply.

5.4.2 Programmable logic controllers (PLCs)

PLCs originated at General Motors Corporation, Detroit, in the late 1960s as a replacement for relay banks which co-ordinated the movements of their assembly lines. They were, and are, chosen because they are smaller than the relay banks, are relatively easily programmed, are fairly easy to troubleshoot, and are rugged, surviving 0–60 °C temperatures and 95% humidity. Early versions used special purpose discrete logic components, but now they tend to be built around microprocessors. Many have a simple programming language based upon ladder logic diagrams (see Chapter 6) so that programs can be written by technicians rather than specialist programmers. Early versions were entirely on–off logic type devices, but modern versions have capabilities for analog input and output, data logging and manipulation, communication with computers and other PLCs, and even PID control laws.

A typical example of PLC use is batch operation weighing and mixing. The following steps would be involved.

1. Weigh bin of material. Strain gauge instrument sends signal to PLC. On satisfying weight requirement, PLC sends OFF signal to valve or screw feeder (or stops sending ON signal).
2. Do step 1 for several bins in parallel. When all have required amount, empty into mixing bin sequentially.
3. Start blending operation.
4. Keep track of feeder conveyors (which often should not be left loaded when not running)

One advantage of the PLC is that when, for different products, the mixture of components must be changed, a new program can be downloaded from the central computer (if there is one) or manually by a technician (for less elaborate setups); no change in hard wiring is needed in either case.

A typical PLC, such as the one pictured in Fig. 5.6(a), is modular and has three module types: the processor module, the I/O modules, and the programming equipment module.

The processor module usually consists of a CPU, memory, and various interface and miscellaneous functions. In one older PLC, the CPU is an 8-bit Intel 8085, while memory is in a 2 kbyte ROM for system monitor and programs and 0.5–2k RAM or EPROM for the user program and intermediate storage. Miscellaneous functions include various timers and counters, and the interfaces are of two types: to programming equipment and (buffered) to an external bus and I/O modules.

In one typical approach, the I/O modules each handle either 32 input or 32 output signals and work with 5 V logic; a fully configured system may have 200–1000 inputs and outputs. Either input or output may be isolated using an optic coupler and the standard signal level is 24 VDC. Interfaces are used to provide 220 V or 20 mA signal levels.

Programming equipment may be expected to consist of at least a plug-in, hand-held calculator type unit capable of simple one-instruction at a time readout and entry. A hand-held ladder logic programmer is shown with the unit in Fig. 5.6(b). VDU terminals are often easier to use for the programming, while downloading from a computer is sometimes possible and usually easier for initial entry; the unit in Fig. 5.6(a) is programmed in that manner using a choice of

(a)

(b)

Figure 5.6 PLCs are commercial special purpose computer systems. We show here (a) a common rack-mounting type of modularized system and (b) a small system with its hand-held programmer.

three languages which are converted to code which is sent to the PLC. A hard copy of the program (i.e. a copy stored in a relatively indestructable medium), usually a printout but occasionally punched tape or cards with older or specialized systems, is often necessary and always desirable.

The basic need in programming of PLCs is to be able to input logical statements of the type 'IF switch 33 open AND switch 91 closed, THEN wait 2.3 s AND after that close switch 14'. It is common for vendors to make one or more special programming languages available for their devices. In particular, a 'language' which allows the programmer to construct logic ladder diagrams (Chapter 8) of the desired logical operations is often supplied, and more conventional-looking computer languages of an assembled or compiled type (e.g. Siemens' STL or various BASIC-like languages, Chapter 6) plus other graph-like languages may be available.

PLCs cost from a few hundred dollars upward. A system configuration including support computers for program generation and down-loading may cost $10 000 up.

5.4.3 Three-term (PID) process controllers

Process controllers of the three-term type implement a control law dating back over a century and which applies to mechanical components, hydraulics, and electronic analog circuits. At its simplest, this very common controller implements the input–output relationship in which the output command $u(t)$ is related to the input signal $e(t)$, so denoted here because it is usually an error signal, by

$$u(t) = K\left(e(t) + \frac{1}{T_i}\int^{t}e(\tau)\,d\tau + T_d\frac{de(t)}{dt}\right)$$

Here the user sets the proportional gain K, the integral or reset time T_i, and the derivative time T_d. Rules of thumb exist for choosing these parameters (see Chapter 8), but the idea is that the first term gives a command proportional to the error from the desired value, the second gives a command proportional to the integral of the error and thereby works to reduce any tendency for the system to develop a steady state offset error, and the third gives a command proportional to the derivative of the error and because of the sign conventions used tends to give more rapid damping of the response oscillations.

The purchased units will usually have a number of features in addition to simply computing the above input–output relationship. In fact, in many respects it is very interesting to consider these as an example of real engineering considerations.

First, the unit will typically display the desired system output (or set point), the actual system output, and the control command being put out by the unit. A vertical display with two needles may well show the first; a second gauge the third. The first gauge may also display, via pointers, the high and low allowable system output values.

It is usual to have manual override of the computed control command. Using a push button the operator can choose to go into a manual control mode and use a button or dial to generate the control command. (This ability to choose leads to terms such as 'hand-auto(matic) units' for these devices.) The computer unit may also check the system output against the preset limits and, if a violation occurs, flash lights or give other warning signals. Figure 4.5 shows the face of an older but typical unit; it is seen again in Fig. 8.5(b), where the unit is opened to show the potentiometers for adjusting the PID gains.

The unit must of course have the capability for the set point and the control parameters to be adjusted. Commonly the former can be done from the instrument face but the latter requires opening the unit.

Communication with a central computer, at least for data logging purposes, may be necessary. The unit may for such purposes have a standard data bus such as IEEE-488 built in.

Finally, there are two operational aspects of the units that are usually allowed for in the real object, but not in the theoretical considerations. These are called **bumpless transfer** and **anti-reset windup**.

Bumpless transfer is desirable to avoid a jump in the command signal when the operator switches from manual mode to automatic mode. The second problem is integrator saturation, called 'reset windup', which occurs when the controller output is limited. Both have engineering solutions (Åström and Wittenmark, 1990).

With all of these built-in functions, the process controller for a single loop (one input and one output) will cost the order of $1000–2000; many are available from manufacturers. The PID function is also available in more general systems.

5.4.4 More general computer control systems

The above functions and more have been implemented in commercial products for which PLC and PID functions may be just a subset. These are pre-assembled systems, and several suppliers assemble and support such systems. The modularity and capability can vary, as can the customizability. It is not unusual to have a basic system consisting of central control console and unit, software, and a few I/O channels starting for around $10 000 and having added capabilities of both logical control and proportional (PID) control varieties. The added units constitute some distribution of the control and cost a few thousand dollars for each unit; these might provide a few tens or a hundred logical (on–off) controls and a few proportional (analog) controls. Examples in use include the Hewlett-Packard 3000 series and the Leeds and Northrup Electromax series systems.

Further capability is obtained with minicomputer-based systems costing from $50 000.

5.4.5 Alternative – in-house design, construction, and programming

In principle, the designer could nearly specify a set of components from different sources to make up a computer system for control. This is particularly so if the basic computer is to be the ubiquitous IBM PC-compatible. A number of manufacturers are prepared to offer 'hardened' versions for industrial application, along with special keyboards and their interfaces and standard level I/O cards to deal with the process. A certain amount of software is also available.

Total cost can be as low as a few thousand dollars; systems engineering studies should show whether this is a good value.

5.5 COMPUTER SYSTEM REQUIREMENTS

A digital computer is by its nature a discrete-event finite state device. This translates into two characteristics of computers as control elements.

1. **Finite word length** Any number will be represented by a finite number of bits, typically 8 to 32 bits. No value will be more accurately represented than $\pm\frac{1}{2}$ of the least significant bit, or

2^{-9} to 2^{-33} of full range.

2. **Finite time** between inputs, between outputs, and to compute responses. The computer has an internal clock running at about 1 MHz to upward of 50 MHz; an instruction takes several clock cycles; an algorithm needs many instructions. For this reason, there will be a sample interval T between inputs and a sample interval T_1 (often the same) between outputs. Neither input nor output can be a continuous function of time.

We survey these two effects in the following sections.

5.5.1 Finite word length

Finite word length affects implementation through four routes: quantization of data on input and output; round-off and truncation errors in arithmetic operations; inexact representations of control law coefficients; and possible non-linear effects such as limit cycles or steady-state errors.

Data quantization

The ADC which converts an incoming analog signal to a digital word is an inherently non-linear device as in Fig. 5.7.

If the data has an n-bit representation, then the least significant bit (lsb) represents 2^{-n} of full scale; an 11-bit-plus-sign ADC has a quantization level for ± 10 V input range of $10*2^{-11}$ V or 4.9 mV. The representation of a signal can be no more accurate than about ± 2.5 mV.

Similarly, the output is fed to a DAC. With a digital word of m bits as input, the output will be in steps of 2^{-m} of full scale. Analog devices can be used to smooth this input from one step to the next, but the intermediate values are interpolations rather than commands.

One usually will choose the ADC with enough bits in the representation that the $\frac{1}{2}$ lsb error is less than noise or other inaccuracies coming from the measuring device: there is little point in having a signal with ± 0.2 V random error represented to an accuracy of ± 0.02 V. The output DAC is usually cheaper than the input ADC, but would rarely need to have more precision than the ADC.

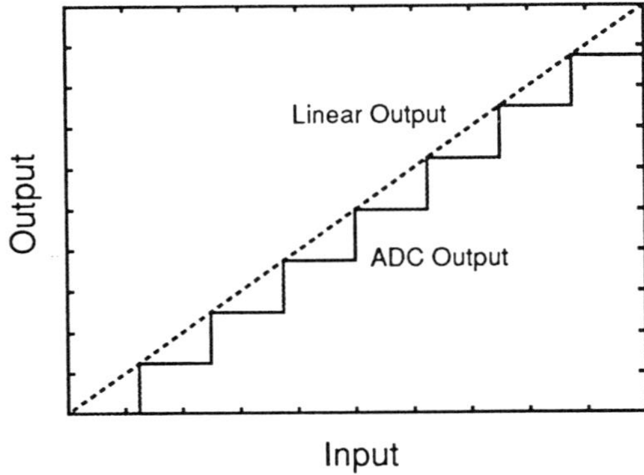

Figure 5.7 Underlying several problems with digital computation is that the representation of a quantity is necessarily non-linear and non-uniquely invertible. This happens right at the conversion stage, when analog to digital conversion is applied, and continues throughout the computation.

Arithmetic errors

Arithmetic errors arise from the fact that multiplication of two numbers with n-bit representations in principle requires $2n$ bits for the product. The product is usually scaled and rounded or truncated back to n bits. The error is (for rounding) up to $\frac{1}{2}$ the least significant bit, or $2^{-(n+1)}$. This error is usually taken as a random number with uniform distribution, and hence will have mean zero and standard deviation $2^{-n}/\sqrt{12}$. These errors have been found to be approximately uncorrelated. Thus a useful model is that multiplication is given by an ideal multiplication operation with additive noise of mean 0 and variance $2^{-2n}/12$.

$$a*b_{\text{actual}} = a*b_{\text{ideal}} + \text{noise}$$

The effect of each noise generator may in principle be propagated to the output, and in linear systems all such effects may be superposed. This is done by Franklin *et al.* (1990) and in the digital signal processing literature, e.g. Roberts and Mullis (1987).

Coefficient representation

Algorithms are usually designed with the expectation that their parameters may be implemented exactly. However, just as in electronic circuits the choice of resistor, capacitor, etc., values is restricted, so in digital computers numbers can only be represented using a finite number of bits. For example, the number $\frac{1}{3} = 0.3333...$ can be represented in binary as 0.1 with one bit, 0.01 with two bits, 0.011 with three bits, 0.0101 with four bits, etc., but all of these are erroneous.

This effect might not be a problem if only one parameter is involved, but multiple parameters rounded independently to a near binary representation can cause changes to the characteristics of the algorithm. Consider the simple algorithm

$$x(k{+}1) = \tfrac{1}{3}\,x(k) + \tfrac{2}{3}\,u(k)$$

The two coefficients can be approximated by one bit, two bits, and so on as

$$x(k{+}1) = \tfrac{1}{2}\,x(k) + \tfrac{1}{2}\,u(k)$$

$$x(k{+}1) = \tfrac{1}{4}\,x(k) + \tfrac{3}{4}\,u(k)$$

$$x(k{+}1) = \tfrac{3}{8}\,x(k) + \tfrac{5}{8}\,u(k)$$

$$x(k{+}1) = \tfrac{5}{16}\,x(k) + \tfrac{11}{16}\,u(k)$$

These have quite different responses.

There is the obvious solution of using more bits, by using double-length arithmetic or a more capable processor, and the not-so-obvious solution of structuring the computation to reduce sensitivity to such problems.

Limit cycles and steady-state errors

Because of quantization, non-linear behaviour is possible from the nominally linear system. For example, a set point reference signal may be slightly different from the measurement to which it is compared. If this difference is small enough, it may be quantized to 0, leading to no corrective command, i.e. to a steady-state offset of up to $\frac{1}{2}$ lsb. Alternatively, the controlled system may 'hunt', oscillating between slight positive error and slight negative error with the control unable to find the exact zero error due to quantization.

5.5.2 Sampling

The digital computer takes a finite time to complete an algorithm computation concerning a data value. Hence it can only consider data every T seconds, for some number T. If we can choose T, or if T is forced by the physics of the problem, the appropriate value must be decided.

Sampling theory

The core idea of sample rate selection is Shannon's Sampling Theorem, which may be stated in many forms and variations, but for our purposes is as follows:

A signal $s(t)$ may be exactly reconstructed from its equally spaced samples $\{s(nT), n=0,\pm1,\pm2,...\}$ if and only if the sampling period T satisfies

$$\frac{\pi}{T} \geq \omega_B$$

where ω_B is the signal bandwidth, for which the signal Fourier transform satisfies

$$S(\omega) = 0 \qquad |\omega| > \omega_B > 0$$

When $T = \pi/\omega_B$, sampling is said to be at the Nyquist rate; a short characterization is that the sampling frequency is twice the highest frequency in the signal.

We make four observations concerning this. The first is the pedantic one that the above applies to a low-pass signal in which $S(\omega) \neq 0$ for almost all $\omega < \omega_B$; the theory is slightly modified for band-pass signals. The second is that in principle an infinite amount of data are needed for the reconstruction. The third is that there are no truly band-limited signals of finite duration.

A more important observation is that the above is concerned with signal reconstruction, which may be irrelevant for applications such as supervisory control of stable systems.

For control systems applications, an alternative viewpoint is that the system is required to respond 'well' to an input with highest significant frequency ω_c. Hence a sampling frequency of at least twice this frequency is necessary.

It should be noted that it is possible to sample too rapidly. This counter-intuitive situation arises primarily because rapid sampling increases the difficulties with numerical/word-length problems. Thus, increasing the frequency of multiplications increases the propagation of round-off errors per second, while coefficients arising in PID controllers may be required to go to 0 or to 1 as $T \rightarrow 0$ (see Chapter 12).

5.5.3 Rules of thumb

There are a number of rules of thumb concerning sampling period or frequency. Among those are (Perdicaris, in Tzafestas (1985)):

1. sample at 10–20 times ω_c;
2. sample period should be at most 0.1 of desired rise time; and
3. choose T small relative to desired closed-loop time constant.

Specific time periods are mentioned by Åström and Wittenmark (1990):

Variable type	*Sampling period*
Flow	$1 - 3\,\mathrm{s}$
Level	$5 - 10\,\mathrm{s}$
Pressure	$1 - 5\,\mathrm{s}$
Temperature	$10 - 20\,\mathrm{s}$

One rule of thumb (Perdicaris in Tzafestas (1985)) for computational word length is to use a word length 4 bits longer than the ADC size chosen, where the latter is presumably chosen consistent with data reliability and noise.

5.6 EXAMPLES

It is difficult to develop an intuition as to the size and speed requirements of programs. Certainly we are becoming accustomed to having hundreds of kilobytes (kbytes) of random access read–write memory (RAM) and read-only memory (ROM) available in microcomputers for personal use (PCs), along with megabytes (Mbytes) of read–write bulk storage on magnetic discs and a promise of gigabytes of read-only or write-once read-many (WORM) bulk storage on optical discs. Common microprocessor CPU units can perform simple instructions such as READ from or WRITE to memory, ADD two integers, or CLEAR a register in tenths to a few microseconds, and special purpose arithmetic units (ALUs) can do multiplication in a similar amount of time. Let us look briefly at a couple of reported examples of relevance to us.

Refai (1986) describes a control law coded for a simple 8-bit microprocessor implementation. The control law was a PID law given by the simple expressions

$$X(n+1) = \omega_d * T - (\omega(n) + \omega(n-1)) * T/2 + X(n)$$

$$e(n) = \omega_d - \omega(n) \tag{5.1}$$

$$u(n+1) = K_p * e(n) + K_i * X(n+1) + K_d * (e(n) - e(n-1))/T$$

where T, K_p, K_i, and K_d are parameters, $\omega(n)$ is the 'present' measurement of the controlled variable, ω_d is the desired value, and $u(n+1)$ is the output command. Although the computation requires several multiplications, the microprocessor used (Motorola M68B00) did not have a multiply instruction and hence a software multiplication was required, i.e. multiplication was performed by shift-and-add instructions, as was (in essence) the division by T.

The task was the sole one performed by the computer; there were no outputs to instruments, no safety monitoring, and no extras (such as anti-reset windup, manual mode, bumpless transfer) as found in commercial units. This bare-bones algorithm then was reported to require 255 bytes of storage and to run one complete cycle (from data read to command output) in about 1 ms.

As a second brief example, we mention the comment by Gressang (1977) concerning adaptive control of an F-8 aircraft. It was found that the Kalman filtering/linear quadratic controller (Chapters 26, 28,

29) for 7–9 states on each of three axes of the aircraft was estimated at that time (1977) to be possible at a rate of about 20 times per second, with filter gain updates at a slower 1/s or 0.2/s rate.

As further examples, we remark that a number of power generation stations use DEC® minicomputers or their equivalents for their basic supervisory computers and have dozens of PLCs in addition. Also, some process industries (such as petrochemical plants) are converting to minicomputer-based systems such as the Honeywell 3000 series.

The message of the above is that direct digital control software is by current standards small, and that the speeds required, by communications standards, are slow. The message not given is that supervisory and monitoring software, as used in control rooms, may be large, complicated, and customized.

5.7 SUMMARY AND FURTHER READING

In this chapter the hardware of computer systems control has been outlined. Included have been descriptions of how the CPU communicates with the outside world, and some prepackaged systems.

We have only touched briefly on any of these, with the intent of giving the systems engineer the flavour of the needs of computer control. Specialists must necessarily study much further into these topics.

Details on the electronics aspects of basic computers can be found in texts such as that of Gibson (1987). A more elaborate discussion oriented toward control systems is in Bennett (1988).

Discussion of word length effects has already been attributed to the digital signal processing literature, such as the classic Rabiner and Gold (1975). A control systems oriented presentation is Williamson (1991).

Control-oriented discussions are presented in journals by Kowalczuk (1989), and Williamson and Kadiman (1989).

6
Computer software

The computer software encodes the control algorithms and the logical commands for the interfacing of the I/O and for the emergency and other routines which the system is expected to perform.

We have already briefly considered how the CPU works. Software (sometimes placed in special unchangeable memory and called **firmware**) is simply the set of bits, loaded into the proper cells in memory, which are operated upon by the hardware. We review languages for that software and emphasize the particular aspects of control systems – namely those associated with real-time operation – which the engineer should keep in mind even when the actual programming details will be done by a specialist programmer.

6.1   SYNOPSIS

Software is the term used for computer programs. When used to implement control algorithms, these have several interlocking aspects:

1. the language used for the encoding;
2. the program which co-ordinates the various tasks, that is, the operating system; and
3. the verification and validation of the system.

The overriding issues in software are that

1. all but the simplest implementations will have multiple tasks to perform, such as algorithm computation, instrument input and command output, and communication with operators and other computers, and
2. the above must be performed in real time, in sometimes unpredictable sequences, with no risk of serious errors. When an error occurs, the computer must not 'crash'.

The user implementing a computer control system will be faced with a number of software choices: algorithms, coding languages, operating systems. The software must be engineered, and in particular validated and verified, to perform the control tasks well and without itself contributing to disasters.

6.2 BASIC PROBLEMS – THE MULTITASKING ENVIRONMENT

A few control system programs are very simple: read data, compute a desired command using some algorithm, output the command, repeat. The program is executed as frequently as possible, so is program-timed – the frequency depends upon the computer's speed. The program starts when the computer is turned on and stops when it is turned off. Many early applications were of this type, and even now the controller for a washing machine or valve need not be complex. It is even possible for such applications to be software timed, i.e. to have the timing controlled by cycling through the software rather than by an external clock circuit. For example, the sequence

1. READ instrument
2. READ set point
3. COMPUTE command
4. OUTPUT command to valve
5. COMPUTE output to display
6. OUTPUT display information
7. CHECK sense lines for supervisor commands
8. EITHER LOAD new supervisory data OR PAUSE an equivalent time
9. UPDATE supervisor data
10. LOAD EITHER standard OR new supervisor response into output
11. END cycle

can be instigated by a real-time clock (RTC) or, if all logic branches are of the same time duration, can be cycled regularly. In either case the computations, etc., might take 50 ms and the desired implementation might be every 75 ms. A cycle could then be instigated by the RTC or, with 'END cycle' expanded using NoOps (No operation – the 'NULL' instruction, taking up time but doing nothing) to 25 ms, the software could simply be in an endless loop.

Such activities as in the above list are called **tasks**, and from the point of view of the computer program they are modules which can be executed independently; although one might request that another be activated, it does not directly call it, as a subroutine would. When there are multiple tasks, then in a real-time system they are related and hence interdependent. The result is that they must communicate with each other and must always run in a certain order and at certain times.

The particular example above is simple and hence straightforward to program because the various tasks are always in the same order and take the same amount of time. If more tasks are required, some are optional, or some (such as causing a printer output) take so long that simply waiting for the task to be completed is inefficient, then a more complicated program is needed.

Even modifying the above so that data are sampled every 0.1 s, the control algorithm is run and an output command given every 0.5 s, the displays are updated every 1.0 s, and the data log is printed every 1.0 min requires a major change in approach. This more complicated program must do the required tasks, interleave tasks to use CPU time effectively, keep tasks in order of importance if there are conflicts for the CPU, etc. Such a program has a structure which occurs frequently in applications of computers – a structure in which there is an overall supervisor plus a number of subroutines specialized for the several tasks. The supervisory program, which ultimately is just another computer program, is called the **operating system.**

When multitasking is done, the various possible tasks must be able to signal when they are ready (e.g. a RTC signals that it is time to take another sample). The signalling is done using the control lines, particularly the sense lines and interrupt lines, and co-ordination is achieved in the CPU by running a supervisory program, called the **scheduler** (usually part of the operating system).

6.2.1 Dealing with the outside world: I/O

It is because of multitasking that control computer problems arise, and it is the input–output properties of these real-time systems which are the main trouble source. This is particularly because the sequencing of tasks, and hence of program instruction execution, may not be fixed.

Control system CPUs must deal with other components and ultimately interact with their outside world. For the system, there are three main I/O sections.

1. **Process I/O** Reading process instruments and outputting actuator commands.
2. **Operator I/O** Reading keyboard and switch inputs from operators and writing data to printers, VDUs, meters, etc.
3. **Computer I/O** Reading and writing data between memory, registers, and ALU, for example.

To handle all of this, the CPU has two options: it may **poll** the devices frequently to see if they are ready to transmit (or receive) data or it may wait for a signal initiated by a device ready for service, i.e. an **interrupt**. In either case, it is the external event which triggers the action.

There are trade-offs between interrupts and sense lines and some off-the-shelf systems use one and some use the other. The essential trade-off is that polling is straightforward but can result in delays in response (some polling systems can take over 500 ms to respond), while interrupts lead to rapid response but possible unforeseen instruction execution sequences. One must decide what it is worth in unpredictability to have rapid responses to inputs such as:

1. alarm inputs;
2. failure indicators, for hardware, power, and transmission failures or errors;
3. override indicators, to allow control panel or other manual inputs to override computer control; and
4. real-time clock, to provide a regularly spaced (in time) signal to the CPU.

6.2.2 Some special software considerations with interrupts

There are several tasks that must be performed when an interrupt occurs.

1. Saving of registers, flags, CPU status, etc., is usually done at least partly by the hardware, so that the program can return to business as usual after the interrupt has been processed.
2. Identification of the source of the interrupt can be by hardware (using effectively a different interrupt signal for each interrupting device – e.g. interrupt vectoring) or by software (which typically must poll devices to see which has sent the interrupt).

3. Establishing priority involves deciding which interrupt gets serviced if several occur simultaneously, which are allowed to interrupt other interrupts, what happens to ignored interrupts (are they eventually forgotten?), etc. Interrupt priority is an important and not necessarily obvious issue.

The above steps may be performed by either software or hardware, depending upon the CPU design. If in software, these tasks are a programming burden, whereas if they are in hardware, the programmer must still be aware of how they affect operation of the algorithms.

We indicate just one of the many problems which must be considered when using interrupts: if the subroutine is re-entered before it is completed, and with completely different data, there is a chance that the results will be muddled or the program will get lost. Suppose, for example, that two registers, R1 and R2, are saved by the interrupt routine when it is entered and restored when it is exited. Then if the main program is interrupted, these cells will contain values A and B, say. Part way into the subroutine, the same registers may contain values a and b when another interrupt of the same type occurs. The subroutine is then entered in the course of this second interrupt response and a is put into R1 and b into R2, *overwriting A and B*. Even if the interrupt returns are unravelled successfully, A and B will be lost and a and b will be processed by the main program. Solutions to this problem include having multiple copies of the subroutine, blocking further interrupts while it is operating, and making it a re-entrant routine (in which all data are saved in stacks).

6.3 SOFTWARE PROGRAMMING

Programming of the computational element may be done in several different ways, or more precisely, at several different levels. It should be remembered that ultimately the CPU executes a sequence of relatively simple instructions (made up of even simpler micro-instructions). The programming language used may consist of more readable commands, but the CPU sees only strings of binary bits, which are voltage settings of, typically, 0 V for binary 0 and 5 V for binary 1. We first indicate one of the principal structural forms for which we must program, the procedure structure with interrupts, and then discuss programming languages.

6.3.1 Program structure – procedures

A simple program will have a simple structure, perhaps such as in Fig. 6.1. Such a program will simply keep running, doing one cycle in whatever time the computer speed allows, and is called software timed.

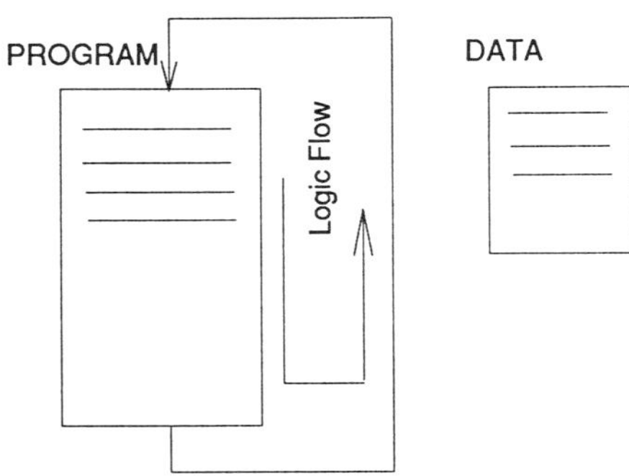

Figure 6.1 A computer program and its data are represented as sequential blocks of storage. A software timed program simply sequences through its list of instructions. The time for one cycle depends upon the computer's speed and the number of instructions.

If a cycle starts when an external real-time clock initiates it, then the structure is more as in Fig. 6.2.

This is of course more natural even for a simple program. A next stage has subroutines, which are used to supply readable structure, to break up the code into manageable segments, and to allow sharing of code segments. This leads to structures such as Fig. 6.3.

The most important variation from our point of view is the possibilities for interrupts posed by the need for multi-tasking. Here, a special subroutine called an interrupt routine may be called at any time (due to an emergency, for example). The impact hardly shows if we impose such a possibility on Fig. 6.3 to create Fig. 6.4.

The attempt in Fig. 6.4 is to indicate that the interrupt may occur at any time and between any two instructions. The software must be so arranged that there will be no problems upon return from the interrupt subroutine, whether it is there to fetch data, to trigger alarms, or simply to update a display or respond to an operator input.

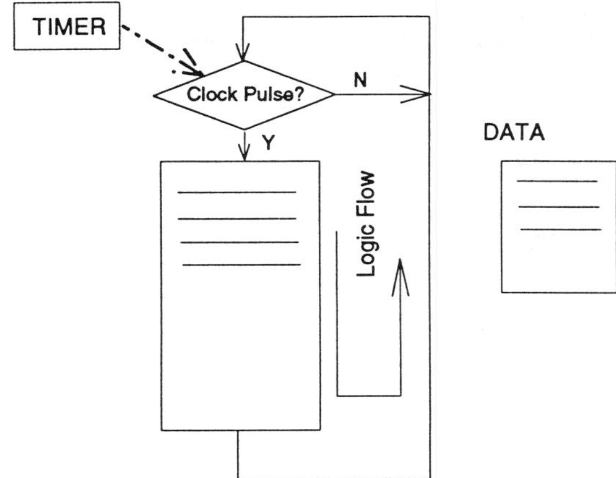

Figure 6.2 An improvement, especially when the sequence is not of fixed duration because of internal branching, is to start the sequence periodically at times determined by a timer.

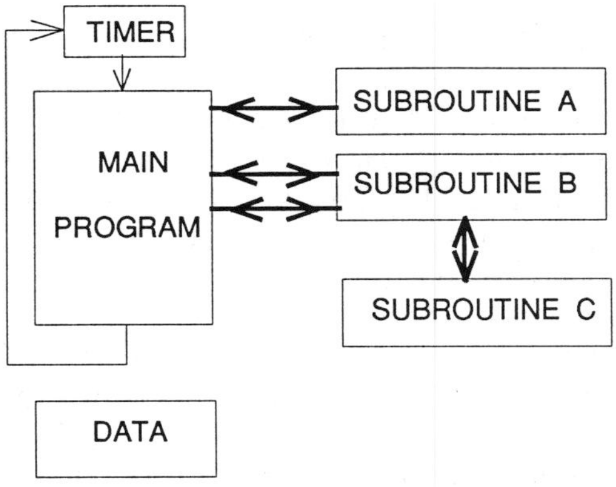

Figure 6.3 A program with subroutines is structurally easier for programmers, and may save memory.

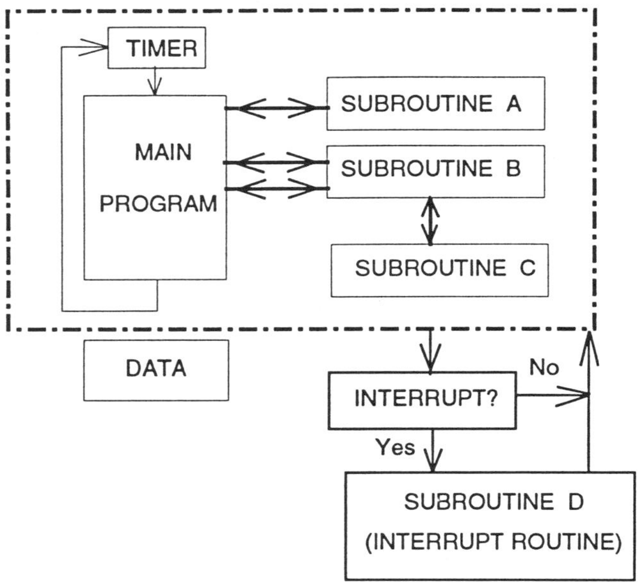

Figure 6.4 An interrupt is a hardware signal causing automatic branching to a special subroutine. In principle, an interrupt might occur anywhere within the standard sequence.

6.3.2 Languages: levels of programming

Machine language consists of the binary sequences referred to above, on which the computer operates directly. The programmer must write his code directly in such sequences of 0s and 1s. Thus the program to add the variable A to variable B and set C to the result requires that the quantities A, B, and C all have memory storage locations known to the programmer, e.g. at locations 10000010, 10000011, and 10000100 (binary) respectively. If the program should then transfer to location 01110101 (i.e. set PC to 01110101), then the program might look like:

MEMORY LOC	CONTENTS
01000111	001010000010
01001000	010010000011
01001001	010110000100
01001010	110101110101

This is an oversimplification in that there are only 4 bits used for the instruction code and the total instruction word has length 12 bits, but the programmer's burden should be clear: the work is tedious and highly error prone. Furthermore, if one of the variables must be reassigned to a different memory location, then finding all of the resulting programming changes will be difficult.

A first step in improving on the above scheme is to use **mnemonics** for the instructions, such as CLA (clear and add) for 0010, ADD for 0100, STO (store) for 0101, and JMP (jump) for 1101. Allowing spaces on the input also helps readability, so that the above might be represented by:

```
CLA        10000010
ADD        10000011
STO        10000100
JMP        01110101
```

This code will require another program to translate it into binary as above, but if the translator is reliable and can be told where in memory to put the instructions, the result should be the same as before.

If a translator program is necessary, it can be used for a few more things. One in particular is **symbolic addressing**, i.e. to tell the translator that *A* as an address means location 10000010, *B* means 10000011, *C* means 10000100, and *NEXT* means 01110101. Then the above becomes

```
CLA        A
ADD        B
STO        C
JMP        NEXT
```

A further improvement in the translator allows it to make some of the **memory management** decisions. If the translator is told to set aside memory cells for variables *A*, *B*, and *C*, and that *NEXT* is a destination location for a transfer, it can keep an appropriate **symbol table** of these labels and their addresses instead of requiring the programmer to do so.

A translator of the type outlined above is a computer program itself, of course. Its inputs are sequences of characters (CLA, *A*, *NEXT*, etc.), and its outputs are binary strings as in the first part of this section. Such a translator is an elementary member of the class of programs called **assemblers**; the input to it is called **assembly language code**, and the output is called **object code**.

The advantages of assembly language programming (over machine code programming) are the relatively easy readability of the input program (compared with a string of binary digits) and the fact that the assembler copes with the memory allocation. Most assemblers also will have some 'macro-instructions', or '**macros**', which the translator recognizes and expands (each time they appear) into machine code on a many-to-one basis. (A simple macro might allow the instruction MADD A, B to represent the two instructions CLA A, ADD B in the above example.) This allows even more programmer convenience.

The disadvantage of writing in assembler code rather than machine code is that the programmer ends up with a computer program, the assembler, between the input alphanumeric source code and the machine instructions. The programmer is usually responsible for the final machine code and hence must either believe the assembler has been properly implemented and used or check the machine code using memory dumps and other methods. On the other hand, programmer errors in coding binary words are avoided, considerable flexibility is gained, and the programmer retains tight control over the final program because most instructions are one-to-one related to the assembly language mnemonic instructions.

The tight control over the program and the cleverness with which programmers can sometimes code algorithms, along with the fact that various parts and ports of the computer are directly accessible to the programmer, means that assembled programs are (in principle) efficient in their usage of both memory and time. For this reason they have in the past been frequently used for real-time computer applications, and are still common for small programs.

A further step along the trail of programmer convenience is to have a more elaborate translator, one to which an instruction such as

```
C := A + B;
```

is meaningful. This instruction, representing the information 'find the sum of the variables A and B (A + B) and place the result (:=) in C, end of instruction (;)' is elementary for programs capable of breaking elaborate character strings such as

```
IF (IN < MAX) THEN C := IN ELSE C := A + B; GOTO NEXT;
```

and

```
FOR I:= 1 TO IN DO C[I] := A[I] + B[I+5-3*K];
```

into appropriate machine code. Such programs are called **compilers,** and the input instructions are in a **compiler language.** There are many such languages, including FORTRAN (FORmula TRANslator) for scientific number-crunching, COBOL (COmmon Business Oriented Language) for business data-keeping and updating of wages and such, Pascal (named for scientist Blaise Pascal) for educating programmers, C (the language which followed A and B) for operating systems programming, and ADA (named for Lord Byron's daughter, Ada, Countess of Lovelace, the 'world's first programmer') for United States of America Department of Defense real-time programming.

Compilers were intended to use the computer to remove some of the tedium of programming, leaving the programmer to be creative and efficient. In just a few characters, a programmer can write a single instruction which when compiled results in several machine instructions being created. Also, the compiler can identify certain types of programming errors, reducing the testing time needed. There are disadvantages, though.

1. The language often keeps that programmer far removed from the actual machine code instructions and their ordering, which can be critical in a real time environment.
2. For small routines at least, a programmer can produce more efficient code (in terms of memory used and/or execution time) than a compiler will generate; this is somewhat debatable for large programs, but is probably true even if not always cost efficient. (It may be better to have slightly inefficient code than to have expensive programmers trying to optimize it.)
3. The more important criticism of compilers is a growth from the one with assemblers: because there is a sophisticated program, namely the compiler, between the source code and the object code, the programmer may make errors in using it. Thus the programmer must be highly skilled with the particular compiler used to avoid, e.g. subtle variable interactions which may produce unexpected results.

Assemblers and compilers take as input a set of character strings called **source code**; their output is usually an entire program of machine code, called **object code**. This object code may be interfaced, or **linked**, with other object codes, typically for subroutines and for standard (library) routines to form the machine code that is finally executable.

A significant variation on this procedure from an operational point of view is the **interpreter**, which decodes and executes the source code on a line-by-line basis but does not save the object code after execution. Hence it translates a given line of instructions each time it is met. There are problems with such an approach, the main ones being

1. slow execution because of the need for translation at execution time, and
2. the possibility that the language will be less flexible or efficient because of the line-by-line approach to translation.

The advantage is that the interpreter may be small in storage requirements and that the source code almost surely needs less memory than the equivalent executable code. The most common interpreted language is BASIC, although it is sometimes compiled and linked; some special purpose languages for PLCs and PCs are also interpreted.

Actual coding of the control computers depends upon the computer system used and its tasks. Special purpose applications on small microprocessors are often coded in an assembly language for reasons involving need for efficient code, lack of a suitable compiler, and relatively small and uncomplicated programs. The code is then assembled for this target machine using a larger (or **host**) machine, a method called **cross-assembly**, and the resulting machine code is then **down-loaded** to the target machine. (A variation of this uses a compiled language and a **cross-compiler**.) Larger machines doing complicated sets of tasks are more likely to be coded using compilers, for the reasons that good compilers are available and their advantages in terms of programmer time needed for coding large programs are more important than potential coding efficiencies from using assembler code.

6.3.3 Special purpose languages

Many commercial devices come with special programming already done, so that the input need only be of special parameters and settings for the particular process. For these devices the input may be considered to be by means of a special purpose programming language. We look briefly at a couple of examples.

Robot programming is often done either with a 'teach' mode in which the arm is controlled manually using a hand-held control terminal and told to remember certain positions, and/or with a proprietary language. Such a language is the VAL language used by Unimation for the PUMA series of robots. Its elements are instructions such as in Table 6.1, and a typical small program string is in Table 6.2.

Table 6.1 Elements of VAL

Command	*Effect*
OPENI	open tool
CLOSEI	close tool
DRAW x, y, z	move arm to coordinates x, y, z
HERE *name*	define present location as '*name*'
MOVE *name*	move arm to location '*name*'

Table 6.2 Sample VAL Program

Instruction	*Comment*
READY	Go to initial position
CLOSEI	Close tool
4	Define this as '4' for loop
MOVE *loc 1*	Move to predefined *loc 1*
DRAW 0, 0, −50	Move −50 in z direction
OPENI	Open tool
DRAW 0, 0, 250	Move 250 in z direction
DRAW 200, 0, 0	Move 200 in x direction
DRAW 0, 0, −200	Move −200 in z direction
HERE *loc 2*	Define this as *loc 2*
DRAW 0, 0, 50	Move 50 in z direction
CLOSEI	Close tool (to grasp)
MOVE *loc 2*	Carry object to *loc 2*
MOVE *loc 3*	Carry object to predefined *loc 3*
MOVE 0, 0, −50	Lower object
OPENI	Release object
SETI *n=n+1*	Increment counter
IF *n<10* THEN 4	Go to 4 if less than 10 times through cycle
HALT	Stop program

Another approach, called **ladder logic programming**, is taken by some makers of PLCs. Here the computer in essence simulates a relay bank plus timers and counters, so the basic elements of the language specify

- input contacts ('relay contacts passing signals into the computer when the relays are actuated by external events')
- output coils ('commands to relays to open or close and thus pass signals to external devices')
- internal coils ('relays for internal logic')
- timers (to count time delays, etc.)
- counters (to count events either internal or external)

and sometimes a few others. Symbols are defined for these, and programming consists of manipulating such symbols on some sort of visual display unit to obtain a desired wiring logic diagram. (An illustration of such symbols and programming is in section 8.2.)

Another class of examples contains the special languages often used for numerically controlled (NC) machines; a typical such language is APT (Automatically Programmed Tool).

6.3.4 Software considerations

Any computer may be coded in machine language, and usually at least primitive assemblers are available; for the smallest systems, the program may sometimes be developed on a larger computer and cross-assembled or cross-compiled for the target system. Commercial systems sometimes provide a suitable real-time operating system combined with an appropriate variation of a compiler. For example, the rapidly developing capabilities of programmable logic controllers (PLCs) mean these can handle both logic and analog controls and hence can be programmed in relay logic language (RLL) and higher level languages such as BASIC and Fortran, developments which are tracked by the trade magazines, e.g. Lodzinsky (1990) and Flynn (1990). For a second example, the IBM 9000 system operates under CCOS, with languages Pascal, BASIC, and Fortran readily available, and, because the CPU is a Motorola 68000, other languages and systems also obtainable.

Software is expensive to write: a good programmer generates the order of 3–10 lines per day of programmed, checked, validated, documented code. This number is somewhat independent of the language level, so it is clear that a language which has many machine instructions generated per line of programmer code has cost advantages over machine languages. Thus, particularly for large systems, programming may out of cost necessity be in a compiled language. The extra hardware – memory and CPU capability – needed may be far cheaper to buy than efficient coding. For this reason we look briefly at some of the newer candidates.

A modern language for control applications should preferably have or support the following attributes.

1. The use of blocks, procedures, functions, etc., to provide structure to the program should feature in the language. Structure will prevent inadvertent sharing of memory cells and will prevent spaghetti-like (and hence possibly wrong) paths through the program. It also helps by ensuring readability of the program.
2. Concurrency of a number of simultaneous asynchronous tasks should be supported. Proper ordering of the tasks (to prevent data output tasks executing before the proper data gathering task, for instance) should be supported.
3. Run-time security is of paramount importance, so the language should have intrinsic features which improve security. Included in these are strong variable typing and exception handling capability.
4. The language should allow reaction to events in real-time.
5. The object code generated should be memory- and time-efficient. Both of these are partly a function of the language and partly of the compiler.
6. It is preferable if the code is portable, so that it may be transferred to other machines. True portability is difficult to obtain, however, because of machine differences and lack of standards.

Preferred features in addition to the above can also be defined.

7. Low-level code facilities, to allow added flexibility for the programmer.
8. Separate compilation of procedures, to help in modularization of the code and in the efficiency of code development.

If the above are taken as requirements, then there are virtually no existing languages with proper compiler supports which meet them.

The closest are the new languages ADA and MODULA–2. ADA is a very large and complex language, and possibly it is unsuited for that reason for implementation at this time; it does in principle meet all of the requirements, however. MODULA–2 is a Pascal derivative (or improvement) in which the independent modules are the most important feature; it does not support exception-handling.

For small programs (a few tens or hundreds of lines of code) a number of possibilities exist; most fail to meet several of the above desired conditions, but all have some use. They tend to be special versions of older or very special languages. Thus, the engineer is likely to meet Concurrent Pascal, BASIC, and FORTH. Process BASIC and Process FORTRAN are other available languages; each is a special version of the obvious regular language. The process versions are not standardized, but of necessity will have language extensions for:

1. interfacing to the operating system, such as communicating the need for periodic task execution, enabling and inhibiting interrupts;
2. process input and output, such as reading an ADC or switch, or commanding a DAC output; and
3. task management, such as activating a task when an interrupt occurs.

For example, since the languages such as BASIC are not normally used in real-time operations, they must be augmented with instructions such as

```
ON EVENT "signal on IEEE–488 bus"      'sets up linkage of receipt
   GOSUB "process instrument data"        of interrupt to processing
EVENT ON                                'to allow event to be recognized
EVENT OFF                               'to block event processing
```

and the compiler and operating system must generate appropriate machine code.

Finally, there is always assembly language and its high-level version, the systems programming language C.

6.4 OPERATING SYSTEMS

In systems of reasonable size, there are a number of tasks which are nearly independent of the application; for example, process plants and aircraft both need operator displays. To avoid reprogramming such software for each application, a supervisor program with appropriate special routines is often used. This **operating system** is the computer program, which is not application specific, but which is the essential interface between the outside world and the computer hardware and also the interface between the applications programs (such as the number crunching routines which give the control output as a function of the instrument inputs). It is found in all but the simplest applications, in which the few functions needed may be coded as part of the applications program.

A typical operating system has components which manage elements such as the system interfaces to the operator, task scheduling, timing, memory allocation, and files. The I/O subsystem takes care of such roles as peripheral device drivers (i.e. special codes for dealing with I/O devices) and queue handling, if outputs (e.g. characters to printers) queue, in such a way that applications programs can seem device independent. The operating system may also support special handling of floating point arithmetic operations, loading of applications and other program segments, and so on. Some of the handling of interrupts may reside with the operating system rather than the applications programs.

6.4.1 Overview

The way to view the operating system is to note first that the computer is required to perform a number of different tasks in the system, e.g.

1.   read instruments for data
2.   output commands to actuators
3.   interface with other computers
4.   accept inputs from operators
5.   output information to operators
6.   log data for future analysis and record keeping
7.   compute algorithms for commands as functions of inputs
8.   do unit conversions for data logging

Associated with these tasks, an operating system must perform at least the following functions.

1. Error handling
2. I/O handling
3. Interrupt handling
4. Scheduling of tasks
5. Suspend, kill, restart tasks
6. Maintain queue of tasks needing execution
7. Resource control
8. Protection
9. Provision of good operator interface
10. Assign/accept user input of task priorities

From a process control point of view, in which events may occur in a non-deterministic order and at variable times, one of the most important roles of the operating system is keeping the tasks in order of importance and not forgetting any of them. Consider the following scenario.

1. The computer is in the process of printing a log of the shift's activities.
2. The real-time clock sends an interrupt. The CPU immediately goes to the interrupt routine. It must remember that it was printing, but call up the program appropriate to the clock interrupt.
3. During step 2 the operator presses a command button. If an important command, this will generate another interrupt which must be processed. Now steps 1 and 2 must both be remembered while a response to the operator is made.
4. After step 3 is completed, the system must work backwards up through step 2 and then continue from step 1.

The above is an ordinary sequence of what might be required, but gives some notion of what the operating system must do in a process control application. The critical thing is timeliness of response, as otherwise a very ordinary batch operating system would suffice.

Although the operating system is a computer program providing services to the applications programs, these services cannot be allowed to be beyond the control of the applications programmer as they are in traditional batch processing systems using their operating systems. Such is the danger of subtle ambiguities and such is the need for rapid response in emergencies and timely response in all circumstances that the programmer must have ultimate control of the I/O devices, physical memory, task priorities, exception processing (such as overflows, which cannot be allowed to simply crash the program), and data integrity.

6.4.2 Buying a RTOS

There are many operating systems around, such as Unix and its clones, MS-DOS and CP/M for microprocessors, etc. Few of these are real-time operating systems. The computer manufacturers, such as Digital Equipment Corporation, Data General, IBM, and Hewlett-Packard have real-time systems for their systems: Intel created IRMX for its microprocessors; IBM produced CSOS for its 9000 microprocessor series; DEC has RT–11 and RSX, etc. Some other systems have been developed as real-time versions of non-real-time operating systems, such as real-time Unix versions, and standards are being developed by IEEE under the name Posix in its Standard 1003 groups.

It is arguable that if the computer is fast enough and the number of possible tasks small enough, an interrupt-handling operating system may be unnecessary. Some small commercial systems simply time-slice the available resources and allocate them to the tasks in turn. This simplifies the possible number of logic paths and reduces the potential for software errors; it relies essentially on sense lines for control line signalling.

6.5   VERIFICATION AND VALIDATION

In an attempt to approach software design and implementation in the same systematic way as other branches of engineering treat hardware, the field known as software engineering is being developed. The approach is to be systematic at each step, from specification through design and coding to testing, maintenance, and updating. The goal is that the software should meet the needs of the purchaser, and an important stage of this is verification and validation (V & V). This in principle, and sometimes in fact, applies to each version of the software, before delivery and after repairs, both in place and pre-delivery. We paraphrase ANSI/IEEE terminology (Thayer, 1990) as

> **Verification** is the process of determining whether or not the products of a given phase of the development cycle fulfil the requirements established during the previous phase. It also is used to refer to the act of such determination, including reviewing, inspecting, testing, checking, etc., the items, processes, documents, and services.
> **Validation** is the process of evaluating the system at the end of the development process to ensure compliance with the system requirements.

These definitions are subtly different, with verification considered to apply to particular activities within the overall development and validation connoting a determination that the entire system is suitable. Sometimes the two are considered together as a single topic 'V & V'. The topic is an important part of software engineering (Francis, 1990) and standards have been developed (Horch, 1987); one book devoted to the question is Quirk (1985).

The overreaching problem is that real-time systems have special requirements. They must be very reliable, because in the real-time environment (such as process control) the consequences of failures can be catastrophic (e.g. in nuclear power plants or aircraft flight control systems) or at least very expensive (as in assembly line control). The environment can make unexpected and conflicting demands on the software (as when alarms go off in an unexpected combination), and the demands may be simultaneous rather than sequential. The system must meet deadlines, in which output commands (for example) must be transmitted every few seconds or milliseconds.

The implications of this can be almost overwhelming if the system is using interrupts, for interrupt response is a branching operation in the software. Hence the path through the execution could in principle be almost any sequence of interrupted procedure calls, i.e. the order of computation is not fixed. Yet the software is expected to be at least fail-safe, and preferably it should be fully recovering or fail-operational. Under these circumstances, the design of the software is very difficult and the verification and validation can seem nearly impossible. Certainly there have been some notable failures in software.

The first thing about V & V is to give the user a chance of designing the software in a sensible and co-ordinated manner. Currently this is seen to be specifications at several different levels, and testing and documentation at each level as it develops. This is simply not a closed book yet; design, as in so many fields, is partly an art form practised by highly talented individuals.

The next thing to do is to code the design well. This will for practical purposes mean coding in a language which can be read by other programmers and into a program which can be changed without using obscure 'patches'. This means ultimately a carefully chosen higher order language for most of the programming, structured programming, and detailed programming standards. Few programming standards specific to real-time systems have evolved; it is felt, however, that programs should be re-initializable, and it should be possible to synchronize routines, to designate portions un-interruptable, and to restrict communications between certain branches.

The verification portions of V & V are performed for each stage in the development process. This may be seen as a common sense management approach in which procedures are applied to ensure that the system specifications reflect the job's functional requirements, the software specifications satisfy the system specifications, the design meets the software specifications, and the code meets the software specifications. Each of these verification stages will have several steps. For example, steps ensuring that the code meets specifications may include analysis of the program, step-by-step walkthrough of the code, formal program-proving methods, and execution.

The validation portion of V & V may overlap heavily with the verification. In fact, sometimes, particularly for small programs created by small groups, little distinction is made between them and V & V is considered a single process. The goal of validation is to demonstrate by operation that the final system meets the user's functional requirements (as opposed to verifying that a step in the development has been properly performed). This will require exercise of the entire actual system if this is feasible. Frequently, the software package will be first shown to execute on a large computer, then on the target computer, then on the target computer in a real-time mode, and finally on the target computer in the actual system and interfaced to the actual devices used. A goal of the testing is to execute each logic branch, each I/O statement, each algorithm, etc., and it should be possible to determine how many times each unit of the program and each logic branch has been tested so that 100% testing can be verified.

Demonstration of functional suitability requires extensive testing, and there should be an *a priori* test plan. The test cases and their nature are an important issue in themselves and fall into two categories: systematic and statistical. The former can be designed to test specific properties of the system, and an expected reaction predicted and compared to the actual reaction. The latter uses randomly generated data (such as interrupt times) and the results must be evaluated.

With systematic testing, a further breakdown of test types is possible: **white box**, in which the program is known to the tester (who may then purposely aim to exercise all paths) and **black box**, in which the tester aims to ensure that all reasonable cases are attempted. In either case, the generation of the test data will require, among other things, a familiarity with the system requirements and environment. In principle a set of expected outputs for each input set should also be prescribed. This all becomes a difficult job, requiring considerable skill and knowledge from the responsible engineers. Finally the tests must be evaluated and documented.

Statistical testing uses randomly generated data and hopes to be content with random answers. Thus measurement data are simulated with a random number generator and the computer output is checked for mean and standard deviation of commands, probability of software failure, or expected number of remaining software errors. The biggest cause of concern is perhaps the timing and interaction of interrupts.

Statistical testing has two possible approaches: pure statistical testing (no failures in N tests $\Rightarrow$? about probability of failure) or indicator cases (with random data, manually verify correct operation and estimate vulnerability). In either case, the hope is that the randomness introduced may find difficulties that systematic testing might miss. For this reason, and because of the sheer number of tests which can be required, statistical testing is often a complement to systematic testing.

For all of the above, the results must be examined to determine whether the program operated as per specifications. This can be the difficult and expensive part of testing. Here systematic testing at least has an expected result, while statistical testing may need some qualitative evaluation. Analysis is clearly necessary, regardless of the formalism used.

6.6   SUMMARY AND FURTHER READING

Software engineering is a problem for all computer software vendors, programmers, and users. Thus it is still being studied by computer scientists. In this chapter we have only touched on some of the issues – language choice, verification and validation, special languages, operating systems – that affect the control systems engineers. The key issue for such an engineer is that the system must operate in real time, probably uses interrupts, and must not contribute further to any abnormalities that occur in operation of the system. This last aspect is the overriding one for any component of the control system.

There are many places where it is possible to learn more than one wants to know about the software issues. A book extending this chapter is Holland's (1983). Languages are just one of the topics covered by Sinha (1986), which includes the work of Ahson (1986). A control system oriented textbook is Bennett (1988).

Mainstream languages such as FORTRAN, ADA, C, etc., usually have a number of textbooks plus specification books associated with them. For example, C is to be found specified in Kernighan and Ritchie (1978) and discussed in such as Schildt (1988).

Special languages, or special versions of general languages, are to be found in manufacturers' documentation. Hence the robot language VAL is in Unimation's documents (1979), although popular descriptions such as Shimano's (1979) also can be found. PLC programming is also in manufacturers' documentation, although the book of Kissell (1986) is very helpful. Manufacturers' documentation is typified by such as IBM's (1983, 1983–4, 1984). An earlier book on NC machine programming is Pressman and Williams (1977).

Software engineering is actively pursued in journals and magazines such as *IEEE Computer* and *IEEE Transactions on Software Engineering*. Extensive coverage in books is just starting to become evident, in texts such as Sommerville (1989). Books on special aspects of software engineering are more common, although the treatment of validation and verification by Quirk (1985) is one of the few of its type.

Operating systems are in a number of texts, including Lister and Eager (1988). A more specialized concentration on real time aspects is given by Allworth (1981) and a critique is presented by Stankovic (1988). A recent textbook is Joseph (1989).

Applications are usually described in journals, such as *IEEE Micro*, *Automatica*, and *IEEE Control Systems Magazine*. Useful information for the non-specialist is frequently to be found in the popular magazine *Byte*. Books such as Sinha (1986) and Sinha (1987) also have examples.

7
Communications

In this chapter we consider the connection of computer control system components at three levels: simple wiring, instruments to computers using digital signalling, and computer networking.

7.1   SYNOPSIS

The control computer must communicate with its sensors and actuators, its operators, and other computers. The CPU, as we saw in Chapter 5, communicates within the computer system by signalling at low voltages along data, address, and control buses. These communications are partly with interface devices for contact with other parts of the control system. These communications devices are of several types and levels of sophistication.

Some aspects of the communication system are simply those of wiring and grounding, with the constraint being that the magnetic fields of the various plant components should not cause excess noise in the communications signal.

Another issue is that instruments and actuators, in their communications with the computer, should send and receive signals easily recognizable by the computer. If the signals have been converted from analog at the source and are digital (as they may be to avoid certain noise problems or to allow time sharing of the wiring) then an agreed protocol must be used for the messages, e.g. RS-232C or IEEE 488.

If computers are to communicate, then they may be networked. Local area networks (LANs) are used for many communications tasks, among them factory communications. Standards are emerging and developing from among several possibilities of physical configurations and protocols. One potential standard is the **manufacturing application protocol (MAP)** and associated **technical office protocol (TOP)**. A likely solution is that present systems will evolve to such a standard but will have several different communications LANs with gateways connecting them.

These all may be used in a single factory, as in Fig. 7.1

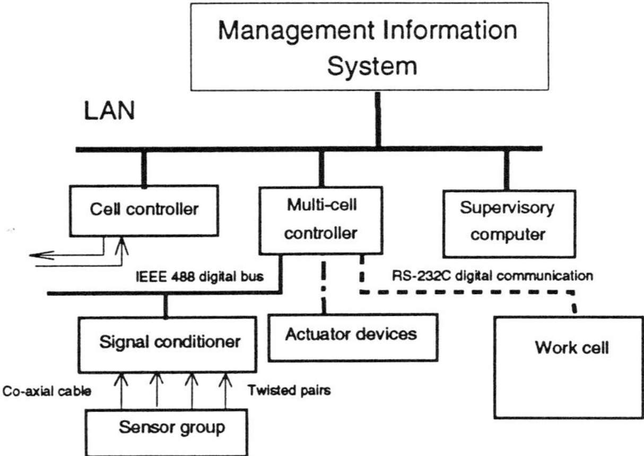

Figure 7.1 A Computer Integrated Manufacturing (CIM) system usually requires communications between operations at various levels, from the lowest Direct Digital Control (DDC) to the Management Information System (MIS). The communications may be done in various ways.

7.2 SENSOR WIRING

The computer system necessarily communicates with the instruments and actuators over wires (or more recently fibre optic cables and occasionally radio links); for a multi-computer configuration, it must also communicate with other computers. In the following sections we look at the problems and principles of wiring and its protection, particularly with regard to sensors.

Interference with signals is always a potential problem, and the engineer's resort is to attempt to plan for it by following established standards. Although cost of materials can be an issue, installation labour costs and their possible trade-off against maintenance and troubleshooting costs are a prime systems engineering consideration.

7.2.1 Wiring, grounding, and shielding

It should go without saying that wiring should be of adequate size for the load carried and should be physically protected from abrasion, accidental cutting, and similar events. Colour coding is necessary, and cabling should be used where possible so that maintenance may be carried out.

Cabling needs are defined by standards issued by agencies such as the SAA in Australia and NFPA in the US. Defining characteristics include size, material, and number of conductors, and requirements are in terms of current capacities, voltage drops allowable, costs, etc. Power cables (such as nominal $1\,mm^2$ copper, approx. AWG #14, suitable for a 1kW motor drawing 10A at 115V) will exhibit resistance of perhaps $0.03\,\Omega/m$ and capacitance of $100\,pF/m$, so that 100m of two-conductor cable will have about $6\,\Omega$ resistance and $0.020\,\mu F$ capacitance; instrument cabling for 4–20mA may be smaller from a current carrying point of view, but will then exhibit (in nominal $0.5\,mm^2$ size) an order of magnitude more resistance and perhaps an unacceptable voltage drop. Also, long cables in particular may be involved in transmission line effects (reflection, etc.) and although 2km cables have been demonstrated with low level ($\pm10\,mV$) signals, even 10m cables can cause problems. The standards also define things such as insulation types and armoured cladding restrictions on number of conductors.

Digital signalling over wires is another possibility, and one that is increasingly feasible with smart instruments and actuators, but here the capacitance and inductance effects may adversely affect transient response.

Part of system wiring is the supply of power to the system devices, including computers, sensors, and actuators. Aspects of wiring to be considered include straightforward properties such as wire size (for which tables of ampacities are available), cabling, protection in the form of fuses and circuit breakers, and grounding. Indirect effects include the electromagnetic interference which power lines may have on sensors, computers, etc. Although power supply is well beyond the scope of this book, an elementary introduction may be found in Webb and Greshock (1990).

Voltage is not an absolute quantity, but a potential difference between two points. For voltages to be meaningful signals (and also for safety in circuits – a matter of lesser importance for us but very important in medical instrumentation) they must all be measured against a stable reference point, called the circuit common. When a circuit is linked to other circuits in a measurement system, the commons are usually connected together to provide the same common for the complete system; the latter may in fact be the universal common – earth ground – and the connections may be to a ground rod, water pipe or power line common. The problem due to not having a common ground is that an extraneous current may flow in the circuit between the two different 'grounds' of the instrument and the computer, yielding an erroneous signal.

The need for all instruments to be connected to a common ground, such as an earthed heavy copper strip (so that there is low resistance to the ground reference point and hence a large current carrying capacity) may impose impractical burdens on the system because of the size of the installation; it is one thing to have a common ground for a single laboratory or room and quite another to have one for a large plant. So, it may be necessary to tolerate some ground loops, with stable grounds being provided for groups of instruments.

An alternative to having voltages ground-referenced is to use differential voltages to carry the signal. The signals are differentially amplified, and since any ground loop voltages now are present in both paths, they will cancel out because they are common mode (common to both inputs). As the inputs to the differential amplification are not identically loaded, the differential amplifiers must have large input impedances, much larger than the source impedances. This helps ensure a high **common mode rejection** (CMR), as does having the source impedances balanced.

Even with the above, it is quite possible to have extraneous signals at the output of long wires to the instruments or actuators, due primarily to capacitive coupling of the instrument cabling to nearby lines. The differences in the amount of pick-up can be reduced substantially by twisting the wire pair carrying the differential signal. This makes the capacitive coupling substantially the same for the two wires of the pair so that high CMR is maintained. Twisted wire pairs are a common and cost-effective way of carrying signals, and are used in many applications, including telephone connections to local exchanges.

A further improvement can be obtained by shielding the signal pair, i.e., surrounding the signal-bearing lines with a conductor and thus inhibiting the capacitive interference. A typical high-quality cable has a foil shield around the signal pair and a copper drain wire. The shield should be connected to the signal ground rather than the amplifier ground, but not both (which would form a ground loop whose currents might induce coupling in the shielded pair). If the instrumentation amplifiers are also shielded by their metal enclosures and chassis, then these enclosures are attached to the cable shield and hence to the signal ground. No commons other than the signal common should be connected to the shield, however, as this would lead to ground loops and defeat the purpose of having differential amplification.

High frequency interference, usually called RF (radio frequency) interference, is also found in many environments. It can be due to such sources as sparks, flash lamps, and laser discharges, and the 'spikes' which can render digital equipment inoperable. Usually the circuit must be enclosed in a metal shield and shielded cable used. RF shielding should be grounded at both ends (to prevent RF reflections), however, in conflict with the conditions for common mode shielding. Thus the best shielding will have two separate shields: an inner shield for low frequency interference and an outer shield, terminated at both ends, for RF interference.

7.2.2 Isolation

Sometimes two circuits must be isolated from each other. A striking example occurs when medical instruments are electrically isolated to prevent any possibility of power supply voltages reaching the patient. Isolation transformers are effective up to about 5 MHz, above which they are ineffective due to stray capacitances.

Excellent isolation can be obtained using opto-isolators, which comprise a light source–detector pair in which the incoming signal flashes the light and the detector senses the flashes and outputs an appropriate signal. No electrical signals need cross the interface. Light sources include neon bulbs and LEDs, while detectors such as photoconductors and photodiodes are frequently used. The method is particularly appropriate for digital signals, as the systems handle high frequencies and discrete signal levels well.

7.2.3 The fibre optic alternative

Many of the above problems can be avoided by using fibre optic technology to carry the signals around the plant. These systems are

- insensitive to electrical noise,
- automatically carrying differential signals (i.e., presumably the sensor and the receiver are electrically at different common levels),
- high speed devices, and
- used for digital signals.

The basic cost can be higher than twisted wire pairs, but the many problems avoided, plus the potential for networking (section 7.4.2) are making this an increasingly attractive alternative.

7.3   COMMUNICATIONS NETWORKS

One of the reasons for using a computer element is that communications of various types in principle becomes rather easy: simply use the address and data lines cleverly. For a number of reasons, including desire for standardization of the interfaces and because the range over which the simple 0–5 V low power signals on the computer's buses can be transmitted is limited, there are several ways of organizing the communications, depending upon the task.

1.   **Modems** (Modulators/Demodulators) for communication between computers – usually one pair at a time – over telephone lines.
2.   **Local area networks** (LANs) for communication among a number of computer equipped elements.
3.   Commonly found **connections and protocols** for interfacing computers with their peripheral devices. These are particularly evident with microprocessors and include RS-232C serial connections, IEEE 488 instrument bus, and Centronics printer ports.

We look at those in the next few sections.

7.3.1 Standard I/O arrangements

The number of ways of connecting devices to the computer would appear to be almost endless. Some instruments are smart, as more and more are becoming, and hence provide some data smoothing, linearization, and A/D conversion, but there is still a demand from users to have straightforward connections with the computer. From this demand have evolved several conventions for instrumentation, of which we look at two. The first is the RS-232C connection, originally defined by the EIA in the US and also called, particularly in Europe, the ISO V24. The alternative is the general purpose interface bus, or GPIB, with the IEEE-488 bus being the defining standard, and occasionally known as the Hewlett-Packard interface bus (HP-IB) in honour of the manufacturer who defined the first version and helped make it a *de facto* standard.

RS-232C, called V24 in Europe, is only semi-standardized in that many implementations extend the definition or make presumptions about the hardware applications. Among the common factors are that the data is sent serially (i.e. one bit at a time) over the connection, only one wire is used for the data, and there is a 25-pin connector at each end of the cable. Of the several response, etc., signals available, in many cases only a few are actually used. The transmission signal levels used are $\pm 10 - \pm 12\,V$, and the receiver is intended to register a mark (or logic 1) on receipt of $+3\,V$ and a space (or 0) on receipt of $-3\,V$.

The range of an RS-232C connection is limited by the noise sensitivity of the data lines, which in some receivers need change by only $1.5\,V$ relative to GND to signal a mark or space and thus false signals may arise in electrically noisy areas. Overall, RS-232C transmission is defined for $15\,m$ range at up to $19\,200\,bits/s$ data rates; special implementations can obtain slower bit rates at ranges of 50–$100\,m$, while personal computer data exchanges tend to be at $115\,kbps$ (kilo bits per second) over 1 or $2\,m$.

To circumvent the noise problem, RS-422 and RS-423 schemes may be used. These are variants of RS-232 in which a pair of wires is used for each of the data links, and the signal is defined as the difference between the two wires. The effect is that noise, which would tend to affect both wires equally (common mode), does not appear in the data. The differences between the two schemes are due to the fact that RS-422 uses only the line difference for signalling and hence may be biased as desired, whereas RS-423 has the data line pair balanced with respect to ground potential. Both have ranges up to $2\,km$ before distortion and noise become problems; RS-422 allows data rates up to $1\,Mbps$ and RS-423 allows up to $100\,kbps$.

There are a number of differences between the RS-232C and GPIB (IEEE 488), including the fact that the latter is capable of carrying 8 bits in parallel, but the most striking one is a philosophical one: GPIB is a bus system in many ways similar to the computer's internal structure of data/address/control lines. The idea is shown in Fig. 7.2. The controller will almost always be the computer, and the devices could be either instruments or actuators. The terminology of the bus has the controller, plus listeners and talkers (who receive and transmit data, respectively).

The bus has 16 signal lines plus several ground return lines. Eight of the signal lines are used for bus operation, three for handshaking (which is the signal and reply which verify that the communication channel is open) and five for bus management. The other eight lines

IEEE 488 Bus

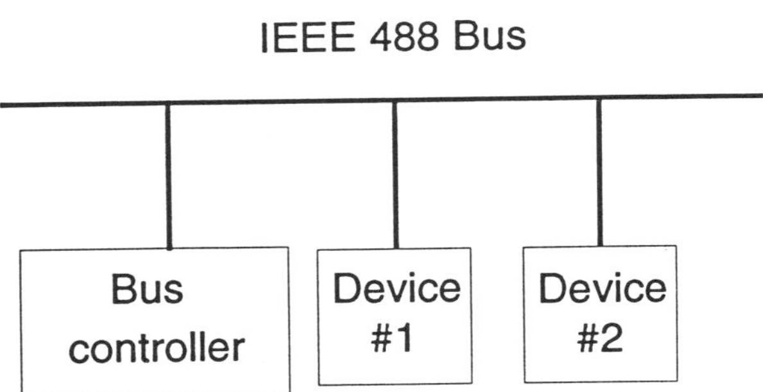

Figure 7.2 The IEEE 488 (GPIB, HPIB) is a bus oriented communication system with a central controller. It is used, for example, to have a data recording system poll the various sensors for data.

are bi-directional data lines and are used to carry both data and address information. The connector is a 24-contact type. Signal levels are TTL (0 V for true and 5 V for false logical). Typical maximum transmission distance is 20 m divided by the number of devices connected, and 2 m cabling lengths are typical. Up to 30 different devices may be addressed, but 15 is a more realistic limit to the system's capability. Short ranges can carry very high data rates: 20 kbytes/s for a few metres or tens of metres using IEEE-488 versions.

In operation, the controller polls the devices to find if they wish to receive, transmit, etc. It then sets up the conversations, one talker and one or more listeners at a time, and commands the data transfer to take place. IEEE 488 does not specify signal coding, but ASCII (American Standard Code for Information Interchange) is commonly used.

We mention the **Centronics connection** only because it is so popular. It is used principally for printers, although of course it can be used for other applications. There are eight data lines, so bytes may be carried in bit-parallel form, plus at least two control lines and several ground lines. The minimal control lines are a data-ready strobe and a receiver-acknowledge signal, used for simple poll-response confirmation, called **handshaking**.

7.4 COMMUNICATION BETWEEN COMPUTERS – NETWORKING OF SYSTEM ELEMENTS

Networking is a very new element on the control system scene; the standard approach has been to have multiple computer elements in a hierarchy with a master–slave relationship. This is evident in multiple computer devices such as robots, where the Unimation PUMAs, for example, have microprocessors controlling each joint and robot control is by a minicomputer supervisor.

Advances in computer networking, the sheer number of computers in a large plant (when counting all the microprocessors in the controllers and smart instruments), and the desire for more management information on activities (CIM – computer integrated manufacturing) have all combined to make pursuit of LAN ideas in manufacturing plants a topic of much interest, with emphasis currently on MAP (manufacturing applications protocol) and SP50 field bus.

Computer networking developed originally because of a desire to save cabling costs in large computer systems with shared resources. Thus printers can now be shared by several computers without running cabling from each computer to each printer; several user terminals (which tend to generate data at relatively low rates) may be attached to a single communication line, etc. When computer systems such as those at universities, comprise a group of terminals, possibly in a separate building from the computer centre, the network concept requires one line from the terminal room to the centre whereas the direct plan requires one line per terminal; the potential cost savings are obvious. The disadvantage of this approach is the need for increased sophistication at the nodes of the network, for each element must now be able to deal with the communication system as well as the computer elements.

Networks are also attractive for control systems, for similar reasons:

- cost savings due to less cabling;
- possible control advantages because of interactions of control elements (without need for a central communication node); and
- flexibility of equipment choices provided that the network contact points are compatible.

There are three properties which characterize a network: the physical topology, the technology of implementation, and the communication protocols.

7.4.1 Topology

There are three building block topologies for a network: star, bus, and ring. An implementation may be purely one or another, or may involve combinations of the three.

The **star** topology has the outlying elements all connected individually to a central element (Fig. 7.3(a)). In a control system, this would have all instruments and controllers connected by individual cables to a central computer, probably in a central control room.

In the **ring** topology, the cable is laid in a closed ring shape as in Fig. 7.3(b). Communications are usually uni-directional, and often two rings with contra-directional data flow are used. The ring's advantage over the bus is in reliability: a single break in the cable will still leave open a communication path between any two elements in the ring provided bi-directional flow can be allowed.

A **bus** topology for the system is quite like the bus layout of the CPU itself. Here (Fig. 7.3(c)) a single cable runs through the plant and the various instruments and controllers time-share usage of the wire. Traffic is usually bi-directional on the bus.

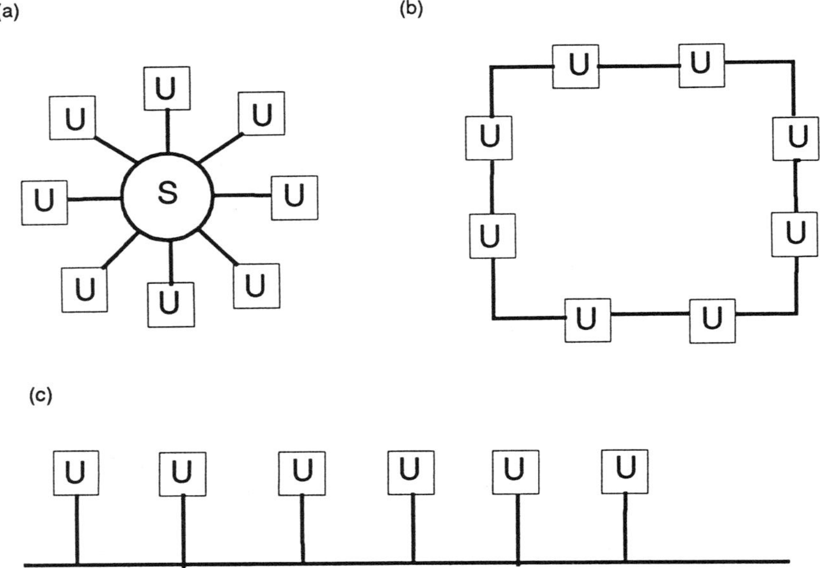

Figure 7.3 Networks can be configured with various topologies. Here we have in pure form (a) star, (b) ring, and (c) bus topologies. S denotes server or supervisor and U denotes user.

For many applications, a combination of star with either bus or ring layouts is useful (Fig. 7.4).   This is particularly appropriate to decentralized control, as control elements and their instruments may be placed in a star layout near their required plant locations (an increasing possibility because of improved hardening of computer and other electronic equipment).   Then only required information is placed on the bus for communication to other control locations or the central control and monitoring room.   A further alternative is the bus–bus topology, using IEEE 488 buses for short distance communication between decentralized controllers plus a long distance line for plant communication.

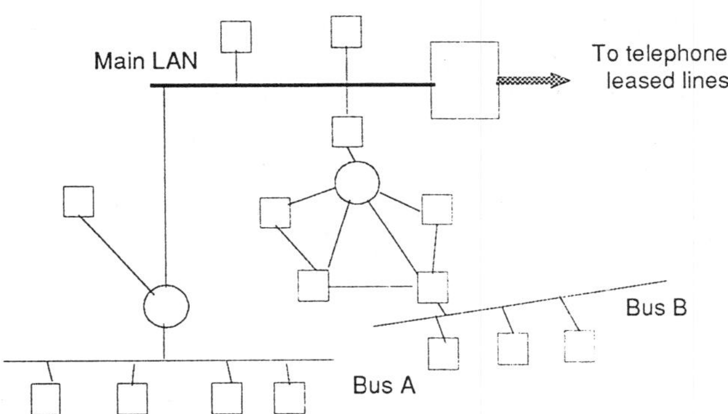

Figure 7.4   There is no technical reason that topologies cannot be mixed, with various buses attached to a master Local Area Network (LAN).   It is even possible to send plant data over the telephone system.   In this diagram, the circles are local LAN or communication supervisors, and the squares represent devices, possibly including inter-network ports, which are supervised.

7.4.2 Technology

The technology of the implementation is a major determinant of communication rates and distances for acceptable signal-to-noise ratios.   Although radio is a possibility (an early network was the Hawaiian Aloha radio communication network), the current choices are basically twisted-wire pairs, coaxial cabling, fibre optics, and occasionally unguided laser light communication.   The interfaces are almost certainly implemented with solid state electronics, and in commercial applications the electronics may be a set of one or two integrated circuit (IC) chips.

The possible communication speeds of the three main contenders are shown in Table 7.1. Most of these are well in excess of the requirements of control systems; 1000 sensors sending their 16 bit data 100 times per second requires only 1.6 Mbps (million bits per second) of capacity plus overhead. Noise problems are likely to at least contribute to making a choice, as industrial plants generate much electrical interference due to motors turning on and off, high voltages in power lines, etc. On these grounds fibre optics are receiving increasing attention for communications over multi-hundred metre distances; coaxial cabling is a very common choice within plants, and twisted wire pairs can be used over short distances (metres).

Table 7.1 Communication media – rates

Medium	Cost (units/m)	Cost (units per connector)	Data rate	Length (km)
Twisted-wire pair	0.1	0	10 Mbps	several
Coaxial cable	1.5	12	10 Mbps	2.5
			50 Mbps	1
Fibre optics	0.5	30	100 Mbps (to 10 Gbps)	100

The tabulated values are typical rather than maxima. Twisted wire pairs constitute an inexpensive but noisy and vulnerable carrier medium. Coaxial cable is very popular, while fibre optics are fast, noise insensitive, and safe in an inflammable or explosive environment; they are still unfamiliar and expensive, however, and are unable to carry electrical power.

7.4.3 Communication protocols

The communication protocols are the rules which determine how the network resources are shared. The first choice to be made is between a centralized controller of the network and distributed control. In fact, it is arguable this is the fundamental determinant of the network data flow. Certainly, control systems can – because of their real-time communication needs, particularly in an emergency situation – have requirements not found with data communications for banks or even telephone conversations between individuals.

Centralized control means just that: there is a single central unit (probably the central computer) which is master of the system. All other units wishing to transmit must gain the permission of the master. The master can also override the slaves. The CPU usually functions in this role within a single computer system, with demands conveyed via interrupts and sense lines and the CPU setting up communication paths such as DMA (direct memory access). The network is philosophically an extended version of the computer bus concept.

In **distributed control** all of the units attached to the network bus or ring are essentially equal, although this is a fairly pure form since priority setting is often possible. The units must contend for the use of the network. Some of the possibilities are CSMA/CD, token rings and buses, and register insertion rings.

CSMA/CD stands for carrier-sensed multiple-access with collision detection. It dates to a radio data communication network (ALOHA) and is analogous to one even when implemented over cable. A node wishing to send a message checks the communication medium to see if it is in use (carrier sense). If so, the node waits. If not, the node begins to transmit its message. If another node has also been waiting and starts to transmit simultaneously, the messages interfere; both transmitters will sense this (collision detection), cease transmission, and retry the transmission some random time later. Available commercially, under such names as Ethernet, these systems give good performance when lightly loaded. A problem is that a maximum delay that a particular message might endure cannot be guaranteed.

In **token-passing networks**, a short data sequence called a token passes from one node to another in sequence on the network. A node receiving the token, but no other node, is allowed to transmit its message if it has one. Upon either not having a message or on completion of its transmission, a node passes the token (i.e. sends the token code) to its successor on the net. The most famous token network is a ring network implemented by Cambridge University and called the Cambridge ring. Token-using networks can be arranged so that maximum delay is restricted (to the time for which all other nodes send one message plus the time for sending a node's own message), but on the average a minimum wait of the time for the token to pass through half the nodes is entailed even when the network is lightly loaded.

Buffer-insertion rings – there seems to be no point in having buffer-insertion buses – use a two-message length buffer at each node.

A node wishing to transmit fills one of its buffers so that it will be ready to transmit if given the opportunity. If a break in transmission by other nodes occurs, the node starts to transmit from its buffer (similar to CSMA). If another message starts to pass through while this node is transmitting, the message is fed into its second buffer and passed on only when it has finished its own transmission, i.e. it saves rather than cancels as in CSMA/CD systems. This system has been implemented so far at a university (Ohio University's DLCN/DDLCN – a variation of the Cambridge ring also using a partial buffer) and has the potential of higher throughput than other schemes. It can be as rapid in response as CSMA/CD; but it does not appear possible to guarantee an upper bound on the delay.

7.4.4 What's happening – MAP/IEEE 802.4 and others

Manufacturing application protocol (MAP)

The evolving standard for control systems applications networks appears to be MAP, which implements a token-passing bus approach to the network. The concepts are defined by IEEE Standard 802.4. In fact, networks are defined in terms of seven layers (see Table 7.2) according to the Open System Interconnection (OSI) reference model of the International Standards Organization (ISO).

Table 7.2 ISO Model

Layer	Description	Layer function
7	Application	Program program
6	Presentation control	Translate format
5	Session control	Make and keep connect
4	Transport end-to-end control	Error check
3	Network control	Message routing
2	Logical link control Medium access control	Intranet error check
1	Physical control	Signalling physical connect

This model is partly the result of liaison between the IEEE and the European Computer Manufacturers' Association (ECMA) begun in

1981. The IEEE standards apply to the bottom two layers of the model. The logical link control falls under IEEE 802.2; medium access control and physical control are covered by 802.3 for CSMA/CD buses, 802.4 for token-passing buses, and 802.5 for token-passing rings. Other standards are under development.

In the case of MAP and the similar proprietary ARCNet, the network is physically of bus structure and operationally a token bus, in which the token passes sequentially from station to station. After a station finishes transmitting data frames, it sends a token-containing frame to its successor. The successor, now in 'possession' of the token, may transmit its data frames if it has any. When finished, or if it has no data, it transmits the token to its successor, and so on. Various procedures, mainly based upon the transmitter of the token listening to see if anything happens, are available for establishing if a successor has failed or gone off-line, for adding new stations to the bus, for setting up priorities for stations, for limiting the number of frames transmitted when in possession of the token, etc.

Standard 802.4 describes three media and transmission methods. These are indicated in Table 7.3; notice that data rates are upward of 1 Mbps.

Table 7.3 Media and Transmission

Data rate	Medium	Drop cable	Signalling
1 Mbps	Coax (RG-6,RG-11 CATV)	25–50 ohm < 350 mm	Manchester Coding
5 & 10 Mbps Baseband	75 ohm semirigid	RG-6 'very short'	FSK direct
1, 5, & 10 Mbps Broadband	CATV		AM three-level

The token-bus system helps provide some of the properties needed in factories: flexibility in topology, with trees, stars, and repeaters; upper bound on response time; reliability checking; possibility of having a master station. Range is dependent upon the speed desired and the number of repeaters, but the order of hundreds of metres is achievable.

Although MAP was developed in part from manufacturers needs, it is in some respects an 'overkill' for the needs on the factory floor.

This may be seen by examining Table 7.4, derived from LeFebvre (1987), which compares functional requirements of the six levels of the National Bureau of Standards model of factory computing systems.

Table 7.4 Functional requirements and suitable communications networks for levels of NBS factory model*

NBS factory level	*Data type (nature)*	*Time scale*	*Possible network*
Corporate Plant	Accounting Inventory Data processing (Batch orientated)	Hours	MAP/TOP Ethernet etc.
Area Cell	Data consolidation ↑ Coordination↓ (Intermittent)	Seconds	MAP MAP
Workstation Equipment (PLCs, PIDs)	Real-time data Direct Digital Control (High Speed Periodic)	msec	mini-MAP proprietary network

* Derived from LeFebvre (1987).

The last two, or maybe three, levels involve communications of a rather specialized type, where the programmable controllers and devices do the actual process control. They operate in real time and need information often, but in small amounts; e.g. each device needs an 8-bit byte of data every few milliseconds (ms). Although MAP is applicable in principle, it appears that not every computing device needs the full generality of MAP. In fact, the Instrumentation Society of America (ISA) has proposed a set of standards for this level which in effect has only layers 1, 2, and 7 of the OSI model. Implemented in a form called Mini-MAP, it is compatible in most respects with MAP. The primary difference, and the reason the devices will not be plug-compatible but will need bridges, is the use of a carrier band. This gives frequency modulated signals rather like frequency shift keying (FSK, in which a different frequency is used briefly for each symbol) superimposed on a carrier. Cable distances are limited to 1.5–4.5 km.

LAN alternatives to MAP

There are competing alternatives to the MAP token-bus systems. One of these is the common Ethernet approach, which some plants are finding adequate in spite of suspicions that its CSMA/CD protocol is not suitable for time-critical communications; one significant advantage is that front offices already have Ethernet installed, so plant-wide networking for such purposes as implementing CIM reduces to connecting the rest of the plant.

A somewhat different approach uses a bus structure similar in concept to the computer's internal bus, for communications. The approach may be called a **field bus**, and essentially has a simple bus such as a twisted wire pair with access controlled by a single supervisory computer. Not atypical is Intel's BITBUS, in which as many as 256 modules are connected to up to 10 km of a twisted-wire pair with communications at 2.4 Mbaud. Concentrators can be used to connect a number (say 16 or so) I/O devices to each module connection, while gateway modules are one bus-connected way to allow communications with parallel input or output devices. The host computer can also, using other buses, networks, or modems, connect to other parts of the system.

Another variation is Europe's SP50 field bus, currently being defined and standardized. The underlying philosophy is to replace the present approach of wiring individual 4–20 mA signal devices to their appropriate sources or sinks, sometimes over long stretches of wiring, with a digital link shared by many devices. Part of the definition effort is devoted to defining and subdividing the OSI physical layer appropriately; the likely outcome will have one protocol for existing systems which derive their power from the field wiring and another protocol to meet requirements of higher performance systems including for example PLCs.

Although serving a communication function, field buses are not considered true networks, but are perhaps best considered a notch down from networks in sophistication. Claimed advantages of this are that they are simple in concept and installation, use existing technology, and, because they are like field wiring rather than a communications system, can be handled by electricians.

7.5 NETWORK ARCHITECTURE – CENTRAL vs. DISTRIBUTED CONTROL

Once it has been decided that a plant will have sensors and actuators in various places, and that these will be computer compatible, then it is possible to connect them either to a centralized control computer at the one extreme or by independently pairing sensors with actuators at the other. The trade-offs include optimality and (perhaps) computer cost-effectiveness in the first case versus fault isolation and short cabling in the second. In fact, because of the ubiquity of microprocessors in loop controllers, PLCs, and smart instruments, and because microprocessors are relatively easily networked, many systems are naturally employing distributed, hierarchical control.

Decomposition of a system into a hierarchical one can be done on several different grounds.

1. **System structure** This can seem very natural to the designer. For example, a pump motor may itself be a controlled device and may in addition be used to maintain a liquid level. Although the level error might in principle be used directly in control of the motor, it is likely that in fact the two loops – level control using the pump and pump speed control – will be nested.
2. **Levels of control** In this approach, direct device control will be the lowest level. Computation of set points for the devices will be a supervisory level. Adaption of the set point computations to changing system characteristics might be a third level in the hierarchy.
3. **Levels of influence** When this is the criterion, fast time constant devices will be at one level and slower ones at a higher level. An example is in control of the aerodynamic surfaces of an aircraft by an autopilot. These will require tight loops and rapid response in the presence of disturbances. The navigational corrections, on the other hand, may be made less rapidly.
4. **Theoretical** Theoretical arguments involving process models can be helpful in determining hierarchies and the controllers to be used at each level. The results are only partial ones, however, although like many theoretical arguments they can help support decisions (see Chapter 34).

7.6   COSTING

Costs are difficult to pinpoint in general. Wiring and shielding may be more of a labour than a materials cost, for example, while differential and isolated amplification are design problems. For standard connection types, the actual computer interfaces can range in cost from $50 or so for RS-232C connections and several hundred dollars per GPIB element through several hundred dollars per computer for a simple CSMA/CD LAN hook-up up to several thousand dollars for a computer modem. For some computers, signal conditioning modules are available which take in raw sensor signals such as those from thermocouples and output RS-232C or IEEE 488 signals; such cost a few hundred dollars per module, where the module may allow several inputs.

7.7   FURTHER READING

Information on wiring and shielding tends to be scattered, although some electronics textbooks will contain it and some issues are covered by standards, or at least by standard practice. Some discussion is in Hunter (1987), and special textbooks such as Webb and Greshock (1990) are helpful. A recent text on interfacing is Derenzo (1990).

Networks are discussed in the communications literature, e.g. Flint (1983) or Gee (1983), with a useful summary in Sloman (1982). One useful text is Tannenbaum (1988).

Details of standards such as RS-232C tend to appear in the computer literature. The ongoing development of networks in industry may be followed in the engineering literature, often under the classification of distributed control (e.g. *I&CS*, Sep. 1990).

Architecture of manufacturing computer systems is often a topic of the control engineering periodicals (see e.g. LeFebvre, 1987) and of texts such as Groover (1987). General theoretical issues are discussed by Larson (1979), and the textbook by Leigh (1988) has a partial presentation.

Developments in a manufacturing environment are most easily followed in the trade journals such as *I&CS*.

8

Control laws without theory

Many control systems can be and have been set up with almost no reliance upon the mathematical theory of control. Many more have been based upon a notion of how the mathematics might turn out but with no explicit reliance upon that tool. This is true even leaving aside the PLCs devoted simply to turning on and off the various machines involved.

Such an approach has a long and partially distinguished history. Early windmills, with their vanes to keep the main propeller pointing to the wind, doubtless had no supporting theory, but only the understanding of some clever inventors of how things worked. (Similar things could be said of boat designers, who learned from their successes – and disasters.) The Watts governor, famed as the first of the explicit control devices, had no theory until Maxwell solved the 'hunting' problem in 1869.

Even today, many systems must do without strong theoretical underpinnings, as the mathematics may be too difficult, too expensive, or irrelevant. More to the point, a number of systems have not been modelled in detail, and hence the manipulation of mathematical models upon which theory is based is not possible.

Such systems are still controlled, however. The control is based upon the engineering artistry of the designer and the skill of the system operators, just as was the case with early aeroplanes. In this chapter we review a few of the schemes in which a system is compensated and tuned through the use of heuristics and tuning. The reader will derive from this a notion of what to do when the theory fails, an appreciation of some real systems and how they are controlled and operated, and an introduction to the heuristics of control.

8.1 SYNOPSIS

The parameters of many control systems may be adjusted to yield good performance, i.e. the control systems may be 'tuned', without recourse to much theory.

1. PLCs are, by their nature as relay logic substitutes, virtually theoryless.
2. A new breed of control devices are self-tuning. Their tuning algorithms were built in by the engineers who designed them.
3. For many of the simpler control laws, and particularly the widely used PID law and its subsets, there are rules of thumb for setting the parameters. These schemes typically involve observing system operation under open-loop or simple closed-loop conditions, estimating parameters of their responses, and inferring satisfactory values of the control law parameters.

These all have their limitations, of course, and few engineers would install them without having a good understanding of the plant.

8.2 PLCs – THE SEMICONDUCTOR-IMPLEMENTED RELAY BANKS

A great deal of what is taught about control systems seems exotic: advanced algorithms for shaping missile trajectories, fancy filters for extracting information from noisy data, methods for moving robot arms efficiently and rapidly, sensitive instruments for monitoring pollution. Underlying most of these are much more mundane tasks: turning equipment on and off, opening and closing valves, checking sensors to be certain they are working, sending alarms when monitored signals go out of range. Process control plants, aircraft, and indeed most controlled systems share this need for simple but important operations.

The tool for handling these simple operations, of which there may be hundreds or thousands in a single factory, is the programmable logic controller (PLC). It is easiest, and also historically correct, to think of the PLC as effectively a large bank of relays implemented in a computer. The great advantage of the PLC over relay banks is its programmability – the implemented logic can be altered by changing a computer program rather than by changing the wire harness of a set of relays.

Implementation of logical commands does not require the engineer to be a theoretician, but a careful and thorough understanding of the plant and its elements is necessary. We have already met much of the PLC's hardware in Chapter 5 and hinted at its software in Chapter 6. In this section we partially reprise those presentations and add some discussion of applications.

8.2.1 The basic scheme

At its simplest, the PLC implements turning on a motor (say) via logic. The simple form of the operation is shown in Fig. 8.1(a), a relay form in Fig. 8.1(b), and a PLC implementation in Fig. 8.1(c).

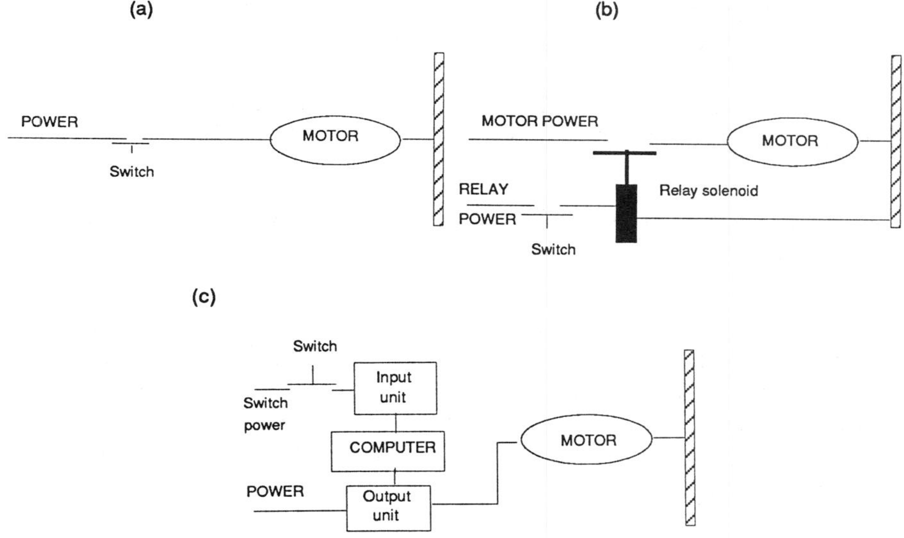

Figure 8.1 The progression in turning a motor on and off from (a) simple switching of the motor power to (c) using a pushbutton to activate a computer which, after suitable logic checks, commands its output unit to allow the power to be connected to the motor.

Because the relays are used in (b), the relay switch need not carry the motor power and could in fact require only the few volts sufficient to power the relay coil. The PLC carries this further: input power is matched to the input unit, and could be an instrument signal (4–20 mA), the computer is basically a low power 0–5 V device, and the output unit can match the output command to the motor requirement, e.g. 120 V, 240 V, or 400 V AC, or perhaps 0–24 V DC. Further, the computer is able to implement logic such as 'turn on motor 2 s after switch is closed provided some other (e.g. safety switch is closed) conditions are met'.

8.2.2 Programming in ladder logic

The PLC turns output commands on and off after making logic checks. In essence, it implements checks on whether contacts are open or closed, runs timers and counters, and does sequencing. The usual logic tests are represented in **logic ladder diagrams** and programming of the computer can be done from those diagrams, although manufacturers may offer other options.

The basic elements available in the PLC and their symbols are shown in Fig. 8.2.

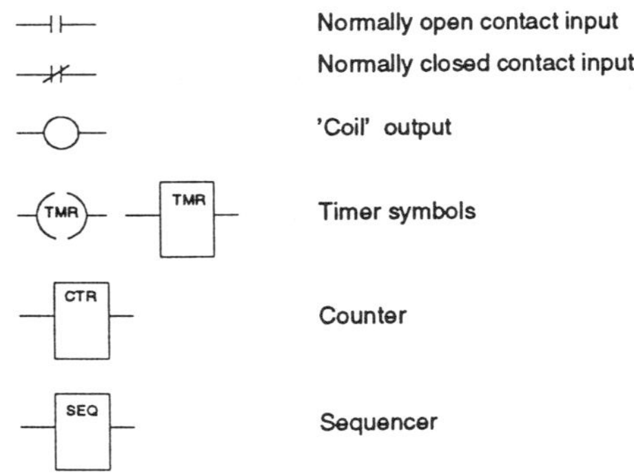

Figure 8.2 A few of the symbol types used in representing PLC programs. These are not standard.

Along with the basic logic functions, PLCs are now available with arithmetic capabilities. These capabilities are sometimes combined with a special purpose auxiliary module to allow continuous (rather than just on–off) output commands. These modules may even have a PID capability (section 8.3), which overlaps into the 'tuning' area and definitely requires more than logical decisions from the engineer or technician setting up the program.

In using the PLC, the logic program – the logical relationship between inputs and outputs – must be determined. There may be a great many instructions of the type

> IF the motor switch is *ON* and the appropriate alarms are *OFF*
> THEN start the motor turn-on sequence.

and the components must be assigned codes (often numbers) that are related to the I/O wiring of the PLC. Thus the command above might translate to:

> IF normally open 0011 is closed
>   AND normally closed 0001 is closed
>   AND normally closed 0003 is closed
> THEN operate sequencer 0201
>   with 2 second steps
>   to close switches 1011 and then 1012 and then 1013
>   to the 'coils' 1111, 1112, 1113 respectively

The corresponding ladder diagram is in Fig. 8.3.

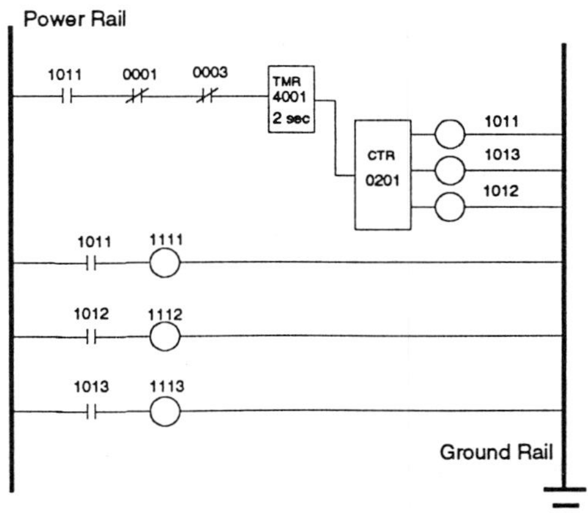

Figure 8.3 A simple PLC program in ladder logic. When pushbutton 1011 is pressed and safety interlocks 0001 and 0003 have not been tripped, then after two seconds a counter is started. This successively trips switches 1011, 1013, 1012 to do the motor start-up sequence.

8.2.3 The programming of PLCs

The standard way of programming the PLCs is to enter the ladder diagram into the computer using a special keyboard and to look at the display on a dedicated device, either a VDU (computer visual display unit) or liquid crystal display. The display will look much like Fig. 8.3, and is said to be easy for technicians to work with. Hard copy (in the form of printer output) is easy to obtain.

Programming of PLCs has also been done in other languages, notably BASIC and its relative FORTRAN and, to a lesser extent, versions of Pascal; one assembly language type approach is that of Siemens' STL. These can be particularly useful if the PLC also allows arithmetic functions and proportional control laws.

8.3   TUNING OF CONTROLLERS

When the control computer's output is in other than a simple logical relationship to the input, there are usually a number of parameters which can be selected. Such selection is called tuning of the controller and is often refined even after theoretical studies, often with the aid of simulators or with the actual system. In some cases, perhaps driven by necessity or convenience, this adjustment is done without theory. This theoryless tuning can be done by special self-tuning controllers, a recent innovation, or by manual means starting with rules of thumb.

8.3.1 Self-tuning controllers

When command signals to the process are functions of errors between desired and actual outputs, the determination of what functions to use and what the parameters of the functions should be is at its core a matter requiring some theory. Long experience with certain function types, plus a certain *ad hoc* intuitive justification has allowed many practitioners to manage with only a modicum of theory. So, instead, they use now-standard functions and 'tune' their parameters to the job at hand.

Since this tuning required manipulating the parameters to achieve a certain desired response pattern to system transients due to start-up, set point changes, or disturbances, it was perhaps inevitable that computers were programmed to do the pattern recognition and parameter manipulation. Controllers using such techniques are adaptive controllers (Chapter 31), although the special nature of their functions and implementations has led them to be described as 'self-tuning'.

The usual function implemented in commercial self-tuners is PID, or three-term, control (see section 8.3.2). The tuning algorithms are mostly proprietary, but would (at least philosophically) be related to the manual tuning algorithms (section 8.3.3). A special subroutine in the control computer may be visualized as overseeing the operation of the PID law and changing the three parameters of that law when the error between desired process output and actual process output shows patterns (such as oscillations) indicating change was necessary.

Claimed advantages of self-tuning, some proven in a research environment and others also established in an industrial environment, include the following.

1. Skilled process engineers are not needed to tune the process initially.
2. Control will be quite good regardless of set point changes or load disturbances.
3. Technicians on start-up can spend their time on tasks other than tuning.
4. The plant is always in tune, so its operation is more efficient. This can result in savings in energy and materials, and improvement has been shown to be substantial in some research programmes.

As commercial self-tuning controllers may even be delivered with 'good guess' initial parameter values, the control engineer may well see this as theoryless control of a device. A responsible engineer, however, might be unwise to install one of these (or any other) black boxes 'blindly' without having some notion as to what it does. Particularly important may be considerations of what happens to the algorithm when things go wrong with the plant. Typical desirable capabilities may be as follows.

1. The ability to find out the current parameters and to manually override them.
2. The ability to enable and disable self-tuning.
3. The ability to instigate a special plant input disturbance-like signal to cause a self-tuning cycle.

Of course, if manual tuning is available and is to be used, the process engineer must have some notion of how to do the tuning. This is the topic of the next section.

Several commercial PID controllers are self-tuning (Clarke, 1986). One of the first of these was the Foxboro 'Exact' self-tuner. While manual tuning and set-up are of course possible the basic self-tuning is based upon the tracking of error transients introduced by load-disturbances (Fig. 8.4) The disturbances are assumed to be of the form of step inputs.

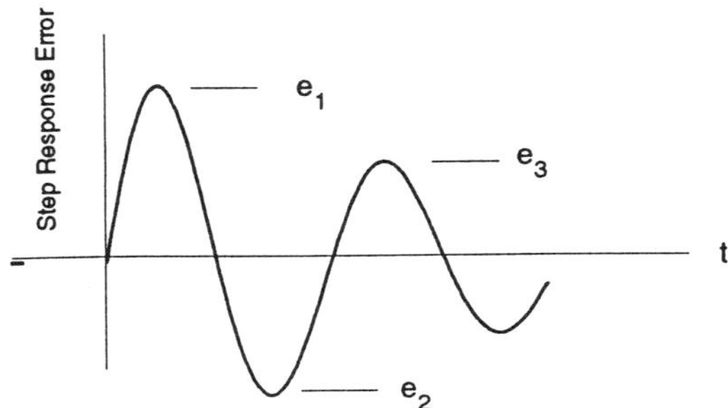

Figure 8.4 A self-tuner may apply a small perturbation and measure the error response. One algorithm uses the peaks e_1, e_2, e_3 to set new PID parameters. After Clarke (1986).

The transient is tracked whenever $|e|$ exceeds a user-set noise band and the algorithm looks for three successive peaks e_1, e_2, e_3 as shown. The computed values

$$\text{Overshoot} = \left| \frac{e_2}{e_1} \right|$$

$$\text{Damping} = \left| \frac{e_3 - e_2}{e_2 - e_1} \right|$$

are compared with user-prescribed values. If the computed quantities are too small, for instance, it is inferred that the gain is too small and the gain is increased; if peaks are indistinct, the integral and derivative time parameters are reduced. There is an elaborate set of interlocks which guide the tuning so that the transients take on a desired form and size; these have not been published in the open literature.

Another self-tuner is that of Turnbull Controls, for which the details have again not been published. This appears to perform a system identification and design a PID controller according to some rule (see Chapter 30 and also section 8.3.2 below).

8.3.2 Heuristics of PID controllers

The standard process control law has a hundred-year plus history, rules of thumb for tuning dating back fifty years, and possibility for implementation with purely mechanical or pneumatic controllers. It is called the three-term, or proportional-plus-integral-plus-derivative (PID), law and has as a special case the proportional-plus-integral (PI) law. The PID rule, relating the controller's output command $u(t)$ to its input signal $e(t)$ as

$$u(t) = K_p \, e(t) + K_i \int^t e(s)\,ds + K_d \, \frac{de(t)}{dt} \tag{8.1}$$

is often built into commercial controllers, with the designer or operator expected to set the three constants K_p, K_i, and K_d according to his application. A popular alternative form of this is

$$u(t) = G_k \left[e(t) + \frac{1}{T_i} \int^t (e(s)\,ds) + T_d \, \frac{de(t)}{dt} \right] \tag{8.2}$$

where T_i is called the integral action time or reset time, and T_d is called the derivative action time or pre-act time.

Because the output depends upon the sum of three terms, these are sometimes called, logically enough, three-term controllers. Other names include loop controllers, PID controllers, and process controllers.

One sampled-data or digital computer version of the above is defined by using a trapezoidal approximation for the integral term and a simple first difference for the derivative, to yield a near-equivalent form

$$i(nT) = i(nT-T) + \left(e(nT) + e(nT-T)\right)/2$$

$$u(nT) = K_p\, e(nT) + K_i T\, i(nT) + K_d\left(e(nT) - e(nT-T)\right)/T \quad (8.3)$$

where T is the time between successive computations (and between samples, and between commands). Suppressing the T in the indexing and incorporating the Ts in the constants yields

$$i(n) = i(n-1) + \left(e(n) + e(n-1)\right)/2$$

$$u(n) = K'_p e(n) + K'_i\, i(n) + K'_d\left(e(n) - e(n-1)\right) \quad (8.4)$$

The equation and notation of (8.2) also have a digitized form.

Such controllers are frequently used in large systems which are not modelled sufficiently for extensive analysis to be done, as often they can be tuned to give satisfactory performance. In the subsections below we develop the heuristics associated with each of the three terms and then present a few of the tuning rules which have been proposed.

It is possible to argue that the various terms of the three-term controller are a good idea – especially since the result is so often satisfactory. In this section we discuss the P-term, then P+I, then P+D. These are of course not derivations but justifications. A more profound question is: why do the simple control laws often work as well as they do?

Proportional control – the P term

Manual tuning has a heuristic motivation which is fruitfully traced to gain insight into the classical approach to control system design. If a plant is not responding accurately to commands sent to it, the first step is usually to introduce a feedback loop, compare the actual output to the desired output, and send a command to overcome the difference. In Fig. 8.5(a) is shown a schematic for this configuration. (We remember that this is in fact an abstraction in which a number of realities have been ignored: desired output is probably a number rather than a signal; actual output is probably a signal which measures output using a transducer; error may be either a number or a signal, but the controller (Fig. 8.5(b)) is probably a digital computer device and hence works with numbers; plant commands may be voltages, hydraulic pressures, or other signal carriers. Our implicit assumption, made explicit here, is that such scaling as is necessary has already been done.)

(a)

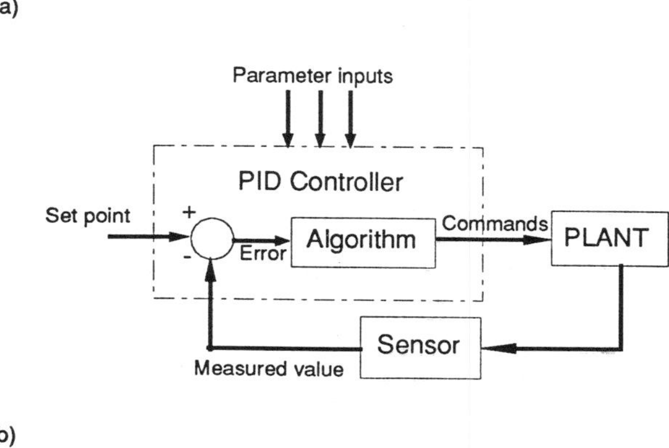

(b)

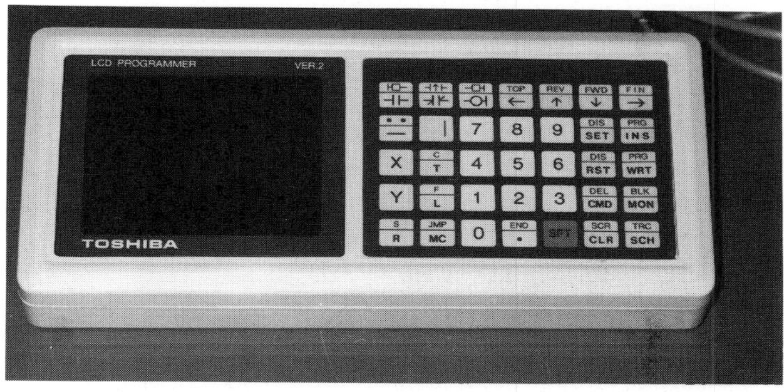

Figure 8.5 The three-term, or PID, or loop controller. (a) schema for inputs, (b) an industrial controller (see Fig. 5.7) opened to show three adjustment dials; the set point may be input using the front dial.

Given the simplifications, we still find this a useful model. To give ourselves a notation, we define

$$\text{desired output} = \text{input} = r(t) \text{ (or in sampled form } r(nT),$$
$$\text{denoted by } r(n))$$

$$\text{actual output} = \text{output} = y(t) \text{ (or } y(nT) \text{ or } y(n))$$

$$\text{error signal} = e(t) \text{ (or } e(nT) \text{ or } e(n))$$

$$\text{control signal} = u(t) \text{ (or } u(nT) \text{ or } u(n)) \tag{8.5}$$

and we remark that the definition of the error has

$$e(t) = r(t) - y(t) \tag{8.6}$$

Now, if $e(t)$ is positive, then $y(t)$ is too small and a control signal is sent to increase $y(t)$, with the opposite if $e(t)$ is negative. Furthermore, if the magnitude of the error is rather large, to have the magnitude of the control signal rather large also seems reasonable. Thus we can argue that a reasonable control law is

$$u(n) = K_p \, e(n) \tag{8.7}$$

As the command is proportional to the error, this is called a proportional (P-type) controller.

The amount of corrective command put out for a given perceived error will depend upon K_p. Heuristically, we will have a large corrective effort, and hopefully rapid response, if K_p is large. Also, we might expect that steady-state error would be small if K_p is large, since there would be a substantial correction for small $e(n)$. On the other hand, if the error signal is noisy, there will be a lot of control effort expended in tracking this noise if K_p is large. Also, if K_p is large and the error can also be large, the command signal might try to overdrive the plant. Responses of a simple system controlled using a proportional control only (also called a gain compensation), are shown in Fig. 8.6. Note that the apparent steady-state error between the actual output and the desired output (=1) decreases with increasing gain K_p, but the oscillation of the transient increases in magnitude at the same time. In fact, for $K_p = 2.5$ the output maintains a sustained oscillation.

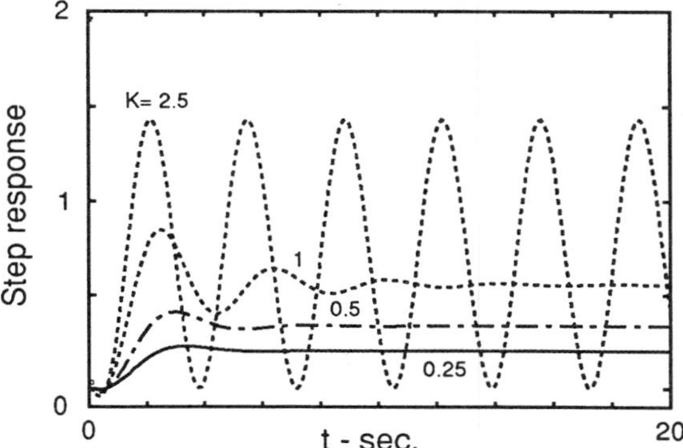

Figure 8.6 Step responses of the simple closed-loop system for various values of proportional gain K. (The examples in this chapter all use the plant transfer function (Chapter 10) defined by $(-s+4)/((s+2)(s^2+2s+2))$

Integral control – the I term

Proportional control will not always do a satisfactory job. In fact the system sometimes acquires a steady-state error. We shall see why later (in Chapter 17) when we look at the mathematics, but the effect is similar to that in which we are sending a voltage to a motor which is supposed to rotate at a certain speed. Here, if the error is zero then the command is zero and the motor will stop and only restart when the error is non–zero. Eventually one hopes for steady offset in the output rotation velocity which yields a steady voltage command which maintains constant speed. For the error to be small in these circumstances, however, K_p must be very large which, if nothing else, will entail extremely large voltages to the motor when it is first switched on and $e(n)$ has significant size.

To overcome the above problem, another term is introduced into the control law. Suppose that $e(t)$ is integrated. Then the integral will be small if the system oscillates but will gradually build up if a steady offset, even a small one, is present. Thus if

$$u(t) = K_p e(t) + K_i \int^t e(s)\,ds \tag{8.8}$$

we might expect the steady-state error to eventually disappear, with the integral retaining the offset information needed for the control to maintain a constant command. The above is called a two-term or proportional-plus-integral (PI-type) controller and is quite common. There are now two parameters to set (K_p and K_i) and tuning can be in stages: first set the rapid transients using K_p and then tune out the steady offsets at a rate determined by K_i. The integral term is sometimes called a 'reset' term and is associated with a minor problem of its own called **reset windup**, in which the integral becomes very large early in the operation (due perhaps to trying to follow big step changes in desired output) and dominates the proportional term. A typical remedy is to reset the integral to zero periodically.

The implementation of the PI law in a digital computer is straightforward, and the reset logic is easy. The implementing equation will usually be derived from

$$u(n) = K'_p e(n) + K'_i \sum^{n} e(i) \tag{8.9}$$

but will probably effectively accumulate the sum in its own register.

$$s(n) = s(n-1) + e(n)$$

$$u(n) = K'_p e(n) + K'_i s(n) \tag{8.10}$$

Another popular variation of this is to compute what is essentially the variation in the control. Thus, noting that

$$s(n) = e(n) + s(n-1)$$

we have

$$u(n) = u(n-1) + K'_i e(n) + K'_p \left(e(n) - e(n-1)\right) \tag{8.11}$$

which has the virtue of maintaining constant control command whenever the error becomes zero.

The influence of I-terms as shown in Fig. 8.7, is that the integral term seems to decrease the steady-state error, but increase the oscillation for the given value of proportional gain.

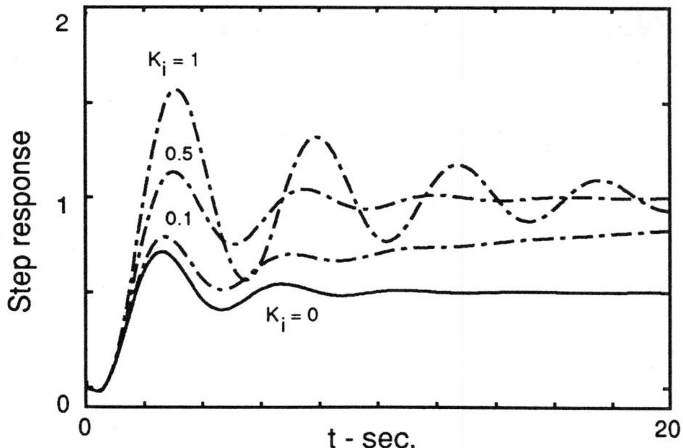

Figure 8.7 Step responses for proportional gain = 1 and various values of integral gain K_i, showing the effect of integral control. Notice that the response eventually approaches 1. (Refer to Fig. 8.6.)

Derivative control – the D term

In use, the above sometimes need large gains to induce rapid enough response. Unfortunately the systems then end up lightly damped, demonstrating overshoot and oscillations of unacceptable levels. For this reason, the controller should contain a 'braking' term, which will act to decrease the magnitude of the command as the error comes close to zero at high speed, but increase the magnitude of the command if the error is diverging from zero. Clearly, then, the quantity of interest is the derivative of the error, and a candidate controller is

$$u(t) = K_p e(t) + K_d \frac{de(t)}{dt} \tag{8.12}$$

Examination of this shows that if $r(t)$ = constant and $y(t) < r$ and increasing, then $e(t) > 0$ and its derivative is negative. Thus $u(t)$ will be smaller (assuming $K_d > 0$) than for the P case, so the system's error will not pass through zero as rapidly as before and in fact as it passes through zero, the command will already be trying to slow the system; hence, the control is proportional to a prediction of future error. Other cases may be argued similarly.

Control laws such as (8.12) are proportional-plus-derivative (PD-type) laws and, although justifiable, can be hard to tune and are less common than PI controllers. The digital implementation of the law typically uses a back difference to approximate the derivative, i.e.

$$u(n) = K_p\, e(n) + K'_d\left(e(n) - e(n-1)\right) \qquad (8.13)$$

Examples of PD control are shown in Fig. 8.8. Starting with a steadily oscillating output with proportional control only, increasing K_d apparently increases the damping of the oscillation.

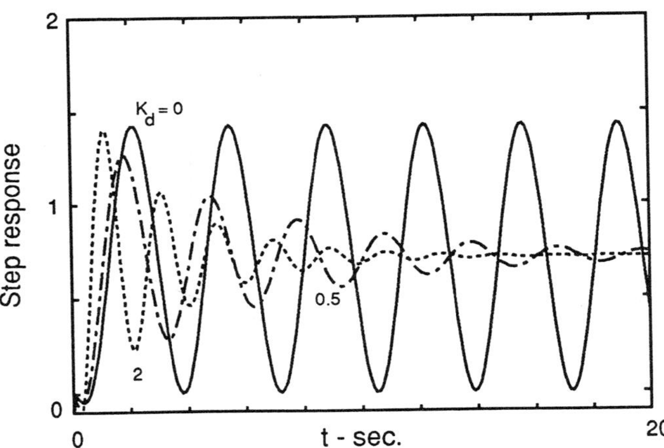

Figure 8.8 The effect of derivative control, demonstrated for various K_d with $K_p = 2.5$. Notice the increased damping. (Refer to Fig. 8.6.)

8.3.3 Rule-of-thumb tuning of PID controllers

A common industrial form uses all three of the above to form the three-term, proportional-integral-derivative, or PID control law. The implementation may be of several forms, but generically is

$$s(n) = s(n-1) + e(n)$$

$$u(n) = K'_p\, e(n) + K'_i\, s(n) + K'_d\left(e(n) - e(n-1)\right) \qquad (8.14)$$

Tuning with three parameters (plus other parameters such as sample rate and integral term reset period to be set) can be tricky and it seems in many cases to be based on perturbation from settings found to be useful in the past. When starting from scratch on systems in which the steps are allowed, two approaches may be used: the first uses the controller in a closed-loop situation, while the second starts from the system's open-loop step response.

Even when well tuned, PID controllers have a problem in that they do not take explicit account of time delays, as opposed to time lags.

Closed-loop tuning

A typical and famous algorithm is that of Ziegler and Nichols (1942), as follows.

Step 1. Operate the system with the loop closed and with K_i and K_d set to 0.

Step 2. Start the system and vary K_p until the system begins a steady oscillation. Set $K_{crit} = K_p$ at oscillation, and let P_{crit} = period of oscillation.

Step 3. For P control, set

$$K_p = K_{crit}/2 \tag{8.15}$$

Step 4. For PI control, set

$$K_p = 0.45 K_{crit}$$

$$K_i = K_p/T_i, \text{ where } T_i = 0.83 P_{crit} \tag{8.16}$$

Step 5. For PID control, set

$$K_p = 0.6 K_{crit}$$

$$K_i = K_p/T_i, \text{ where } T_i = 0.5 P_{crit}$$

$$K_d = K_p T_d, \text{ where } T_d = 0.125 P_{crit} \tag{8.17}$$

An example of the result of this approach appears in Fig. 8.9.

Other methods also exist for rule of thumb parameter settings. In a variation of the above, for example, the initial value of K_p is chosen so that the system exhibits a decaying oscillation in the response, with successive peaks a fraction (say 25%) of the previous peak. Notice that these schemes work on the plant with the control loop closed.

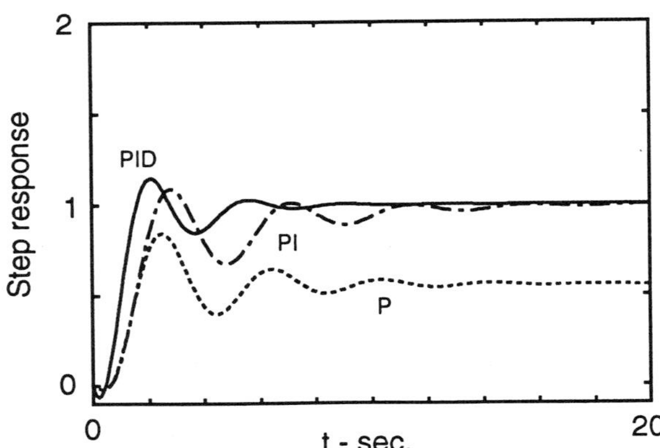

Figure 8.9 The effect of Ziegler–Nichols tuning, demonstrating Proportional only (P), Proportional plus Integral (PI), and Proportional plus Integral plus Derivative (PID) control for the system used in Fig. 8.6.

Open-loop response (reaction curve) based tuning

Another possibility is to find the open-loop plant's response. An example is the transient-response method (Åström and Wittenmark, 1990) as follows.

Step 1. Find the open-loop step response for the system. From this are found the three parameters:

R = the slope of the response,
K = ratio of output change to input change, and
L = the apparent time lag before significant response occurs.

See Fig. 8.10.

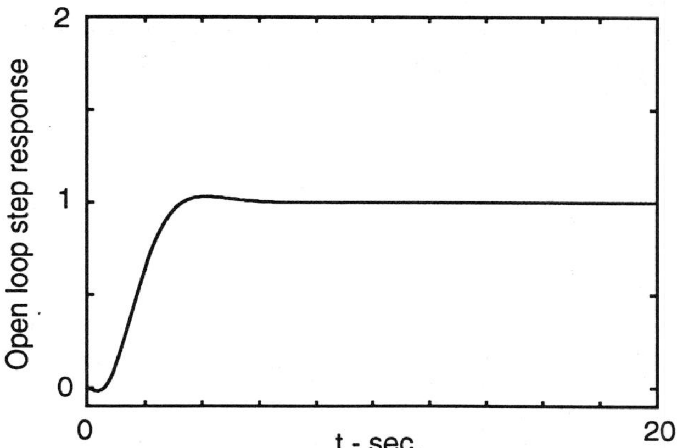

Figure 8.10 The step response of the plant $(-s+4)/((s+2)(s^2+2s+2))$ alone, i.e. the open-loop system. This response is called the reaction curve. Compare with the closed-loop responses in Fig. 8.6.

Step 2. For P control, set

$$G = 1/(KRL) \tag{8.18}$$

Step 3. For PI control, set

$$G = 0.9/(KRL)$$

$$T_i = 3\,L \tag{8.19}$$

Step 4. For PID control, set

$$G = 1.2/(KRL)$$

$$T_i = 2L$$

$$T_d = L/2 \tag{8.20}$$

Using these rules on our example system results in the curves of Fig. 8.11.

This reaction curve method also has variations, including Cohen and Coon, presented by Stephanopoulos (1984) and Clarke (1986).

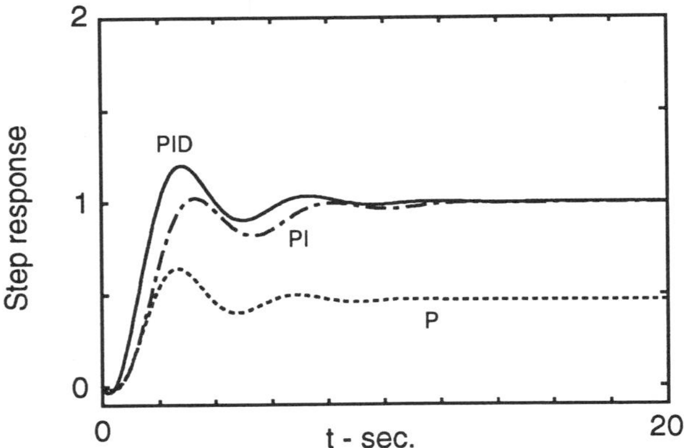

Figure 8.11 Reaction curve tuning, showing P, PI, and PID control for system of Fig. 8.10. Compare with tuning of Fig. 8.9.

8.4 OTHER COMPENSATORS

PID controllers are among a class of control laws, which might be called compensators, whose function is to shape the response of the open-loop system to have desirable properties. In the frequency domain studied in detail in Chapters 15 and 20, they are in essence filters which shape the error signal so that the plant responds 'well'.

The basic form of such compensator elements is the phase-lead phase-lag element

$$\dot{u}(t) = -a_\mathrm{a}\mathrm{u}(t) + K_\mathrm{a}\left(\dot{e}(t) + b_\mathrm{a}\,e(t)\right) \tag{8.21}$$

which has a digital form

$$u(k+1) = -bu(k) + K_\mathrm{d}\left\{e(k+1) + a\,e(k)\right\} \tag{8.22}$$

where the exact characteristics for frequency shaping depend upon a and b (Chapter 20). Setting the parameters requires at least some knowledge of the underlying theory, but some precalculated relationships have been proposed.

One design for the form (8.22) has been proposed (Yeung and Chaid, 1986) which uses frequency response characteristics and which requires the designer to choose two of his parameters:

ω_o: roughly the desired closed-loop $3\,$dB bandwidth, i.e. the frequency at which the gain is down $3\,$dB from the DC value.

ϕ: phase margin desired (typical choice is about 45–55°). Frequency domain parameter related to damping.

Sampling period T must also be chosen. Then a design rule is as follows.

Step 1. Find the open-loop gain $|G(\omega_o)|$ and phase $\text{arc}(G(\omega_o))$ by opening the feedback loop and applying frequency ω_o to the controller input. Measure the outputs.

Step 2. Compute the intermediate parameters

$$\theta = -180° + \phi - \text{arc}(G(\omega_o))$$

$$\alpha = \frac{1 - |G(\omega_o)| \cos\theta}{|G(\omega_o)| \cos\theta - |G(\omega_o)|^2}$$

$$h = \frac{\cos\theta - |G(\omega_o)|}{(2/T)\tan(\omega_o T/2)\sin\theta} \tag{8.23}$$

Step 3. Then the controller parameters are given by

$$a = \frac{1-(2/T)\alpha h}{1+(2/T)\alpha h}$$

$$b = \frac{1-(2/T)h}{1+(2/T)h}$$

$$K_d = \frac{1+b}{1+a} \tag{8.24}$$

PI and PID control gains designed on similar grounds are also suggested. A potential problem for the user is that the choices of bandwidth and phase margin may not be arbitrary, and if any difficulty arises it may be necessary to plunge into the theory or at least to allow a cascade of compensators.

8.5 FURTHER READING

Theoryless control is not a standard topic in textbooks, so a further reading list is somewhat restricted. One cannot properly advise that theory be ignored, but engineers are often well served by these rules of thumb.

PLCs are inherently theoryless. Their use is described in manufacturers' literature. One of the few extensive textbook descriptions is Kissell (1986).

Although theory is absent, it is still possible and desirable to be systematic in planning and programming PLCs. One systematic review and programming approach is described by Pessen (1989).

Self-tuning regulators are a field of their own, and although we will present them along with adaptive control theory, use of such devices without theory is probably best described in manufacturers' literature. Some of the issues are discussed by Åström and Hägglund (1988), who also discuss PID controllers. Another discussion of self-tuners is that of Gawthrop (1987).

For three-term controllers, further reading can be found in manufacturers' literature, in a few texts such as Åström and Wittenmark (1990) and Stephanopoulos (1984), where the latter is notably from the class of chemical process control textbooks, and of course from isolated technical articles such as Ziegler and Nichols (1942). Both Åström and Hägglund (1988), and Bennett (1988) present discussion, including various forms of the implementations and engineering considerations such as **bumpless transfer** (to allow switching from manual to automatic mode without transients) and **anti-reset windup** (to prevent such effects as slow settling due to build-up of the integral term during saturation or set point changes). An engineering look at the topic occasionally appears in the technical magazines; an example is St. Clair (1991).

We also mentioned that there is a class of design rules which require a theoretical background, but for which rules or design tables have been deduced. In such approaches the engineer needs some appreciation of what is happening, but need not perform all of the calculations. One such approach is that of Knowles (1978).

9
Sources of system models

In previous chapters we have virtually ignored the system being controlled. Knowing a temperature was to be controlled, it was implicit that a temperature sensor was to be used to measure the output and a heater of some kind to cause the output to vary, but the mechanism of heat transfer was ignored. It was even suggested that the loop could be tuned by formulae which barely recognized the nature of the controlled process. None of this is completely true, of course, as a good engineer will have a notion of how the system works, and how it reacts to input and he will be influenced by this in many aspects of engineering the system. So, while the previous work did not use mathematical models of the controlled plant, these will be pervasive in what follows. In fact, the use of mathematical models is fundamental to use of control theories.

This chapter considers models, in particular ordinary differential equation models – but first looks at modelling as a scientific and philosophical exercise. We consider a simple and common controlled system, i.e. a DC motor, and its responses in certain types of control configurations with continuous elements. Then we present several examples which show models developed for different fields from differing points of reference: a rocket model from laws of motion, a liquid tank model from thermodynamic principles, a black box model using input–output data, and the nature of modelling of sensor noise and disturbances to the system. Then, because so much control theory is associated with linear systems, we demonstrate how models are linearized and mention a rationale for doing so. Finally, we consider what happens when we sample such systems, i.e. what they look like from within the computer, which sees only samples of the output.

9.1 SYNOPSIS

It is often useful to study models of systems rather than the systems themselves. Commonly used are scale models, analog models, pilot plants, simulations, and mathematical models. All are used in engineering, but the most commonly used in control systems studies are mathematical models.

Of the many types of mathematical models, we deal primarily with differential equation models, exemplified by the equation

$$\tau \frac{d^2\theta}{dt^2} + \frac{d\theta}{dt} = K_m v_a$$

for describing a motor's shaft angle θ as a function of the applied voltage v_a with parameters τ and K_m. This is an approximate model based, as many are, on the laws of physics plus careful engineering approximations.

A common alternative, particularly in computer control, is a model such as

$$y(kT) = 4.5y(kT{-}T) - 3y(kT{-}2T) - 0.5u(kT{-}T)$$
$$- 0.2u(kT{-}2T) - 8e(kT{-}T) - 9e(kT{-}2T)$$

which relates the sampled outputs y of a paper press (paper basis weight) to the command inputs u (steam pressure at drying cylinders) and other inputs e (errors due to noise and disturbances for examples) at times kT, where T is the sample period of the system and k is an indexing integer.

Models are essential for the development and analysis of computer control algorithms, and for this reason the present chapter is devoted to them.

9.2 OVERVIEW

Modelling has been defined as 'a deliberate intelligible cognitive activity aimed at abstracting, and reproducing in some convenient realms of discourse, features of an object or system of interest to the modeller'. The words here are carefully chosen to rule out the trivial, such as assertions that 'we all carry within us models of the world's behaviour', and to emphasize utility by not requiring the inclusion of features not of interest, such as thermodynamic considerations in simple spring-mass systems.

We build models to achieve an understanding of:

1. the inherent nature and characteristics of the system;
2. the results of future changes in the system; and
3. the system's response to external stimuli.

Modelling is generally regarded as indispensable for gaining understanding of the behaviour of complex systems, but it must also be recognized that it is in part an art form, as is much of engineering. Unfortunately for the user, there is no comprehensive, consistent theory of modelling, there is no set of theoretical notions which form an adequate basis for the existing methodologies, and there are no good sets of modelling heuristics nor notions which would lead to such heuristics.

In spite of the above, engineers must have models to work with and, although many subsystems and systems have already been modelled, they must often construct models. To this task, they bring their experience and knowledge plus a great deal of ingenuity; they attempt to be systematic, in a way extending the so-called scientific method – observe the system, formulate a theory, predict new observations, experiment, refine theory if necessary, etc. – and organized, perhaps in the manner advocated by Murthy *et al.* (1990) for mathematical models, who insist that first it is necessary to define the problem to be solved carefully. Then the system must be characterized, a step which requires deep knowledge of its physical aspects and involves considerable idealization and simplification. Next, a mathematical formulation, dictated by both the characterization and by *ad hoc* intuitive considerations, is selected, and the model's variables are related on a one-to-one basis to the variables in the characterization. The resulting mathematical model is analysed, and the variable behaviour is compared to data from the real system. If the data do not agree, according to a pre-defined criterion, then it is necessary to back-track and make changes in the characterization, the mathematical model, or both.

Whether the engineer has been completely systematic or, more likely, a combination of systematic, ingenious, and resourceful in its development, the model must – in the end – be useful. For this, it must be parsimonious, having only the required level of detail. Thus for example a model of a machine may ignore long-term changes in behaviour due to wear and very short-term effects of spikes in the electricity supply to the motor. In addition, the model should be valid in the sense that enough evidence has been obtained to show that it is adequate to the task at hand, i.e. that it correlates with observations of reality and that its theoretical base is adequate.

There are a number of types of models; the overwhelming *caveat* with all of these is that the model is not the real thing, and it may not perform identically.

1. **Scale models** are usually undersized but exactly proportioned models of physical devices such as the aircraft models placed in wind tunnels to show aerodynamic effects. They can be much less expensive and easier to study than the real thing, but have the potential problem that observed effects (such as drag) may not easily scale up to full size.

2. **Analog models** are based upon the notion that two systems show structural similarity in their system models. For example, one physical quantity such as an electrical current in a circuit may be modelled analogously to another quantity such as displacement in a mechanical device or flow of a liquid, and hence the quantities may show similar behaviour. The use of the electrical analogy allows electrical circuits to be constructed which may be operated to show how the mechanical system, say, will operate.

3. **Pilot plant models** are operational systems which are smaller versions of systems to be implemented. They are common in universities, which cannot afford and do not need full scale processes, and for other studies. Arguably similar in some respects to scale models, the results may for various reasons not scale up to full sized processes.

4. **Simulation models** are systems which purport to approximate real system behaviour by the clever use of computer programs. For our purposes they are means of solving mathematical models of systems in which numerical quantities represent real quantities.

5. **Mathematical models** represent the real system by sets of differential and difference equations, with variables assigned to represent real quantities such as voltages, displacements, temperatures, concentrations, etc. The advantage of such models is that the tools of mathematics may be brought to bear to establish stability, expected performance, etc. These properties may be difficult to show conclusively experimentally, a fault shared by the other model forms.

Our primary tools are mathematical models and their expressions in simulation models.

Simulation models

For simulation programs on digital computers, there are a number of special purpose computer languages, although in fact a simulation can be written in a general purpose language, even in machine code. The special languages are usually devoted to certain classes of problems, and their advantages are that they automatically do much of the 'house-keeping' commonly needed by problems of the class they serve. Thus the differential-equation oriented languages will supply numerical integration procedures as part of their implementations, while discrete event languages have a built-in list processing to keep track of event statistics and forthcoming events. SIMULINK, EASY5, SIMNON and PSI are among the languages/programs especially suited to control systems studies, with the last two particularly easy to use for learning purposes.

Simulations have a certain attraction in that they tend to be repeatable, require only a modest computer and a long time (or a large computer for lesser time) plus (considerable) programming. 'Experiments' are usually easily performed on the simulated system – just make another computer run – and models can be very elaborate indeed. The latter feature, with the possibility for consideration of subtle non-linearities and intricate logical configurations, provides some of the incentive for simulation modelling.

The problems with simulation are many and are mostly the obvious ones: the simulation model may be wrong (as any model may be wrong), the implementation may be wrong or misleading, and in any case the simulation allows 'experiments' rather than structural analysis. An elaborate model may require much programming, which may not be cheap, and much computer time, which fortunately is becoming cheaper.

Simulation models will be shown briefly in a later subsection and are used in several places to demonstrate points.

Mathematical models

As soon as we wish to apply a theoretical technique, we need a mathematical model. Partly for this reason and partly for the other reasons listed above for modelling, suitable models are usually part of the knowledge of the computer control systems engineer.

The models used are mostly ordinary differential equations, with difference equations used in computer control. Furthermore, these equations are almost always linear with constant coefficients. That this is often appropriate is perhaps a surprise; this issue is commented upon in section 9.5. If non-linearity is a critical part of the system, we still tend to use linear models to attain insight before using simulation to enable further studies.

9.3 A TYPICAL SMALL PLANT: MODEL DEVELOPMENT AND USE FOR A DC MOTOR

The traditional model presentation in textbooks is of the electric motor, and many examples are motivated by its control. We show such a model development here and include the effects of controlling the motor using some actual devices.

9.3.1 The model

The common mathematical model of a DC motor has

$$\tau \frac{d\omega}{dt} = -\omega + Kv(t)$$

$$\frac{d\theta}{dt} = \omega \tag{9.1}$$

where

θ = shaft angle relative to some reference (radians)

ω = angular velocity (rad/s)

τ = time constant (s)

K = gain coefficient (rad/s/V)

v = input voltage (V)

As we explore this, we shall see that many simplifications have been made. It is not 'correct' because it ignores certain known effects, but it is correct enough for its usual intended application, and thus it is very useful.

We will also add to this model to make a model of a controlled system: we shall assume that the shaft angle is measured using a potentiometer and the rotation rate is measured using a tachogenerator. The resulting closed-loop system will give us considerable insight into the operation of controlled systems.

9.3.2 Derivation of the model

The essence of a direct current (DC) motor is that a magnetic field is created, using either a field current i_f in a coil or a permanent magnet, and a polarized rotatable coil called an armature is placed in this field. The polarization is maintained using a current i_a in the armature, and the magnetic effects cause the armature to rotate. The motor may be either field-controlled (i.e. i_a = constant and input is to the field) or armature controlled (i.e. i_f = constant and input is to the armature), but the latter is much more common because it has good speed control characteristics, whereas the former is preferred where high starting torque is required. The basic configuration and a more detailed representation are in Fig. 9.1.

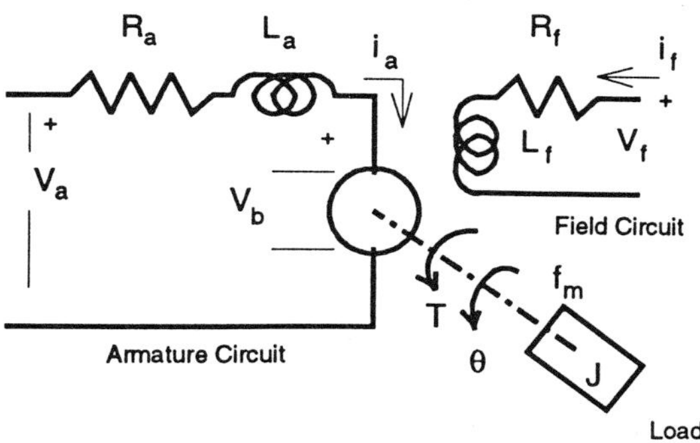

Figure 9.1 A model of an electric motor, showing both mechanical and electrical effects. After Chen (1975).

The motor modelling couples two effects: the mechanical system described by Newton's laws and the electrical system described by Kirchhoff's laws. The assumptions concerning the mechanical system are as follows.

1. The torque applied by the fields is proportional to the two currents, i_f and i_a

$$\text{Torque} = T = K_p i_a i_f$$

$$= K_t i_a \qquad (i_f = \text{constant}) \qquad (9.2)$$

2. The moment of inertia of the armature and any attached system is given by J.

3. The only appreciable friction effect in operation is viscous friction

$$\text{Friction} = f_m \frac{d\theta}{dt} = f_m \omega \tag{9.3}$$

and coulomb and static frictions are neglected.

The assumptions concerning the electrical system are as follows.

1. There is a back electromagnetic force (emf) due to the armature rotation

$$\text{Back emf} = v_b = K_b \omega = K_b \frac{d\theta}{dt} \tag{9.4}$$

2. For voltages within specifications the armature circuit will not saturate. The armature circuit exhibits both resistance R_a and inductance L_a in series so that the circuit is modelled by

$$R_a i_a + L_a \frac{di_a}{dt} + v_b = v_a \tag{9.5}$$

where v_a is the applied voltage input.

Using the above assumptions, we are left with coupled linear constant coefficient ordinary differential equations for the motor.

$$\text{Motor equation: } J \frac{d^2\theta}{dt^2} = \sum (\text{torques})$$

$$= -f_m \frac{d\theta}{dt} + K_t i_a \tag{9.6}$$

$$\text{Circuit equation: } R_a i_a + L_a \frac{di_a}{dt} + v_b = v_a \tag{9.7}$$

Elimination of the armature current then yields

$$JL_a \frac{d^3\theta}{dt^3} + (JR_a + f_m L_a) \frac{d^2\theta}{dt^2}$$

$$+ (f_m R_a + K_t K_b) \frac{d\theta}{dt} = K_t v_a \qquad (9.8)$$

Because L_a is usually quite 'small', this is approximately

$$JR_a \frac{d^2\theta}{dt^2} + (f_m R_a + K_t K_b) \frac{d\theta}{dt} = K_t v_a \qquad (9.9)$$

which is often written

$$\tau \frac{d^2\theta}{dt^2} + \frac{d\theta}{dt} = K_m v_a \qquad (9.10)$$

where

$$\tau = \frac{JR_a}{K_t K_b + f_m R_a} - = \text{DC motor time constant (s)}$$

$$K_m = \frac{K_t}{K_t K_b + f_m R_a} - = \text{DC motor gain constant}$$
$$\text{in ((rad/s)/volt)} \qquad (9.11)$$

We remark that the model was developed carefully from assumptions about the motor and was explicitly simplified by the ignoring of L_a. Thus this model is almost surely 'wrong' in both the assumptions and the simplification, but it is also 'right' in being useful for many tasks and satisfactory for prediction of performance of many systems.

9.3.3 Feedback control of the motor

A typical control task with a motor is to have it produce either a set speed of rotation ω_d or a desired shaft angle θ_d. The latter in particular will almost certainly require feedback control, with the angle measured, the measurement converted to a voltage of level appropriate to control, the measured value compared to the desired value, and the two numbers used to generate a command voltage to the motor. A simple such arrangement is shown in the block diagram of Fig. 9.2.

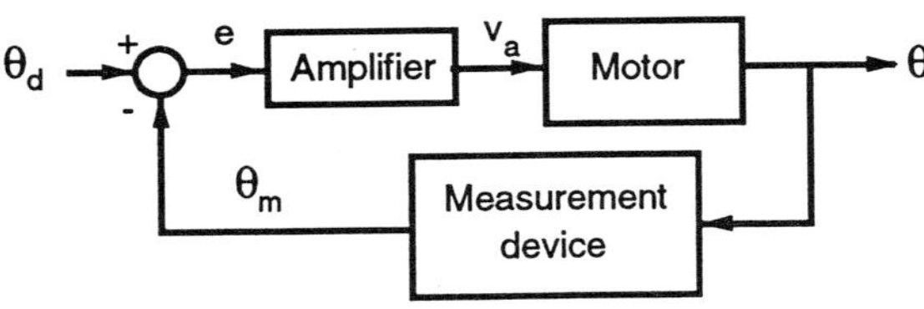

Figure 9.2 An abstraction of the motor model into a block diagram, showing angle feedback.

In this example, the measurement device may be a simple potentiometer, geared to the motor shaft, with output θ_m volts proportional to the shaft angle θ. The desired angle θ_d is also a voltage, perhaps from a manually-set potentiometer with voltage in the same range as the measurer. The voltage differencing can be done with a simple electronic circuit and the result then amplified to a voltage of a level appropriate for driving the motor. In equations, we have

$$\theta_m = K_{pot_1}\, \theta$$

$$\theta_d = K_{pot_2}\, \theta_{ext}$$

$$e = \theta_d - \theta_m$$

$$v_a = K_{amp}\, e \tag{9.12}$$

In these, the amplifier is assumed linear with an amplification of K_{amp} while θ_{ext} is the externally set desired angle in radians (or degrees) and $K_{pot} = K_{pot_1} = K_{pot_2}$ is the appropriate voltage gain per unit angle (say $\pm 15\,\text{V}$ as θ ranges from $-\pi$ to π radians).

With the model now specified in equations (9.8)–(9.11) the task is to design the amplifier gain so that response is 'good'. This task is tackled in several ways in future chapters, but here we note that for this simple situation, we are studying the second-order ordinary linear differential equation

$$\tau \frac{d^2\theta}{dt^2} + \frac{d\theta}{dt} = K_m\,K_{amp}\,K_{pot}\,(\theta_{ext} - \theta) \tag{9.13}$$

which can be written in the form

$$\tau \frac{d^2\theta}{dt^2} + \frac{d\theta}{dt} + K_1\,\theta = K_1\,\theta_{ext} \tag{9.14}$$

and that the design parameter is $K_1 = K_m\,K_{amp}\,K_{pot}$, where K_m and τ are motor parameters.

If the motor speed is also measured, then the system can have a block diagram as in Fig. 9.3.

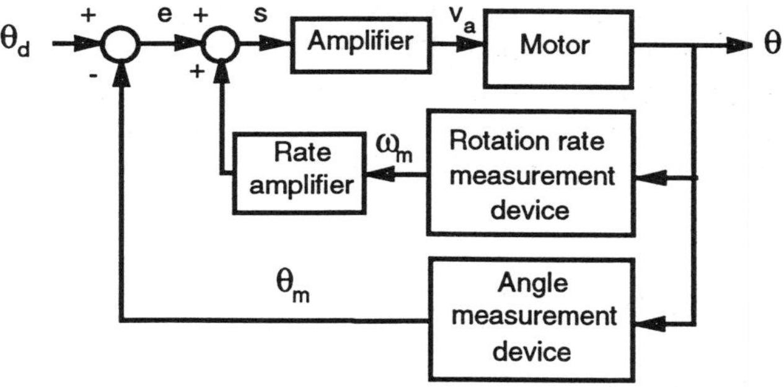

Figure 9.3 The motor block diagram when both shaft angle and rotation speed are measured and used for control.

In this configuration, the speed of rotation is measured by, for example, a tachogenerator giving output $\dot{\theta}_m$ (volts) and this is amplified to give $K_{r,amp}\,\dot{\theta}_m$ (volts). The last is added (or subtracted, if the designer wishes) to the error signal e to give the input s to the amplifier. Hence we now have the system described by

$$\tau \frac{d^2\theta}{dt^2} + \frac{d\theta}{dt} = K_m K_{amp}\left[K_{pot}(\theta_{ext} - \theta) + K_{r,amp} K_{tach} \frac{d\theta}{dt}\right] \quad (9.15)$$

where it is assumed that $\dot{\theta}_m = K_{tach}\, d\theta/dt$ (volts). Rearranging gives the differential equation for analysis as

$$\tau \frac{d^2\theta}{dt^2} + [1 + K_2]\frac{d\theta}{dt} = K_1 \theta = K_1 \theta_{ext} \quad (9.16)$$

Both parameters K_1 and K_2 are to be selected by analysis and implemented by choosing or designing amplifiers and measurement devices.

9.3.4 System studies

In selecting parameters for the motor control problem, the engineer will be presented with the classic choice of slow and steady, or rapid but oscillatory, in addition to the need to avoid instability (Fig. 9.4).

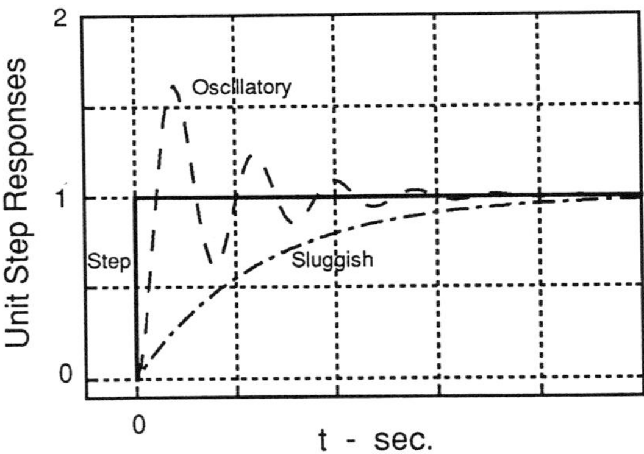

Figure. 9.4 Unit step responses such as might be possible for the motor's shaft angle and many other systems. Ideally, the output would overlie the input step, but a frequent engineering trade-off is between rapid reponse which overshoots the desired value and oscillates, and slower reponse.

All these respond to a sudden change in desired angle θ_{ext}, a so-called step input. The setting of a constant desired angle is called the **set point**; if the system were required to track a changing input $\theta_{ext}(t)$, such a tracking problem would probably require different amplifier gains even though the basic trade-off is the same.

With variable amplifiers, an engineer might well tune this process manually. Analysis, however, will tend to give insight as to the available trade-offs, the achievable goals, and the chance of instability. Hence, modelling and the analysis of models are important parts of control engineering.

Analytical studies may be performed using a number of tools and criteria. We meet conversion to discrete time systems in Chapter 12, design criteria in Chapter 13, and controller analysis and design in most of the theory sections.

Studies can also be done using simulation, which becomes almost a necessity if non-linearity, such as amplifier saturation, is introduced. A SIMNON digital computer simulation program for the motor is shown below.

```
CONTINUOUS SYSTEM motor
STATE omega theta
DER omegadot thetadot
vin=if t < 1 then 0 else 1.
thetadot = omega
omegadot = −omega/tau + ka* vin/tau
tau : 0.1
ka: 2.
END
```

9.4 MATHEMATICAL MODELS FOR CONTROL SYSTEMS

Control systems engineers are unlikely to be required to develop a model from scratch, because they are usually working with systems, or at least components, which are fairly well understood. They should be aware of how those models arise, however, and particularly how to develop 'black-box' models. Thus a systems engineer is often required to have at least a rudimentary understanding of the physics, chemistry, biology, etc., of the systems to be controlled, even if the detailed model is supplied by specialists.

Control systems models range from input–output (black box) models to postulated-form (mathematico-physical, theoretical) models. The usual level is an intermediate one: the system is represented as an interconnection of subsystems on physical grounds and the subsystems are then modelled, using physical laws to supply the form and parameter fitting (grey box) to supply the numerical values. The distinction may be seen by considering the motor in the previous section; mechanical and electrical laws determine the form of the input–output relation for any DC motor, but the actual numbers come from a combination of simple tests and physical arguments.

The black box approach is called **system identification** by control systems engineers; it is an extensive branch of the field which we shall meet in Chapter 30. Done off-line it is truly system identification, or at least parameter identification. Done on-line, identification becomes an important part of adaptive control and self-tuning control, the latter of which is a special case of the former and is a commercial product. Most identification is, in effect, parameter estimation, since most algorithms have strong implicit or explicit assumptions about the system built in.

For examples of mathematical model development, we show the following.

1. On the basis of kinematics and electrical principles, an electrical motor model was developed in section 9.3.
2. A generic response model is mentioned in section 9.4.1.
3. The laws of motion yield models for studying rocket trajectories. A simple such planar model is discussed in section 9.4.2.
4. Conservation of mass and energy give the basis of a model of temperature T and depth h of a liquid in a tank in section 9.4.3.
5. Kinematics via Lagrange theory give a simple robot arm model in section 9.4.4.
6. System identification methods (Chapter 30) give a paper press model mentioned in section 9.4.5.
7. Noise and disturbances are mentioned in section 9.4.6.
8. Physical arguments and knowledge of the system provide models of computer sampling in section 9.4.7.

The above exemplify the types and origins of models of use in control systems studies.

9.4.1 A generic response curve model

Mathematical models may be built in several ways, but most seem to reduce to a combination of physical principles with examination of response curves. Thus, in the approach of simply fitting a response curve, a system may have a sudden step change in input at a time arbitrarily called 0 and have the resulting response of Fig. 9.5. From this, a guess is that the system is roughly modelled as a time delay and a first order response, i.e. that it can be described by the differential equation

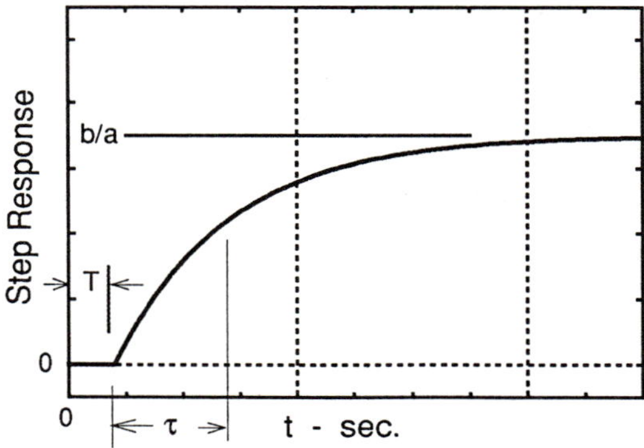

Figure 9.5 For simple modelling, it may be sufficient to get the response to a step at time 0 and note the asymptotic value, *b*, and the delay before response begins, *T*.

$$\frac{\mathrm{d}x}{\mathrm{d}t} = -ax(t) + bu(t-T) \tag{9.17}$$

where *u* represents the input and *x* represents the output. *T* is the time delay, and $\tau = 1/a$ is the time constant. The Laplace transform of this is

$$sX(s) - x(0) = -aX(s) + b\,\mathrm{e}^{-sT}U(s) \tag{9.18}$$

the quantity

$$\frac{X(s)}{U(s)} = \frac{be^{-sT}}{s + a} \tag{9.19}$$

which contains no initial conditions, is the transfer function of the system (whatever it is). Notice that there is no underlying understanding of the system involved, only a guess and a curve fit; this is a type of 'system identification' (Chapter 30) and is very commonly used.

9.4.2 Modelling an exo-atmospheric rocket

Let us consider a rocket launch guidance task and for simplification take a two-dimensional plane approach, considering only vertical and horizontal coordinates. This is not completely unrealistic because space launchers seldom do elaborate cross-plane manoeuvres because of the fuel required. Then the situation is as in Fig. 9.6.

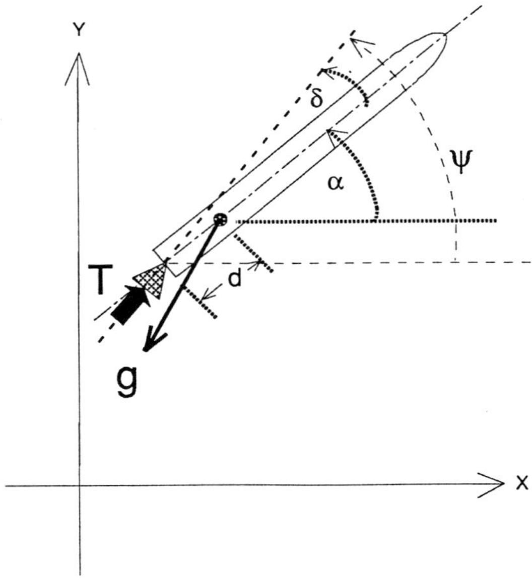

Figure 9.6 Coordinates and variables used in developing an exo-atmospheric rocket flight model.

Taking the coordinates

$$x = \text{horizontal distance (metres)}$$

$$y = \text{vertical distance (metres)}$$

and the definitions

$$\psi = \text{pitch attitude of thrust (rad)}$$

$$u = \text{rate of change of attitude} = \text{command quantity (rad/s)}$$

$$T = \text{rocket thrust (N)}$$

$$g = \text{gravity acceleration (m/s}^2)$$

$$G = \text{gravitational constant (Nm}^2)$$

$$m = \text{vehicle mass (kg)}$$

$$\dot{m} = \text{fuel burning rate (kg/s)}$$

we can apply Newton's law to obtain the equations of motion.

$$\ddot{x} = \text{force in } x\text{-direction/mass}$$

$$= \frac{(-mg + T)_{\text{in } x\text{-direction}}}{m}$$

$$= \frac{-Gx}{m(x^2 + y^2)^{3/2}} + \frac{T\cos\psi}{m} \tag{9.20}$$

and similarly

$$\ddot{y} = \frac{-Gy}{m(x^2 + y^2)^{3/2}} + \frac{T\sin\psi}{m} \tag{9.21}$$

where usually T_{h} is a constant characterizing the motor and

$$T = \begin{cases} 0 & \text{with motor off} \\ T_{\text{h}}(t) & \text{with motor on} \end{cases}$$

$$\dot{m} = \begin{cases} 0 & \text{with motor off} \\ \dot{M}(t) & \text{with motor on} \end{cases}$$

$$\dot{\psi} = u \tag{9.22}$$

One notes that this simplified case is non-linear and time-varying. It involves four differential equations: two of first order and two of second order. The approximations for thrust and mass rate are results of design, the characteristic operation of rocket motors, and observation; actually, they tend to have initial transients and final tail-offs, plus a slight fluctuation during flight.

The model can rapidly become elaborate. The first step is to include the actual rocket vehicle rotational dynamics, i.e. its rotation about its centre of mass: we make the further definitions shown in Fig. 9.6.

Defining the quantities

$$J(m) = \text{moment of inertia about centre of gravity (N$-$m)}$$
$$= \text{function of mass}$$
$$d(m) = \text{distance from c.g. to engine mount (m)}$$
$$= \text{function of mass}$$
$$\alpha = \text{vehicle body attitude angle relative to horizontal}$$
$$= \text{pitch attitude (rad)}$$
$$\delta = \text{angle of thrust relative to vehicle axis (rad)}$$
$$\rho = \text{atmospheric density ($=\rho_0 e^{-kh}$ where ρ_0 is the density at zero altitude and k is a constant)}$$

we may use the torque equation

$$J(m) \frac{\mathrm{d}^2 \alpha}{\mathrm{d}t^2} = \sum (\text{applied torques})$$

$$= -d(m)\,T\sin\delta \tag{9.23}$$

We remark that J and d depend upon m because the centre of gravity of the vehicle shifts toward the motor as the fuel burns. The control equation (9.22) $\dot{\psi} = u$ will now be replaced by a command on the body pitch attitude α, implemented by commanding gyros in the vehicle to change their settings. The result is that

$$\psi = \alpha + \delta$$

$$\delta = C_c\,(\alpha, \alpha_{des})$$

$$\alpha_{des} = N_c\,(x, y, \dot{x}, \dot{y}, \psi) \tag{9.24}$$

in which both the vehicle control function C_c and the navigation function N_c must be designed by the control systems engineering team.

A further elaboration is to consider atmospheric flight, in which case the forces of drag and lift must be accounted for. These are often taken as

$$\text{Drag along axis} = C_d(\beta)\rho A v^2 \qquad \text{(N)}$$

$$\text{Lift perpendicular to axis} = C_L(\beta)\rho A v^2 \qquad \text{(N)} \tag{9.25}$$

where

$$\rho = \text{atmospheric density} \approx \rho_0 e^{-kh}$$

$$A = \text{effective vehicle cross-sectional area (m}^2)$$

$$\beta = \text{angle of attack}$$

$$= \text{angle between body axis and velocity vector}$$

$$= \alpha - \tan^{-1}\left[\frac{\dot{y}}{\dot{x}}\right] \text{ (rad)}$$

$$C_d(\beta), C_L(\beta) = \text{drag and lift coefficients as functions of angle of attack}$$

$$v = \text{speed (or velocity magnitude)}$$

$$= (\dot{x}^2 + \dot{y}^2)^{\frac{1}{2}} \text{ (m/s)}$$

$$h = \text{altitude above reference}$$

$$= (\dot{x}^2 + \dot{y}^2)^{\frac{1}{2}} - h_0 \text{ (m)} \tag{9.26}$$

Since these forces effectively operate through a point called the centre of pressure c_p which is different from the centre of gravity c_g, the distance between them must be defined. Usually the c_p is a fixed function of the vehicle shape and hence is fixed relative to, say, the motor attachment point. Thus we need

$$d_{cp} = \text{distance from rocket motor attachment to } c_p \text{ (m)} \qquad (9.27)$$

A result of all of this is that the model quickly builds to a form such as the following:

$$\ddot{x} = \frac{-Gx}{m(x^2+y^2)^{\frac{3}{2}}} + \frac{T\cos\psi}{m} - \frac{C_d(\beta)\rho A\ v^2\cos\alpha + C_L(\beta)\rho A\ v^2\sin\alpha}{m}$$

$$\ddot{y} = \frac{-Gx}{m(x^2+y^2)^{\frac{3}{2}}} + \frac{T\sin\psi}{m} - \frac{C_d(\beta)\rho A\ v^2\sin\alpha - C_L(\beta)\rho A\ v^2\cos\alpha}{m}$$

$$J(m)\frac{d^2\alpha}{dt^2} = \sum (\text{applied torques})$$

$$= - d(m)\ T\sin\delta + (d_{cp} - d(m))C_L(\beta)\rho Av^2 \qquad (9.28)$$

in which the various quantities were defined above.

In this gradual building of a model, the key tools were Newton's laws of motion. Three dimensions require further refinement, and of course the navigation and attitude control laws must be defined. Elaborate analyses and simulations need even more elaborate models, and the latter will often use much more detailed gravitational models, thrust depending upon external air density, non-constant mass flow rates, etc.

9.4.3 Process control of a tank's contents

Whereas the rocket model above was based upon the laws of motion, the models of chemical processes are based upon conservation of mass and energy and upon the nature of chemical processes. Fundamental are the simple laws governing contents and temperature of a simple tank, as in Fig. 9.7.

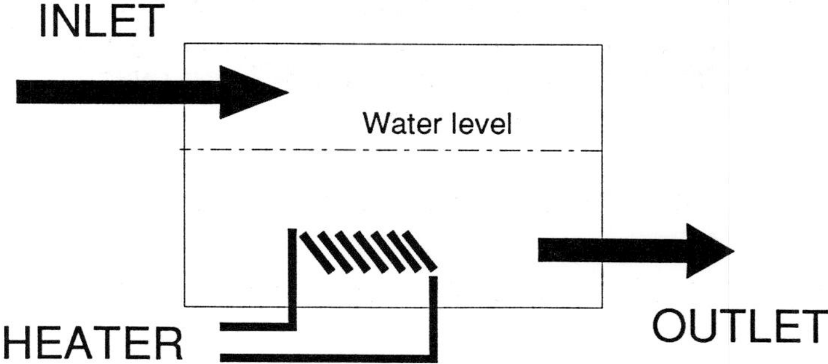

Figure 9.7 A liquid tank, with input of water and heat, and output of heated water, for development of a process model.

Conservation of mass

Taking the contents as liquid and ignoring evaporation yields the expression

$$\frac{d(\text{mass of contents})}{dt} = \text{mass flow in} - \text{mass flow out} \qquad (9.29)$$

so that with a substance of density ρ and a tank cross-section A (m^2) and substance depth h we have, for flow rates F_i and F_o at inlet and outlet respectively,

$$\frac{d(\rho A h)}{dt} = \rho_i F_i - \rho_o F_o \qquad (9.30)$$

or, if the density is constant independent of temperature,

$$A \frac{dh}{dt} = F_i - F_o \qquad (9.31)$$

Conservation of energy

For the energy balance, we consider the energy input and output in the substance and the energy supplied by the heater at rate Q. We assume

$$\text{Energy in substance} \propto c_{\mathrm{p}} \, (T - T_{\mathrm{ref}}) \tag{9.32}$$

where T is actual temperature and T_{ref} is a reference temperature for the approximation. Then we have

$$\frac{\mathrm{d}(\text{energy of tank contents})}{\mathrm{d}t} = \text{energy flow in with substance}$$

$$- \text{energy of outlet substance}$$

$$+ \text{energy input by heater}$$

$$- \text{energy losses}$$

$$\frac{\mathrm{d}(\rho A h\, c_{\mathrm{p}}(T - T_{\mathrm{ref}}))}{\mathrm{d}t} = \rho F_{\mathrm{i}} c_{\mathrm{p}}(T_{\mathrm{i}} - T_{\mathrm{ref}}) - \rho F_{\mathrm{o}} c_{\mathrm{p}}(T - T_{\mathrm{ref}}) + Q - (\text{losses}) \tag{9.33}$$

Using constant values of ρ, T_{ref}, c_{p}, and ignoring losses gives

$$A\, \frac{\mathrm{d}(hT)}{\mathrm{d}t} = F_{\mathrm{i}} T_{\mathrm{i}} - F_{\mathrm{o}} T + \frac{Q}{\rho\, c_{\mathrm{p}}} \tag{9.34}$$

Using the chain rule on the left-hand side and substituting from (9.31) yields

$$A h\, \frac{\mathrm{d}T}{\mathrm{d}t} = F_{\mathrm{i}}\, (T_{\mathrm{i}} - T) + \frac{Q}{\rho\, c_{\mathrm{p}}} \tag{9.35}$$

From a controller's point of view, A, ρ, and c_{p} are parameters, h and T are variables representing the state of the substance in the tank, and F_{o}, F_{i}, and Q are possibly either given or controllable variables. For example, F_{o} and T may be requirements on the tank output imposed by other operating considerations, while the input flow F_{i} and the heater input Q may be manipulated to achieve those ends.

Once again the model can rapidly become more elaborate. In a continuous stirred tank reactor, for example, a simple exothermic reaction A $\rightarrow$ B takes place and the heat is removed by a coolant. Thus we must account for the concentrations of A and B in input and output.

We define

$$V = \text{volume of mixture in tank}$$
$$\rho_i, \rho, \rho_o = \text{densities of input, tank, and output streams}$$
$$F_i, F_o = \text{volumetric flow rates of input and output streams}$$
$$T_i, T_o = \text{temperatures of input and output streams}$$
$$c_{A_i}, c_A, c_{A_o} = \text{concentrations of A in input, tank, and output (moles/vol)}$$
$$c_B, c_{B_o} = \text{concentrations of B in tank and in output (moles/vol)}$$
$$r = \text{rate of reaction per unit volume}$$
$$h_i, h, h_o = \text{specific enthalpy content of input, contents, and output}$$
$$c_p = \text{specific heat capacity of reacting mixture} \tag{9.36}$$

Then total mass balance requires

$$\frac{d(\rho V)}{dt} = \rho_i F_i - \rho_o F_o \pm (\text{mass consumed or generated}) \tag{9.37}$$

Mass balance on component A requires

$$\frac{d(c_A V)}{dt} = c_{A_i} F_i - c_{A_o} F_o - r V \tag{9.38}$$

Energy balance gives approximately for a liquid

$$\text{Energy} = \text{Internal Energy} + \text{Kinetic Energy} + \text{Potential Energy}$$

and for a stationary tank

$$\frac{d(\text{Energy})}{dt} = \frac{d(\text{Internal Energy})}{dt} \approx \frac{d(\text{Enthalpy})}{dt}$$

which with H defined as total enthalpy gives

$$\frac{dH}{dt} = \rho_i F_i h_i(T_i) - \rho_o F_o h_o(T_o) - Q \tag{9.39}$$

Since H is a function of the temperature and composition of the liquid substance

$$H = H(T, c_A V, c_B V)$$

we can find from thermodynamics that

$$\frac{dH}{dt} = \frac{\partial H}{\partial t}\frac{dT}{dt} + \frac{\partial H}{\partial(c_A V)}\frac{d(c_A V)}{dt} + \frac{\partial H}{\partial(c_B V)}\frac{d(c_B V)}{dt}$$

$$= \rho V c_p \frac{dT}{dt} + \tilde{H}_A (c_{A_i} F_i - c_{A_o} F_o - rV) + \tilde{H}_B (0 - c_{B_o} F_o + rV) \quad (9.40)$$

where $\tilde{H}$ denotes a partial molar enthalpy.

Substituting this in (9.39)

$$\rho V c_p \frac{dT}{dt} - \tilde{H}_A (c_{A_i} F_i - c_{A_o} F_o - rV) - \tilde{H}_B (-c_{B_o} F_o + rV)$$

$$+ \rho_i F_i h_i(T_i) - \rho_o F_o h_o(T_o) - Q \quad (9.41)$$

From the definitions of the various concentrations, etc.,

$$\rho_i F_i h_i(T_i) = F_i \left[c_{A_i} \tilde{H}_A(T) + \rho_i c_{P_i}(T_i - T) \right]$$

$$\rho_o F_o h_o(T_o) = F_o \left[c_{A_o} \tilde{H}_A(T_o) + c_{B_o} H_B(T_o) \right] \quad (9.42)$$

Provided the tank is well stirred, the output has the same characteristics as the tank contents, so that $T_o = T$, $\rho_o = \rho$, etc. Hence we have

$$\rho V c_P \frac{dT}{dt} = F_i \rho_i c_{P_i} (T_i - T) + (\tilde{H}_A - \tilde{H}_B)rV - Q \quad (9.43)$$

Introducing the assumptions $\rho_i = \rho$ (the density of the liquid is constant), $r = k_o e^{-E/RT} c_A$ (for the reaction rate dependence on temperature and concentration, with k_o, E and R as parameters), and the notation $J = (\tilde{H}_A - \tilde{H}_B)/(\rho c_P)$ into equations (9.37), (9.38) and (9.43) gives the system model as

$$\frac{dV}{dt} = F_i - F_o$$

$$\frac{dc_A}{dt} = \frac{F_i}{V}(c_{A_i} - c_A) - k_o e^{-E/RT} c_A$$

$$\frac{dT}{dt} = \frac{F_i}{V}(T_i - T) - Jk_o e^{-E/RT} c_A - \frac{Q}{\rho c_p V} \tag{9.44}$$

In this model, k_o, E, R, J and c_p are parameters characterizing the process. Of the variables, various combinations may be control variables to be manipulated, output variables with desired values, and externally affected variables which act as disturbances to impede the achievement of the output values desired. For example, F_o and c_{B_o} (and hence by implication c_{A_o}) might be specified, Q and F_i might be manipulated, and c_{A_i} and T_i might be characteristics of the incoming liquid which vary due to disturbing factors beyond immediate control.

Once again the modelling can be extended to encompass more issues. Here we have used mass balance, energy balance, and thermodynamic processes to give an illustration of the evolution and reasoning in developing system models.

9.4.4 Dynamics of robot arms

The equations associated with robots are complex mostly because of the complicated set of several links, each movable with respect to some others, which comprise them. Entire books are devoted to deriving these equations for specialists, and indeed a number of interesting control tasks are associated with robots: movement of individual links with precision and speed; planning of trajectories of the manipulator; using feedback from sensors such as joint angle measurers, tactile and proximity sensors, vision systems. This does not even mention the systems design tasks involved in selection and installation, design of processes, etc.

We derive the kinetic equations of the simple manipulator of Fig. 9.8 simply to show what the equations of real systems might look like and also to demonstrate the application of lagrangian kinetics (following Snyder, 1985). The system consists of a telescoping radial arm which can rotate at angle θ in the vertical plane. The mass of the telescoping arm and load are modelled as mass m_2 at radius r, and the outer cylinder is modelled as having fixed mass m_1 at fixed radius r_1.

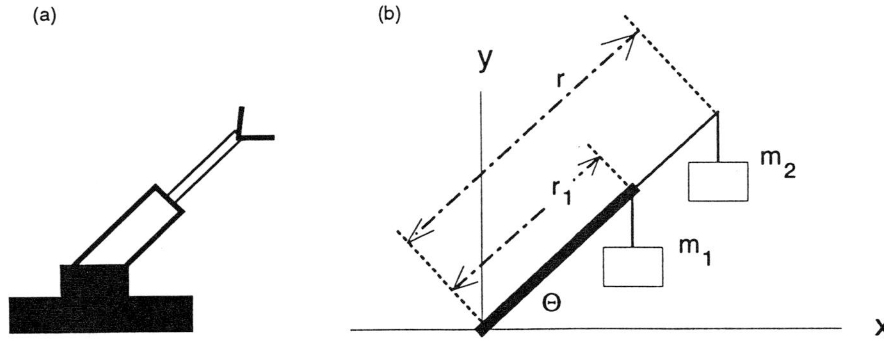

Figure 9.8 A simple planar robotic manipulator in which the arm assembly rotates and the arm itself extends. (a) The robot and (b) a coordinate system and definition of variables.

The lagrangian for this system is

$$L = K - P \tag{9.45}$$

where K is the system's total kinetic energy and P is the total potential energy. If an actuator torque T is controlling a rotary variable, then

$$T = \frac{\partial}{\partial t} \frac{\partial L}{\partial \dot{\theta}} - \frac{\partial L}{\partial \theta} \tag{9.46}$$

while a force F applied in the direction of motion of the prismatic joint in the r-direction gives

$$F = \frac{\partial}{\partial t} \frac{\partial L}{\partial \dot{r}} - \frac{\partial L}{\partial r} \tag{9.47}$$

We must now find L in terms of $r, \dot{r}, r_1, \dot{r}_1, \theta$, and $\dot{\theta}$. We look first at the fixed length outer cylinder. The kinetic energy K_1 is given by

$$K_1 = \tfrac{1}{2} m_1 V_1^2 \tag{9.48}$$

which is easily shown to be

$$K_1 = \tfrac{1}{2} m_1 r_1^2 \dot{\theta}^2 \tag{9.49}$$

Its potential energy P_1 is due to gravity acting upon the mass and is given by

$$P_1 = m_1 g\, r_1 \sin\theta \tag{9.50}$$

For the telescoping arm, which can both rotate at rate $\dot{\theta}$ and extend at rate $\dot{r}$, the expressions are

$$K_2 = \tfrac{1}{2} m_2 r^2 \dot{\theta}^2 + \tfrac{1}{2} m_1 \dot{r}^2 \tag{9.51}$$

$$P_2 = m_2 g r \sin\theta \tag{9.52}$$

From these we find that

$$L = K_1 + K_2 - P_1 - P_2$$

$$= \tfrac{1}{2} m_1 r_1^2 \dot{\theta}^2 + \tfrac{1}{2} m_2 r^2 \dot{\theta}^2 + \tfrac{1}{2} m_1 \dot{r}^2 - m_1 g r_1 \sin\theta$$

$$- m_2 g r \sin\theta \tag{9.53}$$

Computing the partial derivatives as in (9.46) and (9.47) then gives the differential equation model

$$(m_1 r_1^2 + m_2 r^2)\, \ddot{\theta} + 2 m_2 r\, \dot{r}\dot{\theta} + g\,(m_1 r_1 + m_2 r) \cos\theta = T$$

$$m_2 \ddot{r} - m_2 r \dot{\theta}^2 + m_2 g \sin\theta = F \tag{9.54}$$

In a system such as this, the control task will ordinarily be to choose and implement force F and torque T, possibly by commands to electric motors or hydraulic actuators, which will move the manipulator from one position coordinate set (r_i, θ_i) to another set (r_f, θ_f) in a smooth and safe manner. The mass m_2 may be variable as the manipulator picks up and moves objects or fixed because the manipulator is a welder or spray painter unit.

The equations of motion are notably non-linear, and if the desired positions are specified in cartesian coordinates $(x,y) =$ (horizontal,vertical) centred at the rotation axis of the arm, then even the coordinate transformations

$$x = r\cos\theta \qquad r = (x^2 + y^2)^{\frac{1}{2}}$$

$$y = r\sin\theta \qquad \theta = \tan^{-1}\left[\frac{y}{x}\right] \tag{9.55}$$

are non-linear.

The situation is much more complicated in three dimensional motion with an industrial robot and indeed entire books are devoted to deriving the transformations. A standard representation in terms of generalized coordinates in an n-vector q is that

$$D(q)\ddot{q} + C(q,\dot{q})\dot{q} + g(q) = \tau \tag{9.56}$$

where τ is the vector of applied generalized forces. The matrix $D(q)$ is the generalized inertia matrix, $C(q,\dot{q})$ is a matrix containing what are called the Christoffel symbols of the first kind and are results of the centrifugal and Coriolis forces, and $g(q)$ arises in the differentiation of the potential energy terms.

9.4.5 Presses

One of the more famous examples of adaptive control application is the work of Åström *et al.* (1977) on the control of paper making machines. We will meet this again in Chapters 30–31.

The results of this modelling effort gave the surprisingly simple sampled data model

$$y(kT) = -\alpha_1 y(kT-T) - \alpha_2 y(kT-2T) + \beta_1 u(kT-T)$$

$$+ \beta_2 u(kT-2T) + \gamma_1 e(kT-T) + \gamma_2 e(kT-2T) \qquad (9.57)$$

and it was found through recognition methods similar in philosophy to least-squares data fits that (approximately)

$$\alpha_1 = -4.5 \qquad \alpha_2 = 3.0$$
$$\beta_1 = -0.5 \qquad \beta_2 = -0.2$$
$$\gamma_1 = -8.0 \qquad \gamma_2 = -9.0$$

where the quantities involved are $y(kT)$ = paper basis weight = output, $u(kT)$ = steam pressure to drying cylinders = control variable, and $e(kT)$ = 'noise'. T is the sampling period for the computer to measure and command the process, so that kT is the time, $k=0,1,2,...$

Such black box methods have been shown to be quite useful and are at the heart of some commercial controllers which have a self-tuning property.

9.4.6 Modelling of noise and disturbances

An important part of the design of many control systems is allowing for extraneous inputs and for unmodelled plant characteristics. The undesired inputs consist of such effects as sensor noise and random or varying environmental effects such as gusts on aircraft. Unmodelled plant effects can be unknown or known, but ignored, characteristics, such as non-linearities or high frequency vibration modes.

Noise, such as measurement noise, is usually modelled as being a random process with certain known statistical characteristics (Appendix C). The common characteristics assumed for a noise process $\eta(t)$ include stationarity (the properties of the process do not change over time – used in property 2 below), ergodicity (probabilistic properties can be approximated using time averages), and that certain second order properties hold. Thus it is often assumed that the process has zero mean (average) value (properties 1 and 3 below) and that at any instant it has a gaussian (or normal) distribution (property 3). A further common assumption is that the process is white noise, which means among other things that $\eta(t)$ and $\eta(\tau)$ are uncorrelated for $t \neq \tau$ (property 2). In summary the three properties are:

1. $\mathscr{E}[\eta(t)] = 0$

2. $\mathscr{E}[\eta(\tau)\eta^T(t)] = R(t-\tau) = R\delta(t-\tau)$ $R \geq 0$

    $S(\omega) = \mathscr{F}(R(\tau)) = R = \text{constant}$

3. $p(\eta(t)) = N(0,R)$ (9.58)

where $S(\omega)$ is the power spectral density of the process, $\mathscr{F}(\cdot)$ denotes the Fourier transform operation, and $R(\tau)$ is the auto-correlation function.

Many models assume coloured noise rather than white noise. This is based upon either physical arguments (about the auto-correlation of the noise or the frequency band it must use) or on measurements of the noise (which are inferred from measuring known signals). In this case $S(\omega)$ is not constant, and of course then the auto-correlation $R(\tau) \neq 0$ for some $\tau \neq 0$. This situation is often modelled by assuming the noise is the output of a linear system with white noise input. In particular, the output can be 'computed' as the convolution

$$v(t) = \int h(t-\tau)\,\eta(\tau)\,d\tau = h(t)^{**}\,\eta(t) \tag{9.59}$$

and it follows that

$$R_v(\tau) = h(\tau)^{**}\,h(-\tau)^{**}\,R_\eta(\tau)$$

$$S_v(\omega) = |\mathscr{F}(h(t))|^2\,S_\eta(\omega) = |H(\omega)|^2\,S_\eta(\omega) \tag{9.60}$$

Such a model, with a carefully chosen linear system, is often an approximation used in analysis, with the input noise chosen to be white.

Disturbances are modelled in several ways, depending upon the analysis tools available and of course on the physical situation. One way is as a noise, often a low-frequency one. A second way is to suggest that the disturbance is always a special signal, typically a step or ramp, appearing as an input at a particular point of the system; this represents situations such as changes in quality of a process input material or flight into a high altitude jetstream. A third way is to consider all signals of a certain class, such as all those with bounded square integrals. This last we might denote as $d(t) \in D$ where D is a set of signals

$$D = \left\{ d(t) : \int \|d(\tau)\|^2 \, d\tau \leq 1 \right\} \tag{9.61}$$

9.4.7 Sampling of data

When computers are used for control, the data are discretized, i.e. converted to computer approximations for which the accuracy is finite and determined by computer and converter word length; at best, with a 12-bit ADC, the precision is within 1 part in 2^{13}, about 0.0125%, of full range. This effect is usually ignored, simply because other errors in the system (also frequently ignored) contribute 1–2% or more to output variation.

Another problem arises because the computer computes incrementally and sequentially in time. Thus, the data are sampled and the outputs occur only intermittently, albeit usually regularly with period T seconds, where T is typically in the range 0.1–10 s (with the option of much higher and lower values if appropriate). So, between samples the computer must generate commands which apply until the next are computed and output. Both effects must be allowed for in modelling and are discussed further in later chapters.

A summary of what we will need is straightforward. First, we cannot use differential equation models for our analysis, or not for all of our analysis; instead, we use difference equations such as the press model in section 9.4.5. These models work with samples of the signals, as defined by

$$\text{Output}(t) = \begin{cases} \text{Input } (t) & T=nT,\, n=0,\pm1,\pm2,\dots \\ \text{Undefined (or 0)} & t \neq nT \end{cases} \tag{9.62}$$

and sometimes represented mathematically using the Dirac delta function δ as

$$\text{Output}(t) = \Sigma \, \text{Input}(t) \, \delta(t-nT) \tag{9.63}$$

The physical situation and its representation are shown in Fig. 9.9.

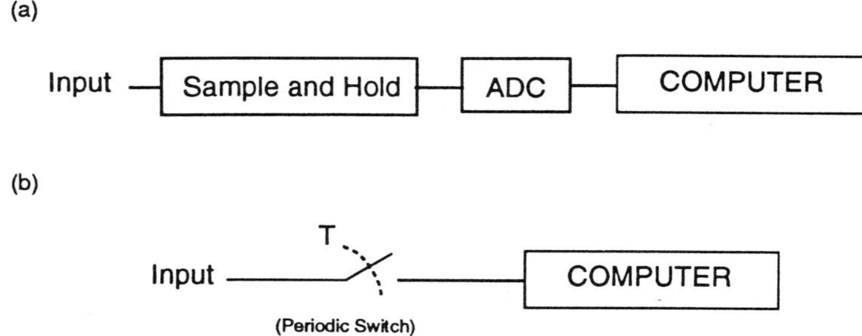

Figure 9.9 Modelling the computer input interface: (a) the devices; and (b) modelling as a periodic switch.

The **zero-order hold** is a component peculiar to computer (and other sampled-data) systems. It is a device with the property

$$\text{Output}(t) = \text{Input}(nT) \qquad\qquad nT \le t < nT+T \qquad\qquad (9.64)$$

used for the output, to approximate the physical device(s) of computer output buffer-DAC combination, as in Fig. 9.10.

Thus if the computer outputs a command $u(kT)$, the plant receives the signal of Fig. 9.11.

Interpreting the computer output as impulses leads to the standard model of the ZOH as having characteristic

$$u_{\text{out}}(t) = u(kT) \qquad\qquad kT \le t < kT+T \qquad\qquad (9.65)$$

The Laplace transform representation of this characteristic is

$$\text{ZOH}(s) = \frac{1 - e^{-Ts}}{s} \qquad\qquad (9.66)$$

9.5 LINEARIZATION

It is surprising that most control theory results apply to linear constant coefficient ordinary differential or difference equation models; exceptions are in the optimal control field (Chapters 26–27), where at

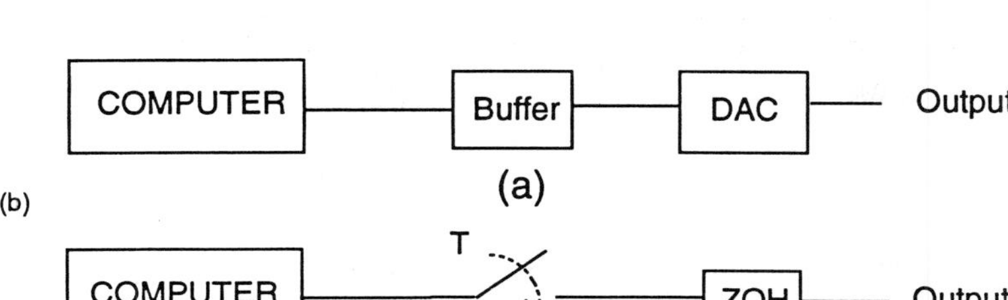

Figure 9.10 Modelling the computer output interface: (a) the devices; and (b) modelling as a switch leading to a zero-order hold (ZOH).

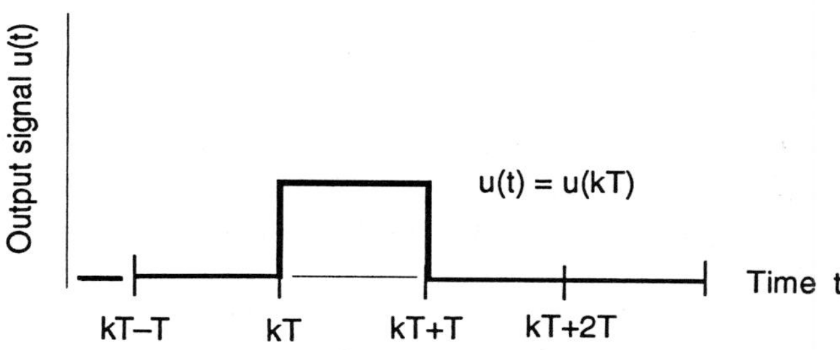

Figure 9.11 The computer's output hardware produces a square voltage pulse when the computer sends an output number.

least the forms of results can be found for non-linear ordinary equations, and some parts of the Lyapunov stability theory (Chapter 16). Yet the world is a non-linear place, exhibiting at least the properties of saturation (a maximum rotational speed for a motor, for example) and, in the case of digital systems, non-linearity due to the minimum resolution with which finite-length computer words can represent numbers.

There are at least two reasons for concentrating on linear constant coefficient ordinary differential/difference equations.

1. Because so much is known about vectors and matrices from linear algebra and from the theory of linear transforms (such as Laplace, Fourier, and z), many useful results are available. Generalizing the system model seldom leads to results amenable to implementation except in special cases.
2. The linear models are a reasonable approximation precisely because we use them in automatic control systems. In what seems a circular argument, by using a control system to force the plant to function 'near' an operating point, we keep deviations from linearity 'small' so that the linear assumption is a reasonable one.

Obtaining a linear model from a non-linear one is, in principle, straightforward – we use a Taylor series expansion of non-linear elements about the proposed operating point and retain only the linear term.

As an example, suppose a system is known to be modelled by the differential equation

$$\ddot{x} = f(x, \dot{x}, u) \tag{9.67}$$

where f is a known and sufficiently smooth function, i.e. its partial derivatives exist and are not too large numerically. Suppose that $x = x_{\text{ref}}(t)$ is known to be the trajectory description when $u_{\text{ref}}(t)$ is applied, so that

$$\ddot{x}_{\text{ref}} = f(x_{\text{ref}}, \dot{x}_{\text{ref}}, u_{\text{ref}}) \tag{9.68}$$

Consider varying $u_{\text{ref}}(t)$ by an amount $\Delta u(t)$, with the result $\Delta x(t)$, so that

$$\Delta \ddot{x} = f(x_{\text{ref}} + \Delta x, \dot{x}_{\text{ref}} + \Delta \dot{x}, u_{\text{ref}} + \Delta u) \tag{9.69}$$

Provided the partial derivatives exist, a Taylor series expansion may be performed, yielding

$$\Delta \ddot{x} = \left(\frac{\partial f}{\partial x}\right)\bigg|_* \Delta x + \left(\frac{\partial f}{\partial \dot{x}}\right)\bigg|_* \Delta \dot{x} + \left(\frac{\partial f}{\partial u}\right)\bigg|_* \Delta u$$

$$+ \left(\frac{\partial^2 f}{\partial x^2}\right)\bigg|_* (\Delta x)^2 + \left(\frac{\partial^2 f}{\partial x \partial \dot{x}}\right)\bigg|_* \Delta x \Delta \dot{x}$$

$$+ \left(\frac{\partial^2 f}{\partial x \partial u}\right)\bigg|_* \Delta x \Delta u + \left(\frac{\partial^2 f}{\partial \dot{x}^2}\right)\bigg|_* \Delta \dot{x}^2 + \cdots \tag{9.70}$$

where $*$ denotes the point $(x_{\mathrm{ref}}(t), \dot{x}_{\mathrm{ref}}(t), u_{\mathrm{ref}}(t))$ for evaluation of the partial derivatives.

Suppose Δu can be made small enough that Δx is small; further suppose that the second partial derivatives are small. Then the above is approximated by

$$\Delta \ddot{x} = a(t)\,\Delta x + b(t)\,\Delta \dot{x} + c(t)\,\Delta u \tag{9.71}$$

where

$$a = \frac{\partial f}{\partial x}\bigg|_* \qquad b = \frac{\partial f}{\partial \dot{x}}\bigg|_* \qquad c = \frac{\partial f}{\partial u}\bigg|_* \tag{9.72}$$

Then under these assumptions, we now have a linear ordinary differential equation in place of the original non-linear ordinary differential equation. If we further have u_{ref} = constant and $\ddot{x}_{\mathrm{ref}} = \dot{x}_{\mathrm{ref}} = 0$, then $a(t)$, $b(t)$, $c(t)$ are constants, which is even simpler.

Cases of such linearization abound, although often the linearization is done at the stage of original model construction. Consider a few examples.

1. For the rocket flight, if the engine burn is a short one, the mass may stay nearly constant – near enough that it can be treated as constant by the guidance algorithm. Similarly, for short periods of time, the altitude may be constant enough that the gravity can be treated as constant in magnitude and the orbital arc short enough that the gravitational direction is also constant. Then the rocket is described by linear ordinary differential equations, even

without the Taylor formalism; in fact, some of the terms are truncated after the constant term in their expansions.

2. Aerodynamic drag and lift forces are proportional to air density and to speed squared, but at near constant altitude and speed linear or even constant, approximations may be good enough. This is particularly true if the objective is to understand an autopilot which will maintain a constant altitude and speed.

3. In mechanical systems, non-linear frictions such as stiction (opposing the initial movement) and speed-squared drag are both present and often ignored.

And finally, if linearization is not good enough, or if it is not known whether it is good enough, what can be done? Usually, the engineer needing answers will use whatever he can learn from the linear model and check it with a simulation. Frequently much design can be done for regions of the operating regime of the plant; for example, helicopters are 'close enough' for linear approximations to apply to within ± 5 km/hr or so of cruising speed of perhaps 85 km/hr.

9.6 MODEL ERRORS

There is little doubt that models will be 'wrong', although for many purposes they will be adequate. Some techniques make implicit allowance for possible modelling errors; for example, one of the justifications for allowing gain margin and phase margin in frequency domain design (Chapters 15 and 20) is to allow for actual device transfer functions being different from the design standard, due perhaps to component tolerances. Other methods admit to explicit allowance for modelling inaccuracies, provided that some model allowing for the inaccuracies is present.

One method for incorporating inaccuracy is to assume that the output depends on the modelled effects plus noise. In such schemes

$$y(t) = F(y(\tau), u(\tau), \tau; \tau \leq t) + \eta(t) \tag{9.73}$$

where F is a known function developed from physical considerations and η is a 'noise' of character which allows for model errors. A small white noise might be used to compensate for unmodelled high frequency effects, for example.

Another approach treats the model as known to within a parameter set θ, with a particular piece of hardware representable as a function

$$y(t) = F(y(\tau), u(\tau), \theta, \tau;\ \tau \le t) \tag{9.74}$$

where $\theta \in \Theta$. One might then design around a particular value of θ, around the average value, or around a worst-case value. Particularly in the latter cases, one might need a characterization of Θ, such as

$$\Theta = \left\{ \theta :\ \|\ \theta\ \| \le 1\ \right\} \tag{9.75}$$

All of these are made more precise in Chapters 10–11 when we start working with specific model representations.

9.7 COMPUTER ASSISTANCE

Computers are used in simulating models, which usually means solving the differential equations describing the system, and in parameter estimation forms of system identification (Chapter 30). Some computer assisted control systems design (CACSD) programs also allow the manipulation of models, such as the building of system models from subsystem models, and the simulation of model responses.

9.8 COMMENTARY

Mathematical models attempt to represent the real system by sets of equations whose variables and parameters are (hopefully) predictors of variables and parameters of the real system. In control systems, the only class available is that of ordinary linear differential equations with constant coefficients and difference equations of the same kind; some randomness is acceptable, provided it is gaussian and not too complicated in its time evolution, but even this variation complicates things. It is indeed fortunate that this restricted class allows us to obtain some quite useful results.

The advantage of mathematical models is best seen by first looking at the alternatives, which only allow us to experiment upon the model and hence indirectly on the system. Then we may miss a revealing experiment because only a finite, usually small, number of tests are available to us. By contrast, the tools of mathematics allow us to analyse the structure of a mathematical model and hence of the system it (hopefully) represents. The example of Maxwell, who demonstrated that Watt's governors needed damping to operate stably and without oscillation – damping lost because of improvements in machinists' abilities to build to close tolerances without friction – is an early demonstration of the value of analysis. One can also envisage a great deal of effort being put into simultaneously controlling several outputs of a process to selected values when analysis might demonstrate that this was impossible without a change in the process (Chapter 22).

The disadvantage of a mathematical model is that it is so seldom computable in a useful form. Computers can be used to convert the equations to numbers, of course, but these may or may not be helpful of themselves. Even if closed form expressions can be found, they may give little insight into the system.

Nevertheless, control theory is essentially applied mathematics, with a whole tool kit of techniques, including graphical ones, which help to make it an enjoyable and useful field. A large fraction of this book is about those techniques.

9.9 FURTHER READING

The question is: where do the mathematical models needed for analysis come from? Our answer has been indirect, through examples and some discussion, but several different approaches have been demonstrated, including Newton's laws of motion, thermodynamics, and Lagrange methods. One observation is that a great deal of knowledge of the process is needed before it can be modelled. Many control engineers will find that they must either use a black box model with an identification algorithm (Chapter 30) or a published model concerning the plant to be controlled. For either of these, the control engineer will find that having basic knowledge of physics (and sometimes chemistry and biology) will be very helpful.

A major discussion of mathematical modelling in a broader context than control systems is in Murthy *et al.* (1990). More immediately devoted to dynamic systems is (McClamroch, 1980).

A few simple models are given without derivation by Åström and Wittenmark (1990, App. A). Dorf (1989, Ch. 2) gives the models of several components, while Franklin *et al.* (1990) present models of several interesting systems. Stephanopoulos (1984, Part II) presents models of chemical processes and was the source of the presentation of section 9.4.2. The modelling of joint movements for robots is covered in a number of texts, including (Paul, 1981) and (Snyder, 1985).

Linearization is done more elaborately in such texts as (Skelton, 1988, Ch. 3). Some studies of non-linear systems were published in the 1950s and 1960s by authors such as Gibson (1963).

Specific examples of applied or studied models of real systems are usually kept within industry, partly because they are too long and complicated to publish in a textbook. Significant studies are occasionally reported. Examples are found in the text of Grimble and Johnson (1988) (Volume 1 has a shape controller for a steel mill; volume 2 has a drill ship positioning system, plus more steel mill problems). Many texts have summaries and leads to real problems. The *IEEE Reprints* and the *IEEE Control Systems* magazine also lead to such models; *Kalman Filtering* (Sorenson, 1985), and *Adaptive Methods for Control Systems Design* (Gupta, 1986) exemplify the former, while the *CACSD Benchmark problems* (August 1987, August 1989, August 1990) lead to others. Solutions to the latter are available from CACSD system producers such as System Control Technology, Inc., Palo Alto, CA, designers of Ctrl-C, illustrating both the problems and CACSD application.

10

Continuous-time system representations

The mathematical model of a plant or process usually comprises a set of differential equations to describe its operations. For convenience, these are frequently converted to other representations and this chapter considers the linear time-invariant systems to which these representations apply.

10.1 SYNOPSIS

The control engineer's 'real world' can usually be described by sets of differential equations, as in Chapter 9. Often useful for both processes and components is the special case of linear ordinary differential equations with constant coefficients, such as

$$y^{(n)} + a_{n-1} y^{(n-1)} + \cdots + a_2 \ddot{y} + a_1 \dot{y} + a_0 y$$

$$= b_m u^{(m)} + b_{m-1} u^{(m-1)} + \cdots + b_1 \dot{u} + b_0 u \tag{10.1}$$

where $y^{(n)}$ is the nth derivative of the function $y(t)$ and $\dot{y}$ and $\ddot{y}$ are the 1st and 2nd derivatives respectively. Alternative representations of the essential aspects of such systems are available in the forms of transfer functions

$$\frac{Y(s)}{U(s)} = \frac{b_m s^m + b_{m-1} s^{m-1} + \cdots + b_1 s + b_0}{s^n + a_{n-1} s^{n-1} + \cdots + a_1 s + a_0} \tag{10.2}$$

or frequency responses

$$\frac{Y(j\omega)}{U(j\omega)} = \frac{b_m (j\omega)^m + b_{m-1}(j\omega)^{m-1} + \cdots + b_1(j\omega) + b_0}{(j\omega)^n + a_{n-1}(j\omega)^{n-1} + \cdots + a_1(j\omega) + a_0}$$

where we use the usual notational conventions that $j = \sqrt{-1}$ and ω is radian frequency. An alternative notation uses the operator $\mathrm{D}^i \equiv \mathrm{d}^i/\mathrm{d}t^i$ to give

$$(\mathrm{D}^n + a_{n-1}\mathrm{D}^{n-1} + \cdots + a_0)\,y(t)$$

$$= (b_m\mathrm{D}^m + b_{m-1}\mathrm{D}^{m-1} + \cdots + b_0)\,u(t)$$

The above are especially useful for the design techniques characterized as **classical** and for SISO systems. The transfer function can be written in a number of different forms, of which (10.2) is the **direct form**; others include **parallel, cascade**, and **continued fraction** forms.

For **modern** approaches and MIMO systems, a set of first-order ordinary differential equations is used (called a state variable model) and in the special case of constant coefficient linear differential equations this model has a matrix representation as

$$\dot{\mathbf{x}} = \mathbf{A}\mathbf{x} + \mathbf{B}\mathbf{u}$$

$$\mathbf{y} = \mathbf{C}\mathbf{x} + \mathbf{D}\mathbf{u}$$

These representations are not unique, although the input–output relationships must be the same for each representation of the same system. It is sometimes convenient to have the matrices in canonical forms; one example is the **controllable canonical form**, which for the system described by (10.1) is

$$\dot{\mathbf{x}} = \begin{bmatrix} 0 & 1 & 0 & \cdots & 0 & 0 \\ 0 & 0 & 1 & \cdots & 0 & 0 \\ \cdot & \cdot & \cdot & & \cdot & \cdot \\ 0 & 0 & 0 & & 1 & 0 \\ 0 & 0 & 0 & \cdots & 0 & 1 \\ -a_0 & -a_1 & -a_2 & \cdots & & -a_{n-1} \end{bmatrix} \mathbf{x} + \begin{bmatrix} 0 \\ 0 \\ \cdot \\ \cdot \\ \cdot \\ 0 \\ 1 \end{bmatrix} \mathbf{u}$$

$$\mathbf{y} = \begin{bmatrix} b_0 & b_1 & \cdots & b_m & 0 & \cdots & 0 \end{bmatrix} \mathbf{x}$$

Others include the **observable canonic form** and the **Jordan canonic form**. These are useful in certain derivations, although ordinarily one attempts to use forms in which the state variables have physical significance.

10.2 TRANSFER FUNCTIONS

Transfer functions allow operation in the frequency- or s-domain of complex numbers, rather than in the time domain. They are usually considered primarily with SISO systems, although at least notationally it is possible to represent MIMO systems. In either case, they apply to linear ordinary differential equations with constant coefficients; we will see the difference equation variation in Chapter 11.

10.2.1 Single-input–single-output (SISO) systems

If the differential equations describing a system are linear, have constant (or highly special) coefficients, and have no time delays, then analysis based upon Laplace transforms is common and convenient. We observe that, in the absence of initial conditions, a variable $x(t)$ with Laplace transform $X(s)$, i.e.

$$x(t) \Leftrightarrow X(s) \equiv \int_0^\infty x(t)\, e^{-st}\, dt$$

also has for integer k

$$\frac{d^k x(t)}{dt^k} \Leftrightarrow s^k X(s) - s^{k-1} x(0+) - s^{k-2} \dot{x}(0+) - \cdots - x^{(k-1)}(0+)$$

where $x(0+)$ is the limit of $x(\varepsilon)$ as $\varepsilon \rightarrow 0$ and $\varepsilon > 0$, and

$$\int_0^t x(\tau)\, d\tau \Leftrightarrow \frac{X(s)}{s}$$

A time delay, which we will have occasion to use, albeit reluctantly for continuous time systems even though they are easily handled for discrete time systems such as those used in computer control studies, has transform effect

$$x(t-\tau) \Leftrightarrow e^{-s\tau} X(s)$$

Using these properties, we see that the Laplace transform variable s can virtually be used as a 'differentiator operator'. For example, the system

$$\frac{d^n x}{dt^n} + a_{n-1} \frac{d^{n-1} x}{dt^{n-1}} + a_{n-2} \frac{d^{n-2} x}{dt^{n-2}} + \cdots + a_1 \frac{dx}{dt} + a_o x$$

$$= b_m \frac{d^m u}{dt^m} + b_{m-1} \frac{d^{m-1} u}{dt^{m-1}} + \cdots + b_1 \frac{du}{dt} + b_o u$$

has, if we define $X(s)$ and $U(s)$ as the Laplace transforms of $x(t)$ and $u(t)$, respectively, a representation (without initial conditions) of

$$\left(s^n + a_{n-1} s^{n-1} + a_{n-2} s^{n-2} + \cdots + a_1 s + a_o\right) X(s)$$

$$= \left(b_m s^m + b_{m-1} s^{m-1} + \cdots + b_1 s + b_o\right) U(s)$$

The ratio of $X(s)$ to $U(s)$ is then called the transfer function from input $u(t)$ to output $x(t)$, e.g.

$$\frac{X(s)}{U(s)} = \frac{b_m s^m + b_{m-1} s^{m-1} + \cdots + b_1 s + b_o}{s^n + a_{n-1} s^{n-1} + a_{n-2} s^{n-2} + \cdots + a_1 s + a_o} \tag{10.3}$$

Roots of the denominator polynomial are the system poles, and roots of the numerator are the system zeros. The transfer function approach has several applications:

* manipulation into special forms,
* solution of the differential equations represented,
* manipulation (such as reduction or elimination of variables) of subsystem models to obtain system models,
* generation of the frequency response.

Transfer function forms

The expression in (10.3) is a **direct form** of the transfer function, but other forms are sometimes useful, particularly if the transfer function represents a filter or control law to be implemented. The basis for two of these is the fundamental theorem of algebra, which allows the numerator and denominator polynomials to be factored to first-order terms, as in

$$\frac{X(s)}{U(s)} = H(s) = K\frac{(s-s_1)\,(s-s_2)\,\cdots\,(s-s_m)}{(s-\lambda_1)\,(s-\lambda_2)\,\cdots\,(s-\lambda_n)}$$

where s_i, $i=1,2,\ldots,m$ are the zeros and λ_j $j=1,\ldots,n$ are the system poles. These may be combined in a variety of orderings to give **cascade** or **series** forms, such as

$$H(s) = H_1(s)\,H_2(s)\,\cdots\,H_k(s)$$

where

$$H_i(s) = K_i\frac{(s-s_{i,1})\,\cdots\,(s-s_{i,m_i})}{(s-\lambda_{i,1})\,\cdots\,(s-\lambda_{i,n_i})} \quad i=1,2,\ldots,k \tag{10.4}$$

Another possibility is **parallel form**, exemplified by

$$H(s) = H_1(s) + H_2(s) + \cdots + H_j(s)$$

with the $H_i(s)$ similar in form to the factors in (10.4). A further possibility is the **continued fraction form**, exemplified by the special case of the Cauer form (used in filter network realizations):

$$H(s) = \cfrac{1}{h_1 + \cfrac{1}{h_2 s + \cfrac{1}{h_3 + \cfrac{1}{h_4 s + \cdots}}}}$$

The utility of the above forms is both analytical and, in some cases, synthetic in that computations of control laws may be organized. The first three structures are shown in Fig. 10.1; the last is met in section 30.2.4.

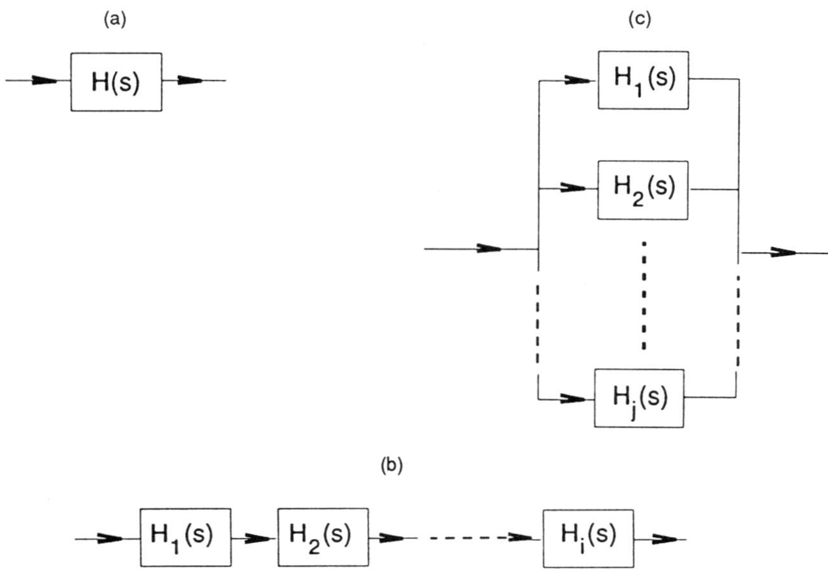

Figure 10.1 The structure of transfer functions. (a) The direct form is a ratio of polynomials. (b) Using partial fraction expansions gives an equivalent form, as does (c) using a product form or cascade decomposition.

Example

The following forms can be easily derived for the system described by the differential equation

$$y^{(3)} + 9\ddot{y} + 23\dot{y} + 15y = \ddot{u} + 6\dot{u} + 8u$$

$$H(s) = \frac{s^2 + 6s + 8}{s^3 + 9s^2 + 23s + 15}$$

$$= \left(\frac{s + 2}{s + 1}\right)\left(\frac{s + 4}{s + 3}\right)\left(\frac{1}{s + 5}\right)$$

$$= \frac{0.375}{s + 1} + \frac{0.25}{s + 2} + \frac{0.375}{s + 3}$$

$$= \cfrac{1}{s + \cfrac{1}{\frac{1}{3} + \cfrac{1}{3s + \cfrac{1}{\frac{1}{6} + \cfrac{1}{12s + \cfrac{1}{(1/30)}}}}}}$$

Solution of differential equations

The obvious use of Laplace transforms is to solve the differential equations represented. Thus, if $G(s)$ is a known transfer function from $u(t)$ to $x(t)$ and if $u(t)$ is a form such that $U(s)$ is readily available, then using the notation $x(t) \Leftrightarrow X(s)$ to mean that $x(t)$ and $X(s)$ form a transform pair in that $X(s)$ is the transform of $x(t)$ and $x(t)$ is the inverse transform of $X(s)$,

$$x(t) \Leftrightarrow X(s) = G(s)U(s)$$

may often be readily found. In control systems, we often seek the response to a unit impulse, i.e. the Dirac delta function $\delta(t)$ (Fig. 10.2(a)) for which

$$\delta(t) = 0, \; t \neq 0$$

$$\int_{-\varepsilon}^{\varepsilon} \delta(\tau)\, d\tau = 1, \; \varepsilon > 0$$

$$\delta(t) \Leftrightarrow 1$$

This special function, one of a class called distributions, is often taken as the limit of a pulse of unit area whose width vanishes. This is shown for a rectangular pulse in Fig. 10.2(a) along with a common graphical symbol for the function.

Another commonly sought response is to the unit step function $\mathscr{U}(t)$, shown in Fig.10.2(b), for which

$$\mathscr{U}(t) = \begin{cases} 0 & t < 0 \\ 1 & t \geq 0 \end{cases}$$

$$\mathscr{U}(t) \Leftrightarrow \frac{1}{s}$$

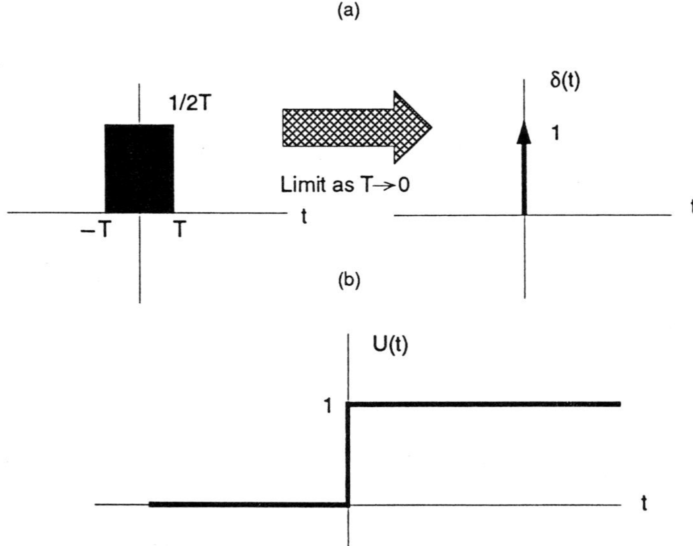

Figure 10.2 (a) Dirac delta function, $\delta(t)$; (b) unit step function, $\mathscr{U}(t)$.

We note that the transform of the impulse response of a system is equal to the transfer function.

The actual solution can usually be found using tables of Laplace transforms, although in general the inverse Laplace transform is needed. A typical case will require that $X(s)$ be expanded in partial fractions to forms available in tables and that the tables be consulted. Hence if

$$X(s) = e^{-s\tau} \frac{b_m s^m + b_{m-1} s^{m-1} + \cdots + b_1 s + b_0}{s^n + a_{n-1} s^{n-1} + \cdots a_1 s + a_0}$$

then we take

$$X(s) = \sum_{i=1}^{k} e^{-s\tau} X_i(s)$$

by partial fractions (out to first-order terms if necessary) and use tables to yield

$$X(t) = \sum_{i=1}^{k} x_i(t-\tau)$$

It is presumed this approach is familiar to readers from their mathematics background.

Example

Consider

$$H(s) = \frac{1}{s(4s+1)}$$

$$= \frac{1}{s} - \frac{4}{4s+1} = H_1(s) + H_2(s)$$

$$H_1(s) = \frac{1}{s} \Leftrightarrow 1 = h_1(t)$$

$$H_2(s) = \frac{-4}{4s+1} \Leftrightarrow e^{-t/4} = h_2(t)$$

then the impulse response of the model is

$$h(t) = h_1(t) + h_2(t)$$

$$= 1 - e^{-t/4}$$

Manipulation of representations

The Laplace transform has a number of properties which make it useful in the study of linear systems: it is a linear operator; the transform of the convolution of two functions equals the product of the transforms of the functions; because of these it is easily shown to be associative, distributive, and (in the case of scalars, but not matrices) commutative. Manifestations of these properties allow manipulation of transfer functions almost algebraically and are very useful in analysing systems. For examples,

a. If $X(s) = G_1(s)U(s)$ and $Y(s) = G_2(s)X(s)$, then $Y(s) = G_2(s)G_1(s)U(s) = G(s)U(s)$. Thus association justifies generation of an intermediate signal $x(t) \Leftrightarrow X(s)$ or the alternative of a single transfer function $G(s) = G_2(s)G_1(s)$.
b. If $G(s) = G_1(s) + G_2(s)$ and $Y(s) = G(s)U(s)$, then $Y(s) = X_1(s) + X_2(s)$ where $X_1(s) = G_1(s)U(s)$ and $X_2(s) = G_2(s)U(s)$. This use of the distribution property allows, for example, the construction of control laws in alternative forms (e.g. three-term controllers as sums of P, I, D terms or as second-order systems).

The most common use is to derive a closed loop transfer function from the transfer functions of several components.
  For the system of Fig. 10.3, we see that

$$Y(s) = G(s)U(s) \qquad\qquad U(s) = C(s)E(s)$$

$$E(s) = R(s) - M(s) \qquad\qquad M(s) = F(s)Y(s)$$

from which a little algebra yields

$$\frac{Y(s)}{R(s)} = \frac{G(s)C(s)}{1 + F(s)G(s)C(s)}$$

as the closed-loop transfer function. The simple manipulations above are common; more complicated versions of the algebra can sometimes be done more easily by looking at and manipulating the patterns of the blocks – an approach called **block diagram algebra** – as found in older textbooks.

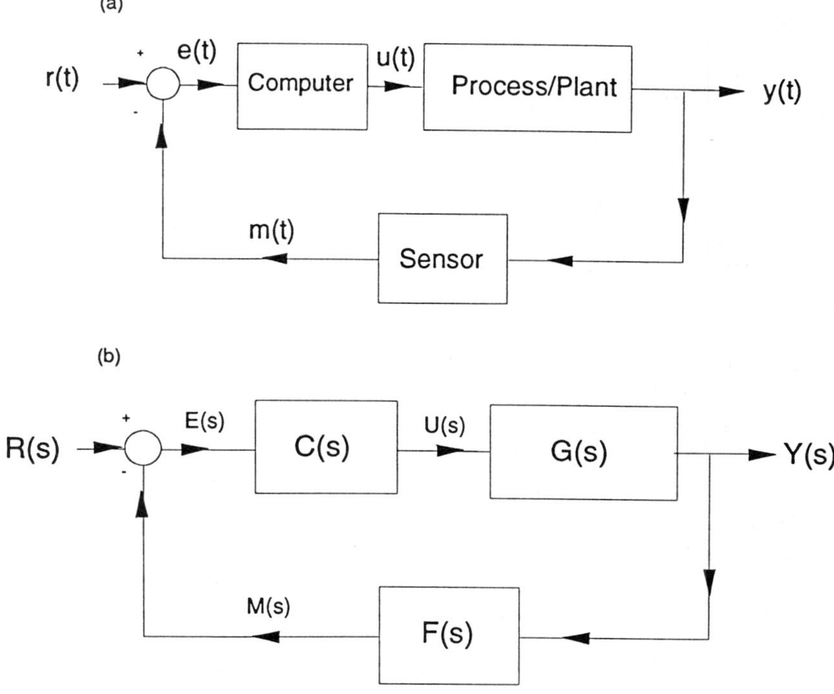

Figure 10.3 Two representations for a control system. (a) Block diagram of the physical situation with variables denoted as indicated. (b) Taking Laplace tranforms allows an equivalent representation.

10.2.2 Multivariable – multi-input–multi-output (MIMO) – transfer functions

For systems with m inputs and n outputs, it is possible to write differential equation sets such as

$$\sum_{k=0}^{n_i} \alpha_{i,k}\, y_i^{(k)} = \sum_{j=1}^{m} \sum_{k=0}^{m_j} \beta_{i,j,k}\, u_j^{(k)} \qquad i=1,\ldots,n$$

with transfer function representations

$$Y_i(s) = \sum_{j=1}^{m} \frac{\sum_{k=0}^{m_j} \beta_{i,j,k}\, s^k}{\sum_{k=0}^{n_i} \alpha_{i,k}\, s^k}\, U_j(s) \qquad i=1,\ldots,n$$

$$= \sum_{j=1}^{m} G_{i,j}(s)\, U_j(s) \qquad i=1,\ldots,n$$

This is more convenient notationally in the matrix form

$$\mathbf{Y}(s) = \begin{bmatrix} Y_1(s) \\ Y_2(s) \\ \vdots \\ Y_n(s) \end{bmatrix} = \begin{bmatrix} G_{1,1}(s) & G_{1,2}(s) & \ldots & G_{1,m}(s) \\ G_{2,1}(s) & G_{2,2}(s) & \ldots & G_{2,m}(s) \\ G_{n,1}(s) & G_{n,2}(s) & \ldots & G_{n,m}(s) \end{bmatrix} \begin{bmatrix} U_1(s) \\ U_2(s) \\ \vdots \\ U_m(s) \end{bmatrix}$$

$$= \mathbf{G}(s)\,\mathbf{U}(s)$$

Since the operators are all linear, this representation can be manipulated almost as if they were scalars. Thus, assuming the dimensions are correct, we may find that series, parallel, and feedback connections may be formed in the expected way. For example, series connection yields

$$\mathbf{Y}(s) = \mathbf{G}(s)\,\mathbf{W}(s) \quad \text{and} \quad \mathbf{W}(s) = \mathbf{H}(s)\,\mathbf{U}(s)$$

$$\Rightarrow \mathbf{Y}(s) = \mathbf{G}(s)\,\mathbf{H}(s)\,\mathbf{U}(s)$$

(but *not* $\mathbf{H}(s)\mathbf{G}(s)\mathbf{U}(s)$ because matrix multiplication is not commutative).

Parallel connection yields that

$$\mathbf{V}(s) = \mathbf{W}(s) + \mathbf{Y}(s)$$

where $\mathbf{W}(s) = \mathbf{H}(s)\,\mathbf{U}(s)$ and $\mathbf{Y}(s) = \mathbf{G}(s)\,\mathbf{U}(s)$

$$\Rightarrow V(s) = [H(s) + G(s)]\, U(s)$$

Finally, feedback connection, defined by Fig. 10.3, has

$$Y(s) = G(s)\,C(s)\,E(s)$$

$$E(s) = R(s) - M(s)$$

$$M(s) = F(s)\,Y(s)$$

and yields

$$Y(s) = [I + G(s)\,C(s)\,F(s)]^{-1}\,G(s)\,C(s)\,R(s)$$

and

$$Y(s) = G(s)\,C(s)\,[I + F(s)\,G(s)\,C(s)]^{-1}\,R(s)$$

where I is the identity matrix. In the basic representation, the matrix $G(s)$ is $n \times m$ and so not necessarily square.

Special forms: the Smith–McMillan form and matrix fraction forms

The notions of system poles and zeros are not so clear for matrices as for SISO systems (where they are simply the roots of the denominator and numerator polynomials, respectively). As part of the determination of this and other questions, it is useful to have a canonical form for the transfer function matrix. The form used is the Smith–McMillan form.

The Smith–McMillan form $M(s)$ of a rectangular matrix $G(s)$ is a rectangular matrix in which the only non-zero elements are those for which the row and column indices are equal. $M(s)$ may be derived from $G(s)$ by elementary row and column operations, and the polynomials making up the rational terms satisfy certain divisibility properties. In particular, the Smith–McMillan form of $G(s)$ is

$$\mathbf{M}(s) = \begin{pmatrix} \mathbf{M}(s) & 0 \\ 0 & 0 \end{pmatrix}$$

$$\mathbf{M}_1(s) = \mathrm{diag}\left\{ \frac{\upsilon_1(s)}{\delta_1(s)}, \frac{\upsilon_2(s)}{\delta_2(s)}, \ldots, \frac{\upsilon_r(s)}{\delta_r(s)} \right\}$$

where r is the normal rank (i.e. rank for almost all s) of $G(s)$, $\upsilon_i(s)$ and $\delta_i(s)$ are co-prime, i.e. they have no common factors, and

$$\left. \begin{array}{l} \upsilon_i(s) \text{ divides } \upsilon_{i+1}(s) \\ \delta_{i+1}(s) \text{ divides } \delta_i(s) \end{array} \right\} i = 1, \ldots, r-1$$

Computation of the Smith–McMillan form follows from first forming

$$\mathbf{G}(s) = \frac{1}{d(s)} \mathbf{P}(s)$$

where $d(s)$ is the least common multiple of the denominators of all the terms of $\mathbf{G}(s)$ and $\mathbf{P}(s)$ is a polynomial matrix. Then $\mathbf{P}(s)$ is manipulated using a sequence of elementary operations; each such operation is representable as a pre- or post-multiplication by a simple matrix called an elementary matrix. The operations and their matrix multiplier equivalents are:

- interchange of two rows/columns – the elementary matrix is an identity of appropriate size in which the required two rows/columns have been interchanged – pre-multiplication interchanges rows while post-multiplication interchanges columns;
- multiplication of a row/column by a constant – the elementary matrix is an identity matrix except that the desired constant is placed on the diagonal at the required row/column; again pre-multiplication affects rows and post-multiplication affects columns;
- addition of a polynomial multiple of one row/column to another row/column – the elementary matrix uses the identity with the polynomial placed in the appropriate row/column – with pre- and post-multiplication effects as above.

The elementary matrices used to pre- or post-multiply $(\mathbf{L}_i, \mathbf{R}_j)$ can be applied to $\mathbf{P}(s)$ to create a matrix $\mathbf{Q}(s)$

$$\mathbf{Q}(s) = \mathbf{L}_k(s)\,\mathbf{L}_{k-1}(s)\,\cdots\,\mathbf{L}_1(s)\,\mathbf{P}(s)\,\mathbf{R}_1(s)\,\mathbf{R}_2(s)\,\cdots\,\mathbf{R}_{L-1}(s)$$

The particular matrix we seek is the Smith form, $\mathbf{S}(s)$, which is given by

$$\mathbf{S}(s) = \begin{pmatrix} \mathbf{S}_1(s) & 0 \\ 0 & 0 \end{pmatrix}$$

$$\mathbf{S}_1(s) = \operatorname{diag}\left\{ \eta_1(s), \eta_2(s), \ldots, \eta_r(s) \right\}$$

where

$$\eta_i(s) = \frac{d_i(s)}{d_{i-1}(s)}$$

with $d_0(s) = 1$, and $d_i(s)$ the greatest common divisor polynomial of all $i \times i$ minors of $\mathbf{P}(s)$, normalized to be monic, i.e. to have the coefficient of the highest point term in s be +1.

Example

As in Maciejowski (1989), consider the two-input–three-output system described by

$$\mathbf{G}(s) = \begin{bmatrix} \dfrac{1}{s^2 + 3s + 2} & \dfrac{-1}{s^2 + 3s + 2} \\[2mm] \dfrac{s^2 + s - 4}{s^2 + 3s + 2} & \dfrac{2s^2 - s - 8}{s^2 + 3s + 2} \\[2mm] \dfrac{s - 2}{s + 1} & \dfrac{2s - 4}{s + 1} \end{bmatrix}$$

$$= \frac{\begin{bmatrix} 1 & -1 \\ s^2 + s - 4 & 2s^2 - s - 8 \\ s^2 - 4 & 2s^2 - 8 \end{bmatrix}}{s^2 + 3s + 2}$$

It is easily shown that $d_0(s) = 1$, $d_1 = 1$, $d_2(s) = s^2 - 4$ and hence the Smith–McMillan form of $\mathbf{G}(s)$ is

$$\mathbf{M}(s) = \frac{\begin{bmatrix} 1 & 0 \\ 0 & s^2 - 4 \\ 0 & 0 \end{bmatrix}}{s^2 + 3s + 2}$$

The elementary matrices in the transformation are presented in

$$\mathbf{M}(s) = \begin{bmatrix} 1 & 0 & 0 \\ 0 & 1 & 0 \\ -s & 0 & 1 \end{bmatrix} \begin{bmatrix} 1 & 0 & 0 \\ -\frac{s}{2} & 1 & 0 \\ 0 & 0 & 1 \end{bmatrix} \begin{bmatrix} 1 & 0 & 0 \\ 0 & \frac{1}{2} & 0 \\ 1 & 1 & -1 \end{bmatrix} \mathbf{G}(s)$$

$$\times \begin{bmatrix} 2 & 0 \\ -1 & 1 \end{bmatrix} \begin{bmatrix} \frac{1}{3} & 0 \\ 0 & 1 \end{bmatrix} \begin{bmatrix} 1 & 1 \\ 0 & 1 \end{bmatrix}$$

$$= \begin{bmatrix} 1 & 0 & 0 \\ -\frac{s}{2} & \frac{1}{2} & 0 \\ 1-s & 1 & -1 \end{bmatrix} \mathbf{G}(s) \begin{bmatrix} \frac{2}{3} & \frac{2}{3} \\ -\frac{1}{3} & \frac{2}{3} \end{bmatrix}$$

Using the Smith–McMillan form, a number of useful definitions may be made.

$$p(s) = \delta_1(s)\,\delta_2(s) \cdots \delta_r(s) = \text{pole polynomial of } \mathbf{G}(s)$$

$$z(s) = v_1(s)\,v_2(s) \cdots v_r(s) = \text{zero polynomial of } \mathbf{G}(s)$$

$$\text{Deg}(p(s)) = \text{McMillan degree of } \mathbf{G}(s)$$

If $(s-s_k)^\rho$ is a factor of $p(s)$ $\{$of $z(s)\}$, then ρ is said to be the multiplicity of the pole $\{$resp. zero$\}$ at s_0.

Zeros of $\mathbf{G}(s)$ are called transmission zeros of the system. Transmission zeros can be observed to be points at which the rank of $\mathbf{G}(s)$ decreases. Their name comes about because they have the property that for particular inputs related to the zero locations, the system will have no output, i.e. they have a transmission blocking property.

An alternative form for a rational transfer function matrix, which makes the notation seem even more analogous to the scalar case, is the **matrix fraction description**. In this, appropriate (and non-unique) matrices of polynomials $\mathbf{N}(s)$ and $\mathbf{D}(s)$ are used to enable representations such as $\mathbf{G}(s) = \mathbf{N}(s)\,\mathbf{D}^{-1}(s)$. This is called the right matrix-fraction description, and $\mathbf{N}(s)$ and $\mathbf{D}(s)$ are called the numerator and denominator matrices, respectively. One way of finding them involves the Smith–McMillan form. In particular, we write

$$\mathbf{G}(s) = \mathbf{L}(s)\,\mathbf{M}(s)\,\mathbf{R}(s)$$

$$= \mathbf{L}(s)\,\mathrm{diag}\left\{\frac{\upsilon_1(s)}{\delta_1(s)}\,,\ \frac{\upsilon_2(s)}{\delta_2(s)}\,,\ \ldots\,,\ \frac{\upsilon_r(s)}{\delta_r(s)}\,,\ 0,\ 0,\ \ldots\,,\ 0\right\}\mathbf{R}(s)$$

where $\mathbf{L}(s)$ and $\mathbf{R}(s)$ are the products of the elementary matrices as before. Then we define

$$\mathbf{N}_\mathrm{M}(s) = \mathrm{diag}\left\{\upsilon_1(s),\ \upsilon_2(s),\ \ldots,\ \upsilon_r(s),\ 0,\ \ldots,\ 0\right\}$$

$$\mathbf{D}_\mathrm{M}(s) = \mathrm{diag}\left\{\delta_1(s),\ \delta_2(s),\ \ldots,\ \delta_3(s),\ 1,\ 1,\ \ldots,\ 1\right\}$$

so that $\mathbf{M}(s) = \mathbf{N}_\mathrm{M}(s)\,\mathbf{D}_\mathrm{M}^{-1}(s)$ and, because elementary matrices are always invertible,

$$\mathbf{G}(s) = \mathbf{L}(s)\left[\mathbf{N}_\mathrm{M}(s)\,\mathbf{D}_\mathrm{M}^{-1}(s)\right]\mathbf{R}(s)$$

$$\mathbf{G}(s) = \left[\mathbf{L}(s)\,\mathbf{N}_\mathrm{M}(s)\right]\left[\mathbf{R}^{-1}(s)\,\mathbf{D}_\mathrm{M}(s)\right]^{-1}$$

$$= \mathbf{N}(s)\,\mathbf{D}^{-1}(s)$$

where the definitions of **N** and **D** are obvious. We remark that it is now clear that a point z is a zero if $\mathbf{N}(z)$ is of lesser rank that $\mathbf{N}(s)$, and a point p is a pole if $\mathbf{D}(p)$ is of decreased rank (and hence not invertible at that point). Notice that similar arguments lead to a left matrix-fraction description:

$$\mathbf{G}(s) = \tilde{\mathbf{D}}^{-1}(s)\,\tilde{\mathbf{N}}(s)$$

10.3 FREQUENCY RESPONSE

A very special input to a system is a sinusoid $B\sin(\omega t)$, and if the system is linear, the post-transient output is $A\sin(\omega t + \phi)$. The effect of the system is represented by the ratio A/B and the phase shift ϕ; the variation of these with ω is the frequency response of the system. Thus when the input is $\sin(\omega t)$, the steady-state output will be of the form $A(\omega)\sin(\omega t + \phi(\omega))$

For a linear constant coefficient system described by transfer function $G(s)$, it is almost always true that the complex function $G(j\omega)$ gives the frequency response:

$$A(\omega) = |G(j\omega)|$$

$$\phi(\omega) = \arg(G(j\omega))$$

or

$$G(j\omega) = A(\omega)\,e^{j\phi(\omega)}$$

This information is usually represented graphically; three different sets of graphs are in common use.

1. Two plots are used: $A(\omega)$ vs ω, called the **amplitude response**, and $\phi(\omega)$ vs ω, called the **phase response**. Frequently the plots are $A(\omega)$ in decibels (i.e. $20\log(A(\omega))$) vs $\log\omega$ (or ω is on a logarithmic scale) and $\phi(\omega)$ in degrees vs $\log\omega$. These are the **Bode plots** for the system when G is an open-loop transfer function.
2. One plot of $\operatorname{Im}(G(j\omega))$ vs $\operatorname{Re}(G(j\omega))$ is made with ω as a parameter. An alternative point of view is that the plot is a polar plot of radius $A(\omega)$ and angle $\phi(\omega)$, with ω as the arc parameter.

This is a portion of the **Nyquist plot** for the system when G is an open-loop transfer function.

3. A plot of $A(\omega)$ in dB is made against $\phi(\omega)$ in degrees. With auxiliary information and when G is an open-loop transfer function, this log (magnitude) – phase plot is the **Nichols' chart** for the system.

These plots can be very useful because the frequency response can for some systems, especially electronic ones, be established experimentally. Many readers will be familiar with the concept of frequency response specifications for audio equipment; the concept here is similar.

The frequency response of a system (e.g. an idealized small motor) with transfer function $H(s) = 1/[s(4s+1)]$ is shown in Fig. 10.4.

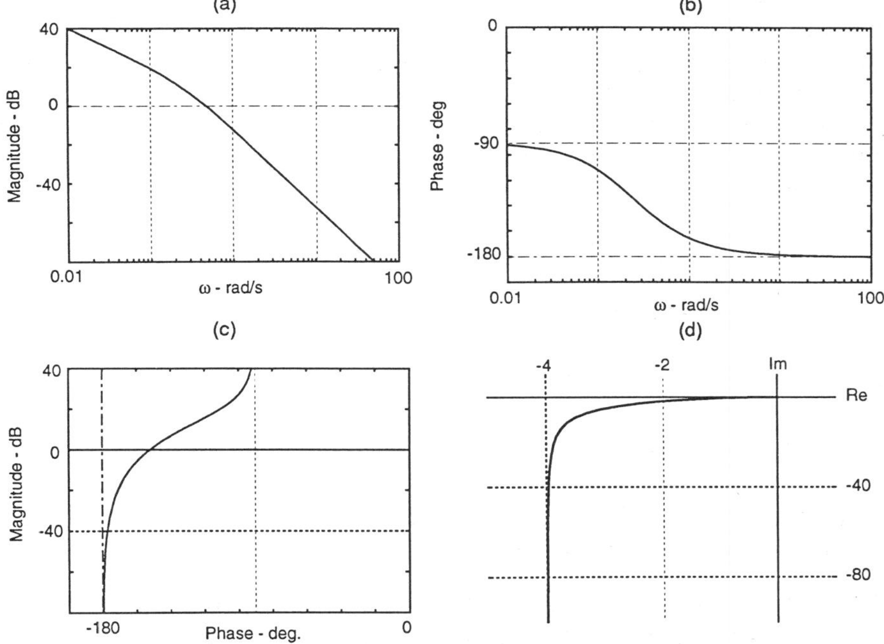

Figure 10.4 Equivalent to the transfer function is the Fourier transform of the impulse response, which if necessary can be found experimentally by measuring the output when sinusoidal inputs are applied. This frequency response has three forms of representation: (a) and (b) are the magnitude and phase of the system (Bode plots), (c) is log-magnitude vs phase (becomes a Nichols plot with additional information), (d) is a polar plot (partial Nyquist plot).

Multivariable Frequency Response

It is quite feasible to plot the frequency responses of the elements of a $G(s)$ matrix, i.e. to compute and plot $G_{ij}(j\omega)$ for input j and output i for frequency ω. In the scalar case, the magnitude of such a term represents a gain at that frequency. To extend this concept of gain to the matrix, it is logical to compute a gain as the magnitude measure of the output divided by the magnitude measure of the input. This is done using induced norms as discussed in Chapters 15, 20 and 33, and Appendix B. In particular, the gain for an input $u(t)$ is defined as

$$\text{Gain} = \frac{\|y(t)\|}{\|u(t)\|}$$

which can be shown to be bounded, when $u(t)$ is a mix of inputs at frequency ω, by the largest and smallest singular values of the matrix $G(j\omega)$. This fact is expressed mathematically by

$$\underline{\sigma}(G(j\omega)) \leq \frac{\|y(t)\|}{\|u(t)\|} \leq \overline{\sigma}(G(j\omega))$$

where the σ are computed using a matrix singular value decomposition (SVD). This of course yields bounds for designing multivariable compensators.

10.4 STATE VARIABLE REPRESENTATION

The notion of the state of a system is fundamental to the class of tools of control systems analysis called 'modern' and can be defined in many ways. Common to all definitions is the following set of notions:

> The **state** of a system is a set of variables which, if known at time t_0, is sufficient to allow the prediction of the same set at any future time $t \geq t_0$ whenever all of the inputs are known for all τ, $t_0 \leq \tau \leq t$. The variables are called **state variables**. The set of all states for a system is called the **state space**.

Note that there is no requirement for the state to be unique or of minimal size, although the latter is often part of the definition. Typically, the state of a SISO system will be of the same size (dimension) as the order of the ordinary differential equation modelling the system.

The notion becomes more precisely defined – when we restrict the discussion to systems of differential equations (and hence implicitly to the physical systems they represent) – to the following:

> The state of a system of equations is any set of variables which, if specified at time t_0, provides the information necessary to solve that set of equations for any $t \geq t_0$ for any specified set of forcing functions.

It is fundamental to the state variable approach to systems studies to assume that the system may be placed in a form in which the variation of the state variables can be described by a coupled set of first-order ordinary differential equations, one for each state variable. This assumption holds for systems of linear ordinary differential equations; it may or may not hold for other, non-linear systems, although it does appear to hold for many systems of interest. In addition, the original model may often be linearized so that a state-space description is possible for the region of linearization.

10.4.1 Obtaining state-space descriptions – linear systems

A set of ordinary linear differential equations may be placed in the vector-matrix form

$$\dot{\mathbf{x}} = \mathbf{A}(t)\,\mathbf{x} + \mathbf{B}(t)\,\mathbf{u}$$

$$\mathbf{y} = \mathbf{C}(t)\,\mathbf{x} + \mathbf{D}(t)\,\mathbf{u}$$

where $\mathbf{x}$ is an n-vector (i.e. an $n \times 1$ matrix) containing the state variables, $\mathbf{u}$ is an m-vector of the inputs to the system, and $\mathbf{y}$ is a p-vector of the system outputs (whether measurement quantities or physical quantities depends upon the system being modelled). The matrices $\mathbf{A}$, $\mathbf{B}$, $\mathbf{C}$, and $\mathbf{D}$ are respectively $n \times n$, $n \times m$, $p \times n$, and $p \times m$; frequently $\mathbf{D}$ is a matrix of zeros and so it is not always shown in the above form. If the original differential equation model has constant coefficients, then the matrices also are constant.

Placing a system of differential equations in the above form is sometimes straightforward, but not infrequently some clever manipulation is needed. We consider some of the possibilities using examples of the forms used. We note, however, that the point of the representation is seldom to convert a simple SISO differential equation system to this form.

For a simple constant coefficient nth order ordinary differential equation

$$\frac{d^n y}{dt^n} + a_{n-1}\frac{d^{n-1}y}{dt^{n-1}} + a_{n-2}\frac{d^{n-2}y}{dt^{n-2}} + \cdots + a_1\frac{dy}{dt} + a_0 y = u \qquad (10.5)$$

we may define the n state variables

$$x_1 = y \qquad x_2 = \dot{y} \qquad x_3 = \frac{d^2 y}{dt^2} \qquad \cdots \qquad x_n = \frac{d^{n-1}y}{dt^{n-1}}$$

so that

$$\dot{x}_1 = x_2 \qquad \dot{x}_2 = x_3 \qquad \cdots \qquad \dot{x}_{n-1} = x_n \qquad (10.6)$$

Then

$$\dot{x} = \frac{d^n y}{dt^n} = -a_{n-1}\frac{d^{n-1}y}{dt^{n-1}} - a_{n-2}\frac{d^{n-2}y}{dt^{n-2}} - \cdots - a_1\frac{dy}{dt} - a_0 y + u$$

$$= -a_0 x_1 - a_1 x_2 - a_2 x_3 - \cdots - a_{n-1}x_n + u$$

Defining

$$\mathbf{x} = \begin{bmatrix} x_1 \\ x_2 \\ \vdots \\ x_n \end{bmatrix} \qquad\qquad \mathbf{u} = [u]$$

$$
A = \begin{bmatrix}
0 & 1 & 0 & \cdots & 0 & 0 \\
0 & 0 & 1 & \cdots & 0 & 0 \\
\vdots & \vdots & \vdots & & \vdots & \vdots \\
0 & 0 & 0 & & 1 & 0 \\
0 & 0 & 0 & \cdots & 0 & 1 \\
-a_0 & -a_1 & -a_2 & \cdots & 0 & -a_{n-1}
\end{bmatrix}
\qquad
B = \begin{bmatrix}
0 \\
0 \\
\vdots \\
0 \\
1
\end{bmatrix}
$$

$$
C = \begin{bmatrix} 1 & 0 & 0 & \cdots & 0 \end{bmatrix}
\qquad
D = \begin{bmatrix} 0 \end{bmatrix}
\tag{10.7}
$$

then gives the desired form with

$$
\dot{x} = Ax + Bu
$$

$$
y = Cx
\tag{10.8}
$$

This is a simple case of a common form called the **controllable canonical form**.

More difficulty arises when derivatives of the control variable appear, as the state variable form does not allow explicit differentiation of an input variable to be another variable. (Thus it is not allowed for **u** to be defined as the vector

$$
u = \begin{bmatrix} u \\ \dot{u} \\ \ddot{u} \end{bmatrix}
$$

for example.) It is usually desired for a SISO system to have the **A** matrix be of the form of (10.7) or a somewhat similar form to be seen below. Even with this proviso, there are several different possible state variable definitions, each giving somewhat different matrices. We consider the differential equation

$$\frac{d^n y}{dt^n} + a_{n-1}\frac{d^{n-1}y}{dt^{n-1}} + a_{n-2}\frac{d^{n-2}y}{dt^{n-2}} + \cdots + a_1\frac{dy}{dt} + a_0 y$$

$$= b_m\frac{d^m u}{dt^m} + b_{m-1}\frac{d^{m-1}u}{dt^{m-1}} + \cdots + b_1\frac{du}{dt} + b_0 u \tag{10.9}$$

and assume $m \le n$. By superposition arguments, we have that if (10.7–8) give the response to an input $u(t)$ for (10.5), then

$$\dot{x} = \mathbf{A}x + \mathbf{B}u$$

$$y = [\, b_0 - b_n a_0 \quad b_1 - b_n a_1 \quad \cdots \quad b_{n-1} - b_n a_{n-1}\,]\,\mathbf{x} + [\, b_n\,]\,\mathbf{u}$$

gives a solution to (10.9). We remark that it is no longer true that the definitions (10.6) apply; e.g. no longer is x_4 the third derivative of the variable y.

An alternative state variable definition may be justified by observing that (10.9) implies

$$y(t) = b_n u(t) + \int(-a_{n-1}y(t_n) + b_{n-1}u(t_n)$$

$$+ \int(-a_{n-2}y(t_{n-1}) + b_{n-2}u(t_{n-1}) \quad + \int(\cdots$$

$$+ \int(-a_0 y(t_0) + b_0 u(t_0))\,dt_0)\,dt_1)\cdots\,)\,dt_{n-1}\,dt_n$$

Defining the variables

$$x_1 = \int(-a_{n-1}y + b_{n-1}u + x_2)\,dt$$

$$x_2 = \int(-a_{n-2}y + b_{n-2}u + x_3)\,dt$$

$$\vdots$$

$$x_k = \int \left(a_{n-k} y + b_{n-k} u + x_{k+1} \right) dt$$

$$\vdots$$

$$x_n = \int a_0 y + b_0 u \, dt \qquad (10.10)$$

and observing that this makes

$$y = b_n u + x_1 \qquad (10.11)$$

gives, with differentiation of (10.10) and substitution of (10.11), the result

$$\mathbf{A} = \begin{bmatrix} -a_{n-1} & 1 & 0 & \cdots & 0 & 0 \\ -a_{n-2} & 0 & 1 & \cdots & 0 & 0 \\ \vdots & \vdots & \vdots & & \vdots & \vdots \\ -a_1 & 0 & 0 & \cdots & 0 & 1 \\ -a_0 & 0 & 0 & \cdots & 0 & 0 \end{bmatrix} \qquad \mathbf{B} = \begin{bmatrix} b_{n-1} - b_n a_{n-1} \\ b_{n-2} - b_n a_{n-2} \\ \vdots \\ b_1 - b_n a_1 \\ b_0 - b_n a_0 \end{bmatrix}$$

$$\mathbf{C} = [\, 1\ 0\ 0 \cdots 0\,] \qquad \mathbf{D} = [\, b_n \,]$$

with

$$\dot{\mathbf{x}} = \mathbf{A}\mathbf{x} + \mathbf{B}\mathbf{u}$$

$$y = \mathbf{C}\mathbf{x} + \mathbf{D}\mathbf{u}$$

This is in fact a commonly seen form called an **observable canonical form**.

Another variation may be defined by proposing a structure and then choosing the coefficients. For example, we propose the following:

$$y = x_1 + \beta_0 u \qquad \dot{x}_1 = x_2 + \beta_1 u \quad \cdots \quad \frac{dx_k}{dt} = x_{k+1} + \beta_k u \quad \cdots$$

$$\frac{dx_n}{dt} = -(\alpha_0 x_1 + \alpha_1 x_2 + \cdots + \alpha_{n-1} x_n) + \beta_n u$$

To find coefficients α_i and β_j, we differentiate y successively to find (following e.g. Wiberg (1971))

$$\begin{bmatrix} \beta_0 \\ \beta_1 \\ \beta_2 \\ \vdots \\ \beta_n \end{bmatrix} = \begin{bmatrix} 1 & 0 & 0 & \cdots & 0 \\ a_{n-1} & 1 & 0 & \cdots & 0 \\ a_{n-2} & a_{n-1} & 1 & \cdots & 0 \\ \vdots & \vdots & \vdots & \vdots & \vdots \\ a_0 & a_1 & \cdot & a_{n-1} & 1 \end{bmatrix}^{-1} \begin{bmatrix} b_n \\ b_{n-1} \\ b_{n-2} \\ \vdots \\ b_0 \end{bmatrix} \qquad (10.12)$$

Using this solution, the state equations become

$$\dot{x} = Ax + Bu$$

$$y = Cx + Du$$

where

$$A = \begin{bmatrix} 0 & 1 & 0 & \cdots & 0 & 0 \\ 0 & 0 & 1 & \cdots & 0 & 0 \\ \vdots & \vdots & \vdots & & \vdots & \vdots \\ 0 & 0 & 0 & & 1 & 0 \\ 0 & 0 & 0 & \cdots & 0 & 1 \\ -a_0 & -a_1 & -a_2 & \cdots & & -a_{n-1} \end{bmatrix} \qquad B = \begin{bmatrix} \beta_1 \\ \beta_2 \\ \beta_3 \\ \vdots \\ \beta_n \end{bmatrix}$$

$$C = [\, 1 \ 0 \ 0 \ 0 \ \cdots \ 0\,] \qquad\qquad D = [\beta_0]\,[b_n]$$

and the β_i solve (10.12).

The above forms of the **A** matrix are called **companion forms** of the system matrix. A very different form is best approached by using the transfer function. For the present system, the transfer function is as given in (10.2), which, by the fundamental theorem of algebra, may be expanded in partial fractions using the roots (here assumed distinct) λ_i, $i = 1, 2, ..., n$, of the denominator. Thus by assumption

$$\lambda_i^n + a_{n-1}\lambda_i^{n-1} + a_{n-2}\lambda_i^{n-2} + \cdots + a_1\lambda_i + a_0 = 0$$

and using this, we may expand (10.2) as

$$\frac{Y(s)}{U(s)} = \frac{c_1}{s-\lambda_1} + \frac{c_2}{s-\lambda_2} + \frac{c_3}{s-\lambda_3} + \cdots + \frac{c_n}{s-\lambda_n} + d_0$$

Putting $G_i(s) = c_i/(s - \lambda_i)$, we see that

$$Y(s) = G_1(s)U(s) + G_2(s)U(s) + \cdots + G_n(s)U(s) + d_0 U(s)$$

$$= X_1(s) + X_2(s) + \cdots + X_n(s) + d_0 U(s)$$

Inversion arguments show that we may define $\dot{x}_1 = \lambda_1 x_1 + c_i u$ $(i = 1, 2, ..., n)$ and have

$$y = x_1 + x_2 + \cdots + x_n + d_0 u$$

In matrix notation, this becomes the **modal form**

$$\dot{\mathbf{x}} = \begin{bmatrix} \lambda_1 & 0 & 0 & \cdots & 0 \\ 0 & \lambda_2 & 0 & \cdots & 0 \\ \vdots & \vdots & \ddots & & \vdots \\ 0 & 0 & 0 & \lambda_{n-1} & 0 \\ 0 & 0 & 0 & \cdots & \lambda_n \end{bmatrix} \mathbf{x} + \begin{bmatrix} c_1 \\ c_2 \\ \vdots \\ c_{n-1} \\ c_n \end{bmatrix} u$$

$$y = [\, 1\ 1\ 1\ \cdots\ 1\,]\mathbf{x} + d_0 u$$

If the roots are not distinct, this approach yields a **Jordan form** for the **A** matrix. As given, roots and coefficients may be complex, and therefore the numbers x_i may be complex.

If the eigenvalues are complex numbers, they necessarily come in pairs. This yields the possibility of using real number representations by rearrangement. Thus the system with eigenvalues $\sigma \pm j\omega$ may be written

$$\dot{\mathbf{x}} = \begin{bmatrix} \sigma & \omega \\ -\omega & \sigma \end{bmatrix} \mathbf{x} + \mathbf{Bu}$$

instead of

$$\dot{\mathbf{z}} = \begin{bmatrix} \sigma+j\omega & 0 \\ 0 & \sigma-j\omega \end{bmatrix} \mathbf{z} + \mathbf{Bu}$$

where the matrices **B** are different in the two representations.

Example

Consider the simple second-order system

$$\ddot{y} + 2\dot{y} + 5y = 3\dot{u} + u$$

Among its alternative representations are

1. $\dfrac{Y(s)}{U(s)} = \dfrac{3s + 1}{s^2 + 2s + 5}$

2. $\dfrac{Y(j\omega)}{U(j\omega)} = \dfrac{3j\omega + 1}{(5 - \omega^2) + 2j\omega}$

3. $\dot{\mathbf{x}} = \begin{bmatrix} 0 & 1 \\ -5 & -2 \end{bmatrix} \mathbf{x} + \begin{bmatrix} 0 \\ 1 \end{bmatrix} u$

$$y = [1 \quad 3]\,\mathbf{x}$$

4. $\quad \dot{\mathbf{z}} = \begin{bmatrix} -2 & 1 \\ -5 & 0 \end{bmatrix} \mathbf{z} + \begin{bmatrix} 3 \\ 1 \end{bmatrix} u$

$\quad y = [1 \quad 0] \, \mathbf{z}$

5. $\quad \dot{\mathbf{w}} = \begin{bmatrix} -1+2j & 0 \\ 0 & -1-2j \end{bmatrix} \mathbf{w} + \begin{bmatrix} (3-j)/2 \\ (3+j)/2 \end{bmatrix} u$

$\quad y = [1 \quad 1] \, \mathbf{w}$

6. $\quad \dot{\alpha} = \begin{bmatrix} -1 & 2 \\ -2 & -1 \end{bmatrix} \alpha + \begin{bmatrix} 0 \\ 1 \end{bmatrix} u$

$\quad y = [-1 \quad 3] \, \alpha$

We remark that representations $3-6$ all have the same input–output relationship, as can be verified, e.g. by computing (as in section 10.4.3)

$$\frac{Y(s)}{U(s)} = \mathbf{C}\,(s\mathbf{I}-\mathbf{A})^{-1}\mathbf{B}$$

This emphasizes the non-unique character of state variable representations.

10.4.2 Multi-input–multi-output (MIMO) linear systems

Compared with SISO systems, provided the systems involved are linear, MIMO systems are just a little more difficult.

The near trivial case is one in which the systems are uncoupled. Suppose, for example, that we have systems modelled by

$$\dot{\mathbf{x}}_i = \mathbf{A}_i \mathbf{x}_i + \mathbf{B}_i \mathbf{u}_i$$

$$\mathbf{y}_i = \mathbf{C}_i \mathbf{x}_i + \mathbf{D}_i \mathbf{u}_i \qquad i = 1,\dots,k$$

where the individual system models were developed as in section 10.4.1. Then defining the concatenated vectors

$$\mathbf{x} = \begin{bmatrix} \mathbf{x}_1 \\ \mathbf{x}_2 \\ \vdots \\ \mathbf{x}_k \end{bmatrix} \qquad \mathbf{u} = \begin{bmatrix} \mathbf{u}_1 \\ \mathbf{u}_2 \\ \vdots \\ \mathbf{u}_k \end{bmatrix} \qquad \mathbf{y} = \begin{bmatrix} \mathbf{y}_1 \\ \mathbf{y}_2 \\ \vdots \\ \mathbf{y}_k \end{bmatrix}$$

and the block-diagonal matrices

$$\mathbf{A} = \begin{bmatrix} \mathbf{A}_1 & \mathbf{0} & \mathbf{0} & \cdots & \mathbf{0} \\ \mathbf{0} & \mathbf{A}_2 & \mathbf{0} & \cdots & \mathbf{0} \\ \mathbf{0} & \mathbf{0} & \mathbf{A}_3 & \cdots & \mathbf{0} \\ \vdots & & & \ddots & \vdots \\ \mathbf{0} & \mathbf{0} & \mathbf{0} & \cdots & \mathbf{A}_k \end{bmatrix} \qquad \mathbf{B} = \begin{bmatrix} \mathbf{B}_1 & \mathbf{0} & \mathbf{0} & \cdots & \mathbf{0} \\ \mathbf{0} & \mathbf{B}_2 & \mathbf{0} & \cdots & \mathbf{0} \\ \mathbf{0} & \mathbf{0} & \mathbf{B}_3 & \cdots & \mathbf{0} \\ \vdots & & & \ddots & \vdots \\ \mathbf{0} & \mathbf{0} & \mathbf{0} & \cdots & \mathbf{B}_k \end{bmatrix}$$

$$\mathbf{C} = \begin{bmatrix} \mathbf{C}_1 & \mathbf{0} & \mathbf{0} & \cdots & \mathbf{0} \\ \mathbf{0} & \mathbf{C}_2 & \mathbf{0} & \cdots & \mathbf{0} \\ \mathbf{0} & \mathbf{0} & \mathbf{C}_3 & \cdots & \mathbf{0} \\ \vdots & & & \ddots & \vdots \\ \mathbf{0} & \mathbf{0} & \mathbf{0} & \cdots & \mathbf{C}_k \end{bmatrix} \qquad \mathbf{D} = \begin{bmatrix} \mathbf{D}_1 & \mathbf{0} & \mathbf{0} & \cdots & \mathbf{0} \\ \mathbf{0} & \mathbf{D}_2 & \mathbf{0} & \cdots & \mathbf{0} \\ \mathbf{0} & \mathbf{0} & \mathbf{D}_3 & \cdots & \mathbf{0} \\ \vdots & & & \ddots & \vdots \\ \mathbf{0} & \mathbf{0} & \mathbf{0} & \cdots & \mathbf{D}_k \end{bmatrix}$$

yields the usual general form

$$\dot{\mathbf{x}} = \mathbf{A}\mathbf{x} + \mathbf{B}\mathbf{u}$$

$$\mathbf{y} = \mathbf{C}\mathbf{x} + \mathbf{D}\mathbf{u}$$

When systems are connected in cascade, so that the output of one is the input to the next, as in Fig. 10.5, then the situation is only slightly more complicated.

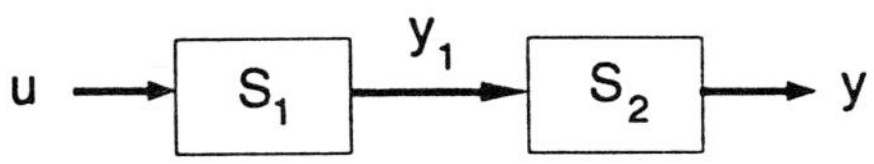

Figure 10.5 Systems in cascade.

In the two-element cascade shown in Fig. 10.5, the two subsystems S_1 and S_2 might have the state-space representations

$$\dot{x}_1 = A_1 x_1 + B_1 u$$

$$y_1 = C_1 x_1 + D_1 u$$

and, using the indicated naming of elements,

$$\dot{x}_2 = A_2 x_2 + B_2 y_1$$

$$y_2 = C_2 x_2 + D_2 y_1$$

Then one possible state-space description of the entire system, with input u and output y, is

$$\dot{x} = \begin{bmatrix} \dot{x}_1 \\ \dot{x}_2 \end{bmatrix} = \begin{bmatrix} A_1 & 0 \\ B_2 C_1 & A_2 \end{bmatrix} x + \begin{bmatrix} B_1 \\ B_2 D_1 \end{bmatrix} u$$

$$y = \begin{bmatrix} D_2 C_1 & C_2 \end{bmatrix} x + \begin{bmatrix} D_2 & D_1 \end{bmatrix} u$$

Example

Consider a motor in cascade with a compensator, satisfying

$$\tau\ddot{\theta} + \dot{\theta} = K_m v \qquad \dot{v} + \alpha v = K_a(\dot{u} + \beta u)$$

respectively. For the motor, define the state $[x_1\ x_2]^T = [\theta\ \dot\theta]^T$ the input as v and the output $y = x_1 = \theta$. Then a state system representation is

$$\dot{\mathbf{x}} = \begin{bmatrix} 0 & 1 \\ 0 & -\dfrac{1}{\tau} \end{bmatrix} \mathbf{x} + \begin{bmatrix} 0 \\ \dfrac{K_m}{\tau} \end{bmatrix} v$$

$$y = [\,1\quad 0\,]\,\mathbf{x}$$

For the compensator control law, let x_3 = state, with input u and output v. Then the (1-dimensional) state representation is

$$\dot{x}_3 = -\alpha x_3 + K_a(\beta-\alpha)u$$

$$v = x_3 + K_a u$$

Combining these yields, with system input u and output θ,

$$\begin{bmatrix} \dot{x}_1 \\ x_2 \\ x_3 \end{bmatrix} = \begin{bmatrix} 0 & 1 & 0 \\ 0 & -\dfrac{1}{\tau} & \dfrac{K_m}{\tau} \\ 0 & 0 & -\alpha \end{bmatrix} \begin{bmatrix} x_1 \\ x_2 \\ x_3 \end{bmatrix} + \begin{bmatrix} 0 \\ \dfrac{K_a K_m}{\tau} \\ K_a(\beta-\alpha) \end{bmatrix} u$$

$$y = [\,1\quad 0\quad 0\,] \begin{bmatrix} x_1 \\ x_2 \\ x_3 \end{bmatrix}$$

We note that x_1 and x_2 retain 'natural' interpretations as shaft angle and rotation rate, but that the rules of state representation (namely, no derivatives of the input) mean that x_3 represents an integral portion of the voltage v rather than all of v.

State-space conversion of multivariable transfer functions

The example on page 248 demonstrated state-space representations of SISO transfer functions, and we saw these were not unique. One of the potential state-space representations for a matrix transfer function $\mathbf{G}(s)$ can be defined as follows. Let $\mathbf{G}_i(s)$ denote the *i*th column of an $n \times m$ transfer function $\mathbf{G}(s)$, so that if $U_i(s)$ is the scalar transform of the scalar control variable $u_i(t)$, we have

$$\mathbf{G}(s) = \sum_{i=1}^{m} \mathbf{G}_i(s)\, U_i(s)$$

Represent $\mathbf{G}_i(s)$ in terms of an n-vector of polynomials $\mathbf{N}_i(s)$ and a scalar polynomial $d_i(s)$, where $d_i(s)$ is the least common denominator of the m denominators in $\mathbf{G}_i(s)$. Thus

$$\mathbf{G}_i(s) = \frac{\mathbf{N}_i(s)}{d_i(s)}$$

where

$$d_i(s) = s^{n_i} + \sum_{j=0}^{n_i-1} d_{ij} s^j$$

$$\mathbf{N}_i(s) = \begin{bmatrix} \upsilon_{i,10} + \upsilon_{i,11} s + \cdots + \upsilon_{i,1n_i-1}\, s^{n_i-1} \\ \upsilon_{i,n0} + \upsilon_{i,n1} s + \cdots + \upsilon_{i,nn_i-1}\, s^{n_i-1} \end{bmatrix}$$

Defining

$$\mathbf{A}_i = \begin{bmatrix} 0 & 1 & 0 & 0 & \cdots & 0 \\ 0 & 0 & 1 & 0 & \cdots & 0 \\ \vdots & \vdots & \vdots & \vdots & & \vdots \\ -d_{i,0} & -d_{i,1} & -d_{i,2} & \cdots & \cdots & -d_{i,n_i-1} \end{bmatrix}$$

$$
\mathbf{B}_1 = \begin{bmatrix} 0 \\ 0 \\ \vdots \\ 0 \\ 1 \end{bmatrix}
\qquad
\mathbf{C}_1 = \begin{bmatrix} \upsilon_{i,10} & \upsilon_{i,11} & \cdots & \upsilon_{i,1n_i-1} \\ \cdot & \cdot & \cdot & \cdot \\ \upsilon_{i,n0} & \upsilon_{i,n1} & \cdots & \upsilon_{i,nn_i-1} \end{bmatrix}
$$

we have that one state-space representation of the system is given by

$$
\mathbf{A} = \begin{bmatrix} \mathbf{A}_1 & 0 & 0 & \cdots & 0 \\ 0 & \mathbf{A}_2 & 0 & \cdots & 0 \\ \vdots & \vdots & & & \vdots \\ 0 & 0 & 0 & \cdots & \mathbf{A}_m \end{bmatrix}
\qquad
\mathbf{B} = \begin{bmatrix} \mathbf{B}_1 & 0 & 0 & \cdots & 0 \\ 0 & \mathbf{B}_2 & 0 & \cdots & 0 \\ \vdots & \vdots & & & \vdots \\ 0 & 0 & 0 & \cdots & \mathbf{B}_m \end{bmatrix}
$$

$$
\mathbf{C} = \begin{bmatrix} \mathbf{C}_1 & \mathbf{C}_2 & \cdots & \mathbf{C}_n \end{bmatrix}
$$

In this, if $M = \sum_{i=1}^{m} n_i$, then $\mathbf{A}$ is an $M \times M$ matrix, $\mathbf{B}$ is $M \times m$, and $\mathbf{C}$ is $n \times M$. This representation is of a type called controllable (Chapter 22), but may not be observable (Chapter 24); a representation which is both controllable and observable is called a minimal realization, and there exist methods, some of which are in CACSD programs, for finding such representations if the ensuing theoretical analysis requires them.

10.4.3 Relationship to Laplace transforms

Having seen a derivation of a state-space representation (the modal form) from a Laplace transform representation, it should be no surprise to find that the reverse can also be done. We need only use notation carefully and realize that

$$X(s) = \mathscr{L}(\mathbf{x}(t)) = \begin{bmatrix} \mathscr{L}(x_1(t)) \\ \mathscr{L}(x_2(t)) \\ \vdots \\ \mathscr{L}(x_n(t)) \end{bmatrix} = \begin{bmatrix} X_1(s) \\ X_2(s) \\ \vdots \\ X_n(s) \end{bmatrix}$$

to see that

$$\dot{\mathbf{x}} = \mathbf{A}\mathbf{x} + \mathbf{B}\mathbf{u}$$

$$\mathbf{y} = \mathbf{C}\mathbf{x} + \mathbf{D}\mathbf{u}$$

has transform

$$s\mathbf{X}(s) - \mathbf{x}(0) = \mathbf{A}\mathbf{X}(s) + \mathbf{B}\mathbf{U}(s)$$

$$\mathbf{Y}(s) = \mathbf{C}\mathbf{X}(s) + \mathbf{D}\mathbf{U}(s)$$

and hence has solution

$$\mathbf{Y}(s) = [\mathbf{C}(s\mathbf{I} - \mathbf{A})^{-1}\mathbf{B} + \mathbf{D}]\mathbf{U}(s)$$

In the common case for which $\mathbf{D} = \mathbf{0}$, expansion of this is useful in that we see, from Cramer's Rule for matrix inversion, that

$$\mathbf{Y}(s) = \frac{\mathbf{C}\,\mathrm{cof}(s\mathbf{I}-\mathbf{A})\mathbf{B}}{\det(s\mathbf{I}-\mathbf{A})}$$

where $\mathrm{cof}(\cdot)$ is the matrix of cofactors and $\det(\cdot)$ is the determinant. The former is a matrix of polynomials, while the latter is a polynomial. From the definitions involved, the zeros of the latter are the eigenvalues of the matrix $\mathbf{A}$ and also the poles of the system represented by the state-space mode.

10.4.4 Changes of variable – similarity transforms

It should be clear from section 10.4.1 that state-space representations are not unique. The state variables are sometimes chosen for physical significance and sometimes for analytical convenience, and occasionally a representation will be both significant and convenient.

The change of representation of a linear constant coefficient system, if we perform one, is called a similarity transformation. To perform one, suppose we have a system described by

$$\dot{x} = Ax + Bu$$

$$y = Cx + Du$$

We change the representation, which could also be considered a change of variable or of linear system basis or of coordinate axes, by defining a new state vector x_T, related to the previous state vector x by an invertible $n \times n$ matrix T, as in

$$x_T = Tx \qquad x = T^{-1}x_T$$

Then clearly we have the state equations for x_T given by

$$\dot{x}_T = TAT^{-1}x_T + TBu$$

$$y = CT^{-1}x_T + Du$$

To see that the transfer function from $u(t)$ to $y(t)$ is unchanged, we may directly compute

$$sX_T(s) = TAT^{-1}X_T(s) + TBu(s)$$

$$Y(s) = CT^{-1}X_T(s) + Du(s)$$

This has solution

$$Y(s) = [CT^{-1}(sI-TAT^{-1})^{-1}TB + D]U(s)$$

$$= [C(sI-A)^{-1}B + D]U(s)$$

Canonical forms

The state–space formulations also have canonical forms, just as there were several standard forms for the transfer functions and one special form (the Bode form) for frequency response functions. These forms can be reached by similarity transformations on systems such as

$$\dot{x} = Ax + Bu$$

$$y = Cx + Du$$

and involve special forms of the **A** matrix plus one or both of **B** and **C**. The actual transformations are in the Appendix B.

The observable canonical forms have

$$A = \begin{bmatrix} & \vdots & I \\ a & \vdots & \cdots \\ & \vdots & O \end{bmatrix} \qquad\qquad C = [\,1\ \ 0\ \ 0\ \cdots\ \ 0\,]$$

or

$$A = \begin{bmatrix} I & \vdots & \\ \cdots & \vdots & a \\ O & \vdots & \end{bmatrix} \qquad\qquad C = [\,0\ \ 0\ \ 0\ \cdots\ \ 1\,]$$

The controllable canonical forms have

$$A = \begin{bmatrix} O & \vdots & I \\ \cdots & \cdots & \cdots \\ & a^T & \end{bmatrix} \qquad\qquad B = \begin{bmatrix} 0 \\ 0 \\ \vdots \\ 0 \\ 1 \end{bmatrix}$$

or

$$A = \begin{bmatrix} & a^T & \\ \cdots & \cdots & \cdots \\ I & \vdots & O \end{bmatrix} \qquad\qquad B = \begin{bmatrix} 1 \\ 0 \\ \vdots \\ 0 \end{bmatrix}$$

In either case, for $n \times n$ matrix $\mathbf{A}$, $\mathbf{I}$ is $(n-1) \times (n-1)$ identity, $\mathbf{0}$ is an $n-1$ vector of zeros, and $\mathbf{a}$ is an n-vector of coefficients from the $\mathbf{A}$ matrix's characteristic equation. Existence of the transformation depends upon observability, resp. controllability, of the $(\mathbf{A};\mathbf{C})$ and $(\mathbf{A};\mathbf{B})$ matrix pairs (Chapters 22 and 24).

Another canonical form uses Jordan blocks, and is a generalized modal decomposition. A typical form, and its importance as a form, is that for MIMO systems it yields

$$\mathbf{A} = \begin{bmatrix} \mathbf{J}_{co} & 0 & 0 & 0 \\ 0 & \mathbf{J}_c & 0 & 0 \\ 0 & 0 & \mathbf{J}_o & 0 \\ 0 & 0 & 0 & \mathbf{J} \end{bmatrix} \qquad \mathbf{B} = \begin{bmatrix} \mathbf{B}_{co} \\ \mathbf{B}_c \\ 0 \\ 0 \end{bmatrix}$$

$$\mathbf{C} = [\, \mathbf{C}_{co} \quad 0 \quad \mathbf{C}_o \quad 0 \,]$$

where the $\mathbf{J}$ are Jordan blocks, and the $\mathbf{B}_i$ and $\mathbf{C}_i$ are suitably non-trivial. The significance of this is the interpretation, in which the system is interpreted as decomposed into controllable and observable (CO subscript), controllable but not observable (C), observable but not controllable (O), and neither controllable nor observable (no subscript) subsystems (Chapters 22 and 24). An alternative decomposition showing the same is

$$\mathbf{A} = \begin{bmatrix} \mathbf{A}_0 & 0 & \mathbf{A}_{1,3} & 0 \\ \mathbf{A}_{2,1} & \mathbf{A}_c & \mathbf{A}_{2,3} & \mathbf{A}_{2,4} \\ 0 & 0 & \mathbf{A}_0 & 0 \\ 0 & 0 & \mathbf{A}_{4,3} & \mathbf{A}_{--} \end{bmatrix} \qquad \mathbf{B} = \begin{bmatrix} \mathbf{B}_{co} \\ \mathbf{B}_c \\ 0 \\ 0 \end{bmatrix}$$

$$\mathbf{C} = [\, \mathbf{C}_{co} \quad 0 \quad \mathbf{C}_o \quad 0 \,]$$

These are useful in decomposing systems, of course, but also in finding representations which are minimal (in size of state vector) and having a given input–output transfer function.

10.4.5 Some state–space descriptions for non-linear systems

For non-linear systems, generation of a state–space description can be difficult. One relatively simple case arises when all derivatives are ordinary, the inputs are not differentiated, and the highest derivative can be isolated easily. Consider for example a system described by

$$\frac{d^n y}{dt^n} = f\left(\frac{d^{n-1}y}{dt^{n-1}}, \frac{d^{n-2}y}{dt^{n-2}}, \ldots, \frac{dy}{dt}, y, u\right)$$

Using the definitions

$$x_1 = y \quad x_2 = \frac{dy}{dt} \quad \cdots \quad x_n = \frac{d^{n-1}y}{dt^{n-1}} \qquad (10.13)$$

we immediately have the representation in terms of first derivatives as

$$\dot{x}_1 = x_2 \quad \dot{x}_2 = x_3 \quad \cdots \quad \dot{x}_{n-1} = x_n$$

$$\dot{x}_n = f(x_n, x_{n-1}, \ldots, x_1, u)$$

$$y = \begin{bmatrix} 1 & 0 & 0 & \cdots & 0 \end{bmatrix} \mathbf{x}$$

or

$$\dot{x} = \hat{f}(\mathbf{x}, u) \text{ and } y = \mathbf{C}\mathbf{x}$$

For systems which seem to exhibit explicit dependence on derivatives of the control variables, the above does not work directly. This is because the required form, even for non-linear systems, is

$$\dot{\mathbf{x}} = f(\mathbf{x}, \mathbf{u}, t) \quad \mathbf{y} = g(\mathbf{x}, \mathbf{u}, t)$$

Here it is to be noted that the input variables $\mathbf{u}$ are not explicitly differentiated.

A sometimes helpful interpretation in these circumstances is to note whether the input variable is a command variable and should have the highest derivative noted as the actual control quantity. If so, then the definition of extra state variables may obviate the apparent difficulty.

For example, consider

$$\frac{d^n y}{dt^n} = f\left(\frac{d^{n-1}y}{dt^{n-1}}, \frac{d^{n-2}y}{dt^{n-2}}, \ldots, \frac{dy}{dt}, y, \frac{d^2 u}{dt^2}, \frac{du}{dt}, u\right)$$

and use the definitions of (10.13) plus the definitions

$$w_1 = u \qquad w_2 = \frac{du}{dt}$$

plus consider the second derivative of u as the control variable, calling it ω. Then we have

$$\dot{w}_1 = w_2 \quad \dot{w}_2 = \omega$$

$$\dot{x}_1 = x_2 \quad \dot{x}_2 = x_3 \quad \cdots \quad \dot{x}_{n-1} = x_n$$

$$\dot{x}_n = f(x_n, x_{n-1}, \ldots, x_1, \omega, w_2, w_1)$$

$$y = [1 \quad 0 \quad 0 \quad \cdots \quad 0]\, \mathbf{x}$$

and $\mathbf{x}$, $\mathbf{w}$ are easily concatenated into a single vector if desired.

Example

For the translational dynamics of the rocket model in Chapter 9, we may define the state vector $\mathbf{x}$ by

$$x_1 = x = \text{horizontal distance}$$

$$x_2 = \dot{x} = \text{horizontal velocity}$$

$$x_3 = y = \text{vertical distance}$$

$$x_4 = \dot{y} = \text{vertical velocity}$$

$$x_5 = \text{mass}$$

$$u_1 = \text{pitch angle}$$

$$u_2 = \text{motor on/off command (1 for on, 0 for off)}$$

$$G = \text{gravitational constant}$$

$$T_h = \text{engine thrust}$$

$$\dot{M} = \text{propellant mass flow rate}$$

Then a state model is given by

$$\dot{x}_1 = x_2$$

$$\dot{x}_2 = \frac{-Gx_1}{x_5(x_1^2 + x_2^2)^{1.5}} + T_h u_2 \frac{\cos(u_1)}{x_5}$$

$$\dot{x}_3 = x_4$$

$$\dot{x}_4 = \frac{-Gx_3}{x_5(x_1^2 + x_2^2)^{1.5}} + T_h u_2 \frac{\sin(u_1)}{x_5}$$

$$\dot{x}_5 = M u_2$$

10.4.6 Finding the system response to input using state-space and transition matrices

There are several choices for finding the system response to a given input. Aside from numerical and simulation methods, the details depend upon the system representation.

- Numerical methods and simulations such as SIMNON, whether analog or digital computer based, tend to work with state descriptions, i.e. sets of first-order differential equations.
- For differential equation representations, the usual methods of solution of differential equations may be used (or attempted).

- For transfer functions, the methods of Laplace transforms are applied.
- For linear state-space models, the methods are those of differential equations but the details are probably unfamiliar to the reader and are presented here.

We consider the model structure

$$\dot{\mathbf{x}} = \mathbf{A}(t)\mathbf{x} + \mathbf{B}(t)\mathbf{u}$$

$$\mathbf{y} = \mathbf{C}(t)\mathbf{x} + \mathbf{D}(t)\mathbf{u} \tag{10.14}$$

with initial conditions

$$\mathbf{x}(t_0) = \mathbf{x}_0$$

To obtain a general solution, we first find what is in effect the free or unforced solution and then derive the particular solution. More precisely, we define the transition matrix $\Theta(t_2, t_1)$ as the solution of the matrix differential equation

$$\frac{d\Theta(t, t_0)}{dt} = \mathbf{A}(t)\,\Theta(t, t_0) \qquad \Theta(t_0, t_0) = \mathbf{I}$$

It is readily shown that the transition matrix, which is unique by the properties of solutions of linear ordinary differential equations, has among its properties the following.

1. Transition property:

$$\Theta(t_2, t_0) = \Theta(t_2, t_1)\,\Theta(t_1, t_0)$$

2. Inversion property:

$$\Theta(t_1, t_0) = \Theta^{-1}(t_0, t_1)$$

3. Determinant property:

$$\det \Theta(t_1, t_0) = \exp\left(\int_{t_0}^{t_1} \text{tr}\,\mathbf{A}(\tau)\,d\tau\right)$$

4. Separation property:

$$\Theta(t_1, t_0) = \Theta(t_1, 0)\,\Theta^{-1}(t_0, 0)$$

$$= \Theta(t_1)\,\Theta^{-1}(t_0)$$

Note the alternative notation allowed by the separation property.

Using the transition matrix, the general solution of the state equations can be written

$$\mathbf{x}(t) = \Theta(t, t_0)\,\mathbf{x}(t_0) + \int_{t_0}^{t} \Theta(t, \tau)\,\mathbf{B}(\tau)\,\mathbf{u}(\tau)\,d\tau \tag{10.15}$$

This may either be checked by direct differentiation (using the fundamental theorem of calculus, the chain rule, and the Leibnitz Rule for differentiation of the integral) or derived using the method of variation of parameters.

From the definition of the transition matrix, it is clear that if the matrix $\mathbf{A}$ is a constant, then

$$\Theta(t_1, t_0) = \Theta(t_1 + \tau, t_0 + \tau) = \Theta(t_1 - t_0, 0)$$

for all τ. For such systems, the single independent variable notation defined by $\Theta(t_1, t_0) \equiv \Theta(t_1, t_0)$, $\Theta(t) \equiv \Theta(t, 0)$ etc. is particularly appropriate and will be used frequently in the sequel.

In the non-time-varying case for which $\mathbf{A}$ is constant, the transition matrix solves

$$\frac{d\Theta(t)}{dt} = \mathbf{A}\,\Theta(t) \qquad \Theta(0) = \mathbf{I}$$

Among the methods for finding the matrix $\Theta(t)$ in the latter case are Laplace transform methods, matrix transformation methods, and numerical methods. Laplace transform methods solve

$$\Theta(t) = \mathscr{L}^{-1}\left\{(s\mathbf{I} - \mathbf{A})^{-1}\right\}$$

which is in fact straightforward because, using Cramer's Rule, the resolvent matrix $(s\mathbf{I} - \mathbf{A})^{-1}$ is rational for each element.

The matrix methods rely upon the matrix interpretation of the first-order differential system (10.14). Thus the solution is by analogy with scalars given by

$$\Theta(t) = e^{\mathbf{A}t}$$

By definition, when matrices are exponents,

$$\Theta(t) = \sum_{i=0}^{\infty} \frac{(\mathbf{A}t)^i}{i!}$$

Using similarity transforms of $\mathbf{A}$, if $\mathbf{A}$ has distinct eigenvalues λ_k, $k = 1, 2, \ldots, n$ (when $\mathbf{A}$ is $n \times n$), then there is a matrix $\mathbf{T}$ (the matrix whose columns are the eigenvectors of $\mathbf{A}$) such that

$$\Lambda = \mathrm{diag}\left(\lambda_1, \lambda_2, \ldots, \lambda_n\right) = \mathbf{T}^{-1}\mathbf{A}\mathbf{T}$$

Using this fact, we compute

$$\Theta(t) = \sum_{i=0}^{\infty} \frac{(\mathbf{T}\Lambda\mathbf{T}^{-1}t)^i}{i!} = \mathbf{T}\left[\sum_{i=0}^{\infty} \frac{(\Lambda t)^i}{i!}\right]\mathbf{T}^{-1} = \mathbf{T}e^{\Lambda t}\mathbf{T}^{-1}$$

But since $\Lambda t = \mathrm{diag}\left(\lambda_1 t, \lambda_2 t, \ldots, \lambda_n t\right)$

$$e^{\Lambda t} = \sum_{i=0}^{\infty} \frac{(\Lambda t)^i}{i!} = \sum_{i=0}^{\infty} \mathrm{diag}\left(\frac{(\lambda_1 t)^i}{i!}, \frac{(\lambda_2 t)^i}{i!}, \ldots, \frac{(\lambda_n t)^i}{i!}\right)$$

$$= \mathrm{diag}\left(e^{\lambda_1 t}, e^{\lambda_2 t}, \ldots, e^{\lambda_n t}\right)$$

Hence we have shown that

$$\Theta(t) = \mathbf{T}\,\mathrm{diag}\left(e^{\lambda_1 t}, e^{\lambda_2 t}, \cdots, e^{\lambda_n t}\right)\mathbf{T}^{-1} = \mathbf{T}e^{\Lambda t}\mathbf{T}^{-1}$$

The third class of methods includes numerical integration and power series expansion of $e^{\mathbf{A}\mathbf{T}}$. We show the second of these, as in Franklin, Powell, and Workman (1990). It amounts to a truncated power series expansion for $e^{\mathbf{A}\mathbf{T}}$. For some K, define Ψ as

$$\Psi = \mathbf{I} + \frac{\mathbf{A}\mathbf{T}}{2!} + \frac{(\mathbf{A}\mathbf{T})^2}{3!} + \cdots + \frac{(\mathbf{A}\mathbf{T})^{K-1}}{K!}$$

and compute it as

$$\Psi = \mathbf{I} + \frac{\mathbf{A}\mathbf{T}}{2}\left(\mathbf{I} + \frac{\mathbf{A}\mathbf{T}}{3}\left(\mathbf{I} + \cdots\left(\mathbf{I} + \frac{\mathbf{A}\mathbf{T}}{K}\right)\right)\cdots\right)$$

Then take

$$e^{\mathbf{A}\mathbf{T}} = \Phi(\mathbf{T}, 0) \approx \mathbf{I} + \mathbf{A}\mathbf{T}\Psi$$

The above techniques fall into two classes: numerical and symbolic. The choice will partly depend upon the needs of the problem. All three approaches can be used also to find the forced solution (10.15), although the input $\mathbf{u}(t)$ must be structured appropriately if closed-form type methods are to be applied.

10.5 NOISE AND DISTURBANCES

For continuous systems, the models of noise can be either in state-space or in transfer functions, but will follow the notion mentioned in section 9.4.6 of white noise driving a linear system. In such cases we have for scalars

$$H(s) = \frac{\displaystyle\sum_{i=0}^{m} \beta_i s^i}{s^n + \displaystyle\sum_{i=0}^{n-1} \alpha_i s^i}$$

$$(D^n + \alpha_{n-1} D^{n-1} + \cdots + \alpha_0)\, w(t) = \eta(t)$$

$$v(t) = (\beta_m D^m + \cdots + \beta_1 D + \beta_0)\, w(t)$$

$$S_\eta(\omega) = R$$

$$S_\upsilon(\omega) = |H(j\omega)|^2 R$$

where $H(s)$ is the transfer function of the linear system and R is the level (variance) of the white noise; D denotes the differentiation operation, and the above avoids differentiation of white noise even though for modelling purposes only this makes little difference. One possible state-space model is

$$\dot{x}(t) = \begin{bmatrix} 0 & 1 & 0 & \cdots & 0 \\ 0 & 0 & 1 & 0 & 0 \\ \vdots & \vdots & \vdots & \vdots & \vdots \\ -\alpha_0 & -\alpha_1 & \cdots & & -\alpha_{n-1} \end{bmatrix} x(t) + \begin{bmatrix} 0 \\ 0 \\ \vdots \\ 1 \end{bmatrix} \eta(t)$$

$$v(t) = \begin{bmatrix} \beta_0 & \beta_1 & \cdots & \beta_m & 0 & \cdots & 0 \end{bmatrix} x(t)$$

and the same result is obtained.

Models of systems with disturbances and noise

A typical representation of a feedback system with measurement noise and output disturbances is shown in Fig. 10.6.

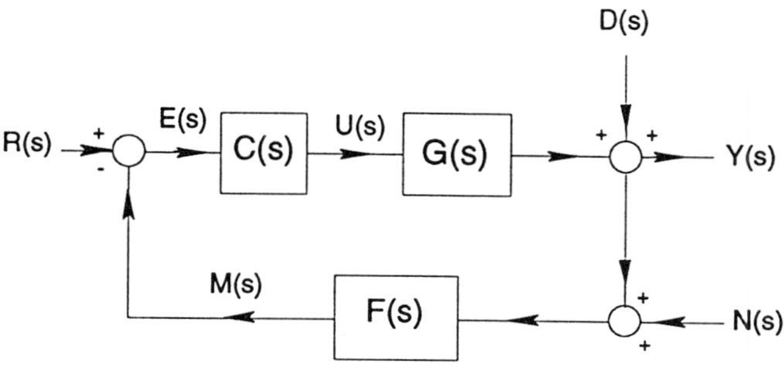

Figure 10.6   Typical closed-loop system block diagram.

Here $\mathbf{G}(s)$ represents the plant transfer function, $\mathbf{C}(s)$ the controller transfer function, $\mathbf{F}(s)$ the sensor transfer function, and $\mathbf{Y}(s)$, $\mathbf{R}(s)$, $\mathbf{D}(s)$, $\mathbf{N}(s)$, $\mathbf{E}(s)$, $\mathbf{U}(s)$, $\mathbf{M}(s)$ are the respective transforms of the output signal, the input reference signal, the disturbance, the (measurement) noise, the error signal, the command signal and the measurement signal. Then transform algebra yields easily

$$\mathbf{E}(s) = \mathbf{R}(s) - \mathbf{F}(s)\,(\mathbf{N}(s) + \mathbf{Y}(s))$$

$$\mathbf{Y}(s) = \mathbf{D}(s) + \mathbf{G}(s)\,\mathbf{U}(s)$$

$$\mathbf{U}(s) = \mathbf{C}(s)\,\mathbf{E}(s)$$

and hence relationships such as

$$\mathbf{E}(s) = (\mathbf{I} + \mathbf{F}(s)\,\mathbf{G}(s)\,\mathbf{C}(s))^{-1}\,(\mathbf{R}(s) - \mathbf{F}(s)\,\mathbf{N}(s) - \mathbf{F}(s)\,\mathbf{D}(s))$$

$$\mathbf{Y}(s) = (\mathbf{I} + \mathbf{G}(s)\,\mathbf{C}(s)\,\mathbf{F}(s))^{-1}\,(\mathbf{G}(s)\,\mathbf{C}(s)\,\mathbf{R}(s) + \mathbf{D}(s))$$

$$- \mathbf{G}(s)\,\mathbf{C}(s)\,\mathbf{F}(s)\,\mathbf{N}(s)$$

where the possibility of multi-input–multi-output transfer functions has been allowed.

The disturbances may be modelled as noise (as above) or as signals such as steps or ramps, e.g. $\mathbf{D}(s) = \varepsilon/s$ or δ/s^2. Alternatively, $\mathbf{D}(s)$ is in robust control theory defined along the lines of $\mathbf{D}(s) \in \mathscr{D}$ where

$$\mathscr{D} = \{ \mathbf{D}(s) : \|\mathbf{D}(j\omega)\|_p < 1 \}$$

where p typically is either 2 or ∞ (see norms in Appendix B or Chapter 33).

Measurement noise $\mathbf{N}(s)$ is ordinarily modelled as noise, but occasionally more like a disturbance in order to represent severe measurement device breakdown.

The usual state-space studies consider only the noise form explicitly, with the linear system model often such as

$$\dot{\mathbf{x}} = \mathbf{Ax} + \mathbf{Bu} + \Gamma\mathbf{v}$$

$$\mathbf{y} = \mathbf{Cx} + \mathbf{w}$$

with input $\mathbf{u}$, output $\mathbf{y}$, state $\mathbf{x}$, and noises $\mathbf{v}$ and $\mathbf{w}$. The noises are assumed to have properties

$$\mathscr{E}[\mathbf{v}] = \mathbf{0} \qquad \mathscr{E}[\mathbf{v}(t)\,\mathbf{v}^\mathrm{T}(\tau)] = \mathbf{Q}\delta(t-\tau)$$

$$\mathscr{E}[\mathbf{w}] = \mathbf{0} \qquad \mathscr{E}[\mathbf{w}(t)\,\mathbf{w}^\mathrm{T}(\tau)] = \mathbf{R}\delta(t-\tau)$$

$$\mathscr{E}[\mathbf{x}(0)] = \mathbf{x}_0 \qquad \mathscr{E}[(\mathbf{x}(0)-\mathbf{x}_0)(\mathbf{x}(0)-\mathbf{x}_0)^\mathrm{T}] = \mathbf{S}$$

In this, $\mathbf{w}$ is clearly intended to represent measurement noise, but the structure is such that if the model is extended so that part of the state vector is used for colouration of the input noise $\mathbf{v}$, then random model errors and disturbances can be represented through $\mathbf{v}$.

Model errors are usually taken in terms of either additive or multiplicative errors in matrices in the representations. Specifically, for transfer function matrices $\mathbf{G}(s)$ in which the nominal or standard model is $\mathbf{G}_0(s)$, the usual approach is to take either multiplicative errors as applying to the output

$$G(s) = (I + \Delta(s))\, G_0(s)$$

or to the input

$$G(s) = G_0(s)\, (I + \Delta(s))$$

or as additive errors

$$G(s) = G_0(s) + \Delta(s)$$

If $\Delta(s)$ contains only a few variable parameters, then the uncertainty is called structured. In much of the recent theoretical development, the only specification on $\Delta(s)$ is of the form

$$\|\Delta(j\omega)\| < L(\omega)$$

for a specified matrix norm and scalar function L. One particular norm used has been the ∞-norm

$$\|\Delta(j\omega)\|_{\infty} = \sup_{\omega} \|\Delta(j\omega)\| < L = \text{constant}$$

Notice that for scalars, the norm becomes the magnitude of the complex number, and the ∞-norm is the maximum value of that magnitude over all frequencies.

In linear state-space models, the above ideas appear also. Here the basic model

$$\dot{x} = Ax + Bu$$

$$y = Cx$$

has its matrices, particularly the A matrix, allowed to vary. An unstructured variation is one for which

$$A = A_0 + \delta A \qquad \|\delta A\| < a$$

while a structured uncertainty tends to be of the form

$$A = A_0 + \sum_{i=1}^{p} \theta_i A_i \quad -1 \le \theta_i \le 1$$

and a random variation has

$$A = A_0 + \sum_{i=1}^{p} w_i A_i$$

where the w_i are scalar noise processes.

Structured uncertainties tend to allow at least the possibility of parameter estimation and hence adaptive control (Chapter 31) in controlling the systems. Unstructured variations force us to consider worst case scenarios and hence conservative designs; these are the main topic of robust control theories (Chapter 33).

10.6 COMPUTER PROGRAMS

Most computer packages should be able to handle computations of the solutions of the models for given inputs. Some, such as SIMNON, do this by simulation. Packages such as MATLAB® will allow conversions of representations from transfer functions to state–space and back; sometimes this is done using canonical forms, since then the coefficients can be written out with little computation.

10.7 FURTHER INFORMATION

Longer discussions of the information in this chapter are found in books such as Dorf (1989), Phillips and Harbor (1991), and Kuo (1987). Wiberg (1971) is fairly thorough concerning the state-space transformations. Multivariable cases are in Maciejowski (1989). Noise is treated in the signal analysis literature and is introduced in Appendix C.

11

Sampled-data system representations

This chapter parallels Chapter 10 for discrete-time systems. Because difference equations are undoubtedly less familiar to students than differential equation methods, we also take a brief look at the characteristics of difference equation time responses.

11.1 SYNOPSIS

The digital computer works in a sampled-data manner, and hence sees the continuous real world through a picket fence, i.e. only at particular instants. This sampling is usually done at regular intervals T (typically a few hundredths of a second to a few seconds). As a result, computer control works with sequences of data samples and commands. The model descriptions are often in terms of difference equations, such as

$$y(kT) + a_1 y(kT-T) + \cdots + a_{n-2} y(kT-nT+2T) + a_{n-1} y(kT-nT+T)$$

$$+ a_n y(kT-nT) = b_0 u(kT) + b_1 u(kT-T) + \cdots + b_m u(kT-mT)$$

$$k = 0, 1, 2, \ldots \tag{11.1}$$

with the alternative representation using the shift operator defined by $q^i x(kT) = x(kT+iT)$ to give

$$(1 + a_1 q^{-1} + a_2 q^{-2} + \cdots + a_n q^{-n})\, y(kT)$$

$$= (b_0 + b_1 q^{-1} + \cdots + b_m q^{-m})\, u(kT) \tag{11.2}$$

The z-transform, which could be considered a discrete analogue of the Laplace transform, proves useful here, giving a discrete-time transfer function

$$\frac{Y(z)}{U(z)} = \frac{b_0 + b_1 z^{-1} + b_2 z^{-2} + \cdots + b_m z^{-m}}{1 + a_1 z^{-1} + a_2 z^{-2} + \cdots + a_n z^{-n}}$$

from which the frequency response can be calculated as

$$\left. \frac{Y(z)}{U(z)} \right|_{z = e^{j\omega T}}$$

Alternative forms of the rational function are of course possible. Models such as (11.1–2) also have state-space models of the form

$$\mathbf{x}(k+1) = \mathbf{A}\mathbf{x}(k) + \mathbf{B}\mathbf{u}(k)$$

$$\mathbf{y}(k) = \mathbf{C}\mathbf{x}(k) + \mathbf{D}\mathbf{u}(k)$$

which have standard canonical forms very similar to those for continuous-time system descriptions (Chapter 10), including controllable, observable, and Jordan forms. For example, an observable canonical form is given by

$$\mathbf{x}(k) = \begin{bmatrix} -a_1 & 1 & 0 & 0 & \cdots & 0 \\ -a_2 & 0 & 1 & 0 & & \\ -a_3 & 0 & 0 & 1 & \cdots & 0 \\ \vdots & \vdots & \vdots & \vdots & \vdots & \vdots \\ -a_{n-1} & 0 & 0 & 0 & \cdots & 1 \\ -a_n & 0 & 0 & 0 & \cdots & 0 \end{bmatrix} \mathbf{x}(k-1) + \begin{bmatrix} b_1 - a_1 b_0 \\ b_2 - a_2 b_0 \\ b_3 - a_3 b_0 \\ \vdots \\ b_n - a_n b_0 \end{bmatrix}$$

$$\mathbf{y}(k) = [\, 1 \quad 0 \quad 0 \quad 0 \quad \cdots \quad 0 \,]\, \mathbf{x}(k) + b_0 \mathbf{u}(k)$$

11.2 TIME RESPONSE OF DIFFERENCE EQUATIONS

Throughout this book, discrete-time systems – either closed- or open-loop – are described by relations of the form

$$y(k) = -a_1 y(k-1) - a_2 y(k-2) - \cdots - a_n y(k-n) + b_0 u(k)$$

$$+ b_1 u(k-1) + \cdots + b_m u(k-m) \tag{11.3}$$

where $y(k)$ is the output at sample number k, i.e. the output at time kT, $u(k)$ is the input at sample number k, i.e. the input at time kT, and $a_1, a_2, ..., a_n$ and $b_0, b_1, ..., b_m$ are known coefficients. There are many variations:

1. the coefficients are unknown;
2. there are multiple inputs, some of them noisy;
3. the coefficients are not constants;
4. the $y(k)$ and $u(k)$ terms do not appear linearly; and/or
5. the representation is in a different form, e.g. matrix form.

However, the above is basic and therefore worth a considerable amount of study concerning the nature of the systems it represents. Equation (11.3) is called a linear, constant coefficient, causal **difference equation**.

Consider a system model (as above, but with $b_0=1$, $b_i=0$, $i=1, 2, ..., m$)

$$y(k) = -a_1 y(k-1) - a_2 y(k-2) - a_3 y(k-3) - \cdots - a_n y(k-n) + u(k)$$

Provided a_n is not zero, this is an nth order difference equation and its state-space representation will be an n-dimensional system. The solutions for a unit pulse at $k = 0$, i.e. for $u(0) = 1$ and $u(k) = 0$ when $k \neq 0$, are given using the roots $\lambda_i, i=1,...,n$ of the **characteristic equation** of the system:

$$\lambda^n + a_1 \lambda^{n-1} + a_2 \lambda^{n-2} + \cdots + a_{n-1} \lambda + a_n = 0 \tag{11.4}$$

With these, the pulse response solutions $y(k)$, $k \geq 0$ satisfy

$$y(k) = c_1 (\lambda_1)^k + c_2 (\lambda_2)^k + \cdots + c_n (\lambda_n)^k \tag{11.5}$$

where the c_i are chosen so that $y(0)=1$, and $y(-i)=0$, $i=1,2,...,n-1$. Because the fundamental theorem of algebra informs us that indeed there are n such roots and that, if the coefficients a_i are real then the

λ_i are either real or in complex conjugate pairs, we know immediately that the solution will have the form of sums of terms of the forms $c_i(r_i)^k$ and $c_j(r_j)^k \cos(\phi_j k + \theta_j)$.

The handling of more complicated inputs is straightforward because of superposition and the observation that a function $u(k)$ starting at time step 0 can be represented as the sum of scaled offset pulses, i.e.

$$u(k) = \sum_{i=0}^{\infty} d_i \delta(k-i) \tag{11.6}$$

where by definition the pulse satisfies $\delta(0) = 1$, $\delta(j) = 0$ for $j \neq 0$ and is the Kronecker δ–function.

Then it is easy to see that if we define $y_i(k)$ as the response to $\delta(k-i)$, $i = 0, 1, \ldots$, so that

$$y_i(k) = -a_1 y_i(k-1) - a_2 y_i(k-2) - \cdots - a_n y_i(k-n) + \delta(k-i)$$

then the response $y(k)$ to an input such as (11.6) is given by

$$y(k) = \sum_{i=0}^{\infty} d_i y_i(k) \tag{11.7}$$

This argument is more easily and more commonly made using the pulse response $h(k)$, i.e. the response of the system to a pulse at step 0. In this case, $h(k) = y_0(k)$. Then the linearity and shift properties of the z transform imply that

$$y_i(k) = y_0(k-i) = h(k-i)$$

so that finally (11.7) becomes the convolution sum

$$y(k) = \sum_{i=0}^{\infty} d_i h(k-i) \tag{11.8}$$

$$= \sum_{j=0}^{\infty} u(j)\, h(k-j) \tag{11.9}$$

Ordinarily, the system is causal, meaning that the output cannot anticipate an input. Causality is interpreted mathematically as $h(j) = 0$, for $j < 0$ and the effect on (11.9) is

$$y(k) = \sum_{j=0}^{k} u(j)\, h(k-j)$$

A change of index variable (replacing i with $k-j$) lets us write (11.8) and (11.9) in general as

$$y(k) = \sum_{j=-\infty}^{k} d_{k-j}\, h(j)$$

$$= \sum_{j=-\infty}^{k} u(k-j)\, h(j)$$

or, if the system is causal,

$$y(k) = \sum_{j=0}^{k} u(k-j)\, h(j)$$

The point of this, besides the justification of the general solution, is to lead to the observation that the system operation will always be characterized by the superposition of terms involving the sums of powers of the roots of the characteristic equation (11.5). These are called the **characteristic roots**, the **poles**, or the **eigenvalues** of the system.

Control laws are chosen to affect the system responses. To see how we can affect this behaviour, consider (11.3) and suppose we choose to do a simple feedback, i.e. to measure the output at each step k and set the controller value $u(k+1)$ proportional to it. (The $k+1$ is to allow computation time between measurement and command.) Then

$$u(k) = Gy(k-1) + u_e(k)$$

where u_e is an external signal, and the closed-loop system is now described by

$$y(k) = -(a_1 - b_0 G)y(k-1) - (a_2 - b_1 G)y(k-2) - \cdots$$

$$- (a_{m+1} - b_m G)y(k-m-1) - \cdots - a_n y(k-n)$$

$$+ b_0 u_e(k) + b_1 u_e(k-1) + \cdots + b_m u_e(k-m)$$

where it has been assumed that $m < n$. From this it can be observed that the characteristic equation (11.4) will become

$$\lambda^n + (a_1 - b_0 G)\lambda^{n-1} + (a_2 - b_1 G)\lambda^{n-2}$$

$$+ \cdots + (a_{m+1} - b_m G)\lambda^{n-m-1} + a_{m+2}\lambda^{n-m-2}$$

$$+ \cdots + a_{n-1}\lambda + a_n = 0$$

This may of course be expected to have characteristic values much different from those of the original system. The combination of art with science for selecting the feedback gain G is one of the problems we will address in the theory sections.

The above represents response which may in principle be found for any input $u(k)$ sequence. One particular input of considerable interest to us is the sinusoidal input

$$u(k) = A \sin(k\phi + \theta) \tag{11.10}$$

in which typically $\phi = \omega T$ because the input is in fact samples taken every T seconds of a signal of frequency ω. By Euler's rules that

$$\sin x = \frac{e^{jx} - e^{-jx}}{2j}$$

$$\cos x = \frac{e^{jx} - e^{-jx}}{2}$$

and the linearity of the response, we see that the crucial element of the input is e^{jx}. Thus let $u(k) = Ae^{j\phi k}$ and assume that

1. any transients have died out and the system is in steady state;
2. the system is stable so that the output is due solely to the input; and
3. the output is of the form $y(k) = Be^{j\phi k}$ where B may be a complex number.

Under these assumptions, we look to see if there is a B such that condition 3 holds. Substituting in (11.3) gives

$$Be^{j\phi k} = -a_1 Be^{j\phi(k-1)} - a_2 Be^{j\phi(k-2)} - \cdots - a_n Be^{j\phi(k-n)}$$

$$+ b_0 Ae^{j\phi k} + b_1 Ae^{j\phi(k-1)} + \cdots + b_m Ae^{j\phi(k-m)}$$

Eliminating the common factor $e^{j\phi k}$ and rearranging gives the result that $y(k) = Be^{j\phi k}$ indeed holds provided that

$$B = \frac{b_0 + b_1 e^{-j\phi} + b_2 e^{-2j\phi} + \cdots + b_m e^{-j\phi m}}{1 + a_1 e^{-j\phi} + a_2 e^{-2j\phi} + \cdots a_n e^{-j\phi n}} A$$

The complex value B/A depends on ϕ, the phase step per time step, and is usually written as such, i.e. as $F(\phi)$ or $F(\omega T)$. As such, it is called the frequency response of the system and is often written in terms of its phase and magnitude.

$$F(\omega T) = |F(\omega T)| \, e^{j\,\mathrm{arc}(F(\omega T))}$$

Returning to the original problem with input (11.10) and using the linearity of the system, we find by little more than substitution that

$$y(k) = |F(\phi)| \, A \sin(k\phi + \theta + \mathrm{arc}(F(\phi)))$$

where we have used the easily shown facts that $|F(\phi)| = |F(-\phi)|$ and $\mathrm{arc}(F(\phi)) = -\mathrm{arc}(F(-\phi))$.

Because almost any periodic input is, by Fourier theory, representable by a sum of sinusoids,

$$u(k) = \sum_{p=-\infty}^{\infty} c_p \exp(j\omega_p Tk)$$

we know that in steady-state operation, a system described by (11.3) has an output

$$y(k) = \sum_{p=-\infty}^{\infty} c_p \exp(j\omega_p Tk)\, F(\omega_p T)$$

Thus it is clear that along with studying the transients inherent in the system due to the roots of the characteristic equation, the steady-state as described by the frequency response $F(\omega T)$ may also be of interest. We shall see this in future sections.

11.3 TRANSFER FUNCTIONS

Transfer functions, and in fact much of discrete-time system analysis, are based upon the z-transform (see Appendix A). For our purposes, we may define the z-transform of a sequence $\{...,y(-2), y(-1),y(0),y(1),y(2),...\}$ as

$$Y(z) = \mathscr{Z}(y(k)) = \sum_{i=-\infty}^{\infty} y(i)\, z^{-i} \tag{11.11}$$

where z is a complex number. This is a linear transformation, with the commonly used shift property that

$$x(k) = y(k-j) \Rightarrow X(z) = z^{-j} Y(z)$$

Tables of transforms for common sequences, as in Appendix 1, are readily available. Thus, one easily finds, either from the definition or from tables, that

$$y(i) = 1,\, i \geq 0 \qquad \Leftrightarrow \qquad Y(z) = \frac{z}{z-1}$$

$$y(i) = a^i,\, i \geq 0 \qquad \Leftrightarrow \qquad Y(z) = \frac{z}{z-a}$$

and so on, where again $\Leftrightarrow$ indicates a transform pair. These have similar appearance to Laplace transforms, and can be manipulated in much the same way.

11.3.1 Basics of use

The solution of difference equations such as (11.3) is possible with z-transforms. Taking the z-transform, using zero initial conditions, and defining $Y(z) = \mathcal{Z}(\{y(k)\})$ and $U(z) = \mathcal{Z}(\{u(k)\})$ we find this becomes

$$Y(z) = -a_1 z^{-1} Y(z) - a_2 z^{-2} Y(z) - \cdots - a_n z^{-n} Y(z)$$

$$+ b_0 U(z) + b_1 z^{-1} U(z) + \cdots + b_m z^{-m} U(z)$$

Hence we find the input–output relationship of (11.3) to be representable by the transform relationship $Y(z) = H(z) U(z)$ in which the expression

$$H(z) = \frac{b_0 + b_1 z^{-1} + b_2 z^{-2} + \cdots + b_m z^{-m}}{1 + a_1 z^{-1} + a_2 z^{-2} + \cdots + a_n z^{-n}} \tag{11.12}$$

is called the discrete time transfer function. This has the same types of applications as the continuous time transfer functions in the Laplace transform variable s, and most of the applications are analogous: solution of difference equations, manipulation of system representations, frequency response calculations.

The expression in (11.12) is called a direct form of the transfer function. Because such forms appear in classical control law designs, it is often important numerically that they be rearranged. Commonly used are the cascade (or series) form and the parallel form. Just as in the continuous time cases of Chapter 10, these are given by $H(z) = H_1(z) H_2(z) \cdots H_k(z)$ where

$$H_i(z) = K_i \frac{(z - s_{i,1}) \cdots (z - s_{i,m_i})}{(z - \lambda_{i,1}) \cdots (z - \lambda_{i,n_i})} \qquad i = 1, 2, \ldots, k$$

and

$$H(z) = H_1(z) + H_2(z) + \cdots + H_j(z)$$

respectively. Expanded fraction forms are also possible, and we will meet the Cauer form in Chapter 30. Mixed forms, such as sums of cascade forms, are of course also possible and sometimes useful.

For a particular input $\{u(k)\}$ for which the z-transform is rational in z, $Y(z)$ is then rational and (in principle) easily inverted. Usually the inputs of interest are the pulse, step, and sometimes ramp input functions, in which $U(z)$ is 1, $(1-z^{-1})^{-1}$, and $Tz^{-1}(1-z^{-1})^{-2}$, respectively. Another interesting input is sinusoidal, which we consider below. The usual solution method is simply to find the expression $Y(z) = G(z)\,U(z)$ and invert it, using tables as with the Laplace transform; alternatives include finding numerical solutions by either long division of the transform functions (in which case $\{y(kT)\}$ is given by the coefficients of z^{-k} in the quotient) or by numerical solution of the difference equation corresponding to the transfer function.

The forms such as in (11.12) are easily combined both additively and multiplicatively to yield transfer functions of composite systems. The manipulations are basically algebraic, although block diagram algebra, presented in some texts (e.g. DiStefano *et al.*, 1976), can give guidance as to manipulations needed.

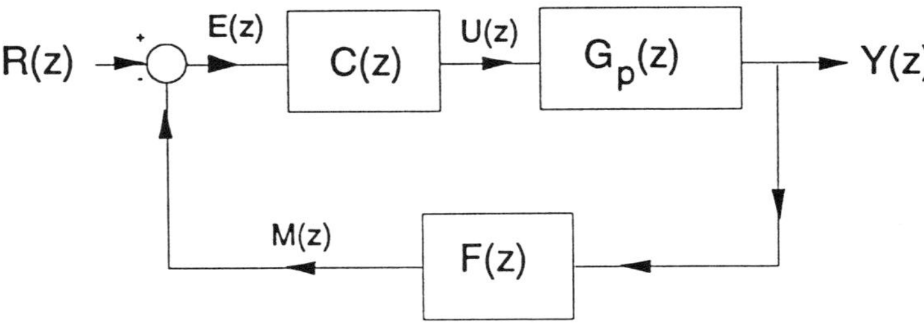

Figure 11.1 A closed-loop system z-transform model without noise or disturbances.

The most common example is for a simple feedback control loop such as in Fig. 11.1, where $C(z)$ is the compensator/controller/ computer transfer function, $G(z)$ the plant or process transfer function, and $F(z)$ the feedback transfer function (of a measurement device, perhaps). $Y(z)$ is the z-transform of the output quantity, $U(z)$ the transform of the control signal, $E(z)$ the transform of the error signal, $M(z)$ the transform of the measurement signal, and $R(z)$ the transform of the input reference signal. We wish to know the transfer function of the system from reference to output.

The algebra of the situation gives, for the various signals,

$$Y(z) = G(z)\,U(z) \qquad U(z) = C(z)\,E(z)$$

$$E(z) = R(z) - M(z) \qquad M(z) = F(z)\,Y(z)$$

By simple algebra, we find that eliminating the intermediate signals $E(z)$, $U(z)$, and $M(z)$ gives

$$Y(z) = G(z)\,C(z)\,[R(z) - F(z)\,Y(z)]$$

This easily rearranges to yield

$$Y(z) = \frac{G(z)\,C(z)}{1 + G(z)\,C(z)F(z)}\,R(z)$$

and from this we may read the system closed-loop transfer function as

$$\frac{Y(z)}{R(z)} = \frac{G(z)\,C(z)}{1 + G(z)\,C(z)\,F(z)}$$

Much more complex systems may be manipulated in the same way, so that the advantage over direct attack on the difference equations is obvious.

Example

A motor under computer control may be modelled (see Chapter 12) to have z-transform model

$$G(z) = \frac{(T + \tau e^{-T/\tau} - \tau)\,z + (\tau - \tau e^{-T/\tau} - T e^{-T/\tau})}{z^2 - (1 + e^{-T/\tau})\,z + e^{-T/\tau}}$$

where T is the sampling period and τ is a time lag parameter characteristic of the motor's speed of response. We multiply through by z^{-2}, which gives

$$G(z) = \frac{(T + \tau e^{-T/\tau} - \tau)\, z^{-1} + (\tau - \tau e^{-T/\tau} - T e^{-T/\tau})\, z^{-2}}{1 - (1 + e^{-T/\tau})\, z^{-1} + e^{-T/\tau}\, z^{-2}}$$

Thus if $Y(z)$ is the transform of the output $y(k)$ and $U(z)$ the transform of the input $u(k)$, we have

$$\left(1 - (1 + e^{-T/\tau})\, z^{-1} + e^{-T/\tau}\, z^{-2}\right) Y(z)$$

$$= \left((T + \tau e^{-T/\tau} - \tau)\, z^{-1} + (\tau - \tau e^{-T/\tau} - T e^{-T/\tau})\, z^{-2}\right) U(z)$$

Distributing $Y(z)$ and $U(z)$, interpreting $z^{-n} Y(z)$ as $y(k-n)$, etc., and a small amount of algebra allow us to write the difference equation model of the motor as

$$y(k) = (1 + e^{-T/\tau})\, y(k-1) - e^{-T/\tau} y(k-2)$$

$$+ (T + \tau e^{-T/\tau} - \tau)\, u(k-1) + (\tau - \tau e^{-T/\tau} - T e^{-T/\tau})\, u(k-2)$$

Clearly the steps may be taken in the opposite direction, from difference equation to transfer function. The difference equation model is easily simulated, even on a programmable calculator. For example, $u(0) = 1$, $u(k) = 0$, $k \neq 0$, and $y(k) = 0$, $k < 1$, allows calculation of the unit pulse response.

11.3.2 Multivariable transfer functions

The sampled data transfer function situation for systems with m inputs and n outputs parallels that for continuous time systems presented in section 10.2.2. Thus, for difference equation sets such as

$$\sum_{k=0}^{n_i} \alpha_{i,k}\, y_i(l-k) = \sum_{j=1}^{m} \sum_{k=0}^{m_j} \beta_{i,j,k}\, u_j(l-k) \quad i = 1, 2, \ldots, n$$

with transfer function representations

$$Y_i(z) = \sum_{j=1}^{m} \frac{\sum_{k=0}^{m_j} \beta_{i,j,k} z^{-k}}{\sum_{k=0}^{n_i} \alpha_{i,k} z^{-k}} U_j(z) \qquad i = 1, 2, \ldots, n$$

$$= \sum_{j=1}^{m} G_{i,j}(z) U_j(z) \qquad i = 1, 2, \ldots, n$$

We use the more convenient matrix form notation

$$\mathbf{Y}(z) = \begin{bmatrix} Y_1(z) \\ Y_2(z) \\ \vdots \\ Y_n(z) \end{bmatrix} = \begin{bmatrix} G_{1,1}(z) & G_{1,2}(z) & \cdots & G_{1,m}(z) \\ G_{2,1}(z) & G_{2,2}(z) & \cdots & G_{2,m}(z) \\ \vdots & & & \vdots \\ G_{n,1}(z) & G_{n,2}(z) & \cdots & G_{n,m}(z) \end{bmatrix} \begin{bmatrix} U_1(z) \\ U_2(z) \\ \vdots \\ U_m(z) \end{bmatrix}$$

$$= \mathbf{G}(z) \mathbf{U}(z)$$

Since the operators are all linear, this representation can be manipulated almost as if they were scalars, just as was done for the continuous time case. We summarize the operations here under the assumption that matrix dimensions are consistent.

Series:

$$\mathbf{Y}(z) = \mathbf{G}(z) \mathbf{W}(z) \qquad \text{and} \qquad \mathbf{W}(z) = \mathbf{H}(z) \mathbf{U}(z)$$

$$\Rightarrow \mathbf{Y}(z) = \mathbf{G}(z) \mathbf{H}(z) \mathbf{U}(z)$$

(but *not* $\mathbf{H}(z) \mathbf{G}(z) \mathbf{U}(z)$ – matrix multiplication is not commutative.)

Parallel: where

$$\mathbf{W}(z) = \mathbf{H}(z) \mathbf{U}(z) \qquad \text{and} \qquad \mathbf{Y}(z) = \mathbf{G}(z) \mathbf{U}(z)$$

$$\Rightarrow \mathbf{V}(z) = \mathbf{W}(z) + \mathbf{Y}(z) \Rightarrow \mathbf{V}(z) = [\mathbf{H}(z) + \mathbf{G}(z)]\,\mathbf{U}(z)$$

Feedback connection:

$$\mathbf{Y}(z) = \mathbf{G}(z)\,\mathbf{E}(z) \quad \mathbf{E}(z) = \mathbf{U}(z) - \mathbf{M}(z) \quad \mathbf{M}(z) = \mathbf{H}(z)\,\mathbf{Y}(z)$$

$$\Rightarrow \mathbf{Y}(z) = [\mathbf{I} + \mathbf{G}(z)\,\mathbf{H}(z)]^{-1}\,\mathbf{G}(z)\,\mathbf{U}(z)$$

and

$$\mathbf{Y}(z) = \mathbf{G}(z)\,[\mathbf{I} + \mathbf{H}(z)\,\mathbf{G}(z)]^{-1}\,\mathbf{U}(z)$$

In the basic representation, the matrix $\mathbf{G}(z)$ is $n \times m$ and so not necessarily square.

Canonical form: Smith–McMillan form

As in the continuous time case (section 10.2), the transfer function $\mathbf{G}(z)$ has a Smith–McMillan form $\mathbf{M}(z)$ defined by

$$\mathbf{M}(z) = \begin{pmatrix} \mathbf{M}_1(z) & 0 \\ 0 & 0 \end{pmatrix}$$

$$\mathbf{M}(z) = \mathrm{diag}\left\{ \frac{v_1(z)}{\delta_1(z)}, \frac{v_2(z)}{\delta_2(z)}, \ldots, \frac{v_r(z)}{\delta_r(z)} \right\}$$

where r is the normal rank (i.e. the rank for almost all z) of $\mathbf{G}(z)$, $v_1(z)$ and $\delta_1(z)$ are co-prime, i.e. they have no common factors and

$$\left. \begin{array}{l} v_i(z) \text{ divides } v_{i+1}(z) \\ \delta_{i+1}(z) \text{ divides } \delta_i(z) \end{array} \right\} i = 1, 2, \ldots, r-1$$

This is clearly the same as for the continuous case with a different argument, and since the manipulations do not depend upon the interpretation of the argument, the computation of a Smith–McMillan form follows the same path as in section 10.2.2.

The alternative form for a rational transfer function matrix called the matrix fraction description is also available using non-unique matrices of polynomials, $\mathbf{N}(z)$ and $\mathbf{D}(z)$, to create representations such as $\mathbf{G}(z) = \mathbf{N}(z)\,\mathbf{D}^{-1}(z)$ for the right matrix-fraction description, and $\mathbf{G}(z) = \tilde{\mathbf{D}}^{-1}(z)\,\tilde{\mathbf{N}}(z)$ for the left matrix-fraction form. These may be derived using the Smith–McMillan form, just as in the continuous time case section 10.2.2.

11.4 FREQUENCY RESPONSE

For the system frequency response, we assume we have a model such as (11.1–2) and that the input is $u(nT) = \mathrm{e}^{\mathrm{j}\omega nT}$. As this has the z-transform

$$\{\mathrm{e}^{\mathrm{j}\omega nT}\} \Leftrightarrow \frac{1}{1 - \mathrm{e}^{\mathrm{j}\omega T}z^{-1}}$$

it is clear that the system output will be

$$\{y(nT)\} = \mathscr{Z}^{-1}\left[H(z)\,\frac{1}{1 - \mathrm{e}^{+\mathrm{j}\omega T}z^{-1}} \right] \tag{11.13}$$

$$= \mathscr{Z}^{-1}\left[\frac{H(\mathrm{e}^{\mathrm{j}\omega T})}{1 - \mathrm{e}^{+\mathrm{j}\omega T}z^{-1}} + \sum_i \frac{[H(z)(z-z_i)]|_{z = z_i}}{(1 - z^{-1}z_i)(1 - \mathrm{e}^{\mathrm{j}\omega T}z_i^{-1})} \right]$$

where the sum is over the poles z_i of $\mathbf{H}(z)$, taken here notationally as distinct but easily generalized, and the expansion is a partial fraction expansion of the function in (11.13). On taking the inverse transform, the terms from the summation will, provided the system is stable (a topic investigated in the theory chapters, especially Chapters 13–14), eventually decay so that for large k,

$$y(kT) = H(\mathrm{e}^{\mathrm{j}\omega T})\,\mathrm{e}^{\mathrm{j}\omega kT}$$

The complex number $H(\mathrm{e}^{\mathrm{j}\omega T})$ then represents the frequency response of the system.

We should particularly notice that $\mathbf{H}(e^{j\omega T})$ is periodic in ωT, repeating for $\omega T = 2k\pi$ for all integer k. A typical response is shown in Fig. 11.2.

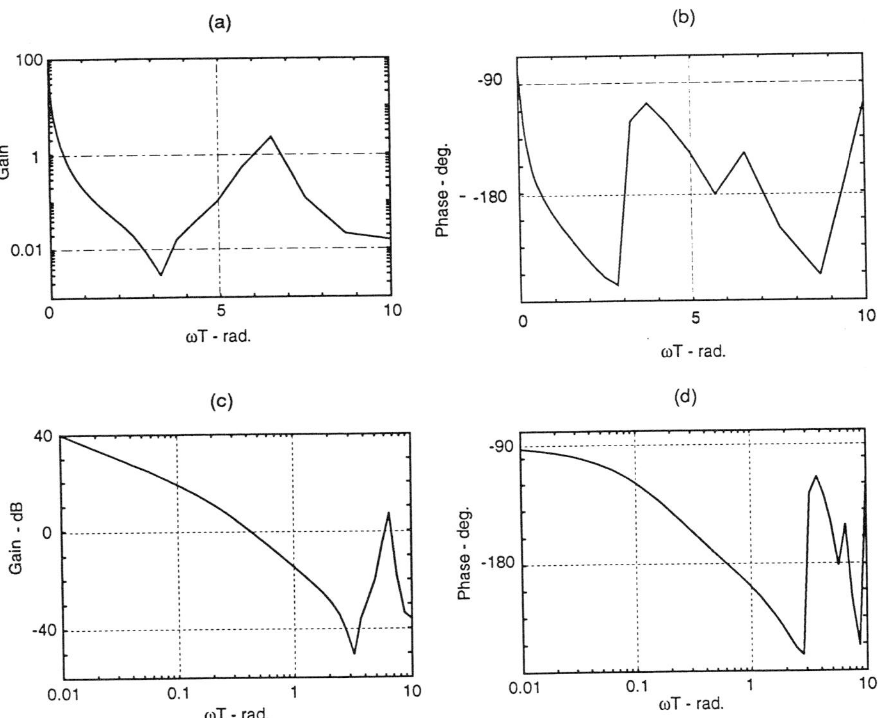

Figure 11.2 Frequency response of a sampled data system. (a) and (b) show gain and phase vs. frequency ω* sample period T. (c) and (d) show the same information plotted against log (ωT). What is striking is that the frequency response is a periodic function.

In the multivariable case, again as in continuous time, we evaluate the matrix transfer function. Thus the frequency response is $\mathbf{G}(e^{j\omega T})$, and the matrix gain is bounded by the largest and smallest singular values:

$$\underline{\sigma}(\mathbf{G}(e^{j\omega T})) \le \frac{\|\mathbf{y}(t)\|}{\|\mathbf{u}(t)\|} \le \overline{\sigma}(\mathbf{G}(e^{j\omega T}))$$

where $\mathbf{u}(t)$ is any input vector with components at frequency ω and sampling period T, and $\mathbf{y}(t)$ is the corresponding output vector.

11.5 STATE-SPACE REPRESENTATIONS

11.5.1 Introduction

In this section we explore the use of simple state-space representations, particularly those for linear constant coefficient systems. We are concerned with representations of the form

$$\mathbf{x}(k+1) = \mathbf{A}(k)x(k) + \mathbf{B}(k)\mathbf{u}(k)$$

$$\mathbf{y}(k) = \mathbf{C}(k)x(k) + \mathbf{D}(k)\mathbf{u}(k) \tag{11.14}$$

where $\mathbf{A}$ is $n \times n$, $\mathbf{B}$ is $n \times m$, $\mathbf{C}$ is $p \times n$, $\mathbf{D}$ is $p \times m$, and $\mathbf{x}$ is the n-dimensional state vector, $\mathbf{u}$ is the m-dimensional input or control vector, and $\mathbf{y}$ is the p-dimensional output or measurement vector. For many of our examples, $m = 1$ and $p = 1$ and the system is said to be an n-dimensional SISO system.

A very useful alternative time-domain representation to (11.1) and (11.3), i.e. to

$$y(k) = -\sum_{i=1}^{n} a_i y(k-i) + \sum_{i=0}^{L} b_i u(k-i) \tag{11.15}$$

is given by the matrix form of (11.14) where in this case x is an n-vector, $\mathbf{B}$ is an $n \times 1$ matrix (sometimes denoted as a vector $\mathbf{b}$), $\mathbf{C}$ is a $1 \times n$ matrix (notation $\mathbf{c}^{\mathrm{T}}$), $\mathbf{D}$ is 1×1 (notation d and a scalar), and $\mathbf{A}$ is an $n \times n$ matrix. The vector $\mathbf{x}$ is the state vector of the system, y and u are the scalar output and input respectively, and $\mathbf{A}, \mathbf{B}, \mathbf{C}, \mathbf{D}$ (or notationally $\mathbf{A}, \mathbf{b}, \mathbf{c}^{\mathrm{T}}$, and d) are constants. We should stress the dimensional relationships:

1. if there is one control variable, then $m = 1$;
2. if there are n steps of delay (and $n \geq L$), then n is the length of the state vector;
3. if y is a scalar output, then $p = 1$; and
4. if $b_0 = 0$, then the 1×1 matrix $\mathbf{D}(=d)= 0$.

The matrices are not unique for a given input–output relationship (11.15). The two types of relationships are equivalent, of course, but will be seen in the sequel to be useful or convenient for different types of studies.

In this section we simply look at some of the state–space representations equivalent to (11.15). For this, it is convenient to define the shift operator q by

$$q^k x(j) = x(j+k) \quad k = \text{integer}$$

Using this notation, which is almost like the z-transform, (11.15) becomes

$$y(k) = -\sum_{i=1}^{n} a_i q^{-i} y(k) + \sum_{i=0}^{L} b_i q^{-i} u(k)$$

We stress that at this point this is only used for notational convenience and motivating clarity; we do not yet use any special properties of the expression.

11.5.2 Special forms

The special forms with which we commonly deal include several very special ones (where the matrices have zeros and ones in certain places) called canonical forms) which (because almost any linear constant coefficient system can be placed into a canonical form) can be treated with some quite powerful tools.

For a highly instructive case, we choose $b_1 = 1$ and $b_i = 0$, $i \neq 1$, so that (11.1) becomes

$$y(k) = \sum_{i=1}^{n} a_i q^{-i} y(k) + u(k-1)$$

Then defining state variables x_i by

$$x_1(k) = y(k) \qquad x_2(k) = y(k-1) \quad \cdots \quad x_n(k) = y(k-n+1)$$

we can easily show

$$x_2(k) = x_1(k-1) \quad x_3(k) = x_2(k-1) \quad \cdots \quad x_n(k) = x_{n-1}(k-1)$$

so that

$$x_1(k) = y(k) = -\sum_{i=1}^{n} a_i\, y(k-i) + u(k-1)$$

$$= -a_1 x_1(k-1) - a_2 x_2(k-1) - \cdots - a_n x_n(k-1) + u(k-1)$$

$$y(k) = x_1(k)$$

or in matrix form

$$\mathbf{x}(k+1) = \begin{bmatrix} -a_1 & -a_2 & -a_3 & \cdots & \cdots & -a_n \\ 1 & 0 & 0 & \cdots & & 0 \\ 0 & 1 & 0 & \cdots & & 0 \\ 0 & 0 & 1 & \cdots & & 0 \\ \vdots & \vdots & \vdots & & & \vdots \\ 0 & & & 1 & 0 & 0 \\ 0 & 0 & 0 & 0 & 1 & 0 \end{bmatrix} \mathbf{x}(k) + \begin{bmatrix} 1 \\ 0 \\ 0 \\ 0 \\ \vdots \\ 0 \\ 0 \end{bmatrix} u(k)$$

$$y(k) = [1 \quad 0 \quad 0 \quad \cdots \quad 0]\, \mathbf{x}(k)$$

To generalize this to cases where $b_1 \neq 0$, $i \geq 0$, is straightforward when we observe that superposition still applies, e.g. that if the input sequence is delayed by q then so is the output sequence. Thus, if $\{y(k)\}$ is the response to $\{u(k-1)\}$, then $\{b_2 y(k-1) + b_j y(k-j+1)\}$ is the response to $\{b_2 u(k-2) + b_j u(k-j)\}$. More generally, the response of (11.11) is represented, for $b_0 \neq 0$, by

$$\mathbf{x}(k+1) = \begin{bmatrix} -a_1 & -a_2 & -a_3 & \cdots & \cdots & -a_n \\ 1 & 0 & 0 & \cdots & & 0 \\ 0 & 1 & 0 & \cdots & & 0 \\ 0 & 0 & 1 & \cdots & & 0 \\ \vdots & \vdots & \vdots & & & \vdots \\ 0 & & & 1 & 0 & 0 \\ 0 & 0 & 0 & 0 & 1 & 0 \end{bmatrix} \mathbf{x}(k) + \begin{bmatrix} 1 \\ 0 \\ 0 \\ 0 \\ \vdots \\ 0 \\ 0 \end{bmatrix} u(k)$$

$$y(k) = [\, b_1 - a_1 b_0 \quad b_2 - a_2 b_0 \quad b_n - a_n b_0 \,] x(k) + b_0 u(k) \qquad (11.16)$$

This is a common, canonical form and is called the **phase canonical** form, the **controllable canonical** form, or the first canonical form.

An alternate matrix form may be written by grouping the delays. Then (11.1) becomes

$$y(k) = b_0 u(k) + (b_1 u(k-1) - a_1 y(k-1) + (b_2 u(k-2)$$
$$- a_2 y(k-2) + (b_3 u(k-3) - \cdots + (b_n u(k-n)$$
$$- a_n y(k-n)) \cdots)))$$

Using the shift operator q makes the next step seem more obvious. Using it, we write

$$y(k) = b_0 u(k) + q^{-1}(b_1 u(k) - a_1 y(k) + q^{-1}(b_2 u(k)$$
$$- a_2 y(k) + q^{-1}(\cdots q^{-1}(b_n u(k) - a_n y(k)) \cdots)))$$

Now define

$$x_n(k) = q^{-1}(b_n u(k) - a_n y(k))$$
$$= b_n u(k-1) - a_n y(k-1)$$
$$x_{n-1}(k) = b_{n-1} u(k-1) - a_{n-1} y(k-1) + x_n(k-1)$$

$$x_2(k) = b_2 u(k-1) - a_2 y(k-1) + x_3(k-1)$$

$$x_1(k) = b_1 u(k-1) - a_1 y(k-1) + x_2(k-1)$$

$$y(k) = b_0 u(k) + x_1(k)$$

Substituting the last of these for $y(k-1)$ and rearranging gives

$$x_1(k) = -a_1 x_1(k-1) + x_2(k-1) + (b_1 - a_1 b_0) u(k-1)$$

$$x_2(k) = -a_2 x_1(k-1) + x_3(k-1) + (b_2 - a_2 b_0) u(k-1)$$

$$\vdots$$

$$x_n(k) = -a_{n-1} x_1(k-1) + (b_n - a_n b_0) u(k-1)$$

$$y(k) = x_1(k) + b_0 u(k)$$

or in matrix form

$$\mathbf{x}(k) = \begin{bmatrix} -a_1 & 1 & 0 & 0 & \cdots & 0 \\ -a_2 & 0 & 1 & 0 & & \\ -a_3 & 0 & 0 & 1 & \cdots & 0 \\ \vdots & & & & & \\ -a_{n-1} & 0 & 0 & 0 & \cdots & 1 \\ -a_n & 0 & 0 & 0 & \cdots & 0 \end{bmatrix} \mathbf{x}(k-1) + \begin{bmatrix} b_1 - a_1 b_0 \\ b_2 - a_2 b_0 \\ b_3 - a_3 b_0 \\ \vdots \\ b_n - a_n b_0 \end{bmatrix} u(k-1)$$

$$y(k) = [\, 1 \quad 0 \quad 0 \quad 0 \quad \cdots \quad 0 \,] \, \mathbf{x}(k) + b_0 u(k) \tag{11.17}$$

This is the second, or **observable, canonical** form. Like the first form, it can be written down directly from the original difference equation.

Example

The motor under computer control in the example of section 11.3.1 had

$$G(z) = \frac{(T + \tau e^{-T/\tau} - \tau)z + (\tau - \tau e^{-T/\tau} - T e^{-T/\tau})}{z^2 - (1 + e^{-T/\tau})z + e^{-T/\tau}}$$

One possible state-space model of this is

$$\mathbf{w}(k+1) = \begin{bmatrix} w_1(k+1) \\ w_2(k+1) \end{bmatrix} = \begin{bmatrix} 1 + e^{-T/\tau} & -e^{-T/\tau} \\ 1 & 0 \end{bmatrix} \begin{bmatrix} w_1(k) \\ w_2(k) \end{bmatrix} + \begin{bmatrix} 1 \\ 0 \end{bmatrix} u(k)$$

$$y(k) = \begin{bmatrix} T + \tau e^{-T/\tau} - \tau & \tau - \tau e^{-T/\tau} - T e^{-T/\tau} \end{bmatrix} \begin{bmatrix} w_1(k) \\ w_2(k) \end{bmatrix}$$

Other forms are of course possible. We stress that the inputs and outputs, $u(k)$ and $y(k)$ respectively, are fixed by the system definition, as is their relationship. The model's internal variables, $w_1(k)$ and $w_2(k)$, may or may not be samples of physical variables describing the motor.

11.5.3 Combinations of state representations

Many control systems are configured with serial connections of components, as in Fig. 11.3.

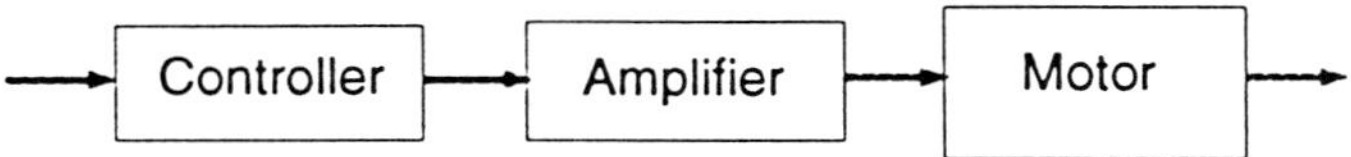

Figure 11.3 Serial (or cascade) connection of components for modelling.

Often the individual components have well-known models and the system is to be modelled while retaining information about the components. This is easily done with state-space notation by suitable concatenation of the vectors and matrices. Hence if

$$y(k) = -\sum_{i=1}^{n_1} a_i y(k-i) + \sum_{i=1}^{m_1} b_i e(k-i)$$

$$w(k) = -\sum_{j=1}^{n_2} c_j w(k-j) + \sum_{j=1}^{m_2} d_j y(k-j)$$

then using the first canonic form of each gives the obvious representation

$$\mathbf{x}(k) = \begin{bmatrix} \mathbf{x}_1(k) \\ \mathbf{x}_2(k) \end{bmatrix}$$

with

$$\mathbf{x}_1(k) = A_1\,\mathbf{x}_1(k-1) + B_1\,e(k-1)$$

$$y(k) = C_1\,\mathbf{x}_1(k)$$

$$\mathbf{x}_2(k) = A_2\,\mathbf{x}_2(k-1) + B_2\,y(k-1)$$

$$w(k) = C_2\,\mathbf{x}_2(k)$$

so that, eliminating $y(k)$,

$$\mathbf{x}(k) = \begin{bmatrix} A_1 & 0 \\ \hdotsfor{2} \\ B_2 C_1 & A_2 \end{bmatrix}\mathbf{x}(k-1) + \begin{bmatrix} B_1 \\ \cdots \\ 0 \end{bmatrix}e(k-1)$$

$$w(k) = [\,0 \quad C_2\,]\,\mathbf{x}(k)$$

Expanded, these matrices yield the following form:

$$\mathbf{x}(k) = \left[\begin{array}{ccccc|ccccc}
-a_1 & -a_2 & \cdots & -a_{n-1} & -a_n & 0 & 0 & \cdots & 0 & 0 \\
1 & 0 & \cdots & 0 & 0 & 0 & 0 & \cdots & 0 & 0 \\
\vdots & \vdots & \vdots & \vdots & \vdots & \vdots & \vdots & \vdots & \vdots & \vdots \\
0 & 0 & \cdots & 1 & 0 & 0 & 0 & \cdots & 0 & 0 \\
\hline
b_1 & b_2 & \cdots & b_{m_1} & b_0 & -c_1 & -c_2 & \cdots & -c_{n-1} & -c_n \\
0 & 0 & \cdots & 0 & 0 & 1 & 0 & \cdots & 0 & 0 \\
\vdots & \vdots & \vdots & \vdots & \vdots & \vdots & \vdots & \vdots & \vdots & \vdots \\
0 & 0 & \cdots & 0 & 0 & 0 & 0 & \cdots & 1 & 0
\end{array}\right]\mathbf{x}(k-1) + \begin{bmatrix} 1 \\ 0 \\ \vdots \\ 0 \\ 0 \\ 0 \\ \vdots \\ 0 \end{bmatrix}e(k-1)$$

$$w(k) = \begin{bmatrix} 0 & 0 & \cdots & d_1 & d_2 & \cdots & d_{m_2} & \cdots & 0 \end{bmatrix}\mathbf{x}(k)$$

A special case arises when the input is more delayed than the output in the difference equation, i.e. when in

$$y(k) = -\sum_{i=1}^{n} a_i y(k-i) + \sum_{i=0}^{L} b_i u(k-i)$$

we have $L > n$. This case has some analogies to having higher order derivatives on the input than on the output in a differential equation, and for that reason modellers often seek one of the representations we have already seen. In fact, it is easy in a computer to store the back values $\{u(k-1), u(k-2),..., u(k-L)\}$, so it is sometimes convenient to have a state representation for this. This is easily done if we define

$$w_1(k) = u(k-1) \quad w_2(k) = u(k-2) \ \cdots \ w_L(k) = u(k-L)$$

i.e. $w_i(k) = w_{i-1}$ $(i=2,3,...,L)$. Using matrix notation, this is

$$w(k+1) = \begin{bmatrix} 0 & 0 & 0 & \cdots & 0 \\ 1 & 0 & 0 & & 0 \\ 0 & 1 & 0 & \vdots & 0 \\ \vdots & \vdots & \vdots & 0 & \vdots \\ 0 & 0 & 0 & 1 & 0 \end{bmatrix} w(k) + \begin{bmatrix} 1 \\ 0 \\ 0 \\ \vdots \\ 0 \\ 0 \end{bmatrix} u(k) \qquad (11.18)$$

Shown below is one of possible 'fixes'. Starting with a variation on the form (11.16), but using (11.18) as part of the driving function, we can work with the concatenated vectors

$$\mathbf{z}(k) = \begin{bmatrix} \mathbf{w}(k) \\ \mathbf{x}(k) \end{bmatrix}$$

and the definitions

$$
\mathbf{A} = \begin{bmatrix}
0 & 1 & 0 & 0 & \cdots & & 0 \\
0 & 0 & 1 & 0 & & & \\
0 & 0 & 0 & 1 & 0 & & \\
\vdots & \vdots & \vdots & \vdots & & \ddots & \vdots \\
0 & 0 & 0 & 0 & & & 1 \\
-a_n & -a_{n-1} & -a_{n-2} & & \cdots & -a_2 & -a_1
\end{bmatrix}
$$

$$
\mathbf{B} = \begin{bmatrix} b_0 & b_1 & \cdots & b_{L-1} & b_L \end{bmatrix}
$$

$$
\mathbf{W} = \begin{bmatrix}
0 & 0 & 0 & \cdots & & 0 \\
1 & 0 & 0 & \cdots & & 0 \\
0 & 1 & 0 & & & 0 \\
& & & \cdots & \vdots & \vdots \\
\vdots & \vdots & \vdots & & 0 & 0 \\
0 & 0 & 0 & \cdots & 1 & 0
\end{bmatrix}
$$

to obtain the representation

$$
\mathbf{z}(k) = \left[\begin{array}{c|c}
\mathbf{W} & 0 \\ \hline
0 & \\ \hline
B & A
\end{array}\right] \mathbf{z}(k-1) + \begin{bmatrix} 1 \\ 0 \\ 0 \\ \vdots \\ 0 \end{bmatrix} u(k-1)
$$

$$
y(k) = \begin{bmatrix} 0 & 0 & 0 & \cdots & 0 & | & 0 & 0 & 0 & \cdots & 1 \end{bmatrix} \mathbf{z}(k) + b_0 u(k)
$$

The dimension of the state is $n+L$. Even if $L \le n$, we know that we can use a state of dimension n – as in all of our representations so far. This wastes resources (space in computer memory and execution time) and explains the search for more efficient representations as found in the literature.

Representation of multivariable systems follows immediately from the above. For example, one representation of an $n \times m$ matrix transfer function $\mathbf{G}(z)$ assumes $\mathbf{G}_i(z)$ denotes the ith column so that if

$U_i(z)$ is the scalar transform of the scalar control variable $\{u_i(nT)\}$, we have

$$Y(z) = \sum_{i=1}^{m} G_i(z) U_i(z)$$

Now we represent $G_i(z)$ in terms of an n-vector of polynomials $N_i(z)$ and a scalar polynomial $d_i(z)$, where $d_i(z)$ is the least common denominator of the m denominators in $G_i(z)$, so that

$$G_i(z) = \frac{N_i(z)}{d_i(z)}$$

and the denominator has the form

$$d_i(z) = z^{n_i} + \sum_{j=0}^{n_i-1} d_{i,j} z^j$$

$$N_i(z) = \begin{bmatrix} v_{i,10} + v_{i,11}\, z + \cdots + v_{i,1n_i-1}\, z^{n_i-1} \\ \vdots \quad \vdots \quad \vdots \quad \vdots \\ v_{i,n0} + v_{i,n1}\, z + \cdots + v_{i,nn_i-1}\, z^{n_i-1} \end{bmatrix}$$

Letting the matrices A_i, B_i, C_i be defined by

$$A_i = \begin{bmatrix} 0 & 1 & 0 & 0 & \cdots & 0 \\ 0 & 0 & 1 & 0 & \cdots & 0 \\ \vdots & \vdots & \vdots & \vdots & & \vdots \\ -d_{i,0} & -d_{i,1} & -d_{i,2} & \cdots & \cdots & -d_{i,n_i-1} \end{bmatrix}$$

$$B_i = \begin{bmatrix} 0 \\ 0 \\ \vdots \\ 0 \\ 1 \end{bmatrix} \qquad C_i = \begin{bmatrix} v_{i,10} & v_{i,11} & \cdots & v_{i,1n_i-1} \\ \vdots & \vdots & \vdots & \vdots \\ v_{i,n0} & v_{i,n1} & \cdots & v_{i,nn_i-1} \end{bmatrix}$$

we have that one state-space representation of the system is given by

$$
\mathbf{A} = \begin{bmatrix} \mathbf{A}_1 & 0 & 0 & \cdots & 0 \\ 0 & \mathbf{A}_2 & 0 & \cdots & 0 \\ \vdots & \vdots & & & \vdots \\ 0 & 0 & 0 & \cdots & \mathbf{A}_m \end{bmatrix} \qquad \mathbf{B} = \begin{bmatrix} \mathbf{B}_1 & 0 & 0 & \cdots & 0 \\ 0 & \mathbf{B}_2 & 0 & \cdots & 0 \\ \vdots & \vdots & & & \vdots \\ 0 & 0 & 0 & \cdots & \mathbf{B}_m \end{bmatrix}
$$

$$
\mathbf{C} = [\, \mathbf{C}_1 \;\; \mathbf{C}_2 \;\; \cdots \;\; \mathbf{C}_n \,]
$$

In this, if $M = \sum_{i=1}^{m} n_i$, then $\mathbf{A}$ is an $M \times M$ matrix, $\mathbf{B}$ is $M \times m$, and $\mathbf{C}$ is $n \times M$. This may not be the best possible representation, depending upon the application.

11.5.4 Applications

The use of matrices lends a certain elegance to the representation, but it has more important ramifications than simply easing the notation problem since solution is, in principle, straightforward. Observe that for a system with $\mathbf{x}(0) = \mathbf{x}_0$ given and

$$
\mathbf{x}(k) = \mathbf{A}\mathbf{x}(k-1) + \mathbf{B}\mathbf{u}(k-1)
$$

$$
\mathbf{y}(k) = \mathbf{C}\mathbf{x}(k)
$$

the solution is

$$
\mathbf{x}(k) = \mathbf{A}^k \mathbf{x}_0 + \sum_{i=0}^{k-1} \mathbf{A}^i \mathbf{B}\mathbf{u}(k-i-1) \tag{11.19}
$$

with $\mathbf{y}(k)$ following immediately. In fact, this generalizes quite nicely and if matrix methods are used the computations can be simplified.

For the time-varying representation

$$\mathbf{x}(k+1) = \mathbf{A}(k)\mathbf{x}(k) + \mathbf{B}(k)\mathbf{u}(k)$$

$$\mathbf{y}(k) = \mathbf{C}(k)\mathbf{x}(k) + \mathbf{D}(k)\mathbf{u}(k)$$

it is convenient to define the transition matrix $\theta(k,j)$ by

$$\theta(k,j) = \mathbf{A}(k-1)\theta(k-1,j) \qquad \theta(j,j) = 0 \qquad k > j$$

and write the solution as

$$\mathbf{x}(k) = \theta(k,j)\mathbf{x}(j) + \sum_{i=j}^{k} \Theta(k,i+1)\mathbf{B}(i)\mathbf{u}(i)$$

When $\mathbf{A}$ and $\mathbf{B}$ are constants, then

$$\theta(k,j) = \mathbf{A}^{k-j}$$

and solutions as in (11.19) result.

Note that this discrete time transition matrix has the limited transition property $\theta(k,j) = \theta(k,i)\theta(i,j)$, $k \geq i \geq j$, but would not have other transition matrix properties (see section 10.4.6) unless the $\mathbf{A}(k)$ matrices are invertible (which they usually are if $\mathbf{A}$ models a physical continuous time system).

11.5.5 Change of variable and similarity transformations

The choice of state variables must not affect the input–output relationships of the system – otherwise our designs would be representation dependent – and it is convenient to choose states which are 'meaningful' to us.

Consider

$$\mathbf{x}(k) = \mathbf{A}\mathbf{x}(k-1) + \mathbf{B}\mathbf{u}(k-1)$$

$$\mathbf{y}(k) = \mathbf{C}\mathbf{x}(k)$$

with $\mathbf{x}(0) = \mathbf{x}_0$ given, and try the similarity transform $\mathbf{w}(k) = \mathbf{P}\mathbf{x}(k)$.

The system description then is

$$w(k+1) = \mathbf{PAP^{-1}} w(k) + \mathbf{PB} u(k)$$

$$y(k) = \mathbf{CP^{-1}} w(k)$$

For any k, we can solve to find

$$w(k) = (\mathbf{PAP^{-1}})^k w(0) + \sum_{i=1}^{k} (\mathbf{PAP^{-1}})^{i-1} \mathbf{PB} u(k-i)$$

which with $w(0) = \mathbf{P}x(0)$ gives

$$w(k) = \mathbf{PA}^k x(0) + \mathbf{P} \sum_{i=1}^{k} \mathbf{A}^{i-1} \mathbf{B} u(k-i)$$

$$= \mathbf{P}x(k)$$

and gives output

$$y(k) = \mathbf{CP^{-1}} w(k) = \mathbf{CP^{-1}} \mathbf{P} x(k) = \mathbf{C}x(k)$$

Thus y has the same dependence on $\mathbf{u}$ as in the original representation.

11.5.6 The standard canonical forms

One important implication of the above is that we may use canonical forms for the system representation when this is convenient, as by doing so we are not affecting the system behaviour. The important canonical forms are the controllable or phase variable form (11.16), the observable form (11.17), and the Jordan form (analogous to 10.6) which often becomes a diagonal form. These may be obtained in general using the methods of Appendix B, which are in fact matrix transformations independent of the model involved.

11.5.7 The z-transform of state equations

The z-transform of state equations follows simply enough in principle from the operations of the scalar z on vectors and matrices, i.e. the fact (definition) that

$$z\,\mathbf{A} = z\,\{a_{ij}\} = \{za_{ij}\}$$

Using this fact and the shift property on the state equations

$$\mathbf{x}(k+1) = \mathbf{A}\mathbf{x}(k) + \mathbf{B}u(k)$$
$$\mathbf{y}(k) = \mathbf{C}\mathbf{x}(k) + \mathbf{D}u(k)$$

yields

$$z\,\mathbf{X}(z) - z\,\mathbf{x}(0) = \mathbf{A}\mathbf{X}(z) + \mathbf{B}\mathbf{U}(z)$$
$$\mathbf{Y}(z) = \mathbf{C}\mathbf{X}(z) + \mathbf{D}\mathbf{U}(z)$$

which rearranges to give a solution

$$\mathbf{X}(z) = [z\mathbf{I} - \mathbf{A}]^{-1}\ [z\,\mathbf{x}(0) + \mathbf{B}\mathbf{U}(z)]$$
$$\mathbf{Y}(z) = \{\,\mathbf{C}\,[z\mathbf{I}-\mathbf{A}]^{-1}\mathbf{B} + \mathbf{D}\,\}\,\mathbf{U}(z) + z\,\mathbf{C}\,[z\mathbf{I}-\mathbf{A}]^{-1}\mathbf{x}(0)$$

This can give an alternative and useful demonstration of the similarity transform argument of section 11.5.5. We note that the original system has z-transform relationship

$$\mathbf{Y}(z) = \mathbf{C}(z\mathbf{I}-\mathbf{A})^{-1}\mathbf{B}\mathbf{U}(z) + z\mathbf{C}\,(z\mathbf{I}-\mathbf{A})^{-1}\mathbf{x}_0$$

while the changed coordinate system has

$$\mathbf{Y}(z) = \mathbf{C}\mathbf{P}^{-1}(z\mathbf{I}-\mathbf{P}\mathbf{A}\mathbf{P}^{-1})^{-1}\mathbf{P}\mathbf{B}\mathbf{U}(z)$$
$$+ z\mathbf{C}\mathbf{P}^{-1}(z\mathbf{I}-\mathbf{P}\mathbf{A}\mathbf{P}^{-1})^{-1}\mathbf{P}\mathbf{x}_0$$

$$= \mathbf{CP}^{-1}(\mathbf{P}(z\mathbf{I}-\mathbf{A})\mathbf{P}^{-1})^{-1}\mathbf{PBU}(z)$$

$$+ z\mathbf{CP}^{-1}(\mathbf{P}(z\mathbf{I}-\mathbf{A})\mathbf{P}^{-1})^{-1}\mathbf{Px}_0$$

$$= \mathbf{C}(z\mathbf{I}-\mathbf{A})^{-1}\mathbf{BU}(z) + z\mathbf{C}(z\mathbf{I}-\mathbf{A})^{-1}\mathbf{x}_0$$

11.5.8 Transfer function to state–space conversions

We have already seen many examples of transfer function to state–space transformations (section 11.5.2) and the multivariable case (section 11.5.3). The coefficients from transfer functions can easily be read into the observable and controllable canonical forms, so algorithmic conversion is possible.

11.6 REPRESENTING NOISE, DISTURBANCES, AND ERRORS

11.6.1 Modelling of noise and disturbances

For discrete time systems, as with continuous time systems, the models of noise can be in either state-space or transfer functions, but will follow the notion (see section 9.4.6) of white noise driving a linear system and parallels the discussion of section 10.5. The mathematics are somewhat easier for discrete time systems because the noise is taken as a sequence of random numbers. In such cases we have for scalars

$$H(z) = \frac{\displaystyle\sum_{i=0}^{m} \beta_i z^{-i}}{1 + \displaystyle\sum_{i=0}^{n-1} \alpha_i z^{-1}}$$

$$v(k) = -\sum_{i=1}^{n} \alpha_i \upsilon(k-i) + \sum_{j=0}^{m} \beta_j \eta(k-j)$$

with input noise uncorrelated

$$\mathscr{E}[\eta(k)\eta(j)] = R\,\delta(k-j)$$

$$S_\eta(e^{j\omega T}) = R$$

$$S_v(e^{j\omega T}) = |H(e^{j\omega T})|^2\,R$$

where $H(z)$ is the transfer function of the linear system and R is the level (variance) of the white noise. One possible state-space model is

$$\mathbf{x}(k+1) = \begin{bmatrix} 0 & 1 & 0 & \cdots & 0 \\ 0 & 0 & 1 & 0 & 0 \\ \vdots & \vdots & \vdots & \vdots & \vdots \\ -\alpha_0 & -\alpha_1 & & \cdots & -\alpha_{n-1} \end{bmatrix} \mathbf{x}(k) + \begin{bmatrix} 0 \\ 0 \\ \vdots \\ 1 \end{bmatrix} \eta(k)$$

$$v(k) = [\,0 \quad \cdots \quad 0 \quad \beta_m \quad \cdots \quad \beta_1\,]\,\mathbf{x}(k)$$

and the same result is obtained.

A typical representation of a feedback system with measurement noise and output disturbances was shown in Fig. 10.6, and the discrete-time case in Fig. 11.4 is the exact analog, as are the definitions of the transfer functions and variable transforms.

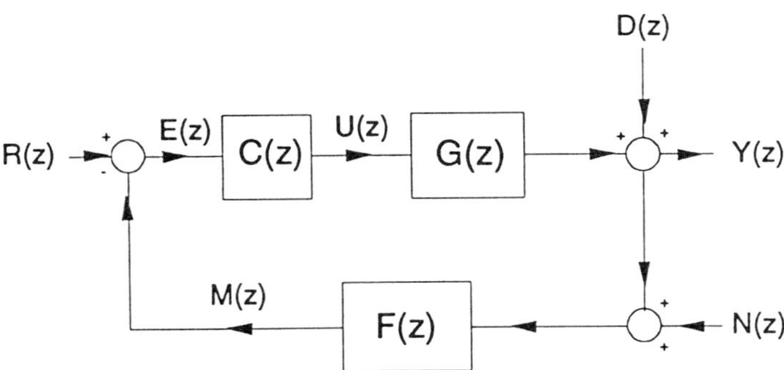

Figure 11.4 Typical z-transform system representation. See Fig.10.6.

Thus $\mathbf{G}(z)$ represents the plant transfer function, $\mathbf{C}(z)$ is the controller transfer function, $\mathbf{F}(z)$ represents the measurement device, and $\mathbf{Y}(z)$, $\mathbf{R}(z)$, $\mathbf{D}(z)$, $\mathbf{N}(z)$, $\mathbf{E}(z)$, $\mathbf{U}(z)$, $\mathbf{M}(z)$ are the respective transforms of the output signal, the input reference signal, the disturbance, the (measurement) noise, the error signal, the command signal and the measurement signal. Then transform algebra yields easily

$$\mathbf{E}(z) = \mathbf{R}(z) - \mathbf{F}(z)(\mathbf{N}(z) + \mathbf{Y}(z))$$

$$\mathbf{Y}(z) = \mathbf{D}(z) + \mathbf{G}(z)\mathbf{U}(z)$$

$$\mathbf{U}(z) = \mathbf{C}(z)\mathbf{E}(z)$$

and hence relationships such as

$$\mathbf{E}(z) = (\mathbf{I} + \mathbf{F}(z)\mathbf{G}(z)\mathbf{C}(z))^{-1}(\mathbf{R}(z) - \mathbf{F}(z)\mathbf{N}(z) - \mathbf{F}(z)\mathbf{D}(z))$$

$$\mathbf{Y}(z) = (\mathbf{I} + \mathbf{G}(z)\mathbf{C}(z)\mathbf{F}(z))^{-1}\mathbf{G}(z)\mathbf{C}(z)\mathbf{R}(z)$$

$$+ (\mathbf{I} + \mathbf{G}(z)\mathbf{C}(z)\mathbf{F}(z))^{-1}\mathbf{D}(z)$$

$$- (\mathbf{I} + \mathbf{G}(z)\mathbf{C}(z)\mathbf{F}(z))^{-1}\mathbf{G}(z)\mathbf{C}(z)\mathbf{F}(z)\mathbf{N}(z)$$

where the possibility of MIMO transfer functions has been allowed.

The disturbances may be modelled as noise (as above) or as signals such as steps or ramps, e.g.

$$\mathbf{D}(z) = \frac{\varepsilon}{1 - z^{-1}} \qquad \text{or} \qquad \mathbf{D}(z)\frac{\delta T z^{-1}}{(1 - z^{-1})^2}$$

Alternatively, $\mathbf{D}(z)$ is in robust control theory defined along the lines of $\mathbf{D}(z) \in \mathscr{D}$

$$\mathscr{D} = \left\{ \mathbf{D}(z) : \|\mathbf{D}(e^{j\omega T})\|_p < 1 \right\}$$

where p typically is either 2 or ∞ as in Appendix B.

Measurement noise $\mathbf{N}(z)$ is ordinarily modelled as noise, but occasionally more like a disturbance in order to represent severe measurement device breakdown.

State-space studies usually consider only the noise form explicitly, with the linear system model often such as

$$\mathbf{x}(k+1) = \mathbf{A}\mathbf{x}(k) + \mathbf{B}\mathbf{u}(k) + \mathbf{\Gamma}\mathbf{v}(k)$$

$$\mathbf{y}(k) = \mathbf{C}\mathbf{x}(k) + \mathbf{w}(k)$$

with input $\mathbf{u}$, output $\mathbf{y}$, state $\mathbf{x}$, and noises $\mathbf{v}$ and $\mathbf{w}$ with

$$\mathscr{E}[\mathbf{v}] = \mathbf{0} \qquad \mathscr{E}[\mathbf{v}(k)\,\mathbf{v}^{\mathrm{T}}(j)] = \mathbf{Q}\delta(k-j)$$

$$\mathscr{E}[\mathbf{w}] = \mathbf{0} \qquad \mathscr{E}[\mathbf{w}(k)\,\mathbf{w}^{\mathrm{T}}(j)] = \mathbf{R}\delta(k-j)$$

$$\mathscr{E}[\mathbf{x}(0)] = \mathbf{x}_{o} \qquad \mathscr{E}[(\mathbf{x}(0) - \mathbf{x}_{o})\,(\mathbf{x}(0) - \mathbf{x}_{o})^{\mathrm{T}}] = \mathbf{S}$$

As in the continuous time case, $\mathbf{w}$ is intended to represent measurement noise, but the structure is such that if the model is extended so that part of the state vector is used for 'coloration' of the input noise $\mathbf{v}$, then random model errors and disturbances can be represented through $\mathbf{v}$.

11.6.2 Model errors

Model errors are usually taken in terms of either additive or multiplicative errors in matrices in the representations, and the results are just like those of section 10.5 except that the z-transform appears in place of the Laplace transform and the state-space difference equations appear in place of the state differential equations.

11.7 COMPUTER PROGRAMS

Virtually the same programs used in Chapter 10 can be used for many of the calculations for sampled data system analysis.

11.8 FURTHER INFORMATION

Although almost all of this chapter could be considered analogous to Chapter 10 with a change of polynomial variable, we have somewhat belaboured the point because of some students' lack of familiarity with discrete time models, difference equations, and *z*-transforms. Most of the information in this chapter is presented in standard textbooks. For examples, longer discussions are found in books such as Franklin, Powell, and Workman (1990), Kuo (1980), and Ogata (1987). Noise modelling is in various engineering communications and random processes texts, such as Papoulis (1977), and multivariable representations are in Maciejowski (1989).

12

Conversions of continuous time to discrete time models

To a computer, a plant looks like a discrete time system even though usually it is well defined for continuous time. In addition, the computer issues its commands at discrete times even if the original control law design was based on differential equations. For these reasons, it is necessary to be able to convert continuous time representations to equivalent discrete time representations.

12.1 SYNOPSIS

There are two ways to approach discrete time control:

- find a continuous time controller and approximate it, or
- design a sampled data controller using a discrete time model of the plant.

The conditions under which the approximation method applies are indicated in Fig. 12.1.

If a control law (e.g. a PID law) has already been designed then it may be satisfactory to convert it for use in the digital computer. In the typical situation, a differential equation or its transfer function is known for the law and the digital computer expression is to approximate it. For example, if the control law has transfer function $C_a(s)$, the sampled data transfer function $C_d(z)$ is formed in one of the following ways.

1. Substitution:

$$C_d(z) = C_a(s)\,|_{s=f(z)} = C_a(f(z))$$

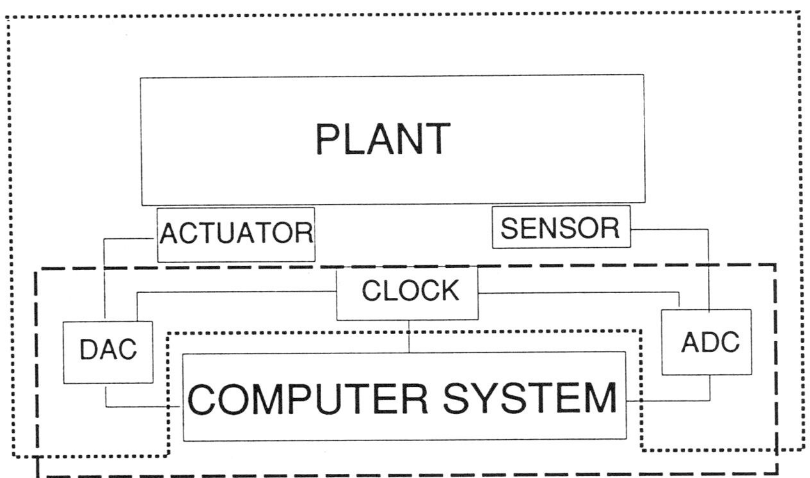

Figure 12.1 Simple system configuration. Digital aspects can be considered by either taking the digital equivalent of the plant system (dotted outline) or by deriving the computer system (dashed outline) from a continuous time controller.

where the choices for $f(z)$ must be rational in z and common choices include

(a) $\qquad f(z) = \dfrac{z - 1}{Tz}$

(b) $\qquad f(z) = \dfrac{2}{T} \dfrac{z - 1}{z + 1}$

2. Pole-zero mapping: If $C_a(s)$ has poles at b_i, $i = 1, 2, ..., n$ and zeros at a_j, $j = 1, 2, ..., m$, then $C_d(z)$ is taken as rational with poles at $e^{b_i T}$, $i = 1, 2, ..., n$ and zeros at $e^{a_j T}$, $j = 1, 2, ..., m$. Gain adjustment may be necessary.

3. $C_d(z)$ may be chosen so that the controller's output is invariant at the command times kT when the input is a chosen function, typically an impulse or a step.

When the plant is to be modelled such that only outputs at discrete times are necessary and the inputs are to be piecewise constants, then although the methods above are occasionally used, it would seem proper that for linear time invariant systems the choice be restricted. When the original model is a transfer function $H_a(s)$, the most appropriate model is the impulse invariant transformation

$$H_d(z) = \frac{z-1}{z} \mathscr{Z}_{eq}\left(\frac{H_a(s)}{s}\right)$$

When the continuous time model is a state-space model such as

$$\dot{x}(t) = Ax(t) + Bu(t)$$

$$y(t) = Cx(t)$$

a straightforward argument gives the sampled data version

$$x(kT+T) = e^{AT}x(kT) + \left[\int_0^T e^{A(T-\tau)}\,d\tau\right]Bu(kT)$$

$$y(kT) = Cx(kT)$$

12.2 CONVERSIONS OF CONTINUOUS TIME CONTROL LAWS

Often a plant has been operating quite satisfactorily with analog controllers but replacement with digital forms is desired to allow logic decisions, communications with central computers, unit conversions for data logging, etc. In these circumstances, the engineer has to replace the analog control element with a digital element as in Fig. 12.2, and there are several ways of achieving this.

PID controllers can be converted using time domain approximations, for example,

$$u(t) = K_p e(t) + K_i \int_{-\infty}^{t} e(\tau)\,d\tau + K_d \frac{de(t)}{dt}$$

becomes

$$u_{\text{d}}(kT) = K_{\text{p}}\,e(kT) + K_{\text{i}}T \sum_{j=-\infty}^{k} e(jT) + \frac{K_{\text{d}}}{T}\,(e(kT)-e(kT-T))$$

$$u(t) = u_{\text{d}}(kT) \qquad kT \le t < kT+T \qquad\qquad (12.1)$$

where the actual calculation may be rearranged for convenience. The second equation of (12.1) is actually implemented by the hardware of the buffer and DAC, i.e. the zero-order hold (ZOH).

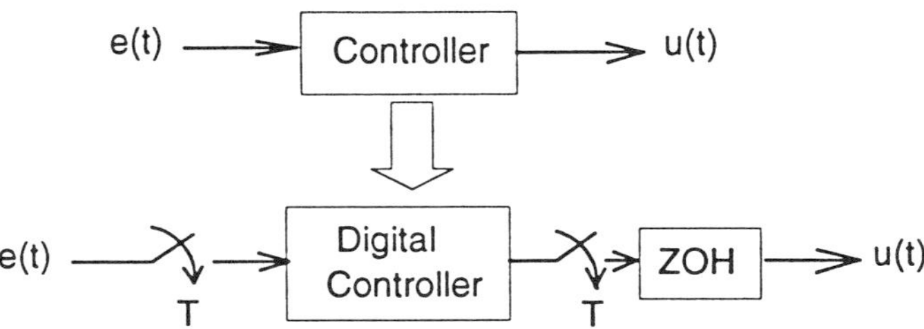

Figure 12.2 Desired replacement of an analog controller by a digital controller with ZOH.

Alternative methods are well known in the digital signal processing literature and are usually discussed in relation to the conversion of filter designs; the problem is inherently different from that of modelling the controlled system, for which we use impulse invariant designs below. There we need a good model for a physical situation, with the model applying only at sample instants. Here we are looking for useful approximations for one type of algorithm to be replaced by another. Let us review some of the alternatives.

The problem is essentially one of converting a continuous-time control law characterized by a transfer function

$$G_{\text{a}}(s) = \frac{b_m s^m + b_{m-1} s^{m-1} + \cdots + b_0}{a_n s^n + a_{n-1} s^{n-1} + \cdots + a_0}$$

where usually $n \ge m$ and $a_n \neq 0$. The object is to find a digital computer control law, characterized by its transfer function

$$G_d(z) = \frac{d_0 z^n + d_1 z^{n-1} + \cdots + d_n}{z^n + c_1 z^{n-1} + \cdots + c_n}$$

which in some way approximates $G_a(s)$. The methods break into three classes.

1. Often a substitution of $f(z)$ for s is made and a heuristic argument is used for the conversion.
2. Coefficients are found such that for a given input $r(t)$, the two systems have approximately the same outputs at the sample instants. Hence, if

$$u(t) = \mathcal{L}^{-1}\left[G_a(s)R(s)\right]$$

with $R(s) = \mathcal{L}[r(t)]$, and

$$u_d(kT) = \mathcal{Z}^{-1}\left[G_d(z)R(z)\right]$$

with $R(z) = \mathcal{Z}\{r(kT)\}$, then it is desired that

$$u_d(kT) \approx u(kT)$$

3. Coefficients d_i and c_j are found such that the frequency responses of the two systems are approximately the same, i.e. if the input signal is sinusoidal in both the continuous- and discrete-time cases, then the outputs should also be sinusoidal and should be similar in both magnitude and phase at the sample instants. This translates to a desire that

$$G_d(e^{j\omega T}) \approx G_a(j\omega)$$

We consider some of the possibilities below.

12.2.1 Substitutions and their heuristics

The heuristic method is often quite appealing and will work with fast enough sampling on some systems. A typical argument observes that, when the sampling period is T,

$$\frac{dx(t)}{dt}\bigg|_{t=nT} \approx \frac{x(kT) - x(kT-T)}{T}$$

and also that

$$\int_{-\infty}^{kT} x(\tau)\,d\tau \approx \int_{-\infty}^{kT-T} x(\tau)\,d\tau + T\,x(kT-T)$$

or alternatively, if $I(kT)$ denotes the integral at time sample kT

$$I(kT) = I(kT-T) + Tx(kT)$$

Furthermore, because the Laplace transform operator s corresponds to differentiation (and s^{-1} to integration), whereas the z-transform operator corresponds to a forward shift, the above approximations lead to

$$s = \frac{1 - z^{-1}}{T} \tag{12.2}$$

as a reasonable substitution for conversion purposes.

Another heuristic argument notes that the original transfer function has poles at p_i, $i=1,2,\ldots,n$ and zeros at q_j, $j=1,2,\ldots,m$. One mapping from the stable area of the Laplace transform (the left-half plane) to the stable area of the z-transform (within the unit circle) is

$$z = e^{sT} \tag{12.3}$$

Using this to map the poles p_i and zeros q_j of the continuous time transfer function to corresponding poles and zeros of the sampled data transfer function, gives

$$G_{\mathrm{d}}(z) = K\frac{(1 - z^{-1}e^{q_1 T})\,(1 - z^{-1}e^{q_2 T})\,\cdots\,(1 - z^{-1}e^{q_m T})}{(1 - z^{-1}e^{p_1 T})(1 - z^{-1}e^{p_2 T})\,\cdots\,(1 - z^{-1}e^{p_n T})}$$

where usually K is selected such that $G_{\mathrm{d}}(1) = G_{\mathrm{a}}(0)$. An alternate version of this has

$$G_{\mathrm{d}}(z) = K\frac{(z - e^{q_1 T})\,(z - e^{q_2 T})\,\cdots\,(z - e^{q_m T})\,N(z)}{(z - e^{p_1 T})\,(z - e^{p_2 T})\,\cdots\,(z - e^{p_n T})}$$

where $N(z) = (z + 1)^{m_1}(z)^{m_2}$ and $m_1 + m_2 = n - m \geq 0$ are terms to represent poles at ∞. This pole-zero mapping technique has had variable success.

A third substitution can be derived in several ways, but results in the bilinear or **Tustin transformation**. Notice that (12.3) also gives the form

$$sT = \ln z$$

Series expansion yields

$$\ln z = 2\left[\frac{z-1}{z+1} + \frac{1}{3}\left(\frac{z-1}{z+1}\right)^3 + \frac{1}{5}\left(\frac{z-1}{z+1}\right)^5 + \cdots\right] \tag{12.4}$$

Hence truncation gives the approximation

$$s \approx \frac{2}{T}\frac{z-1}{z+1} = \frac{2}{T}\frac{1-z^{-1}}{1+z^{-1}} \tag{12.5}$$

which is used for the substitution. We remark that this substitution also maps the left-hand real plane of s into the unit circle in the z plane. Alternatively, it can be derived as a trapezoidal integration approximation to $1/s$ interpreted as an integration operator.

12.2.2 Invariant transformations

Invariant transformations are those in which a selected continuous time function, when sampled, gives the same samples as a discrete time system. For example, it might be desired that a discrete time system have the same step response as the step response of the continuous time system to which it is an approximation. In general, we select $G_d(z)$ as 'equivalent' to $G_a(s)$ by forcing, for reference signal $r(t)$ with Laplace transform $R(s)$ and z transform of its samples $R(z)$, that

$$G_d(z)\,R(z) = \mathscr{Z}\,[\text{samples of}\,\mathscr{L}^{-1}(G_a(s)\,R(s))]$$

$$\equiv \mathscr{Z}_{eq}\,[G_a(s)\,R(s)]$$

Where $\mathscr{Z}_{eq}$ is a notation for the transformation indicated (Appendix A). Notice that different inputs may lead to different system approximations.

The invariant response criterion is necessary for the study of control systems design. Occasionally impulse invariance is desired for control law conversion, so that $R(s) = 1$ and $R(z) = 1$. In this circumstance, we have $G_d(z)$ selected so that

$$G_d(z) = \mathscr{Z}_{eq}\,(G_a(s)) \tag{12.6}$$

Among the alternatives is step invariance, in which

$$G_d(z)\,\frac{z}{z-1} = \mathscr{Z}_{eq}\left[\frac{G_a(s)}{s}\right]$$

$$G_d(z) = \frac{z-1}{z}\,\mathscr{Z}_{eq}\left[\frac{G_a(s)}{s}\right]$$

This same result is given if the square pulse input

$$u(t) = \begin{cases} 1 & 0 \le t < T \\ 0 & \text{otherwise} \end{cases}$$

is used. This represents the ZOH model of computer output with buffer and DAC (see Chapter 9) and yields $R(z) = 1$ and $R(s) = (1-e^{-Ts})/s$ so that

$$G_d(z) = \mathscr{Z}_{eq}\left[(1 - e^{-Ts})\frac{G_a(s)}{s}\right]$$

$$= \frac{1-z}{z}\,\mathscr{Z}_{eq}\left[\frac{G_a(s)}{s}\right]$$

12.2.3 Frequency response

Sometimes control law designs are made using frequency responses, so we can use this as a criterion for the conversion. Heuristic methods seldom come with guarantees about the response. In the impulse invariant case of (12.6) signal processing texts (such as Rabiner and Gold, 1975) show that

$$T\, G_{\mathrm{d}}(e^{j\omega T}) = \sum_{k=-\infty}^{\infty} G_{\mathrm{a}}\left(j\omega + jk\frac{2\pi}{T}\right)$$

so that for T sufficiently small and $-\pi < \omega T < \pi$,

$$T\, G_{\mathrm{d}}(e^{j\omega T}) \approx G_{\mathrm{a}}(j\omega)$$

(By T 'sufficiently' small, we need $G_{\mathrm{a}}(j\omega + jk(2\pi/T)) \approx 0$ for $k \neq 0$.) The bilinear (or Tustin) transformation (12.5) yields

$$G_{\mathrm{d}}(z) = G_{\mathrm{a}}\left(\frac{2}{T}\frac{1-z^{-1}}{1+z^{-1}}\right)$$

Direct substitution shows that the frequency response is

$$G_{\mathrm{d}}(e^{j\omega T}) = G_{\mathrm{a}}\left(j\frac{2}{T}\tan\left(\frac{\omega T}{2}\right)\right)$$

so that for small ωT

$$G_{\mathrm{d}}(e^{j\omega T}) \approx G_{\mathrm{a}}(j\omega)$$

as desired. No guarantees about the pulse response are given by this approach, but the frequency response is quite good for rapid sampling.

Example

To show the effects of the three techniques, a typical analog transfer function is converted: the lag or lead network form

$$G(s) = \frac{\dfrac{s}{a} + 1}{\dfrac{s}{b} + 1}$$

converts as

$$G_1(z) = \frac{b}{a} \frac{1 + Ta - z^{-1}}{1 + Tb - z^{-1}}$$

using the heuristic substitution (12.2). Using the bilinear substitution (12.4) yields

$$G_2(z) = \frac{b}{a} \frac{(2 + aT) - z^{-1}(2 - aT)}{(2 + bT) - z^{-1}(2 - bT)}$$

while making the transfer function step-response/square pulse invariant gives

$$G_3(z) = \frac{\dfrac{b}{a} - z^{-1}\left[e^{-bT} - 1 + \dfrac{b}{a}\right]}{1 - e^{-bT}z^{-1}}$$

The results of these substitutions are compared, for differing values of sample period T, in Fig. 12.3. Numerical values used were $a = 50$ and $b = 10$, and sampling periods T were 0.1 s and 0.01 s.

It is interesting to remark that all of the step responses appear reasonably close to that of the continuous time system, although only $G_3(z)$ is exactly correct at the sample instants. On the other hand, a lag network is supposedly designed for its frequency response properties, and here the $T = 0.01$ case, particularly the bilinear transformation, appears close to the original gain while the $T = 0.1$ case is noticeably distorted.

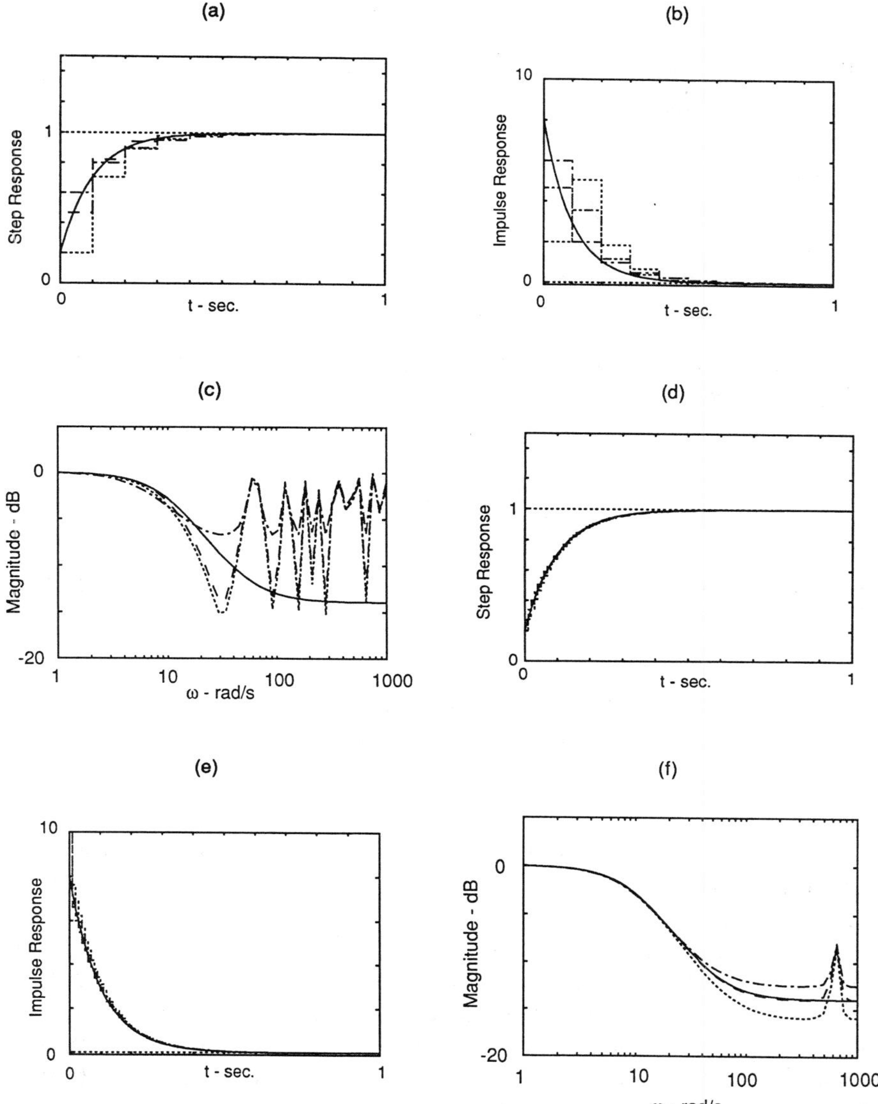

Figure 12.3 Examples of responses from various controller approximations. In all figures, G_1 is $-\cdot$, G_2 is $--$, G_3 is $\cdots$, and the original analog element is the solid line. (a) – (c) are respectively step, impulse, and frequency responses for sample period $T = 0.1$ s; (d) – (f) are the same for $T = 0.01$ s.

12.3 RELATION OF DISCRETE TO CONTINUOUS TRANSFER FUNCTIONS FOR SAMPLING OF CONTINUOUS TIME SYSTEMS

The problem with the above methods is that, while satisfactory for conversions of existing laws, they are inexact in various ways as representations of the computer's view of the plant. The hardware in a simple feedback system (Fig. 12.4), in effect, has samplers placed as in Fig. 12.5.

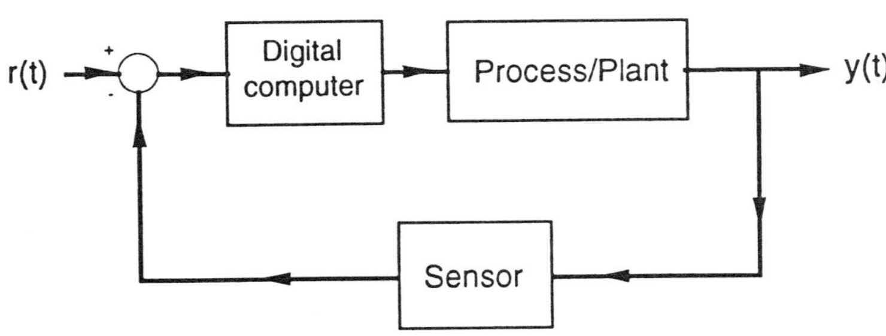

Figure 12.4 Simple feedback controller with analog signal differencing.

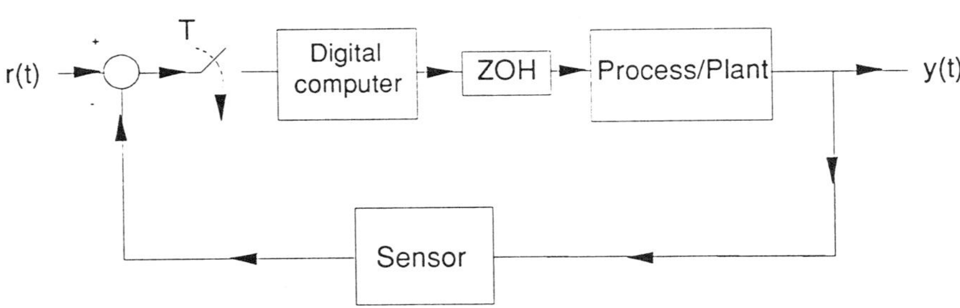

Figure 12.5 Elaboration of Fig. 12.4 with sampler and output ZOH.

The sampling of the error signal at the computer's input is obvious. The output sampler and ZOH are a model of the output buffer and DAC. The Laplace transform model of the buffer-DAC hardware is

$$\text{ZOH}(s) = \frac{1 - e^{-Ts}}{s}$$

Using this model and taking transforms (recognizing that the computer's operation has no identifiable continuous time variable) yields a model such as given in Fig. 12.6.

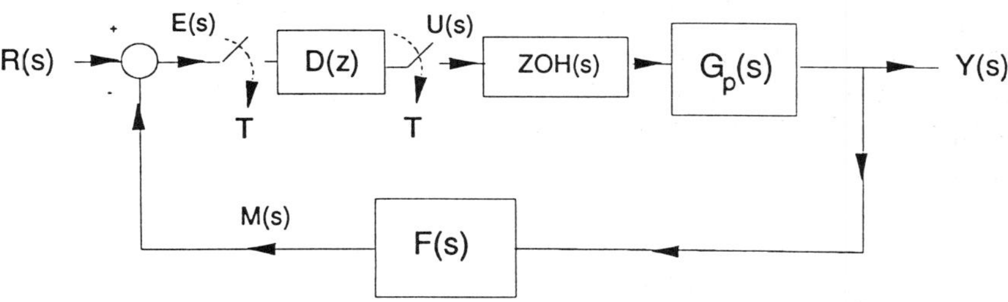

Figure 12.6 Transform model of Fig. 12.5.

In fact, this mixture of Laplace and z-transforms does not lend itself to analysis, particularly when the delay term e^{-Ts} in the ZOH is also considered. Thus it is necessary to work with one or the other of the transforms, and the one of choice is usually the z-transform. All Laplace transforms must be 'converted' to their sampled values. Then, because the z-transform is additive but not multiplicative, i.e.

$$\mathscr{Z}_{eq}\left\{\mathscr{L}(x(t)) + \mathscr{L}(y(t))\right\} = \mathscr{Z}_{eq}\left\{\mathscr{L}(x(t))\right\} + \mathscr{Z}_{eq}\left\{\mathscr{L}(y(t))\right\}$$

$$\mathscr{Z}_{eq}\left\{\mathscr{L}(x(t))\right\}\mathscr{L}\{(y(t))\} \neq \mathscr{Z}_{eq}\left\{\mathscr{L}(x(t))\right\} * \mathscr{Z}_{eq}\left\{\mathscr{L}(y(t))\right\}$$

we must convert with care: segments between samplers must be converted in a block, not factor by factor.

We consider the system in more detail. If we are interested in the time history of the output $y(t)$ for a given set of computer control 'pulses', we can find that the input to the ZOH-plant is

$$u(t) = \sum_k u(kT)\,\delta(t - kT)$$

$$U(s) = \sum_k u(kT)\,e^{-kTs}$$

The plant is a cascade of the zero-order hold

$$\text{ZOH}(s) = \frac{1 - e^{-Ts}}{s}$$

with the plant transfer function $G_p(s)$. Hence

$$G_{p0}^{*}(z) = \mathcal{Z}_{eq}\left[(1 - e^{-Ts})\,\frac{G_p(s)}{s}\right]$$

$$= (1 - z^{-1})\,\mathcal{Z}_{eq}\left[\frac{G_p(s)}{s}\right]$$

From this we argue that

$$Y(z) = (1 - z^{-1})\,\mathcal{Z}_{eq}\left[\frac{G_p(s)}{s}\right]U(z)$$

In the configuration of Fig. 12.6 the situation for feedback analysis is made more complicated. We observe that

$$e(kT) = \text{samples } (r(t) - m(t))$$

$$= r(kT) - m(kT)$$

$$E(z) = R(z) - M(z)$$

$$m(kT) = \mathcal{Z}^{-1}\left[(1 - z^{-1})\,\mathcal{Z}_{eq}\left[\frac{G_p(s)\,F(s)}{s}\right]U(z)\right]$$

$$M(z) = (1 - z^{-1})\,\mathcal{Z}_{eq}\left[\frac{G_p(s)\,F(s)}{s}\right]U(z)$$

$$U(z) = D(z)\,E(z)$$

where usually $D(z)$ is to be chosen, as it is the computer algorithm. Algebra then allows us to determine that

$$E(z) = \frac{R(z)}{1 + D(z)\,(1 - z^{-1})\,\mathscr{Z}_{eq}\left[\dfrac{G_p(s)\,F(s)}{s}\right]}$$

which gives the system transfer function

$$\frac{Y(z)}{R(z)} = \frac{(1 - z^{-1})\,\mathscr{Z}_{eq}\left[\dfrac{G_p(s)}{s}\right]D(z)}{1 + D(z)\,(1 - z^{-1})\,\mathscr{Z}_{eq}\left[\dfrac{G_p(s)\,F(s)}{s}\right]}$$

The alternative symbols

$$G^*(z) = (1 - z^{-1})\,\mathscr{Z}_{eq}\left[\dfrac{G_p(s)}{s}\right]$$

$$GF^*(z) = (1 - z^{-1})\,\mathscr{Z}_{eq}\left[\dfrac{G_p(s)\,F(s)}{s}\right]$$

allow this to be written as

$$\frac{Y(z)}{R(z)} = \frac{D(z)\,G^*(z)}{1 + D(z)\,GF^*(z)}$$

and the special case of perfect sensor $(F(s) = 1)$ becomes

$$\frac{Y(z)}{R(z)} = \frac{D(z)\,G^*(z)}{1 + D(z)\,G^*(z)}$$

Example

A motor with transfer function

$$G_p(s) = \frac{1}{s(\tau s + 1)}$$

with a perfect sensor of shaft angle has (using tables)

$$G_p^*(z) = \tau \, \frac{(\alpha - 1 + e^{-\alpha})\, z^{-1} + (1 - e^{-\alpha})\, z^{-2}}{(1 - z^{-1})\,(1 - e^{-\alpha} z^{-1})}$$

where $\alpha \equiv T/\tau$. $D(z)$ can be various choices, including the simplest of proportional control with $D(z) = K$ with gain K to be determined (using, for example, root locus methods as in Chapter 18).   The closed-loop transfer function in this case is

$$\frac{K\tau\,((\alpha - 1 + e^{-\alpha})\, z^{-1} + (1 - e^{-\alpha} - \alpha e^{-\alpha})\, z^{-2})}{(1 - z^{-1})(1 - e^{-\alpha} z^{-1}) + K\tau\,((\alpha - 1 + e^{-\alpha})\, z^{-1} + (1 - e^{-\alpha} - \alpha e^{-\alpha})\, z^{-2})}$$

and K is to be chosen to give 'good' performance (see Chapter 13).

12.4 DISCRETE TIME SAMPLING OF CONTINUOUS SYSTEMS IN STATE–SPACE

Working in the time domain with continuous time state–space models is, in principle, straightforward.   In this section we first show the form needed and then show how the computations can be performed.
  Consider the continuous time model

$$\dot{\mathbf{x}} = \mathbf{G}\mathbf{x} + \mathbf{H}\mathbf{u}$$

$$\mathbf{y} = \mathbf{J}\mathbf{x} \tag{12.7}$$

We recall from Chapter 10 the $n \times n$ transition matrix $\Theta(t, \tau)$, which is the solution of

$$\frac{\mathrm{d}\Theta(t, \tau)}{\mathrm{d}t} = \mathbf{G}\Theta(t, \tau) \qquad \Theta(\tau, \tau) = \mathbf{I}$$

Using the transition matrix, the solution to a differential equation of form (12.7) with initial condition $x(t_0) = x_0$ is

$$\mathbf{x}(t) = \Theta(t, t_0)\, \mathbf{x}_0 + \int_{t_0}^{t} \Theta(t, \tau)\, \mathbf{H}\mathbf{u}(\tau)\, d\tau$$

To discretize the above, we note that when the matrices are not time-varying then Θ is stationary. This means that

$$\Theta(t, \tau) = \Theta(t-\tau, 0) = \Theta(t-h, \tau-h)$$

Using this and the fact that time intervals of interest are of length T, if $\mathbf{u}(t) = \text{constant} = \mathbf{u}(kT)$ for $kT \le t < kT + T$, we have

$$\mathbf{x}(kT + T) = \Theta(T, 0)\, \mathbf{x}(kT) + \int_{0}^{T} \Theta(T, \tau)\, d\tau\, \mathbf{H}\mathbf{u}(kT)$$

which, with the definitions

$$\mathbf{A} = \Theta(T, 0)$$

$$\mathbf{B} = \int_{0}^{T} \Theta(T, \tau)\, d\tau\, \mathbf{H}$$

$$\mathbf{C} = \mathbf{J}$$

gives the standard form

$$\mathbf{x}(k+1) = \mathbf{A}\mathbf{x}(k) + \mathbf{B}\mathbf{u}(k)$$

$$\mathbf{y}(k) = \mathbf{C}\mathbf{x}(k)$$

Chapter 10 suggests several ways of calculating the transition matrix Θ, including numerical integration, using matrix transformations, using numerical series, and Laplace transforms.

The matrix similarity transforms yield, for $\mathbf{G} = \mathbf{M}\boldsymbol{\Lambda}\mathbf{M}^{-1}$, and having distinct eigenvalues λ_i, $i = 1, 2, \ldots, n$, that

$$\mathbf{A} = e^{GT} = \mathbf{M}\, \text{diag}\, (e^{\lambda_1 T}, e^{\lambda_2 T}, \ldots, e^{\lambda_n T})\, \mathbf{M}^{-1}$$

$$\mathbf{B} = \mathbf{M} \operatorname{diag}\left(\frac{e^{\lambda_1 T} - 1}{\lambda_1}, \frac{e^{\lambda_2 T} - 1}{\lambda_2}, \cdots, \frac{e^{\lambda_n T} - 1}{\lambda_n}\right) \mathbf{M}^{-1} \mathbf{H}$$

The Laplace transforms yield

$$\mathbf{A} = \mathscr{L}^{-1}\{ (s\mathbf{I} - \mathbf{G})^{-1} \}$$

$$\mathbf{B} = \mathscr{L}^{-1}\left\{ \frac{(s\mathbf{I} - \mathbf{G})^{-1}}{s} \right\} \mathbf{H}$$

A numerical method is

$$\psi = \mathbf{I} + \frac{\mathbf{GT}}{2} + \frac{(\mathbf{GT})^2}{6} + \cdots + \frac{(\mathbf{GT})^k}{(k + 1)!}$$

$$\mathbf{A} = \mathbf{I} + \mathbf{GT}\psi$$

$$\mathbf{B} = \mathbf{T}\psi\mathbf{H}$$

where we emphasize the series for ψ is finite, with typical value of k of about 10–12.

An alternative and approximate method works through transform approximation. In particular, we have from (12.7) that

$$\mathbf{x}(t + T) = \int_{t_0}^{t+T} (\mathbf{Gx}(\tau) + \mathbf{Hu}(\tau))\, d\tau$$

$$= \mathbf{x}(t) + \int_{t}^{t+T} (\mathbf{Gx}(\tau) + \mathbf{Hu}(\tau))\, d\tau$$

$$\approx \mathbf{x}(t) + (\mathbf{Gx}(t) + \mathbf{Hu}(t))\, T$$

$$= (\mathbf{I} + \mathbf{TG})\, \mathbf{x}(t) + \mathbf{THu}(t)$$

or using a trapezoidal rather than an Euler integration approximation

$$\mathbf{x}(t + T) \approx \mathbf{x}(t) + \frac{T}{2}\,(\mathbf{G}\,[\mathbf{x}\,(t + T) + \mathbf{x}\,(t)\,] + \mathbf{H}\,[\mathbf{u}\,(t + T) + \mathbf{u}(t)\,])$$

$$\approx (\mathbf{I} - \frac{T}{2}\,\mathbf{G})^{-1}\left[(\mathbf{I} + \frac{T}{2}\,\mathbf{G}\,)\,\mathbf{x}(t) + \frac{T}{2}\,\mathbf{H}\,[\mathbf{u}\,(t + T) + \mathbf{u}\,(t)\,]\right]$$

This can of course lead to adequate, albeit not theoretically elegant, approximations.

A newly suggested transformation is the δ-operator (Goodwin *et al.*, 1992) given by

$$\delta\mathbf{x}(t) = \begin{cases} \dfrac{d\mathbf{x}(t)}{dt} & T = 0 \\[2ex] \dfrac{\mathbf{x}\,(t + T) - \mathbf{x}(t)}{T} & T \neq 0 \end{cases}$$

and having effects such as

$$e^{AT} = \mathbf{I} + \mathbf{A}T$$

$$s = \frac{z - 1}{T}$$

This is claimed to be particularly appropriate for small sampling periods T.

12.5 BETWEEN-SAMPLE BEHAVIOUR

The behaviour of the system between samples depends upon the system. Inversion of $\mathscr{Z}_{eq}$, for example, to give a Laplace transform and hence a continuous time function is invalid simply because the continuous function yielding a given set of samples is not unique; the method may yield one of the possible time functions. Analysis is possible, however, provided that the original system is well known and the input to it is defined for all time.

The situation for state space is particularly straightforward. Here we use transition matrices as in

$$\mathbf{x}(t) = \phi(t, kT)\, \mathbf{x}(kT) \int_{kT}^{t} \phi(t, \tau)\, d\tau\, \mathbf{B}\mathbf{u}(kT)$$

$$\mathbf{y}(t) = \mathbf{C}\mathbf{x}(t) + \mathbf{D}\mathbf{u}(kT)$$

With transforms, the situation is more complicated, as one must use the modified z-transform, which is beyond the scope of this book. Its concept is straightforward, as it involves a ZOH(s) in cascade with the plant $\mathbf{G}_p(s)$, but with the response to an impulse sampled at $kT + \delta T$, $k = 0, 1, \ldots$ and $0 \le \delta < 1$. Thus we have

$$\mathbf{y}(t) = \mathcal{L}^{-1}\left[(1 - e^{-Ts})\, \frac{\mathbf{G}_p(s)}{s} \right]$$

as usual, but

$$\mathbf{Y}(z, \delta) = \mathcal{Z}\left\{ y(t)|_{t=kT+\delta} \right\}$$

Choosing $\delta \ne 0$ gives us the value of the output between sample instants kT, $k = 0, 1, 2, \ldots$. For example, the modified z-transform equivalent of the motor example of section 12.3 can be shown to be

$$G_p^*(z, \delta) = \frac{T}{z(z-1)} + \tau\frac{\delta\alpha - 1}{z} + \tau\frac{(z-1)e^{-\delta\alpha}}{z(z - e^{-\alpha})}$$

One notes that $G_p^*(z)$ is not given simply by $\lim_{\delta \to 0} G_p^*(z, \delta)$, and detailed examination of the definition of the modified transform shows why: the nature of the sensitivity of the first sample to its exact location in time.

12.6 COMPUTER SUPPORT

It is relatively straightforward to implement several of the methods of this chapter for computation on a personal computer. Commercial programs such as MATLAB® and Ctrl-C® are programmable allowing formula-type conversions and transition matrix calculations, and they can certainly compute and plot impulse, step, and frequency responses. Simulation programs also are instructive.

12.7 SUMMARY AND FURTHER READING

The issue of conversion of continuous time transfer functions to sampled data transfer functions is a standard one in the digital signal processing literature, such as the classic Rabiner and Gold (1975) and most other basic textbooks. Application of these methods to conversion of control laws is to an extent *ad hoc*, but usually successful.

Plant representation for the purposes of digital control law design is usually addressed, more or less well, in textbooks on digital control systems. The situation with constant coefficient state-space representation is apparently settled, but the advice with respect to transfer functions varies. Textbooks discussing the topic include Kuo (1980) and Franklin, Powell and Workman (1990).

The modified z-transform is not standard in elementary textbooks, but some do address it. Among those who do are Kuo (1980) and Ogata (1987), and the former has a table applying to a few such cases.

A discussion of some of the issues and a different candidate substitution are presented by Goodwin *et al.* (1992).

13

System performance indicators

Closed-loop systems are expected to give 'good' performance. In this chapter we introduce some of the properties used to measure performance.

13.1 SYNOPSIS

Design of any system requires a specification, implicitly or explicitly, of the desired performance.

1. The fundamental requirement of a controlled system is that it be stable, and so that is the first item we review. Two important definitions are BIBO stability and Lyapunov stability.
2. There are number of classical servomechanism step response characteristics which are used as performance indicators and specifications. We define these characteristics, such as percent overshoot and rise time, and discuss their relationships in section 13.3. Included are the notions of bandwidth and frequency response.
3. Single performance indices such as minimum time to traverse a trajectory are very useful for some system specifications such as space vehicles and robot arm motion. These are ultimately different in nature from the classical servomechanism indicators; some are also readily amenable to mathematical optimal control techniques. These indices are the subject of section 13.4.
4. Many design techniques require that the controlled system have special properties called structural properties. The most important of these are controllability and observability, which have matrix tests and are given in section 13.5 along with the notions of sensitivity and robustness.

13.2 STABILITY

The property of stability is fundamental to good systems and so the pursuit of control laws which assure stable operation is the *sine qua non* of the control engineer's art: without stablity the rest of the problem is irrelevant. Being stable means the system does not oscillate or diverge from desired operating conditions under any reasonable conditions, but the exact encoding of this notion in terms suitable for the analysis of systems has led to many mathematical definitions. In this section we concentrate on two notions: bounded-input–bounded-output (BIBO) stability and Lyapunov stability. Methods for testing stability of constant coefficient systems are met in Chapters 14–16.

13.2.1 Definitions

Definition 1 A system with input sequence $\{u(t)\}$ and output sequence $\{y(t)\}$, where either or both may be vectors, is said to be **BIBO stable** if, for any bounded input, the output is bounded. Thus if all elements of the input sequence satisfy $\|u(t)\| \leq U < \infty$, then in a BIBO system $\|y(t)\| \leq Y < \infty$, where the $\| \cdot \|$ denotes a scalar function (i.e. a metric or norm, see section B.13) measuring the size of the argument and U and Y are scalars.

We remark that an integrator is not BIBO stable, since

$$\int_0^t x(\tau)\, d\tau \to \infty \quad \text{as } t \to \infty$$

even for the simple bounded input $x(t) = 1$.

We notice that this definition refers to a stable system, and hence, for several reasons, is apt to be misleading. For example, an aircraft may have both easily maintained and stable flight configurations and unmaintainable configurations (such as a stall, for instance); it is difficult to call the system stable, but straightforward to call certain flight regimes stable or unstable. For such cases, an alternative point of view refers only to stable operating conditions, rather than stable systems. We start here by assuming that the inputs are known (perhaps they are test inputs of a certain type) so that

$$x(k) = f_d(x(k-1); k) \tag{13.1}$$

or in continuous time

$$\dot{x} = f_c(x,t)$$

is an adequate state-space description of the system. Then we define an equilibrium operating condition or equilibrium point.

Definition 2 An **equilibrium state** of a system (13.1) is a vector $\mathbf{x}_e$ such that $\mathbf{x}_e = f_d(\mathbf{x}_e; k)$ for all k (or $\dot{\mathbf{x}}_e = 0 = f_c(\mathbf{x}_e; t)$ in continuous time).

Using this, the notion of stability is related to an equilibrium state or point.

Definition 3 An equilibrium state $\mathbf{x}_e$ of (13.1) is said to be stable in the sense of Lyapunov (stable i.s.L.) or **Lyapunov stable** if, for any given $\delta > 0$ there is an $\varepsilon > 0$ (which may depend upon δ) such that for any disturbance $\mathbf{d}$, $\|\mathbf{d}\| < \varepsilon$, for which the initial conditions $\mathbf{x}(p) = \mathbf{x}_e + \mathbf{d}$ apply, then the solution $\mathbf{x}(k)$, $k \geq p$, has the property $\|\mathbf{x}(k) - \mathbf{x}_e\| \leq \delta$. The relevant property in continuous time is that if $\|\mathbf{x}(t_0) - \mathbf{x}_e\| < \varepsilon$, then a stable equilibrium has $\|\mathbf{x}(t) - \mathbf{x}_e\| \leq \delta$ for all $t \geq t_0$.

If, eventually, the error becomes almost zero we have the notion of asymptotic stability of the equilibrium state.

Definition 4 An equilibrium state $\mathbf{x}_e$ is **asymptotically stable** in the sense of Lyapunov if it is stable i.s.L. and in addition,

$$\|\mathbf{x}(k) - \mathbf{x}_e\| \to 0 \qquad \text{as } k \to \infty$$

or

$$\|\mathbf{x}(t) - \mathbf{x}_e\| \to 0 \qquad \text{as } t \to \infty$$

Finally, because the i.s.L. stability is defined only in terms of infinitesimals, we need terminology to describe the situation if the disturbances are allowed to be 'large'.

Definition 5 An equilibrium state $\mathbf{x}_e$ is **globally** (asymptotically) stable i.s.L., or (asymptotically) stable **in the large**, if it is (asymptotically) stable i.s.L. and ε and δ are unbounded (except that $\varepsilon \leq \delta$).

The notions of Lyapunov stability are shown in Fig. 13.1.

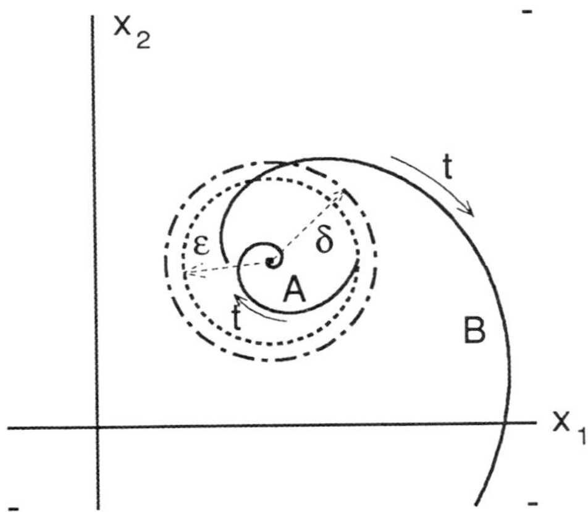

Figure 13.1 The ideas of Lyapunov stability. Trajectory A stays within radius δ of the equilibrium point and is stable i.s.L., while B diverges from the disc and is unstable.

13.2.2 Relative stability

Since a BIBO unstable system will eventually have unbounded output for bounded input, and many unstable systems will have a growing oscillation for a pulse input, and since we find experimentally that such unstable behaviour is often close in parameter values to stable behaviour, we sometimes look for measures of 'nearness to going unstable'. Typically, slow sluggish responses are associated with imperturbability, whereas rapid, perhaps oscillatory, responses are associated with being susceptible to instability in the face of equipment parameter variations. Figure 13.2 shows responses illustrating these features.

The notions are related to having poles/eigenvalues close to the stability boundaries (i.e. to the imaginary axis for continuous time systems and to the unit circle for sampled data systems), but the usual measures of relative stability are gain and phase margins (defined in the next section) and are encountered when using frequency domain techniques (Chapters 15 and 20). Indicators of response

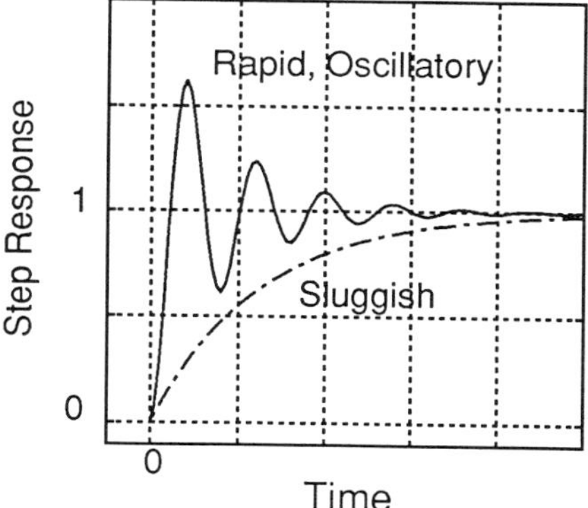

Figure 13.2 Illustrating rapid but oscillatory step response and sluggish response. Choice is often an engineering trade-off.

characteristics are the pole locations for linear time invariant systems; these are discussed in Chapter 19.

13.3 CLASSICAL RESPONSE PERFORMANCE INDICATORS

From the classical point of view there are two problems to solve: achieving satisfactory response to commands and ignoring disturbances – and there are two approaches to handling them.

Good response means that the error between desired and actual output in steady state is small (preferably zero) and that the transient response in attaining that small error is rapid and suitable in form. The transient response is usually characterized in terms of several parameters of the system's response to a unit step in its input, as in Fig. 13.3.

The details of the parameter definitions vary, but a common set is as follows: the initial value is taken as 0, and the final value is taken as y_f which is assumed constant.

1. t_r, the rise time, is the time to rise from 10% of y_f to 90% of y_f (the 10–90% rise time).
2. The percent overshoot (PO) is the percentage by which the peak value M_{pk} of the time response exceeds y_f, so that

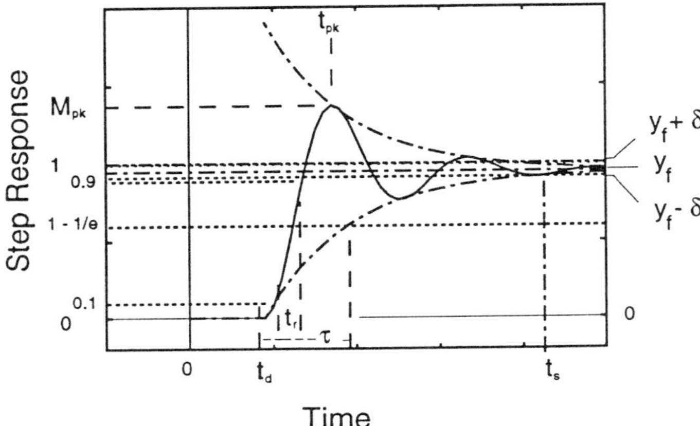

Figure 13.3 Showing response characteristics to input step applied at time 0: delay time t_d, 10–90% rise time t_r, time constant τ, δ% settling time t_s, peak time t_{pk}, overshoot M_{pk}, and steady-state value y_f (implying steady-state error $1 - y_f$).

$$PO = \frac{(M_{pk} - y_f)}{y_f} \times 100\%$$

3. Settling time t_s is the time for the system output to settle to within a fraction $\pm \delta$ of the final value, so that $y(t)$, $t > t_s$, lies between $y_f - \delta$ and $y_f + \delta$. Often δ is taken as 5% or 2% of either y_f or of the input value, which is 1.0 in the case of the unit step.
4. Peak time t_{pk} is the time after the input step is applied that the peak value M_{pk} is attained.
5. Steady-state error e_{ss} is the error between the desired value (usually 1.0 for the unit step) and the actual value (y_f) of the output once the error becomes essentially constant (assuming it does).
6. Delay time t_d is the time from application of an input to start of response by the system. To gain a consistent definition, sometimes t_d is taken as the time from application of input to 10% (say) of final value.
7. Time constant τ is the parameter associated with the response envelope $(1 - e^{-t/\tau})\,y_f$ of the error; in the absence of oscillations, it is the time after the start of the response to reach 63% of the final value.

Some indicators are defined in terms of the open- or closed-loop frequency responses, as in Figs 13.4 and 13.5.

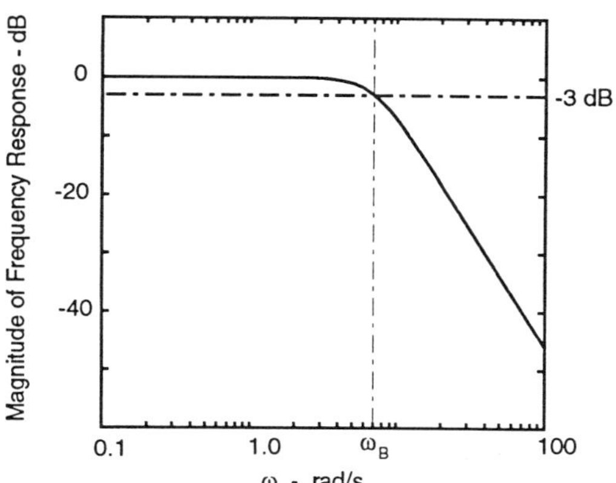

Figure 13.4 Definition notion for 3dB bandwidth ω_B.

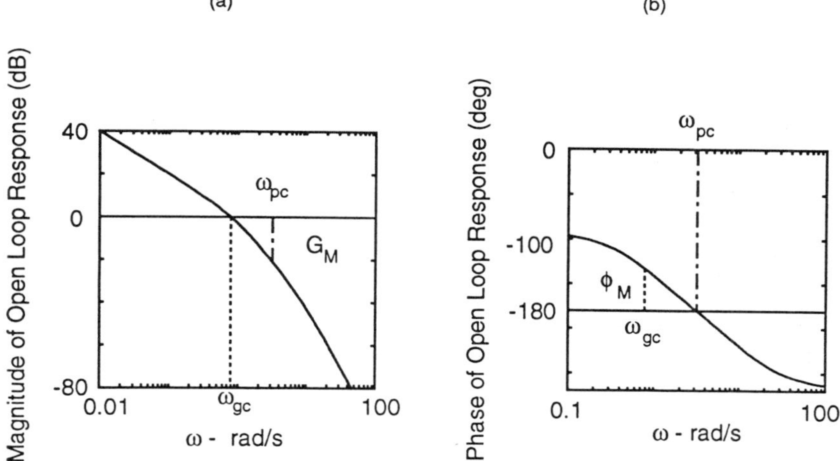

Figure 13.5 Demonstrating gain crossover frequency ω_{gc}, phase crossover frequency ω_{pc}, gain margin G_M, and phase margin ϕ_m on frequency response plots.

8. Bandwidth ω_B is the frequency (for a system such as most control systems, i.e. low-pass) at which the magnitude of the frequency response has fallen to $\frac{1}{2}$ (or $- 3\,\text{dB}$ down) of the steady value. It indicates the speed of the system's response to input: larger bandwidth is associated with faster response.
9. Phase margin ϕ_m and gain margin G_M are defined for the open-loop response of a system and indicate the relative stability (or closeness to instability) and the damping of the response of the system. More of either is roughly associated with a more sluggish response. Phase margin is the phase above $-\pi$ of the transfer function when the magnitude is 1 ($= 0\,\text{dB}$). Gain margin is the additional gain needed to reach unit magnitude when the phase is $-\pi$. The frequencies at which the margins are measured are the gain and phase crossover frequencies ω_{gc} and ω_{pc}, respectively. These are shown in Fig. 13.5.

The control engineer's task is to design for 'good' dynamic and steady-state response in spite of parameter variations while 'ignoring' noise and load disturbances. The above parameters are indicators of suitable performance. Typical specifications are:

$$G_M = 6\text{--}10\,\text{dB} \qquad \phi_m = 35\text{--}50°$$

$$\text{PO} < 15\% \qquad e_{ss} < 0.1\%$$

There are many ways of attempting the achievement of the above control system properties. Approaches range from *ad hoc* and heuristic to sophisticated and mathematically 'optimum'.

The above have been defined relative to desired outputs. The same numbers concerning dynamic response affect disturbed outputs, but in those cases the desired response is still zero error relative to desired input: it is preferred that the output not reflect disturbances, but, if it does so, the return to normal should be 'rapid'.

13.4 MODERN PERFORMANCE INDICATORS

The various response indicator parameters above are ultimately determined by engineering trade-offs by the engineer who tunes the control system, often using a control law of pre-specified type, such as a PID law. A quite different approach is to define a single function which is optimized mathematically. The resulting 'optimal' control may or may not be 'good', depending upon a number of factors, not least of which is the appropriateness of the indicator function.

13.4.1 Weighted error approaches

One older approach with much intuitive appeal considers the error between desired and actual response and attempts to minimize some chosen function of this error. The error $e(t)$ between actual and desired responses is shown in Fig. 13.6(a); the aim is to keep the error as 'small' as possible. Positive and negative errors might cancel each other, so a possible function uses the absolute value (i.e. ignoring the sign), e.g.

$$\text{IAE} = \text{integral absolute error} = \int_0^T |e(\tau)| \, d\tau$$

which would appear to give small errors on average if the controller can be chosen to make this – or its summation equivalent in discrete time control – take its minimum value.

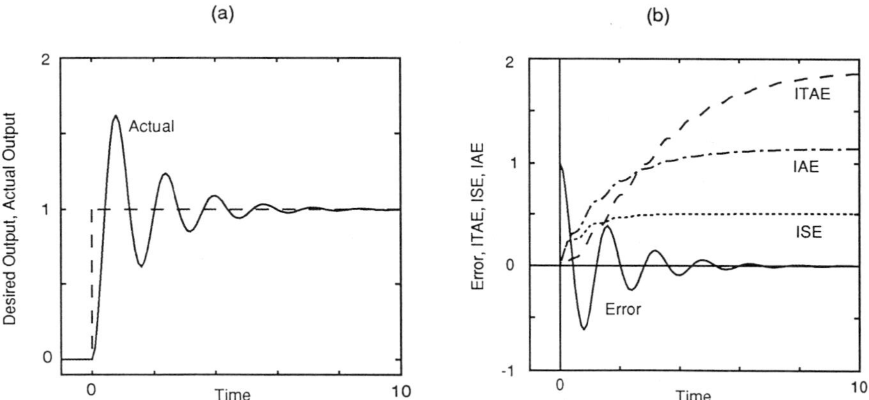

Figure 13.6 (a) Desired and actual step responses; (b) error in step response and the accumulation with time of ISE, ITAE, and IAE performance criteria. These can be used as minimization criteria in optimizing performance.

Unfortunately absolute values can be difficult to work with mathematically. Furthermore, one might wish for the cost function to penalize occasional large errors much more than common small ones. One solution to both of these problems is to use the function

$$\text{ISE} = \text{integral squared error} = \int_0^T e^2(\tau)\,d\tau$$

A further refinement is to penalize errors which occur late in time more than early transients. Explicit dependence of the functions on time can be implemented easily with the following forms:

$$\text{ITAE} = \text{integral time absolute error} = \int_0^T |e(\tau)|\,\tau\,d\tau$$

$$\text{ITSE} = \text{integral time squared error} = \int_0^T e^2(\tau)\,\tau\,d\tau$$

Figure 13.6(b) shows some of these measures for the step response of Fig. 13.6(a).

There are many variations on the above which might have heuristic justification. An important one is to use the ISE form but let T become large; this is a regulator problem of an almost classic type, but for mathematical naturalness it needs division by T to remain bounded. This yields the form

$$\text{MSE} = \text{mean square error} = \lim_{T \to \infty} \frac{1}{T} \int_0^T e^2(\tau)\,d\tau$$

which has been found useful, e.g. when disturbances have the characteristics of random noise.

13.4.2 Modern control theory and optimal control

Optimal control theory approaches the problem of determining the best control in a quite different manner from the above. Usually it is presented in a state-space framework, and a fairly general formulation of the problem in differential equation form follows.

Suppose a system with vector control commands $\mathbf{u}$ and state $\mathbf{x}$ is described by

$$\dot{\mathbf{x}} = \mathbf{f}(x, \mathbf{u}, \mathbf{t}) \qquad \mathbf{x}(0) = \mathbf{x}_0 \text{ given} \tag{13.2}$$

Then find a control law $\mathbf{u}(\mathbf{x}, t)$ if possible, or a control history $\mathbf{u}(t)$, such that the scalar functional

$$J(u) = \int g(\mathbf{x}(\tau), \mathbf{u}, \tau) \, d\tau$$

takes on a minimum value and the relationship (13.2) holds. Typically a boundary relationship $\mathbf{x}(T) = \mathbf{x}_f$ must also be met as a constraint on the control.

The most common forms of $J(u)$ are the minimum time control, in which

$$J(u) = \int_0^T d\tau = T$$

is to be minimized, usually subject to the additional control constraint such as a bound on a metric $\|u(t)\| \leq U$, and the quadratic problem in which the form

$$J(u) = \mathbf{x}^T(T) \, \mathbf{S} \, \mathbf{x}(T) +$$

$$\int_0^T \{ \mathbf{x}^T(\tau) \, \mathbf{Q}(\tau) \, \mathbf{x}(\tau) + \mathbf{u}^T(\tau) \, \mathbf{R}(\tau) \, \mathbf{u}(\tau) \} \, d\tau \tag{13.3}$$

is to be minimized, where the matrices $\mathbf{Q}$ and $\mathbf{S}$ are positive semidefinite and $\mathbf{R}$ is positive definite (section B.5).

A common minimum time control problem is that of finding a control law for the attitude control thrusters of a space satellite such that the satellite is rotated from an initial state $\mathbf{x}_0$ to a final orientation $\mathbf{x}_f$ in minimum time while satisfying the laws of dynamics, where the latter are modelled by a function (13.2). The quadratic functional may be viewed as a mathematical generalization of the ISE function of the previous section.

Although the minimum time criterion may seem obvious, the quadratic form (13.3) is less so and is worthy of comment. We first note that using matrices simply extends the ideas of scalars. Then we consider a scalar form of (13.3). This is

$$J(u) = \int_0^T (qx^2(\tau) + ru^2(\tau))\,d\tau \qquad q \geq 0, r > 0$$

If $x(t)$ is an error, then clearly this expression is penalizing error and control input, with relative weightings r and q. Using these parameters, we are able to trade-off control input for small error and vice versa. In many instances, u^2 is proportional to power (as in voltage^2 or current^2) and its integral then is proportional to energy. The version

$$J(u) = \int_0^T u^2(\tau)\,d\tau$$

is then a minimum control energy problem.

The optimal control formulation is in principle general, and many other problem statements may be set up. One, for example, is the minimum fuel problem, in which

$$J(u) = \int_0^T |u(\tau)|\,d\tau$$

It is so named because the control is often proportional to the fuel used, for example, in an aerospace situation.

Comments

Modern control theory is a mathematician's approach to control problems. The engineer must still develop an appropriate criterion, and it is noteworthy that the mathematical criteria do not include monetary cost. Furthermore, the fit of the solvable problems to the actual situations can be difficult. One should recall that the optimal control theory has its greatest successes in aerospace problems, where minimum time (for an airplane to climb to altitude) and minimum fuel (for a satellite to change orbit) are relevant and important.

On the other hand, fitting $\mathbf{Q}$ and $\mathbf{R}$ matrices to real problems is often an elusive goal when one desires only that rise time and overshoot be nicely traded off.

13.5 OTHER INDICATORS OF STRUCTURE AND PERFORMANCE

Other properties of the system, also important to designers, which we will address are the 'structural' properties of state-space descriptions, i.e. those properties associated with the nature of the interactions rather than the exact numbers.

13.5.1 Controllability and observability

Two properties of the linear state-space system descriptions, often needed in proofs about existence of certain types of controllers, etc., are controllability and observability. Loosely stated, these are respectively indicators of whether the model can be driven to a precisely specified state in finite time and whether complete knowledge of the state can be extracted from a finite set of measurements, but it should be recognized that they are ultimately technical terms. Their primary use is in tests to show whether some design techniques we will meet later, especially pole placement in Chapter 23 and state estimation in Chapter 25, can be expected to be applicable.

The definitions of the terms – we remark that they are considered together because they are duals of each other in the linear algebraic sense – help clarify their roles and also why they are technical terms.

Definition A system is said to be **controllable** if any initial state $x(t_0)$ at any initial time t_0 can be moved to any other desired state x_f, $x(t_f) = x_f$, in a finite time interval $\tau = t_f - t_0$ by applying an admissible control function $u(t)$, $t_0 \leq t \leq t_f$.

Definition A system is said to be **observable** if any initial state $x(t_0)$ can be determined after a finite time interval $t - t_0$ from a measurement history $Y(t) = \{y(\tau), t_0 \le \tau < t\}$ and the control variable history $U(t) = \{u(\tau), t_0 \le \tau < t\}$. Furthermore, given the usual uniqueness of solution arguments, $x(t)$, $t \ge t_0$, can also be determined.

It is quite possible to have states which are controllable but not observable, observable but not controllable, both controllable and observable, or neither controllable nor observable. This decomposition can in fact be made explicit, as we will see in a Chapters 23 and 25.

Now let us emphasize what the terminology allows. A pen recorder may well be such that any pen position and speed can be attained, but the acceleration is not also specifiable; such a system, while not controllable in the definition sense, may be quite adequate and be easy to design control laws for. Similarly, the fact that a radar target's roll rate cannot be extracted from the radar tracking data – making the system not observable – does not necessarily make for a poor radar. Systems which are observable/controllable may be easier to design for because more design theorems apply, but systems which are not observable/controllable are not necessarily inadequate.

The above have been only the main definitions. Variations are sometimes met in more advanced theory. Among these are output controllability, controllability and observability of particular states, and the concept of state reachability. For linear time invariant systems the distinctions are usually not important. Tests for controllability are given in Chapter 22, while those for observability are in Chapter 24.

13.5.2 Sensitivity

Sensitivity of a quantity is defined as the fractional change in that quantity per fraction change in an independent quantity. Ideally, we seek highly sensitive reaction to input and low sensitivity to disturbances and parameter errors.

The usual definition of the sensitivity of T with respect to changes in G concerns the ratio of the fractional change in each and is of the form

$$S_T^G \equiv \lim_{\Delta G \to 0} \frac{(\Delta T)/T}{(\Delta G)/G}$$

which can also be written when the limits exist as

$$S_T^G = \frac{d(\ln T/\ln G)}{dG}$$

$$= \frac{G}{T}\frac{dT}{dG}$$

Elaboration is given, for example, by DiStefano *et al.* (1976).

The above notions have recently become prominent again because of the development of robust control theory. One of the standard control system configurations is shown in Fig. 13.7.

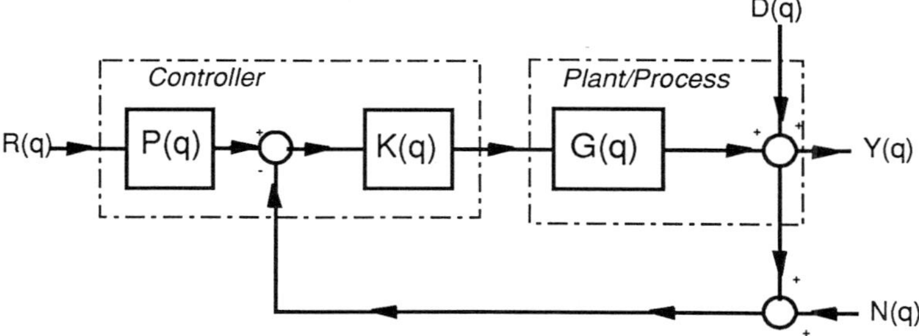

Figure 13.7 System block diagram model showing disturbances affecting the plant, noise affecting the feedback signal, with a possible controller configuration.

Straightforward algebra of the transforms shows that

$$\mathbf{Y}(q) = [\mathbf{I} + \mathbf{G}(q)\,\mathbf{K}(q)]^{-1}\mathbf{D}(q) - [\mathbf{I} + \mathbf{G}(q)\,\mathbf{K}(q)]^{-1}\mathbf{G}(q)\,\mathbf{K}(q)\,\mathbf{N}(q)$$

$$+ [\mathbf{I} + \mathbf{G}(q)\,\mathbf{K}(q)]^{-1}\mathbf{G}(q)\,\mathbf{K}(q)\,\mathbf{P}(q)\,\mathbf{R}(q) \qquad (13.4)$$

Then

$$\mathbf{S}(q) = [\mathbf{I} + \mathbf{G}(q)\,\mathbf{K}(q)]^{-1}$$

is called the **sensitivity function** and

$$\mathbf{T}(q) = [\mathbf{I} + \mathbf{G}(q)\,\mathbf{K}(q)]^{-1}\mathbf{G}(q)\,\mathbf{K}(q)$$

is called the **closed-loop transfer function**. Also,

$$\mathbf{F}(q) = \mathbf{I} + \mathbf{G}(q)\,\mathbf{K}(q)$$

is called the **return difference** and $\mathbf{P}(q)$ is the **prefilter**. Using these, (13.4) becomes

$$\mathbf{Y}(q) = \mathbf{S}(q)\,\mathbf{D}(q) - \mathbf{T}(q)\,\mathbf{N}(q) + \mathbf{T}(q)\,\mathbf{P}(q)\,\mathbf{R}(q) \qquad (13.5)$$

We may note that from the definitions

$$\mathbf{S}(q) + \mathbf{T}(q) = \mathbf{I}$$

But it is clear from (13.5) that we want both $\mathbf{S}(q)$ and $\mathbf{T}(q)$ to be 'small' (which for matrices means that their respective metrics – see Appendix B.13 – are near zero) so that the disturbances and noise have little effect on the output. (If $\mathbf{T}(q)$ is small but we choose $\mathbf{P}(q) = \mathbf{T}^{-1}(q)$, then $\mathbf{Y}(q) = \mathbf{R}(q)$, which is desired.) Usually we are then involved in a trade-off; one common such trade-off assumes that disturbances are low frequency events and noise is at high frequency, so that $\mathbf{S}(q)$ and $\mathbf{T}(q)$ can be small at the appropriate frequencies.

These are consistent with the classical definitions, of course. For scalar systems, for example, it is easy to show that the input transfer function with $P(q) = 1$ is

$$T(q) = \frac{G(q)\,K(q)}{1 + G(q)\,K(q)}$$

while the disturbance transfer function is

$$S(q) = \frac{1}{1 + G(q)\,K(q)}$$

Then going through the formalism of the definitions, we find that

$$S_T^G = \frac{1}{1+K(q)\,G(q)}$$

$$S_S^G = \frac{-\,G(q)\,K(q)}{1\,+\,G(q)\,K(q)}$$

Thus we might choose K large to make $S(q)$ small and $T(q) \approx 1$. However, each 1% error in G may then make for a small change in $T(q)$ (which is what feedback control is partly about), but also a -1% change in $S(q)$ (which is already small).

In SISO systems, the frequency response ideas are straightforward. In MIMO or multivariable systems, the notions need more refinement. This is particularly so in defining the meaning of 'small' and 'large' when often not all elements of the matrices **S** and **T** behave the same way, and it leads to seemingly esoteric mathematics simply to express the ideas. These issues are partly addressed in section B.13 and Chapters 15 and 33.

13.5.3 Robustness

Related to sensitivity is the idea of robustness. 'Robust' is generally taken to mean 'the system still works satisfactorily if a large modelling error has been made in doing the design or if a large disturbance occurs'. The mathematical treatment of this concept, as opposed to the heuristic treatment, requires explicit statements of the nature of the errors and the performance requirements, and the approach treats the robust control problem as the problem of analysing and designing accurate control systems for plants which may contain significant uncertainties. These uncertainties must be defined as to type and size, and the notion of good performance must be specified. The uncertainties are described as follows.

For the state-space model such as

$$\mathbf{x}(k+1) = \mathbf{A}\mathbf{x}(k) + \mathbf{B}\mathbf{u}(k)$$

$$\mathbf{y}(k) = \mathbf{C}\mathbf{x}(k)$$

uncertainties are modelled relative to a nominal value $\mathbf{A}_0$ as in

$$A = A_0 + \delta A$$

or

$$A = A_0 (I + \Delta)$$

which are additive and multiplicative uncertainties, respectively. If they are constrained using matrix norms (section B.13) in a form such as $\| \delta A \| \leq l_a$ or $\| \Delta \| \leq l_m$, the uncertainties are said to be **unstructured**. If the individual elements of δA or Δ are constrained, as in

$$\delta A = \sum_i \alpha_i A_i$$

where A_i is known and $-1 \leq \alpha_i \leq 1$, or

$$\delta A x = \sum_i x_i G_i w$$

where w is a vector noise process, then the uncertainties are respectively called **structured** and **stochastic**.

For the transfer matrix model $G(q)$, uncertainty may be modelled relative to a nominal model $G_0(q)$ as in

$$G(q) = (I + \Delta(q)) G_0(q)$$

where $\| \Delta(\omega) \| < l_m(\omega)$, or

$$G(q) = G_0(q)(I + \Delta(q))$$

where $\| \Delta(\omega) \| < l_m(\omega)$, or

$$G(q) = G_0(q) + \delta G(q)$$

where $\| \delta G(\omega) \| < l_a(\omega)$.

The first two models are input and output multiplicative errors, respectively, and the third is an additive error. As shown, the uncertainties are unstructured; if only certain elements of Δ and δG are variable, then the uncertainties have structure.

The sources of uncertainty are potentially many and varied: unmodelled dynamics (perhaps high frequency dynamics), neglected non-linearities, and parameter variations due to factors such as temperature and age.

A robustly stable system is then one which is stable for all uncertainties of the allowed type. A system shows robust performance if it meets specified levels of performance for all allowed uncertainties.

13.5.4 System type

In characterizing a feedback system, some linear systems will be stable plus have the ability to track a polynomial of up to order $n-1$ with zero steady-state, i.e. post-transient, error and a polynomial of order n with a finite non-zero error – briefly the system is of type n. Thus a type 1 system, the most important for many applications, will have zero steady-state response to a step input (order 0 polynomial) and finite error in response to a ramp (constant velocity, or order 1 polynomial) input. We will meet system type in more detail in Chapter 17.

13.5.5 Reachability

The idea of reachability is essentially that of controllability as presented here. It came about in part because of more restricted definitions of controllability, in particular a definition which called a system controllable if the state $x_f = 0$ could be obtained in finite time using some control; in this case, the term reachable was used for a state other than the origin which could be obtained in finite time and for a system in which all states are reachable.

13.6 SUMMARY AND FURTHER READING

This chapter has been concerned primarily with defining the terms associated with desirable properties of systems: stability, performance indicators and indices, structural properties. In classical analysis and design, the essence of a system's response characteristics is due to its pole (or eigenvalue) locations, provided that fixed gains are used. Hence the history of classical control system design is one of obtaining 'good' response by using feedback and choosing gains so that the poles are appropriately placed.

The typical requirements on the system are that it be stable, that the step response exhibit acceptable overshoot and damping, and that the steady-state errors be 'small'. These problems have been classically attacked in three different manners:

1. by suggesting a form for the control law, e.g. proportional, and then placing its parameters using trial and error tuning methods;
2. by suggesting a form for the control law and then choosing its parameters after examining the locus of the pole locations as the parameters vary;
3. by examining the system's frequency response and then using compensators and choosing parameters to obtain a desired type of frequency response.

We meet all of these in Chapters 14–20.

In modern control systems design, the work is primarily with state-space models. Pole locations are obtained with a fairly general structure (i.e. with state feedback combined with, if necessary, a state observer, giving in effect a high order and specialized compensator) and many elegant results are available. An alternative is optimal control, in which the control law is a mathematical result rather than a designer input. Modern control approaches dominate Chapters 22–33. Design involves

*   specification of performance index,
*   solution for the form of the required control law, and
*   manipulation of the parameters of the index to obtain 'good' response characteristics.

Most of the topics covered here are in standard theory textbooks, such as Kuo (1980), Franklin *et al.* (1990), Dorf (1989), Phillips and Harbor (1991), Ogata (1987), etc. One text concerned with modern sensitivity approaches and matrix norms is Maciejowski (1989).

This book also discusses most of the topics. Thus BIBO stability is covered in Chapters 14–15, and Lyapunov stability in Chapter 16. Applications of classical performance criteria will be seen in Chapters 17 and 20, while optimal control is the subject of Chapters 26–27. Controllability is discussed in Chapter 22 and observability in

Chapter 24. Much of this material has also been presented in a similar manner in Åström and Wittenmark (1990). The robustness ideas and some of the classical multivariable ideas are covered in Chapters 15 and 33, and those include references in addition to the above Maciejowski (1989).

14

BIBO stability and simple tests

BIBO stability of constant coefficient linear systems, whether described by differential or difference equations, is determined by the pole locations of the closed-loop systems. These poles are, by definition, the roots of the denominator polynomial in transfer function representations and of the characteristic equation of the **A** matrix in state-space representations. The poles must lie in the left-half plane for continuous time systems and within the unit circle for discrete time systems. The straightforward way of checking this is to compute the poles. An alternative that is easy and can lead to other insights is to process the coefficients of the denominator polynomial of the transfer function, which is the same as the determinant of the state-space dynamics matrix. This chapter demonstrates those tests and shows how they may be used in three different ways.

1. To check whether a system is stable, the test is applied to the characteristic polynomial describing the system in question.
2. To find the range of a parameter – often a gain K – such that the system is stable, the system denominator polynomial is used with K as an unknown. The array then leads to a set of inequalities which will define the stability-ensuring values of K.
3. To check speed of response, we check whether the poles are well away from the stability line ($\mathrm{Re}(s) = 0$ for continuous time and $|z| = 1$ for discrete time).

Applications 2 and 3 can be combined to give parameter values yielding fast response.

14.1 SYNOPSIS

For BIBO stability of a linear system, the integral of its impulse response (or the sum of its response pulses in the sampled data case) must be finite. When the system is described by constant coefficients, this requirement leads to the fact that the transfer function poles must lie in the left-half of the s plane or within the unit circle in the z plane, as appropriate. For testing, one may either:

- compute the poles; or
- infer the pole locations by examining the characteristic polynomial, in which case the determination of whether the roots of a polynomial of degree n, such as

$$p^n + a_1 p^{n-1} + a_1 p^{n-2} + \cdots + a_{n-1} p + a_n = 0$$

lie in the left-half plane or inside the unit circle can be performed by processing an array derived from the coefficients. The first criteria are checked by Routh testing and the second are evaluated using Jury testing.

Although the tests give only a yes–no type answer concerning stability, when unknown parameters are part of the arrays, the values of those parameters which yield stability can be determined. Furthermore, changes of variable allow the tests to be used to test against selected half planes other than the left-half plane and selected circles of radius other than unity.

14.2 BIBO STABILITY TESTS FOR LINEAR SYSTEMS

For a linear constant coefficient differential equation, the modelling sections have shown that, in the absence of initial conditions, the output Laplace transform is

$$Y(s) = H(s) U(s) \tag{14.1}$$

and the inverse transform for causal systems is

$$y(t) = \int_0^t h(t-\tau) u(\tau)\, d\tau = \int_0^t h(\tau)\, u(t-\tau)\, d\tau \tag{14.2}$$

If the input $u(t)$ is bounded, so that $|u(t)| \leq U < \infty$, then

$$|y(t)| \leq \left| \int_0^t h(\tau)\, d\tau \right| U \leq \int_0^t |h(\tau)|\, d\tau\, U$$

and hence the output $y(t)$ is bounded if the impulse response $h(t)$ is bounded and converges to 0. If $h(t)$ is unbounded, then clearly an input $u(t)$ can be found such that $y(t)$ is unbounded.

Theorem The linear system described by (14.1)–(14.2) is **BIBO stable** iff the input response $h(t)$ is integrally bounded:

$$\int_0^\infty |h(\tau)|\, d\tau < \infty$$

Now, $H(s)$ is the transform of the impulse response; expanding $H(s)$ in partial fractions gives

$$H(s) = \sum_{i=1}^k \sum_{j=1}^{m_i} \frac{\alpha_{ij}}{(s - \lambda_i)^j} \qquad n = \sum_{i=1}^k m_i$$

where the λ_i are the (complex) poles of $H(s)$, α_{ij} are (complex) coefficients, and m_i is the multiplicity of pole λ_i. Inverting this gives

$$h(t) = \sum_{i=1}^k \sum_{j=1}^{m_i} \alpha_{ij} t^{j-1} \exp(\lambda_i t))$$

We then can compute

$$\int_0^\infty |h(\tau)|\, d\tau = \int_0^\infty \left| \sum_{i=1}^k \sum_{j=1}^{m_i} \alpha_{ij} t^{j-1} \exp(\lambda_i t) \right| dt$$

$$\leq \sum_{i=1}^k \sum_{j=1}^{m_i} \int_0^\infty | \alpha_{ij} t^{j-1} \exp(\lambda_i t)|\, dt$$

This will be bounded and its terms will approach 0 asymptotically if the real part of $\lambda_i < 0$ for all $i = 1, 2, ..., k$.

Theorem   A system modelled by an ordinary differential equation with constant coefficients is **BIBO stable** iff, for all poles λ_i of the transfer function, $\mathrm{Re}(\lambda_i) < 0$, i.e. if all system poles lie in the left-half of the s plane.

Similar arguments for difference equations give

$$h(nT) = \sum_{i=1}^{k} \sum_{j=1}^{m_i} \alpha_{ij} (nT)^{j-1} (\lambda_i)^n$$

and the definition below.

Theorem   A system modelled by a difference equation with constant coefficients is **BIBO stable** if, for all poles λ_i of the transfer function, $|\lambda_i| < 1$, i.e. iff all system poles lie within the unit circle in the z plane.

Alternative demonstrations can be done using state-space models. For difference equations with constant coefficients, we have the model

$$\mathbf{x}(k+1) = \mathbf{A}\mathbf{x}(k) + \mathbf{B}\mathbf{u}(k)$$

$$\mathbf{y}(k) = \mathbf{C}\mathbf{x}(k) + \mathbf{D}\mathbf{u}(k)$$

with solution

$$\mathbf{y}(k) = \mathbf{C}\mathbf{x}(k) + \mathbf{D}\mathbf{u}(k)$$

$$= \mathbf{C}\mathbf{A}^k \mathbf{x}(0) + \sum_{i=0}^{k-1} \mathbf{C}\mathbf{A}^{k-i-1} \mathbf{B}\mathbf{u}(i) + \mathbf{D}\mathbf{u}(k)$$

If a similarity transform to Jordan form is done, so that $\mathbf{A} = \mathbf{T}\mathbf{J}\mathbf{T}^{-1}$, and if we assume (for convenience) that the eigenvalues of $\mathbf{A}$ are distinct, so that $\mathbf{J} = \mathrm{diag}(\lambda_1, \lambda_2, ..., \lambda_n)$, then this becomes

$$\mathbf{y}(k) = \mathbf{CT}\,\mathrm{diag}\,(\lambda_1^k, \lambda_2^k, \ldots, \lambda_n^k)\,\mathbf{T}^{-1}\mathbf{x}(0) + \mathbf{D}\mathbf{u}(k)$$

$$+ \sum_{i=0}^{k-1}\mathbf{C}\,\mathbf{T}\,\mathrm{diag}\left(\lambda_1^{k-i-1}, \lambda_2^{k-i-1}, \ldots, \lambda_n^{k-i-1}\right)\mathbf{T}^{-1}\mathbf{B}\mathbf{u}(i)$$

Clearly, to have $|\mathbf{y}(k)|$ bounded as k becomes large, if $|\mathbf{u}(j)|$ is bounded for all j, then it is necessary and sufficient that $|\lambda_i| < 1$ for all $i = 1, 2, \ldots, n$. Since the eigenvalues of the $\mathbf{A}$ matrix are the system poles, this is an alternative proof of the statement above.

The parallel of the above for the continuous time state-space model uses the transition matrix $\mathbf{e}^{\mathbf{A}t}$, the similarity transformation

$$\Lambda = \mathrm{diag}(\lambda_1, \lambda_2, \ldots, \lambda_n) = \mathbf{T}^{-1}\mathbf{A}\mathbf{T}$$

and the fact that $\mathbf{e}^{\mathbf{A}t} = \mathbf{T}\,\mathbf{e}^{\Lambda t}\mathbf{T}^{-1}$ with $\mathbf{e}^{\Lambda t} = \mathrm{diag}\,(e^{\lambda_1 t}, e^{\lambda_2 t}, \ldots, e^{\lambda_n t})$ to argue that BIBO stability is equivalent to $\mathrm{Re}(\lambda_i) < 0$ for all $i = 1, 2, \ldots, n$.

It is clear that a candidate system may be investigated by simply computing the pole locations. Many computer programs have little difficulty (except for numerical problems) in doing this. An alternative which allows some design work is to work on the characteristic polynomial without factoring it, as in in the next sections.

14.3 LEFT-HALF PLANE TESTS FOR CONTINUOUS TIME SYSTEMS

For continuous time systems described by linear constant coefficient ordinary differential equations, we are dealing with transfer functions $H(s)$ which are rational in s, i.e. are the ratio of polynomials in s

$$H(s) = \frac{b_0 s^m + b_1 s^{m-1} + b_2 s^{m-2} + \cdots + b_m}{s^n + a_1 s^{n-1} + a_2 s^{n-2} + \cdots + a_{n-1}s + a_n}$$

and where by assumption one usually has $m \leq n$. If the system has a state-space description

$$\dot{\mathbf{x}} = \mathbf{A}\mathbf{x} + \mathbf{B}\mathbf{u}$$

$$\mathbf{y} = \mathbf{C}\mathbf{x}$$

we are interested in the polynomial

$$\det(s\mathbf{I} - \mathbf{A}) = s^n + a_1 s^{n-1} + a_2 s^{n-2} + \cdots + a_{n-1} s + a_n$$

Whether all of the roots of this polynomial lie in the open left-half plane can be determined by examining the coefficients. This can be done with a Routh test.

For the usual test, consider the polynomial

$$D(s) = a_0 s^n + a_1 s^{n-1} + \cdots + a_{n-1} s^1 + a_n$$

Then form the $(2+[n/2])$ column, $n+1$ row array

$$
\begin{array}{c|ccccccc}
s^n & a_0 & a_2 & a_4 & a_6 & \cdots & 0 \\
s^{n-1} & a_1 & a_3 & a_5 & a_7 & \cdots & 0 \\
\hline
s^{n-2} & c_0 & c_1 & c_2 & c_3 & \cdots & 0 \\
s^{n-3} & d_0 & d_1 & d_2 & d_3 & 0 & \\
\vdots & & & & & & \\
s & e_0 & e_1 & 0 & & & \\
s^0 & f_0 & 0 & & & & \\
\end{array}
$$

where each row after the second is formed from the two above it using a simple determinant pattern, e.g.

$$c_0 = \frac{-\begin{vmatrix} a_0 & a_2 \\ a_1 & a_3 \end{vmatrix}}{a_1}$$

$$c_i = \frac{-\begin{vmatrix} a_0 & a_{2i+2} \\ a_1 & a_{2i+3} \end{vmatrix}}{a_1} \qquad i = 0, 1, 2, \ldots,$$

and continuing

$$d_i = \frac{-\begin{vmatrix} a_1 & a_{2i+3} \\ c_0 & c_{i+1} \end{vmatrix}}{a_1} \qquad i = 0, 1, 2, \ldots,$$

etc. Using this array, and in particular its first column, we may consider the location of the system poles, i.e. of the zeros of the polynomial used to generate the array.

1. If
   (a) all of the coefficients a_i, $i = 0, 1, 2, \ldots, n$, are positive, and
   (b) there are no sign changes in the first column of the above array

   then the polynomial $D(p)$ has no right-half plane zeros.

2. If 1(a) holds but there are sign changes, then
   (a) if the elements of the first column are all non-zero, then the number of zeros in the right-half plane is equal to the number of sign changes in the first column, but
   (b) if a first column element is zero and the remainder of that row is also zero, an auxiliary row is generated and the test continued – the use of an auxiliary computation implies that either there is a pair of poles lying on the boundary (i.e. the imaginary axis) or two poles placed symmetrically about the origin on the real axis; the former case will have no sign changes associated with it, while the latter will have one such change – while
   (c) if a first column element is zero and the remainder of the row is not all zeros, the zero is replaced by a small positive element ε and the test is continued, with $\varepsilon \rightarrow 0$ and signs examined as in step 1.

The auxiliary row is formed from the row immediately above it. This is done as the derivative of the auxiliary equation formed from the coefficients of the row above, so that if the row with index k is all zeros, we form the auxiliary row starting with s^k as follows.

$$
\begin{array}{ccccc}
s^{k+1} & h_0 & h_1 & h_2 & h_3 & \cdots \\
s^k & (k+1)h_0 & (k-1)h_1 & (k-3)h_2 & (k-5)h_3 & \cdots
\end{array}
$$

In this case the auxiliary equation is

$$h_0 s^{k+1} + h_1 s^{k-1} + h_2 s^{k-3} + \cdots = 0$$

Example

1. Consider the polynomial

$$D(s) = 1.425s^3 + 4.825s^2 + 1.675s + 0.075$$

The coefficients are all seen to be positive, so the roots might all be in the left-half plane. We form the Routh array.

$$
\begin{array}{l|ll}
s^3 & 1.425 & 1.675 \\
s^2 & 4.825 & 0.075 \\
\hline
s & 5.5965 & 0 \\
1 & 0.0870 &
\end{array}
$$

Since all first column elements are positive, they have the same sign. Hence all roots of $D(s)$ lie in the left-half plane. If $D(s)$ is the denominator of a closed-loop transfer function of a continuous time system, then that system is stable – all of its poles have negative real parts.

2. Now consider $D(s) = s^3 + s^2 + s + 1$. The array starts as

$$
\begin{array}{l|ll}
s^3 & 1 & 1 \\
s^2 & 1 & 1 \\
\hline
s & 0 & 0 \\
1 & ? &
\end{array}
$$

The s^1 row is replaced by the auxiliary, i.e. the derivative of $s^2 + 1$, to form

$$
\begin{array}{c|cc}
s^3 & 1 & 1 \\
s^2 & 1 & 1 \\
\hline
s & 2 & 0 \\
1 & 1 &
\end{array}
$$

There are three roots of the polynomial, as it is third order. It has no right-half plane roots because the first-column signs do not change, and it has two roots which satisfy the (second-order) auxiliary equation, which in this example means two imaginary roots.

14.4 DISCRETE TIME SYSTEMS

When the system is described by linear constant coefficient difference equations, the region for testing becomes the unit circle, but the approach is essentially similar to that above. These systems have difference equation models

$$
y(k) = -\sum_{i=1}^{n} a_i y(k-i) + \sum_{i=0}^{m} b_i u(k-i)
$$

or transfer function models

$$
\frac{Y(z)}{U(z)} = \frac{b_0 + b_1 z^{-1} + b_2 z^{-2} + \cdots + b_m z^{-m}}{1 + a_1 z^{-1} + a_2 z^{-2} + \cdots + a_n z^{-n}}
$$

or state-space models

$$
\mathbf{x}(k+1) = \mathbf{A}\mathbf{x}(k) + \mathbf{B}\mathbf{u}(k)
$$

$$
\mathbf{y}(k) = \mathbf{C}\mathbf{x}(k) + \mathbf{D}\mathbf{u}(k)
$$

where the latter has characteristic polynomial

$$
\det(z\mathbf{I} - \mathbf{A}) = z^n + a_1 z^{n-1} + a_2 z^{n-2} + \cdots + a_n
$$

For this there are several tests, plus a transformation which allows use of the Routh test.

14.4.1 The Jury–Schur–Cohn test: Raible form

Suppose a system has characteristic equation

$$D(z) = 0 = a_0 z^n + a_1 z^{n-1} + a_2 z^{n-2} + \cdots + a_{n-1} z + a_n \qquad (14.3)$$

where we assume that $a_0 > 0$. Then form the array

$$
\begin{array}{cccccc}
a_0 & a_1 & a_2 & a_3 \cdots & a_{n-1} & a_n \\[4pt]
a_n & a_{n-1} & a_{n-2} & \cdots \ a_1 & a_0 & \alpha_n = \dfrac{a_n}{a_0} \\[6pt]
\hline
{}^{n-1}a_0 & {}^{n-1}a_1 & {}^{n-1}a_2 & \cdots \ {}^{n-1}a_{n-1} & & \\[4pt]
{}^{n-1}a_{n-1} & {}^{n-1}a_{n-2} & {}^{n-1}a_{n-3} & \cdots \ {}^{n-1}a_0 & & \alpha_{n-1} = \dfrac{{}^{n-1}a_{n-1}}{{}^{n-1}a_0} \\[6pt]
\hline
\vdots & & & & & \\[4pt]
\hline
{}^{0}a_0 & & & & &
\end{array}
$$

where for $k = n, n-1, \ldots, 1$ and $i = 0, 1, \ldots, k-1$

$$
{}^{k-1}a_i = {}^{k}a_i - \alpha_k \, {}^{k}a_{k-i} \qquad\qquad \alpha_k = \frac{{}^{k}a_k}{{}^{k}a_0}
$$

$$
{}^{n}a_i = a_i \qquad\qquad\qquad i = 0, 1, \ldots, n
$$

It can be observed that the first two rows are the coefficients in forward and reverse order, the third is obtained by subtracting the second multiplied by α_n from the first, the fourth is the third in reverse order, the fifth is the result of subtracting the fourth multiplied by α_{n-1} from the third, the fifth is the fourth in reverse order, etc.

By Jury's theorem (Åström and Wittenmark, 1990), if $a_0 > 0$, then the characteristic equation has all of its roots inside the unit circle if $^k a_0 > 0$, $k = 0, 1, \ldots, n-1$. Furthermore, if a $^k a_0 \neq 0$ for all k in that range, the number of roots outside the unit circle equals the number of negative $^k a_0$.

Example

We consider the stability of the system described in part by

$$\mathbf{x}(k+1) = \begin{bmatrix} 0 & 1 & 0 & 0 & 0 \\ 0 & 0 & 1 & 0 & 0 \\ 0 & 0 & 0 & 1 & 0 \\ 0 & 0 & 0 & 0 & 1 \\ \dfrac{-9}{256} & \dfrac{3}{16} & \dfrac{-7}{64} & \dfrac{-1}{2} & 1 \end{bmatrix} \mathbf{x}(k) + \mathbf{B}\mathbf{u}(k)$$

Its characteristic equation is the polynomial

$$D(z) = z^5 - z^4 + \frac{1}{2} z^3 + \frac{7}{64} z^2 - \frac{3}{16} z + \frac{9}{256}$$

and it is this which is analysed. The resulting array is

1	−1	0.5	0.1094	−0.1875	0.0352
0.0352	−0.1875	0.1094	0.5	−1	1

$\alpha_5 = 0.0352$

0.9988	−0.9934	0.4962	0.0918	−0.1523
− 0.1523	0.0918	0.4962	−0.9934	0.9988

$\alpha_4 = -0.1525$

1.0220	−1.0074	0.4205	−0.0597
− 0.0597	0.4205	−1.0074	1.0220

$\alpha_3 = -0.0584$

1.0185	−0.9828	0.3616
0.3616	−0.9828	1.0185

$\alpha_2 = 0.3526$

0.8980	−0.6363
−0.6363	0.8980

$\alpha_1 = -0.7086$

0.4471

As all of the $^i a_0$, that is, all of the leading elements of the odd numbered rows, are positive, the roots of the system are all within the unit circle. (In fact, separate computation shows that the roots are at

$$z = \mp\frac{1}{2}, \frac{1}{4}, \frac{3}{4}, \exp(\mp j \cos^{-1} 0.5)$$

which are indeed within the unit circle.)

14.4.2 Jury test: Jury form

The above is the Raible tabulation for the Jury test. An alternative form of the same test on (14.3) uses the array below, in which the coefficients come in row pairs, with the second row of each pair derived from the pair above it by forming a determinant of the first column and the column next to the right of the coefficient in question; the odd row above this new row is the new row in the reverse direction.

a_n	a_{n-1}	a_{n-2}			a_1	a_0
a_0	a_1	a_2	$\cdots$		a_{n-1}	a_n
b_{n-1}	b_{n-2}	b_{n-3}		$\cdots$	b_1	b_0
b_0	b_1	b_2		$\cdots$	b_{n-2}	b_{n-1}
c_{n-2}	c_{n-3}	c_{n-4}	$\cdots$	c_1	c_0	
c_0	c_1	c_2	$\cdots$	c_{n-3}	c_{n-2}	

$$\vdots$$

p_3	p_2	p_1	p_0
p_0	p_1	p_2	p_3
q_2	q_1	q_0	

where

$$b_k = \begin{vmatrix} a_n & a_{n-1-k} \\ a_0 & a_{k+1} \end{vmatrix} = a_n a_{k+1} - a_0 a_{n-(k+1)} \qquad k = 0, 1, 2, \ldots, n-1$$

$$c_k = \begin{vmatrix} b_{n-1} & b_{n-(k+2)} \\ b_0 & b_{k+1} \end{vmatrix} = b_{n-1} b_{n+1} - b_0 b_{n-1-(k+2)} \quad k = 0, 1, 2, \ldots, n-2$$

$$\vdots$$

$$q_k = p_3 p_{k+1} - p_0 p_{3-(k+1)} \qquad\qquad k = 0, 1, 2$$

Then the criterion is that the system is stable if all of the following hold:

1. $|a_n| < |a_0|$

2. $D(1) > 0$

3. $D(-1)*(-1)^n > 0$

4. $|b_{n-1}| > |b_0|$

    $|c_{n-2}| > |c_0|$

    $$\vdots$$

    $|q_2| > |q_0|$

14.4.3 Refinements

If we wish to determine whether roots lie exactly on the unit circle (and hence lead to constant amplitude oscillations), we may refine the test. We notice first that the Raible form is indeterminate as to number when a test coefficient $^k a_0 = 0$. Furthermore, the array cannot be continued (because the corresponding α will be undefined). To get around this, we can test whether roots are inside the circle of radius $(1+\varepsilon)$ for small positive and negative ε; see also section 14.5 below. In principle, we replace z by $(1+\varepsilon)z'$ and check how many roots are outside the circle defined by $\|z'\| = 1$, i.e. outside $\|z\| = (1+\varepsilon)$. In operation, we replace the coefficients a_i with $(1+\varepsilon)^{n-i}a_i$ and approximate this (because ε is 'small') by $(1+(n-i)\varepsilon)a_i$. The test array is then built with ε carried as a symbol. Since we would seldom be so unlucky as to have the test array now be indeterminate, we could look at the number of coefficients which are negative for $\varepsilon < 0$ and for $\varepsilon > 0$: the number which are negative in both instances are the number outside the original unit circle, and the difference between the number in the two instances is the number on the original unit circle. Examples are given by Kuo (1980), among others.

14.4.4 Route test of *w*-transformed equation

The above are two variations of a direct test for poles inside the unit circle. A third possibility is to transform the discrete time transfer function into a form to which the Routh test is applicable. The particular change used is the bilinear form, which sets z to

$$z = \frac{1+w}{1-w} \tag{14.4}$$

because it maps the left-half of the w-plane (i.e. $\mathrm{Re}(w) < 0$) into the unit circle ($\|z\| < 1$) and the right-half plane into $\|z\| > 1$, with the imaginary axis into $\|z\| = 1$. The inverse transformation is also bilinear, using the substitution

$$w = \frac{z-1}{z+1} \tag{14.5}$$

This is frequently called the **w-transform**, although it is rarely used as a true transform; sometimes the factor $2/T$ appears multiplying the right-hand side in (14.5), with a corresponding change in (14.4) and then the expression is similar to that used for the Tustin transformation in Chapter 12. When applied to an expression such as

$$D(z) = a_0 z^n + a_1 z^{n-1} + \cdots + a_n$$

(14.4) yields

$$D_w(w) = \{a_0(1+w)^n + a_1(1-w)(1+w)^{n-1} + \cdots$$

$$+ a_{n-1}(1-w)^{n-1}(1+w) + a_n(1-w)^n\}/(1-w)^n \tag{14.6}$$

We are interested in the number of unstable zeros of $D(z)$, which is the number of unstable zeros of $D_w(w)$, where the former is the number of zeros outside the unit circle and the latter is the number of zeros in the right-half plane. Rearranging the numerator of (14.6) yields the polynomial of interest, i.e.

$$D''(w) = b_0 w^n + b_1 w^{n-1} + \cdots + b_{n-1} w + b_n$$

It is this polynomial which is tested using the Routh test of section 14.3.

14.5 RELATIVE STABILITY

The tests in this section are, at their core, yes/no tests: a system of equations does or does not have its poles inside the unit circle or in the left-half plane, as appropriate. It is quite possible, however, to modify the procedures slightly so as to test:

- whether the poles are inside a circle other than the unit circle in the case of Jury tests; or
- whether the poles are to the left of a line other than $s = 0$ in the case of Routh tests.

In the first case, the Jury test can be performed on the polynomial $D(z_r) = D(z/r)$. If $D'(z_r) = D(rz_r)$ has all roots inside $z_r = 1$, then $D(z)$ has all roots inside $z = r$. Hence the method is to replace z by rz_r in $D(z)$ and test whether z_r is inside the unit circle.

In the second case, we see that the Routh test can be performed on the polynomial $D'(p) = D(p+a)$. If $D'(p)$ has all zeros in the left-half plane (i.e. with $\text{Re}(p) < 0$), then $D(s)$ has all zeros to the left of $s = a$, (i.e. with $\text{Re}(s) < a$). The method is to replace s by $p+a$ and test whether all zeros of $D'(p)$ lie in the left-half plane.

The procedure is straightforward: perform the indicated substitution to obtain new coefficients of the polynomial and then apply the Jury or Routh test as appropriate to the system.

Example

Suppose that a unity feedback system of standard form has compensator $C(s) = K$, an unknown parameter, feedback instrument transfer function $F(s) = 1$, and plant transfer function

$$G(s) = \frac{0.075}{1.425s^3 + 4.825s^2 + 1.675s + 0.075}$$

We wish to determine whether the poles decay at least as fast as e^{-t} for $K = 10$. First we find the closed-loop transfer function as

$$H(s) = \frac{KG(s)}{1 + KG(s)}$$

$$= \frac{0.075K}{1.425s^3 + 4.825s^2 + 1.675s + 0.075 + 0.075K}$$

For all transients to decay at least as rapidly as e^{-t}, the poles must lie to the left of $s = -1 + j\omega$, i.e. the roots of the denominator polynomial must have real part less than -1. We know that the denominator polynomial is

$$D(s) = 1.425s^3 + 4.825s^2 + 1.675s + 0.825$$

because we have chosen $K = 10$. We substitute $p - 1$ for s, since then $\mathrm{Re}(p) < 0 \Rightarrow \mathrm{Re}(s) < -1$.

$$D(p-1) = 1.425p^3 + 0.55p^2 - 3.70p + 2.55$$

We see immediately, because the coefficients are not all positive, that $\mathrm{Re}(p) < 0$ is not possible. Hence, the transients cannot decay faster than e^{-t} for $K = 10$.

Other transformations allow other regions to be tested. For example, the replacement $z = r(z_t + \alpha)$, where α is a real number, applied with a Jury-type test allows us to determine whether z_t is inside a circle of radius r centred at $(\alpha, 0)$ in the complex plane.

14.6 DESIGN FOR STABILITY

As mentioned several times, the tests themselves give only checks on
stability and tell nothing about how to make an unstable system stable.
Clever use, however, allows a certain amount of design to be done.
This has two steps, which are in fact independent. First the designer
introduces one or more parameters, perhaps the implementation
parameters of some controller, into the system transfer function.
Then the tests are set up and the resulting algebra examined to
determine which values of the parameters, if any, will result in poles
inside the region of interest (e.g. unit circle for stability, radius r
circle for transient speed of decay). The procedure is straightforward
provided the algebra can be solved. This last is a severe restriction
with problems of appreciable size.

Example 1

For the example problem of the previous section, we now wish to
determine the range of K for which the system is stable. Stability is
indicated if the poles, i.e. the roots of the denominator polynomial,
are in the left-half plane. To determine this, we apply the Routh test.
   First we see that for all the coefficients to be positive we must have
$K > -1$. Then we set up the array as follows:

s^3	1.425	0.675
s^2	4.825	$0.075\,(1+K)$
s	$5.5965 - 0.075K$	0
1	$0.0155\,(1+K)\,(5.5965-0.075K)$	

   For the first column to have no sign changes, we need all elements
positive.   From this it follows that the system is stable for
$K < (5.5965/0.075 =)$ 74.62 and then, with the s term positive, the s^0
row requires $K > -1$ (which we already knew from the coefficient
test). Thus, the system is stable provided that $-1 < K < 74.62$.

Example 2

Consider the system described by

$$\mathbf{x}(k+1) = \begin{bmatrix} 0 & 1 \\ 0 & -\frac{1}{2} \end{bmatrix} \mathbf{x}(k) + \begin{bmatrix} 0 \\ 1 \end{bmatrix} \mathbf{u}(k)$$

for which a control law

$$\mathbf{u}(k) = [\, K \ \ 0 \,] \mathbf{x}(k)$$

is to be designed. Then we evaluate the resulting characteristic equation, which is

$$D(z) = 0 = z^2 + 0.5z + K$$

Then $a_0 = 1 > 0$ and we may apply the basic test. The array is

$$
\begin{array}{ccc}
1 & 0.5 & K \\
K & 0.5 & 1
\end{array}
$$

$$\alpha_2 = K$$

$$
\begin{array}{cc}
1 - K^2 & 0.5(1 - K) \\
(1 - K)/2 & 1 - K^2
\end{array}
$$

$$\alpha_1 = 0.5/(1 + K)$$

$$(1 - K)(1 + K) - (1 - K)/4(1 + K)$$

Our conditions are that

$$1 - K^2 > 0$$

$$1 - K^2 - (1 - K)/4(1 + K) > 0$$

The first condition is satisfied if $-1 < K < 1$. The second requires, provided the first holds, that

$$4(1 + K)(1 - K^2) - (1 - K) > 0$$

or

$$(1-K)(1+2K)(3+2K) > 0$$

which needs $-0.5 < K < 1$. Combining these, the system for which the above is the characteristic equation is stable iff $-0.5 < K < 1$.

As an alternative, the Jury form requires

1. $|K| < 1$
2. $D(1) = 1.5 + K > 0$
3. $D(-1) = 0.5 + K > 0$
4. Not applicable: $n = 2$

This is clearly satisfied by $-0.5 < K < 1$.

Finally, for a comparison of methods, we also use the modified Routh method for the same problem. Making the bilinear w-substitution in $D(z)$ yields

$$D_w(w) = (1+w)^2 + 0.5\,(1-w)(1+w) + K(1-w)^2$$

$$= (0.5+K)w^2 + (2-2K)\,w + (1.5+K)$$

The Routh array is

w^2	$0.5+K$	$1.5+K$
w	$2-2K$	0
1	$1.5+K$	

For no sign changes, all of the first column must have the same sign, yielding $K > -0.5$, $K < 1$, $K > -1.5$, respectively, or $-0.5 < K < 1$, or else $K < -0.5$, $K > 1$, $K < -1.5$, respectively, which is inconsistent. The result is seen to be the same as that given by the Jury test, but the substitution is tedious.

14.7 COMPUTER ASSISTANCE

The tests presented here are easily encoded for numerical evaluation, and students should find it easy to write such programs to determine the utility and characteristics of the array tests. For stability determination, most computer assisted design software will include root finders, so that poles may be calculated directly. For studies of parameter changes, root locus and root contour methods (Chapter 18) are also usually available.

14.8 SUMMARY AND FURTHER READING

The stability of linear constant coefficient systems is determined by the locations of their closed-loop poles. These may be computed directly, or the general regions of their locations may be inferred from Routh-type tests which process arrays developed from the characteristic equations. The tests are straightforward to perform, especially with a computer. They may be extended to test regions with circles other than the unit circle and half planes other than the negative half plane by simple algebraic transformations. They also have limited uses for finding regions, if they exist, for which parameters will yield desirable properties. An important fault is that, if the test fails, it gives little indication of what to do about the situation.

Routh tests and Hurwitz variations date from the late 1800s; Jury's unit circle tests date from the 1950s. The tests are standards in most control systems textbooks. For examples, Routh tests and some variations appear in texts such as (Dorf, 1989) and (Phillips and Harbor, 1991). The basic Jury form is in most digital control system texts, including (Franklin *et al.*, 1990) and (Ogata, 1987), while the Raible form is included in (Kuo, 1980).

15

Nyquist stability theory

One of the now classical methods of testing closed-loop stability as a function of plant model and of control loop structure is based upon complex variable theory, with the particular application due to Nyquist. The idea is subtle for beginners but is basically straightforward. It has been applied to basic stability testing, to certain non-linear systems, to multivariable systems, and to developing the concepts of relative stability.

15.1 SYNOPSIS

Nyquist theory treats a transfer function $F(s)$ (or $F(z)$) as a mapping from the s (resp. z) plane to the complex plane. When the independent variable traverses a closed path, then the mapping will encircle the origin a number of times N equal to the net number of poles and zeros enclosed by the closed path. With a number of modifications and interpretations according to application, this simple idea can be used for a number of stability tests. The basic approach works as follows.

1. The contour is taken as enclosing the unstable region of the independent variable plane. In discrete time systems the basic contour follows the curve $e^{j\theta}$, as θ ranges from 0 to 2π, while in continuous time systems the contour forms a D-shaped figure with the vertical segment following the imaginary axis and the arc 'enclosing' the right-half plane. In particular the contour traverses $s = j\omega$ for $-R \leq \omega \leq R$ and the semicircle $Re^{j\theta}$ for $\pi/2 \geq \theta \geq -\pi/2$, with $R \to \infty$; the traversal is clockwise and usually starts at the origin.
2. The function F is taken as $1 + k\,G(q)$; then the function $G(c)$ is plotted for c on the appropriate contour.
3. For any chosen value of k, the number N of clockwise encirclements of $-1/k$ by the mapping is counted.
4. The number P of poles of $G(q)$ inside the contour is determined.

5.   Then, the number Z of zeros of $1 + k\,G(q)$ inside the contour, and hence the number of unstable poles of the closed-loop system for which $1 + k\,G(q) = 0$ is the characteristic equation, is given by $Z = P + N$.

Extensions may be made relatively easily. For example, the notion of relative stability arises by considering, for a value of k for which the system is stable, the magnitude or phase change needed in $G(q)$ for the system to become unstable. Multivariable problems involve simultaneous consideration of the various scalar input–output transfer functions, while certain static non-linearities in otherwise linear systems may be examined in terms of 'variable $-1/k$ points'. In principle, compensator design can also be done using the Nyquist plots.

15.2 THE UNDERLYING BASIS – CAUCHY'S THEOREM

The underlying theory is that of complex functions of complex variables, so we will consider here a variable q, taken later to be s or z as appropriate. Then the function $F(q)$ defines a mapping from the q coordinate system to the F coordinates, where both systems have real and imaginary coordinates. A line between two points q_0 and q_1 in the q plane will map into a line between $F(q_0)$ and $F(q_1)$ in the F plane, as in Fig. 15.1, provided that $F(\cdot)$ is a 'nice' function; this means basically that it is an analytic function with continuous derivative along the line, which is a condition easily met by linear transfer functions at all but a small number of points.

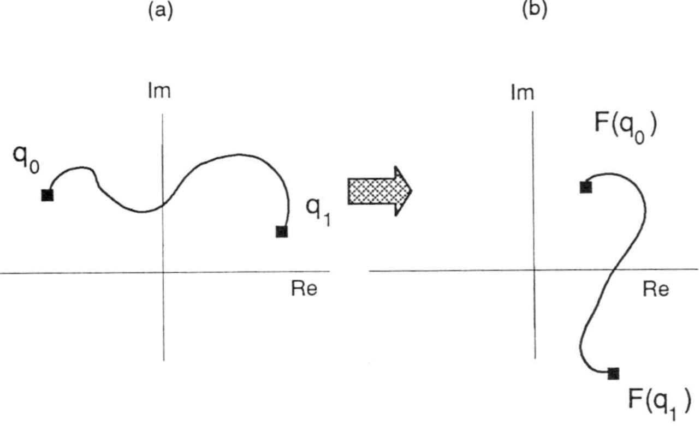

Figure 15.1 Continuous mapping from one complex plane to another.

Furthermore, a closed contour from q_0 to q_1 and back to q_0 via q_2 will map a closed figure from $F(q_0)$ to $F(q_1)$ and back to $F(q_0)$ via $F(q_2)$, as in Fig. 15.2. The mapping may be one-to-one, as shown in the figure, but this is not always the case; a many-to-one function will result in the mapping's crossing itself.

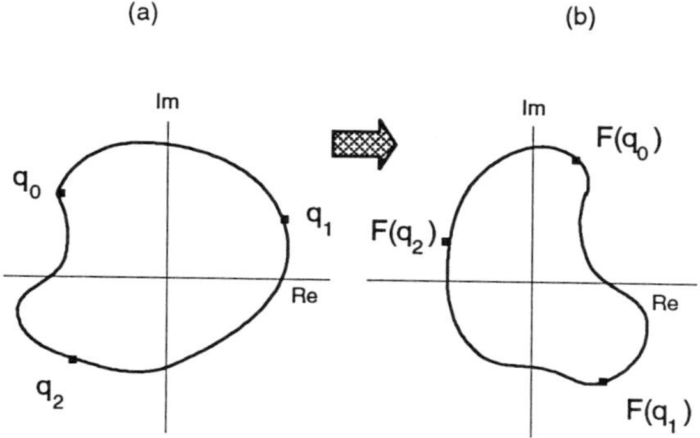

Figure 15.2 Mapping of a closed contour from one complex plane to another.

Now consider an arbitrary closed contour Γ in the q-plane. We use the following definitions.

Enclosed A point P is enclosed by a contour Γ if P is encircled by Γ in either the clockwise or counter-clockwise direction.

Encircled A point P is encircled n times in a clockwise direction by a contour Γ if, when a point S on the contour moves from a starting point q_0 around the contour and back to the starting point, an observer on P of S turns n times (net) in the clockwise direction while facing S. Thus a vector from P to S goes through $-360°n$ as S traverses Γ. The encirclement is in the counter-clockwise direction if the angle is $+360°n$. The point is not encircled if the angle is $0°$.

Then, the following theorem may be stated:

Theorem (Cauchy) Let a contour Γ encircle a region C of the q-plane in a clockwise direction so that Γ encloses P poles and Z zeros of a function $F(q)$ which is analytic (basically, is 'nice') in C. Then the origin of the $F(q)$ plane is encircled N times in the clockwise direction by the mapping of Γ where $N = Z - P$.

We remark that encirclement N times in the clockwise direction is the same as encirclement $-N$ times in the counter-clockwise direction. A hypothetical case illustrating the theorem with $Z = 1$, $P = 2$, and $N = -1$ is shown in Fig. 15.3.

(a)   (b)

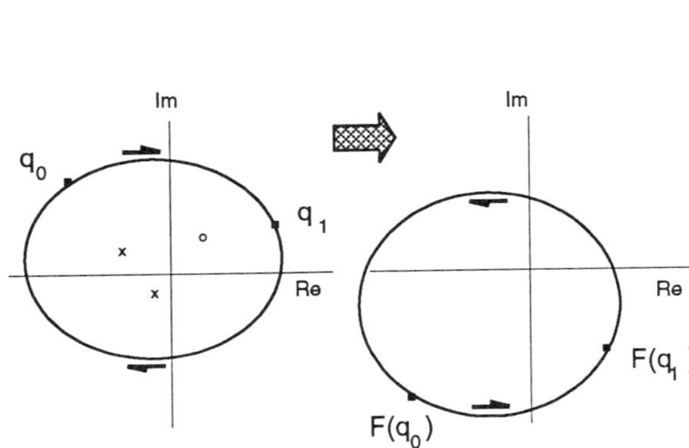

Figure 15.3 Mapping in which two poles and one zero of the function $F(q)$ are encircled by the closed contour, leading to -1 encirclements of the origin. Illustrating the Cauchy Theorem.

The following corollary involves a simple coordinate translation of the above.

Corollary If the function $F(q)$ is of the form $F(q) = a + G(q)$, then the mapping $G(q)$ of the contour Γ will encircle the point $(-a)$ in the $G(q)$ plane N times, where $N = Z - P$ and Z and P are defined as above.

15.3 APPLICATION TO DETERMINING STABILITY OF LINEAR CONTROL SYSTEMS – NYQUIST CONTOURS AND CURVES

The application of the above involves the choice of functions $F(\cdot)$ to evaluate and of the contour Γ. For a system of form as in Fig. 15.4, the closed-loop transfer function is given by

$$\frac{Y(q)}{R(q)} = \frac{C(q)\,G(q)}{1 + C(q)\,G(q)\,H(q)}$$

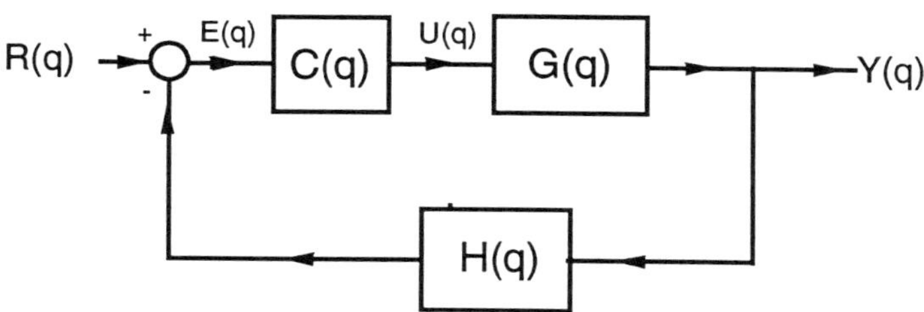

Figure 15.4 A standard closed-loop system model, where q may be either s or z.

The key to stability for such a system is the location of the closed-loop poles, i.e. of the zeros of the denominator. For this reason, we take

$$F(q) = 1 + C(q)\,H(q)\,G(q)$$

and determine, using the Cauchy Theorem, how many zeros of this function lie in the right-half plane for continuous time representations or outside the unit circle in the discrete time case.

Continuous time case

1. Define the Nyquist contour Γ as the semidisc with centre at the origin, radius $\rightarrow \infty$, and enclosing the right-half plane. Thus, take the three-arc contour defined by

    (a) $s = j\omega$ ω increases for 0 to R
    (b) $s = Re^{j\theta}$ θ decreases from $\pi/2$ to $-\pi/2$
    (c) $s = -j\omega$ ω increases for $-R$ to 0

and let R become very large, i.e. $R \rightarrow \infty$ (Fig. 15.5(a)).

2. Evaluate $C(s)H(s)G(s)$ along the contour, thus obtaining the Nyquist plot.
3. Count the number P of enclosed (i.e. right-half plane) poles of the open-loop transfer function $C(s)H(s)G(s)$.
4. Count the number N of clockwise encirclements of the point $(-1,0)$ by the Nyquist plot.
5. Then the number Z of unstable poles of the closed-loop system is given by $Z = P + N$.

Special cases, such as poles lying on the imaginary axis, are accommodated using simple tricks in the definition of the Nyquist contour, in particular by bypassing the point by a semicircle of radius ε with $\varepsilon \to 0$.

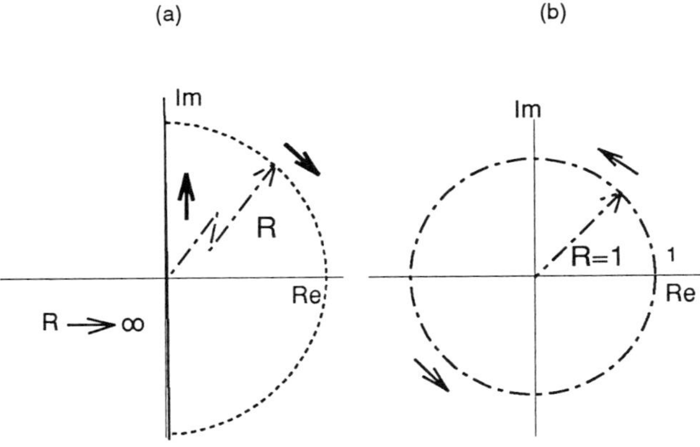

Figure 15.5 The basic Nyquist contours: (a) for the continuous time system enclosing the right half plane; and (b) for the discrete time system, 'enclosing' the region outside the unit disc.

Discrete time case

The Nyquist contour Γ is defined by the unit circle, i.e. by

$$z = \mathrm{e}^{\mathrm{j}\phi} \qquad\qquad \phi = 0 \to 2\pi$$

and this is taken as 'enclosing' the region of the z-plane for which $\| z \| > 1$ (Fig. 15.5(b)). Variations of this definition exist for those who are uncomfortable with this enclosure of an unbounded region.

An important variation of this scheme has the denominator of the general form

$$F(q) = 1 + KG(q) \tag{15.1}$$

Here, one can plot $G(q)$ for the appropriate contour and consider encirclements of $-1/K$ for any K. By doing so, the range of values of K for which the closed-loop system is stable can be established.

Examples

Nyquist plots are demonstrated in Fig. 15.6 for the continuous time system

$$G(s) = \frac{4(s+1)}{s\left(\dfrac{s^2}{625} + 0.8\dfrac{s}{25} + 1\right)\left(0.3s^2 + s - 1\right)}$$

and the discrete time system

$$G(z) = \frac{0.05z^{-1}}{(1 - 0.9z^{-1})\,(1 - 0.5z^{-1})}$$

In the first case, the contour is that of Fig. 15.5(a) modified in the vicinity of the origin to be $\rho e^{j\theta}$, $-\pi/2 < \theta < \pi/2$ to avoid having the contour pass through 0 and hence to keep the path Γ such that the function $G(z)$ is analytic for all points on it. Since ρ is taken as small, $G(\rho e^{j\theta})$ will have large magnitude here. The Nyquist plot (Fig. 15.6(a)) shows the regions of the real axis (and hence $-1/K$ in (15.1)) for which N, and hence Z, is constant; P is of course constant and equal to 1.

The Nyquist plot of $G(z)$ is determined from $G(e^{j\phi})$ for $0 \le \phi < 2\pi$ and is shown in Fig. 15.6(b). Indicated are the regions of encirclement of $-1/K$ and the resulting values of Z, the number of closed-loop unstable poles of the closed-loop system. Notice that one value of K for which Z changes is $K = 57$, while the other value is $K = -1$; the axis is crossed by the Nyquist plots at $(-1/57, 0)$ and $(1, 0)$ respectively.

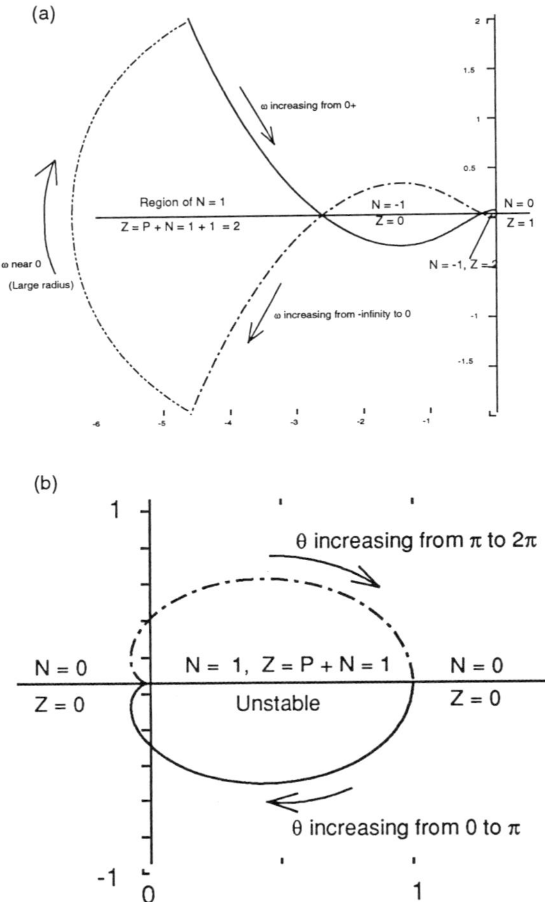

Figure 15.6 Nyquist plots for text examples. (a) Close-up to show detail near the origin for the continuous system, showing regions of various N (and hence Z) values. $Z = 0$ only for $-2.5 < -1/K < -0.4$. (b) Full Nyquist plot for discrete time system. Here the regions for which $Z = 0$ are those for which $-1/K < -0.0175$ or $-1/K > 1$.

15.4 THE MULTIVARIABLE EXTENSION

In the above, the system was SISO. The MIMO, or multivariable, extension, involves intricacies which we will not dwell upon. Some of the ideas which are important to later work, however, are presented here.

Rather than a scalar function, let $\mathbf{G}(q)$ be a matrix of dimension $n \times n$ and the system as in Fig. 15.7, where in particular the choice $\mathbf{C}(q) = \mathbf{K}$ has been made.

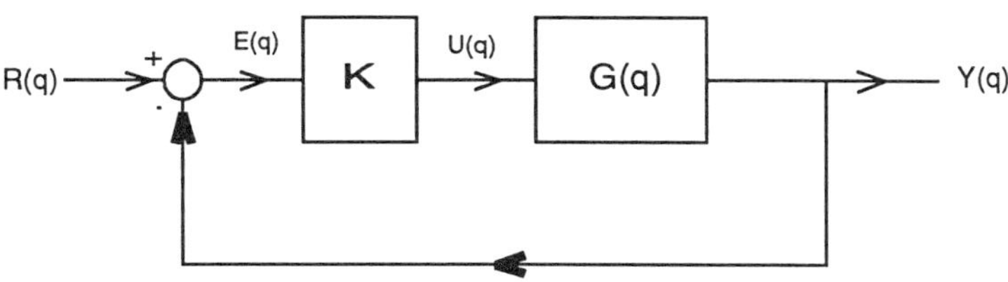

Figure 15.7 Very simple multivariable system feedback configuration. $\mathbf{K}$ and $\mathbf{G}(q)$ are matrices, and $\mathbf{R}(q)$ and $\mathbf{Y}(q)$ are vectors.

In general, $\mathbf{K}$ will be $n \times n$, and we will attempt to establish system stability limits for various $\mathbf{K}$. It is easily shown that

$$\mathbf{Y}(q) = [\mathbf{I} + \mathbf{G}(q)\,\mathbf{K}]^{-1}\,\mathbf{G}(q)\,\mathbf{K}\mathbf{R}(q)$$

It can also be shown that $\mathbf{Y}(q)$ is exponentially stable (for pulse input, and hence for all inputs), iff $\det[\mathbf{I} + \mathbf{G}(q)\,\mathbf{K}]$ has no poles in the unstable region (right-half plane or outside the unit circle, as appropriate to the system). As the determinant is a scalar function, it is straightforward to treat it in the same manner as the function $F(q)$ used in the scalar function Cauchy arguments. Thus in the usual manner,

N = number of clockwise encirclements of origin by $\det[\mathbf{I} + \mathbf{G}(q)\,\mathbf{K}]$

P = number of unstable poles of the Smith–McMillan form of $\mathbf{G}(q)$ (see Chapters 10 and 11), which is the number of unstable roots of $p(q)$ when $\mathbf{G}(q) = \mathbf{N}(q)/p(q)$ for $\mathbf{N}(q)$ a matrix of polynomials and $p(q)$ is a scalar polynomial.

Z = number of zeros of $[\mathbf{I} + \mathbf{G}(q)\,\mathbf{K}]$ enclosed by the Nyquist contour

and the result is that

$$N = Z - P$$

We could in principle draw the Nyquist plots for each matrix $\mathbf{K}$ of interest. This is not an attractive option, for obvious reasons: the ability to determine a range of stable K is one of the attractions of the Nyquist SISO theory. To examine this problem partially we look first at the special case $\mathbf{K} = k\mathbf{I}$. Now we are interested in Nyquist plots associated with $\det [\mathbf{I}+k\mathbf{G}(q)]$. Since a determinant of a matrix is equal to the product of that matrix's eigenvalues, we instead use the eigenvalues (perhaps more accurately called eigenvalue functions, since they will depend upon q) of $[\mathbf{I}+k\mathbf{G}(q)]$. But the eigenvalues of $[\mathbf{I}+k\mathbf{A}]$ are respectively equal to 1 + eigenvalues of $[k\mathbf{A}]$ which in turn equal $1+k$ eigenvalues of $[\mathbf{A}]$. Thus we first find n functions $\lambda_i(q)$ for which

$$\det [\lambda_i(q)\mathbf{I} + k\mathbf{G}(q)\,] = 0 \qquad i = 1, 2, ..., n$$

Then

$$\det [\mathbf{I} + k\mathbf{G}(q)] = \prod_i (1 + k\lambda_i(q))$$

and hence

$$\Delta \arg \det [\mathbf{I} + k\mathbf{G}(q)] = \sum_i \Delta \arg [\,1 + k\lambda_i(q)\,]$$

Thus we can count the number N by counting the number of clockwise encirclements of the origin by the n functions $[1 + k\lambda_i(q)]$, of the point $(-1,0)$ by the functions $k\lambda_i(q)$, or the point $(-1/k, 0)$ by the functions $\lambda_i(q)$. We remark that the individual functions, because they are not necessarily rational, may not form closed curves as the Nyquist contour is traversed, but that the set of functions will be closed.

The next case to consider has $\mathbf{K} = \text{diag}\{k_1, k_2, ..., k_n\}$. To examine it, we need more definitions and background. The key ideas, however, are those of Gershgorin bands and Nyquist arrays. The latter are the Nyquist plots of the elements $g_{ij}(s)$ of $\mathbf{G}(s)$. The former are motivated by the following result.

Theorem (Gershgorin) The eigenvalues of a complex $n \times n$ matrix $\mathbf{Z} = \{z_{ij}\}$ lie in the union of the n circles c_i, where c_i = circle centred at z_{ii} with radius

$$r_i = \sum_{\substack{j \neq i \\ j=1}}^{n} \| z_{ji} \|$$

A matrix $\mathbf{G}(q)$ for which the Gershgorin bands exclude the origin is said to be **diagonally dominant**, and a system with narrow bands is one which closely resembles a set of non-interacting SISO systems.

In application, we plot the Nyquist array of a square $\mathbf{G}(q)$ by:

1. plotting (separately) each $g_{ij}(q)$ for q on the appropriate Nyquist contour; and
2. on the plots $g_{ii}(q)$, plotting for all (actually for sufficient) points $q_c \in$ contour (e.g. $q_c = j\omega_c$ for selected ω_c) the circles centred at $g_{ii}(q_c)$ with radius

$$\sum_{\substack{j \neq i \\ j=1}}^{n} \| g_{ji}(q_c) \|$$

The union of the circles forms a band around the plot $g_{ii}(q)$. Thus we have n Gershgorin bands, one for each element of the diagonal of the Nyquist array.

Example

A Gershgorin band is illustrated in Fig. 15.8 for the system defined by

$$\mathbf{G}(s) = \begin{bmatrix} \dfrac{1.5}{s^3 + 2s^2 + 2s + 1} & x \\[2ex] \dfrac{1}{4s + 1} & x \end{bmatrix}$$

where the x denote elements irrelevant to the computation.

Finally, we relate this to the diagonal $\mathbf{K}$ feedback gain using a result due to Rosenbrock.

Theorem (Rosenbrock, 1970) For $\mathbf{G}(q)$ square and $\mathbf{K} = \text{diag}\{k_1, k_2, \ldots, k_n\}$, suppose

$$\left| g_{ii}(q) + \frac{1}{k_i} \right| > \sum_{\substack{j \neq i \\ j=1}}^{n} |g_{ij}(q_c)| \qquad i = 1, 2, \ldots, n$$

for all q on the Nyquist contour, and let the Gershgorin band associated with g_{ii} encircle the point $(-1/k_i, 0)$ a net of N_i times in the clockwise direction. Then the negative feedback system characterized by $[\mathbf{I} + \mathbf{G}(q)\mathbf{K}]^{-1}$ has Z zeros enclosed by the Nyquist contour, where

$$Z = P + \sum_{i=1}^{n} N_i$$

and P is the number of poles of $\mathbf{G}(q)$ enclosed by the contour.

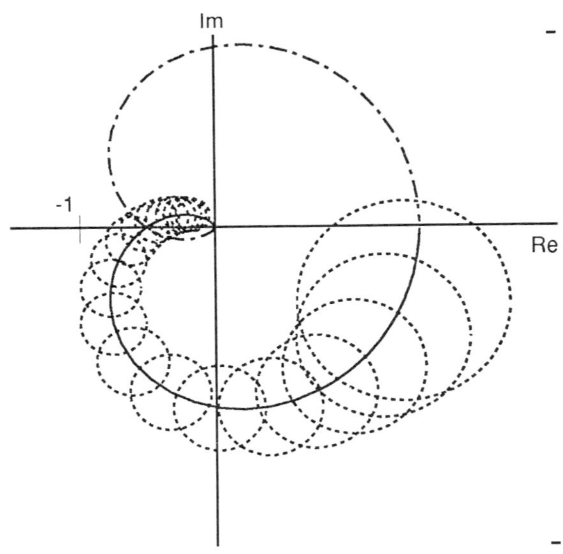

Figure 15.8 Gershgorin circles for the text example.

Using the above, it is clear that the requirement is that

$$P = -\sum_{i=1}^{n} N_i$$

for closed-loop stability.

The above discussion has avoided several issues, including the necessity in the last case for the transfer function to be diagonally dominant and for the Gershgorin bands not to enclose the points in question.

15.5 NON-LINEARITIES AND LIMIT CYCLES

One of the interesting applications of Nyquist ideas is to simple, static non-linearities such as relays, saturation elements, etc. The basic idea is that the non-linearity is approximated to first-order by a linear term in which the gain depends upon the amplitude, but not the instantaneous value, of the input. In this section we look briefly at an example of the approach (called the **describing function**) and show the Nyquist ideas.

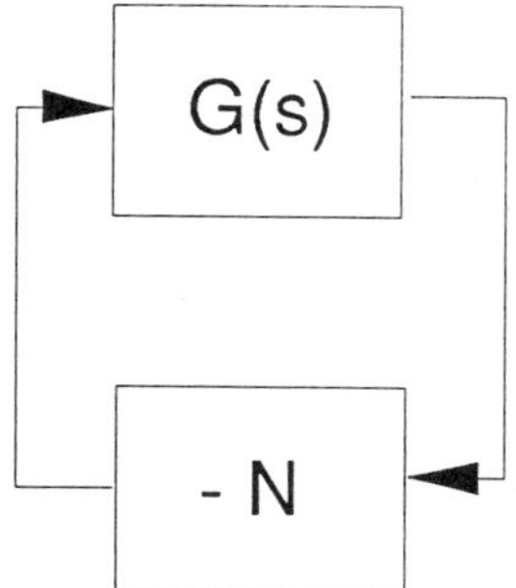

Figure 15.9 Common non-linear system configuration for studies of non-linear stability. *N* is a static non-linearity (such as a relay).

More precisely, consider the autonomous system of Fig. 15.9. The object is to determine whether the system can sustain an unforced oscillation, or **limit cycle**. To this end, let $e = E \sin \omega t$ and assume that *N* can be modelled as

$$u = N(e) = K_{eq}(E)\, e + f_d(e)$$

where $K_{eq}(E)$ is a gain dependent upon the parameter E and $f_d(e)$ contains higher harmonics of the input. Assuming $G(s)$ is low-pass in nature, it may be reasonable to say that

$$G(j\omega) f_d(E \sin \omega t) \approx 0$$

Then the question is whether the assumed oscillation e is sustained, i.e. whether

$$e = -G(j\omega)\, N(e) = -G(j\omega)\, K_{eq}(E)\, e$$

An alternative way of saying this is that we must determine whether

$$\frac{1}{K_{eq}(E)} + G(j\omega) = 0$$

for some ω and E. To see this, we can plot the two functions

$$G(j\omega) \qquad \text{for all } \omega$$

$$-1/K_{eq}(E) \qquad \text{for all } E$$

and see whether they intersect. The first function is clearly a Nyquist plot; the second is a 'variable gain'. Hence, we see whether for some E the point $-1/K_{eq}(E)$ intersects, for some ω, $G(j\omega)$. The ω is of course the frequency of oscillation, and the critical E is the amplitude of oscillation; whether the oscillation is a stable one is determined by examining the stability of perturbed values of K_{eq}.

Example

An ideal relay, with

$$N(e) = \begin{cases} M & E \sin \omega t > 0 \\ -M & E \sin \omega t < 0 \end{cases}$$

can be shown to have equivalent gain (or describing function)

$$K_{eq}(E) = \frac{4M}{\pi E}$$

Suppose the relay is switching a simple amplifier plus motor combination, with shaft angle the variable of interest. In this case, a typical transfer function will have the form

$$G(s) = \frac{10}{s(s + 1)(s + 10)}$$

The two plots, of $G(s)$ for $s \in \Gamma$, and $-1/K_{eq}(E)$ for varying E, are shown in Fig. 15.10, with the regions of various number of encirclements of $-1/K$ values and their resulting Z values also shown.

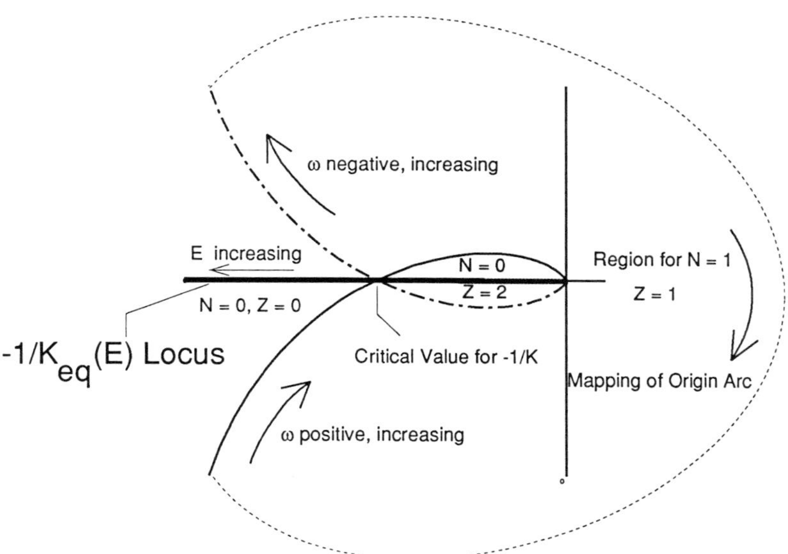

Figure 15.10 Text example system Nyquist plot showing images of both $G(s)$ and of the non-linearity, as represented by $-1/K_{eq}(E)$. The values of Z indicate that E will decrease for large values and will increase for small values, suggesting that there is a possible steady oscillation at the frequency and magnitude defined by the intersection of the two curves.

It is easy to show that the Nyquist plot crosses the negative real axis for $\omega = \sqrt{10}$ with amplitude $1/11$. This will occur for $K_{eq}(E) = 11$, which means $M/E = 11\pi/4$. This point is therefore a potential oscillation point. Consider further if the amplitude E should decay slightly for any reason. Then K_{eq} will increase and the point $-1/K_{eq}$ will move to the right, where it will be encircled twice by the $G(s)$ plot and hence be unstable. This instability means the amplitude will increase toward the critical point. Similar arguments show that too large an amplitude will decay. Hence, the analysis indicates a possible stable oscillation at $\sqrt{10}$ rad/s with amplitude $E = 4M/11\pi$.

15.6 NYQUIST PLOTS AND CLOSED-LOOP RESPONSE

For particular closed-loop structures, it is possible by using graphical methods to extract from the Nyquist plot the closed-loop frequency response. This is done using a set of loci called M-circles and a second set called N-circles for magnitude and phase, respectively. These are used with the Nyquist plot in a manner which we shall indicate below.

Consider any point w in the complex plane. We can compute the magnitude and phase of the complex number

$$\frac{w}{1 + w} = M e^{j\phi}$$

where

$$M = \left|\frac{w}{1 + w}\right| \quad \text{and} \quad \phi = \arg\left\{\frac{w}{1 + w}\right\}$$

Writing $x = \text{Re}(w)$ and $y = \text{Im}(w)$, so that $w = x + jy$, yields

$$M^2 = \frac{x^2 + y^2}{(1 + x)^2 + y^2} \qquad \phi = \tan^{-1}\left\{\frac{y}{x^2 + x + y^2}\right\} \tag{15.2}$$

Alternatively, we can find all w yielding a particular value of M or of N. Consider first the loci for constant M. For $M = 1$, (15.2) easily yields the vertical line $x = -\frac{1}{2}$ (and y arbitrary). For $M \neq 1$, the expression may be rearranged to yield

$$\left(x + \frac{M^2}{M^2 - 1}\right)^2 + y^2 = \left(\frac{M}{M^2 - 1}\right)^2$$

This is easily seen to be the equation of a circle centred at

$$\left(-\frac{M^2}{M^2 - 1}, \ 0\right)$$

with radius

$$\left|\frac{M}{M^2 - 1}\right|$$

Similarly, if $N = \tan \phi$, we can find the locus of all w which yield a given ϕ from

$$\left(x + \frac{1}{2}\right)^2 + \left(y - \frac{1}{2N}\right)^2 = \left(\frac{N^2 + 1}{4N^2}\right)$$

Here, the circles are centred at

$$\left(-\frac{1}{2}, \ \frac{1}{2N}\right)$$

and have radius

$$\left(\frac{N^2 + 1}{4N^2}\right)^{\frac{1}{2}}$$

Now consider plotting $G(j\omega)$ in a complex coordinate system containing M- and N-circles as in Fig. 15.11. For the particular closed-loop system (unity feedback and $G(s)$ representing the entire forward path transfer function, including selected gains), the closed-loop transfer function is

$$T(s) = \frac{G(s)}{1 + G(s)}$$

and the frequency response is

$$T(j\omega) = \frac{G(j\omega)}{1 + G(j\omega)}$$

Then it is obvious that if for frequency ω_t, $G(j\omega_t)$ intersects a particular M-circle M_t and a particular N-circle N_t, then it must be true that

$$T(j\omega_t) = \frac{G(j\omega_t)}{1 + G(j\omega_t)} = M_t \exp(j \tan^{-1} N_t)$$

This shows that we can obtain the closed-loop transfer function graphically for this particular feedback control system structure. More important, perhaps, is the fact that the circles indicate the closed-loop characteristics and hence can be considered design constraints. This will partly be addressed in Chapter 20.

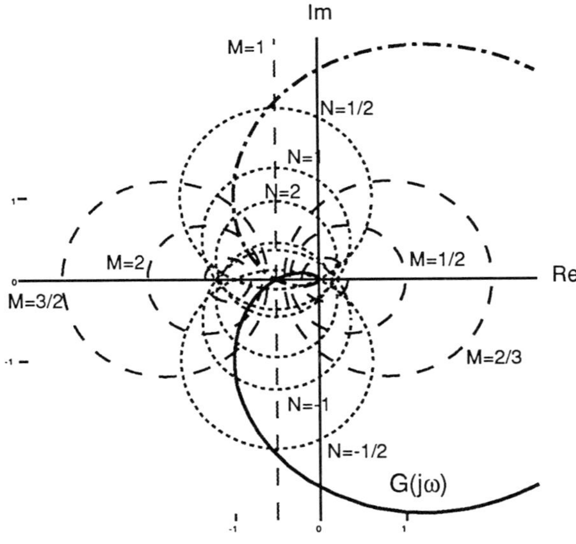

Figure 15.11 A Nyquist plot of a function $G(s)$ along with M-circles ($M = 1/3$, $1/2$, $2/3$, 1, $3/2$, 2, 3) and N-circles ($N = \pm 1/2$, ± 1, ± 2).

15.7 RELATIVE STABILITY

For many purposes it is not only desirable to know that a system is stable, but how close it is to instability. For example, if components are far enough out of specification, a nominally satisfactory system may become unstable if we do not allow for the errors. Also, it turns out that systems 'close to unstable' often display a highly oscillatory response, while those which are 'highly stable' may be sluggish. Thus we like to have reasonable measures of 'closeness to instability'; these are usually called measures of relative stability.

The measures associated with Nyquist plots and representations derived from them are gain margin and phase margin. These are best understood by referring to Fig. 15.12 where a typical portion of a Nyquist plot is shown.

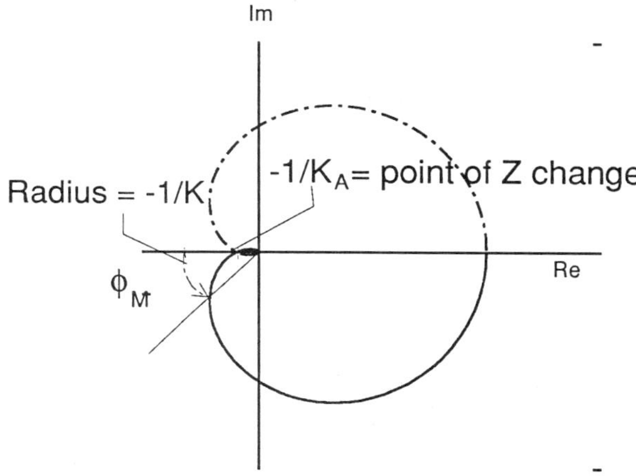

Figure 15.12 Illustrating phase margin ϕ_m and gain margin K_A/K on a Nyquist plot.

Presumably, the region to the left of the intersection of the curve $G(q)$, $q \in \Gamma$ = Nyquist contour, with the negative real axis is known to be stable and we are considering the point $(a, 0)$, $a = -1/k$, in the stable region. The intersection of $G(q)$ with the axis is at $(b, 0)$ and occurs for $q = q_b$. If we were to increase the gain to $k_a = -1/b$, the system would be unstable (because $1 + k_b G(q_a) = 0$). The **gain margin** is defined as the multiplicative increase needed from the candidate gain before the system becomes unstable, i.e. the change in gain before the point $(-1/k, 0)$ is encircled enough times that $Z > 0$. In this case

$$G_\mathrm{m} = \frac{k_\mathrm{a}}{k} = \frac{a}{b} \quad \text{or} \quad G_\mathrm{m,db} = 20\log\left(\frac{k_\mathrm{a}}{k}\right)$$

Alternative representations, including those for which the gain has been included in $G(q)$ so that we are considering encirclements of $(-1, 0)$ for stability, include

$$G_\mathrm{m,db} = -20\log\left[G(q_\mathrm{b})\right] \qquad \text{for } k \text{ 'included'}$$

$$G_\mathrm{m} = \frac{-1}{kG(q_\mathrm{b})} \qquad \text{for explicit candidate } k$$

Gain margin has one interpretation as 'the amount gain should change before the system becomes unstable'; similarly, phase margin is 'the amount the phase of $G(q)$ should change before the closed-loop system becomes unstable'. To define it, let q_a be the value of q on the Nyquist contour Γ for which

$$|G(q_\mathrm{a})| = \frac{1}{k}$$

Then the **phase margin** ϕ_m is defined as

$$\phi_\mathrm{m} = \arg\{G(q_\mathrm{a})\} + 180°$$

Once again, the minor adjustment needed if $G(q)$ incorporates all gains and $(-1, 0)$ is the test point is obvious.

We remark that typical 'good' values of the margins are

$$\phi_\mathrm{m} \approx 40\text{–}50°$$

$$G_\mathrm{m} \approx 6\text{–}12\,\mathrm{dB}$$

Some texts define the margins using the above equations but allowing initially unstable systems to be considered. Thus a system might have −30° of phase margin, indicating that phase must be decreased by 30° before stability is attained.

15.8 DESIGN USING NYQUIST PLOTS

The underlying nature of design using Nyquist plots is straightforward: one chooses compensation elements which reshape the Nyquist plot to yield desired properties. These compensation elements are typically cascades of basic elements of the form

$$C(q) = K \frac{\dfrac{q}{a} + 1}{\dfrac{q}{b} + 1}$$

This has the effect of increasing the magnitude and phase of the open-loop transfer function $C(q)\,G(q)$ at the frequency a (i.e. at $s \rightarrow j\omega$, $\omega = a$, or at $z \rightarrow \exp(j\omega T)$, $\omega T = \ln \alpha T$) and decreasing them in the same manner at b. Design involves selecting the form of the elements and then their parameters. This is done using a different representation in Chapter 20.

Example

Figure 15.13 shows the Nyquist diagram for a plant such as a motor,

$$G(s) = \frac{1}{s(0.2s + 1)}$$

It is seen that for $K = 10$, $G_m = \infty$ and $\phi_m \approx 40°$. The same figure also shows the motor with the compensator

$$C(s) = \frac{\dfrac{s}{6} + 1}{\dfrac{s}{60} + 1}$$

for which the shape is different and the phase margin has increased to

75°. These numbers were established graphically; computation shows the respective phase margins are 38.6° and 77°.

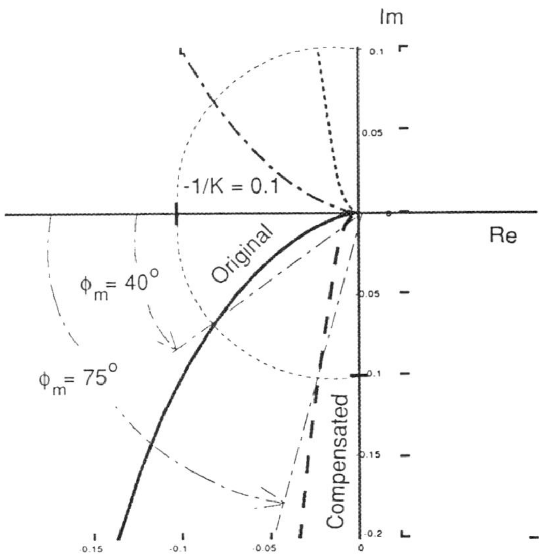

Figure 15.13 The original and compensated Nyquist plots for the text example.

15.9 NYQUIST AND OTHER PLOTS

The Nyquist plots are, except for loops 'at infinity', the frequency responses of the open-loop transfer functions given by $G(j\omega)$ in the continuous time models and $G(e^{j\omega t})$ in sampled data models. We have seen that these plots indicate stability and relative stability, which suggests that they might be the basis of designing closed-loop systems, particularly when M- and N-circles are also considered.

This is partly true, in that Nyquist theory arguably provides the theoretical underpinning of possible design methods. In actual design work, however, it turns out that other graphical representations are more convenient for some tasks.

We recall that the Nyquist plot is of complex numbers in the complex plane as a frequency parameter (ω or ωT) varies. Alternatives to this representation include those of Bode and Nichols.

Bode In Bode plots, we make two plots:

1. $|G(q)|_{dB}$ (= $20\log G(q)$) versus $\log \omega$, called the Bode magnitude plot; and
2. $\arg G(q)$ versus $\log \omega$, called the Bode phase plot.

Nichols In Nichols plots, we graph $|G(q)|_{dB}$ versus $\arg G(q)$, as parameter ω varies, i.e. a log magnitude versus phase plot, on a plot on to which the M- and N-circles have been mapped (and become distorted from circles).

Because they are primarily design tools rather than stability testing methods, these are left to later chapters. In particular, Bode plots are discussed in Chapter 20. Nyquist, Bode, and log magnitude versus phase plots are shown together in Chapter 10.

15.10 COMPUTER AIDS

Computer aids such as MATLAB® and Ctrl-C will generate the mapping of the contour Γ for finite points on the basic contour which are not poles of the open-loop transfer function. In addition, they may be programmed to provide Gershgorin bands. Hence, they generate the details of the plots but not necessarily the number N of encirclements. Within this limitation, they usually provide the Nyquist plots rapidly and easily, and can easily generate the Bode and log magnitude versus phase (for Nichols charts) plots.

15.11 FURTHER READING

Nyquist theory for SISO continuous time systems is found in many elementary textbooks. This, along with brief mentions of non-linear applications and sampled data systems are to be found, for example, in Phillips and Harbor (1991), while more extensive coverage of continuous time systems is in DiStefano, Stubberud and Williams (1976). Digital control texts usually build on the continuous time theory; examples are Franklin, Powell and Workman (1990) and Åström and Wittenmark (1990). The multivariable extensions are presented by Maciejowski (1989).

16
Lyapunov stability testing

The second or direct method of Lyapunov is entirely different from pole analysis in philosophy, nature and detail, although there are a few overlaps for linear time-invariant systems.

The approach is based on consideration of a generalized energy content of the system in question. Suppose we can define a generalized energy such that (a) the energy is minimized at an equilibrium point and (b) when the state is not at the equilibrium point, the energy must decrease. Then if the system is disturbed from equilibrium, its energy must go up and then start decreasing. If it keeps decreasing, it must obviously go to its minimum. But the minimum energy configuration is the equilibrium point, and hence the state must approach the equilibrium point. Thus the equilibrium point must be stable.

A more intuitive point of view for some people is to consider a mathematical description of the shape of the famous 'bowl' of explanations of stability used in elementary physics classes. In this viewpoint a stable 'marble' must always roll down the side of the bowl, and in doing so must necessarily return to the lowest point, which is the centre, or equilibrium point.

The formalization of the above notions is the subject of the Lyapunov theory, and their application is a tool of some assistance to the control engineer. A major problem with using the approach is that we cannot always find an appropriate generalized energy.

16.1 SYNOPSIS

The core idea of Lyapunov methods is built around the Lyapunov function and state-space representation: the Lyapunov function is a scalar function V of the state vector $\mathbf{x}$ such that:

1. $V(\mathbf{x}) \geq 0$ for $\mathbf{x}$ in the region $\mathscr{R}$ of interest.
2. $V(\mathbf{x})$ is rather like a measure of distance of $\mathbf{x}$ from an equilibrium point $\mathbf{x}_e$ and gets larger as $\mathbf{x}$ gets further from that

point (within $\mathscr{R}$).

3. For the equilibrium point considered, $V(\mathbf{x}(t))$ does not increase with time as the system operates (for stability i.s.L.) and in fact decreases as time goes on (for asymptotic stability i.s.L.).

4. If for the equilibrium point considered and $V(\mathbf{x})$ as in (1) and (2) above, $V(\mathbf{x}(t))$ increases with time, then the equilibrium point is unstable.

In section 16.2 we meet the basic ideas expressing this for discrete time systems. Section 16.3 shows design applications, and then in section 16.4 we meet some of the results for continuous time systems.

16.2 THE BASIC IDEAS OF LYAPUNOV'S DIRECT METHOD

The definitions associated with Lyapunov stability were presented in Chapter 13. For operating with the definition, we are concerned with a system of equations which generate a state sequence $\{y(k)\}$ according to dynamics

$$\mathbf{y}(k+1) = f(\mathbf{y}(k);k) \qquad k=0,1,\dots \qquad \mathbf{y}(0) = \mathbf{y}_0 \qquad (16.1)$$

where $\mathbf{y}$ is the n-vector describing the state. We study an equilibrium point $\mathbf{y}_e$, a point such that

$$\mathbf{y}_e = f(\mathbf{y}_e;k) \text{ for all } k$$

Let $V(\mathbf{x})$ be a scalar function of the n-vector which is positive definite in a region $\mathscr{R}$ of the origin, i.e.

$$V(\mathbf{x}) > 0 \qquad \mathbf{x} \in \mathscr{R}, \mathbf{x} \neq \mathbf{0}$$

$$V(\mathbf{0}) = 0 \qquad (16.2)$$

Then if for the equilibrium point $\mathbf{y}_e$ of (16.1) we have

$$V(\mathbf{y}(k+1) - \mathbf{y}_e) \leq V(\mathbf{y}(k) - \mathbf{y}_e) \qquad (16.3)$$

for all $k = 0, 1, 2, \ldots$, and if $(\mathbf{y}(0) - \mathbf{y}_e) \in \mathscr{R}$, i.e. $\mathbf{y}_0$ belongs to a suitable region near $\mathbf{y}_e$, it must be that the equilibrium point is stable i.s.L. To show the dynamics explicitly, we rewrite this as

$$V(f(\mathbf{y}(k);k) - \mathbf{y}_e) \le V(\mathbf{y}(k) - \mathbf{y}_e)$$

In fact, the theorem states that the equilibrium point is stable i.s.L. if there exists a function V for which (16.2) and (16.3) hold. The underlying idea is represented in Fig. 16.1, which shows contours of a function $V(\mathbf{x})$ near an equilibrium point $\mathbf{x}_e$ and a possible trajectory $\mathbf{x}(t)$ for which $V(\mathbf{x}(k))$ is non-increasing.

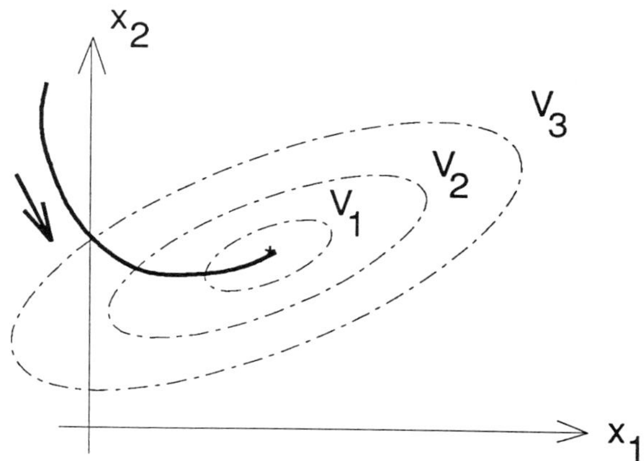

Figure 16.1 Contours of a Lyapunov function, with $V_1 < V_2 < V_3$, and a stable i.s.L. trajectory evolving over time to smaller V values.

Example

Consider the scalar dynamics

$$y(k+1) = ay(k) \qquad y(0) = y_0$$

Let $V(x) \equiv x^2$, which obviously is positive definite. We notice that $y_e = 0$ is an equilibrium point of the dynamics. We also note that the pole of the system is at a, so the system is BIBO stable for $|a| < 1$.

Calculating (16.3) for this example yields

$$V(f(y(k);k)) = (ay(k))^2 = a^2 y(k)^2$$

$$= a^2 V(y(k))$$

Thus we have stability i.s.L. of the equilibrium point 0 provided $a^2 \leq 1$, and asymptotic stability i.s.L if $a^2 < 1$.

Application: Linear system stability

We generalize the above scalar system to use matrices and vectors. Thus let

$$x(k+1) = Ax(k)$$

and let $V(\mathbf{x})$ be defined with the positive definite matrix $\mathbf{P}$ as

$$V(\mathbf{x}) = \mathbf{x}^T \mathbf{P} \mathbf{x}$$

Applying this as per the theorem, we find

$$V(\mathbf{x}(k+1)) - V(\mathbf{x}(k)) = \mathbf{x}^T(k+1)\mathbf{P}\mathbf{x}(k+1) - \mathbf{x}^T(k)\mathbf{P}\mathbf{x}(k)$$

$$= \mathbf{x}^T(k)\{\mathbf{A}^T\mathbf{P}\mathbf{A} - \mathbf{P}\}\mathbf{x}(k)$$

$$= -\mathbf{x}^T(k)\mathbf{Q}\mathbf{x}(k)$$

where the definition of $\mathbf{Q}$ is obvious. Then the origin is asymptotically stable i.s.L. if for any positive definite $\mathbf{P}$ there exists a positive definite $\mathbf{Q}$ for which

$$-\mathbf{x}^T\mathbf{Q}\,\mathbf{x} \leq 0$$

holds. In fact further theory tells us the origin $\mathbf{x} = \mathbf{0}$ of

$$x(k+1) = Ax(k)$$

is asymptotically stable i.s.L. if for any $\mathbf{Q} > 0$ there exists some $\mathbf{P} > 0$ for which

$$\mathbf{A}^{\mathrm{T}}\mathbf{P}\mathbf{A} - \mathbf{P} = -\mathbf{Q}$$

This is called a Lyapunov equation and the computation of a $\mathbf{P}$ given some $\mathbf{Q}$ (often taken as $\mathbf{Q} = \mathbf{I}$) and the checking of the result to see if $\mathbf{P} > 0$ is best done using some of the computer aided control system design programs.

Application: Non-linear system stability

The method applies readily to non-linear systems, at least in principle. We see this in another example, which also introduces the notion of a region of stability.

Example

Consider the scalar system with dynamics

$$y(k+1) = y(k)^3$$

which obviously has one equilibrium point at the origin, $y = 0$, and another at $y = -1$. To evaluate the stability of the former point, we consider the function

$$V(y) = y^2$$

which is obviously non-negative and is zero only at the origin. The change in V is computed directly as

$$\Delta V(k) = V(y(k+1)) - V(y(k))$$

$$= y(k+1)^2 - y(k)^2$$

$$= y(k)^6 - y(k)^2$$

$$= y(k)^2 (y(k)^4 - 1)$$

This is non-positive for

$$|y(k)| \le 1$$

and hence the equilibrium point is stable, with region of stability $\mathscr{R} = \{y : |y| \le 1\}$. It is asymptotically stable for strict inequalities in the expressions for y.

The equilibrium point at $y = -1$ may be shown to be unstable by using the fact mentioned in comment 1 at the end of this section and the function $V(y) = (y+1)^2$.

Application: Transient response estimation

The Lyapunov technique can sometimes be used to estimate the speed of transient response. As usual, the idea is straightforward, but the application in general can be difficult.

Let the equilibrium state $y_e = 0$ be known to be asymptotically stable i.s.L. Define

$$K = \max_{\mathbf{x} \in \mathscr{R}} \frac{V(\mathbf{x})}{-\Delta V(\mathbf{x})}$$

Then it follows that

$$\Delta V(\mathbf{x}(k)) = V(\mathbf{x}(k+1)) - V(\mathbf{x}(k)) \le \frac{-1}{K} V(\mathbf{x}(k))$$

and hence

$$V(\mathbf{x}(k+L)) \le \left(1 - \frac{1}{K}\right)^{L} V(\mathbf{x}(k))$$

This can be used to estimate, e.g. the number of steps L required to reach 1% of an initial value.

In the special case of a linear system, it can be shown that if $V(\mathbf{x}) = \mathbf{x}^T \mathbf{P} \mathbf{x}$ and $\Delta V(\mathbf{x}) = -\mathbf{x}^T \mathbf{Q} \mathbf{x}$, then

$$K = \max_{x} \frac{x^T P x}{x^T Q x} = \text{largest eigenvalue of } Q^{-1}P$$

Comments

The problem with all this is that for real systems it becomes virtually impossible to find functions V for which either stability or instability can be demonstrated with the resulting ΔV.

There are a number of variations of the basic theorems. We mention only two.

1. The equilibrium point is unstable if there exists a $V(y) \geq 0$ for which $\Delta V > 0$.
2. The equilibrium point is asymptotically stable i.s.L. if (a) it is stable i.s.L. and (b) $V > 0$ and $\Delta V < 0$ for $y \neq y_e$.

16.3 DESIGN USING LYAPUNOV'S DIRECT METHOD

Certain types of design may be done using Lyapunov stability theory. We will meet in section 30.3.5 an example in the design of adaptive observers, in which an algorithm is made a stable estimator of a system's parameters, and in section 32.3.3 an example of convergence analysis for learning control. Most Lyapunov-based design is concerned with finding a class of controllers, often a set of parameters for a controller with specified structure, so that a system will be guaranteed stable. This can sometimes be modified so that the system is 'very stable', as we will see in examples in this section.

We will restrict the discussion to linear systems, for as usual most results are for them.

Consider the state–space n-vector system with state feedback matrix $\mathbf{K}$ to be chosen, i.e. let

$$\mathbf{x}(k+1) = \mathbf{A}\mathbf{x}(k) + \mathbf{B}\mathbf{u}(k)$$

$$\mathbf{u}(k) = \mathbf{K}\mathbf{x}(k)$$

so that we are to choose $\mathbf{K}$ in

$$\mathbf{x}(k+1) = [\mathbf{A} + \mathbf{B}\mathbf{K}]\mathbf{x}(k)$$

Choose the positive definite function (Appendix B5) $V(\mathbf{x})$ as

$$V(\mathbf{x}) = \mathbf{x}^T \mathbf{P} \mathbf{x}$$

Then

$$V(\mathbf{x}(k+1)) - V(\mathbf{x}(k)) = \mathbf{x}(k+1)^T \mathbf{P} \mathbf{x}(k+1) - \mathbf{x}^T(k)\mathbf{P}\mathbf{x}(k)$$

$$= \mathbf{x}(k)^T [\mathbf{A} + \mathbf{B}\mathbf{K}]^T \mathbf{P} [\mathbf{A} + \mathbf{B}\mathbf{K}] \mathbf{x}(k) - \mathbf{x}^T(k)\mathbf{P}\mathbf{x}(k)$$

$$= - \mathbf{x}^T(k)\mathbf{Q}\mathbf{x}(k)$$

where the definition of $\mathbf{Q}$ is obvious. Then the origin is stable i.s.L. provided $\mathbf{Q} \geq 0$, i.e. that $\{\mathbf{P} - [\mathbf{A}+\mathbf{B}\mathbf{K}]^T \mathbf{P}[\mathbf{A}+\mathbf{B}\mathbf{K}]\}$ is positive semidefinite. Another way to put this is that the origin is a Lyapunov stable equilibrium point for a given $\mathbf{K}$ provided that, for some positive definite matrix $\mathbf{P}$, the matrix $\mathbf{Q}$ defined from

$$[\mathbf{A} + \mathbf{B}\mathbf{K}]^T \mathbf{P} [\mathbf{A} + \mathbf{B}\mathbf{K}] - \mathbf{P} = -\mathbf{Q} \tag{16.4}$$

is positive semidefinite.

We may proceed further than this: we may consider making the system optimum in the sense that the V function decreases as rapidly as possible. To do this, we want to choose $\mathbf{K}$ to minimize

$$\Delta V(k) = -\mathbf{x}^T(k)\mathbf{Q}\mathbf{x}(k)$$

which means we must in some sense maximize $\mathbf{Q}$.

Since the expression in (16.4) is quadratic, we must simply take a gradient equal to zero. This means that we need (Appendix B) for the optimum $\mathbf{K}$, call it $\mathbf{K}^0$, that

$$\mathbf{0} = \mathbf{B}^T \mathbf{P} \mathbf{A} + \mathbf{B}^T \mathbf{P} \mathbf{B} \mathbf{K}^0$$

or

$$\mathbf{K}^0 = -(\mathbf{B}^T \mathbf{P} \mathbf{B})^{-1} \mathbf{B}^T \mathbf{P} \mathbf{A}$$

16.4 CONTINUOUS TIME SYSTEMS

The theory for continuous time systems is analogous to the above. In particular consider systems described by

$$\dot{\mathbf{x}} = f(\mathbf{x}) \qquad \mathbf{x}(0) = \mathbf{x}_0$$

Here one seeks a scalar function $V(\mathbf{x}(t))$ such that

$$V(\mathbf{x}) \geq 0 \qquad \mathbf{x} \in \mathscr{R}$$

Then the point $\mathbf{x} = \mathbf{0}$ is said to be stable i.s.L. provided that

$$\dot{V}(\mathbf{x}(t)) \leq 0 \qquad \mathbf{x} \in \mathscr{R}$$

Example

For the linear system

$$\dot{\mathbf{x}} = \mathbf{A}\mathbf{x}$$

$$\mathbf{y} = \mathbf{C}\mathbf{x}$$

we consider the function

$$V(\mathbf{x}) = \mathbf{x}^T \mathbf{P}\mathbf{x} \qquad \mathbf{P} \geq 0$$

and find that the origin is stable i.s.L. provided that

$$\dot{V}(\mathbf{x}) = \mathbf{x}^T \mathbf{P}\mathbf{A}\mathbf{x} + \mathbf{x}^T \mathbf{A}^T \mathbf{P}\mathbf{x} = -\mathbf{x}^T \mathbf{Q}\mathbf{x} \leq 0$$

or

$$\mathbf{P}\mathbf{A} + \mathbf{A}^T \mathbf{P} = -\mathbf{Q} \leq 0$$

which is called the continuous time Lyapunov equation.

Other examples as for the discrete time system models will also yield results similar to those cases.

16.5 LYAPUNOV'S FIRST METHOD

There is also a first method of Lyapunov. This works with linear or linearized systems and is notable for its characterization of the equilibrium points as, e.g. nodes (for which the eigenvalues of the linear form are real), foci (for which they are complex) and vortices (for which they are imaginary). Nodes and foci can be either stable or unstable; a node with one stable and one unstable eigenvalue is called a saddle point. Usually pictured for two-dimensional systems, as in Fig. 16.2, these are helpful more for forming mental images of system behaviour than for theoretical insights.

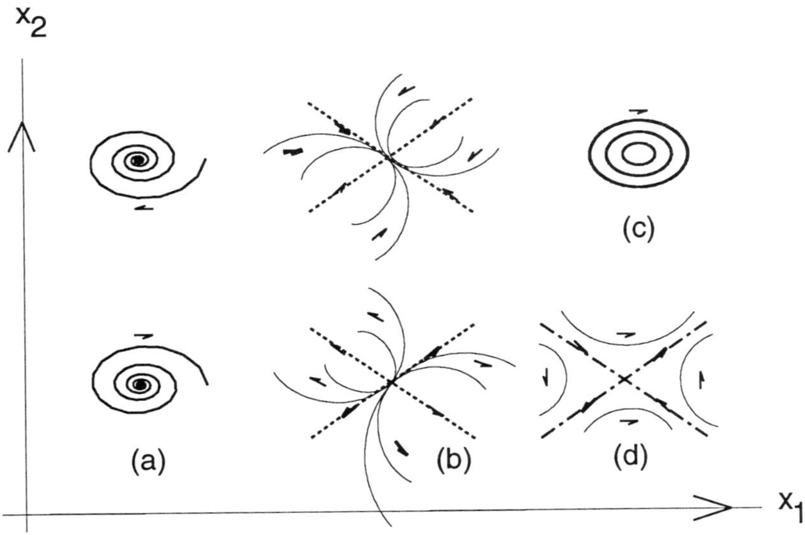

Figure 16.2 Showing equilibrium points in the phase plane. The top row are stable, and the bottom are unstable: (a) focus; (b) node; (c) centre; and (d) saddle point.

The form considered is

$$\ddot{y} - (\lambda_1 + \lambda_2)\dot{y} + \lambda_1 \lambda_2 y = 0$$

with the cases:

1. λ_i, $i = 1, 2$ real
      Both $> 0 \Rightarrow$ unstable node
      Both $< 0 \Rightarrow$ stable node
      Opposite signs $\Rightarrow$ saddle point
2. $\text{Re}(\lambda_i) = 0$, $i = 1, 2 \Rightarrow$ centre or vortex
3. $\lambda_i = $ conjugate pair.
      $\text{Re}(\lambda_i) < 0 \Rightarrow$ stable focus
      $\text{Re}(\lambda_i) > 0 \Rightarrow$ unstable focus

Use of the first method for non-linear systems simply involves linearizing about each equilibrium point in turn and computing the eigenvalues of the resulting linear system. This will indicate stability or lack thereof for a small region $\mathscr{R}$ about the equilibrium point provided the eigenvalues are not on the stability boundary. The method tells nothing about larger regions, a fault shared by all linearization approaches.

16.6 COMPUTER ASSISTANCE

Lyapunov theory tends to be application specific except for linear constant coefficient systems. For these, there has been considerable study of algorithms for the solution of the Lyapunov equations

$$\mathbf{PA} + \mathbf{A}^T\mathbf{P} = -\mathbf{Q} \leq \mathbf{0}$$

for continuous time systems and

$$\mathbf{A}^T\mathbf{PA} - \mathbf{P} = -\mathbf{Q}$$

for discrete time systems. These arise in a variety of ways in the study of control systems, including modelling (minimal and balanced realizations, see e.g. Moore (1981)) and system noise covariance calculations. Systems such as MATLAB® have algorithms for solving these equations.

16.7 COMMENTS AND FURTHER READING

The ideas of Lyapunov's direct method, sometimes called the 'second method', are straightforward. The difficulty comes in finding suitable functions $V(\mathbf{x})$, for the fact that we cannot find such a function does not prove one does not exist. Occasionally, it is possible to find a $V(\mathbf{x}) \geq 0$ such that $\dot{V}(\mathbf{x}) \geq 0$, in which case instability is demonstrated, but in many practical cases we either cannot find $V(\mathbf{x})$ or cannot prove that $\dot{V}$ does not change sign; either way, we are left in a limbo with nothing proven.

Lyapunov methods sometimes seem to be going out of fashion, as not all recent introductory level textbooks mention them (perhaps because of the difficulties mentioned above). Shinners (1992) discusses both the first and second methods. Other texts with some coverage are Kuo (1980) and Ogata (1987). The functions sometimes appear in the study of algorithm convergence in, for example, adaptive methods (Chapter 31), and this often seems one of the more common uses of the technique.

17

Steady-state response: error constants and system type

The interest of the engineer is in two aspects of the system response: steady-state response and transient response. The former is the output as $t \rightarrow \infty$, while the latter is the portion of the output which dies away after a short time. Provided that the system is asymptotically stable, the transients will indeed die out. This section is concerned with aspects of the steady-state error.

17.1 SYNOPSIS

In studying SISO control systems, the basic system block diagram is an interconnection of linear systems so that using their transfer functions leads to the form of Fig. 17.1.

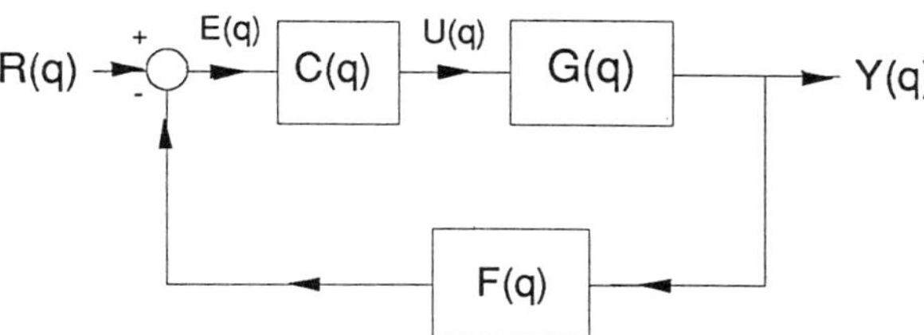

Figure 17.1 Repeat of transfer function block diagram model of typical SISO system.

For this it is easy to derive that, whether q is the Laplace transform variable s or the z transform variable z,

$$Y(q) = \frac{G(q)\,C(q)}{1 + G(q)\,C(q)\,F(q)}\,R(q) = H(q)\,R(q)$$

while the error between input and measured output is given by

$$E(q) = R(q) - F(q)\,Y(q) = \frac{1 + G(q)\,C(q)\,[F(q) - 1]}{1 + G(q)\,C(q)\,F(q)}\,R(q)$$

For a stable system in which $e(t) \to$ constant $= e_{ss}$ we have for Laplace transforms and continuous time systems

$$e_{ss} = \lim_{t \to \infty} e(t) = \lim_{s \to 0} sE(s)$$

and for z-transforms and discrete time systems

$$e_{ss} = \lim_{nT \to \infty} e(nT) = \lim_{z \to 1} (1 - z^{-1})E(z)$$

Hence the steady-state errors are easy to find once an input is specified. This specification is usually done for step, ramp, and parabolic inputs, and it turns out that the errors are closely related to the system property called **system type**.

A unity feedback system (i.e. one in which $F(q) = 1$) has type L if the open-loop transfer functions have the form

$$G(s)\,C(s) = \frac{N(s)}{s^L}$$

or

$$G(z)\,C(z) = \frac{N(z)}{(z - 1)^L}$$

where $N(q)$ is a rational polynomial not containing factors of the indicated denominator form. A type L system has the property of producing zero error in following an input polynomial form of degree $L-1$, finite error when the polynomial has degree L, and infinite error when the degree is greater than L as shown in Table 17.1.

Table 17.1 Properties of type-L systems

L	K_{pos}	K_{vel}	K_{acc}	$e(\infty)\vert_{step}$	$e(\infty)\vert_{ramp}$	$e(\infty)\vert_{acc}$
0	finite	0	0	finite	∞	∞
1	∞	finite	0	0	finite	∞
2	∞	∞	finite	0	0	finite
3	∞	∞	∞	0	0	0
>3	∞	∞	∞	0	0	0

In the following sections we discuss and justify this for continuous time and discrete time systems separately.

17.2 DISCRETE TIME PROBLEMS

We consider the system of Fig. 17.1 for the discrete time case and choose the unity feedback case $F(z) = 1$, corresponding to perfect measurement with no transients in the instrument. Evaluation of the error in terms of the input is straightforward.

$$E(z) = \frac{R(z)}{1 + C(z)\,G(z)}$$

At this point we use the fact that the steady-state limit of $e(t)$ as $t \to \infty$ is given by the z-transform theory as follows.

$$\lim_{t \to \infty} e(t) = e(\infty) = \lim_{z \to 1} \frac{z - 1}{z} E(z)$$

$$= e_{ss} = \lim_{z \to 1} \left[\frac{z - 1}{z} \frac{R(z)}{1 + C(z)\,G(z)} \right]$$

Hence, given a system $G(z)$, the final steady-state error e_{ss} will depend upon the input $R(z)$ and on the compensator $C(z)$. The typical input chosen for consideration, because it amounts to a set point or constant reference value, is the step, $r(t) = A$, $t \geq 0$, $r(t) = 0$, $t < 0$. For this, we already know

$$R(z) = \frac{Az}{z - 1}$$

and thus

$$e(\infty)\big|_{\text{step}} = \lim_{z \to 1} \frac{A}{1 + C(z)\,G(z)}$$

Defining the position error constant K_{pos} by

$$K_{\text{pos}} = \lim_{z \to 1} C(z)\,G(z)$$

gives

$$e_{\text{ss}}\big|_{\text{step}} = \frac{A}{1 + K_{\text{pos}}}$$

Usually, we want e_{ss} to be 'small' and preferably zero, which means that K_{pos} must be 'large'. Since $G(z)$ and $C(z)$ are rational, this means that the denominator of the product should have a factor which tends to 0 as $z \to 1$. If $G(z)$ has no such factor, we may as part of our design introduce one, so that $C(z)$ takes the form

$$C(z) = \frac{G_{c1}(z)}{(z - 1)^L}$$

where $G_{c1}(z)$ is rational with no poles or zeros at $z = 1$, and L is an integer greater than or equal to 1. We remark that the PID controller has z-transform transfer function (in the PI form with $K_d = 0$)

$$\frac{U(z)}{E(z)} = \frac{(K_i + K_p)\,z - K_p}{(z - 1)}$$

and thus will have the correct form to yield zero steady-state error to a constant input, just as our heuristic arguments of Chapter 8 suggested, provided that $G(z)$ does not have a zero at $z = 1$.

The other inputs for which error constants are sometimes computed are the ramp and parabola functions, representing constant velocity and acceleration respectively and constituting tests of the system's tracking ability.

For the ramp input, $r(t) = At, t > 0$, the z-transform is

$$R(z) = \frac{ATz}{(z-1)^2}$$

where T is the sampling interval, so that

$$e_{ss}\bigg|_{\text{ramp}} = \lim_{z \to 1} \left[\frac{z-1}{z} \; \frac{ATz}{(z-1)^2 \left[1 + C(z)G(z)\right]} \right]$$

$$= \lim_{z \to 1} \frac{AT}{(z-1)\left[1 + C(z)G(z)\right]}$$

$$= \lim_{z \to 1} \frac{AT}{(z-1)C(z)G(z)}$$

Defining the velocity error constant

$$K_{\text{vel}} = \lim_{z \to 1} \frac{z-1}{T} C(z)G(z) \tag{17.1}$$

gives

$$e_{ss}\bigg|_{\text{ramp}} = \frac{A}{K_{\text{vel}}} \tag{17.2}$$

Finally, the acceleration function $r(t) = \dfrac{At^2}{2} U(t)$, with z-transform

$$R(z) = \frac{AT^2\, z(z + 1)}{2(z - 1)^3}$$

results in the same manner in

$$e_{ss}\Big|_{\text{parab}} = \frac{A}{K_{\text{acc}}}$$

where the acceleration error constant K_{acc} is given by

$$K_{\text{acc}} = \frac{1}{T^2}\lim_{z \to 1}\ (z - 1)^2\, C(z)\, G(z)$$

The error constants are indicators of the closed-loop system's ability to track inputs of a given shape. Although it is preferable that the error be zero, certainly a specification will have that e_{ss} be finite for a class of inputs, which immediately implies that the applicable error constant (K_{pos} or K_{vel}, usually) be non-zero, possibly finite, and preferably infinite (i.e. unbounded). Let us look briefly at the implications of this.

Example

Suppose that it is specified that a unit ramp should be followed with no more that 2% (say) error with $T = 0.01\,\text{s}$. Hence from (17.2) we need

$$0.02 > e_{ss}\Big|_{\text{ramp}} = \frac{1}{K_{\text{vel}}}$$

so that $K_{\text{vel}} > 50$ is required. From (17.1) we then have

$$50 < K_{\text{vel}} = \lim_{z \to 1}\frac{(z - 1)}{0.01}\, C(z)\, G(z)$$

From this it follows that $G(z)\, C(z)$ must have at least one pole at $z = 1$. Further, the constraint is easily met if there is more than one such pole. If there is only one such pole, so that

$$C(z)\,G(z) = \frac{[C(z)\,G(z)]'}{z - 1} \tag{17.3}$$

where $[C(z)\,G(z)]'$ has no poles at $z = 1$, then we also need

$$0.5 < [C(1)\,G(1)]'$$

We note that with the form (17.3), $K_{pos} = \infty$ and $K_{acc} = 0$, i.e. a system able to track a ramp with finite error can follow steps with no error but will have unbounded errors for constant accelerations. We can continue this argument, requiring zero error for a ramp and finding $C(z)\,G(z)$ needs at least two poles at $z = 1$, etc. Also, other inputs may be considered. To summarize, we define

$$C(z)\,G(z) = \frac{[C(z)\,G(z)]'}{(z - 1)^L} \tag{17.4}$$

with the rational numerator having no poles or zeros at $z = 1$, as the form for a 'type-L' system. It is clear that if $G(z)\,C(z)$ has the form of (17.4) then:

1. for a **step input**, $e_{ss} = 0$ and $K_{pos} = \infty$ provided $L \geq 1$ – for all L, e_{ss} is finite but usually non-zero;
2. for a **ramp input**, $e_{ss} = 0$ and $K_{vel} = \infty$ provided $L \geq 2$ – for $L = 1$, e_{ss} is finite but usually non-zero and it is infinite for $L = 0$; and
3. for a **parabolic input**, $e_{ss} = 0$ and $K_{acc} = \infty$ provided $L \geq 3$ – for $L = 2$, e_{ss} is finite but usually non-zero and it is unbounded for $L < 2$.

Hence parameter L, called the **system type**, is a parameter which immediately indicates whether steady-state errors after asymptotically stable responses are zero, potentially finite, or always infinite.

The above extends readily but unhelpfully to more complicated systems. The situation to beware of is of misleading analogies. Hence, one must be careful of simple extensions of the above section to a system as in Fig. 17.1 with $F(z) \neq 1$. Then although the pseudo-error $e'(t) = r(t) - f(t)**y(t)$ satisfies the form used above, to wit,

$$E'(z) = \frac{R(z)}{1 + F(z)\,H(z)\,C(z)}$$

we must point out that the 'error' is no longer input minus output. In fact, if error is to have the usual meaning of $r(t) - y(t)$, then

$$E(z) = \frac{(1 + F(z)\,H(z)\,C(z) - H(z)\,C(z))}{1 + F(z)\,H(z)\,C(z)}R(z)$$

Here, although it is certainly possible to compute $e(\infty)$ for various $R(z)$, the resulting implications for designing $C(z)$ are not so clear as before except when $F(1) = 1$.

Example

Consider a small motor which is to be used in position control, so that the output shaft angle is to take on a commanded value. Further, assume that the shaft angle is measured by an attached potentiometer or shaft encoder, so that there is effectively no delay in reading the angle. How elaborate must the controller be?

Discussion A simple motor model, including a zero-order hold for the computer output command/motor input amplifier command, is given by

$$G_{\mathrm{p}}(z) = K_{\mathrm{p}}\frac{(\alpha - 1 + e^{-\alpha})\left(z + \dfrac{(1 - e^{-\alpha} - \alpha e^{-\alpha})}{(\alpha - 1 + e^{-\alpha})}\right)}{(z - 1)\,(z - e^{-\alpha})}$$

where $\alpha = T/\tau$ is the ratio of sampling period T to motor time constant τ, and K_{p} is the system gain from input voltage to output angle.

Clearly the motor system is type 1, so that zero error to a step command should be achievable with a proportional control law. On the other hand, if the motor is to track an input angle increasing at a constant rate, the control law should be of type 1 or higher. A PI controller should handle the job. Small, difficult to remove, errors to constant inputs may show up in real systems for various reasons, however; among these reasons are stiction in the rotation and finite brush size and number which make the motor not quite linear in its responses.

17.3 THE CONTINUOUS TIME CASE

The arguments for continuous time use Laplace transform transfer functions and the appropriate limits, but are otherwise exactly parallel to those above. We quickly review them.

It is straightforward to show that, for a system as in Fig. 17.1, the error quantity $e(t) = y(t) - r(t)$ has transfer function relationship

$$E(s) = (1 - H(s))R(s)$$

$$= \frac{1 + (F(s) - 1)G(s)C(s)}{1 + F(s)G(s)C(s)}R(s)$$

In the unity feedback case, which is the one for which most results apply, $F(s) = 1$ and this becomes

$$E(s) = \frac{R(s)}{1 + G(s)C(s)}$$

and we may consider what happens in steady-state provided this is stable.

Thus for a step input

$$r(t) = \begin{cases} A & t \geq 0 \\ 0 & t < 0 \end{cases}$$

with transform $R(s) = A/s$, we have

$$e_{ss} = \lim_{t \to \infty} e(t) = \lim_{s \to 0} s\, E(s) = \lim_{s \to 0} \frac{R(s)s}{1 + G(s)C(s)}$$

$$= \lim_{s \to 0} \frac{A}{1 + G(s)C(s)}$$

If we define the position error constant K_{pos} as

$$K_{pos} = \lim_{s \to 0} G(s)C(s)$$

then

$$e_{ss} = \frac{A}{1 + K_{pos}}$$

Similarly we may evaluate with ramp inputs

$$r(t) = At\, U(t) \iff R(s) = As^{-2}$$

to find

$$e_{ss} = \lim_{s \to 0} \frac{R(s)s}{1 + G(s)C(s)} = \lim_{s \to 0} \frac{A}{s + G(s)C(s)s}$$

$$= \lim_{s \to 0} \frac{A}{sG(s)C(s)}$$

With the velocity error constant K_{vel} defined as

$$K_{vel} = \lim_{s \to 0} sG(s)C(s)$$

this is

$$e_{ss} = \frac{A}{K_{vel}}$$

for the position error due to an input with constant rate of change. Continuing, we find

$$K_{acc} = \lim_{s \to 0} s^2 G(s)C(s)$$

with

$$e_{ss} = \frac{A}{K_{acc}}$$

when the input is the ramp $r(t) = A\ t^2/2\ U(t) \Leftrightarrow A\ s^{-3}$. It is clear that if $G(s)\,C(s)$ has the form

$$G(s)\,C(s) = \frac{N(s)}{s^L}$$

for some rational $N(s)$ for which s is not a factor of either numerator or denominator, then the parameter L immediately indicates whether steady-state errors of asymptotically stable responses are zero, potentially finite, or always infinite. This parameter is the system type for continuous time systems.

The above results are for unity feedback systems. For $F(s) \neq 1$, we have

$$e_{ss} = \lim_{t \to \infty} e(t) = \lim_{s \to 0} sE(s)$$

$$= \lim_{s \to 0} \frac{1 + (F(s) - 1)\,G(s)\,C(s)}{1 + F(s)\,G(s)\,C(s)} sR(s)$$

for which the error can be computed, but for which rules about system types and error constants are less reliable.

17.4 COMPUTER AIDS

System type is read off from the transfer function and is not amenable to, or requires, computer assistance. Steady-state errors can be established from the definitions and error constants or by simulation.

17.5 SUMMARY OF PROPERTIES

If we define L as the number of open-loop system poles at the origin in the continuous time case and at $z = 1$ in the sampled data case, then we can refer to L as the system type. In terms of response to certain types of inputs, this type indicates what steady-state position error can be expected for a stable system. This was shown in Table 17.1.

The pattern there is an obvious one, as should be the design element. For example, if the system is to track step inputs such as set points with zero error in a unity feedback situation, it is necessary that the compensator be such that the system is type 1 or higher; the implementation of this is probably with an integrator element. Other requirements, such as finite error in tracking a parabola (constant acceleration input for a positioning system), have analogous solutions. These are requirements on the structure of the compensator; so far, parameter values have not been set.

In addition, we have defined parameters K_{pos}, K_{vel}, and K_{acc} which can be used as numerical indicators of the final error values. One example is the position error constant K_{pos}, defined by

$$K_{pos} = \lim_{z \to 1} C(z)\, G(z)$$

and which is used as in

$$e_{ss}\Big|_{step} = \frac{A}{1 + K_{pos}}$$

The other error constants and the continuous time case are similar.

We should remember (a) that the above is not the only way of computing steady-state errors and (b) that knowing nothing about a system except its type number is indicative of the value of the steady-state error for various inputs.

17.6 FURTHER READING

System type and error constants are standard in teaching and may be seen in most elementary texts, such as Dorf (1989) and Phillips and Harbor (1991) for continuous time and Kuo (1980) and Franklin *et al.* (1990) for discrete time systems.

18

Root locus methods for analysis and design

Root locus methods allow study of the changes in the roots of a polynomial as a parameter varies: their implicit but important message is that these roots vary in a regular manner as a parameter varies. Root locus methods are applied to the denominator polynomials of closed-loop transfer functions and hence indicate the movement of system poles as a parameter (typically a control compensator parameter such as a gain) varies. The techniques are independent of whether the system is discrete time or continuous time, but the criteria for good and bad poles depend upon that fact (Chapter 19).

Although the root locus plots are easily generated by computers and in fact become very tedious to create manually for systems of order greater than 6 or 8, it is worthwhile to understand the fundamentals of sketching them so that design is other than pure trial and error. For this reason, we consider first the basics of sketching root locus plots, and then discuss design using the plots.

18.1 SYNOPSIS

The root locus diagram (sometimes in the general cases called a root contour) plots the closed-loop poles of a system as a parameter varies. In the paradigm situation, this is a plot of the poles of the transfer function of the form

$$G(q) = \frac{\dfrac{N_1(q)}{D(q)}}{1 + K\dfrac{N(q)}{D(q)}} \tag{18.1}$$

and more particularly the zeros of the denominator

$$1 + K \frac{N(q)}{D(q)}$$

as the parameter K varies, when N and D are polynomials in q which do not contain K. The calculations are easily done by a computer in several different ways. Design methods, however, sometimes require that the control law $C(q)$ be such that this be modified to

$$1 + KC(q) \frac{N(q)}{D(q)} = 1 + K \frac{N'(q)}{D'(q)}$$

As knowing how to choose $C(q)$ is something of an engineering art form, sketching methods are presented in section 18.2 and design, which really is repeated analysis for any but simple parameter choices, is the topic of section 18.3.

The locus shows all possible pole locations as the parameter varies; the issue of 'good' choices of pole locations is left to Chapter 19.

18.2 SKETCHING OF ROOT LOCUS PLOTS

18.2.1 Basics

The rules for sketching root locus plots apply to closed-loop transfer functions placed in the form (18.1) where $N_1(q)$, $N(q)$ and $D(q)$ are polynomials in q, and K is a real constant, frequently a gain constant, which is varied to give the locus. In this, q may be either the Laplace transform variable s or the z-transform variable z. Further, because the interest is in the poles of $G(q)$, it is the zeros of the denominator which are of interest, i.e. the system poles are complex numbers q_p, dependent upon K, such that

$$1 + K \frac{N(q_p)}{D(q_p)} = 0$$

For the purposes of drawing the locus (and for some of the frequency domain methods of later sections), the two polynomials are presumed to be available in factored form as

$$\frac{N(q)}{D(q)} = \frac{(q - z_1)\,(q - z_2)\,\cdots\,(q - z_m)}{(q - p_1)(q - p_2)\,\cdots\,(q - p_n)} \tag{18.2}$$

where the z_i are zeros and the p_j are poles of the numerator and denominator, respectively, and are complex numbers; a gain factor within N/D is ordinarily incorporated into the locus parameter K. It is presumed that $n \geq m$.

The main rules are summarized here.

Rule 1: Basic Rule

Any point q_k lies on the locus for some value of K, $K > 0$, if for some integer ℓ

$$\arg\left[\frac{N(q_k)}{D(q_k)}\right] = 180° \pm \ell \times 360°$$

and for some $K < 0$ if

$$\arg\left[\frac{N(q_k)}{D(q_k)}\right] = \pm \ell \times 360°$$

The value of K is given by

$$K = -\frac{D(q_k)}{N(q_k)}$$

Rule 2

The locus has $\max(n, m)$ branches and the closed-loop system has one pole (zero of the denominator polynomial) from each branch. Each starts at one of the poles at $K = 0$ and goes to one of the zeros as $|K| \to \infty$. If $n > m$, then $n - m$ zeros are interpreted as being 'at infinity', while if $m > n$ then $m - n$ poles are 'at infinity'. We remark that having many zeros 'at infinity' is more common with analog than with digital systems.

Rule 3

The $n - m$ zeros 'at infinity' are reached by loci converging onto $n - m$ asymptotes, provided $n \neq m$. These asymptotes form angles ϕ_i

(measured counter-clockwise positive from the real axis of the q-plane) of

$$
\phi_i = \begin{cases} \dfrac{(2i + 1) \times 180^\circ}{n - m} & \text{for } K > 0 \\[2em] \dfrac{2i \times 180^\circ}{n - m} & \text{for } K < 0 \end{cases}
$$

where $i = 0, 1, 2, \ldots, n - m - 1$. The asymptotes intersect at the point on the real axis with coordinates $(\sigma_c, 0)$, with

$$
\sigma_c = \frac{\displaystyle\sum_{i=1}^{n} p_i - \sum_{k=1}^{m} z_k}{n - m}
$$

Rule 4
A branch of the locus will lie on the real axis to the left of an odd number of poles and/or zeros for $K > 0$ and to the left of an even number for $K < 0$. For these purposes, 0 is considered an even number.

Rule 5
The locus is symmetric about the real q–axis.

Rule 6
The locus may exhibit points on the real axis at which closed-loop poles transition between real pairs and complex values as K varies. These are called breakaway points $q_{\text{breakaway}}$ if the locus branches from the real axis as K increases and breakin points q_{breakin} if the locus arrives at the real axis as K increases; the former occur between adjacent real open-loop poles (including perhaps poles at infinity), and the latter between real open-loop zeros. The locations of such points, whether breakaway or breakin, are given by the real solutions q_b of

$$
\sum_{i=1}^{n} \frac{1}{q_b - p_i} = \sum_{k=1}^{m} \frac{1}{q_b - z_k}
$$

The computation of this location can be performed several different ways. It is often possible to guess roughly the location of q_b, as the above calculation and several variant forms of computation are quite tedious. One alternative arises if the poles and zeros are real, for then the above denominators are distances which are easily determined graphically. Another approach (to be used in the example below) exploits the observation that the breakaway is the point for which K is a maximum on the real axis (or K is a minimum for a breakin point). Ordinarily the breakaway or breakin forms an angle perpendicular to the real axis.

Rule 7

The locus leaves a complex pole p_i of multiplicity L at an angle given by

$$\Theta_{p_i} = 180° + \arc\left\{\frac{N(q)\,(q - p_i)^L}{D(q)}\right\}_{q=p_i}$$

where the cancellation of factors is assumed, and arrives at a complex zero z_i of multiplicity M at an angle given by

$$\Theta_{z_i} = 180° - \arc\left\{\frac{N(q)}{D(q)\,(q - z_i)^M}\right\}_{q=z_i}$$

The 180° angles used become 0° when considering $K < 0$.

Usually for sketching purposes, Rule 1 should be kept in mind, but Rules 2–6 are crucial, while Rule 7 is a refinement. Even Rule 6 need not be computed exactly for sketching purposes.

The algorithm for root locus diagrams, then, is as follows.

1. Place the transfer function in form (18.1). Determine the number and locations of the zeros and poles of the rational function of q in the denominator, i.e. find representation (18.2) of the denominator.
2. Place the poles (using crosses) and zeros (using circles) on the graph paper.
3. Compute the centroid q_c for the asymptotes and draw in the asymptote lines, which all start from the point $(q_c, 0)$ in the q-plane.

4. Start sketching locus branches in the upper half plane (the lower half will be a mirror image). First place lines on the real axis. Draw arrowheads (showing K increasing in magnitude) as in Rule 3. If two arrowheads point at each other (or away from each other), then guess (optionally compute) a breakaway point (breakin point) and sketch in the perpendiculars.
5. Fair in lines with the asymptotes. For complex zeros and poles either guess or (preferably) compute the arrival (departure) angles.
6. Connect the various line segments. Be sure that each pole has a locus branch going to some zero.

The above comes with experience and the use of a good computer program.

Example: root locus manual technique

As an illustration of the algorithm for root locus diagrams, we consider the rather artificial example

$$G(z) = \frac{1}{(z - 0.25)(z - 0.75)(z + 0.5)}$$

and the locus as K varies of the closed-loop poles, that is, the poles of

$$H(z) = \frac{KG(z)}{1 + KG(z)}$$

To do this, we need only examine the zeros of the denominator. Applying the root locus approach, we first mark the poles and zeros of the denominator function $G(z)$: we see there are ($n =$) 3 poles at $q_p = 0.75, 0.25, -0.5$ and ($m =$) 0 zeros, as shown in Fig. 18.1(a). There are three branches to the loci: one starting at each of the three poles – these go to three open-loop zeros 'at infinity'.

Next we find the centroid of the poles and zeros and the asymptote angles. The centroid is real, as it must be because all complex poles and zeros are paired, and is given by

$$\sigma_c = \frac{0.75 + 0.25 - 0.5 - \{0 \text{ for sum of zeros}\}}{3 - 0} = \frac{1}{6}$$

The angles are

$$\phi = \frac{(2k + 1)\,180°}{3 - 0} = 60°, 180°, 300° = \frac{\pi}{3}, \pi, \frac{5\pi}{3}$$

as shown in Fig. 18.1(b) where the asymptote at π is the negative real axis.

Rule 4 is illustrated in Fig. 18.1(c), in which the segments to the left of -0.5 and between 0.25 and 0.75 must lie on branches of the locus.

Finally, the breakaway point which must lie between 0.25 and 0.75 is determined. It could reasonably be estimated as being half way between those poles, i.e. at $z = 0.5$, but was here calculated. The calculation 'trick' used was to calculate the value of K which would give a value of closed-loop pole on the real axis in the range $\{0.25, 0.75\}$. The largest such K is that 'just prior to' breakaway. In particular, we calculate

$$(\sigma - 0.25)(\sigma - 0.75)(\sigma + 0.5) + K = 0$$

for K as σ varies. A simple search, as illustrated by Table 18.1, is usually sufficient for sketching purposes. Clearly the largest value of K yielding a real pole in the interval examined is $K_{\text{breakaway}} \approx 0.063448$, giving $z_{\text{breakaway}} \approx 0.530 \pm 0j$.

Table 18.1 K corresponding to particular real axis values σ of s

σ	K
0.5	0.0625
0.6	0.05775
0.4	0.04725
0.55	0.06300
0.52	0.063342
0.53	0.063448
0.525	0.06342187
0.535	0.06341963
0.529	0.06344711
0.531	0.063446700
etc.	

The breakaways are sketched in Fig. 18.1(d), and the poles for $K = 1.094$ (at $z = -1 + 0j$ and at $z = 0.75 \pm 0.79j$ and respective magnitudes of 1.0 and 1.09) are shown also. The latter were found by direct calculation. This figure also shows the final root locus. The arrowheads indicate the directions of traversing the locus as K increases.

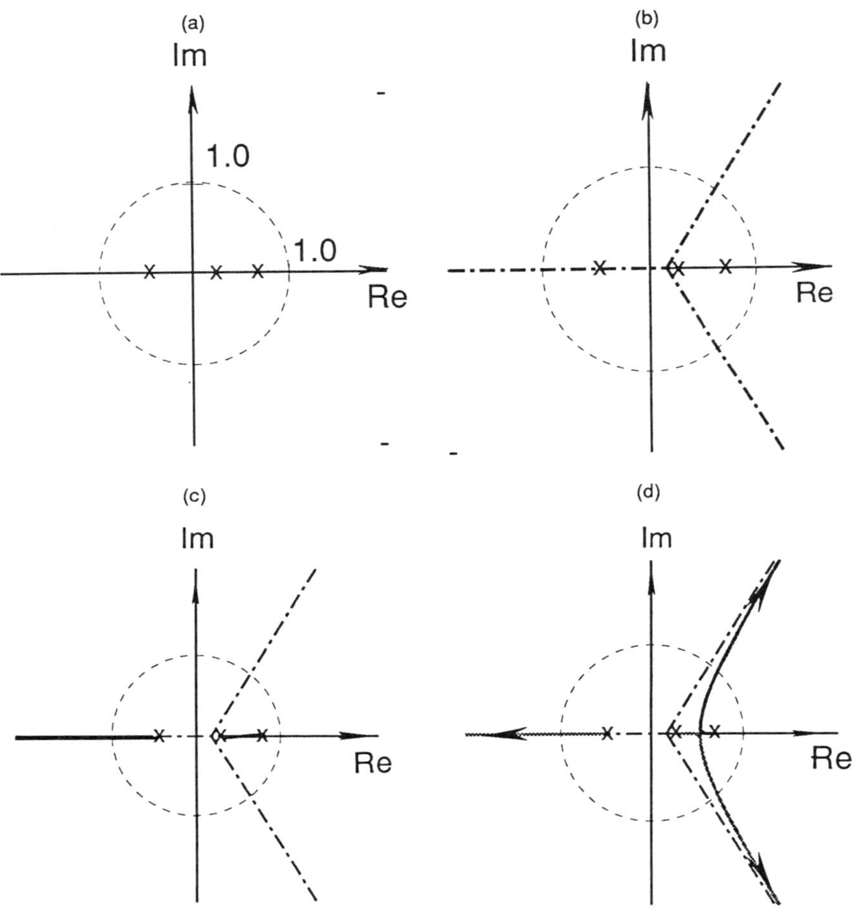

Figure 18.1 Development of a root locus diagram example for a discrete time system, showing the stability boundary (unit circle). (a) Placement of poles. (b) Sketching of asymptotes. (c) Finding locus segments on real axis. (d) Completion of diagram with breakaway point and asymptotic loci.

The root locus manual technique may also be used to find general patterns of loci: general form for two real poles, two complex poles, three poles in various combinations of real and complex plus one zero, etc. Some of this is done in tables presented by Phillips and Harbor (1991, Table 7.6). Many examples are in elementary textbooks.

We should observe, however, that a very typical discrete time system will lead to a form of z-transform such as

$$G(z) = \frac{b_1 z^{-1} + b_2 z^{-2} + b_3 z^{-3} + \cdots + b_m z^{-m}}{1 + a_1 z^{-1} + a_2 z^{-2} + \cdots + a_n z^{-n}}$$

with $b_1 \neq 0$ and $n \geq m$. In this case we see that the system will have n poles and $n-1$ zeros ($n-m$ of them at the origin). Then there will be one asymptote out to π and all other parts of the locus running from finite poles to finite zeros. Figure 18.2 shows the locus for a system with 4 poles and 3 zeros.

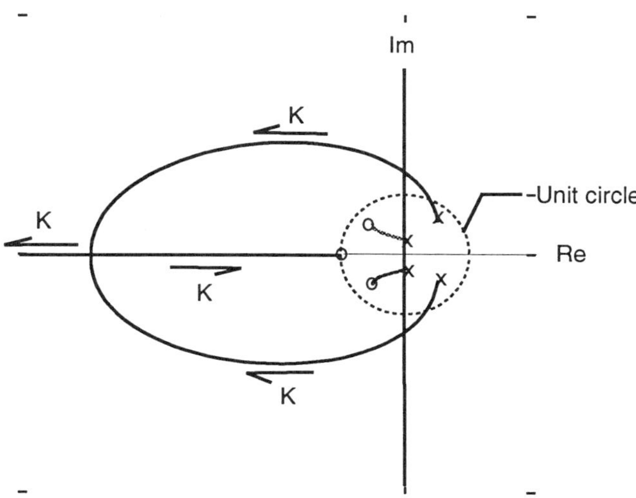

Figure 18.2 Typical 4-pole 3-zero root locus diagram showing unit circle stability boundary.

18.2.2 Variations

The treatment of $K<0$ is often seen as a variation of the basic method, but the essentials have already been noted above. Two other, somewhat related, variations also arise:

1. treatment of a parameter other than gain; and
2. treatment of multiple parameters.

The first variation is easy for a computer program based upon root computation, as it does not care which parameter is varying. For sketching methods based upon the above, a rearrangement of the transfer function is needed.

Suppose that the denominator has the form

$$1 + \frac{N_1(q) + KN_2(q)}{D_1(q) + KD_2(q)}$$

Simple rearrangement puts it in the form

$$\frac{\{N_1(q) + D_1(q)\} + K\{N_2(q) + D_2(q)\}}{D_1(q) + KD_2(q)}$$

and because we are only interested in zeros of this, we may concentrate upon

$$1 + K\frac{N_2(q) + D_2(q)}{N_1(q) + D_1(q)}$$

which is of the required form. This approach is sometimes called a 'root contour' method.

Example

Consider the simple example

$$H(s) = \frac{K\dfrac{1 + \alpha}{(s + \alpha)}}{1 + K\dfrac{1 + \alpha}{(s + \alpha)}}$$

The design parameter is taken to be α (which might, for example, be related to a hardware time constant) and K is preset to 0.5. Then the poles of $H(s)$ are the zeros of

$$D'(s) = 1 + \alpha \, \frac{1.5}{s + 0.5}$$

This function may be studied in the usual manner.

When several parameters must be chosen, the root locus approach becomes difficult to implement, for the approach by its nature is concerned with a single parameter and a certain structure. To make fruitful use of root loci, one approach is to start with all parameters set to zero and work sequentially through them, choosing a good value for each before proceeding to the next. Each parameter in the sequence is evaluated using a root contour. A variation is to place the successive parameters on the one plot, and to evaluate the successive ones for several values of the first. We best see this with an example.

Example

We are to choose the gain and a zero location for a controller to be used with the system

$$G_{\mathrm{p}}(z) = \frac{1}{(z - 0.25)\,(z - 0.75)\,(z + 0.5)}$$

Thus we are to choose α and β in $C(z) = \beta z + \alpha$ and are concerned with $1 + C(z)\,G_{\mathrm{p}}(z)$.

Setting $\beta = 0$, we examine the roots as α varies; the resulting diagram was in Fig. 18.1. We might choose values α_1 and α_2 for further study, plotting new root loci as functions of β as in Fig. 18.3(a). An alternative is to show many loci together as in Fig. 18.3(b), which shows varying α starting a family of loci in β.

It is clear from the example that multiple parameter cases can place a considerable burden on the designer in trading off the design choices to be made. On the other hand, patterns which are useful may become apparent.

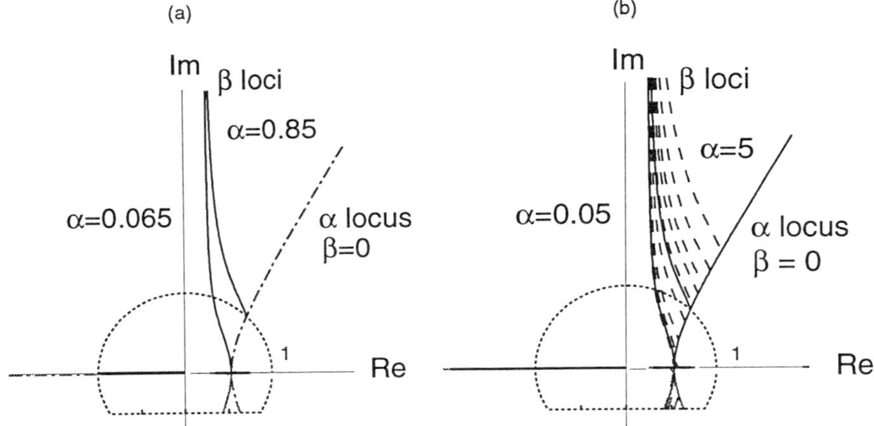

Figure 18.3 Development of plots representing two-parameter design: (a) showing the basic α locus and β loci for two values of α; and (b) the family of β loci as α changes from 0.05 to 5.

18.3 DESIGN IN ROOT LOCUS

Design with root loci is a matter primarily of design by repeated analysis, except of course for the obvious selection of a single parameter.

The situation when the desirable root set is unattainable using only K becomes one in which the control law must be changed. Sometimes it is obvious what must be done to change the locus: for example, one may guess at a new open-loop pole or pole set (or zero or set of zeros) and then repeat the single variable design. Thus, a PID controller introduces two zeros and a pole along with the gain constant; any one of these may be a design parameter. Alternatively, one may try to look at two parameters simultaneously. Either approach requires a bit of artistry on the part of the design engineer.

Example

Let us consider the selection of a single parameter by root locus methods. In particular, we study the proportional control of a simple motor; the object is to choose the gain to obtain 'good' response. The motor model is taken as

$$G(z) = \frac{K_1(z + b)}{(z - 1)(z - e^{-aT})}$$

where

$$K_1 = \frac{aT - 1 + e^{-aT}}{a}$$

$$b = \frac{1 - e^{-aT} - aT\,e^{-aT}}{aT - 1 + e^{-aT}}$$

We choose the motor time constant as $\tau = 1\,\mathrm{s}$ and select sampling frequency as 10 samples/s, so that $T = 0.1$ and $aT = 0.1$. Then

$$K_1 = 0.048 \quad \text{and} \quad b = 0.967$$

and hence

$$G(z) = \frac{0.048\,(z + 0.967)}{(z - 1)(z - 0.905)}$$

Using the root locus approach, we first plot the diagram labelled P control in Fig. 18.4. We also calculate a few numerical values, in particular,

$$z_{\text{breakaway}} \approx 0.952 \qquad\qquad \text{where } K \approx 0.024$$

$$z_{\text{breakin}} \approx -2.87 \qquad\qquad \text{where } K \approx 158$$

$$\|z\| \approx 1 \text{ at } \omega T \approx 30° = \pi/6 \qquad\qquad \text{where } K \approx 2.05$$

We return to this in Chapter 19, but at this point we comment that the alternative choices of K are not attractive: either slow frequency of response for small K or slow decay of transients for larger K.

Let us reconsider the design and in particular consider using a PI controller, digitized as

$$C(z) = K_p + K_i \frac{z}{z-1} = \frac{K\,(z-c)}{z-1}$$

where K and c are to be chosen. A quick root locus sketch shows this is unlikely to help. We try instead a PD controller of the form

$$C(z) = K_p + K_d \frac{z-1}{z} = K \frac{(z-c)}{z}$$

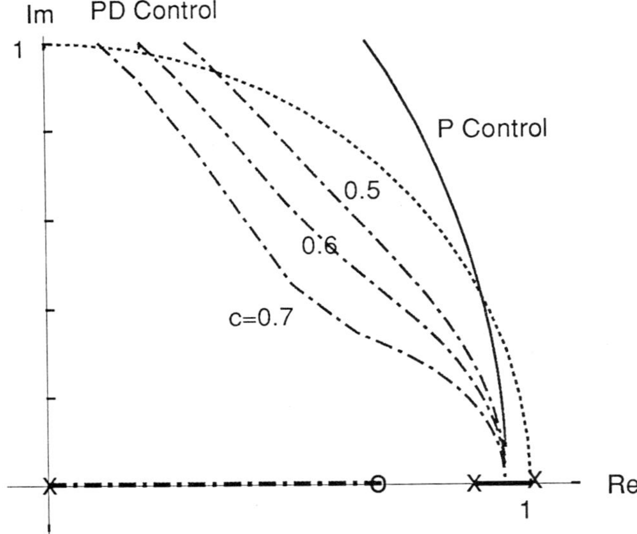

Figure 18.4 Two parameter loci. The effect of various zero locations (PD control) on the basic locus. Example design with varying zero location.

Sketches of the resulting root locus diagrams for various $c = K_d/K$ with $K = K_p + K_d$ as the root locus parameter are in Fig. 18.4. By choosing $c \approx 0.5$ or greater, the root locus is made to bend into the unit circle. We choose $c = 0.7$ from which the K value of $K \approx 5$ ($\Rightarrow$ $K_p = 1.5$ and $K_d = 3.5$) yields poles at $z_p \approx 0.28$, $0.69 \pm 0.32j$. The reasons for these choices come in Chapter 19. Figure 18.5 shows step responses of some of the compensators considered.

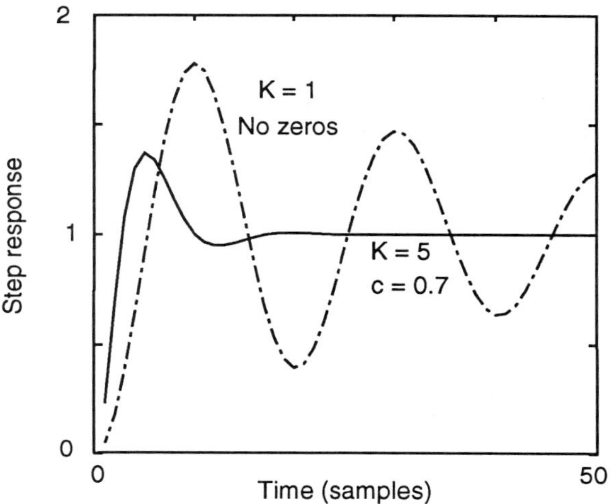

Figure 18.5 Step responses for P control and PD control, with parameters chosen from the root loci of Fig. 18.4.

18.4 COMPUTER AIDS

The above is in principle not difficult to do on a computer, and algorithms tend to be of two varieties.

1. Compute the roots of $D(q) + K\,N(q)$ as K ranges over some selected range. Attempt to build loci by connecting the appropriate roots in each set. Thus for an nth order system, for each K compute the n roots and then attempt to connect the corresponding roots from each set to form the plot of the loci.
2. Search for points z satisfying the angle criterion and 'near' a point already on the locus. Connect the new point to the old and continue. Start each branch of the locus at a pole and end it at a zero.

Both methods have problems, usually associated with branching points, such as breakaway and breakin points, of the loci. Either method needs careful programming and use to avoid ambiguities and errors in the plots. For this reason, one is well advised to have at least some knowledge of the expected appearance of the locus.

The first option is easy for a personal computer and is used in some commercial packages such as MATLAB®; it presents difficulties in showing fine detail unless the user is prepared to interactively adjust the parameter set to be plotted or to mentally fill in gaps which may appear in the loci between calculated roots. The second option iterates on the basic angle rules while searching over the complex plane for candidate locus points; the programming can be tricky.

18.5 SUMMARY

We have considered only the essence of root locus plotting so that the engineer will be able to develop notions of the effects of design choices. Computers make it easy to generate approximate root locus plots for any given parameter variation, and the methods are independent of whether the system under consideration is continuous time or discrete time.

18.6 FURTHER READING

Root locus plots are standard in most introductory level textbooks. Root locus diagrams themselves are actually independent of the system studied – only the interpretation of good and bad poles depends on whether the system is discrete time or continuous time. Hence most textbooks (e.g. Dorf (1989), Kuo (1980), Franklin *et al.* (1990), Phillips and Harbor (1991), etc.) can help. For drill, the student might look at such as (DiStefano *et al.*, 1976).

19

Desirable pole locations

It is always required that system characteristic values, also known as transfer function poles and dynamics matrix eigenvalues, be stable, meaning that they should be in the left-half plane for continuous time systems and within the unit circle for sampled data systems. It is also true, however, that some pole values may yield more desirable system responses than other values. We explore this issue in this chapter.

19.1 SYNOPSIS

There are two ways to set the poles of linear systems: by direct selection in the complex plane and indirectly by optimization methods.

We study the direct evaluations in section 19.2 for both continuous and sampled data systems. The characteristic values λ of linear constant coefficient systems are either real or in complex conjugate pairs. The usual representation for continuous time systems has

$$\lambda = -\sigma \quad \text{or} \quad \lambda_{1,2} = -\zeta\omega_n \pm j\omega_n \sqrt{1-\zeta^2}$$

for real and complex poles, respectively, where ζ is the damping coefficient and ω_n is the undamped natural frequency. Usually if one considers the traditional performance indicators, then speed of response requires that if possible σ or $\zeta\omega_n$ be 'large', while reasonable trade-offs on overshoot and speed of response lead to ζ in the vicinity of 0.5–0.6.

Similarly, for sampled data systems the poles are of the form

$$\lambda = r \quad \text{or} \quad \lambda_{1,2} = re^{\pm j\phi}$$

with stability requiring $|r| < 1$, speed of transient decay higher when $|r| \ll 1$, and speed of oscillation increasing as ϕ ranges from 0 toward π.

When there are many poles, as in a high-dimensional system, then sometimes the response is dominated by (i.e. looks most like) a particular pole or pole pair: a pole pair much closer to the origin in a continuous time system, or closer to the unit circle in a sampled data system, than other poles. This pair is then used for design to meet classical performance measures, using approximate rules of thumb such as:

1.   ratio of successive oscillations $\approx \delta = r^{2\pi/\phi}$;
2.   oscillation period $= 2\pi/\omega_d \approx 2\pi T/\phi$;
3.   percent overshoot to step response $\approx 100\sqrt{\delta}$;
4.   2% settling time $\approx T \ln(0.02)/\ln(r)$; and
5.   time to peak of overshoot $\approx \pi T/\phi$.

Here T is the sample period. Similar approximations for continuous time systems are:

1.   the ratio of successive peaks $\approx \exp(-2\pi\zeta/(1-\zeta^2)^{\frac{1}{2}})$;
2.   period of oscillation $= 2\pi/\omega_{osc} = 2\pi(\omega_n(1-\zeta^2)^{\frac{1}{2}})^{-1}$;
3.   percent overshoot $= 100\ \exp(-\pi\zeta/(1-\zeta^2)^{\frac{1}{2}})$
     $\qquad\qquad\qquad \approx 100\ (1-\zeta/0.6) \quad \zeta < 0.6$;
4.   2% settling time $t_s \approx 4\tau = 4/\zeta\omega_n$;
5.   peak time $t_p = \pi/(\omega_n\ (1-\zeta^2)^{\frac{1}{2}})$; and
6.   rise time $t_r \approx 2.5/\omega_n$.

An alternative point of view is to choose poles to minimize a particular criterion such as ISE (integral squared error) or a trade-off of error and control effort as in the linear quadratic regulator (LQR) problem. This gives the appearance of objective design but requires freedom in choosing the design parameters.

19.2   PLACEMENT OF POLES IN THE COMPLEX PLANE FOR DISCRETE TIME SYSTEMS

For a linear discrete time system, there will be a transfer function

$$G(z) = \frac{Y(z)}{U(z)} = \frac{\displaystyle\sum_{i=0}^{m} b_i z^{-i}}{1 + \displaystyle\sum_{i=1}^{n} a_i z^{-i}} \tag{19.1}$$

relating the output sequence $\{y(k)\}$ to the input sequence $\{u(k)\}$.

From the fundamental theorem of algebra (that any polynomial may be factored into a product of terms linear in the variable, or terms at most quadratic if all coefficients are real) it follows that the above is a sum of terms of the forms such as

$$K\,z^{-j} \qquad j=0,1,\dots \tag{19.2}$$

$$K\,\frac{(1-p)\,z^{-1}}{1-p\,z^{-1}} \tag{19.3}$$

$$K\,G_c\,\frac{z^{-1}\,(1-z_0\,z^{-1})}{1-2r\cos\phi\;z^{-1}+r^2\,z^{-2}} \tag{19.4}$$

where:

$$G_c = \frac{1-2r\cos\phi+r^2}{1-z_0}$$

and the complex zero form

$$K\,G_c\,\frac{1-2r_0\cos\phi_0\,z^{-1}+r_0^2\,z^{-2}}{1-2r\cos\phi\;z^{-1}+r^2\,z^{-2}} \tag{19.5}$$

with G_c now given by

$$G_c = \frac{1-2r\cos\phi+r^2}{1-2r_0\cos\phi_0+r_0^2}$$

All of the parameters $K, p, z_0, r,$ and ϕ are real and have the following interpretations: K is a gain, p is a real pole, z_0 is a real zero, and in (19.4) there is a complex pole located at $(r\cos\phi, r\sin\phi)$, i.e. a pole pair of magnitude r and angle $\pm\,\phi$ to the real axis. G_c is chosen so that when $K = 1$, a unit input gives – eventually – a unit output for a stable system. In (19.5) both a complex zero pair at $(r_0\cos\phi_0, r_0\sin\phi_0)$ and a complex pole pair as in (19.4) are present.

Complex zeros with a single real pole and stand-alone poles (or zeros) are also possible in digital compensators.

Since the transfer function can be written in a partial fraction expansion, it follows that the output is the sum of outputs from individual sections of the above types, each with the same input.

The term in (19.2) is simple and uninteresting, since the output is simply a delayed multiple of the input; it gets more interesting when several such terms are considered together, as this constitutes what is called a finite-duration impulse response (FIR) filter in the signal processing literature, and it is discussed in detail there, including design for noise removal and signal emphasis.

The first-order term (19.3) is more interesting *per se*. It may be interpreted as relating input sequences $\{u(k)\}$ to output sequences $\{y(k)\}$ as in (19.1) and may be implemented as

$$y(k+1) = p\, y(k) + K(1-p)\, u(k)$$

This system's response for a unit pulse input is

$$y(k+1) = K(1-p)p^k \qquad k=0,1,\ldots,$$

when $y(0) = 0$. The unit step response is

$$y(k) = K(1-p^{k+1})$$

One should note that these converge if $|p|<1$, and the pulse response, unlike continuous time first-order systems, oscillates between + and − errors if $p < 0$.

Most interesting is the quadratic type term in (19.4), and we look at an example to study its characteristics.

Example

We look at (19.4) with $K = 1$ and consider the effect of varying r, ϕ, and z_0 relative to a nominal case defined by

$$r = 0.8$$

$$\phi = 0.5 \text{ rad}$$

$$z_0 = 0$$

Direct inverse transformation shows that the decay rate of the oscillation depends heavily on r, the frequency on ϕ, and that z_0 particularly affects the initial direction and magnitude of the response. The latter appears to have a heuristic interpretation related to a derivative type of term, and partly because of the gain term G_c, it can contribute a large output when the input is a pulse, step, or other rapidly changing signal. This contribution can be seen in Fig. 19.1.

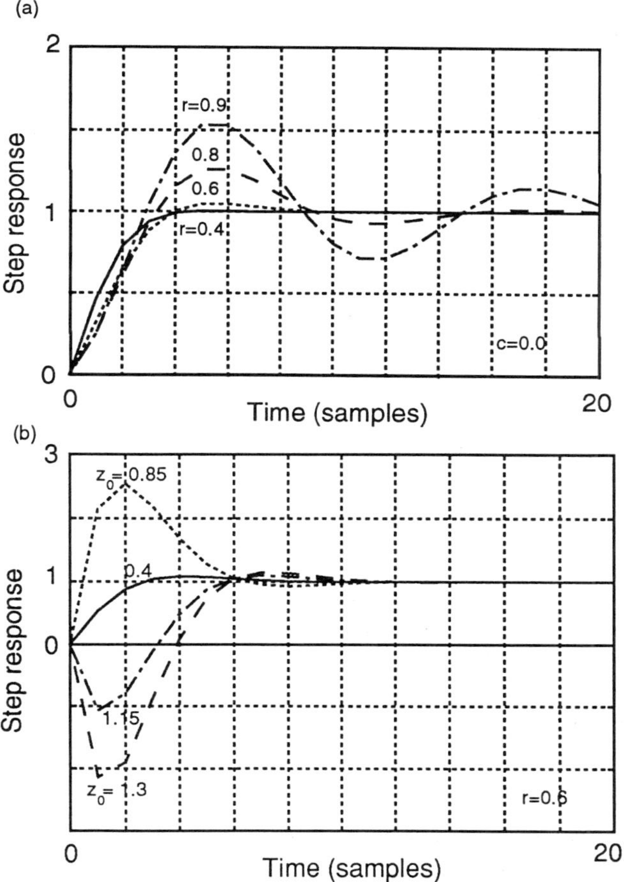

Figure 19.1 Effects of the parameters in quadratic discrete time systems: (a) effect of distance r of poles from origin; (b) effect of zero location z.

In spite of this occasional dominance by the zeros, it is possible to identify desirable regions for the poles. We consider how the poles affect the solution.

Expanding (19.1) to first-order terms yields the form for $G(z)$ as

$$G(z) = \sum_{i=1}^{n} \frac{a_i}{1 - z_{p_i} z^{-1}}$$

where a_i and z_{p_i} may be complex; for convenience we assume that no factors of $G(z)$ are repeated in the original form (19.1). Then the inverse transform yields

$$g(kT) = \sum_{i=1}^{n} a_i z_{p_i}^{k}$$

The individual terms of the expansion are time-varying of the form

$$g_i(kT) = a_i z_{p_i}^{k} = a_i \| z_{p_i} \|^k \exp(jk \arc(z_{p_i})) \tag{19.6}$$

Hence, the magnitude of $g(kT)$ develops as the kth power of the pole with the largest magnitude (which necessarily must be less than or equal to one for stability). This 'dominant pole' then can have its magnitude be a design constraint. Very roughly, if error magnitude after p steps is to be less than q times the initial magnitude, then we need

$$\| z_p \|_{\max}^{p} e_0 < q e_0$$

or

$$\| z_p \|_{\max}^{p} < q$$

Example

The ten-step error magnitude of a system response is to be less than 0.01 of the original value. Then

$$\| z_p \|_{\max} < (0.01)^{1/10} = 0.63$$

Circles of constant z are useful additions to the root locus diagram, as in Fig. 19.2(a).

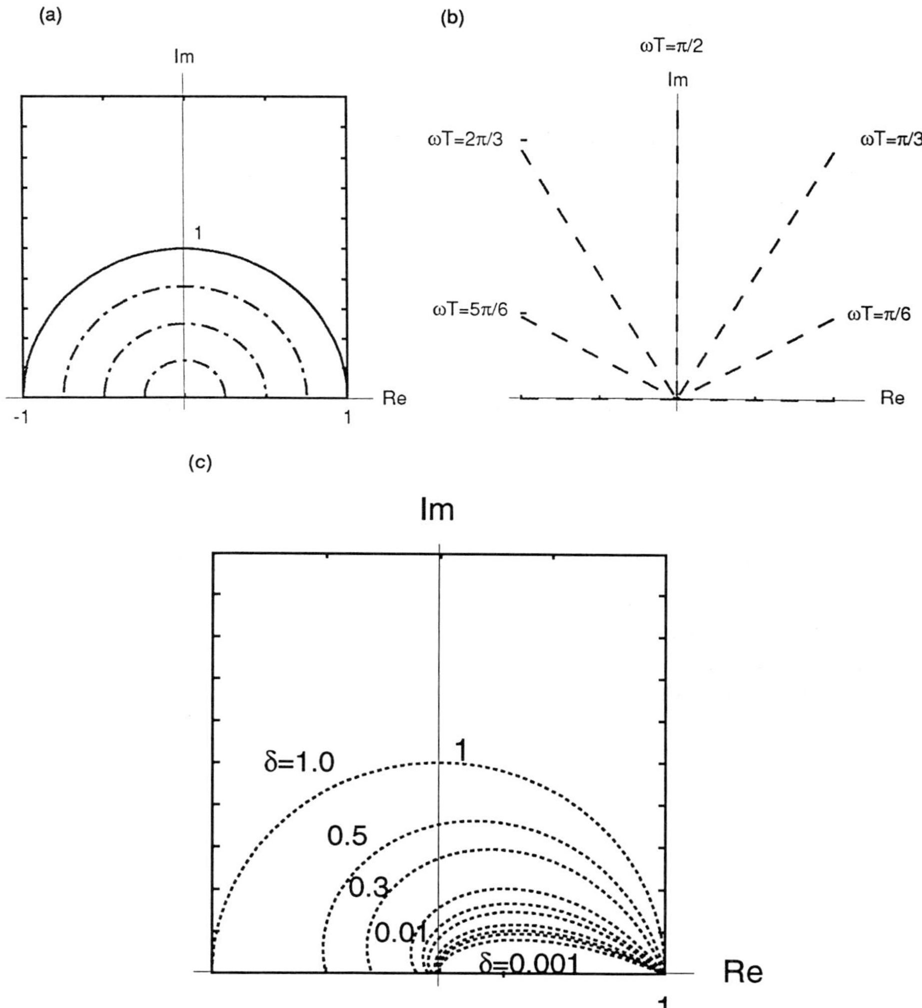

Figure 19.2 Development of graphical display of pole influence representations: (a) distance from origin ($\Rightarrow$ speed of decay); (b) angle relative to real axis ($\Rightarrow$ oscillation frequency); and (c) damping indicator.

There is also the question of oscillation frequency to consider. From (19.6) it is clear that the angle of a term will change by one cycle (i.e. an angle of 2π) in p steps if

$$p \, \text{arc}'(z_{pj}) = 2\pi \qquad\qquad (19.7)$$

where

$$\text{arc}'(a) = \text{arc}(a) \qquad\qquad \text{if } \text{arc}(a) \leq \pi$$

$$= \text{arc}(a^*) \qquad\qquad \text{otherwise}$$

This gives us a criterion for choosing or evaluating pole angles. In particular, for example, if a response is to oscillate through one cycle in p steps, (19.7) suggests how to place the pole angle. Lines of constant cycle frequency are indicated in Fig. 19.2(b).

Example

The response of the preceding example should go through one oscillation in 12 steps. Then we place the dominant pole so that

$$\text{arc}'(z_{p\max}) = \pi/6 = 30°$$

Using this with the magnitude criterion used above,

$$z_p = 0.63 \, e^{j30°} = 0.63\cos 30° + 0.63j \sin 30°$$

$$= 0.55 + 0.32j$$

Since complex poles must necessarily come in pairs, $0.55 - 0.32j$ will also be a pole of this system.

More important, or at least more accessible and understandable, may be a requirement to understand the damping of a term. By this we will mean the ratio of successive peaks in an oscillatory response. Since a single term contributes

$$a \left(\| z_p \| \, e^{j \, \text{arc}(z_p)} \right)^k$$

to the response, it contributes one oscillation in the number of steps k for which $k = 2\pi/\mathrm{arc}(z_p)$. We define the damping δ as the ratio of successive peaks, i.e. the ratio defined by

$$\delta = \frac{a \, \| z_p \|^{k+q}}{a \, \| z_p \|^{q}}$$

where q and $(k + q)$ are indices of times yielding oscillation peaks in the error. Using this and the defining relation (19.7) for oscillation period yields

$$\delta = \| z_p \|^{2\pi/\mathrm{arc}(z_p)}$$

This also gives us a criterion for design and evaluation of pole locations.

Example

The dominant response should oscillate once in each 10 steps and should have a ratio of successive peaks of 0.3. Where should the pole lie?

By performing a calculation like that in the previous example, we may find $\mathrm{arc}(z_p) = 36°$. Then

$$0.3 = \| z_p \|^{10}$$

yields

$$\| z_p \| = 0.887$$

and hence

$$z_p = 0.72 \pm 0.52j$$

where again both members of a complex conjugate pair must be present if one member is.

Spirals of constant δ may be drawn and appear in Fig. 19.2(c). A form combining all indicator lines is in Fig. 19.3.

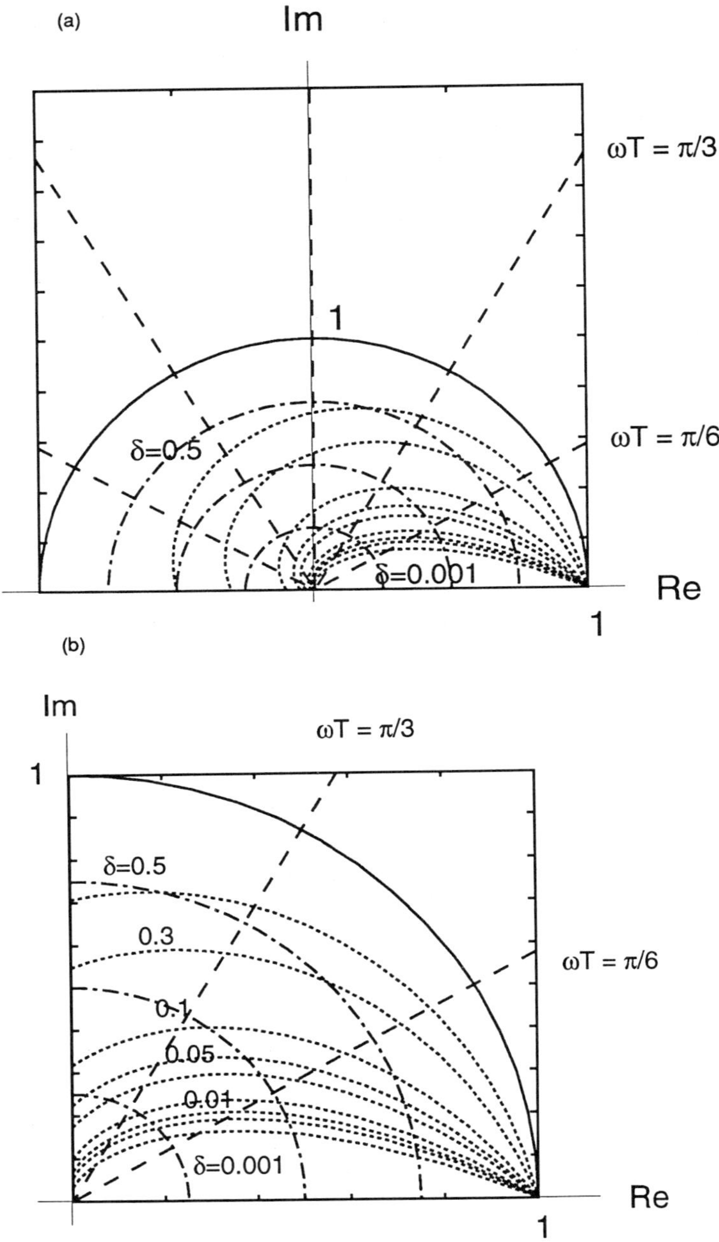

Figure 19.3 Root locus pole placement form: (a) upper half of unit circle; and (b) first quadrant.

The above design 'rules' are definitive only if we indeed have dominant poles, and then only when the terms due to other poles have essentially died out. Nevertheless, they always will help bound the actual performance. More difficult to separate are the rise time and percent overshoot, which may well happen while several poles are still making their presence felt. To study this, we must assume dominance plus make an assumption about the presence or absence of zeros. (Results here are of the type which may make initial rules of thumb, but should be checked for the actual system – perhaps by simulation.)

We have already seen time constant effects as indicated by $r = \|z_p\|$ and oscillation frequency effects as indicated by ϕ. We now look at rise time and peak overshoot indicators. The effects of zeros will show in a phase shift θ, so that errors are of the form

$$e(nT) = Kr^{nT} \cos(\phi_1 n + \theta)$$

This response has an envelope given by the exponential term. If $\theta = 0$, then the maximum error after time $= 0$ will for a stable system occur at approximately the first time n_p for which $|\cos(\mathrm{arc}(\mathrm{pole}))| = 1$, which is when $n_p = \pi/\mathrm{arc}(\mathrm{pole})$ and hence the overshoot will be

$$\| z_p \|^{(\pi/\mathrm{arc}(\mathrm{pole}))}$$

This of course presumes n_p is an integer; otherwise a 'nearest integer' situation may prevail.

When $\theta \neq 0$, then $n_p = (\pi - \theta)/\mathrm{arc}(\mathrm{pole})$ and

$$\mathrm{overshoot} \approx \| z_p \|^{(\pi-\theta)/\mathrm{arc}(z_p)}$$

We note that the expression $\| \mathrm{pole} \|^{(\pi/\mathrm{arc}(\mathrm{pole}))}$ is the square root of the ratio δ used earlier. Hence we may parameterize the percent overshoot using δ. The phase shift θ is related to the zero location $z_0 = c$, and either θ or c may be used as a parameter. The effects are shown in Fig. 19.4, which resulted from simulation.

In summary, we have the following approximate relations to classical performance indicators:

1. Oscillation period $= 2\pi/\omega_d \approx 2\pi T/\mathrm{arc}(z_p)$
2. Percent overshoot to step response $\approx 100\sqrt{\delta}$

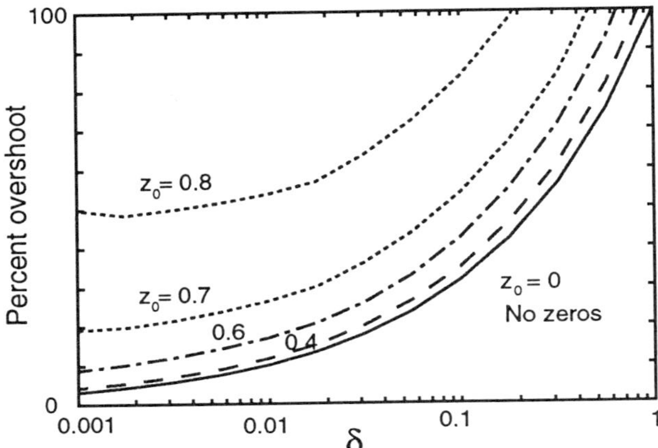

Figure 19.4 The effect of damping parameter δ on percent overshoot for various zero locations c for discrete time systems.

3.   2% settling time $\approx 4\,\tau \approx \dfrac{T\ln(0.02)}{\ln(\|z_p\|)}$

4.   Time to peak of overshoot $\approx \dfrac{\pi T}{\mathrm{arc}(z_p)}$

5.   Ratio of successive oscillations $\approx \delta = \|z_p\|^{2\pi/\mathrm{arc}(z_p)}$

With these parameters in mind, it is possible to select some pole locations. This is particularly true if there are only two poles, or if one pole or pole pair clearly dominates the response of the system. The latter, incidentally, also shows up if a pole and a zero are 'close' together, for then the pole's term in the transient response will have a small coefficient and hence make a small contribution to the response.

19.3   POLE LOCATIONS FOR CONTINUOUS TIME SYSTEMS

The situation is similar for continuous time systems. We first consider systems described by linear coefficient constant ordinary differential equations and hence having transfer functions such as

$$G(s) = \frac{b_m s^m + b_{m-1} s^{m-1} + \cdots + b_1 s + b_0}{a_n s^n + a_{n-1} s^{n-1} + \cdots + a_1 s + a_0}$$

Usually $n \geq m$ and all coefficients are real. The denominator can be expanded into factors of the form

$$(s + \sigma)$$

$$(s^2 + 2\zeta\omega_n s + \omega_n^2)$$

and hence $G(s)$ has a partial fraction expansion

$$G(s) = G_1(s) + G_2(s) + \cdots + G_k(s)$$

in which the denominator of each $G_i(s)$ has only one such factor and the numerator order is less than or equal to that of the denominator. For this reason the system response is a superposition of responses due to the individual terms, and we are justified in considering only the responses of typical such terms.

Linear factors When $G_i(s)$ has the form

$$\frac{a}{s + \sigma}$$

then it is easy to see that the unit step response with zero initial conditions is

$$y(t) = \frac{a}{\sigma}(1 - e^{-\sigma t})$$

Clearly all responses are of the same shape and having σ large means that the transient will decay very rapidly until $y(t)$ takes on its steady-state value a/σ. There is no overshoot, and all speed related indicators (rise time, settling time, delay time) are faster for large σ than they are for small σ. An alternative notation has

$$\tau = \text{time constant} = \frac{1}{\sigma}$$

Quadratic factors The situation becomes more interesting for quadratic factors such as

$$\frac{as + b}{\left(s^2 + 2\zeta\omega_n s + \omega_n^2\right)}$$

where ζ, $0 \le \zeta \le 1$, is called the **damping ratio** and ω_n is called the **undamped natural frequency**. The factor has its poles at

$$-\zeta\omega_n \pm j\omega_n\sqrt{1 - \zeta^2} \qquad 0 \le \zeta \le 1$$

If we take $a = 0$ and $b = \omega_n^2$ and normalize the time axis for various ζ we can find unit step responses as in Fig. 19.5.

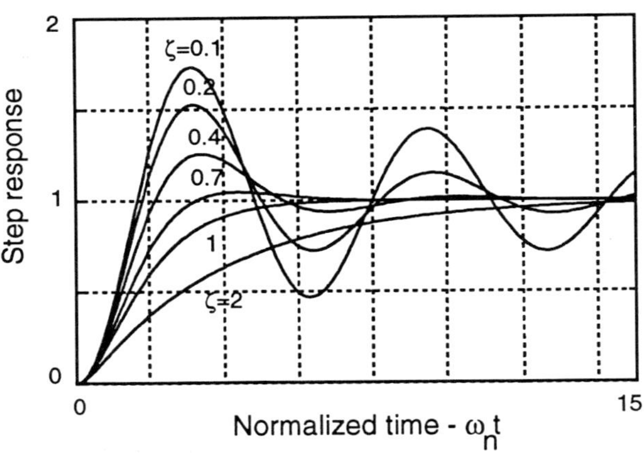

Figure 19.5 Effect of damping ratio ζ on step response for continuous time systems.

It is straightforward to show that the error is of the form

$$e(t) \approx \frac{\exp(-\zeta\omega_n t)}{\sqrt{1-\zeta^2}} \cos\left(\omega_n\sqrt{1-\zeta^2}\, t + \tan^{-1}\left[\frac{\sqrt{1-\zeta^2}}{\zeta}\right]\right)$$

Various parameters are of interest and can be computed; these are defined in Chapter 13. The time constant τ is sometimes used and is defined here as

$$\tau = \frac{1}{\zeta\omega_n}$$

Other parameters specific to the second-order system with complex poles are as follows.

1. Peak time $t_p = \dfrac{\pi}{\omega_n\sqrt{1-\zeta^2}}$

2. Percent overshoot $= 100\,\exp\dfrac{-\pi\zeta}{\sqrt{1-\zeta^2}}$

$$\approx 100\left(1 - \frac{\zeta}{0.6}\right) \quad \zeta < 0.6$$

3. Settling time t_s. 2% settling time is approximately after 4 time constants:

$$t_s \approx 4\tau = \frac{4}{\zeta\omega_n}$$

4. The ratio of successive peaks is roughly

$$r \approx \exp\left(\frac{-2\pi\zeta}{\sqrt{1-\zeta^2}}\right)$$

5. Rise time $t_r \approx \dfrac{2.5}{\omega_n}$

From these one trade-off is immediately obvious, as it was on the graphs: large ζ means small overshoot but late peak time. It is common to consider ζ the order of 0.5–0.6 as being a reasonable choice. Graphs of some of the functions and the effects of a zero at $s = -a$ are given in Figs 19.6 and 19.7, respectively.

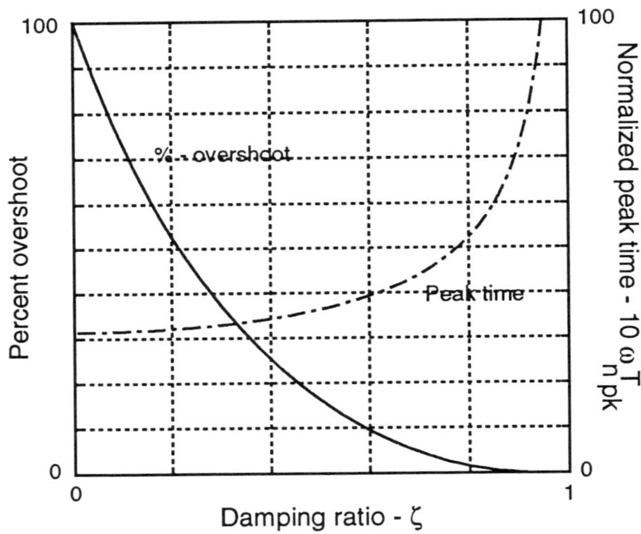

Figure 19.6 Effect of damping ratio ζ on percent overshoot and peak time for continuous time systems.

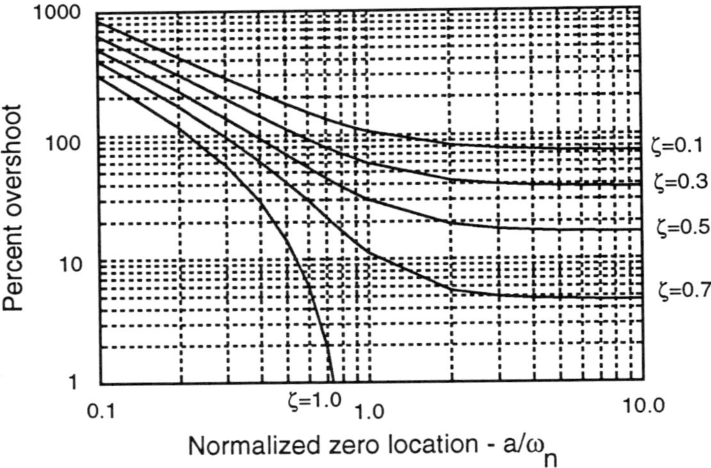

Figure 19.7 Effect of damping ratio ζ on percent overshoot as a function of zero location for continuous time systems.

The above define the response characteristics of a pole or pole pair in terms of standard representation parameters. They are conveniently placed upon a root locus diagram for guidance in choosing poles. To this end we notice that for a particular pole

ω_n = distance from origin

= undamped natural frequency

$\zeta = \cos(\phi)$ where ϕ is the angle of pole with negative real axis

$\zeta\omega_n = \tau^{-1}$

= $-$ real part of pole

ω_d = actual oscillation frequency

= imaginary part of pole

= $\omega_n \sqrt{1-\zeta^2}$

One can place (slanted) lines of constant ζ, circles of constant undamped frequency ω_n, vertical lines representing constant values of τ (since $\text{Re}(s) = -1/\tau$), and horizontal lines corresponding to constant oscillation frequency ω_d, as appropriate to represent specifications. Such guidelines are demonstrated in Fig. 19.8. The design steps involve plotting a root locus on such a form and using it to choose the closed-loop poles.

With the root locus diagram in hand, the engineer attempts to choose a 'good' value K'' of the available parameter K. Good is taken to mean a value such that the resulting poles are those of a closed-loop system with rapid response in which transients are small and die out rapidly. This translates to requirements as follows.

1. K'' must be such that all poles are in the left-half plane. This is to yield a stable system.
2. It is preferred, for reasons of speed of response, that poles be as far to the left as possible. A requirement is often one that transients will decay faster than $e^{-\sigma t}$, which necessitates $\text{Re(poles)} < -\sigma < 0$ for a specified σ.
3. Overshoot in response to step inputs should be at a reasonable level. This often means that the most important complex poles should have damping coefficient ζ in the range $0.4 < \zeta < 0.7$.

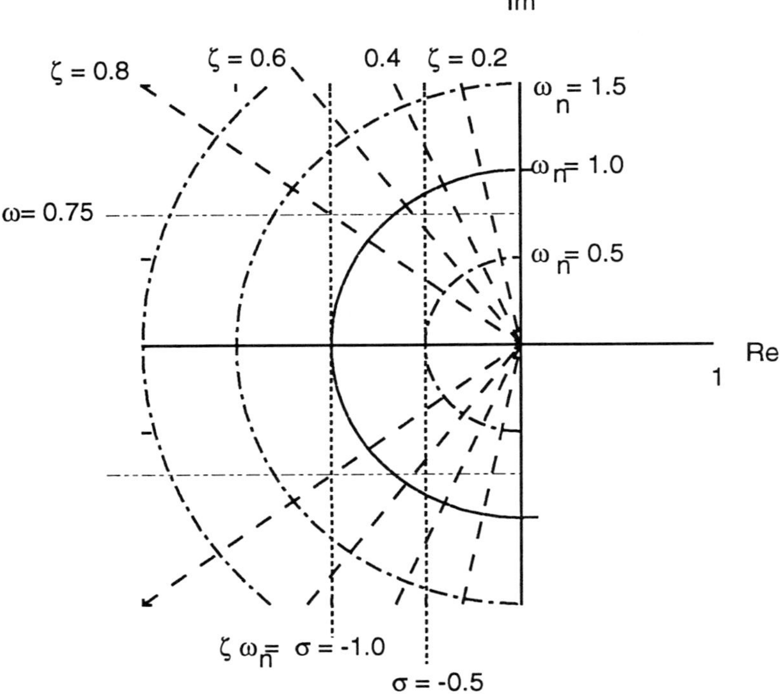

Figure 19.8 Root locus form for continuous time systems, showing circles (·—·—·—·—) of constant natural frequency ω_n (except $\omega_n = 1$ is solid), slanted lines (– – – – –) of constant damping ratio ζ, vertical lines of constant $\sigma = -\zeta\omega_n = -1/\tau$, and horizontal lines (··—··—··—) of constant damped frequency.

Example

An example of $G(s)$ is shown overlaid with indications of the regions $\sigma \le -1.25$ and $0.4 < \zeta < 0.7$ in Fig. 19.9. K'' should be chosen if possible so that poles will be in the desired region, i.e. $0.47 < K'' < 1.36$.

With the above in mind, it is possible to attempt to meet specifications on response. If they cannot be met, then it is necessary to introduce more complicated elements into the compensator $C(s)$.

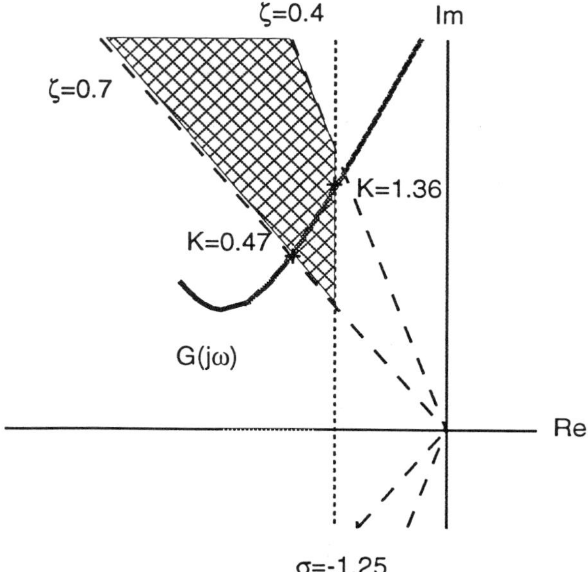

Figure 19.9 Root locus plot showing region of desirable poles. It is possible here to select a gain such that the poles are in that region.

19.4 DESIGN FOR DESIRED POLES

If specific properties are sought, this translates into a certain specificity in pole location. Our choices then are to modify root locus plots so that the desired poles are obtained, to use a design method in which all poles can be specified either precisely or to be within a certain region (Chapter 23).

19.4.1 Design in root locus

Design with root loci is a matter primarily of design by repeated analysis, except of course for the obvious selection of a single parameter.

Example

Let us consider again the example of section 18.3 and the selection of
a single parameter by root locus methods. The root locus plots of the
proportional control and the PD control with zero at $c = 0.7$ is
overlaid with the parameter plot Fig. 19.3 to yield Fig. 19.10.

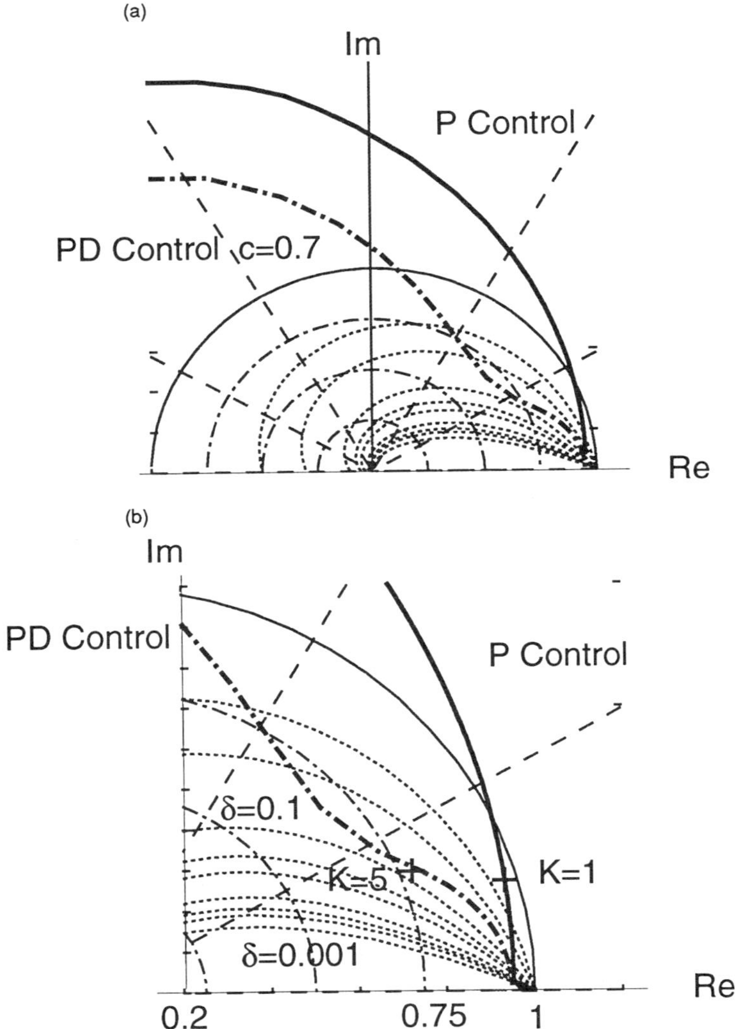

Figure 19.10 Selecting the attainable closed-loop pole for the example
system of section 18.3: (a) overview; and (b) close-up of portion of first
quadrant.

Examining the plot makes the trade-off clear: increasing K leads to faster response oscillations (larger ωT) but slower damping (smaller δ) and slower transient decay (larger pole magnitude). One possible design chooses $\delta = 0.1$, for which $\omega T \approx \pi/24$ and from the graph $z_p \approx 0.9 \pm 0.2j$ and $K \approx 0.05$. It may be predicted that the system with this gain will be stable, will oscillate at around one cycle each 48 samples (i.e. about every 4.8 s), and will have oscillations reduced by 90% each cycle. Results for the chosen values $K = 1$ for P control and $K = 5$ for PD control were shown in Fig. 18.5.

For this simple system, all of the numbers can be calculated exactly, but this defeats some of the purpose of using the graphical method.

The situation when the desirable root set is unattainable becomes one in which the control law must be changed (section 18.3) or the desires must be modified. Either approach requires a bit of artistry on the part of the design engineer.

19.4.2 Design for regions of the complex plane

The above examples, and many in later chapters on pole placement techniques, involved the selection of particular poles. In fact, what is needed is good response, which means only that poles must be such that the response has particular properties. The latter, in fact, then means that poles must be within a region Γ of the complex plane, with exact choice being influenced by other considerations. One suggestion for sampled data systems is that Γ be the intersection of the regions defined by $r \leq r_c$ and $\delta \leq \delta_c$ for choices such as $r_c = 0.4$ and $\delta_c = 10^{-4}$; to simplify the mathematics, this can often be taken as the circle

$$\Gamma = \left\{ z \;\middle|\; \|z - \frac{r_c}{2}\| \leq \frac{r_c}{2} \text{ where } r_c < 0.5 \right\}$$

Design algorithms for such regions are considered in Chapter 23.

19.5 ALTERNATIVE POLE SPECIFICATION – OPTIMAL POLES

An indirect specification of the poles results from using criteria such as 'optimal' response and doing numerical optimization. Two of these methods are indicated in this section.

19.5.1 Prototype design

One technique which serves to specify all of the poles is that of prototype design. For this section, we let

$$\hat{s} = \frac{s}{\omega_0}$$

where ω_0 is roughly (sometimes exactly) a bandwidth parameter and hence can be selected, e.g. using rise time specifications. The idea here is that all of the settable poles are chosen according to some criterion, usually representing some optimality argument.

For example, the closed-loop poles for a zero-less system (able to have zero steady-state error for a step input) of third order using a criterion that the ITAE (integral time absolute error)

$$\mathrm{ITAE} = \int t\,|e(t)|\,dt$$

should be minimized have been found numerically to be at (Franklin, 1985) $\hat{s} = -0.7081, -0.521 \pm 1.068j$.

These can be mapped to sampled data system poles if necessary. It is worth remarking that a one-zero closed-loop system will not have the same poles, nor will a system designed to a different criterion. The latter is clearly illustrated in the different characteristic equations found by D'Azzo and Houpis (1968), when they considered five different criteria, including ITAE and 5% settling time.

Tables 19.1 and 19.2 show pole locations for these criteria, and Figs 19.11 and 19.12 show some normalized step responses.

Table 19.1 Characteristic equations for optimum ITAE response

$$\hat{s} + 1$$
$$\hat{s}^2 + 1.4\,\hat{s} + 1$$
$$\hat{s}^3 + 1.75\,\hat{s}^2 + 2.15\,\hat{s} + 1$$
$$\hat{s}^4 + 2.1\,\hat{s}^3 + 3.4\,\hat{s}^2 + 2.7\,\hat{s} + 1$$
$$\hat{s}^5 + 2.8\,\hat{s}^4 + 5.0\,\hat{s}^3 + 5.5\,\hat{s}^2 + 3.4\,\hat{s} + 1$$
$$\hat{s}^6 + 3.25\,\hat{s}^5 + 6.6\,\hat{s}^4 + 8.6\,\hat{s}^3 + 7.45\,\hat{s}^2 + 3.95\,\hat{s} + 1$$

*Derived from D'Azzo and Houpis 1968, 1988

Table 19.2 Characteristic equations for minimum 5% settling time

$$\hat{s} + 1$$
$$\hat{s}^2 + 1.4\ \hat{s} + 1$$
$$\hat{s}^3 + 1.55\ \hat{s}^2 + 2.10\ \hat{s} + 1$$
$$\hat{s}^4 + 1.6\ \hat{s}^3 + 3.15\ \hat{s}^2 + 2.45\ \hat{s} + 1$$
$$\hat{s}^5 + 1.57\ \hat{s}^4 + 4.05\hat{s}^3 + 4.10\ \hat{s}^2 + 3.02\ \hat{s} + 1$$
$$\hat{s}^6 + 1.45\ \hat{s}^5 + 5.1\ \hat{s}^4 + 5.3\ \hat{s}^3 + 6.25\ \hat{s}^2 + 3.426\hat{s} + 1$$

*Derived from D'Azzo and Houpis 1968, 1988

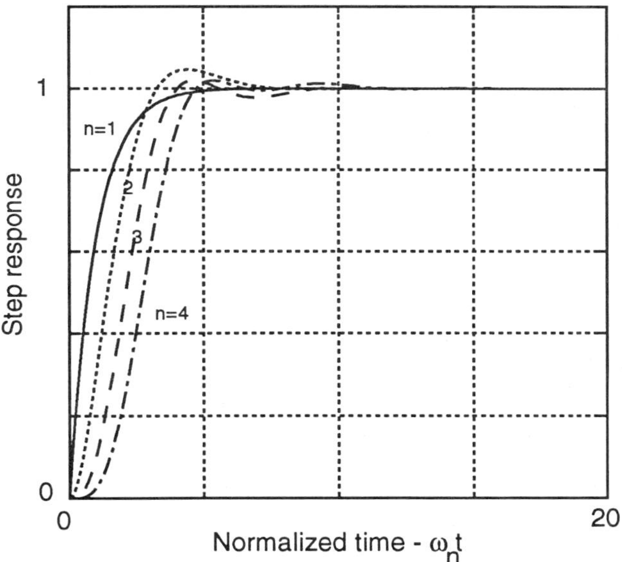

Figure 19.11 Step responses of nth order systems with poles chosen to optimize the ITAE criterion, for $n = 1, 2, 3, 4$.

It is possible to consider minimizing other performance indices such as integral square error (ISE) and integral absolute error (IAE) (Chapter 13). This is done in texts such as Dorf (1989) and D'Azzo and Houpis (1968, 1988). An alternative is to make the system explicitly have one of the well-known low-pass filter transfer functions of known type; one of the possibilities is a Butterworth characteristic (the transfer function is very flat for low frequencies). This is presented by the text of Franklin and Powell (1980) and the paper of Franklin (1985).

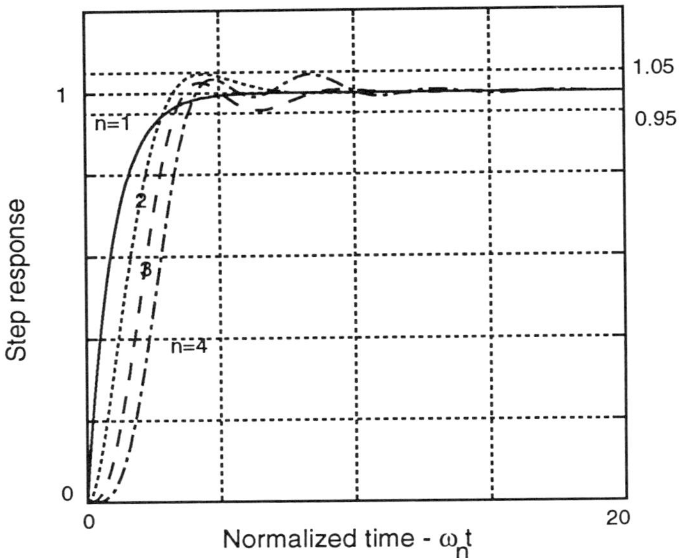

Figure 19.12 Step responses when poles are optimized to minimize settling time, for $n = 1, 2, 3, 4$.

The above may in principle be converted to sampled data system forms using impulse invariant response conversions or may be recalculated by going back to the original works (Graham and Lathrop, 1953).

19.5.2 Optimal pole locations

We shall meet optimal control theory in Chapters 26–27, but let us summarize a result here which is relevant. We have already met criteria ITAE and settling time, and we indicated that ISE and IAE criteria are also minimized by choice of pole locations. Consider the quadratic criterion

$$J = \sum_{k=0}^{N} \{ \mathbf{x}^{\mathrm{T}}(k)\,\mathbf{Q}\mathbf{x}(k) + \mathbf{u}^{\mathrm{T}}(k)\,\mathbf{R}\mathbf{u}(k) \} \tag{19.8}$$

for the usual linear system

$$\mathbf{x}(k+1) = \mathbf{A}\mathbf{x}(k) + \mathbf{B}\mathbf{u}(k)$$

If the sum is from 0 to ∞, then this is called the linear quadratic regulator (LQR) problem of optimal control theory. It turns out that the minimum value for (19.8) is attained when the control $\mathbf{u}(k)$ is chosen to be the value, which we denote by $\mathbf{u}^*(k)$ to indicate optimality, computed from

$$\mathbf{u}^*(k) = \mathbf{K}\mathbf{x}(k)$$

where $\mathbf{K}$ is the solution of

$$\mathbf{K} = -[\mathbf{R} + \mathbf{B}^{\mathrm{T}}\mathbf{P}\mathbf{B}]^{-1}\,\mathbf{B}^{\mathrm{T}}\mathbf{P}\mathbf{A}$$

$$\mathbf{P} = \mathbf{Q} + \mathbf{A}^{\mathrm{T}}\mathbf{P}\mathbf{A} - \mathbf{A}^{\mathrm{T}}\mathbf{P}\,[\mathbf{R} + \mathbf{B}^{\mathrm{T}}\mathbf{P}\mathbf{B}]^{-1}\mathbf{B}^{\mathrm{T}}\mathbf{P}\mathbf{A}$$

In continuous time, the corresponding result has $\mathbf{u}^*(t) = \mathbf{K}\mathbf{x}(t)$ where

$$\mathbf{K} = -\mathbf{R}^{-1}\mathbf{B}^{\mathrm{T}}\mathbf{P} \quad \text{and} \quad \mathbf{A}^{\mathrm{T}}\mathbf{P} + \mathbf{P}\mathbf{A} + \mathbf{Q} - \mathbf{P}\mathbf{B}\mathbf{R}^{-1}\mathbf{B}^{\mathrm{T}}\mathbf{P} = 0$$

The closed-loop poles will then be given by the characteristic equation of $(\mathbf{A} + \mathbf{B}\mathbf{K})$. The derivations of this are given in Chapter 26.

The point here is that a systematic way of arriving at a pole set is possible, provided that the weighting matrices $\mathbf{Q}$ and $\mathbf{R}$ can be determined. The latter is the engineering problem, and it is one that has not really been solved in general: there is no clear relation between $\mathbf{Q}$, $\mathbf{R}$, and various traditional criteria even though it is known that for any $\mathbf{K}$ there is a corresponding optimal control problem.

19.6 SUMMARY

We first comment upon what can and cannot be done with pole choice alone. In evaluating the system, a common response waveform to examine is the unit step response, which may be taken as indicating how rapidly and accurately the system will take up a new set point. We discussed this in Chapter 13 and suggested various criteria, such as rise time and percent overshoot, for the response. By choosing poles, we choose the decay rate of the envelope of the response and the frequency of oscillation of the error. This has a bearing on all of the indicators, but ultimately is the main arbiter only of the settling time; all of the other factors can be strongly, even critically, affected also by zero locations.

Given the above disclaimers, it is still reasonable to attempt to choose poles. This is particularly because they get us started in a design, which might nevertheless be tuned further, e.g. using simulation methods. When we choose poles, their relation to response should be kept in mind, and that was really the point of sections 19.2 and 19.3, along with giving a preliminary guide to 'good' locations.

If optimization is allowed, there is implicitly a feeling that many poles can be set, at least in the methods presented in section 19.5. One also has a certain disquiet about those. Considering that an SISO system with identity observer (Chapter 25) and state feedback (Chapter 23) will have us placing up to $2n$ poles for a system with n state variables. and that 'good' response requires perhaps five specifications ($T_d, T_r, T_s, M_p, e_{ss}$ for some signal form) to be met plus 'good' disturbance rejection, all without excessive sensitivity to modelling errors and with a reasonable cost, it can seem overly restrictive to require us to choose so many poles. It may well be that good design can be done with only a few parameters chosen. One attempts then to place all poles within a region of the complex plane, rather than at particular points. Techniques for this are not yet well established, but one approach is mentioned in Chapter 23.

19.7 FURTHER READING

Pole choice in root locus design methods is given in various texts, including Dorf (1989) and Franklin, Powell, and Workman (1990). D'Azzo and Houpis (1968) is one of the earlier textbooks which elaborates somewhat upon prototype-based design and related methods. See also D'Azzo and Houpis (1988). A recent unifying paper concerning the pole location region Γ is (Haddad and Bernstein, 1992).

20
Bode diagrams for frequency domain analysis and design

The classical methods known as frequency domain techniques have their origin with electrical engineers, who rely extensively on representations of signals as sums of sinusoids in their modelling and analysis. It has seemed natural for them to carry such ideas with them into control systems analysis and synthesis, often with considerable success. Less successful has been the direct use of the methods for sampled data control systems (although indirect use characterized by conversion of continuous system designs is possible). This has been for several reasons: the approximations that allow sketching do not always apply, the compensators of most common use are not so relevant for digital control systems, and well established relationships between step responses, pole locations, and frequency responses seem to hold only roughly for sampled data systems. Nevertheless having at least some knowledge of frequency domain methods is fundamental, and for that reason we review them here.

20.1 SYNOPSIS

Interestingly enough, and perhaps curiously, it is not the closed-loop frequency response which is usually used in frequency domain analysis. Rather, the design methods are used for studying the denominator of the closed-loop frequency response of certain structured systems. Typically we have a form

$$G(q) = \frac{C(q)\,G_{\mathrm{p}}(q)}{1 + C(q)\,F(q)\,G_{\mathrm{p}}(q)}$$

for the closed-loop transfer function, and we wish to study and design compensators $C(q)$ (where q can be either z or s at this stage) for the form

$$1 + C(q)\,G(q) = 1 + C(q)\,F(q)\,G_\mathrm{p}(q) = 1 + C(q)\,\frac{N(q)}{D(q)}$$

in which a gain compensator, $C(q) = K$, is only a special but important case. To do this, we make one or more of three types of graphs of the frequency response: the Nyquist (polar) plot, the Bode (magnitude and phase versus frequency) plots, and the Nichols (log magnitude versus phase) plot.

This frequency response information is available experimentally for some plants using an experimental set-up as in Fig. 20.1.

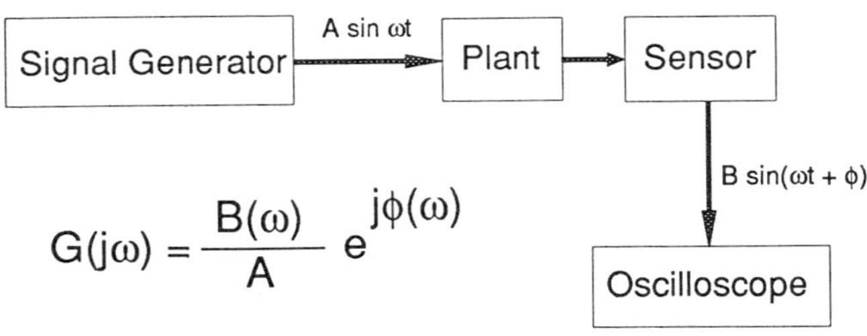

Figure 20.1 Equipment configuration for experimental determination of frequency response.

The Bode plots seem particularly familiar because of their common use for representing frequency response data for such consumer items as audio systems. We concentrate upon them here because they are easy to sketch and to perform rough designs upon. Furthermore they give an indication of stability margin for stable systems.

Serious users will probably, after learning how to make sketches and preliminary designs, use computers to generate the necessary plots and even the final designs. They will also pursue the alternative representations, Nyquist diagrams (Chapter 15) and Nichols charts, which we only mention here.

20.2 THE BODE PLOTS

The Bode plot representations of frequency response data are useful partly because they are easy to sketch and perform a basis for rough designs. They are particularly applicable to continuous time systems, as the sketching property does not carry over directly to sampled data systems. Hence we present first continuous time models and then show how the techniques may be applied to sampled data models.

20.2.1 Continuous time systems and associated sketches

The starting place for Bode plots is the transfer function

$$H(s) = \frac{b_0 s^m + b_1 s^{m-1} + b_2 s^{m-2} + \cdots + b_m}{s^n + a_1 s^{n-1} + a_2 s^{n-2} \cdots + a_{n-1} s + a_n}$$

where by assumption one usually has $m \leq n$.

The Bode plots have two characteristics, both resulting from the choice of quantities graphed, which make sketching them easy.

1. Because of the ordinate choices (log (magnitude) and arc()), the plots can be reduced to additions of simple terms.
2. Because of the abscissa choice ($\log \omega$) along with the ordinate choices, it turns out that straight line asymptotic approximations are fairly accurate for each term.

To see this we consider the typical $H(j\omega)$. This will have the factored form

$$H(j\omega) = \frac{N(j\omega)}{D(j\omega)} = K' \frac{(j\omega - z_1)(j\omega - z_2) \cdots (j\omega - z_m)}{(j\omega)^k (j\omega - p_1)(j\omega - p_2) \cdots (j\omega - p_n)}$$

By combining complex conjugate poles (and zeros) and manipulating the result, this is placed in the Bode form

$$H(j\omega) = \frac{K_B\left(\dfrac{j\omega}{a_1}+1\right)\left(\dfrac{j\omega}{a_2}+1\right)\cdots\left(\dfrac{j\omega}{a_{m1}}+1\right)}{(j\omega)^k\left(\dfrac{j\omega}{b_1}+1\right)\left(\dfrac{j\omega}{b_2}+1\right)\cdots\left(\dfrac{j\omega}{b_{n1}}\right)} \times$$

$$\frac{\left(1-\left(\dfrac{\omega^2}{\omega_{a1}}\right)+j2\zeta_{a1}\dfrac{\omega}{\omega_{a1}}\right)\left(1-\left(\dfrac{\omega^2}{\omega_{a1}}\right)+j2\zeta_{a2}\dfrac{\omega}{\omega_{a2}}\right)\cdots\left(1-\left(\dfrac{\omega^2}{\omega_{am2}}\right)+j2\zeta_{am2}\dfrac{\omega}{\omega_{am2}}\right)}{\left(1-\left(\dfrac{\omega^2}{\omega_{b1}}\right)+j2\zeta_{b1}\dfrac{\omega}{\omega_{b1}}\right)\left(1-\left(\dfrac{\omega^2}{\omega_{b1}}\right)+j2\zeta_{b2}\dfrac{\omega}{\omega_{b2}}\right)\cdots\left(1-\left(\dfrac{\omega^2}{\omega_{bn2}}\right)+j2\zeta_{bn2}\dfrac{\omega}{\omega_{bn2}}\right)}$$

From this we have that the magnitude and angle can be given by

$$20\log\|H(j\omega)\| = 20\log|K_B| - 20k\log\omega$$

$$+ \sum_{i=1}^{m_1} 20\log\left\|\dfrac{j\omega}{a_i}+1\right\| - \sum_{i=1}^{n_1} 20\log\left\|\dfrac{j\omega}{b_i}+1\right\|$$

$$+ \sum_{j=1}^{m_2} 20\log\left\|1-\left(\dfrac{\omega}{\omega_{aj}}\right)^2 + j2\zeta_{aj}\dfrac{\omega}{\omega_{aj}}\right\|$$

$$- \sum_{j=1}^{n_2} 20\log\left\|1-\left(\dfrac{\omega}{\omega_{bj}}\right)^2 + j2\zeta_{bj}\dfrac{\omega}{\omega_{bj}}\right\|$$

$$\arc(H(j\omega)) = -k\dfrac{\pi}{2} + \sum_{i=1}^{m_1}\arc\left(\dfrac{j\omega}{a_i}+1\right) - \sum_{i=1}^{n_1}\arc\left(\dfrac{j\omega}{b_i}+1\right)$$

$$+ \sum_{j=1}^{m_2}\arc\left(1-\left(\dfrac{\omega}{\omega_{aj}}\right)^2 + j2\zeta_{aj}\dfrac{\omega}{\omega_{aj}}\right)$$

$$- \sum_{j=1}^{n_2}\arc\left(1-\left(\dfrac{\omega}{\omega_{bj}}\right)^2 + j2\zeta_{bj}\dfrac{\omega}{\omega_{bj}}\right) + \arc(K_B)$$

In the above, all of the coefficients ω, a, b, ζ are real numbers and $m_1 + 2m_2 = m$ and $n_1 + 2n_2 + k = n$.

We notice that the additive property is apparent. Consider now the magnitude of a typical first-order term. We see that

$$20 \log \left\| \frac{j\omega}{a_i} + 1 \right\| \approx \begin{cases} 0 & \omega \ll a_i \\ 20 \left\| \frac{\omega}{a_i} \right\| & \omega \gg a_i \end{cases}$$

The second expression can be written

$$20 \log \omega - 20 \log \| a_i \|$$

which is linear in $\log \omega$ with slope 20 dB per decade (factor of 10 increase in ω), also called equivalently 6 dB per octave (doubling of ω). These asymptotes appear in Fig. 20.2(a), along with the exact magnitude of the term. Notice that as a denominator term this is subtracted rather than added.

(a)

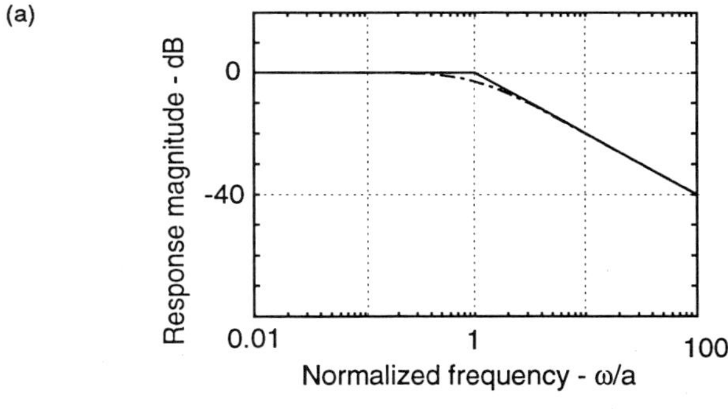

(b)

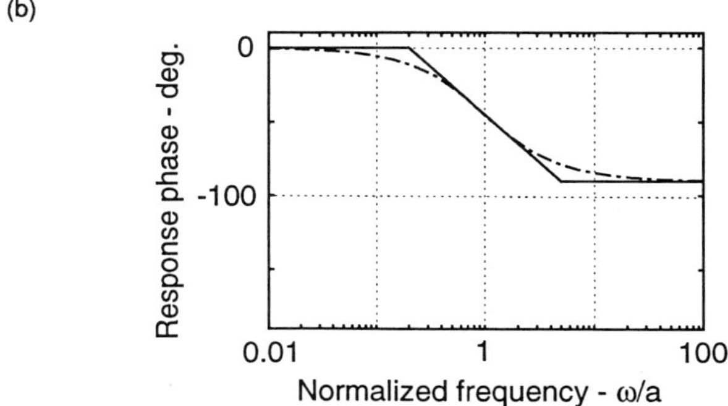

Figure 20.2 Frequency response of first-order pole: (a) magnitude; and (b) phase. Straight line asymptotes are solid lines; actual response is dotted.

The angle of this term is given by

$$\text{arc}\left(\frac{j\omega}{a_i} + 1\right) \approx \begin{cases} 0 & \omega << a_i \\ 45° & \omega = a_i \\ 90° & \omega >> a_i \end{cases}$$

This is shown in Fig. 20.2(b), along with a commonly used interpolating straight line approximation.

Similar arguments for the second-order terms give the following:

$$20\log\left\| 1 - \left(\frac{\omega}{\omega_{aj}}\right)^2 + j2\zeta_{aj}\frac{\omega}{\omega_{aj}}\right\| = \begin{cases} 0 & \omega << \omega_{aj} \\ 40\log\omega - 40\log\omega_{aj} & \omega >> \omega_{aj} \end{cases}$$

$$\text{arc}\left(1 - \left(\frac{\omega}{\omega_{aj}}\right)^2 + j2\zeta_{aj}\frac{\omega}{\omega_{aj}}\right) \approx \begin{cases} 0 & \omega << \omega_{aj} \\ 90° & \omega = \omega_{aj} \\ 180° & \omega >> \omega_{aj} \end{cases}$$

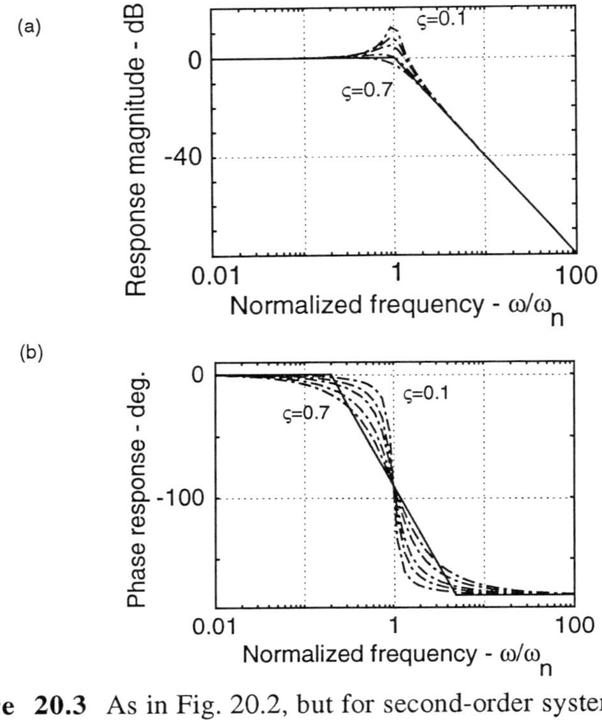

Figure 20.3 As in Fig. 20.2, but for second-order systems with ζ from 0.1 to 0.7.

Such terms are shown in Fig. 20.3, including both exact expressions and the asymptotes. Notice that the asymptotes depend only on ω_{aj}, while the exact expressions depend also on the damping coefficient ζ.

The term $(j\omega)^k$ in the denominator is easy: the log-magnitude $20k\log\omega$ is clearly linear in $\log\omega$, and the angle is a multiple of $-\pi/2$, as is easily seen in the complex plane (Fig. 20.4).

The Bode gain K_B is the final type of term of interest. Its magnitude is given by $20\log\|K_B\|$ and its angle is either $0°$ for a positive gain or $180°$ for a negative gain.

Using the asymptotes, it is easy and straightforward to find the approximate frequency response of a transfer function $G_D(s)$: one places the function $G_D(j\omega)$ in Bode form, plots the gains and phases of the individual factors, and graphically adds and subtracts the terms as appropriate to their presence in the numerator or denominator, respectively.

(a)

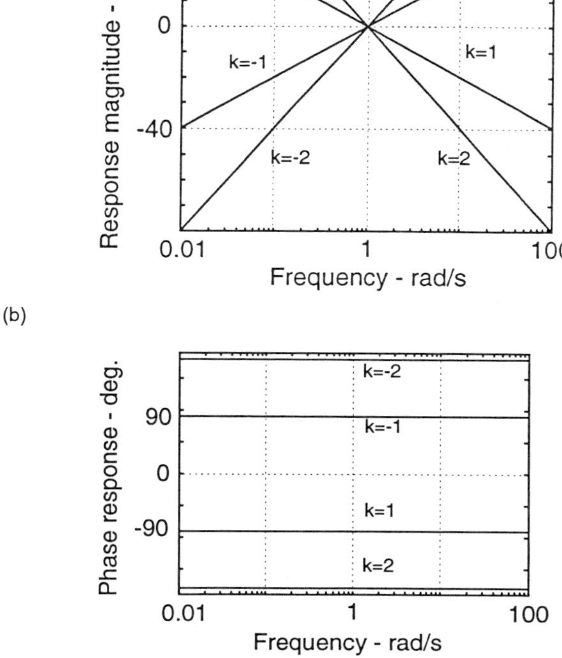

Figure 20.4 As in Fig. 20.2, but for terms $1/j\omega^k$.

Examples

Consider the system of Fig. 20.5 which is to be studied using frequency response methods.

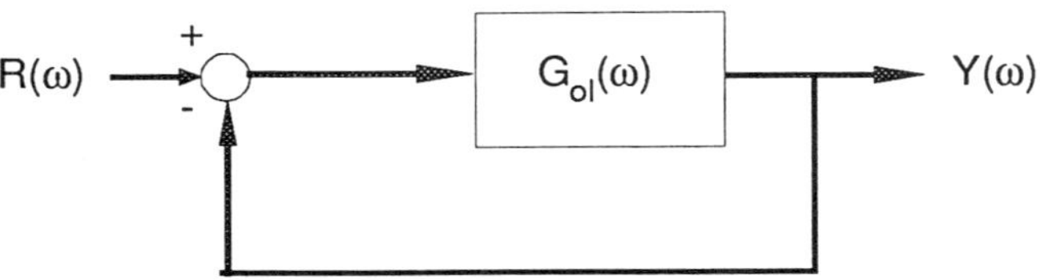

Figure 20.5 The basic system configuration for Bode analysis. G_{ol} is known and may include both plant and controller models.

The open-loop system frequency transfer function is

$$G_{ol}(\omega) = \frac{5\,(j\omega + 7)}{(j\omega)\,((j\omega)^2 + 3.6\,(j\omega) + 9)}$$

which is first placed into Bode form as

$$G_{ol}(\omega) = \frac{\dfrac{35}{9}\left(1 + \dfrac{j\omega}{7}\right)}{(j\omega)\left(1 - \dfrac{\omega^2}{9} + 1.2\,\dfrac{j\omega}{3}\right)}$$

The asymptotic magnitude and phase approximations and their sums are shown in Fig. 20.6. Although shown as a point by point sum, most users will quickly develop the knack of working from left to right, with break points up and down for zeros and poles, respectively, as the critical frequencies (here 7 and 3) are passed.

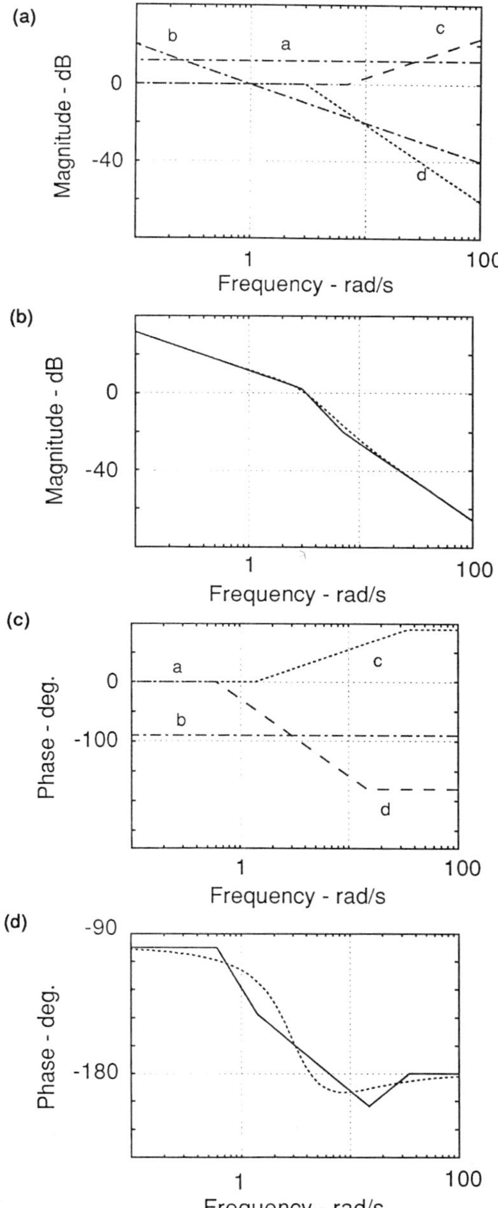

Figure 20.6 Construction of system Bode plots using asymptotes:
(a) approximations to separate terms for magnitude response;
(b) sum of approximations and (dotted) actual response;
(c) and (d) are repeats of (a) and (b) for phase.

20.2.2 Application of Bode approximation methods to digital control systems

To apply the above to digital systems, we have two possibilities: sample rapidly enough that approximations will work, or use a w-transformation to obtain the required form. This is only necessary for manual design, in which the easy sketching allowed by the asymptotic approximations greatly simplifies the task, but it is also helpful in computer aided design, because the required variations are easily visualized.

Let us elaborate upon the above by assuming that we are concerned with a closed-loop system with denominator in the form

$$1 + C(z)\,H(z) = 1 + C(z)\,\frac{N(z)}{D(z)}$$

We expand both numerator and denominator polynomials into their first- and second-order terms (so that all coefficients are real).

$$C(z)\,\frac{N(z)}{D(z)} = K_b\,z^L\,\frac{(1-z_1 z^{-1})(1-z_2 z^{-1})\;\cdots}{(1-p_1 z^{-1})(1-p_2 z^{-1})\;\cdots}$$

$$\times\;\frac{(1+2r_{z1}\cos\phi_{z1}\,z^{-1} + r_{z1}^2\,z^{-2})\,(1+2r_{z2}\cos\phi_{z2}\,z^{-1} + r_{z2}^2 z^{-2})}{(1+2r_{p1}\cos\phi_{p1}\,z^{-1} + r_{p2}^2 z^{-2})\;\cdots}$$

where there are presumably n possibly complex poles and m such zeros, with $n \geq m$. Then we see that the magnitude of this expression can be given from

$$\log|\,H(e^{-j\omega T})\,| = \log|\,K_b\,|$$

$$+ \log|\,1 - z_1\,e^{-j\omega T}\,| + \log|\,1 - z_2\,e^{-j\omega T}\,| + \cdots$$

$$+ \log|\,1 + 2\,r_{z1}\,\cos\phi_{z1}\,e^{-j\omega T} + r_{z1}^2\,e^{-2j\omega T}\,| + \cdots$$

$$- \log|\,1 - p_1\,e^{-j\omega T}\,| - \log|\,1 - p_2\,e^{-j\omega T}\,| - \cdots$$

$$- \log|\,1 + 2\,r_{p1}\,\cos\phi_{p1}\,e^{-j\omega T} + r_{p1}^2\,e^{-2j\omega T}\,| - \cdots$$

and the angle is

$$\text{arc}\left(H(e^{-j\omega T})\right) = \text{arc}(K_b) + \text{arc}(e^{jL\omega T})$$

$$+ \text{arc}\left(1 - z_1 e^{-j\omega T}\right) + \text{arc}\left(1 - z_2 e^{-j\omega T}\right) + \cdots$$

$$+ \text{arc}\left(1 + 2 r_{z1} \cos\phi_{z1} e^{-j\omega T} + r_{z1}^2 e^{-2j\omega T}\right) + \cdots$$

$$- \text{arc}\left(1 - p_1 e^{-j\omega T}\right) - \text{arc}\left(1 - p_2 e^{-j\omega T}\right) - \cdots$$

$$- \text{arc}\left(1 + 2 r_{p1} \cos\phi_{p1} e^{-j\omega T} + r_{p1}^2 e^{-2j\omega T}\right) - \cdots$$

As the addition is simple, we need in principle only to know how to handle generic terms evaluated on the unit circle. Unfortunately, the terms are not linear in the frequency, so sketching is not possible except for small angles ωT. We see this in Fig. 20.7; particularly notable is the periodicity of the response, i.e. that the frequency response is periodic in ωT with period 2π. This reminds us that sampled data systems are not intrinsically low-pass filters in nature; an input of high frequency may lead with sampling to an output of low frequency unless precautions are taken. This effect is called 'aliasing' and preventive measures are taken either in the form of a guard filter or in knowledge that the sampling is fast enough that high frequency disturbances are so small as to be negligible.

It is possible to compute the frequency response with a calculator and thereby gain some of the visualization benefits provided by Bode diagrams. The additivity property remains, so simple controller designs can be quickly established. If ωT is 'small enough', which usually means that $\omega T < 0.1$, a condition usually associated with rapid sampling compared to the frequencies of interest, then the approximation $e^{j\omega T} \approx 1 + j\omega T$ may be used to obtain forms linear in $\log \omega$ and the methods of the previous section may be applied to obtain sketches.

When the rapid-sampling assumption about sampled data systems is not justified, then to reap the sketching benefits of Bode methods a further transformation is sometimes made. This is the so-called w-transformation, a change of variable which we met in Chapter 14, and which may be motivated as an approximation to e^w or by arguments about mappings. In either case, the transformation is given by the substitution

$$z = \frac{1 + w}{1 - w}$$

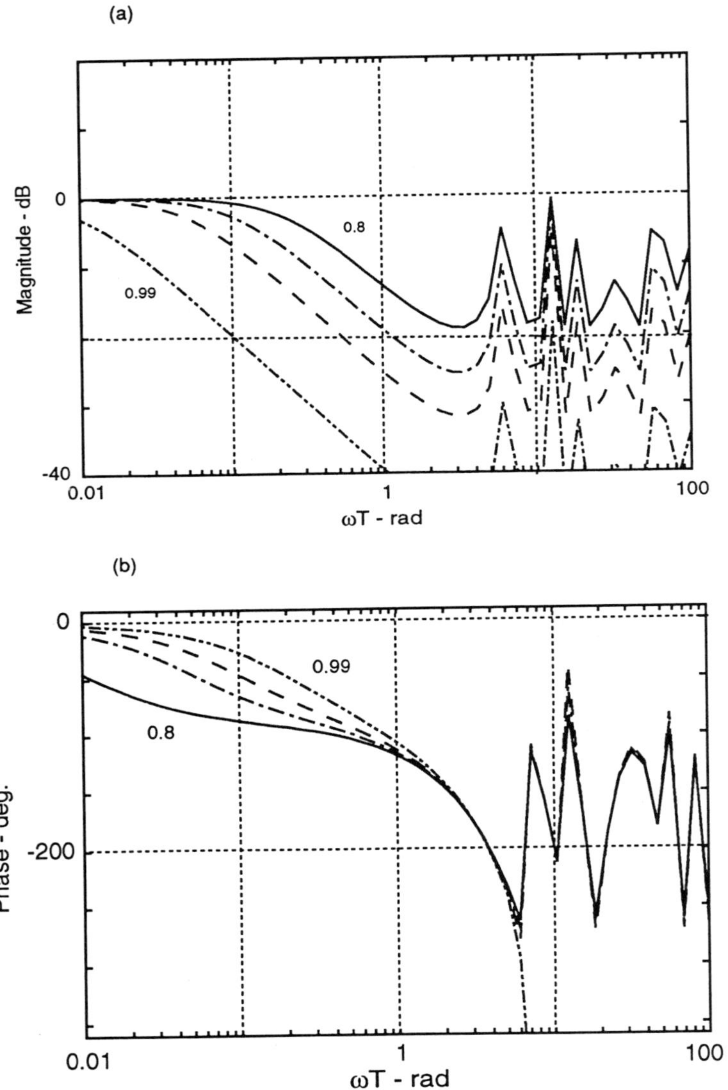

Figure 20.7(a) and (b) Bode plots for discrete system terms: (a) magnitude; and (b) phase for a first-order term, showing effect of pole at 0.8, 0.9, 0.95 or 0.99.

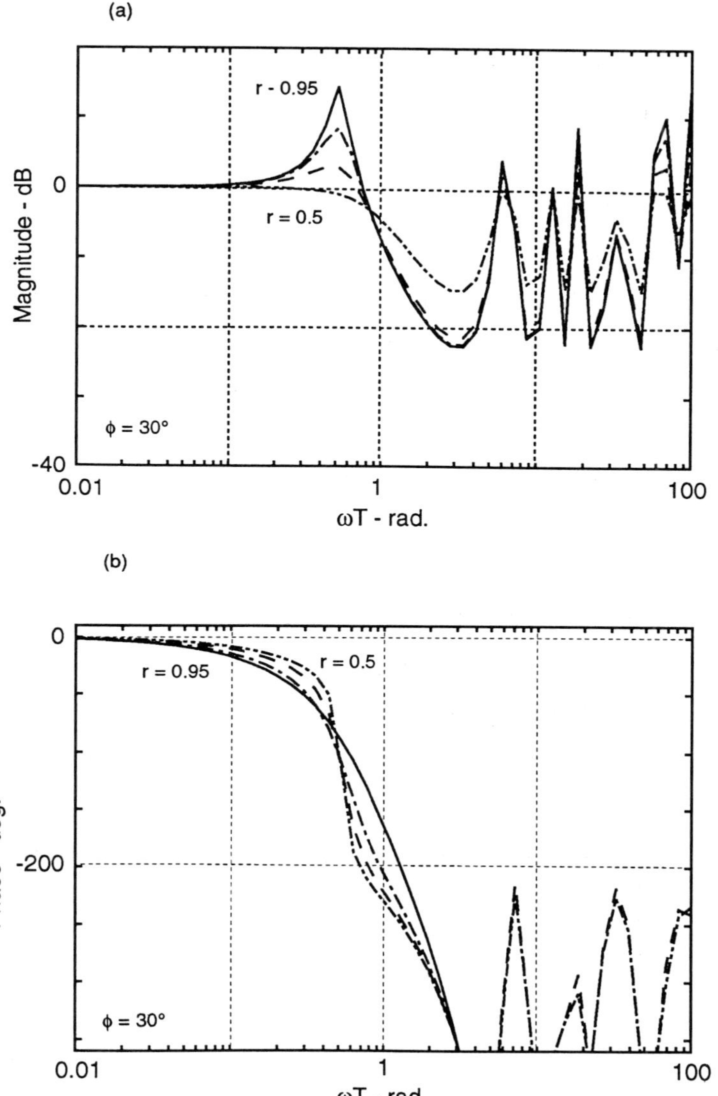

Figure 20.7(c) and (d) Bode plots for discrete system terms: (c) magnitude and (d) phase for a second-order term, showing effect of pole pair with $\phi = 30°$ and $r = 0.5, 0.8, 0.9,$ or 0.95. Noteworthy is the unlikelihood of using asymptotes successfully except for small T values.

or a scaled version thereof in which the scaling uses $2/T$. The reverse (or inverse) transformation is then

$$w = \frac{z - 1}{z + 1}$$

To use the transformation we substitute

$$\hat{H}(w) = H(z)\Big|_{z = \frac{1+w}{1-w}}$$

to obtain the form

$$\hat{H}(w) = K_b \frac{\displaystyle\prod_{i=1}^{m1}\left(1 + \frac{w}{a_i}\right) \prod_{i=1}^{m2}\left(1 + 2\zeta_{ni}\left(\frac{w}{\omega_{ni}}\right) + \left(\frac{w}{\omega_{ni}}\right)^2\right)}{\displaystyle w^l \prod_{l=1}^{n1}\left(1 + \frac{w}{b_i}\right) \prod_{l=1}^{n2}\left(1 + 2\zeta_{di}\left(\frac{w}{\omega_{di}}\right) + \left(\frac{w}{\omega_{di}}\right)^2\right)}$$

If we evaluate this for the pseudo-frequency ω_w by the substitution $w = j\omega_w$ and take log magnitude and argument, we obtain the same forms as appeared in the continuous time case.

It is interesting to compare the pseudo-frequency ω_w with the real frequency ω in using this approach. Since we have set z to $e^{j\omega T}$ and w to $j\omega_w$, we have that the frequency is warped

$$j\omega_w = \frac{e^{j\omega T} - 1}{e^{j\omega T} + 1} = j \tan \frac{\omega T}{2}$$

or

$$\omega_w = \tan \frac{\omega T}{2} \qquad \omega = \frac{2}{T} \tan^{-1} \omega_w$$

Example

For this example, we return to the motor model used in the examples in sections 18.3 and 19.4.1. First we perform the w substitution. With

$$H(z) = \frac{0.048 \ (z + 0.967)}{(z - 1) \ (z - 0.905)}$$

we find

$$\hat{H}(w) = H\left(\frac{1 + w}{1 - w}\right) = \frac{0.048 \left(\frac{1 + w}{1 - w} + 0.967\right)}{\left(\frac{1 + w}{1 - w} - 1\right)\left(\frac{1 + w}{1 - w} - 0.905\right)}$$

$$= \frac{0.048 \ (1.967 + 0.033w) \ (1 - w)}{2w \ (0.095 + 1.905w)}$$

In Bode form, this is

$$\hat{H}(w) = \frac{0.497 \ (1 - w) \ (1 + w/59.6)}{w(1 + w/0.050)}$$

The frequency response at the pseudo-frequency ω_w is given by

$$\hat{H}(j\omega_w) = \frac{0.497 \ (1 - j\omega_w) \ (1 + j\omega_w/59.6)}{j\omega_w(1 + j\omega_w/0.050)}$$

The asymptotic Bode plots are shown in Fig. 20.8. Notice that the numerator term $(1 - j\omega_w)$ contributes $-\pi/2$ to phase rather than the $+\pi/2$ usually contributed by numerator terms; this is due to the minus

(a) (b)

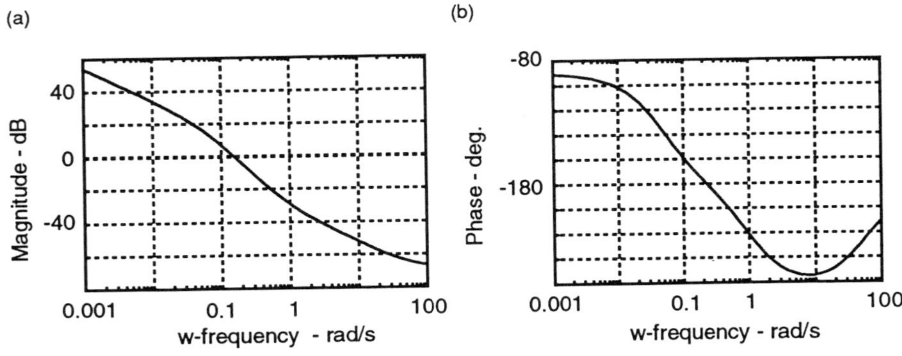

Figure 20.8 (a) Magnitude and (b) phase in terms of *w*-frequency ω_w for text example.

sign and is called a non-minimum phase[1] situation in the literature.

The *w*-transformation makes the Bode approach very easy to use because of the asymptotes, but the actual substitution to obtain $\hat{H}(w)$ is tedious and error-prone. Obtaining the coefficients using a simple computer algorithm is easy of course, but if a computer is available, one might as well design directly with the actual frequencies.

20.3 APPLICATIONS OF THE BODE DIAGRAMS

The Bode diagram sketches have several uses for the student.

1. When experimental data is obtained for a system, a fit of the data can be used to identify the system, i.e. to estimate its parameters.
2. Measures of relative stability can be read directly from the sketches.
3. Approximate compensator designs can be developed using the sketches.

Since the first of these is nearly obvious and will be met later (in Chapter 30), we elaborate on the second and third topics in this section.

---

[1] A transfer function with all poles and zeros in the stable region, either left-half plane or unit disc as appropriate, is said to have minimun phase. It can be shown that the phase of a transfer function with all zeros in the stable region has less displacement from 0° as ω goes from 0 to ∞ than any transfer function with the same magnitude but with one or more zeros outside that region. This becomes a particular problem for design techniques which effectively cancel plant zeros with compensator poles, as the compensator is then unstable.

20.3.1 Gain and phase margins

We recall that we are interested first and foremost in having a stable system. As we argued in discussing the Nyquist stability criterion (Chapter 15), this means we want no frequencies ω_0 such that the transfer function

$$H(p) = \frac{C(p)\,G(p)}{1 + C(p)\,F(p)\,G(p)}$$

has, for $p = j\omega_0$ (or $p = e^{j\omega_0 T}$ for the discrete time case)

$$1 + C(j\omega_0)\,F(j\omega_0)\,G(j\omega_0) = 1 + G_D(j\omega_0) = 0$$

Hence we do not want $G_D(j\omega_0) = -1$, i.e. we do not want $\mathrm{arc}(G_D(j\omega_0)) = -\pi$ and $|G_D(j\omega_0)| = 1$ for the same frequency ω_0.

From this observation we introduced two related measures of 'closeness to instability': phase margin and gain margin.

Definition   The **phase margin** ϕ_m is the angle by which the actual phase lag $\mathrm{arc}(G_D(j\omega_{gc}))$ exceeds $-\pi$ at the frequency ω_{gc} at which $|G_D(j\omega_{gc})| = 1$. ω_{gc} is called the **gain crossover frequency.**

Definition   The **gain margin** G_M is the amount by which the gain $|G_D(j\omega_{pc})|$ is too small to lead to instability at the frequency ω_{pc} for which $\mathrm{arc}(G_D(j\omega_{pc})) = -\pi$. ω_{pc} is called the **phase crossover frequency.** It is often expressed in decibels, so that

$$G_M = -20\log|G_D(j\omega_{pc})|\,\mathrm{dB}$$

Both of these are easily found from the Bode diagrams. They are shown for a typical system in Fig. 20.9.

It turns out that a system 'near to instability' shows faster response with less damping than one 'far from instability'. An engineering design trade-off is speed of response against rapid settling (accuracy of response) and robustness in the face of parameter errors, and nominal desirable values are often taken as

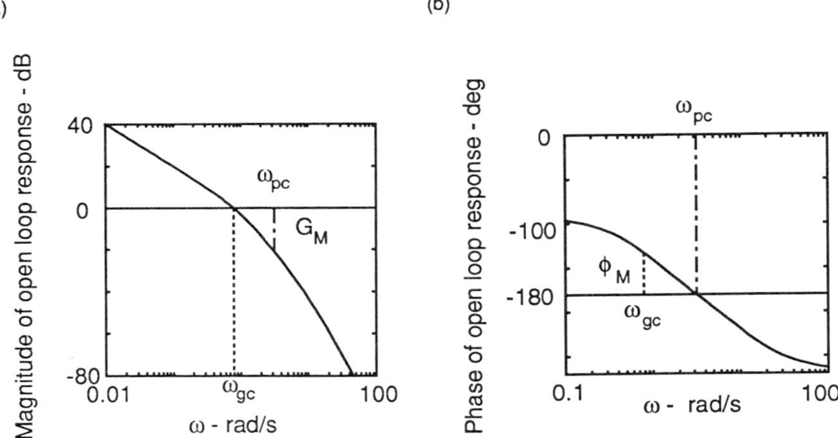

Figure 20.9 Illustration showing (a) gain margin G_M and (b) phase margin and ϕ_m – and gain and phase crossover frequencies ω_{gc} and ω_{pc} for each.

$$G_M \approx 6\text{--}10\,\text{dB}$$

$$\phi_m \approx 45\text{--}55°$$

For second-order poles the fact that phase margin is related to damping coefficient (as in Fig. 20.10) is argued as follows.

For unit step response of a second-order system with no zeros, various texts (such as Dorf (1989) and Saucedo and Schiring (1968)) show that the relationship between phase margin ϕ_m and damping coefficient ζ is

$$\phi_m = \tan^{-1}\left\{\frac{2\zeta}{((4\zeta^4 + 1)^{\frac{1}{2}} - 2\zeta^2)^{\frac{1}{2}}}\right\}$$

and the percent overshoot PO is

$$PO = 100\exp\left(\frac{-\pi\zeta}{(1 - \zeta^2)^{\frac{1}{2}}}\right)$$

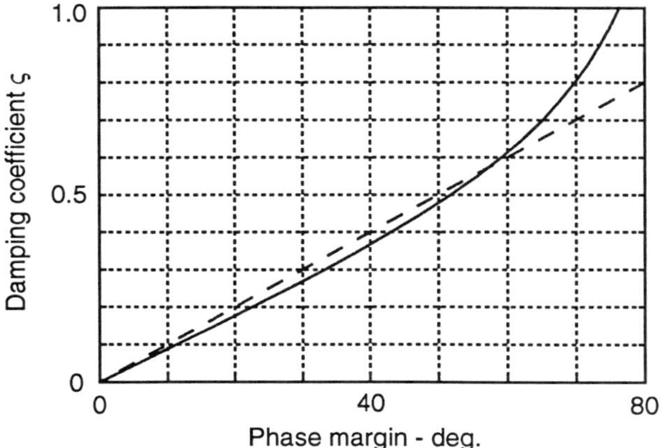

Figure 20.10 Relationship of damping coefficient to percent overshoot of step response, showing also linear approximation (dashed).

Using the definitions in Chapter 19 (root location) the corresponding approximations for discrete time systems are

$$PO = 100\ \delta^{\frac{1}{2}}$$

and

$$\phi_m = \tan^{-1}\left\{\frac{-2\alpha}{((5\alpha^4 + 2\alpha^2 + 1)^{\frac{1}{2}} - 2\alpha^2)^{\frac{1}{2}}}\right\}$$

where $\alpha = \ln(\delta)/2\pi$. An approximation is that

$$\phi_m \approx 100\zeta \sim \frac{-100\ \alpha}{(1 + \alpha^2)^{\frac{1}{2}}}$$

From these (Fig. 20.11), it is possible to design approximately for a particular PO or one of its relatives (settling time, damping coefficient ζ, damping ratio δ) by choosing the phase margin.

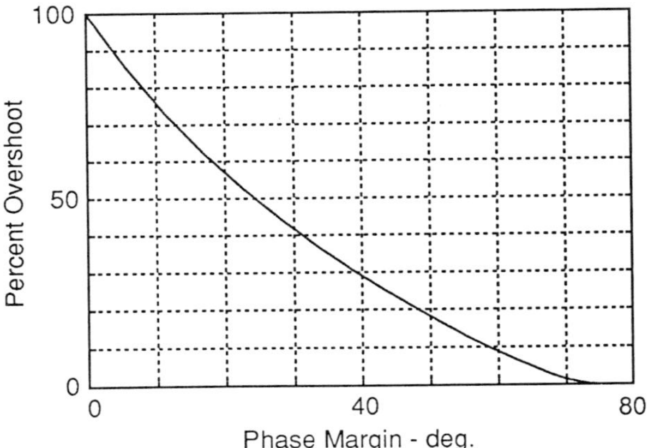

Figure 20.11 Relationship of percent overshoot to phase margin for simple second-order systems.

20.3.2 Design of compensators

With the above in mind, the designer first generates the Bode plots for the basic open-loop system and evaluates the gain and phase margins. If these are unsatisfactory, the compensator design process begins. This usually works with three elements:

1. gain (in an amplifier, for example);
2. lead networks, i.e. circuits for which

$$C(s) = \frac{\frac{s}{a} + 1}{\frac{s}{b} + 1} \qquad a < b$$

3. lag networks, for which

$$C(s) = \frac{\frac{s}{a} + 1}{\frac{s}{b} + 1} \qquad a > b$$

The lead and lag networks have the same form of transfer functions, but different parameter characteristics.

The above elements, which can be cascaded if necessary to obtain the required compensation, have their origin in electrical circuits. Each has an associated passive circuit (i.e. one made up of resistors, capacitors, and inductors) which will yield the necessary form, and for this reason the above set dominates the classical electrical engineering approach to control law design.

A fourth classical approach, the PID or three-term controller, has transfer function

$$C(s) = K_{\mathrm{p}} + \frac{K_{\mathrm{i}}}{s} + K_{\mathrm{d}}s$$

This is common in process control applications and can be implemented with pneumatic and hydraulic controllers. If one allows for the fact that true differentiators are difficult to construct by replacing this with the different model

$$C(s) = K_{\mathrm{p}} + \frac{K_{\mathrm{i}}}{s} + \frac{K_{\mathrm{d}}s}{\dfrac{s}{b} + 1}$$

one finds that rearrangement gives

$$C(s) = K \frac{\left(\dfrac{s}{a} + 1\right)\left(\dfrac{s}{c} + 1\right)}{s\left(\dfrac{s}{b} + 1\right)}$$

which is of the general form of an amplifier cascaded with a pair of lead/lag type elements.

We indicate the nature of lead/lag compensator design. One first notes the following effects of the basic elements on a set of Bode diagrams of a plant or plant plus partial compensator.

1. Gain compensation moves the Bode gain plot up and down without changing its shape. The Bode phase plot is unaffected.
2. A lead network increases the gain between $\omega = a$ and $\omega = b$ until, above $\omega = 5b$ or so, it causes a constant gain increase of

$20\log(b/a)$ dB. The phase plot receives an upward hump between about $\omega = a/10$ and $\omega = 10b$, with the maximum increase being at $\omega = \sqrt{(ab)}$ and the angle at this point equal to $\tan^{-1}(b-a)\sqrt{(2ab)}$ (Fig. 20.12).

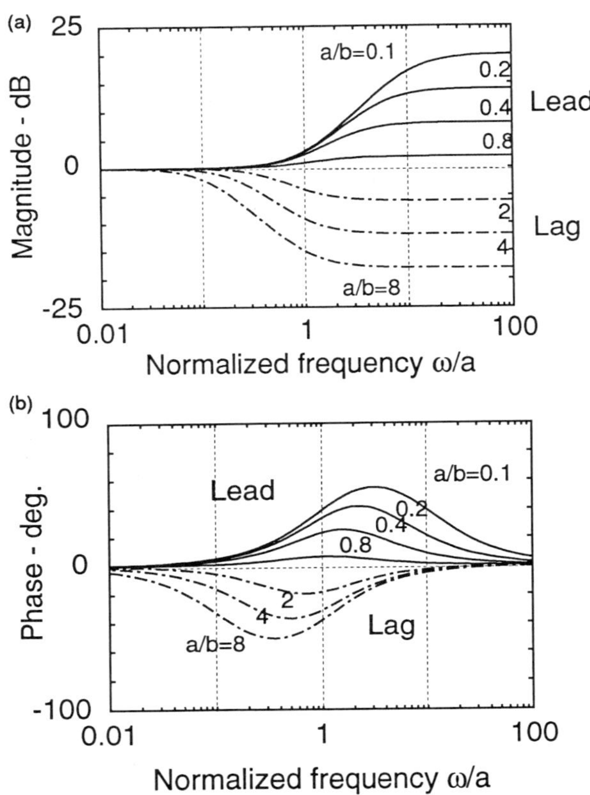

Figure 20.12 Gain (a) and phase (b) of lead and lag networks as parameters vary, for continuous time systems.

3. A lag network decreases the gain between $\omega = a$ and $\omega = b$ until, above $\omega = 5b$ or so, it causes a constant gain decrease of $20\log(b/a)$ dB. The phase plot receives a downward hump between about $\omega = b/10$ and $\omega = 10a$, with the maximum decrease being at $\omega = \sqrt{(ab)}$ and the angle at this point equal to $\tan^{-1}(a-b)\sqrt{(2ab)}$ (Fig. 20.12).

4. Digital implementations of the same idea have the form

$$C(z) = K\frac{z - a}{z - b}$$

with a and b both typically in the interval $(0, 1)$. Here there will be phase lead for $1 > a > b > 0$ and lag for $0 < a < b < 1$. The gain K makes a third parameter which may be used, e.g. to provide a particular steady-state error level (provided the system type is appropriate). It is quite possible and straightforward to generate curves of phase and gain versus frequency for various values of a and b; the normalized cases with $C(e^{j\omega T})$ are shown in Fig. 20.13.

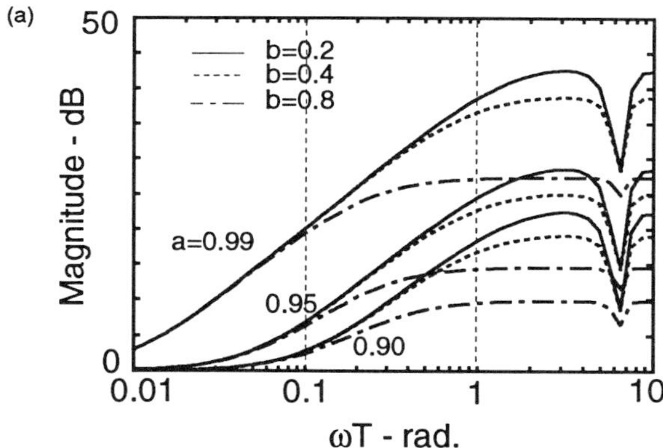

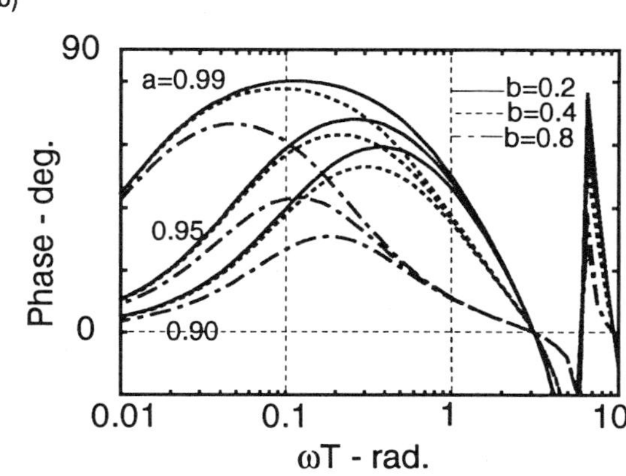

Figure 20.13 Gain (a) and phase (b) of certain first-order digital compensators as parameters a = zero location and b = pole location vary.

As an alternative to the above, the *w*-transform allows us to work with the forms

$$\hat{G}_c(j\omega_w) = \frac{1 + \dfrac{j\omega_w}{a}}{1 + \dfrac{j\omega_w}{b}}$$

to have lag ($a < b$) and lead ($b > a$) elements as in Fig. 20.12. This can be considerably easier to use than the *z*-transform model.

The art of the designer is to manipulate these to meet specifications, either explicit or implicit (e.g. to obtain 'good' performance). These may include (one or more of) the following:

1. having prescribed steady-state error – this usually determines minimum 0-frequency gain and/or system type;
2. having prescribed or implied gain and/or phase margins;
3. having prescribed bandwidth.

General design rules are fraught with exceptions, but one is inclined to make the following statements.

1. Increased gain compensation decreases steady-state errors and stability margins and increases bandwidth.
2. Lag compensation decreases bandwidth (and hence slows response) and relative stability margins. Lag filter implementations are associated with decreased steady-state errors. One might think of lag as similar in effect to integral control.
3. Lead compensation increases bandwidth (and hence yields faster response) and relative stability margins, usually with marginal effects on steady-state errors. Intuitively, lead is rather like derivative control.

Example

For this example we again return to the motor,

$$H(z) = \frac{0.048\,(z + 0.967)}{(z - 1)\,(z - 0.905)}$$

with *w*-form

$$\hat{H}(w) = \frac{0.497\,(1 - w)\,(1 + w/59.6)}{w(1 + w/0.050)}$$

Phase margin is chosen to be 40°, and velocity constant ≥ 50 is needed. Treating the velocity constant first, we find (Chapter 17)

$$K_{\text{vel}} \equiv \lim_{z \to 1} \frac{z - 1}{T}\, C(z)H(z)$$

$$= \lim_{w \to 0} \frac{2w}{T}\, \hat{C}(w)\hat{H}(w)$$

$$= \frac{0.994}{T} \lim_{w \to 0} \hat{C}(w)$$

With $T = 0.1$, we need $\hat{C}(0) > 5$. Imposing $\hat{C}(0) = 5 \approx 14\,\text{dB}$ gives a gain crossover frequency of $\omega_w = 0.36$, at which the phase is $-192°$. We must now choose between a lag compensator, which will need to bring the gain crossover back to about 0.05, or a lead compensator which will increase the phase by 52° plus enough to compensate for the shift it will bring to the crossover frequencies. In either case we need the fact from the compensator effects above that for lead/lag elements

$$\frac{1 + w/a}{1 + w/b}$$

the maximum phase is at $\omega_w = \sqrt{ab}$ and the phase contribution is

$$\phi = \tan^{-1}\left(\frac{b}{a}\right)^{\frac{1}{2}} - \tan^{-1}\left(\frac{a}{b}\right)^{\frac{1}{2}}$$

so that

$$\tan \phi_{\text{max}} = \frac{b + a}{2\,\sqrt{(ab)}}$$

We choose to design a lead filter. Then with $\omega_w = 1$ as the guess for the new crossover frequency and with the choice $\phi = 75°$ as hopefully providing enough margin change (we remark that this is almost the maximum a single lead can provide), we obtain approximately $a = 0.2$ and $b = 7.8$. The results of this are shown in Fig. 20.14, and are noteworthy in being only moderately successful.

As seems too often the case with pure lead of this amount, the crossover frequency has shifted by enough to dilute some of the added phase effect. In fact, we have gone from a gain margin of about 6 dB and a phase margin of 9° in the original system to about 4 dB and 35° in the new one with the improved velocity constant. The result is

$$\hat{C}(w) = 5 \frac{1 + \dfrac{w}{0.2}}{1 + \dfrac{w}{7.8}}$$

which leads after the inverse w substitution to

$$\hat{C}(z) = 195 \frac{1.2z - 0.8}{8.8z + 6.8}$$

A better design would have used another stage of lead. A single stage of lag compensation would have met specifications, although not necessarily have constituted a 'good' compensator, because of low bandwidth. With the lag, we might select that the gain crossover frequency should be 0.045 and design the lag so that the original phase response is unaffected in this area, which could be shown to lead to a very small b/a ratio and a small difference of phase from $-180°$ at low frequencies.

The above simple example shows some of the difficulties with oversimplifying classical design methods. The traditional remedy is to cascade several lead and lag compensators.

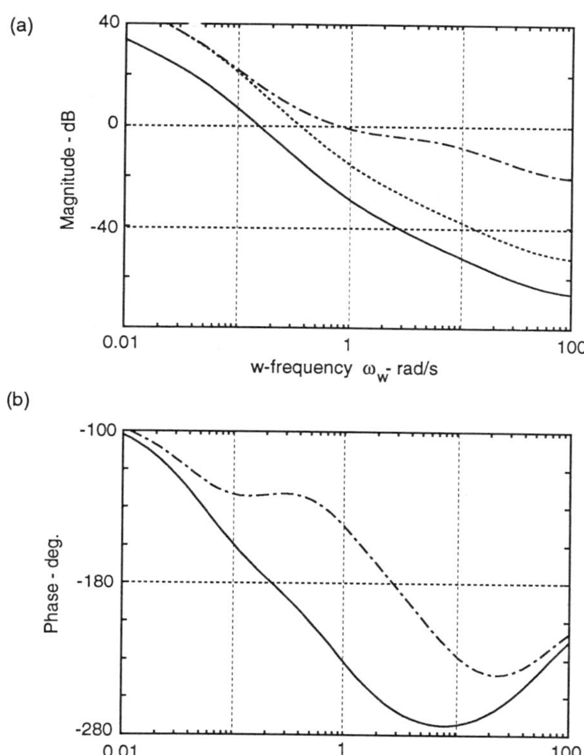

Figure 20.14 Example of continuous system design with lead network: (a) gain; and (b) phase. Original system frequency response is solid line; effect of gain change only is dotted; compensated frequency response is dot-dash.

20.4 VARIATIONS

20.4.1 Other frequency domain representations

Two other graphical representations of frequency domain models are in use: the Nyquist diagram (Chapter 15) and the Nichols chart. Each is a single plot of the frequency response with frequency as a parameter.

The Nyquist diagram (Fig. 20.15(a)) plots $\mathrm{Im}(H(j\omega))$ versus $\mathrm{Re}(H(j\omega))$ for $-\infty < \omega < \infty$ when $H(j\omega)$ is the continuous time frequency response, and plots $\mathrm{Im}(H(e^{j\phi}))$ versus $\mathrm{Re}(H(e^{j\phi}))$ for $-\pi < \phi < \pi$ when $H(e^{j\phi})$ is the sampled data frequency response. It is

(a)

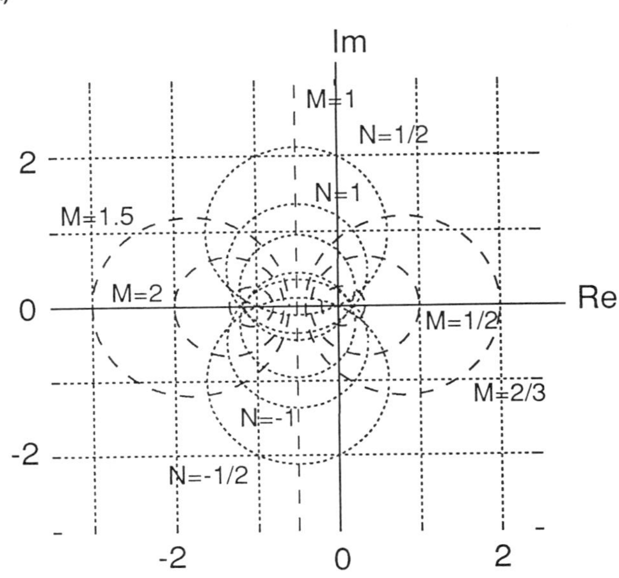

(b)

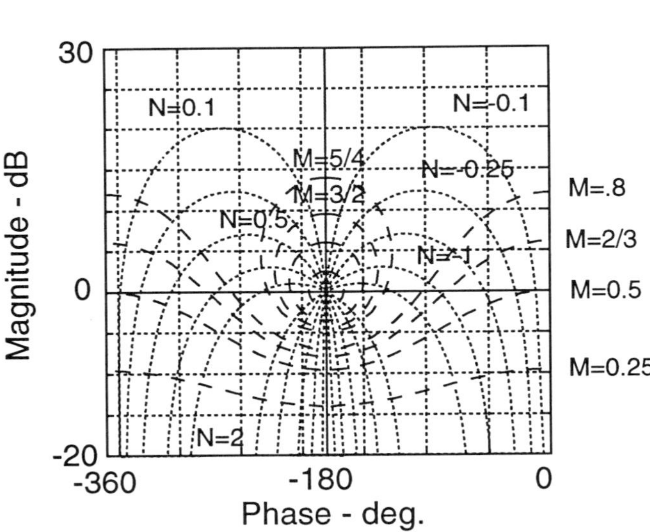

Figure 20.15 The alternatives: (a) Nyquist chart with *M*- and *N*-circles; and (b) Nichols chart (log-magnitude versus phase with loci of constant closed-loop magnitude and phase).

difficult to sketch and poorly scaled for detailed design, but shows gain and phase margins and design concepts readily.

The Nichols chart (Fig. 20.15(b)) shows a plot of $|H|$ versus arc(H) with ω or ϕ as a parameter, depending upon whether continuous time systems or sampled data systems are under consideration; it also has the M- and N-circles (of constant closed-loop magnitude and phase, respectively, in a unity feedback configuration; see Chapter 15) mapped onto the graph. Margins and co-ordinated design of the gains and phases of compensators are relatively easy, as is representation of experimental data; sketching, however, is not straightforward.

20.4.2 Multi-dimensional transfer functions

When considering MIMO systems described by transfer function matrices, it is not always best to consider the effects on an element by element basis. One alternative is to note that we frequently consider the magnitude of the open-loop gain, i.e. $|G(j\omega)|$. In multi-dimensions, this magnitude is interpreted as a matrix norm (Appendix B), in particular an induced norm. Thus it is natural to look at

$$\| \mathbf{G} \| = \sup_{\mathbf{R}} \frac{\| \mathbf{Y} \|}{\| \mathbf{R} \|} = \sup_{\mathbf{R}} \frac{\| \mathbf{GR} \|}{\| \mathbf{R} \|}$$

where $\mathbf{R}$ is the input and $\mathbf{Y}$ is the output. It turns out that

$$\underline{\sigma}(\mathbf{G}(j\omega)) \leq \frac{\| \mathbf{G}(j\omega)\, \mathbf{R}(j\omega) \|}{\| \mathbf{R}(j\omega) \|} \leq \bar{\sigma}(\mathbf{G}(j\omega))$$

where $\bar{\sigma}(\cdot)$ and $\underline{\sigma}(\cdot)$ denote the largest and smallest singular values, and are computed using a singular value decomposition. The induced norm is

$$\| \mathbf{G}(j\omega) \| = \sup_{\mathbf{R}(j\omega)} \frac{\| \mathbf{G}(j\omega)\, \mathbf{R}(j\omega) \|}{\| \mathbf{R}(j\omega) \|} = \bar{\sigma}(\mathbf{G}(j\omega))$$

The design of $\mathbf{G}(j\omega)$ such that low frequency gain is high and high frequency gain is low might be specified in terms of requirements that

$$\underline{\sigma}(\mathbf{G}(j\omega)) > l(j\omega) \qquad \omega < \omega_l$$

$$\overline{\sigma}(\mathbf{G}(j\omega)) < h(j\omega) \qquad \omega > \omega_h$$

Some aspects of actually doing such designs are met in Chapter 33.

Example

We consider the system

$$\mathbf{G}(s) = \begin{bmatrix} \dfrac{1}{s(0.2s + 1)} & \dfrac{2s + 1}{s(s + 1)\,(0.2s + 1)} \\[2em] \dfrac{\dfrac{s}{1.2} + 1}{(0.5s + 1)\left(\dfrac{s^2}{9} + 1.2\,\dfrac{s}{3} + 1\right)} & \dfrac{1}{\left(\dfrac{s^2}{9} + 1.2\,\dfrac{s}{3} + 1\right)} \end{bmatrix}$$

The upper and lower singular value limits (in fact there are only two singular values) and the frequency responses of the individual terms are shown in Fig. 20.16.

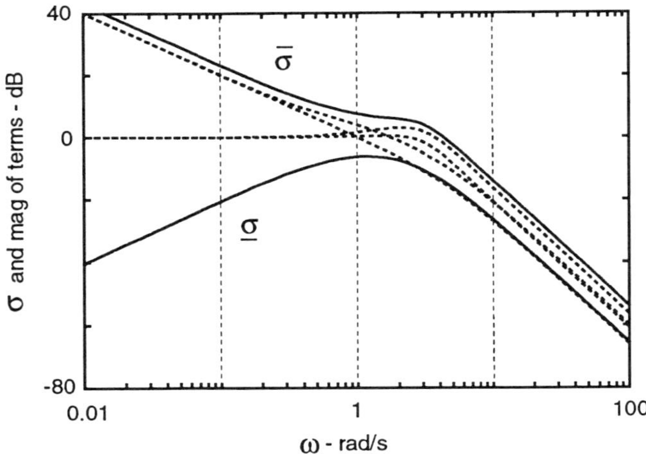

Figure 20.16 The multivariable case: example showing magnitude plots of individual terms and the upper and lower singular value bounds.

20.5 COMPUTER AIDED DESIGN

One would expect almost any computer design aid program to allow some frequency manipulations. Given a model, for example, MATLAB® will generate frequency response data and plot it in any of the three approaches of this chapter.

Formulae and graphs have been published for the design of lead/lag compensators, and these can also be put into computers if desired.

20.6 SUMMARY AND FURTHER READING

Bode analysis should be viewed as straightforward mathematically and quite possible experimentally. It is well covered in elementary texts such as Dorf (1989) and drill problems are in DiStefano *et al.* (1976).

Of the alternative frequency domain analysis methods, Nyquist theory, which is sometimes considered their theoretical basis, was dealt with in Chapter 15, and is in the above textbooks. The pursuit of Nichols methods is more common in older texts than more recent ones; one more recent one is D'Azzo and Houpis (1988). Multivariable methods were seen in Chapter 15 and will be met again in Chapter 34; one text is Maciejowski (1989).

Elementary design is typically done by introducing compensator structures with known generic effects and then iterating their parameters. Electrical engineering text books such as DiStefano *et al.* (1976) do this with lead–lag compensators, while three-term PID controllers are in the chemical engineering process control texts such as Stephanopoulos (1984). More advanced approaches called loop shaping are introduced in Doyle *et al.* (1992). Design is an art form rather than being purely algorithmic. This is particularly so for classical approaches such as those in this chapter; optimal control (Chapters 26–27) and robust control design (Chapter 33) methods shift the artistry from element choice to selection of appropriate weighting functions. In any case, only practice and experience will make the designer competent with any design method.

21

A special control law: deadbeat control

Most of the methods we have considered so far have their origins in linear control laws for systems described by linear constant coefficient differential equations, with sampled data control studied by adapting those theories. Even in the nominally linear control law realm, however, it is possible to develop a special and interesting performance property with discrete time control. In particular, it is possible in principle to achieve a zero error to an input after finite time with a linear control law; this contrasts with continuous time control, which can only asymptotically provide zero error, and follows from the fact that computer control commands are piecewise constant in nature. A response which quickly reaches zero error at the sampling instants and has little ripple between samples is called a **deadbeat response**. In this chapter we develop both transfer function oriented and state–space oriented approaches to design of the control laws, called deadbeat controllers, which yield such response.

21.1 SYNOPSIS

A deadbeat controller has the property that for a given input type, such as a step, the error between input and output,

$$e(kT) = y_{in}(kT) - y(kT)$$

will always show $e(kT) = 0$ $(k > n)$ for a particular n which depends upon the plant.

Such controllers can be designed using transfer functions or state–space arguments. In the first case one first finds an appropriate closed-loop transfer function $G_{cl}(z)$ and then generates, from it and the plant model, a control law $C(z)$ which yields it.

When using state–space arguments, we in effect use the controller in a state feedback configuration $u(kT) = Kx(kT)$ (see Chapter 23) to place all of the closed-loop eigenvalues at 0.

The use of deadbeat controllers can be controversial. Design requires excellent knowledge of the system model and is highly tuned to particular inputs.

21.2 TRANSFER FUNCTION APPROACHES TO DESIGN

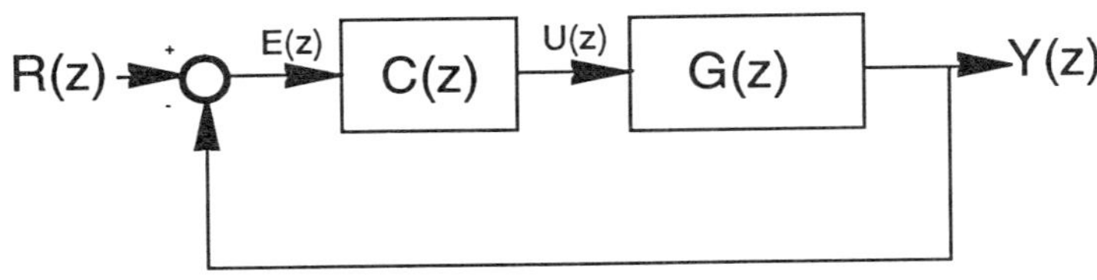

Figure 21.1 Basic digital control system block diagram for examination of compensators for deadbeat response.

Consider a control system with compensator $C(z)$ and plant $G(z)$ in a unity feedback configuration (Fig. 21.1). The closed-loop transfer function $G_{cl}(z)$ is

$$\frac{Y(z)}{R(z)} = G_{cl}(z) = \frac{C(z)\,G(z)}{1 + C(z)\,G(z)}$$

We notice that if we can choose $G_{cl}(z) = 1/z$ and if the input $R(z)$ is a unit step, $R(z) = z/(z-1)$, then

$$Y(z) = G_{cl}(z)\,R(z) = 1/(z-1) = z^{-1} + z^{-2} + \cdots$$

which gives $E(z) = 1 = R(z) - Y(z)$, i.e. $e(nT) = 1$, $n = 0$, and $e(nT) = 0$, $n > 0$. Thus the response is deadbeat, at least at the sampling instants.

Not all responses are deadbeat: we must specify the inputs to which the response is to be deadbeat. If the input is a polynomial in t of order $N-1$, then its transform is of the form

$$R(z) = \frac{A(z)}{(1 - z^{-1})^N}$$

where $A(z)$ is a polynomial in z^{-1} with no zeros at $z = 1$. If the system has unity feedback, then

$$E(z) = R(z) [1 - G_{cl}(z)]$$

and what we want is

$$\lim_{n \to \infty} e(nT) = \lim_{z \to 1} (1 - z^{-1}) E(z)$$

$$= \lim_{z \to 1} \frac{A(z)(1 - G_{cl}(z))}{(1 - z^{-1})^{N-1}}$$

$$= 0$$

For this to hold, $(1 - G_{cl}(z))$ must have $(1 - z^{-1})^N$ as a factor, i.e. it should have the form

$$1 - G_{cl}(z) = (1 - z^{-1})^N F(z) \tag{21.1}$$

where $F(z)$ is a polynomial in z^{-1} and has no zeros at $z = 1$. Thus $G_{cl}(z)$ must be of the form

$$G_{cl}(z) = \frac{z^N - (z-1)^N F(z)}{z^N} \tag{21.2}$$

i.e. the closed-loop transfer function must have a characteristic equation with all poles at the origin. The error will have

$$E(z) = A(z) F(z)$$

which is clearly a finite polynomial in z^{-1}, yielding from the definition of the z transform that $e(nT)$ is non-zero only for finite duration.

Physical realizability, in the sense of causal relationships of input and output, requires (Kuo, 1980) that if the plant pulse response $G(z)$ shows a delay of k samples, then the closed-loop pulse response must show at least the same delay. Thus if

$$G(z) = g_k z^{-k} + g_{k+1} z^{-k-1} + \cdots$$

then

$$G_{cl}(z) = c_n z^{-n} + c_{n+1} z^{-n-1} + \cdots \tag{21.3}$$

with $n \geq k$. We remark that k is given by the pole excess, i.e. the number of poles minus the number of zeros of $G(z)$. Subject to this constraint, the design procedure is as follows.

1. Choose the input to be tracked. This yields $A(z)$ and more particularly N.
2. Choose $G_{cl}(z)$ satisfying (21.3) and also (21.2). $F(z)$ is fairly arbitrary for a stable minimum phase plant (all poles and zeros inside unit circle) for which the pole excess is zero; Kuo (1980) discusses the alternative.
3. Then $C(z)$ is calculated from rearranging the standard unity feedback relationship as

$$C(z) = \frac{1}{G(z)} \frac{G_{cl}(z)}{1 - G_{cl}(z)} \tag{21.4}$$

We remark that (21.4) shows that part of the controller consists of cancellation of the poles and zeros of the plant; if the plant is not accurately known, this can be a problem.

Example (Kuo, 1980, p. 518)

Let $G(z) = 1/(z^2 - z - 1)$ and require deadbeat response to a unit step.

Step 1 Because of the unit step requirement, $N = 1$ and $A(z) = 1$.

Step 2 Since $G(z) = z^{-2} + 2z^{-3} + 3z^{-4} + \cdots$, $G_{cl}(z)$ must start no sooner than $k = 2$. Choose $G_{cl}(z) = z^{-2}$. Then $F(z)$ can be derived from (21.1) as

$$F(z) = \frac{1 - G_{cl}(z)}{1 - z^{-1}} = 1 + z^{-1}$$

Step 3 The control law is now

$$C(z) = \frac{1 - z^{-1} - z^{-2}}{z^{-2}} \times \frac{z^{-2}}{1 - z^{-2}}$$

$$= \frac{1 - z^{-1} - z^{-2}}{1 - z^{-2}}$$

Notice that we have in effect cancelled the plant's poles with the compensator.

To emphasize the fact that the response is tuned to the polynomial of order N, we consider a different input $R_1(z)$. Then we might have the form

$$R_1(z) = \frac{B(z)}{(1 - z^{-1})^K}$$

with $B(z)$ having neither zeros nor poles at $z = 1$ and with K an integer. Then

$$E(z) = B(z) F(z) (1 - z^{-1})^{N-K}$$

for which the control will be deadbeat only for $K \le N$ and $B(z)$ a polynomial in z; the limiting error could be 0 if $B(z)$ is rational and $K < N$.

21.3 STATE–SPACE APPROACHES

It is instructive to consider the issue of pole placement in state–space. We take only a preliminary look at two results, with more detail becoming obvious in Chapters 22, 23, and 27.

We take the simple view that the object is to drive the system to the state $\mathbf{x}_f = 0$ and the model is

$$\mathbf{x}(k+1) = \mathbf{A}\mathbf{x}(k) + \mathbf{B}\mathbf{u}(k) \qquad \mathbf{x}(0) = \mathbf{x}_0 \tag{21.5}$$

We choose the control law

$$\mathbf{u}(k) = \mathbf{K}\,(\mathbf{x}_f - \mathbf{x}(k))$$

Then if $\mathbf{e}(k) = \mathbf{x}_f - \mathbf{x}(k)$ with $\mathbf{x}_f = 0$ we have

$$\mathbf{e}(k+1) = \mathbf{A}\mathbf{e}(k) - \mathbf{B}\mathbf{K}\mathbf{e}(k) = (\mathbf{A} - \mathbf{B}\mathbf{K})\,\mathbf{e}(k)$$

Since

$$\mathbf{e}(k) = (\mathbf{A} - \mathbf{B}\mathbf{K})^k\,\mathbf{e}(0)$$

and since a matrix must satisfy its characteristic equation, we have, for any λ which is an eigenvalue of $(\mathbf{A} - \mathbf{B}\mathbf{K})$ that

$$\lambda^n + a_1\lambda^{n-1} + a_2\lambda^{n-2} + \cdots + a_n = 0$$

and hence

$$(\mathbf{A} - \mathbf{B}\mathbf{K})^n + a_1(\mathbf{A} - \mathbf{B}\mathbf{K})^{n-1} + \cdots + a_n\,(\mathbf{A} - \mathbf{B}\mathbf{K}) = 0$$

If the eigenvalues λ_i of $(\mathbf{A} - \mathbf{B}\mathbf{K})$ are all 0, this means

$$(\mathbf{A} - \mathbf{B}\mathbf{K})^n = 0$$

Thus, if we can choose $\mathbf{K}$ such that the eigenvalues of $(\mathbf{A} - \mathbf{B}\mathbf{K})$ are all 0, then $\mathbf{e}(n) = 0$ for any initial condition.

For an alternative approach, notice that the solution of (21.5) is

$$\mathbf{x}(k) = \mathbf{A}^k\mathbf{x}_0 + \sum_{j=0}^{k-1}\mathbf{A}^{k-j-1}\mathbf{B}\mathbf{u}(j)$$

If $\mathbf{u}$ is a scalar, this can be written

$$\mathbf{x}(k) - \mathbf{A}^k \mathbf{x}_0 = [\ \mathbf{A}^{k-1}\mathbf{B} \ \vdots \ \mathbf{A}^{k-2}\mathbf{B} \ \vdots \ \dots \ \vdots \ \mathbf{B}\ ] \begin{bmatrix} u(0) \\ u(1) \\ \vdots \\ u(k-1) \end{bmatrix}$$

If $k = n$, the right-hand side matrix is square. If it can be inverted, then

$$\begin{bmatrix} u(0) \\ u(1) \\ \vdots \\ u(n-1) \end{bmatrix} = [\ \mathbf{A}^{n-1}\mathbf{B} \ \vdots \ \mathbf{A}^{n-2}\mathbf{B} \ \vdots \ \dots \ \vdots \ \mathbf{B}\ ]^{-1} [\mathbf{x}(n) - \mathbf{A}^n \mathbf{x}_0]$$

for any chosen $\mathbf{x}(n)$.

The above gives a sequence of controls to be applied open-loop to go from $\mathbf{x}_0$ to $\mathbf{x}(n) = \mathbf{x}_f$ in n steps. Provided $\mathbf{x}_f = 0$, we can use linear algebra arguments to obtain a closed-loop control. For those, we note that if $\mathbf{A}^{-1}$ exists (the approach can be used if it does not exist, but is not quite so straightforward), then we can write

$$\mathbf{x}_0 = \mathbf{A}^{-n} [\ \mathbf{A}^{n-1}\mathbf{B} \ \vdots \ \mathbf{A}^{n-2}\mathbf{B} \ \vdots \ \dots \ \vdots \ \mathbf{B}\ ] \begin{bmatrix} u(0) \\ u(1) \\ \vdots \\ u(n-1) \end{bmatrix}$$

Clearly, we can write any $\mathbf{x}$, including $\mathbf{x}(k)$, as a linear combination of the columns of the r.h.s. matrix, i.e.

$$\mathbf{x}(k) = [\ \mathbf{A}^{-1}\mathbf{B} \ \vdots \ \mathbf{A}^{-2}\mathbf{B} \ \vdots \ \dots \ \vdots \ \mathbf{A}^{-n}\mathbf{B}\ ]\ \mathbf{z}$$

for some vector $\mathbf{z}$ with elements z_i, $i = 1, 2, \dots, n$. If we choose $u(i) = -z_1$ at each stage, then we can show that $\mathbf{x}(n) = 0$. To see this, we observe that for arbitrary $\mathbf{x}(0)$, we can take $u(\cdot) = -z(\cdot)$ to find that

$$\mathbf{x}(1) = \mathbf{A}\mathbf{x}(0) - \mathbf{B}z_1(0)$$

$$= \mathbf{A}(\mathbf{x}(0) - \mathbf{A}^{-1}\mathbf{B}z_1(0))$$

$$= \mathbf{A}([\, 0 \;\vdots\; \mathbf{A}^{-2}\mathbf{B} \;\vdots\; \ldots \;\vdots\; \mathbf{A}^{-n}\mathbf{B}\,]z(0))$$

$$= [\, \mathbf{A}^{-1}\mathbf{B} \;\vdots\; \mathbf{A}^{-2}\mathbf{B} \;\vdots\; \ldots \;\vdots\; \mathbf{A}^{-n+1}\mathbf{B} \;\vdots\; 0\,]z(1)$$

Hence, $\mathbf{x}(1)$ cannot have a component in the $\mathbf{A}^{-n}\mathbf{B}$ direction. Repeating the approach, with $u(1) = -z_1(1)$, will eliminate the $\mathbf{A}^{-n+1}\mathbf{B}$ component in the representation of $\mathbf{x}(2)$, etc. Hence $\mathbf{x}(n)$ is necessarily 0 by simply continuing the argument for n stages. The feedback law is

$$u(k) = -\text{ component of }\mathbf{A}^{-1}\mathbf{B}\text{ in the representation of }\mathbf{x}(k)$$

$$= -\left\{ [\, \mathbf{A}^{-1}\mathbf{B} \;\vdots\; \mathbf{A}^{-2}\mathbf{B} \;\vdots\; \ldots \;\vdots\; \mathbf{A}^{-n}\mathbf{B}\,]^{-1}\mathbf{x}(k) \right\}_{\text{First element}}$$

$$= -[\, \mathbf{A}^{-1}\mathbf{B} \;\vdots\; \mathbf{A}^{-2}\mathbf{B} \;\vdots\; \ldots \;\vdots\; \mathbf{A}^{-n}\mathbf{B}\,]^{-1}_{1\text{st row}}\,\mathbf{x}(k)$$

$$= \mathbf{K}^{\mathrm{T}}\mathbf{x}(k)$$

Example

The example of the previous section has among its state–space representations the following:

$$\mathbf{x}(k+1) = \begin{bmatrix} 1 & 1 \\ 1 & 0 \end{bmatrix}\mathbf{x}(k) + \begin{bmatrix} 0 \\ 1 \end{bmatrix}u(k)$$

$$\mathbf{y}(k) = [\, 1 \quad 0\,]\,\mathbf{x}(k)$$

We find

$$\mathbf{A}^{-1} = \begin{bmatrix} 0 & 1 \\ 1 & -1 \end{bmatrix}$$

$$[\,\mathbf{A}^{-1}\mathbf{B} \;\vdots\; \mathbf{A}^{-2}\mathbf{B}\,]^{-1} = \begin{bmatrix} 2 & 1 \\ 1 & 1 \end{bmatrix}$$

and hence

$$u(k) = - [\, 2 \quad 1 \,]\, \mathbf{x}(k)$$

It is easy to show that the closed-loop eigenvalues are 0. Also, by direct calculation, if $\mathbf{x}(0) = [\, \alpha \quad \beta \,]^T$, then $\mathbf{x}(1) = [\, \alpha+\beta \quad -\alpha-\beta \,]^T$ and $\mathbf{x}(2) = [\, 0 \quad 0 \,]^T$.

Example

We present an arbitrary example showing the characteristic of deadbeat response. Let the system be randomly chosen using the rand(·) command in MATLAB® as

$$\mathbf{x}(k+1) = \begin{bmatrix} 0.8886 & 0.5911 & 0.4154 & 0.1537 & 0.4985 \\ 0.2332 & 0.8460 & 0.5373 & 0.5717 & 0.9554 \\ 0.3063 & 0.4121 & 0.4679 & 0.8024 & 0.7483 \\ 0.3510 & 0.8414 & 0.2872 & 0.0331 & 0.5546 \\ 0.5133 & 0.2693 & 0.1783 & 0.5344 & 0.8907 \end{bmatrix} \mathbf{x}(k)$$

$$+ \begin{bmatrix} 0.0269 \\ 0.7098 \\ 0.9379 \\ 0.2399 \\ 0.1809 \end{bmatrix} u(k)$$

$$\mathbf{x}(0) = \begin{bmatrix} 0.1304 \\ 0.0910 \\ 0.2746 \\ 0.0030 \\ 0.4143 \end{bmatrix}$$

$$\mathbf{y}(k) = [1 \ 0 \ 0 \ 0 \ 0] \, \mathbf{x}(k)$$

The coefficients are random numbers, and the resulting poles are at 2.5782, $0.4082 \pm 0.1855j$, $-0.1342 \pm 0.1393j$. Using Ackermann's formula (see Chapter 22 and Appendix B) in MATLAB® to place all of the five poles of $(\mathbf{A} - \mathbf{BK})$ at $z = 0$ gives the gain

$$\mathbf{K} = [\; 1.6585 \quad 1.7921 \quad 1.0839 \quad 1.3599 \quad 2.5802\;]$$

(Numerically, in fact, this places the poles at 0.0013, $0.0004 \pm 0.0013\mathrm{j}$, $-0.0011 \pm 0.0008\mathrm{j}$, an indication of the possibility of numerical problems when using CAD methods.) Figure 21.2 shows the response with the control $\mathbf{u}(k) = -\mathbf{K}\,\mathbf{x}(k)$ to the initial condition $\mathbf{x}(0)=0.25$. Note that $\mathbf{y}(k) = 0$, $k \geq 5$, with the implication, easily verified with the simulation, that $\mathbf{x}(k) = 0$ for $k \geq 0$.

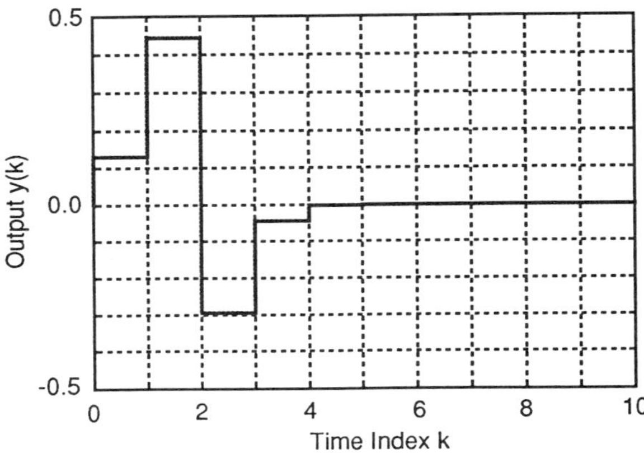

Figure 21.2 Example of deadbeat response to initial error for fifth-order system.

The modifications to take a state representation to $\mathbf{x}_f \neq 0$ in n stages are tricky, but in principle straightforward; maintaining $\mathbf{x} = \mathbf{x}_f \neq 0$ cannot generally be done. One way to do the former is to choose $\mathbf{K}$ such that $(\mathbf{A}-\mathbf{BK})^n = 0$ as above, and then find $\mathbf{u}_{\mathrm{ref}}(k)$, $k=0, 2, ..., n-1$ such that

$$\mathbf{x}_f = \sum_{i=0}^{n-1} (\mathbf{A} - \mathbf{BK})^{n-1-i}\, \mathbf{B}\mathbf{u}_{\mathrm{ref}}(i) \tag{21.6}$$

and finally apply the control

$$\mathbf{u}(k) = -\mathbf{K}\mathbf{x}(k) + \mathbf{u}_{\text{ref}}(k)$$

Since the closed-loop system has

$$\mathbf{x}(k+1) = (\mathbf{A} - \mathbf{B}\mathbf{K})\mathbf{x}(k) + \mathbf{B}\mathbf{u}_{\text{ref}}(k)$$

with solution

$$\mathbf{x}(n) = (\mathbf{A} - \mathbf{B}\mathbf{K})^n \, \mathbf{x}(0) + \sum_{i=0}^{n-1}(\mathbf{A} - \mathbf{B}\mathbf{K})^{n-1-i}\,\mathbf{B}\mathbf{u}_{\text{ref}}(i)$$

the choices of $\mathbf{K}$ and $\mathbf{u}_{\text{ref}}$ (21.6) yield $\mathbf{x}(n) = \mathbf{x}_{\text{f}}$. If $\mathbf{x}_{\text{f}} = 0$, the choice $\mathbf{u}_{\text{ref}}(k) = 0$ is the obvious choice, and in fact this choice is required if $(\mathbf{A} + \mathbf{B}\mathbf{K}; \mathbf{B})$ is controllable (Chapter 22) and the control is a scalar. We have implicitly assumed the latter in all of the above. Maintaining $\mathbf{x}(n+i) = \mathbf{x}_{\text{f}}$ is difficult with this law unless $(\mathbf{I} - \mathbf{A} + \mathbf{B}\mathbf{K})\mathbf{x}_{\text{f}} = \mathbf{B}\mathbf{u}$ is possible for some $\mathbf{u} = \mathbf{u}_{\text{ref}}$, in which case the above control law continues to work satisfactorily.

21.4 PROBLEMS AND FIXES

There are a number of problems with deadbeat design.

1. The controller is tuned for one particular input (a unit step, say) and may be inappropriate for other inputs.
2. The controller is based upon cancelling plant poles and zeros (the $1/G(z)$ factor in $C(z)$). If these are not known precisely, the controller will not be deadbeat. Hence, the design is not robust.
3. Related to (2) is the fact that if the plant has zeros outside the unit circle, then the controller will have poles outside that circle, i.e. the controller will be an unstable subsystem.

Kuo (1980, p.520) considers problem 3 in some detail. The solution is essentially one of incorporating the suspect poles and zeros into constraints on the allowable $G_{\text{cl}}(z)$.

Problem 1 is treated by Kuo (1980, p.526) by modifying the deadbeat constraint. In particular, the desired response is changed by the incorporation of a new pole at $z = c$, $-1 < c < 1$. The system is no longer deadbeat, and c is used to 'tune' the controller.

Problem 2 is a mathematical problem (the control law is infinitely sensitive to parameter errors) but perhaps not a 'real' problem. Åström and Wittenmark (1990) claim that the difficulties are overemphasized and that deadbeat control is worth considering.

21.5 COMPUTER AIDS

The design is straightforward with state–space, because then it is a special case of pole placement. For this reason, Ackermann's formula (Appendix B, Chapter 22) can help with design, and several CACSD packages contain pole placement algorithms of that or similar type. Simulation programs can be used for testing.

21.6 SUMMARY, EVALUATION, AND FURTHER READING

Deadbeat control is unique to computer control systems in that analog linear control laws do not provide that capability. It seems worth considering in an application, but a designer should always beware of controllers based upon pole-zero cancellation.

References include Kuo (1980), whose approach we followed in section 21.2, and Åström and Wittenmark (1990). Alternative derivations can be associated with controllability studies (see Chapter 22) and pole placement (Chapter 23).

22

Controllability

Two properties of the linear state–space system descriptions which are often needed in proofs about existence of certain types of controllers and state estimators are controllability and observability. Although these have important and easily visualized interpretations in terms of the natural language definitions of the words, it should be recognized that they are ultimately technical terms, used as shorthand to summarize properties of the system which allow certain types of controllers and state estimators to be designed. In particular, a system which is not controllable is not 'uncontrollable' in the natural language sense, nor is a not observable system 'unobservable'.

In this chapter we investigate controllability; observability is covered in Chapter 24.

22.1 SYNOPSIS

Definition A system is said to be **controllable** if any initial state x_0 at any initial time t_0 can be moved to any other desired state x_f, $x(t_f) = x_f$, in a finite time interval $\tau = t_f - t_0$ by applying an admissible control function $u(t) = u(t; t_0, x_0, x_f)$, $t_0 \leq t \leq t_f$.

From this definition, the following is the most commonly quoted result.

For a discrete time system modelled by the linear time invariant n-dimensional state difference equation

$$\mathbf{x}(k+1) = \mathbf{A}\mathbf{x}(k) + \mathbf{B}\mathbf{u}(k)$$

$$\mathbf{y}(k) = \mathbf{C}\mathbf{x}(k) + \mathbf{D}\mathbf{u}(k) \tag{22.1}$$

or the continuous time system modelled by

$$\dot{\mathbf{x}} = \mathbf{A}\mathbf{x} + \mathbf{B}\mathbf{u}$$

$$\mathbf{y} = \mathbf{C}\mathbf{x} + \mathbf{D}\mathbf{u} \qquad (22.2)$$

the system is **state controllable** iff (*if* and only *if*)

$$\text{rank } [\mathbf{B} \vdots \mathbf{A}\mathbf{B} \vdots \dots \vdots \mathbf{A}^{n-2}\mathbf{B} \vdots \mathbf{A}^{n-1}\mathbf{B}] = n \qquad (22.3)$$

Because the test applies to the matrices, we sometimes say the matrix pair $(\mathbf{A}; \mathbf{B})$ is controllable if (22.3) holds.

22.2 DEFINITIONS

Definitions are crucial to understanding the topic of controllability. The following are starting points.

Definition   A state x_0 is **controllable at time** t_0 if it can be transferred, in finite time starting at time t_0, to the origin $x_0 = 0$ by some allowable control function $u(t, x_0)$, $t_0 \le t \le t_0 + \tau < \infty$.

Definition   A state x_T is **reachable at time** t_f if the origin $x_0 = 0$ can be transferred to it, in finite time starting at time t_0, by some allowable control function $u_r(t, x_t)$, $t_0 \le t \le t_f < \infty$.

We choose to consider controllability as incorporating both controllability to the origin and reachability from the origin by a generalizing of the term. This is quite acceptable for linear constant coefficient systems and therefore all that is of interest to us here.

Definition   A state x_0 is **controllable** if, for any t_0 and any x_f, there is an admissible control function $u(t)$, which may depend upon x_0, x_f, t_0, such that $x(t)$ is driven from $x(t_0) = x_0$ to $x(t_f) = x_f$ during a finite time interval $t_0 \le t \le t_f$, $t_f - t_0 < \infty$. More succinctly, a state x_0 is controllable if any state is reachable from it in a finite time.

Definition   A **system is controllable** if all states x_0 are controllable.

Alternative terminology sometimes calls a system which is controllable by the term **completely controllable**. The above can be combined to give the definition in the synopsis. A system which does not meet the restrictions of the definitions is said to be **not controllable**. In fact, there are a number of variations in the development of controllability and reachability definitions, but for our purposes with linear constant coefficient systems the technical differences are operationally slight.

We emphasize that a system may not be controllable in the sense of the definition and yet may be quite adequate and be easy to design control laws for. In particular, a system which is not controllable is not necessarily uncontrollable in the common sense of the word. Systems which are controllable are often easier to design for because more design theorems apply, but other systems which are not controllable are often quite useful.

22.3 CONTROLLABILITY TESTS FOR CONSTANT COEFFICIENT SYSTEMS

In this section we present the standard tests and canonical decomposition for the system described by

$$\mathbf{x}(k+1) = \mathbf{A}\mathbf{x}(k) + \mathbf{B}\mathbf{u}(k)$$

$$\mathbf{y}(k) = \mathbf{C}\mathbf{x}(k) + \mathbf{D}\mathbf{u}(k)$$

where $\mathbf{A}$ is assumed to be $n \times n$, $\mathbf{B}$ is $n \times m$, and $\mathbf{C}$ is $p \times n$. $\mathbf{D}$ is obviously then $p \times m$.

The above system is controllable if and only if the $n \times nm$ matrix $\mathbf{Q}_c$, given by

$$\mathbf{Q}_c = [\, \mathbf{B} \,\vdots\, \mathbf{AB} \,\vdots\, \mathbf{A}^2\mathbf{B} \,\vdots\, \ldots \,\vdots\, \mathbf{A}^{n-1}\mathbf{B} \,]$$

has full rank, i.e. rank $(\mathbf{Q}_c) = n$. The demonstration of this is as follows. The solution of the above system is

$$\mathbf{x}(k+q) = \mathbf{A}^k\,\mathbf{x}(q) + \sum_{i=0}^{k-1} \mathbf{A}^{k-i-1}\,\mathbf{B}\mathbf{u}(q+i)$$

If both $\mathbf{x}(k+q)$ and $\mathbf{x}(q)$ can be arbitrary, then the above holds iff

$$\Delta \mathbf{x} = \sum_{i=0}^{k-1} \mathbf{A}^{k-i-1}\,\mathbf{B}\mathbf{u}(q+i) \qquad\qquad (22.4)$$

can take on any (vector) value for some k and some choice of $\mathbf{u}(j)$, $j=0,1,...,k-1$. The largest useful value of k is n, because of the implications of the Cayley–Hamilton theorem (see Appendix B) that

$$\mathbf{A}^j = \sum_{i=0}^{n-1} \alpha_i\,\mathbf{A}^{n-i-1}$$

for some scalars α_i and any $j \geq 0$. In particular, (22.4) with $k \geq n$ can be rewritten as a sum with $k=n$. Hence we have the possibility of system controllability iff

$$\Delta \mathbf{x} = \sum_{i=0}^{n-1} \mathbf{A}^{n-i-1}\mathbf{B}\mathbf{u}(i)$$

can be made arbitrary by choice of $\mathbf{u}(i)$. Rewriting this in matrix form, we see that, given any $\Delta \mathbf{x}$, we need to be able to solve

$$\Delta \mathbf{x} = [\,\mathbf{B} \,\vdots\, \mathbf{AB} \,\vdots\, \mathbf{A}^2\mathbf{B} \,\vdots\, ... \,\vdots\, \mathbf{A}^{n-1}\mathbf{B}\,] \begin{bmatrix} u(n-1) \\ u(n-2) \\ \vdots \\ u(0) \end{bmatrix}$$

for the controls $\{\mathbf{u}(i)\}$. This means that the matrix $\mathbf{Q}_c$ must have an invertible submatrix, which requires the matrix to have full rank. As the dimension is $n \times nm$, this means that

$$\text{rank}\,[\,\mathbf{B} \,\vdots\, \mathbf{AB} \,\vdots\, \mathbf{A}^2\mathbf{B} \,\vdots\, ... \,\vdots\, \mathbf{A}^{n-1}\mathbf{B}\,] = n$$

is required for system controllability. (See Appendix B for tests of rank.) If $m = 1$, then it is necessary to use all of the test matrix.

If the system is controllable, there exists a smallest number ν such that

$$\text{rank } [\, \mathbf{B} \,\vdots\, \mathbf{AB} \,\vdots\, \mathbf{A}^2\mathbf{B} \,\vdots\, \cdots \,\vdots\, \mathbf{A}^{\nu-1}\mathbf{B} \,] = n$$

This number $\nu \le n$ is called the **controllability index** of the controllable system and has the interpretation of being the number of stages generally necessary to transit from one state $\mathbf{x}_0$ to any other state $\mathbf{x}_f$.

An alternative viewpoint and test can be derived by considering the **canonical decomposition**. The system may, by similarity transformation $\mathbf{T}$ (i.e. $\mathbf{x} = \mathbf{T}\mathbf{z}$), be placed in the form

$$\mathbf{z}(k+1) = \Lambda\mathbf{z}(k) + \mathbf{T}^{-1}\,\mathbf{B}\mathbf{u}(k)$$

$$\mathbf{y}(k) = \mathbf{C}\mathbf{T}\mathbf{z}(k) + \mathbf{D}\mathbf{u}(k)$$

where Λ is in Jordan form and is diagonal if the eigenvalues of $\mathbf{A}$ are distinct. With this decomposition, the transformed state vector $\mathbf{z}$ has elements z_i, $i=1,2,\ldots,n$. Then the result is that z_i is controllable if each Jordan block has at least one non-zero row of $\mathbf{T}^{-1}\mathbf{B}$ associated with it and the non-zero rows corresponding to identical Jordan blocks are not identical; if Λ is diagonal with distinct eigenvalues, then z_i is controllable if the ith row of $\mathbf{T}^{-1}\mathbf{B}$ is not zero. The system is controllable if and only if z_i is controllable for each i.

A similar result holds for observability (Chapter 24), so it turns out that the state is decomposable to elements which are controllable, observable, both, or neither.

Example

Let $n=3$, $m=2$, $p=1$, with

$$\mathbf{A} = \begin{bmatrix} 1 & 0 & 0 \\ 0 & 1 & 1 \\ 1 & 1 & 0 \end{bmatrix} \quad \mathbf{B} = \begin{bmatrix} 0 & 0 \\ 1 & 0 \\ 0 & 1 \end{bmatrix}$$

$$\mathbf{C} = [0 \ 1 \ 1] \quad \mathbf{D} = [0 \ 0]$$

Then by direct computation

$$
\mathbf{Q_c} = \begin{bmatrix} 0\,0 & 0\,0 & 0\,0 \\ 1\,0 & 1\,1 & 2\,1 \\ 0\,1 & 1\,0 & 1\,1 \end{bmatrix}
$$

By inspection, rank $\mathbf{Q_c} = 2$ (two linearly independent rows; the first is zero and is therefore dependent). Since $n = 3$, the system is not controllable.

For the canonical decomposition, the eigenvalues of $\mathbf{A}$ are $\lambda_1 = 1$ and $\lambda_{2,3} = (1 \pm \sqrt{5})/2$, which are distinct so that $\Lambda = \mathrm{diag}(\lambda_1, \lambda_2, \lambda_3)$. A transformation matrix is then

$$
\mathbf{T} = \begin{bmatrix} 1 & 0 & 0 \\ -1 & (1+\sqrt{5})/2 & (1-\sqrt{5})/2 \\ 0 & 1 & 1 \end{bmatrix}
$$

for which the inverse is

$$
\mathbf{T}^{-1} = \begin{bmatrix} 1 & 0 & 0 \\ 1/\sqrt{5} & 1/\sqrt{5} & \dfrac{\sqrt{5}-1}{2\sqrt{5}} \\ -1/\sqrt{5} & -1/\sqrt{5} & \dfrac{1+\sqrt{5}}{2\sqrt{5}} \end{bmatrix}
$$

From this we compute the matrix product $\mathbf{T}^{-1}\mathbf{B}$.

$$
\mathbf{T}^{-1}\mathbf{B} = \begin{bmatrix} 0 & 0 \\ 1/\sqrt{5} & \dfrac{\sqrt{5}-1}{2\sqrt{5}} \\ -1/\sqrt{5} & \dfrac{1+\sqrt{5}}{2\sqrt{5}} \end{bmatrix}
$$

Because the first row of $\mathbf{T}^{-1}\mathbf{B}$ is zero, z_1 is not controllable.

22.4 CONTROLLABILITY AND A CONTROLLER

To see how the above tests evolved and also to generalize the results, we examine the time-varying case of

$$\mathbf{x}(k+1) = \mathbf{A}(k)\mathbf{x}(k) + \mathbf{B}(k)\mathbf{u}(k)$$

$$\mathbf{y}(k) = \mathbf{C}(k)\mathbf{x}(k) + \mathbf{D}(k)\mathbf{u}(k) \tag{22.5}$$

where $\mathbf{A}(k)$ is $n \times n$, $\mathbf{B}(k)$ is $n \times m$, $\mathbf{C}(k)$ is $p \times n$, $\mathbf{D}(k)$ is $p \times m$. Consider first the influence of the control on an arbitrary $\mathbf{x}(0) = \mathbf{x}_0$. We see that

$$\mathbf{x}(1) = \mathbf{A}(0)\,\mathbf{x}_0 + \mathbf{B}(0)\,\mathbf{u}(0)$$

$$\mathbf{x}(2) = \mathbf{A}(1)\,\mathbf{x}(1) + \mathbf{B}(1)\,\mathbf{u}(1)$$

$$= \mathbf{A}(1)[\mathbf{A}(0)\mathbf{x}_0 + \mathbf{B}(0)\mathbf{u}(0)] + \mathbf{B}(1)\mathbf{u}(1)$$

$$= \mathbf{A}(1)\mathbf{A}(0)\mathbf{x}_0 + \mathbf{A}(1)\mathbf{B}(0)\mathbf{u}(0) + \mathbf{B}(1)\mathbf{u}(1)$$

and so

$$\mathbf{x}(k) = \mathbf{A}(k-1)\mathbf{A}(k-2) \cdots \mathbf{A}(1)\mathbf{A}(0)\mathbf{x}_0$$

$$+ \mathbf{A}(k-1) \cdots \mathbf{A}(1)\mathbf{B}(0)\mathbf{u}(0)$$

$$+ \mathbf{A}(k-1) \cdots \mathbf{A}(2)\mathbf{B}(1)\mathbf{u}(1)$$

$$+ \cdots + \mathbf{A}(k-1)\mathbf{B}(k-2)\mathbf{u}(k-2)$$

$$+ \mathbf{B}(k-1)\mathbf{u}(k-1)$$

$$= \left[\prod_{j=0}^{k-1} \mathbf{A}(j) \right] \mathbf{x}_0 + \sum_{j=0}^{k-1} \left[\prod_{i=j+1}^{k-1} \mathbf{A}(i) \right] \mathbf{B}(j)\mathbf{u}(j)$$

or

$$\mathbf{x}\,(k) - \left[\prod_{j=0}^{k-1} \mathbf{A}\,(j)\right]\mathbf{x}_0 = \sum_{j=0}^{k-1}\left[\prod_{i=j+1}^{k-1}\mathbf{A}\,(i)\right]\mathbf{B}(j)\,\mathbf{u}(j)$$

For the left-hand side to have arbitrary $\mathbf{x}_0$, arbitrary $\mathbf{x}_f$ and $\mathbf{x}(k) = \mathbf{x}_f$ for some finite k, the essence of controllability, means that for some k, the matrix relation

$$[\mathbf{A}(k-1)\cdots\mathbf{A}(1)\mathbf{B}(0) \,\vdots\, \mathbf{A}(k-1)\cdots\mathbf{A}(2)\mathbf{B}(1) \,\vdots\, \cdots \,\vdots\, \mathbf{A}(k-1)\mathbf{B}(k-2) \,\vdots\, \mathbf{B}(k-1)]\,\mathbf{U}$$

$$= \text{arbitrary vector} = \mathbf{x}_f - \mathbf{A}(k-1) \,\cdots\, \mathbf{A}(0)\mathbf{x}_0 \qquad (22.6)$$

where

$$\mathbf{U} = \begin{bmatrix} \mathbf{u}(0) \\ \mathbf{u}(1) \\ \vdots \\ \mathbf{u}(k-1) \end{bmatrix}$$

must hold for some choice of $\mathbf{U}$. For this to happen, the matrix on the left must have at least one set of columns whose linear combinations (by choosing elements of $\mathbf{U}$) give any possible columns of n numbers. In linear algebraic terms, the matrix must have n linearly independent columns, or its columns must 'span the space' of n-vectors. In another set of words, the property

$$\text{rank}\,[\,\mathbf{A}(k-1)\,\cdots\,\mathbf{A}(1)\mathbf{B}(0)\,\vdots\,\ldots\,\vdots\,\mathbf{A}(k-1)\mathbf{B}(k-2)\,\vdots\,\mathbf{B}(k-1)\,] = n$$

for some k, is necessary and sufficient for controllability starting at time 0. Because the matrices $\mathbf{A}$ and $\mathbf{B}$ vary, the number of steps k may vary also if the initial time is different.

If $\mathbf{A}(k) = \mathbf{A} = $ constant and $\mathbf{B}(k) = \mathbf{B} = $ constant, the above becomes a requirement that

$$\text{rank}\,[\,\mathbf{A}^{k-1}\mathbf{B}\,\vdots\,\ldots\,\vdots\,\mathbf{A}\mathbf{B}\,\vdots\,\mathbf{B}\,] = n$$

for some k and then the result of the previous section for the basic test applies.

We remark that it is not only the index that is important here. If the system is controllable, then we can actually compute the control necessary for a transition. We look at (22.6) and expand $\mathbf{U}$, which generally is an m-vector, to

$$
\mathbf{U} = \begin{bmatrix} u_1(0) \\ u_2(0) \\ u_3(0) \\ \vdots \\ u_m(0) \\ u_1(1) \\ u_2(1) \\ \vdots \\ u_m(k-1) \end{bmatrix}
$$

We argue that the test for controllability means that the left-hand side matrix in (22.6) has an invertible submatrix. If it is controllable, we can solve for the control by setting all elements of $\mathbf{U}$ to zero except those associated with this submatrix, and then determine the others from a matrix inversion. Notice that the matrix in question is of dimension $n \times km$. Since it has rank n by assumption of controllability, we form the submatrix $\mathbf{S}_A$ by choosing columns $1 \leq i_1 < i_2 < \cdots < i_n \leq n$ from it such that $\mathbf{S}_A$ has rank n also, i.e. choose n linearly independent columns of the matrix. Then

$$
\begin{bmatrix} U_{i_1} \\ U_{i_2} \\ \vdots \\ U_{i_n} \end{bmatrix} = \mathbf{S}_A^{-1} (\mathbf{x}_f - \mathbf{A}(k-1) \cdots \mathbf{A}(0)\mathbf{x}_0)
$$

This is easily formalized by using a selection matrix $\mathbf{S}$. In particular, $\mathbf{S}$ is a $km \times n$ matrix with n non-zero columns, and with each non-zero column containing a different one of the elementary unit vectors. Under these conditions, $\mathbf{SS}^T$ contains an identity matrix, and straightforward algebra yields that $\mathbf{Q}_c\mathbf{SS}^T\mathbf{x} = \Delta$ is solved in our sense by $\mathbf{x} = \mathbf{S}[\mathbf{Q}_c\,\mathbf{S}]^{-1}\Delta$.

The situation is easy if the matrices are constant and $m = 1$, for then the controllability matrix $\mathbf{Q}_c$ is square and taking of submatrices is unnecessary.

Example

Consider the simple example with dynamics equation

$$\mathbf{x}(k+1) = \begin{bmatrix} 0 & 1 & 0 \\ 0.5 & 0.5 & 0 \\ 0 & 0 & 1 \end{bmatrix} \mathbf{x}(k) + \begin{bmatrix} 0 & 0 \\ 1 & 0 \\ 0 & 1 \end{bmatrix} \mathbf{u}(k)$$

The controllability matrix is given by

$$\mathbf{Q}_c = \begin{bmatrix} 0.5 & 0 & 1 & 0 & 0 & 0 \\ 0.75 & 0 & 0.5 & 0 & 1 & 0 \\ 0 & 1 & 0 & 1 & 0 & 1 \end{bmatrix}$$

which has rank 3. Using only the first element of the control vector yields rank $= 2$, and using only the second yields rank $= 1$. The controllability index can be seen to be 2. As we are seeking rapid control, we will choose to do a two-step command sequence, so that we have

$$\mathbf{x}(2) - \mathbf{A}^2 \mathbf{x}_0 = \mathbf{Q}_c' \mathbf{U}$$

with

$$\mathbf{Q}_c' = \begin{bmatrix} 1 & 0 & 0 & 0 \\ 0.5 & 0 & 1 & 0 \\ 0 & 1 & 0 & 1 \end{bmatrix}$$

We now choose $i_1 = 1$, $i_2 = 2$, $i_3 = 3$ to obtain the selection matrix $\mathbf{S}$ as

$$S = \begin{bmatrix} 1 & 0 & 0 \\ 0 & 1 & 0 \\ 0 & 0 & 1 \\ 0 & 0 & 0 \end{bmatrix}$$

and

$$\mathbf{Q}'_c \, \mathbf{S} = \begin{bmatrix} 1 & 0 & 0 \\ 0.5 & 0 & 1 \\ 0 & 1 & 0 \end{bmatrix}$$

for which the inverse is

$$[\mathbf{Q}'_c \mathbf{S}]^{-1} = \begin{bmatrix} 1 & 0 & 0 \\ 0 & 0 & 1 \\ -0.5 & 1 & 0 \end{bmatrix}$$

Then for any final condition $\mathbf{x}_f$ and initial condition $\mathbf{x}_0$ the control given by

$$\begin{bmatrix} u_1(0) \\ u_2(0) \\ u_1(1) \\ u_2(1) \end{bmatrix} = \mathbf{S}\,[\mathbf{Q}'_c \mathbf{S}]^{-1}(\mathbf{x}_f - \mathbf{A}^2 \mathbf{x}_0) = \begin{bmatrix} 1 & 0 & 0 \\ 0 & 0 & 1 \\ -0.5 & 1 & 0 \\ 0 & 0 & 0 \end{bmatrix}(\mathbf{x}_f - \mathbf{A}^2 \mathbf{x}_0)$$

should be satisfactory. The same operations can be done with $\mathbf{Q}_c$ and a 6×3 matrix $\mathbf{S}$ to yield the same result.

22.5 OTHER GENERALIZATIONS

The above discussions have been of a restricted subset of the possibilities of controllability. We mention certain variations in this section.

22.5.1 Output controllability

The controllability discussion above has in fact been concerned with what is called 'state controllability', because it is concerned with reaching a state which may be specified. One variation is called 'output controllability', which is concerned with whether it is possible to move the system's output from one condition y_0 to another y_f. The conditions for output controllability are less restrictive than those for state controllability, because the output dimension p is no higher than the dimension n of the state, provided the C matrix in (22.5) has full rank p. The above tests carry over in the obvious manner. For example, when $A(k)$, $B(k)$, and $C(k)$ are constants A, B, C, respectively, the output is controllable iff the matrix

$$\mathbf{C}\,\mathbf{Q}_c = [\ \mathbf{B} \vdots \mathbf{CAB} \vdots \mathbf{CA^2B} \vdots \ldots \vdots \mathbf{CA}^{n-1}\mathbf{B}\ ]$$

has rank p.

22.5.2 Continuous time system models

The concepts also apply to continuous time systems, with the result that the system described by

$$\dot{\mathbf{x}}(t) = \mathbf{A}\mathbf{x}(t) + \mathbf{B}\mathbf{u}(t)$$

$$\mathbf{y}(t) = \mathbf{C}\mathbf{x}(t) + \mathbf{D}\mathbf{u}(t)$$

with dimensions that $\mathbf{x}$ is an n-vector, $\mathbf{u}$ is an m-vector, $\mathbf{y}$ is a p-vector, is (state-) controllable iff

$$\text{rank}\ \mathbf{Q}_c = \text{rank}\ [\ \mathbf{B} \vdots \mathbf{AB} \vdots \mathbf{A^2B} \vdots \ldots \vdots \mathbf{A}^{n-1}\mathbf{B}\ ] = n$$

The tests are clearly the same as those in the discrete time case. Proofs appear in a number of places, e.g. (Graupe, 1972, Ch. 2), and in Chapter 24 for the related observability test.

22.5.3 Controllability of linear combinations of state variables

It is worth commenting that if the rank of one of the test matrices is not the maximum value, all is not lost. First, it is quite possible that we can adequately control a 'not controllable' system; we choose not to call it an uncontrollable system. Alternatively, the tests may indicate the need for another command input. The second piece of information is that the rank of the matrices really tells us the number of different linear combinations of the state we can control or observe: for instance, if rank $Q_c = 1$, there is one linear combination

$$f_1 x_1 + f_2 x_2 + \cdots + f_n x_n = f_0$$

which we can make take on an arbitrary value of f_0, although the coefficients f_i, $i = 1, 2, \ldots, n$, are fixed by the system properties. The coefficients can be found, if desired, by doing a canonical decomposition, finding which z, is controllable, and then doing the inverse transformation. More particularly, perform the similarity transformation $x = Tz$ and do a canonical decomposition. If z is controllable, then the linear combination $[\, T^{-1}\,]_i x$ can be driven to an arbitrary value, where $[\,]_i$ denotes the ith row of the matrix.

22.5.4 Time-varying systems

For the time-varying system

$$x(k+1) = A(k)\, x(k) + B(k)\, u(k)$$

we use the transition matrix defined by

$$\phi(k,j) = \phi(k,j+1)\, A(j)$$

$$\phi(k,k) = I$$

For this system, we may define a matrix, called the controllability gramian, as

$$W(k_0, k_1) = \sum_{j=k_0}^{k_1-1} \phi(k_1, j+1) B(j) B^T(j) \phi^T(k_1, j+1) \qquad (22.7)$$

and show, in a manner similar to that used for the observability matrix in section 24.3.3, that the system is controllable on the interval $[k_0, k_1]$ iff rank $W(k_0, k_1) = n$.

Furthermore, a feedback control which will steer the system from $x(k_0)$ to an arbitrary x_f is given by

$$u(k) = B^T(k)\, \phi^T(k_1, k+1)\, W^{-1}(k_0, k_1)\, (x_f - \phi(k_1, k_0)\, x(k_0))$$

$$(22.8)$$

The latter is seen by noting that, using the transition matrix,

$$x(k) = \phi(k, k_0)\, x(k_0) + \sum_{j=k_0}^{k-1} \phi(k, j+1)\, B(j)\, u(j)$$

$$x(k_1) = \phi(k_1, k_0)\, x(k_0) + \sum_{j=k_0}^{k_1-1} \phi(k_1, j+1)\, B(j)\, B^T(j)\, \phi^T(k_1, j+1)$$

$$*\; W^{-1}(k_0, k_1)\, (x_f - \phi(k_1, k_0)\, x(k_0))$$

$$= \phi(k_1, k_0)\, x(k_0)$$

$$+ W(k_0, k_1)\, W^{-1}(k_0, k_1)\, (x_f - \phi(k_1, k_0)\, x(k_0))$$

$$= x_f$$

The above applies in a straightforward manner if the system is time invariant, for then $\phi(k, j) = A^{k-j}$ and (22.7) becomes

$$W(k_0, k_1) = \sum_{j=k_0}^{k_1-1} A^{k_1-j-1} B B^T (A^{k_1-j-1})^T$$

with $k_1 - k_0 \leq n - 1$ because of the Cayley–Hamilton result.

The continuous time result is as usual similar, with the gramian defined by

$$W(t_0, t_f) = \int_{t_0}^{t_f} \phi(t_f, \tau)\, B(\tau)\, B^T(\tau)\, \phi^T(t_f, \tau)\, d\tau$$

and in the time-invariant case given, after change of variables, by

$$W(t_0, t_0 + T) = \int_0^T \exp(A\tau)\, BB^T \exp(A^T\tau)\, d\tau$$

The gramian test on rank $[W]$ is sometimes more useful than the direct form check on rank $[Q_c]$ because of better numerical properties.

22.5.5 Minimum energy control from controllability

On an interval $[k_0, k_1]$ the control to reach a given point x_1 may not be unique. In this case, we may decide to use an optimization criterion to choose among the candidates. One such candidate is the weighted energy control, in which the criterion to be minimized is

$$J(u) = \sum_{k=k_0}^{k_1-1} u^T(j)\, R(j)\, u(j)$$

where $R(j)$ is a symmetric positive definite weighting matrix. We define the modified controllability gramian matrix

$$W_R(k_0, k_1) = \sum_{j=k_0}^{k_1-1} \phi(k_1, j+1)\, B(j)\, R^{-1}(j)\, B^T(j)\, \phi(k_1, j+1)$$

Then the optimal control is given (Klamka, 1991) by

$$u(k) = R^{-1}(k)\, B^T(k)\, \phi^T(k_1, k+1)\, W_R^{-1}(k_0, k_1)(x_f - \phi(k_1, k_0)\, x(k_0))$$

The fact that this control allows x_f to be reached follows just as in the basic case in (22.8). The fact that it yields a minimum for the cost

J requires considerably more algebra (but not calculus or elaborate optimization arguments).

22.5.6 Controllability of non-linear systems

General controllability results for non-linear systems are difficult to come by, simply because there is no general way of writing the solutions of non-linear differential equations, and controllability theory tends to rely heavily on solution structure. Some local results have been obtained by linearization. Thus the system

$$\mathbf{x}(k+1) = \mathbf{f}(k, \mathbf{x}(k), \mathbf{u}(k))$$

for which the known nominal input sequence $\{\mathbf{u}_{\text{nom}}(k)\}$ yields a nominal state trajectory $\{\mathbf{x}_{\text{nom}}(k)\}$, can be studied by examining the deviations

$$\Delta x(k) = x(k) - x_{\text{nom}}(k)$$

and

$$\Delta u = u(k) - u_{\text{nom}}(k)$$

If these deviations are small enough, then we consider the approximation given by the linearized system

$$\Delta x(k+1) = \frac{\partial f(k, x(k), u(k))}{\partial x}\bigg|_{x_{\text{nom}}(k), u_{\text{nom}}(k)} x(k)$$

$$+ \frac{\partial f(k, x(k), u(k))}{\partial u}\bigg|_{x_{\text{nom}}(k), u_{\text{nom}}(k)} u(k)$$

$$= A(k)\,\Delta x(k) + B(k)\,\Delta u(k)$$

Another approach studies the quasi-linear case

$$\mathbf{x}(k+1) = \mathbf{A}(k, \mathbf{x}(k), \mathbf{u}(k))\,\mathbf{x}(k) + \mathbf{B}(k, \mathbf{x}(k), \mathbf{u}(k))\,\mathbf{u}(k)$$

$$\|\mathbf{A}(k, \mathbf{x}(k), \mathbf{u}(k))\| \leq \alpha_k \qquad \|\mathbf{B}(k, \mathbf{x}(k), \mathbf{u}(k))\| \leq \beta_k$$

This is reviewed by Faradzhev *et al.* (1986).

The special non-linear case of bilinear systems, such as

$$\mathbf{x}(k+1) = \mathbf{A}\left(\mathbf{I} + \sum_{j=1}^{m} u_j(k)\,\mathbf{B}_j\right)\mathbf{x}(k) + \mathbf{B}_0\,\mathbf{u}(k)$$

where $\mathbf{A}$, $\mathbf{B}_j$, are all $n \times n$ matrices, the u_j are scalar elements of the m-vector $\mathbf{u}$, and $\mathbf{B}_0$ is $n \times m$, has been examined by researchers with some results, summarized by Klamka (1991).

22.6 COMPUTER TESTING

Computer testing is straightforward, needing only a capacity to test matrix rank. Numerical problems can be present, however, because of potential ill-conditioning of the controllability matrix $\mathbf{Q}_c$. Some packages, therefore, test the gramian form.

22.7 FURTHER READING

Observability and controllability as concepts and structural indicators are usually traced back to Kalman (1961), and the ideas for linear system models are now found in most standard textbooks. The material of much of this chapter has also been presented in a similar manner in the textbook by Åström and Wittenmark (1990).

A good exposition of the results for controllability of linear systems, including time-varying systems, constrained control variables, and the minimum energy controller design above, is presented by Klamka (1991).

For non-linear systems, the problem quickly becomes very difficult. One old discussion is by Gershwin and Jacobson (1971). A survey including issues of non-linear systems and limited control is given by Faradzhev *et al.* (1986).

23

Controller design by pole placement

In classical methods of control law design, a structure for the controller is introduced by the designer and then several parameters within that structure are chosen to yield a response which meets specifications. Design work is usually done with the transfer function, either in a complex plane (as in root locus) or in the frequency domain (Nyquist–Bode–Nichols). The mathematics of pole placement design are as useful and interesting as many of the results, and therefore this chapter considers many variations on the basic problem: the essential result is that under certain conditions the poles of the closed-loop system may be placed at arbitrary locations of the designer's choice.

In this chapter, we examine state–space methods of pole placement design for linear constant coefficient systems of equations (and hence for the physical systems which those systems purportedly model). We start with the basic ideas, assuming the entire state is measured and that the system is controllable, proceed to a variation of the solution when the entire state is not available and output feedback is necessary, and continue to the problem of stabilizing non-controllable systems. As an alternative, we look at transfer function approaches to pole placement.

23.1 SYNOPSIS

When there is enough freedom available to the control law designer, then the poles of a closed-loop system may be arbitrarily placed through choice of control law parameters. This may be done with either state–space or transfer function representations.

In a state–space model

$$\mathbf{x}(k+1) = \mathbf{A}\mathbf{x}(k) + \mathbf{B}\mathbf{u}(k)$$

$$\mathbf{y}(k) = \mathbf{C}\mathbf{x}(k)$$

the use of state feedback $\mathbf{u}(k) = -\mathbf{K}\mathbf{x}(k)$ is sometimes of interest to provide reference cases and is sometimes implementable, as in the case of a motor with both shaft position and rotation velocity measured. In such cases, all $n = \dim(\mathbf{x})$ system poles can be placed arbitrarily by selecting $\mathbf{K}$ such that $(\mathbf{A} - \mathbf{B}\mathbf{K}) = \mathbf{F}$ has its eigenvalues at the desired pole values. This is possible provided that the system is controllable, i.e. that the controllability matrix $\mathbf{Q}_\mathrm{c}$, defined in Chapter 22, has full rank, i.e.

$$\mathrm{rank}\,[\ \mathbf{B} \mid \mathbf{A}\mathbf{B} \mid \mathbf{A}^2\mathbf{B} \mid \dots \mid \mathbf{A}^{n-1}\mathbf{B}\ ] = n$$

If the system is not controllable, it may still be possible to select some poles and have a stable system. When $\dim(\mathbf{u}) = 1$ and the system is controllable, one convenient design algorithm uses Ackermann's formula for the gain $\mathbf{K}$,

$$\mathbf{K} = [\ 0\ \ 0\ \dots\ 0\ \ 1\ ]\,[\ \mathbf{B} \mid \mathbf{A}^2\mathbf{B} \mid \mathbf{A}\mathbf{B} \mid \dots \mid \mathbf{A}^{n-1}\mathbf{B}\ ]^{-1}\,\alpha_\mathrm{des}(\mathbf{A})$$

where $\alpha_\mathrm{des}(\lambda)$ is the desired characteristic polynomial and $\alpha_\mathrm{des}(\mathbf{A})$ is the matrix polynomial defined by substituting the matrix $\mathbf{A}$ in place of the scalar λ. It is sometimes possible to design an output feedback law $\mathbf{u}(k) = -\mathbf{K}\mathbf{y}(k)$ so that p of the poles may be selected, where $p \geq \max(\dim(\mathbf{y}), \dim(\mathbf{u}))$. Also of interest is the possibility of stabilizing a system which is not controllable.

The alternative to state–space arguments is provided by considering that the control law transfer function

$$U(z) = \frac{T(z)}{R(z)}\,U_\mathrm{c}(z) - \frac{S(z)}{R(z)}\,Y(z)$$

should be chosen so that the plant transfer function

$$H(z) = \frac{Y(z)}{U(z)} = \frac{N(z)}{D(z)}$$

becomes, upon feedback loop closure, equal to a desired (or model) closed-loop transfer function

$$H_m(z) = \frac{Y(z)}{U_c(z)} = \frac{N_m(z)}{D_m(z)}$$

with perhaps some additional constraints to ensure, for instance, that the system rejects disturbances.

Finally, for the cases in which selecting all of the eigenvalues is more restrictive on the designer than is necessary to meet performance requirements, researchers are seeking algorithms which place the eigenvalues within a region rather than at precise locations. One such technique is briefly outlined.

23.2 THE STATE–SPACE APPROACH

The application of state–space methods in a feedback control paradigm is not always clear. Our attitude will be more general than the usual structure, as in Fig. 23.1. Here the plant and the measurement system are linear, but so far the processing of the measurements and of the input signals has not been specified. The system is taken as defined by

$$\mathbf{x}(k+1) = \mathbf{A}\mathbf{x}(k) + \mathbf{B}\mathbf{u}(k)$$

$$\mathbf{y}(k) = \mathbf{C}\mathbf{x}(k)$$

where the first is the dynamics equation and the second is the measurement equation.

In the course of this section we will gradually make more elaborate assumptions about the nature of the matrices and functions involved.

23.2.1 The simplest case

The basic case in design has the entire state fed back in the control of the system, with the control variable being a scalar. We explore this problem before proceeding to advanced versions.

In terms of Fig. 23.1, the basic case has $\hat{\mathbf{x}}(n) = \mathbf{x}(n)$ (called an identity observer) and $\mathbf{u}(n) = \mathbf{K}(\mathbf{r}(n) - \mathbf{x}(n))$, with $\mathbf{K}$ a gain to be determined. We build slowly to the general case by first having $u(n)$ be scalar and $\mathbf{A}$ being of controllable canonical form. Then we generalize $\mathbf{A}$, and finally we increase the dimension of $\mathbf{u}(n)$.

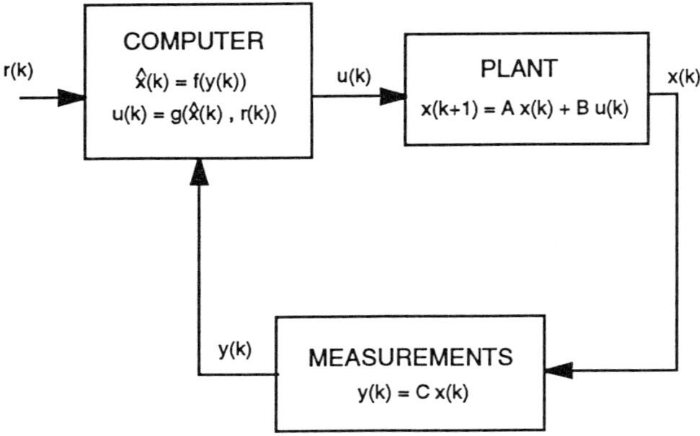

Figure 23.1 State–space control system representation for linear system and measurements with possibly non-linear controller.

Thus to start we consider the case $\mathbf{C} = \mathbf{I}$, so that the state is measured exactly, the plant has a single input (i.e. $m = 1$) and the dynamics are in controllable canonical form. Hence

$$\mathbf{x}(k+1) = \begin{bmatrix} -a_1 & -a_2 & \cdots & \cdots & & -a_n \\ 1 & 0 & 0 & 0 & \cdots & 0 \\ 0 & 1 & 0 & 0 & \cdots & 0 \\ & \vdots & & & & \vdots \\ 0 & 0 & 0 & 0 & 1 & 0 \end{bmatrix} \mathbf{x}(k) + \begin{bmatrix} 1 \\ 0 \\ 0 \\ \vdots \\ 0 \end{bmatrix} u(k) \qquad (23.1)$$

which has characteristic equation

$$\lambda^n + a_1 \lambda^{n-1} + a_2 \lambda^{n-2} + \cdots + a_n = 0$$

and poles which are the roots of this equation. We note that this system is obviously controllable from trivial computation of the controllability matrix. We choose the control law structure

$$u(n) = -\mathbf{K}\mathbf{x}(n) + g(\mathbf{r}(n))$$

If we wish for the poles to lie at $\lambda_1, \lambda_2, \ldots, \lambda_n$, this is the same as requiring the characteristic equation to be

$$\lambda^n + \alpha_1 \lambda^{n-1} + \alpha_2 \lambda^{n-2} + \cdots + \alpha_n = 0$$

$$= (\lambda - \lambda_1)(\lambda - \lambda_2) \cdots (\lambda - \lambda_n)$$

By direct substitution in (23.1) it is clear that choosing gain $\mathbf{K}$ such that

$$\mathbf{K} = [\, \alpha_1 - a_1 \quad \alpha_2 - a_2 \quad \cdots \quad \alpha_{n-1} - a_{n-1} \quad \alpha_n - a_n \,] \tag{23.2}$$

will give the closed-loop state equation description

$$\mathbf{x}(k+1) = \begin{bmatrix} -\alpha_1 & -\alpha_2 & \cdots\cdots\cdots & -\alpha_n \\ 1 & 0 & 0 \ 0 \ \cdots & 0 \\ 0 & 1 & 0 \ 0 \ \cdots & 0 \\ & \vdots & & \vdots \\ 0 & 0 & 0 \ 0 \ 1 & 0 \end{bmatrix} \mathbf{x}(k) + \begin{bmatrix} 1 \\ 0 \\ 0 \\ \vdots \\ 0 \end{bmatrix} g(\mathbf{r}(k))$$

Our design method is clear: choose the poles as real numbers and complex conjugate pairs (so all coefficients are real), form the desired characteristic equation, then form $\mathbf{K}$ as in (23.2).

Example

Consider the simple system

$$\mathbf{x}(k+1) = \begin{bmatrix} 1.5 & -0.54 \\ 1 & 0 \end{bmatrix} \mathbf{x}(k) + \begin{bmatrix} 1 \\ 0 \end{bmatrix} u(k)$$

with poles at -0.9, -0.6. We wish to have poles at $\pm\, 0.5$, giving a characteristic equation $\lambda^2 - 0.25 = 0$. Choosing

$$\mathbf{K} = [\, 0 + 1.5 \quad -0.25 - 0.54 \,] = [\, 1.5 \quad -0.79 \,]$$

gives

$$u(k) = [\, -1.5 \quad 0.79 \,]\, \mathbf{x}(k) + u_{\text{ref}}(k)$$

and

$$\mathbf{x}(k+1) = \left\{ \begin{bmatrix} 1.5 & -0.54 \\ 1 & 0 \end{bmatrix} + \begin{bmatrix} -1.5 & 0.79 \\ 0 & 0 \end{bmatrix} \right\} \mathbf{x}(k) + \begin{bmatrix} 1 \\ 0 \end{bmatrix} u_{\text{ref}}(k)$$

$$= \begin{bmatrix} 0 & 0.25 \\ 1 & 0 \end{bmatrix} \mathbf{x}(k) + \begin{bmatrix} 1 \\ 0 \end{bmatrix} u_{\text{ref}}(k)$$

which clearly has the required poles.

23.2.2 Single input systems of general form

The more general system $\mathbf{x}(n+1) = \mathbf{A}\mathbf{x}(n) + \mathbf{b}u(n)$ is easily handled using similarity transformations. This is because similarity transformations do not affect eigenvalues. Thus let $\mathbf{P}$ define a change in coordinates

$$\mathbf{w}(n) = \mathbf{P}\mathbf{x}(n)$$

Then

$$\mathbf{w}(n+1) = \mathbf{P}\mathbf{A}\mathbf{P}^{-1}\mathbf{w}(n) + \mathbf{P}\mathbf{b}u(n) \tag{23.3}$$

We remember that

$$\det(\mathbf{P}\mathbf{A}\mathbf{P}^{-1} - \lambda\mathbf{I}) = \det(\mathbf{P})\det(\mathbf{A} - \lambda\mathbf{I})\det(\mathbf{P}^{-1})$$

$$= \det(\mathbf{A} - \lambda\mathbf{I})$$

so that the characteristic equation is not affected by the transformation. We select $\mathbf{P}$ such that the system is in one of the controllable canonical forms (23.1) and then choose $\mathbf{K}$ as in section

23.2.1, giving a control law

$$u(n) = [\, a_1 - \alpha_1 \;\; \cdots \;\; a_{n-1} - \alpha_{n-1} \;\; a_n - \alpha_n \,]\, \mathbf{w}(n) + g(\mathbf{r}(n)) \qquad (23.4)$$

Remembering the initial transformation, the control law for the original representation is

$$u(n) = -\, \mathbf{K}^{\#} \mathbf{x}(n) + g(\mathbf{r}(n))$$

$$= [\, a_1 - \alpha_1 \;\; a_2 - \alpha_2 \;\; \cdots \;\; a_n - \alpha_n \,]\, \mathbf{P}\, \mathbf{x}(n) + g(\mathbf{r}(n))$$

To find $\mathbf{P}$, we compute the controllability matrix for a transformed system (23.3)

$$\mathbf{Q}_{\mathrm{cp}} = [\, \mathbf{PB} \;\vdots\; (\mathbf{PAP}^{-1})\mathbf{PB} \;\vdots\; \ldots \;\vdots\; (\mathbf{PAP}^{-1})^{n-1}\mathbf{PB} \,]$$

and note that this has the form

$$\mathbf{Q}_{\mathrm{cp}} = \mathbf{P}\,[\, \mathbf{B} \;\vdots\; \mathbf{AB} \;\vdots\; \ldots \;\vdots\; \mathbf{A}^{n-2}\mathbf{B} \;\vdots\; \mathbf{A}^{n-1}\mathbf{B} \,] = \mathbf{P}\mathbf{Q}_{\mathrm{c}}$$

From this form follows the relationship $\mathbf{P} = \mathbf{Q}_{\mathrm{cp}}\mathbf{Q}_{\mathrm{c}}^{-1}$. Thus our algorithm is as follows.

1. Compute the characteristic equation of $\mathbf{A}$.
2. Compute the controllability matrix $\mathbf{Q}_{\mathrm{c}}$ of the original system ($\mathbf{x}(n+1) = \mathbf{A}\mathbf{x}(n) + \mathbf{b}u(n)$) and the controllability matrix of the controllable form (23.1), which is easy because we can simply write down the canonical form from the characteristic equation.
3. Compute $\mathbf{P} = \mathbf{Q}_{\mathrm{cp}}\,\mathbf{Q}_{\mathrm{c}}^{-1}$. Note that the invertability of $\mathbf{Q}_{\mathrm{c}}$ implies that the system must be controllable for this inverse to exist and hence for us to design the controller.
4. Compute the desired characteristic equation and then the gain $\mathbf{K}^{\#}$ from (23.4).

Example

We demonstrate the above with the system

$$\mathbf{x}(k+1) = \begin{bmatrix} 1 & 0.3 \\ -0.2 & 0.5 \end{bmatrix} \mathbf{x}(k) + \begin{bmatrix} 0.5 \\ 1 \end{bmatrix} u(k)$$

We find that

$$\mathbf{Q}_c = \begin{bmatrix} 0.5 & 0.8 \\ 1 & 0.4 \end{bmatrix} \qquad \mathbf{Q}_c^{-1} = \begin{bmatrix} -2/3 & 4/3 \\ 5/3 & -5/6 \end{bmatrix}$$

and the characteristic equation is

$$\lambda^2 - 1.5\,\lambda + 0.56 = 0$$

Then with or without writing out the canonical form system, we have

$$\mathbf{Q}_{cp} = \begin{bmatrix} 1 & 1.5 \\ 0 & 1 \end{bmatrix}$$

and hence

$$\mathbf{P} = \begin{bmatrix} 11/6 & 1/12 \\ 5/3 & -5/6 \end{bmatrix} \qquad \mathbf{P}^{-1} = \begin{bmatrix} 1/2 & 1/20 \\ 1 & -11/10 \end{bmatrix}$$

If we choose to have the poles at ± 0.5, then

$$u(k) = [\,-1.5 + 0 \quad 0.56 + 0.25\,]\,\mathbf{P}\,\mathbf{x}(k) + u_{\text{ref}}(k)$$

$$= [\,-1.4 \quad -0.8\,]\,\mathbf{x}(k) + u_{\text{ref}}(k)$$

Using this control, the closed-loop system has the apparent state–space model

$$\mathbf{x}(k+1) = \begin{bmatrix} 0.3 & -0.1 \\ -1.6 & -0.3 \end{bmatrix}\mathbf{x}(k) + \begin{bmatrix} 0.5 \\ 1 \end{bmatrix} u_{\text{ref}}(k)$$

This is easily shown to have the correct poles.

Manipulation of the transformations and restriction to the special case of $m = \dim(\mathbf{u}) = 1$ allows Ackermann's formula to be used. This is given by

$$K = [\, 0 \; 0 \; ... \; 0 \; 1 \,] \, Q_c^{-1} \, [\, A^n + \alpha_1 A^{n-1} + \alpha_2 A^{n-2} + \cdots + \alpha_n I \,] \quad (23.5)$$

and is proven in Appendix B.

Example

We repeat the above example using Ackermann's formula. Since it is desired that the characteristic equation be $\lambda^2 - 0.25 = 0$, we have $\alpha_1 = 0$ and $\alpha_2 = -0.25$. Hence we easily compute

$$A^2 + \alpha_1 A + \alpha_2 = \begin{bmatrix} 0.69 & 0.45 \\ -0.3 & -0.6 \end{bmatrix}$$

Q_c^{-1} was computed in the previous example, and we use that to find

$$[\, 0 \quad 1 \,] \, Q_c^{-1} = [\, 1.6667 \quad -0.8333 \,]$$

Finally, completing the computation of (23.5)

$$K = [\, 0 \quad 1 \,] \, Q_c^{-1} \, [\, A^2 + \alpha_1 A + \alpha_2 \,] = [\, 1.4 \quad 0.8 \,]$$

Except for the inversion of the controllability matrix, this is clearly a quite straightforward computation to perform.

The above is fairly direct and simple. We must remember, however, that we have not addressed several problems, including where to place all those eigenvalues; this problem was mentioned in Chapter 19, and a possible solution approach is introduced in section 23.7.

23.2.3 Multiple input systems

We now consider the general case $x(k+1) = Ax(k) + Bu(k)$ where the control $u(k)$ is an m-vector. If there is a column B_i of the matrix B such that $(A; B_i)$ is a controllable pair, then we can obviously do a scalar design, using only the ith component of u in the command, as in the preceding section. The result of course is

$$\mathbf{u}(j) = - \begin{bmatrix} 0 \\ \vdots \\ 0 \\ K \\ 0 \\ \vdots \\ 0 \end{bmatrix} \mathbf{x}(j) + \mathbf{g}(\mathbf{r}(j)) \tag{23.6}$$

If $(\mathbf{A}; \mathbf{B})$ is controllable only when more than one column of $\mathbf{B}$ is used, however, then our design task is somewhat more difficult. The method usually presented is to find a scalar sequence $u_s(k) = \mathbf{q}^T\mathbf{x}(k)$ and an allocation vector $\mathbf{p}$ of dimension $m \times 1$, such that

1. $(\mathbf{A}; \mathbf{Bp})$ is controllable;
2. $(\mathbf{A} - \mathbf{Bpq}^T)$ has its eigenvalues at the required pole locations; and then let
3. $\mathbf{u}(k) = - \mathbf{pq}^T \mathbf{x}(k) + \mathbf{g}(\mathbf{r}(k))$

As presented by Kuo (1980), the choice of $\mathbf{p}$ is done first and is nearly arbitrary, whereas Willems and Mitter (1971) present a constructive method. We look first at the former, then the latter, and then make comments.

23.2.4 Semi-arbitrary allocation

Following Kuo, choose the m-vector $\mathbf{p}$ arbitrarily but such that $(\mathbf{A}; \mathbf{Bp})$ is a controllable pair, and let $\mathbf{b}_p$ denote the n-vector $\mathbf{Bp}$. Design, as in section 23.2.2, a gain vector $\mathbf{q}$ such that the system

$$\mathbf{x}(k+1) = \mathbf{A}\,\mathbf{x}(k) + \mathbf{b}_p\,u_s(k)$$

has eigenvalues in the desired locations. Then clearly the law

$$u_s(k) = - \mathbf{q}^T\mathbf{x}(k)$$

gives

$$\mathbf{x}(k+1) = (\mathbf{A} - \mathbf{b}_p\,\mathbf{q}^T)\,\mathbf{x}(k)$$

with the proper poles. But from the notation, this is

$$\mathbf{x}(k+1) = (\mathbf{A} - \mathbf{Bpq}^{\mathrm{T}}) \, \mathbf{x}(k)$$

and is the same result as choosing

$$\mathbf{u}(k) = -\mathbf{pq}^{\mathrm{T}}\mathbf{x}(k)$$

$$= -\mathbf{Kx}(k)$$

in the original problem.

It seems as if almost any allocation vector $\mathbf{p}$ will do in the above, and we observe that allocation vector is an appropriate name since in fact the design of the feedback is for a scalar control relationship $u_s(k)$, which is allocated among the components of the commands $\mathbf{u}(k)$ using $\mathbf{p}$. Notice that $\mathbf{K}$ has rank 1 in this scheme – there has been no need for complicated control relationships.

23.2.5 Systematic construction

Compared to the above, the method of Willems and Mitter (1971) is similar philosophically but more systematic. They do the following.

1. Construct an $m \times n$ matrix $\mathbf{L}$ such that $(\mathbf{A} - \mathbf{BL}; \mathbf{b})$ is controllable for some arbitrary $\mathbf{b} = \mathbf{Bp}$ where $\mathbf{p}$ is arbitrary but non-trivial.
2. Find a gain $\mathbf{q}$ such that the eigenvalues of $(\mathbf{A} - \mathbf{BL} + \mathbf{Bpq}^{\mathrm{T}})$ are as desired.
3. Let the feedback control law be

$$\mathbf{u}(k) = (-\mathbf{L} + \mathbf{pq}^{\mathrm{T}}) \, \mathbf{x}(k)$$

The key part of the approach is the construction of a matrix $\mathbf{L}$, which proceeds as follows. Let $\mathbf{B}_i$ denote the ith column of $\mathbf{B}$ and let k_1 be the largest index such that

$$\mathbf{Q}_1 = [\, \mathbf{B}_1 \mathrel{\vdots} \mathbf{AB}_1 \mathrel{\vdots} \ldots \mathrel{\vdots} \mathbf{A}^{k_1-1}\mathbf{B}_1 \,]$$

has full rank. If this rank is less than n, then let k_2 be the largest index for which

$$\mathbf{Q}_2 = [\ \mathbf{Q}_1 \ \vdots \ \mathbf{B}_2 \ \vdots \ \mathbf{A}\mathbf{B}_2 \ \vdots \ ... \ \vdots \ \mathbf{A}^{k_2-1}\mathbf{B}_2\ ]$$

has full rank. If this rank is less than n, let k_3 be the largest index for which

$$\mathbf{Q}_3 = [\ \mathbf{Q}_2 \ \vdots \ \mathbf{B}_3 \ \vdots \ \mathbf{A}\mathbf{B}_3 \ \vdots \ ... \ \vdots \ \mathbf{A}^{k_3-1}\mathbf{B}_3\ ]$$

has full rank. Continue this until, for some $r \le m$ and sequence $k_1, k_2, ..., k_r$, we reach

$$\mathbf{Q}_r = [\ \mathbf{Q}_{r-1} \ \vdots \ \mathbf{B}_r \ \vdots \ \mathbf{A}\mathbf{B}_r \ \vdots \ ... \ \vdots \ \mathbf{A}^{k_r-1}\mathbf{B}_r\ ]$$

which has rank n. By the method of construction, $\mathbf{Q}_r$ will be of dimension $n \times n$ and non-singular. Let $\mathbf{Q} = \mathbf{Q}_r$.

Continuing, we now define the selection matrix $\mathbf{S}$, an $m \times n$ matrix, as the matrix with all zeros except for a 1 in the $(j+1)$st row of the r_jth column, where

$$r_j = \sum_{i=1}^{j} k_i \qquad\qquad j = 1, 2, ..., r-1$$

In matrix terms

$$\mathbf{S}_{r_j} = \begin{cases} \mathbf{e}(j+1) & \text{for } r_j = \sum_{i=1}^{j} k_i \quad j = 1, 2, ..., r-1 \\[2mm] 0 & \text{otherwise} \end{cases}$$

where $\mathbf{e}(j)$ is the natural unit vector with jth element equal to 1 and all others equal to 0. We remark that $\Sigma_{i=1}^{r} k_i = n$ from the nature of the construction.

Using $\mathbf{S}$ as defined above, we finally define $\mathbf{L} = \mathbf{S}\mathbf{Q}^{-1}$ for which it can be shown that $(\mathbf{A} - \mathbf{B}\mathbf{L}; \mathbf{B}_i)$ is a controllable pair for any non-zero $\mathbf{B}_i$. We then can safely form an allocation vector $\mathbf{p}$ so that, provided $\mathbf{B}\mathbf{p}$ is non-zero, we can place the eigenvalues of $(\mathbf{A} - \mathbf{BL} + \mathbf{Bp}\mathbf{q}^{\mathrm{T}})$ as we wish with choice of the gain vector $\mathbf{q}$ for the system

$$\mathbf{x}(k+1) = (\mathbf{A} - \mathbf{BL})\,\mathbf{x}(k) + \mathbf{Bp}\,w(k)$$

i.e.

$$w(k) = \mathbf{q}^{\mathrm{T}}\mathbf{x}(k)$$

and hence

$$\mathbf{u}(k) = (-\mathbf{L} + \mathbf{pq}^{\mathrm{T}})\,\mathbf{x}(k)$$

Observations

We now make some observations about both of the above schemes. First, the underlying problem allows us to choose the elements of an $n \times m$ matrix, i.e. to choose nm scalars. We are using this freedom to place n scalars $-$ the eigenvalues of the closed-loop system. Neither of the schemes presented uses the other freedom available. Finally, the Willems and Mitter approach could be used with a 'guessed' $\mathbf{L}$ matrix, since design only needs $(\mathbf{A} - \mathbf{BL}; \mathbf{B}_i)$ controllable for some $\mathbf{B}_i$ whereas their approach allows a more general $\mathbf{b} = \mathbf{Bp}$ for the allocation.

In both schemes, the allocation vector $\mathbf{p}$ can be chosen nearly arbitrarily. It turns out that the actual choice will define things like the closed-loop zeros and the gain magnitudes, neither of which is explicitly addressed by the design methods. One might think that this constitutes a waste of design flexibility.

Example

We consider the system

$$\mathbf{x}(k+1) = \begin{bmatrix} 1 & 1 & 0 \\ 0 & 1 & 0 \\ 0 & 0 & 1 \end{bmatrix} \mathbf{x}(k) + \begin{bmatrix} 0 & 0 \\ 1 & 0 \\ 0 & 1 \end{bmatrix} \mathbf{u}(k)$$

It is easy to show that this is controllable, but that both elements of the control are necessary to obtain this controllability. The eigenvalues are all at $z = 1$. The controllability matrix for $\mathbf{B}_1$ is

$$\begin{bmatrix} 0 & 1 & 2 \\ 1 & 1 & 1 \\ 0 & 0 & 0 \end{bmatrix} \quad \text{so that} \quad \mathbf{Q}_1 = \begin{bmatrix} 0 & 1 \\ 1 & 1 \\ 0 & 0 \end{bmatrix}$$

This has rank 2 and hence $k_1 = 2$. Using this with the controllability matrix for $\mathbf{B}_2$, which is

$$\begin{bmatrix} 0 & 0 & 0 \\ 0 & 0 & 0 \\ 1 & 1 & 1 \end{bmatrix}$$

gives $k_2 = 1$ and hence $r = 2$. The $\mathbf{Q}_2$ matrix is

$$\mathbf{Q}_2 = \begin{bmatrix} 0 & 1 & 0 \\ 1 & 1 & 0 \\ 0 & 0 & 1 \end{bmatrix}$$

Then we may show from the definitions that

$$\mathbf{Q}_2^{-1} = \begin{bmatrix} -1 & 1 & 0 \\ 1 & 0 & 0 \\ 0 & 0 & 1 \end{bmatrix} \quad \mathbf{S} = \begin{bmatrix} 0 & 0 & 0 \\ 0 & 1 & 0 \end{bmatrix} \quad \mathbf{L} = \begin{bmatrix} 0 & 0 & 0 \\ 1 & 0 & 0 \end{bmatrix}$$

The system for our design work, if we select $\mathbf{b} = \mathbf{B}_1$, is

$$\mathbf{x}(k+1) = (\mathbf{A} - \mathbf{B}\mathbf{L})\,\mathbf{x}(k) + \mathbf{b}\mathbf{u}(k)$$

$$= \begin{bmatrix} 1 & 1 & 0 \\ 0 & 1 & 0 \\ -1 & 0 & 1 \end{bmatrix} \mathbf{x}(k) + \begin{bmatrix} 0 \\ 1 \\ 0 \end{bmatrix} \mathbf{u}(k)$$

This has controllability matrix

$$\mathbf{Q} = \begin{bmatrix} 0 & 1 & 2 \\ 1 & 1 & 1 \\ 0 & 0 & -1 \end{bmatrix}$$

We may now proceed as with any scalar input case to find the transformation to controllable form, etc. The transformation $\mathbf{z} = \mathbf{Tx}$ is

$$\mathbf{T} = \begin{bmatrix} 0 & 0 & 1 \\ 1 & 0 & -1 \\ 2 & 1 & -1 \end{bmatrix}$$

The present system has its eigenvalues all at +1 and its characteristic equation is

$$\lambda^3 - 3\lambda^2 + 3\lambda - 1 = 0$$

We would like the poles to be at, say, $0.5 \pm 0.5j$, -0.25, yielding a desired characteristic equation of

$$\lambda^3 - 0.75\lambda^2 + 0.25\lambda + 0.125 = 0$$

The resulting gain for the transformed system is $[-1.125 \ 2.75 \ -2.25]$, so that the gain for the transformed system is

$$\mathbf{K} = [\,-1.125 \quad 2.75 \quad -2.25\,] \, \mathbf{T}$$

$$= [\,-1.75 \quad -2.25 \quad 0.625\,]$$

Our control law then is

$$\mathbf{u}(k) = \left[\, -\mathbf{L} + \begin{bmatrix} \mathbf{K} \\ \mathbf{0} \end{bmatrix} \,\right] \mathbf{x}(k) + \mathbf{u}_{\mathrm{ref}}(k)$$

$$= -\mathbf{K}^{\#}\mathbf{x}(k) + \mathbf{u}_{\text{ref}}(k)$$

$$= \begin{bmatrix} -1.75 & -2.25 & 0.625 \\ -1 & 0 & 0 \end{bmatrix}\mathbf{x}(k) + \mathbf{u}_{\text{ref}}(k)$$

It can be shown (using computer programs, for example) that $\mathbf{x}(k+1) = [\mathbf{A} - \mathbf{BK}^{\#}]\mathbf{x}(k) + \mathbf{Bu}_{\text{ref}}(k)$ has the desired characteristic equation.

23.3 OUTPUT FEEDBACK BASED UPON STATE–SPACE ARGUMENTS

Since the entire state is rarely measured, it seems obvious that a useful design method would use feedback of only the measured quantities, i.e. that the control should be

$$\mathbf{u}(j) = \mathbf{Ky}(j) + \mathbf{r}(j)$$

$$= \mathbf{KCx}(j) + \mathbf{r}(j)$$

It turns out that, if $\mathbf{C}$ has rank p, then at least $n_{\text{e}} = \max(p, m)$ eigenvalues of the system can be arbitrarily placed. The problem is that the other $n - n_{\text{e}}$ eigenvalues may or may not be suitable, so although we present the methods of placement here for completeness, it seems preferable to use a dynamic feedback, as in the next section.

In this section, we first present a simple design method for a simple problem and then examine the method of Davison (1970) for more generality.

23.3.1 Scalar control with a canonical form

Consider the special case in which the system has controllable canonical form and the measurements are of p elements of the state vector. Then

$$\mathbf{x}(j+1) = \begin{bmatrix} 0 & \vdots & \mathbf{I} \\ -a_1 & -a_2 & \dots & -a_n \end{bmatrix}\mathbf{x}(j) + \begin{bmatrix} 0 \\ 1 \end{bmatrix}\mathbf{u}(j)$$

$$\mathbf{y}(j) = [\; \dots \; \mathbf{e}_1 \; \dots \; \mathbf{e}_2 \; \dots \; \mathbf{e}_3 \; \dots \; \mathbf{e}_p \;]\,\mathbf{x}(j)$$

where the p natural unit vectors $\mathbf{e}_i$ are in columns $1 \le r_i \le n$ and without loss of generality we take these as ordered with $1 \le r_1 < r_2 < \cdots < r_p \le n$. If we have

$$\mathbf{u}(j) = k_1 \mathbf{x}_{r_1}(j) + k_2 \mathbf{x}_{r_2}(j) + \cdots + k_p \mathbf{x}_{r_p}(j)$$

then (taking for illustrative purposes only that $r_1 > 1$ and $r_p < n$)

$$\mathbf{x}(j+1) = \begin{bmatrix} 0 & \vdots & \mathbf{I} \\ -a_1 \ldots & -a_{r_1}+k_1 \ldots & -a_{r_p}+k_p \ldots -a_n \end{bmatrix} \mathbf{x}(j) \tag{23.7}$$

is the closed-loop expression and the characteristic equation is of the form

$$\lambda^n + a_n \lambda^{n-1} + \cdots + (a_{r_p} - k_p)\lambda^{r_p-1} + \cdots (a_{r_1} - k_1)\lambda^{r_1-1} + \cdots + a_1 = 0 \tag{23.8}$$

We can in principle insert the desired p eigenvalues $\lambda_1, \lambda_2, \ldots, \lambda_p$ into (23.8) to create a set of p simultaneous linear equations in the p unknowns $k_1, k_2, \ldots, k_p$. Thus solve the set

$$\begin{bmatrix} 1 & \lambda_1 & \lambda_1^2 & \cdots & \lambda_1^n \\ 1 & \lambda_2 & \lambda_2^2 & \cdots & \lambda_2^n \\ & & & \vdots & \\ 1 & \lambda_p & \lambda_p^2 & \cdots & \lambda_p^n \end{bmatrix} \begin{bmatrix} a_1 \\ a_2 \\ \vdots \\ a_n \\ 1 \end{bmatrix} = \begin{bmatrix} \lambda_1^{r_1-1} & \cdots & \lambda_1^{r_p-1} \\ \lambda_2^{r_1-1} & \cdots & \lambda_2^{r_p-1} \\ & \ddots & \\ \lambda_p^{r_1-1} & \cdots & \lambda_p^{r_p-1} \end{bmatrix} \begin{bmatrix} k_1 \\ k_2 \\ \vdots \\ k_p \end{bmatrix} \tag{23.9}$$

The r.h.s. matrix is $p \times p$ and will be invertible except when we ask for repeated eigenvalues, 0 eigenvalues with all $r_i \ne 1$, or have some other exceptional case. In effect we are setting p of n coefficients in the characteristic equation, with the other $n - p$ due to the original system dynamics. This will force p eigenvalues to have required values but place the other $n - p$ beyond our control.

Example

Consider the system described by

$$\mathbf{x}(j+1) = \begin{bmatrix} 0 & 1 & 0 & 1 \\ 0 & 0 & 1 & 0 \\ 0 & 0 & 0 & 1 \\ 1 & -\frac{1}{2} & 1 & \frac{3}{2} \end{bmatrix} \mathbf{x}(j) + \begin{bmatrix} 0 \\ 0 \\ 0 \\ 1 \end{bmatrix} \mathbf{u}(j)$$

$$\mathbf{y}(j) = \begin{bmatrix} 0 & 1 & 0 & 0 \\ 0 & 0 & 1 & 0 \end{bmatrix} \mathbf{x}(j)$$

It is desired that the $n - p = 2$ settable poles be placed at $\pm\frac{1}{4}$ using the measurement feedback law

$$\mathbf{u}(j) = k_1 \, \mathbf{y}_1(j) + k_2 \, \mathbf{y}_2(j)$$

Inspection of the $\mathbf{C}$ matrix reveals that $r_1 = 2$, $r_2 = 3$. Applying this to (23.7) or using direct computation shows that we are placing two eigenvalues of

$$\begin{bmatrix} 0 & 1 & 0 & 1 \\ 0 & 0 & 1 & 0 \\ 0 & 0 & 0 & 1 \\ 1 & -\frac{1}{2}+k_1 & 1+k_2 & \frac{3}{2} \end{bmatrix}$$

The resulting characteristic equation is

$$\lambda^4 + \tfrac{3}{2}\lambda^3 - (1 + k_2)\lambda^2 - (k_1 - \tfrac{1}{2})\lambda - 1 = 0$$

Forcing this to have the roots $\pm\frac{1}{4}$ can be done in several ways. Using the formula (23.9) gives

$$
\begin{bmatrix} 1 & -1/4 & 1/16 & -1/64 & 1/256 \\ 1 & 1/4 & 1/16 & 1/64 & 1/256 \end{bmatrix}
\begin{bmatrix} -1 \\ 1/2 \\ -1 \\ -3/2 \\ 1 \end{bmatrix}
= \begin{bmatrix} -1/4 & 1/16 \\ 1/4 & 1/16 \end{bmatrix}
\begin{bmatrix} k_1 \\ k_2 \end{bmatrix}
$$

for which the solution is $k_1 = 13/32$, $k_2 = -271/16$. The resulting system eigenvalues are $\pm\frac{1}{4}$, $\frac{3}{4} \pm 3.8002\mathrm{j}$. Notice that two of these are the desired eigenvalues, but the other two are artefacts which in this case are unstable.

23.3.2 General systematic design of output feedback

The pole placement may be approached more systematically by using Davison's approach. Here we have the usual general formulation

$$
\mathbf{x}(j+1) = \mathbf{A}\mathbf{x}(j) + \mathbf{B}\mathbf{u}(j)
$$

$$
\mathbf{y}(j) = \mathbf{C}\mathbf{x}(j)
$$

where state $\mathbf{x}$, control $\mathbf{u}$, and measurement $\mathbf{y}$ are n-, m-, and p-dimensional vectors respectively. It is assumed for ease of presentation that $\mathbf{A}$ has distinct eigenvalues. Following Davison (1970) we place p eigenvalues arbitrarily using the following method.

1. Find the similarity transformation $\mathbf{T}$ which places $\mathbf{A}$ into diagonal form

$$
\Lambda = \mathbf{T}^{-1}\mathbf{A}\mathbf{T}
$$

2. Denote $\mathbf{B} = [\, \mathbf{B}_1 \,\vdots\, \mathbf{B}_2 \,\vdots\, \ldots \,\vdots\, \mathbf{B}_m \,]$ and create an allocation vector $\mathbf{q}$ with $q_1 = 1$ and such that the indicator vector

$$
\zeta = \mathbf{T}^{-1}\mathbf{B}\mathbf{q}
$$

has non-zero elements by choosing $\mathbf{q}_k$, $k=2,3,\ldots,m$, recursively such that, for $j = 1, 2, \ldots, n$, the jth elements, $\{\cdot\}_j$ satisfy

$$\{\mathbf{T}^{-1}\mathbf{B}_1 + \mathbf{q}_2\mathbf{T}^{-1}\mathbf{B}_2 + \cdots + \mathbf{q}_{k-1}\mathbf{T}^{-1}\mathbf{B}_{k-1}\}_j + \mathbf{q}_k\{\mathbf{T}^{-1}\mathbf{B}_k\} \neq 0$$

whenever either

$$\{\mathbf{T}^{-1}\mathbf{B}_1 + \mathbf{q}_2\mathbf{T}^{-1}\mathbf{B}_2 + \cdots + \mathbf{q}_{k-1}\mathbf{T}^{-1}\mathbf{B}_{k-1}\}_j \neq 0$$

or

$$\{\mathbf{T}^{-1}\mathbf{B}_k\}_j \neq 0$$

3. Let $\mathbf{K}^* = \mathbf{q}\mathbf{K}^{\mathrm{T}}$ and determine the gain $\mathbf{K}$ such that

$$\mathbf{x}(j+1) = \mathbf{A}\mathbf{x}(j) + \mathbf{B}\mathbf{q}\mathbf{w}(j)$$

$$\mathbf{y}(j) = \mathbf{C}\mathbf{x}(j)$$

$$\mathbf{w}(j) = \mathbf{K}^{\mathrm{T}}\mathbf{y}(j) \tag{23.10}$$

has p required eigenvalues, as follows.

(a) Place (23.10) into controllable canonical form using the transformation matrix $\mathbf{P}$ as in section 23.2.2, i.e. let $\eta(j) = \mathbf{P}\mathbf{x}(j)$, $\mathbf{y}(j) = \mathbf{C}\mathbf{P}^{-1}\eta(j)$, etc.
(b) We observe that

$$\mathbf{K}^{\mathrm{T}}\mathbf{C}\mathbf{P}^{-1} = [\, d_1 \quad d_2 \quad \ldots \quad d_n \,]$$

gives a characteristic equation

$$\lambda^n - (a_n + d_n)\lambda^{n-1} - \cdots - (a_1 + d_1) = 0$$

where the d_i are not necessarily independent (since $\mathbf{K}^{\mathrm{T}}$ is $1 \times p$).

Then for the actual algorithm, for the desired eigenvalues $\lambda_1, \lambda_2, .., \lambda_p$ we define the Vandermonde matrix

$$
\mathbf{L} = \begin{bmatrix}
1 & 1 & 1 & \cdots & 1 \\
\lambda_1 & \lambda_2 & \lambda_3 & & \lambda_p \\
\lambda_1^2 & \lambda_2^2 & \lambda_3^2 & \cdots & \lambda_p^2 \\
\vdots & \vdots & \vdots & \vdots & \vdots \\
\lambda_1^{n-1} & \lambda_2^{n-1} & \lambda_3^{n-1} & \cdots & \lambda_p^{n-1}
\end{bmatrix}
$$

$$
\boldsymbol{\delta}^{\mathrm{T}} = [\, \delta_1 \quad \delta_2 \quad \cdots \quad \delta_p \,]
$$

where $\delta_i = \lambda_i^n - a_n \lambda_i^{n-1} - a_{n-1} \lambda_i^{n-2} - \cdots - a_1$ is the 'error' in the original characteristic equation.

(c)   Define the $p \times p$ matrix $\mathbf{S}$ by $\mathbf{S} = \mathbf{C}\mathbf{P}^{-1}\mathbf{L}$. Then $\mathbf{K}^{\mathrm{T}} = \boldsymbol{\delta}^{\mathrm{T}}\mathbf{S}^{-1}$. If $\mathbf{S}$ is singular, the desired eigenvalues λ_i are to be perturbed slightly until it is non-singular.

4.   Finally, $\mathbf{K}^* = \mathbf{q}\mathbf{K}^{\mathrm{T}}$ as required.

Example

As an example, we use two simple systems with first-order dynamics whose sum of outputs is the only measurement. Thus

$$
\mathbf{x}(k+1) = \begin{bmatrix} 0.9 & 0 \\ 0 & 0.6 \end{bmatrix} \mathbf{x}(k) + \begin{bmatrix} 1 & 0 \\ 0 & 1 \end{bmatrix} \mathbf{u}(k)
$$

$$
\mathbf{y}(k) = [\, 1 \quad 1 \,]\, \mathbf{x}(k)
$$

This system is controllable and observable and in fact is in diagonal form already. Hence the transformation $\mathbf{T}$ is the identity matrix $\mathbf{I}$, $\mathbf{T}^{-1} = \mathbf{I}$, and

$$
\mathbf{T}^{-1}\mathbf{B}_1 = \begin{bmatrix} 1 \\ 0 \end{bmatrix} \qquad \mathbf{T}^{-1}\mathbf{B}_2 = \begin{bmatrix} 0 \\ 1 \end{bmatrix} \qquad \mathbf{q} = \begin{bmatrix} 1 \\ \theta \end{bmatrix} \quad \text{where } \theta \neq 0
$$

We choose $\theta = 1$. The system for pole placement becomes

$$\mathbf{x}(k+1) = \begin{bmatrix} 0.9 & 0 \\ 0 & 6 \end{bmatrix} \mathbf{x}(k) + \begin{bmatrix} 1 \\ 1 \end{bmatrix} \mathbf{w}(k)$$

with $\mathbf{w}(k) = \mathbf{K}^{\mathrm{T}}\mathbf{y}(k)$. As is usual, we find

$$\mathbf{P} = \mathbf{Q}_{\mathrm{p}}\mathbf{Q}_c^{-1} = \begin{bmatrix} 10/3 & -10/3 \\ 3 & -2 \end{bmatrix} \qquad \lambda^2 - 1.5\lambda + 0.54 = 0$$

Then we compute, if we choose to have a pole at 0.3,

$$\mathbf{L} = \begin{bmatrix} 1 \\ 0.3 \end{bmatrix}$$

$$\delta_1 = (0.3)^2 - 1.5\,(0.3) + 0.54 = 0.18$$

$$\mathbf{S} = \begin{bmatrix} 1 & 1 \end{bmatrix} \begin{bmatrix} -3/5 & 1 \\ -9/10 & 1 \end{bmatrix} \begin{bmatrix} 1 \\ 0.3 \end{bmatrix} = [-0.9] \quad \mathbf{S}^{-1} = [-1/0.9]$$

Then $\mathbf{K} = [0.18/(-0.9)] = [-0.2]$. Finally

$$\mathbf{K}^{\#} = \mathbf{q}\mathbf{K}^{\mathrm{T}} = \begin{bmatrix} -0.2 \\ -0.2 \end{bmatrix}$$

With $\mathbf{u}(k) = \mathbf{K}^{\#}\mathbf{y}(k) + \mathbf{u}_{\mathrm{ref}}(k) = \mathbf{K}^{\#}\mathbf{C}\mathbf{x}(k) + \mathbf{u}_{\mathrm{ref}}(k)$, it can be shown that the poles are at 0.3, 0.8. It should be emphasized that a different choice of θ in $\mathbf{q} = [1 \quad \theta]^{\mathrm{T}}$ would have yielded a different second pole.

We should notice that the allocation vector is once again virtually arbitrary and, as claimed in the example, the choice of this will affect the location of the unplaced poles.

23.4 THE STATE–SPACE APPROACH TO STABILIZATION OF NON-CONTROLLABLE SYSTEMS

A controllable system can always be stabilized, since its poles may be placed as we wish, namely, inside the unit circle. The question arises as to whether a system which is not controllable can be stabilized, and the answer is 'sometimes'. We investigate this using a variation of Wonham's (1967) arguments.

The system, as usual, is taken as described by

$$\mathbf{x}(k+1) = \mathbf{A}\mathbf{x}(k) + \mathbf{B}\mathbf{u}(k)$$

but in this instance we assume that $(\mathbf{A}; \mathbf{B})$ is not controllable, so that

$$\mathbf{Q}_c = [\ \mathbf{B} \ \vdots \ \mathbf{AB} \ \vdots \ \mathbf{A}^2\mathbf{B} \ \vdots \ \ldots \ \vdots \ \mathbf{A}^{n-1}\mathbf{B}\ ]$$

does not have full rank $(= n)$. We let $\mathbf{a}(\mathbf{A})$ be the minimal polynomial of $\mathbf{A}$, i.e. the polynomial of least degree for which

$$\mathbf{a}(\mathbf{A}) = \mathbf{0}$$

This will have the same factors as the characteristic polynomial of $\mathbf{A}$, but some of the factors may have lower powers. Now factor $\mathbf{a}(\cdot)$ into a product of two polynomials, one with zeros inside the unit circle, called $\mathbf{a}_-$, and one called $\mathbf{a}_+$ with zeros on or outside the unit circle. Then $\mathbf{a}(\lambda) = \mathbf{a}_-(\lambda)\,\mathbf{a}_+(\lambda)$.

Letting

$$\mathbf{E}_+ = \{\ \mathbf{x} \mid \mathbf{a}_+(\mathbf{A})\mathbf{x} = 0\ \}$$

$$\mathbf{E}_- = \{\ \mathbf{x} \mid \mathbf{a}_-(\mathbf{A})\mathbf{x} = 0\ \}$$

$$\mathbf{Q}_R = \{\ \mathbf{x} \mid \text{For some } \alpha,\ \mathbf{x} = \mathbf{Q}_c\,\alpha;\ \text{i.e. } \mathbf{x} \text{ in range of } \mathbf{Q}_c\}$$

$$\mathbf{P}_\perp = \text{projection matrix of } \mathbf{x} \text{ onto } \mathbf{E}_+$$

$$\mathbf{C}_R = \text{matrix such that } \mathbf{A}\mathbf{P}_\perp + \mathbf{P}_\perp\mathbf{B}\mathbf{C}_R \text{ has desired eigenvalues}$$

then the results are as follows:

1. $(\mathbf{A};\mathbf{B})$ is stabilizable iff $\mathbf{E}_+$ is a subset of $\mathbf{Q}_R$;
2. $(\mathbf{A}\,\mathbf{P}_\perp;\mathbf{P}_\perp\mathbf{B})$ is controllable; and
3. $\mathbf{u}(k) = \mathbf{C}_R\,\mathbf{P}_\perp\mathbf{x}(k)$ is a stabilizing control yielding the desired eigenvalues, where the number setable equals the rank of the controllability matrix $\mathbf{Q}_c$.

To see how this comes about, we must consider the projection of a general $\mathbf{x}(k)$ onto $\mathbf{E}_+$ and $\mathbf{E}_-$, with orthogonal components $\mathbf{y}(k)$ and $\mathbf{z}(k)$, respectively.

In fact, as seen by considering the definitions and the section on projection (where $\mathbf{a}_{++}(\mathbf{A})$ equals full rank row submatrix of $\mathbf{a}_+\,(\mathbf{A})$),

$$\mathbf{P}_\perp = \mathbf{I} - \mathbf{a}_{++}(\mathbf{A})^T\,[\ \mathbf{a}_{++}(\mathbf{A})\,\mathbf{a}_{++}(\mathbf{A})^T\ ]^{-1}\,\mathbf{a}_{++}(\mathbf{A})$$

Then we quickly find

$$\mathbf{y}(k+1) = \mathbf{P}_\perp\mathbf{x}(k+1) = \mathbf{P}_\perp(\mathbf{A}\mathbf{x}(k) + \mathbf{B}\mathbf{u}(k))$$

$$= \mathbf{P}_\perp\mathbf{A}\,(\mathbf{y}(k) + \mathbf{z}(k)) + \mathbf{P}_\perp\mathbf{B}\mathbf{u}(k)$$

Now note that if $\mathbf{y}$ belongs to $\mathbf{E}_+$, then $\mathbf{A}\mathbf{y}$ does also, since $\mathbf{a}_+(\mathbf{A})\mathbf{A}\mathbf{y} = \mathbf{A}\mathbf{a}_+(\mathbf{A})\mathbf{y}$ when $\mathbf{a}_+$ is a polynomial. Hence $\mathbf{P}_\perp\mathbf{A} = \mathbf{A}\mathbf{P}_\perp$, and $\mathbf{P}_\perp\mathbf{A}\mathbf{P}_- = 0$ (where $\mathbf{P}_- = \mathbf{I} - \mathbf{P}_\perp$). Using this information,

$$\mathbf{y}(k+1) = \mathbf{A}\mathbf{P}_\perp\mathbf{y}(k) + \mathbf{P}_\perp\mathbf{B}\mathbf{u}(k)$$

$$\mathbf{z}(k+1) = \mathbf{A}\mathbf{P}_-\mathbf{z}(k) + \mathbf{P}_-\mathbf{B}\mathbf{u}(k)$$

Since $\mathbf{P}_-$ is also $\mathbf{A}$-invariant, $\mathbf{A}\mathbf{P}_-$ is stable.

The controllability matrix for the $\mathbf{y}$-subspace is

$$\mathbf{Q}_y = [\ \mathbf{P}_\perp\mathbf{B}\ \vdots\ \mathbf{A}\mathbf{P}_\perp\mathbf{P}_\perp\mathbf{B}\ \vdots\ \ldots\ \vdots\ (\mathbf{A}\mathbf{P}_\perp)^{n-1}\mathbf{P}_\perp\mathbf{B}\ ]$$

$$= [\ \mathbf{P}_\perp\mathbf{B}\ \vdots\ \mathbf{P}_\perp\mathbf{A}\mathbf{B}\ \vdots\ \ldots\ \vdots\ \mathbf{P}_\perp\mathbf{A}^{n-1}\mathbf{B}\ ]$$

because $\mathbf{P}_\perp\mathbf{P}_\perp = \mathbf{P}_\perp$ and $\mathbf{P}_\perp\mathbf{A} = \mathbf{A}\mathbf{P}_\perp$, etc.

Now we see that, iff $\mathbf{E}_+\subset\mathbf{Q}_R$, then $\mathbf{E}_+=\mathbf{P}_\perp\mathbf{Q}_R$. But then if $\mathbf{y}\in\mathbf{E}_+$, there exists $\mathbf{q}$ such that $\mathbf{y} = \mathbf{P}_\perp \mathbf{Q}_R \, \mathbf{q}$, i.e. $\mathbf{y}$ is in the space spanned by

$$[\, \mathbf{P}_\perp \mathbf{Q}_R \,] = [\, \mathbf{P}_\perp \mathbf{B} \,\vdots\, \mathbf{P}_\perp \mathbf{AB} \,\vdots\, \dots \,\vdots\, \mathbf{P}_\perp \mathbf{A}^{n-1} \mathbf{B} \,]$$

Hence by comparing to $\mathbf{Qy}$ we see that any $\mathbf{y}$ in $\mathbf{E}_+$ is controllable with the pair $(\mathbf{AP}_\perp; \mathbf{P}_\perp \mathbf{B})$.

Choose $\mathbf{C}_R$ such that $\mathbf{AP}_\perp + \mathbf{P}_\perp \mathbf{BC}_R$ has desired eigenvalues (we can choose $n_\perp$ eigenvalues, where $n_\perp = \mathrm{rank}(\mathbf{P}_\perp)$). Then letting $\mathbf{C} = \mathbf{C}_R \mathbf{P}_\perp$ and choosing $\mathbf{u}(k) = \mathbf{Cx}(k)$ gives

$$\mathbf{y}(k+1) = \mathbf{AP}_\perp \mathbf{y}(k) + \mathbf{P}_\perp \mathbf{BC}(\mathbf{y}(k) + \mathbf{z}(k))$$

$$= (\mathbf{AP}_\perp + \mathbf{P}_\perp \mathbf{BC}_R)\, \mathbf{y}(k)$$

$$\mathbf{z}(k+1) = \mathbf{A}\, \mathbf{P}_- \mathbf{z}(k) + \mathbf{P}_- \mathbf{BC}(\mathbf{y}(k) + \mathbf{z}(k))$$

$$= \mathbf{AP}_- \mathbf{z}(k) + \mathbf{P}_- \mathbf{B}\, \mathbf{C}_R \mathbf{y}(k)$$

Written in matrix notation,

$$\begin{bmatrix} \mathbf{y}(k+1) \\ \mathbf{z}(k+1) \end{bmatrix} = \begin{bmatrix} \mathbf{AP}_\perp + \mathbf{P}_\perp \mathbf{BC}_R & \mathbf{0} \\ \mathbf{P}_- \mathbf{BC}_R & \mathbf{AP} \end{bmatrix} \begin{bmatrix} \mathbf{y}(k) \\ \mathbf{z}(k) \end{bmatrix}$$

this is clearly stable. Intuitively, the part which is not controllable is $\{\mathbf{z}(k)\}$, and this is stable by itself; the part which is controllable is $\{\mathbf{y}(k)\}$, and we placed the poles of this part and kept them from interfering with the stability of $\{\mathbf{z}(k)\}$.

Example

Consider the trivial system

$$\mathbf{x}(k+1) = \begin{bmatrix} 1.5 & 1 \\ 0 & -0.5 \end{bmatrix} \mathbf{x}(k) + \begin{bmatrix} 1 \\ 0 \end{bmatrix} \mathbf{u}(k)$$

which has poles at -0.5 and 1.5 and, with controllability matrix

$$\mathbf{Q}_c = \begin{bmatrix} 1 & 1.5 \\ 0 & 0 \end{bmatrix}$$

having rank 1, is not controllable. The minimum polynomial is the characteristic polynomial

$$a(\lambda) = (\lambda + 0.5)\,(\lambda - 1.5)$$

and hence

$$a_+(\lambda) = \lambda - 1.5$$
$$a_-(\lambda) = \lambda - 0.5$$

From the definitions,

$$\mathbf{E}_+ = \{\mathbf{x} \mid \mathbf{x}_2 = 0\}$$
$$\mathbf{E}_- = \{\mathbf{x} \mid \mathbf{x}_2 = -2\mathbf{x}_1\}$$
$$\mathbf{Q}_R = \{\mathbf{x} \mid \mathbf{x}_2 = 0\}$$

We see that $\mathbf{Q}_R = \mathbf{E}_+$, so stabilizability is expected. Continuing,

$$\mathbf{a}_+(\mathbf{A}) = \begin{bmatrix} 0 & 1 \\ 0 & 2 \end{bmatrix} \qquad \mathbf{a}_{++}(\mathbf{A}) = \begin{bmatrix} 0 & 1 \end{bmatrix} \qquad \mathbf{P}_\perp = \begin{bmatrix} 1 & 0 \\ 0 & 0 \end{bmatrix}$$

From this we compute

$$\mathbf{A}\mathbf{P}_\perp = \begin{bmatrix} 1.5 & 0 \\ 0 & 0 \end{bmatrix} \qquad \mathbf{P}_\perp \mathbf{B} = \begin{bmatrix} 1 \\ 0 \end{bmatrix}$$

We choose to place one pole (rank of $\mathbf{P}_\perp$ is 1) at -0.25 using $\mathbf{C}_R$ in $(\mathbf{A}\mathbf{P}_\perp + \mathbf{P}_\perp \mathbf{B}\mathbf{C}_R)$. Since we are in this case seeking to place a pole of

$$\begin{bmatrix} 1.5 + c_1 & c_2 \\ 0 & 0 \end{bmatrix}$$

we see that the choice

$$\mathbf{C_R} = [\, -1.75 \quad c_2 \,]$$

is suitable for any c_2, so that the final gain is

$$\mathbf{C} = \mathbf{C_R}\,\mathbf{P_\perp} = [\, -1.75 \quad 0 \,]$$

Using this gain in the control loop, it is easy to compute that

$$\mathbf{x}(k+1) = \mathbf{A}\mathbf{x}(k) + \mathbf{B}\mathbf{u}(k)$$

$$\mathbf{u}(k) = \mathbf{C}\mathbf{x}(k) + \mathbf{u}_{\mathrm{in}}(k)$$

has

$$\mathbf{x}(k+1) = \begin{bmatrix} -0.25 & 1 \\ 0 & -0.5 \end{bmatrix} \mathbf{x}(k) + \begin{bmatrix} 1 \\ 0 \end{bmatrix} \mathbf{u}(k)$$

with poles at –0.25, –0.5. We notice that the problem had an obvious solution $\mathbf{C} = [-1.75 \; c_2]$ where the second component is arbitrary; the method used here was an 'overkill' demonstrating that the approach may be unduly restricting.

23.5 POLE AND ZERO PLACEMENT WITH TRANSFER FUNCTIONS

Pole placement can be performed with transfer function approaches as well as with state–space methods. A little reflection concerning root locus methods (Chapter 18) will perhaps indicate that the entire complex plane is potentially coverable by selection of compensator parameters.

23.5.1 General approach

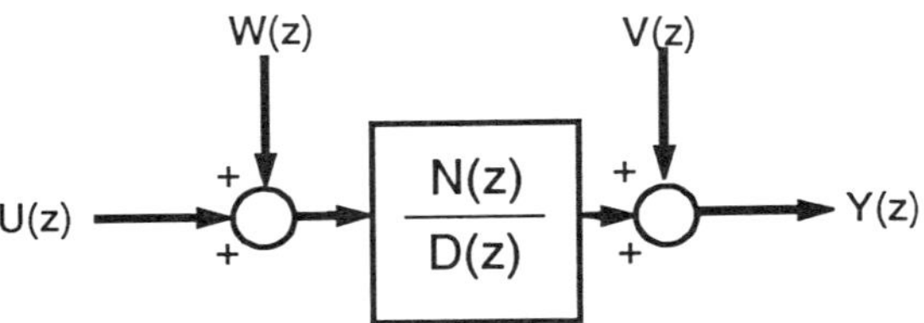

Figure 23.2 Transfer function plant representation, with (transforms of) dynamics noise w and measurement noise v.

The transfer function approach starts with a system model as in Fig. 23.2, in which the plant transfer function is modelled as

$$H(z) = \frac{Y(z)}{U(z)} = \frac{N(z)}{D(z)}$$

with $N(z)$ and $D(z)$ being polynomials in the z-transform variable (and forward shift operator) of degrees $\deg(N(z))$ and $\deg(D(z))$, respectively. By assumption $\deg(N(z)) \leq \deg(D(z))$, so that the system is causal.

It is desired that a controller be designed, with

$$U(z) = \frac{T(z)}{R(z)} U_c(z) - \frac{S(z)}{R(z)} Y(z)$$

so that the overall system transfer function

$$H_m(z) = \frac{Y(z)}{U_c(z)} = \frac{N_m(z)}{D_m(z)} \tag{23.11}$$

from command input to system output takes on the desired, or model, value $H_m(z)$. It is assumed that $\deg(N_m(z)) \leq \deg(D_m(z))$.

The design task is to specify the polynomials $T(z)$, $R(z)$, and $S(z)$ so that the requirements are met. It is required for causality of the resulting control law that

$$\deg(T(z)) \le \deg(R(z))$$

$$\deg(S(z)) \le \deg(R(z)) \tag{23.12}$$

and common choices are

$$\deg(T(z)) = \deg(S(z)) = \deg(R(z))$$

when the computer is fast enough to give no computational delay and

$$\deg(R(z)) = 1 + \deg(T(z)) = 1 + \deg(S(z))$$

when there is computational delay. Further constraints may be that disturbances have 'small' effects, a problem we address shortly.

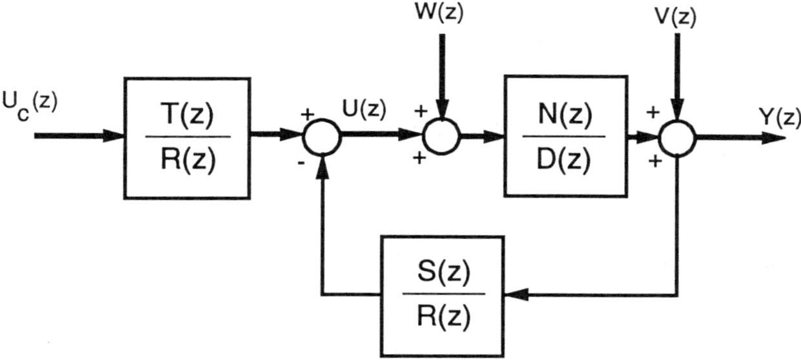

Figure 23.3 Closed-loop control system model for plant as in Fig. 23.2 and control law defined by $T(z)$, $S(z)$, and $R(z)$.

With the above control, the closed-loop system appears as in Fig. 23.3. From this it is easy to see that the closed-loop transfer function is given by

$$\frac{Y(z)}{U_c(z)} = \frac{N(z)\,T(z)}{D(z)\,R(z) + N(z)\,S(z)} \tag{23.13}$$

and the disturbance noise transfer functions are

$$\frac{Y(z)}{W(z)} = \frac{N(z)R(z)}{D(z)R(z) + N(z)S(z)} \tag{23.14}$$

$$\frac{Y(z)}{V(z)} = \frac{R(z)D(z)}{D(z)R(z) + N(z)S(z)} \tag{23.15}$$

The design requirement is that (23.13) be equal to the required transfer function (23.11), and usually it is also desired that (23.14) and (23.15) be made 'small'. The latter is typically approached by making the loop gain

$$H_{\mathrm{lg}}(z) = \frac{S(z)N(z)}{R(z)D(z)}$$

be large for certain values of frequency ω in $H_{\mathrm{lg}}(e^{j\omega T})$. For example, if there is to be no error to constant disturbance values, then the zero-frequency loop gain $H_{\mathrm{lg}}(e^{j0T})$ can be made very large by imposing

$$R(z) = (z - 1)^K R'(z)$$

for some positive integer K. This is in fact integral control. Similarly, to block out noise at a frequency ω_n (such as power line interference), one might make $H_{\mathrm{lg}}(e^{j\omega_n T})$ large by giving $S(z)/R(z)$ a notch filter action; such filter design is only mentioned here, as it is the topic of digital signal processing courses and their textbooks.

23.5.2 The open-loop control solution

An obvious partial solution to the basic problem is pole-zero cancellation: choose $S(z) = 0$ and

$$\frac{T(z)}{R(z)} = \frac{N_{\mathrm{m}}(z)D(z)}{D_{\mathrm{m}}(z)N(z)}$$

This yields the system of Fig. 23.4, which is an open-loop system. Although having the required transfer function, it will be disturbance sensitive because the loop gain is zero. Furthermore, if either $N(z)$ or $D(z)$ has roots outside the unit circle (meaning the system is either non-minimum phase or unstable, respectively), then the overall system will have unstable elements which are uncorrected.

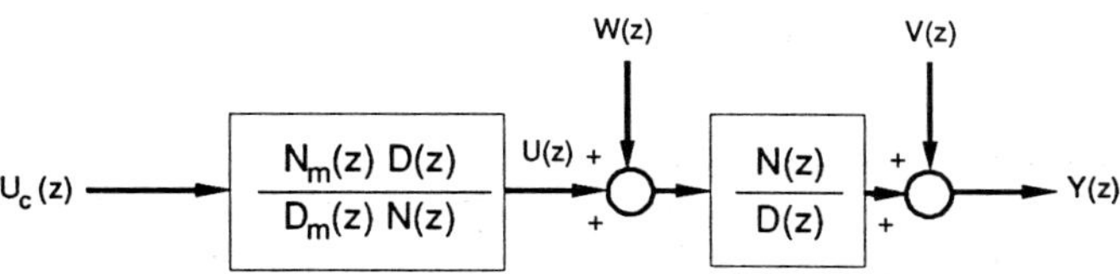

Figure 23.4 Open-loop system resulting from choice $S(z) = 0$.

23.5.3 Standard feedback control

The choice $T(z) = S(z)$ yields standard feedback control, for then with $e(k) = u_c(k) - y(k)$ we have

$$U(z) = \frac{S(z)}{R(z)} E(z)$$

The transfer function requirement yields

$$\frac{Y(z)}{U_c(z)} = \frac{N(z)\,S(z)}{D(z)R(z) + N(z)\,S(z)} = \frac{N_m(z)}{D_m(z)}$$

This requires

$$\frac{S(z)}{R(z)} = \frac{D(z)\,N_m(z)}{N(z)(D_m(z) - N_m(z))}$$

and yields the system of Fig. 23.5. The loop gain will depend upon the choices of $N_m(z)$ and $D_m(z)$. Because the controller has $N(z)$ in its denominator, the controller will be unstable if the system is not minimum phase; the plant system may be unstable, but the feedback should account for this to create a system which is stable.

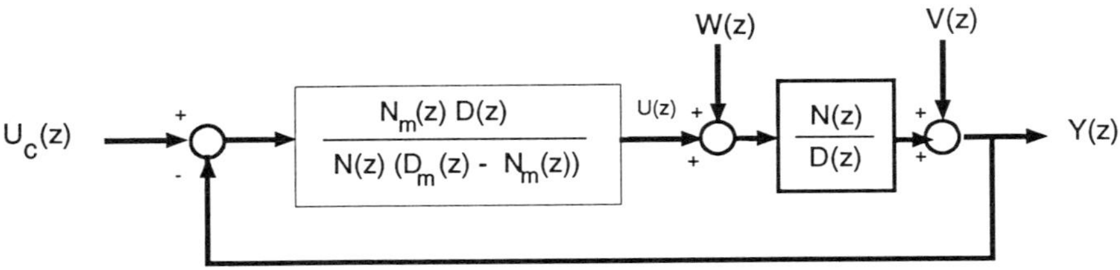

Figure 23.5 Compensation of feedback error to obtain a required closed-loop transfer function.

23.5.4 More general solutions

Since it is characteristic of the controllers that they will cancel factors of $N(z)$ not appearing in $N_m(z)$, any such factors which lie outside the unit circle plus any others not to be cancelled should be retained. Formally this means that $N(z)$ should be factored as

$$N(z) = N^+(z)N^-(z)$$

where $N^-(z)$ contains the 'protected' factors such as those with roots outside the unit circle and $N^+(z)$ contains the remaining factors. For uniqueness, $N^+(z)$ is chosen to be a monic polynomial (the coefficient of its highest power of z is 1). Using this, we must have

$$N_m(z) = N^-(z)N_m'(z)$$

Keeping this in mind, we have the requirement from (23.13) that $R(z)$, $T(z)$, and $S(z)$ be found such that

$$\frac{N(z)T(z)}{D(z)R(z) + N(z)S(z)} = \frac{N_m(z)}{D_m(z)} \tag{23.16}$$

To maintain reasonable dimensional relationships – the above indicates

$$\deg(N(z)) + \deg(T(z)) = \deg(N_{\mathrm{m}}(z))$$

– some cancellation must take place. In particular, $N^+(z)$ must divide both numerator (obviously) and denominator on the left-hand side of (23.16). This gives

$$\frac{N^+(z)N^-(z)T(z)}{D(z)N^+(z)R'(z) + N^+(z)N^-(z)S(z)} = \frac{N^-(z)N_{\mathrm{m}}'(z)}{D_{\mathrm{m}}(z)}$$

which reduces to

$$\frac{T(z)}{D(z)R'(z) + N^-(z)S(z)} = \frac{N_{\mathrm{m}}'(z)}{D_{\mathrm{m}}(z)} \tag{23.17}$$

This gives the design equations

$$T(z) = N_{\mathrm{m}}'(z)$$

$$D(z)R'(z) + N^-(z)S(z) = D_{\mathrm{m}}(z)$$

$$R(z) = R'(z)N^+(z)$$

This is still awkward dimensionally, as it requires

$$\deg(R'(z)) \le \deg(D_{\mathrm{m}}(z)) - \deg(D(z))$$

for example. Because of the causality constraints (23.8), this can easily leave us with too few parameters for adequate design of the control law. To circumvent this difficulty, we introduce an extra design polynomial $F(z)$, arguably related to observers (Chapter 25), into (23.17) as

$$\frac{T(z)}{D(z)R'(z) + N^-(z)S(z)} = \frac{N_{\mathrm{m}}'(z)F(z)}{D_{\mathrm{m}}(z)F(z)}$$

This means the design is a solution of

$$T(z) = N'_m(z)\,F(z)$$

$$D(z)\,R'(z) + N^-(z)\,S(z) = D_m(z)\,F(z)$$

$$R(z) = R'(z)\,N^+(z) \tag{23.18}$$

A good choice for the new factor appears to be $F(z) = z^L$, $L \geq 0$, but its net effect on the final design is to allow more complexity (more terms) in the difference equation for $u(k)$.

Using (23.18), the real problem is the second equation, which is of a form called a Diophantine equation. We can find a unique solution to it provided that (see Åström and Wittenmark, (1990)

$$\deg\,(R'(z)) < \deg\,(N^-(z))$$

$$\deg S(z) < \deg D(z)$$

$$\deg D_m(z) - \deg N_m(z) \geq \deg D(z) - \deg N(z) \tag{23.19}$$

$$\deg F(z) \geq 2\deg D(z) - \deg D_m(z) - \deg\,(N^+(z)) - 1$$

Using these polynomial dimensional relationships, the general Diophantine solution methods may prove unnecessary; algebraic simultaneous linear equations may suffice. We summarize the algorithm before showing this in an example.

Algorithm

$N(z)$ and $D(z)$ are of course presumed known, and the result is the control law

$$U(z) = \frac{T(z)}{R(z)}\,U_c(z) - \frac{S(z)}{R(z)}\,Y(z)$$

Step 1   Choose $N_m(z)/D_m(z)$, the desired transfer function, and $F(z)$, the 'observer' function.

Step 2   Factor the numerator polynomials, so that

$$N(z) = N^-(z)\,N^+(z) \qquad N_\mathrm{m}(z) = N^-(z)\,N_\mathrm{m}'(z)$$

where $N^-(z)$ contains the zeros protected from cancellation and $N^+(z)$ is monic.

Step 3 Solve the Diophantine equation

$$D(z)\,R'(z) + N^-(z)\,S(z) = F(z)\,D_\mathrm{m}(z)$$

for $R'(z)$ and $S(z)$. Use a solution for which

$$\deg\,(S(z)) < \deg\,(N(z))$$

$$\deg\,(R'(z)) = \deg\,(F(z)) + \deg\,(D_\mathrm{m}(z)) - \deg\,(D(z))$$

Step 4 Complete the control law by taking S(z) from step 3 and

$$R(z) = N^+(z)\,R'(z) \qquad T(z) = N_\mathrm{m}'(z)\,F(z)$$

Example

A straightforward example from Åström and Wittenmark (1990) will demonstrate what is involved with the technique. A model of a simple motor is

$$H(z) = \frac{K(z - b)}{(z - 1)(z - a)}$$

where the input is a command voltage and the output quantity is shaft angle. It is desirable that this behave as

$$H_\mathrm{m}(z) = \frac{G(z - b)}{z^2 + p_1 z + p_2})$$

where gain G and pole parameters p_1 and p_2 are selected to meet design requirements on the system performance, and we have chosen not to cancel the motor's zero. Hence

$$N^+(z) = 1 \qquad N^-(z) = K(z - b) \qquad N'_m(z) = G/K$$

It is clear that

$$\deg\left(D_m(z)\right) - \deg\left(N'_m(z)\right) = 2 \geq \deg\left(D(z)\right) - \deg\left(N^+(z)\right) = 2$$

and that we need $\deg(F(z)) \geq 1$. This can be met by choosing $\deg(F(z)) = 1$, and hence from (23.19)

$$\deg\left(R(z)\right) = 1 \qquad \deg\left(S(z)\right) = 1$$

We choose the monic forms (i.e. the forms in which the highest power in the polynomial has coefficient $+1$) for $F(z)$ and $R(z)$, as constants will eventually cancel. In particular, choose

$$F(z) = z - \alpha \qquad S(z) = s_1 z + s_0 \qquad R(z) = z + r_0$$

Thus, we need

$$D(z)R(z) + N(z)S(z) = F(z)D_m(z)$$

which on substitution gives

$$(z - 1)(z - a)(z + r_0) + K(z - b)(s_1 z + s_0)$$

$$= (z - \alpha)(z^2 + p_1 z + p_2)$$

This may be solved to give

$$r_0 = -b + \frac{(b - \alpha)(b^2 + p_1 b + p_2)}{(b - a)(b - 1)}$$

$$s_0 = -a(1 - \alpha)d_1 + (a - \alpha)d_2$$

$$s_1 = (1 - \alpha)d_1 - (a - \alpha)d_2$$

where

$$d_1 = \frac{G}{K\,(1-b)\,(1-a)}$$

$$d_2 = \frac{(a^2 + ap_1 + p_2)}{K(a-b)(1-a)}$$

We also find

$$T(z) = (z - \alpha)\,\frac{G}{K}$$

The final control law is

$$u(k) = -\,r_0\,u(k-1) + \frac{G}{K}\,(u_c(k) - \alpha u_c(k-1)) - s_1\,y(k) - s_0\,y(k)$$

We note that fixing G, p_1, and p_2 fixes the system's closed-loop response. All of the control parameters, however, depend upon the zero at α in the function $F(z)$, and hence this will be a parameter in the control system operation even though it does not affect the input/output characteristic of the system.

23.6 CONTINUOUS TIME SYSTEMS IN STATE–SPACE

For continuous time state–space descriptions such as

$$\dot{x} = Ax + Bu$$

$$y = Cx$$

the use of state feedback

$$u(t) = -\,Kx(t) + u_{\text{ref}}(t)$$

or of output feedback

$$u(t) = -Gy(t)$$

can be done exactly as for discrete time systems. The major difference is that chosen poles for continuous time systems will be in the left-half plane rather than inside the unit circle.

23.7 PLACING POLES IN DESIRABLE REGIONS

The above schemes all place some number of poles (n in the case of an n-dimensional system using state feedback) at or very near precise pre-chosen values. For large systems, this can mean the designer must choose many poles when it is only required that the system meet a few performance specifications: there are more degrees of freedom than the problem requires. Since in fact the requirements can often be met simply by ensuring that all of the poles are in a certain region of the complex plane, there has been considerable interest in design methods which place poles within a chosen region.

Typical discrete-time system specifications may, as argued in Chapter 19, require that the poles lie within the intersection of a circle around the origin (for transient decay rate), a sector of the unit circle (for speed of response) and a curve of exponential decay (for damping). Although the general region definition of this type cannot yet be handled directly, approximations based on known results may be satisfactory. One such simple case is given below.

We refer to Fig. 23.6 and see that a possible region definition is approximated by a circle centred at $(\alpha, 0)$ with radius r. Then for a system such as $\mathbf{x}(k+1) = \mathbf{A}\mathbf{x}(k) + \mathbf{B}\mathbf{u}(k)$ it has been shown (Furuta and Kim, 1987) that the control law $\mathbf{u}(k) = -\mathbf{K}\mathbf{x}(k)$, where

$$\mathbf{K} = (r^2\mathbf{R} + \mathbf{B}^{\mathrm{T}}\mathbf{P}\mathbf{B})^{-1}\,\mathbf{B}^{\mathrm{T}}\mathbf{P}(\mathbf{A} - \alpha\mathbf{I})$$

and $\mathbf{P}$ solves the Riccati equation

$$\mathbf{P} = \frac{(\mathbf{A} - \alpha)^{\mathrm{T}}}{r}\,\mathbf{P}\,\frac{(\mathbf{A} - \alpha)}{r} + \mathbf{Q}$$

$$- \frac{(\mathbf{A} - \alpha)^{\mathrm{T}}}{r}\,\mathbf{P}\mathbf{B}(r^2\mathbf{R} + \mathbf{B}^{\mathrm{T}}\mathbf{P}\mathbf{B})^{-1}\mathbf{B}^{\mathrm{T}}\mathbf{P}\,\frac{(\mathbf{A} - \alpha\mathbf{I})}{r}$$

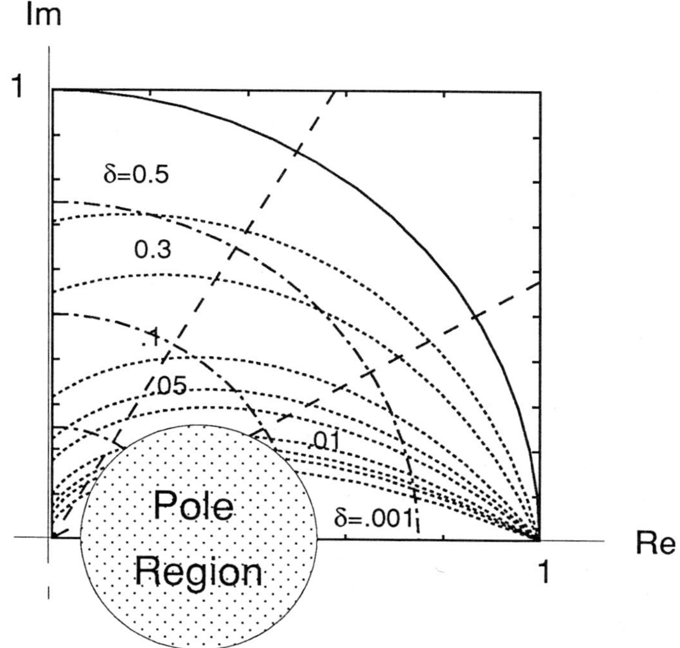

Figure 23.6 Region of definition for desired poles.

yields eigenvalues of $(\mathbf{A} - \mathbf{BK})$ within the defined disc. In this, the positive definite matrix $\mathbf{R}$ and the positive semi-definite matrix $\mathbf{Q}$, where $\mathbf{Q} = \mathbf{D}^{\mathrm{T}}\mathbf{D}$ for arbitrary $\mathbf{D}$ for which $(\mathbf{A};\mathbf{D})$ is controllable, are parameters of the design and are available to use for 'tuning' of the pole selection within the region.

The tuning aspect comes about because the above control law is precisely that which gives optimal feedback regulation (see Chapter 26) for the system

$$\mathbf{w}(k+1) = \frac{(\mathbf{A} - \alpha\mathbf{I})}{r}\,\mathbf{w}(k) + \frac{\mathbf{B}}{r}\,\mathbf{v}(k)$$

where the optimization criterion is to minimize

$$\mathbf{J} = \sum_{k=0}^{\infty} [\mathbf{w}^{\mathrm{T}}(k)\,\mathbf{Q}\,\mathbf{w}(k) + \mathbf{v}^{\mathrm{T}}(k)\,\mathbf{R}\,\mathbf{v}(k)]$$

Example

Let the system be defined by

$$
\mathbf{x}(k+1) = \begin{bmatrix} 1 & 1 & 0 \\ 0 & 1 & 1 \\ 1 & 0 & 1 \end{bmatrix} \mathbf{x}(k) + \begin{bmatrix} 0 & 0 \\ 1 & 0 \\ 1 & 1 \end{bmatrix} \mathbf{u}(k)
$$

The desired region for the poles is chosen to be the circle centred at $(\alpha, 0) = (0.3, 0)$ with radius $r = 0.25$.

Using a CACSD program such as the MATLAB® function *dlqr*, the following cases are easily computed.

1. $\mathbf{Q} = \mathbf{I}_3$, $\mathbf{R} = \mathbf{I}_2$.

$$
\mathbf{K} = \begin{bmatrix} 0.4626 & 1.3416 & 0.9983 \\ 0.2641 & -1.6801 & -0.2882 \end{bmatrix}
$$

$$
\text{eig}\,(\mathbf{A} - \mathbf{BK}) = 0.3120 \pm 0.0402\text{j}, \ 0.3242
$$

2. $\mathbf{Q} = \mathbf{I}_3$, $\mathbf{R} = 100\mathbf{I}_2$.

$$
\mathbf{K} = \begin{bmatrix} 0.4368 & 1.3090 & 0.9957 \\ 0.0748 & -1.8675 & -0.2726 \end{bmatrix}
$$

$$
\text{eig}\,(\mathbf{A} - \mathbf{BK}) = 0.3157 \pm 0.0679\text{j}, \ 0.3365
$$

The same design equations may be used for the selection of poles for the continuous time system $\dot{\mathbf{x}} = \mathbf{Ax} + \mathbf{Bu}$ for a similar disc of radius r and centre $(\alpha, 0)$, but of course the disc would lie in the left-half plane rather than inside the unit circle. This is because the algorithm is one for selecting $\mathbf{K}$ to give eigenvalues of $(\mathbf{A} - \mathbf{BK})$ within a region, but the algorithm is independent of the meaning of the various matrices; the interpretation in terms of the optimization criterion is similar to, but not the same as, that for the discrete time problem.

23.8 COMPUTER AIDS

Among computer aids, state–space designs, based as they are upon vector–matrix manipulations, are often straightforward to program. Pole placement is no exception. Packages such as MATLAB® have standard functions which will compute the gain needed for pole placement via state feedback. They will also assist in computing the gains needed for regional pole placement.

23.9 SUMMARY

One of the obvious things to do when there are many elements of the state available for feedback is to use all of them in a systematic manner to achieve the desirable system response. In this section we have worked through various algorithms for instances when controls are scalars, controls are vectors, feedback is the state, feedback is a subset of the state, and the system is to be stabilized.

The question of where to place all those poles was partly addressed in Chapter 19. One is likely to wish only that the poles all be in a region Γ but research is continuing on straightforward (as opposed to search) methods for doing this. This may be just as well, as the placement methods do not of themselves guarantee implementable control laws with appropriately small gains and commands and low enough complexity relative to the control task.

Rarely are all the elements of the state available for state feedback control. Rather than use output feedback, however, the designer may prefer as a first-cut method to assume full state feedback for the controller design and use a state estimator (such as an observer, Chapter 25) to supply its input. (See also Chapter 29.)

23.10 FURTHER READING

References have been cited within this chapter, but the first, basic methods are in many textbooks, including Ogata (1987). For transfer function approaches, one good source is Åström and Wittenmark (1990). Franklin *et al.* (1990) are among those who concentrate more on state–space methods.

One of the recent research articles which summarizes and generalizes results such as those in section 23.7 on placement within regions is by Haddad *et al.* (1992).

24

Observability

Like controllability, observability has important and easily visualized interpretations in terms of the natural language definitions of the word, but it should be recognized ultimately as a technical term. In particular, a system which is not an observable system is not 'unobservable'. In this section we look at some of the tests and concepts for observability. The reader may note the striking similarity to the Chapter 22 section on controllability; this is not quite accidental, as the two concepts are mathematical duals of each other. To avoid near repetition, some results for continuous time systems are emphasized along with standard discrete time system results.

24.1 SYNOPSIS

The definition of observability is a technical one. A system is said to be **observable** if any initial state $x(t_0)$ can be determined after a finite time interval $t - t_0$ from a measurement history $Y(t) = \{y(\tau), t_0 \leq \tau < t\}$ and, in the case of a forced system, the control variable history $U(t) = \{u(\tau), t_0 \leq \tau < t\}$. Furthermore, given the usual uniqueness of solution arguments, $x(t)$, $t \geq t_0$, can also be determined.

For systems modelled as the discrete time form

$$\mathbf{x}(k+1) = \mathbf{A}\mathbf{x}(k) + \mathbf{B}\mathbf{u}(k)$$

$$\mathbf{y}(k) = \mathbf{C}\mathbf{x}(k) \tag{24.1}$$

or the continuous time form

$$\dot{\mathbf{x}} = \mathbf{A}\mathbf{x} + \mathbf{B}\mathbf{u}$$

$$\mathbf{y} = \mathbf{C}\mathbf{x} \tag{24.2}$$

where $\mathbf{x}$ is an n-vector, the standard test is that the system is observable, also stated as the matrix pair $(\mathbf{A};\mathbf{C})$ is observable, iff the observability test matrix $\mathbf{Q}_o$, given by

$$\mathbf{Q}_o = \begin{bmatrix} \mathbf{C} \\ \mathbf{C}\mathbf{A} \\ \mathbf{C}\mathbf{A}^2 \\ \vdots \\ \mathbf{C}\mathbf{A}^{n-1} \end{bmatrix}$$

has full rank, i.e. rank$[\,\mathbf{Q}_o\,] = n$. If $\mathbf{Q}_o$ has full rank, then the smallest integer v such that

$$\text{rank} = \begin{bmatrix} \mathbf{C} \\ \mathbf{C}\mathbf{A} \\ \mathbf{C}\mathbf{A}^2 \\ \vdots \\ \mathbf{C}\mathbf{A}^{v-1} \end{bmatrix} = n$$

is called the **observability index**.

24.2 DEFINITIONS

The notion of system observability is built from state observability much as was true for system controllability. Hence we have the following definitions.

Definition A system **state** $\mathbf{x}(t_0)$ at some given t_0 is **observable** if knowledge of the input $\mathbf{u}(t)$ and output $\mathbf{y}(t)$ over a finite time segment $t_0 \le t \le t_0 + \tau < \infty$ allows the determination of $x(t_0)$.

Because of the uniqueness of solutions of difference and differential equations, if we know the state at some instant and know the inputs from that instant onward, we know the state at later instants. Hence we have the equivalent definition of state observability given below.

Definition A **state** $\mathbf{x}(t)$ of a system is **observable** if knowledge of the input $u(\tau)$ and measurements $\mathbf{y}(\tau)$ over a finite time segment $-\infty < t_0 \le \tau \le t < \infty$ allows the determination of $\mathbf{x}(t)$. If all states $\mathbf{x}(t)$ are observable, the system is said to be observable, or **completely observable**. If observability depends upon t_0, the state is **observable at t_0**. If the system (or state) is observable when $u(\tau) \equiv 0$, it is said to be **zero-input observable**.

The above can for most purposes be combined into a single operating definition. A system is said to be **observable** if any state $\mathbf{x}(t_0)$ with t_0 arbitrary can be determined after a finite time interval $t - t_0$ from a measurement history $Y(t) = \{\mathbf{y}(\tau), t_0 \le \tau < t\}$ and, in the case of a forced system, the control variable history $U(t) = \{\mathbf{u}(\tau), t_0 \le \tau < t\}$. Furthermore, given the usual uniqueness of solution arguments, $\mathbf{x}(t)$, $t \ge t_0$, can also be determined.

It may be seen from examining the definitions above and in Chapter 22 that it is quite possible to have states which are controllable but not observable, observable but not controllable, both controllable and observable, or neither controllable nor observable. This decomposition can in fact be made explicit, using a canonical decomposition.

Consider what the terminology allows: the fact that a radar target's roll rate cannot be extracted from the radar tracking data – making the system not observable – does not necessarily make for a poor radar system. Systems which are observable are often easier to design for because more design theorems apply, but other systems which are not observable are often quite useful.

24.3 TESTS FOR CONSTANT COEFFICIENT SYSTEMS

Tests for observability are defined primarily for linear systems, because they are so highly structured. Time-invariant linear systems, i.e. those with constant coefficients, are particularly easy to test.

24.3.1 Sampled data systems

In this section we present the standard tests and canonical decomposition for the system described by

$$\mathbf{x}(k+1) = \mathbf{A}\mathbf{x}(k) + \mathbf{B}\mathbf{u}(k)$$

$$\mathbf{y}(k) = \mathbf{C}\mathbf{x}(k) + \mathbf{D}\mathbf{u}(k)$$

where $\mathbf{A}$ is assumed to be $n \times n$, $\mathbf{B}$ is $n \times m$, and $\mathbf{C}$ is $p \times n$. $\mathbf{D}$ is obviously then $p \times m$. We recall that a solution of this is

$$\mathbf{x}(k+k_0) = \mathbf{A}^k \mathbf{x}(k_0) + \sum_{i=0}^{k-1} \mathbf{A}^{k-i-1} \mathbf{B} \mathbf{u}(i+k_0)$$

If we take measurements $\mathbf{y}(k_0), \ldots, \mathbf{y}(k_0+k)$, then

$$\mathbf{y}(k_0) = \mathbf{C}\mathbf{x}(k_0) + \mathbf{D}\mathbf{u}(k_0)$$

$$\mathbf{y}(k_0+1) = \mathbf{C}\mathbf{A}\mathbf{x}(k_0) + \mathbf{C}\mathbf{B}\mathbf{u}(k_0) + \mathbf{D}\mathbf{u}(k_1)$$

$$\vdots$$

$$\mathbf{y}(k_0+k) = \mathbf{C}\mathbf{A}^k\mathbf{x}(k_0) + \mathbf{C}\sum_{i=0}^{k-1}\mathbf{A}^{k-i-1}\mathbf{B}\mathbf{u}(i+k_0) + \mathbf{D}\mathbf{u}(k_0+k)$$

This may be rewritten as

$$\mathbf{C}\mathbf{x}(k_0) = \mathbf{y}(k_0) - \mathbf{D}\mathbf{u}(k_0)$$

$$\mathbf{C}\mathbf{A}\mathbf{x}(k_0) = \mathbf{y}(k_0+1) - \mathbf{C}\mathbf{B}\mathbf{u}(k_0) - \mathbf{D}\mathbf{u}(k_1)$$

$$\vdots$$

$$\mathbf{C}\mathbf{A}^k\mathbf{x}(k_0) = \mathbf{y}(k_0+k) - \mathbf{C}\sum_{i=0}^{k-1}\mathbf{A}^{k-i-1}\mathbf{B}\mathbf{u}(i+k_0) - \mathbf{D}\mathbf{u}(k_0+k)$$

The right-hand side of these is known from taking the measurements $\mathbf{y}(j+k_0)$, $j=0,1,\ldots,k$, and knowing the commands $\mathbf{u}(j+k_0)$, $j=0,1,\ldots,k-1$. Hence, we have

$$\begin{bmatrix} \mathbf{C} \\ \mathbf{C}\mathbf{A} \\ \mathbf{C}\mathbf{A}^2 \\ \vdots \\ \mathbf{C}\mathbf{A}^{k-1} \end{bmatrix} \mathbf{x}(k_0) = \text{known values}$$

and we can solve for $x(k_0)$ provided its coefficient matrix has an invertible submatrix. For this to be true, since the matrix has n columns and pk rows, it needs to have n linearly independent rows. Thus we need that

$$\text{rank} \begin{bmatrix} C \\ CA \\ CA^2 \\ \vdots \\ CA^{k-1} \end{bmatrix} = n$$

for some k. The smallest such k, if a k exists for a given system, is called the **observability index**, and it indicates the number of vector measurements needed to find $x(k_0)$. The maximum value of k we need to check is n, because a square matrix of dimension n (such as A) is a linear combination of matrices A^i where $0 \leq i < n$ (by the Cayley–Hamilton theorem; see Appendix B).

The above arguments are clearly independent of k_0, and hence the basic test for observability of such a system is as follows.

Observability The system (24.1) is **observable** iff the matrix Q_o, given by

$$Q_o = \begin{bmatrix} C \\ CA \\ CA^2 \\ \vdots \\ CA^{n-1} \end{bmatrix}$$

has full rank, i.e. rank $Q_o = n$

Canonical decomposition The system may, by similarity transformation T (i.e. $x = Tz$), be placed in the form

$$z(k+1) = \Lambda z(k) + T^{-1} Bu(k)$$

$$y(k) = CTz(k) + Du(k)$$

where Λ is in Jordan form and is diagonal if the eigenvalues of A are distinct. With this decomposition, the transformed state vector z has elements z_i, $i = 1, 2, ..., n$. Then the result for the diagonal case with distinct eigenvalues is that, for each i, z_i is observable if the ith column of CT is not zero; for Jordan blocks a generalization of this requirement is obvious.

Combining this with the result of Chapter 22 concerning the rows of $T^{-1}B$, we see that the state is decomposable to elements which are controllable, observable, both, or neither. This is a canonical decomposition. The tests are modified in the obvious manner if the Jordan form has off-diagonal terms, as only one column of CT or row of $T^{-1}B$ need be non-zero for each coupled state set for observability or controllability, respectively.

Example

We continue the example of section 22.2, in which $n = 3$, $m = 2$, $p = 1$, with

$$A = \begin{bmatrix} 1 & 0 & 0 \\ 0 & 1 & 1 \\ 1 & 1 & 0 \end{bmatrix} \quad B = \begin{bmatrix} 0 & 0 \\ 1 & 0 \\ 0 & 1 \end{bmatrix} \quad C = [\, 0 \ 1 \ 1 \,] \quad D = [\, 0 \ 0 \,]$$

Then by direct computation

$$Q_o = \begin{bmatrix} 0 & 1 & 1 \\ 1 & 2 & 1 \\ 2 & 3 & 2 \end{bmatrix}$$

By inspection, rank $Q_o = 3$. Since $n = 3$, the system is observable.

Using the transformation matrix T from that example, we find that

$$CT = \begin{bmatrix} -1 & \dfrac{3 + \sigma}{2} & \dfrac{3 - \sigma}{2} \end{bmatrix}$$

where $\sigma = \sqrt{5}$. No columns of $\mathbf{CT}$ are zero, so all elements of $\mathbf{z}$ are observable – the same as finding that the system is observable.

In the decomposition, then, z_1 is observable but not controllable; z_2 and z_3 are observable and controllable.

24.3.2 Basic ideas in the time-varying case

The same thing can be done when the system is time-varying as was done for the controllability test. Again consider the system

$$\mathbf{x}(k+1) = \mathbf{A}(k)\mathbf{x}(k) + \mathbf{B}(k)\mathbf{u}(k)$$

$$\mathbf{y}(k) = \mathbf{C}(k)\mathbf{x}(k) + \mathbf{D}(k)\mathbf{u}(k)$$

Ignoring (actually setting to zero) the inputs for simplicity, we consider a set of measurements $\mathbf{y}(0),...,\mathbf{y}(k)$ when the initial state is $\mathbf{x}(0) = \mathbf{x}_0$. Then

$$\mathbf{y}(0) = \mathbf{C}(0)\mathbf{x}_0$$

$$\mathbf{y}(1) = \mathbf{C}(1)\mathbf{x}(1) = \mathbf{C}(1)\mathbf{A}(0)\mathbf{x}_0$$

$$\mathbf{y}(2) = \mathbf{C}(2)\mathbf{x}(2) = \mathbf{C}(2)\mathbf{A}(1)\mathbf{A}(0)\mathbf{x}_0$$

$$\vdots$$

$$\mathbf{y}(k) = \mathbf{C}(k)\left[\prod_{j=0}^{k-1}\mathbf{A}(j)\right]\mathbf{x}_0 \tag{24.3}$$

Then it is clear that if this system can be solved for $\mathbf{x}_0$ for some k, we have observability, because if we know $\mathbf{x}_0$, then we can determine $\mathbf{x}(k)$ for any k. To find this solution for $\mathbf{x}_0$, we need only that the $(k+1)p \times n$ matrix

$$\mathbf{Q}_{ok} = \begin{bmatrix} \mathbf{C}(0) \\ \mathbf{C}(1)\,\mathbf{A}(0) \\ \mathbf{C}(2)\,\mathbf{A}(1)\,\mathbf{A}(0) \\ \vdots \\ \mathbf{C}(k)\,\mathbf{A}(k-1)\,\mathbf{A}(k-2)\,...\,\mathbf{A}(0) \end{bmatrix}$$

should have an invertible submatrix for some k. This means the rank of the matrix should be n, i.e. it should have n linearly independent rows, its rows should span the space, etc.

The arguments follow those of the controllability case in Chapter 22. If $C(k) = C$ = constant and $A(j) = A$ = constant, the test immediately reduces to that of the previous subsection.

We remark that we can get $\mathbf{x}_0$ explicitly by using an $n \times n$ submatrix, so that from (24.3) we get the form, using a selection matrix S with dimension $n \times np$ and columns which select the independent rows of $\mathbf{Q}_{ok}$,

$$\mathbf{x}_0 = [S\ \mathbf{Q}_{ok}]^{-1}\ [S\ Y]$$

where $S\ Y$ is the appropriate submatrix (subvector) of

$$Y = \begin{bmatrix} \mathbf{y}(0) \\ \mathbf{y}(1) \\ \vdots \\ \mathbf{y}(k) \end{bmatrix}$$

24.3.3 The continuous time case

It turns out that, just as was claimed for controllability, the basic test applies to the constant coefficient continuous time system described by (24.2). Rather than simply claim that result, we examine continuous time here with the understanding that, because of duality, many of the methods and a variation of the results apply to controllability.

The essential results can be derived for unforced linear time-varying systems, as known inputs are simply subtracted out as above. In particular, consider the system

$$\dot{\mathbf{x}} = A(t)\,\mathbf{x}$$

$$\mathbf{y} = C(t)\,\mathbf{x}$$

with transition matrix $\phi(t, \tau)$. Then this system is observable on the interval (t_0, t_1) if

1. for any $\lambda \neq 0$, there exists some t in the interval (which may depend on λ) such that

$$\mathbf{C}(t)\,\phi(t, t_0)\,\lambda \neq 0 \qquad\qquad (24.4)$$

or equivalently,
2. the observability gramian matrix $\mathbf{M}(t_0, t_1)$ defined as

$$\mathbf{M}(t_0, t_1) = \int_{t_0}^{t_1} \phi^{\mathrm{T}}(\tau, t_0)\,\mathbf{C}^{\mathrm{T}}(\tau)\,\mathbf{C}(\tau)\,\phi(\tau, t_0)\,\mathrm{d}\tau$$

is non-singular.

The proof of the above is more intricate, but instructive. Notice that

$$\mathbf{x}(t) = \phi(t, t_0)\,\mathbf{x}_0$$

$$\mathbf{y}(t) = \mathbf{C}(t)\,\phi(t, t_0)\,\mathbf{x}_0$$

by definition of transition matrices. Now if there is a λ violating condition 1, then $\mathbf{x}(0) = \lambda$ gives no output and hence is not observable. On the other hand, if condition 1 holds, then $\mathbf{C}(t)\,\phi(t, t_0)\,\mathbf{x}_0$ is not identically 0, and hence $\mathbf{y}(t)$ is not zero on the interval. Therefore

$$\eta(t, t_0) = \int_{t_0}^{t_1} \phi^{\mathrm{T}}(\tau, t_0)\,\mathbf{C}^{\mathrm{T}}(\tau)\,\mathbf{y}(\tau)\,\mathrm{d}\tau$$

$$= \int_{t_0}^{t_1} \phi^{\mathrm{T}}(\tau, t_0)\,\mathbf{C}^{\mathrm{T}}(\tau)\,\mathbf{C}(\tau)\,\phi(\tau, t_0)\,\mathrm{d}\tau\,\mathbf{x}_0$$

$$= \mathbf{M}(t_0, t_1)\,\mathbf{x}_0$$

$$\neq 0$$

Since this is non-zero for arbitrary x_0, M must have full rank and therefore be invertible. The explicit solution is

$$x_0 = M^{-1}(t_0, t_1)\, \eta(t_1, t_2) \tag{24.5}$$

Thus, if condition 1 holds, then the system is observable and M is non-singular. Now if M is singular, (24.5) is impossible, so the system must not be observable.

If the system is observable, then $y(t) \not\equiv 0$ for $x_0 \neq 0$. In this case

$$0 < \int_{t_0}^{t_1} y^T(\tau)\, y(\tau)\, d\tau = x_0^T M(t_0, t_1)\, x_0$$

This inequality for arbitrary x_0 ensures that M has full rank and hence that M^{-1} must exist. This concludes the basic demonstration of the truth of the above assertions.

Let us now specialize the result to the time-invariant case. We consider

$$\dot{x} = A x$$

$$y = C x$$

Since in this case $\phi(t, t_0) = e^{At}$, (24.4) means that we have observability iff $C e^{At} \lambda \not\equiv 0$ for any λ. Since A satisfies its characteristic equation $P(A) = A^n + a_1 A^{n-1} + \cdots + a_{n-1} A + a_n I = 0$, we have $CP(A)\, e^{At}\, \lambda = 0$. Expanding the polynomial and noticing that

$$A^k e^{At} = \frac{d^k e^{At}}{dt^k}$$

means that

$$\frac{d^n C e^{At} \lambda}{dt^n} + a_1 \frac{d^{n-1} C e^{At} \lambda}{dt^{n-1}} + \cdots + a_n\, C e^{At} \lambda = 0$$

For this to hold with $Ce^{At}\lambda \neq 0$, it must be that one of the initial conditions of the differential equation is non-zero. Hence we need

$$\left. \frac{d^k\, Ce^{At}\lambda}{dt^k} \right|_{t=0} = CA^k\lambda \neq 0$$

for at least one k. This means that, for any λ,

$$\begin{bmatrix} C \\ CA \\ CA^2 \\ \vdots \\ CA^{n-1} \end{bmatrix} \lambda \neq 0$$

This is equivalent to requiring the matrix to have full rank n. Essentially the same argument can be used to develop the test for continuous time controllability of time invariant systems, Chapter 22.

24.4 OTHER GENERALIZATIONS AND COMMENTS

The definitions of observability apply for non-linear systems, but just as there are no standard non-linear systems, there are no standard tests.

It is worth commenting that if the rank of one of the test matrices is not the maximum value, all is not lost. First, it is quite possible that we can find out what we wish to know about a 'not observable' system. Alternatively, the tests may indicate the need for another measurement sensor. The second piece of information is that the rank of the matrix Q_o really tells us the number of different linear combinations of the state we can observe: for instance, if rank $Q_o = 1$, there is one linear combination

$$f_1 x_1 + f_2 x_2 + \cdots + f_n x_n = f_0$$

of which we can make measurements, although the coefficients f_i, $i = 1, 2, \ldots, n$, are fixed by the system properties. The coefficients can be found, if desired, by doing a canonical decomposition, finding which z_i is observable, and then doing the inverse transformation. More particularly, perform the similarity transformation $\mathbf{x} = \mathbf{T}\mathbf{z}$ and perform a canonical decomposition. If z_i is observable, then the linear combination $[\ \mathbf{T}^{-1}\ ]_i\ \mathbf{x}$ can be observed, where $[\ ]_i$ denotes the ith row of the matrix.

A second observation is that the observability test is clearly similar to the controllability test. Comparison with the tests of Chapter 22 show that if (24.1) is observable, then the system modelled by

$$\mathbf{z}(k+1) = \mathbf{A}^T\mathbf{z}(k) + \mathbf{C}^T\mathbf{w}(k)$$

$$\mathbf{v}(k) = \mathbf{B}^T\mathbf{z}(k) \tag{24.6}$$

is controllable; also, if the latter is observable, then the former is controllable. Again, the notation occasionally used is that $(\mathbf{A};\mathbf{C})$ is observable iff $(\mathbf{A}^T;\mathbf{C}^T)$ is controllable; also, $(\mathbf{A};\mathbf{B})$ is controllable iff $(\mathbf{A}^T;\mathbf{B}^T)$ is observable. The system (24.6) is said to be dual to (24.1), whether or not either is controllable or observable. A similar definition of dual system holds for continuous time systems, and the same relationships between observability and controllability hold.

24.5 COMPUTER AIDS

Computer testing for time-invariant cases is, since they are essentially a transpose of those for controllability, easy to do; mathematical aid programs such as MATLAB® and Ctrl-C have no difficulty doing such tests. The time-varying case is more difficult because it is necessary to check whether there exist any t_0 and τ for which observability holds; inability to find such a time interval does not mean it does not exist. Some programs check the gramian rather than the observability matrix.

24.6 FURTHER READING

The concepts of observability, like those of controllability, are usually traced back to Kalman (1961), but the ideas are also found in most standard textbooks. The material of this chapter has also been presented in a similar manner in the textbook of Åström and Wittenmark (1990).

25

State observers

State estimators are algorithms for deriving state estimates from measurements. Based upon proper manipulation of system models, they allow, for example, the estimation of state variables such as accelerations from position measurements. Applications can be found in instrumentation, flight reconstruction, and in providing inputs to state feedback control laws. The most famous state estimators are the Luenberger observer and the Kalman filter. The basic form of the former does not explicitly consider the possibility of noise on the measurements and it is in many respects related to pole-placement controllers. Hence it is reasonably accessible theoretically and we introduce it at this point. We concentrate on discrete time formulations simply because implementation is likely to be on digital computers, but analogous results hold for continuous-time observers.

25.1 SYNOPSIS

In the state estimation problem, a dynamic system is described by the n-dimensional state vector $\mathbf{x}$, input m-vector $\mathbf{u}$, output p-vector $\mathbf{y}$, and noise vectors $\mathbf{v}$ and $\mathbf{w}$, as

$$\mathbf{x}(k+1) = f(k, \mathbf{x}(k), \mathbf{u}(k), \mathbf{w}(k))$$

$$\mathbf{y}(k) = g(k, \mathbf{x}(k), \mathbf{v}(k))$$

The immediate object of the estimation is to find an estimate $\hat{\mathbf{x}}(k)$ of $\mathbf{x}(k)$ which is 'good' in some sense. The estimate is to be derived only from measurements, knowledge of the input commands, and perhaps *a priori* information. Most state estimators require a good system model.

This problem generally is very difficult. Important special cases are well understood, however. In particular, the linear constant coefficient case

$$\mathbf{x}(k+1) = \mathbf{A}\mathbf{x}(k) + \mathbf{B}\mathbf{u}(k)$$

$$\mathbf{y}(k) = \mathbf{C}\mathbf{x}(k)$$

has a straightforward class of solutions: observers. These were originally defined by Luenberger in the following terms: a system $\mathbf{S}_0$ which

1. is intended to approximate the state vector $\mathbf{x}$ of another system $\mathbf{S}$ by means of a vector $\hat{\mathbf{x}}$,
2. has as its inputs the inputs and available outputs of the latter system, and
3. has a state vector $\mathbf{z}$ which is linearly related to the approximation, i.e. $\mathbf{z} = \mathbf{T}\,\hat{\mathbf{x}}$,

is called an **observer**. For the special case in which $\mathbf{T} = \mathbf{I}$, called the **identity observer**, the answer is easily shown, provided $\mathbf{S}_1$ is observable, to be

$$\hat{\mathbf{x}}(k+1) = (\mathbf{A} - \mathbf{G}\mathbf{C})\hat{\mathbf{x}}(k) + \mathbf{G}\mathbf{y}(k) + \mathbf{B}\mathbf{u}(k) \tag{25.1}$$

where the observer gain $\mathbf{G}$ can be chosen such that the eigenvalues of the matrix $\mathbf{F} = \mathbf{A} - \mathbf{G}\mathbf{C}$ are as the designer wishes. The resulting error $\mathbf{e}(k) = \mathbf{x}(k) - \hat{\mathbf{x}}(k)$ satisfies $\mathbf{e}(k) = \mathbf{F}^k\,\mathbf{e}(0)$.

More advanced observers, including those for time-varying systems, and those of lower order than the nth-order system indicated by (25.1), are presented later in this section. Explicit allowance for noise is deferred until Chapter 28.

25.2 THE BASIC LUENBERGER OBSERVER

The idea of observers arose with Luenberger in the mid 1960s, shortly after the publication of Kalman's and Bucy's works on state estimation in explicit noise (Chapter 28). We study observers in the following sections. Students will undoubtedly note many parallels with pole-placement controllers.

25.2.1 The identity observer

The core idea of the observer is as follows. Consider a computational structure

$$\hat{\mathbf{x}}(k+1) = \mathbf{F}\hat{\mathbf{x}}(k) + \mathbf{G}\mathbf{y}(k) + \mathbf{H}\mathbf{u}(k) \tag{25.2}$$

We wish to find matrices $\mathbf{F}, \mathbf{G}, \mathbf{H}$ such that, as k becomes large, $\hat{\mathbf{x}}(k) \to \mathbf{x}(k)$ for the system with state $\mathbf{x}(k)$ given by

$$\mathbf{x}(k+1) = \mathbf{A}\mathbf{x}(k) + \mathbf{B}\mathbf{u}(k)$$

$$\mathbf{y}(k) = \mathbf{C}\mathbf{x}(k) \tag{25.3}$$

assuming we know $\mathbf{u}(k)$ and $\mathbf{y}(k)$.

Since we want the error to be small, we consider it directly. Thus let $\mathbf{e}(k) = \mathbf{x}(k) - \hat{\mathbf{x}}(k)$. Then from (25.2–3) we find

$$\mathbf{e}(k+1) = \mathbf{A}\mathbf{x}(k) + \mathbf{B}\mathbf{u}(k) - \mathbf{F}\hat{\mathbf{x}}(k) - \mathbf{G}\mathbf{y}(k) - \mathbf{H}\mathbf{u}(k)$$

$$= \mathbf{F}\mathbf{e}(k) + (\mathbf{A} - \mathbf{G}\mathbf{C} - \mathbf{F})\mathbf{x}(k) + (\mathbf{B} - \mathbf{H})\mathbf{u}(k) \tag{25.4}$$

Since we would like the error to be independent of the $\mathbf{x}(k)$ and $\mathbf{u}(k)$ sequences, which may be arbitrary, we choose

$$\mathbf{H} = \mathbf{B}$$

$$\mathbf{F} = \mathbf{A} - \mathbf{G}\mathbf{C}$$

Since $\mathbf{G}$ is arbitrary, this imposes a form but not numerical values on $\mathbf{F}$. Further, we now have

$$\mathbf{e}(k+1) = \mathbf{F}\mathbf{e}(k) = \mathbf{F}^k \mathbf{e}(0)$$

Therefore if $\mathbf{G}$ is chosen so that the eigenvalues of $\mathbf{F}$ are inside the unit circle, $\mathbf{e}(k) \to 0$. This is a problem similar to the pole-placement problem for control laws (Chapter 23). It turns out that $\mathbf{G}$ can be chosen to place the poles of $\mathbf{F}$ arbitrarily provided $(\mathbf{A};\mathbf{C})$ is observable, i.e. that the matrix

$$\mathbf{Q}_o = \begin{bmatrix} \mathbf{C} \\ \mathbf{CA} \\ \mathbf{CA}^2 \\ \vdots \\ \mathbf{CA}^{n-1} \end{bmatrix} \tag{25.5}$$

has full rank: rank $(\mathbf{Q}_o) = n$. Poles are often placed so that they are closer to the origin, and hence are associated with faster responses, than those of the system being observed; this typically means that the largest magnitude of an observer pole is taken as considerably smaller than the smallest system pole. This can be overdone, however, as very fast response from the observer means it may follow too faithfully the (ignored) noise of the measurements.

Example

An extremely simple example is given by the system

$$\mathbf{x}(k+1) = \begin{bmatrix} 1 & 1 \\ 0 & 1 \end{bmatrix} \mathbf{x}(k) + \begin{bmatrix} 0 \\ 1 \end{bmatrix} u(k)$$

$$y(k) = x_1(k) = \begin{bmatrix} 1 & 0 \end{bmatrix} \mathbf{x}(k)$$

The observer is to be of the form (25.2), for which

$$\mathbf{F} = \mathbf{A} - \mathbf{GC} = \begin{bmatrix} 1 & 1 \\ 0 & 1 \end{bmatrix} - \begin{bmatrix} g_1 \\ g_2 \end{bmatrix} \begin{bmatrix} 1 & 0 \end{bmatrix}$$

The characteristic equation of $\mathbf{F}$ is

$$\lambda^2 + (g_1 - 2)\lambda + (1 - g_1 + g_2) = 0$$

If we choose to place the poles at $\pm\frac{1}{2}$, then the desired characteristic equation is

$$(\lambda - 0.5)\,(\lambda + 0.5) = \lambda^2 - 0.25 = 0$$

Matching coefficients then gives $g_1 = 2$ and $g_2 = 0.75$ and the observer is

$$\hat{\mathbf{x}}(k+1) = \begin{bmatrix} -1 & 1 \\ -0.75 & 1 \end{bmatrix} \hat{\mathbf{x}}(k) + \begin{bmatrix} 2 \\ 0.75 \end{bmatrix} y(k) + \begin{bmatrix} 0 \\ 1 \end{bmatrix} u(k)$$

Simulation results for this are shown in Fig. 25.1. The choices $\{u(0)\} = \{0\}$, $\mathbf{x}(0) = [\ 10\ \ 1\ ]^{\mathrm{T}}$, $\hat{\mathbf{x}}(0) = [\ 0\ \ 0\ ]^{\mathrm{T}}$ were made arbitrarily.

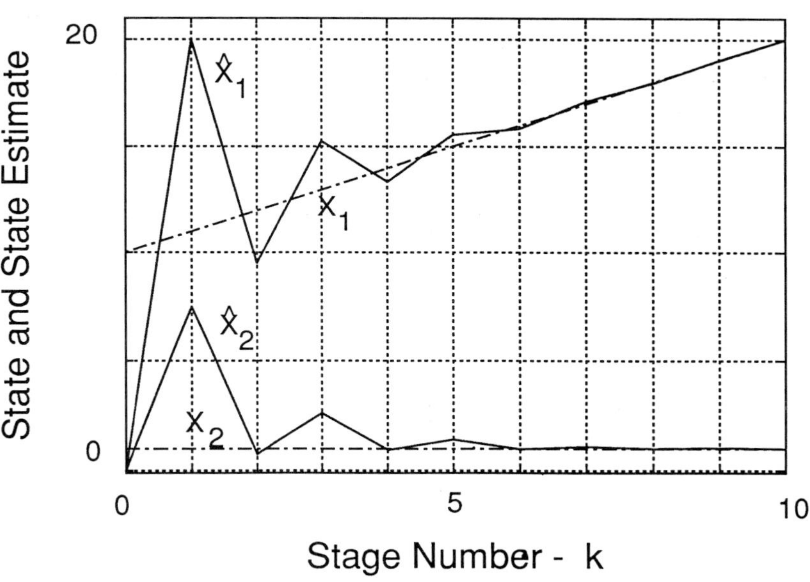

Figure 25.1 Actual system state (dashed line) and estimator state (solid line) for identity observer example.

It is interesting to rearrange the observer into the form

$$\hat{\mathbf{x}}(k+1) = \mathbf{A}\hat{\mathbf{x}}(k) + \mathbf{B}u(k) + \mathbf{G}(y(k) - \mathbf{C}\hat{\mathbf{x}}(k))$$

This has the interpretation

next observer estimate =

{prediction based on old estimate and command}

+ gain × {measurement − measurement predicted by old estimate}

which makes the observer appear an entirely sensible thing to do, and it turns out that many state estimators for linear systems may be put in this form.

25.2.2 General design equations for identity observers

The fairly general case of identity observer design is analogous to that for pole-placement controllers, and involves using a similarity transformation.   We concentrate on the system with scalar measurements

$$\mathbf{x}(k+1) = \mathbf{A}\mathbf{x}(k) + \mathbf{b}u(k)$$

$$y(k) = \mathbf{c}^{\mathrm{T}}\mathbf{x}(k)$$

and characteristic equation

$$\lambda^n + a_n\lambda^{n-1} + \cdots + a_1 = 0$$

for which the observer is

$$\hat{\mathbf{x}}(k+1) = \mathbf{F}\hat{\mathbf{x}}(k) + \mathbf{G}y(k) + \mathbf{H}u(k)$$

subject to

$$\mathbf{H} = \mathbf{b} \qquad\qquad \mathbf{F} = \mathbf{A} - \mathbf{G}\mathbf{c}^{\mathrm{T}}$$

The desired characteristic equation of $\mathbf{F}$ is

$$\lambda^n + d_1\lambda^{n-1} + \cdots + d_n = 0$$

We define the similarity transformation (change of coordinates) $\mathbf{P}$ by its result, which is the companion form of $\mathbf{A}$ given by

$$\mathbf{PAP^{-1}} = \begin{bmatrix} 0 & 0 & \cdots & 0 & -a_n \\ 1 & 0 & 0 & 0 & -a_{n-1} \\ 0 & 1 & 0 & 0 & -a_{n-2} \\ \vdots & \vdots & \ddots & \vdots & \vdots \\ 0 & 0 & \cdots & 1 & -a_1 \end{bmatrix}$$

and

$$\mathbf{c^T P^{-1}} = [\, 0 \ \ 0 \ \ 0 \ \cdots \ 0 \ \ 1 \,]$$

We can choose $\mathbf{PG} = [\, k_n \ \ k_{n-1} \ \ k_{n-2} \ \ ... \ \ k_1 \,]^T$ so that finally the original system is transformed to the canonical form, called an **observable canonical form**,

$$\mathbf{w}(k+1) = \begin{bmatrix} 0 & 0 & \cdots & 0 & -a_n \\ 1 & 0 & 0 & 0 & -a_{n-1} \\ 0 & 1 & 0 & 0 & -a_{n-2} \\ \vdots & \vdots & \ddots & \vdots & \vdots \\ 0 & 0 & \cdots & 1 & -a_1 \end{bmatrix} \mathbf{w}(k) + \mathbf{Pb}u(k)$$

$$y(k) = [\, 0 \ \ 0 \ \ 0 \ \cdots \ 0 \ \ 1 \,] \mathbf{w}(k)$$

The observer for this system will have

$$\mathbf{z}(k+1) = (\mathbf{PAP^{-1}} - \mathbf{PGc^T P^{-1}}) \, \mathbf{z}(k) + \mathbf{Pb}u(k) + \mathbf{PG}y(k)$$

which because of its special form has

$$\mathbf{z}(k+1) = \begin{bmatrix} 0 & 0 & \cdots & 0 & -a_n - k_n \\ 1 & 0 & 0 & 0 & -a_{n-1} - k_{n-1} \\ 0 & 1 & 0 & 0 & -a_{n-2} - k_{n-2} \\ \vdots & \vdots & \ddots & \vdots & \vdots \\ 0 & 0 & \cdots & 1 & -a_1 - k_1 \end{bmatrix} \mathbf{z}(k) + \begin{bmatrix} k_n \\ k_{n-1} \\ k_{n-2} \\ \vdots \\ k_1 \end{bmatrix} y(k) + \mathbf{P}bu(k)$$

Since the a_i are coefficients of the characteristic equation of the original system and the $a_i + k_i$ are clearly the observer's coefficients, we simply choose $k_i = -a_i + d_i$, where the d_i are the desired coefficients. Then for our original observer,

$$\mathbf{PG} = [\, k_n \,\ldots\, k_2 \, k_1 \,]^{\mathrm{T}} \qquad \text{or} \qquad \mathbf{G} = \mathbf{P}^{-1}[\, k_n \,\ldots\, k_2 \, k_1 \,]^{\mathrm{T}}$$

To find $\mathbf{P}$ explicitly, we observe that if $\mathbf{Q}_o$ is the original observability matrix and $\mathbf{Q}_{op}$ is the transformed system's matrix, then we have

$$\mathbf{Q}_o = \begin{bmatrix} \mathbf{c}^{\mathrm{T}} \\ \mathbf{c}^{\mathrm{T}}\mathbf{A} \\ \mathbf{c}^{\mathrm{T}}\mathbf{A}^2 \\ \vdots \\ \mathbf{c}^{\mathrm{T}}\mathbf{A}^{n-1} \end{bmatrix}$$

with $\mathbf{Q}_{op}$ having the explicit form

$$\mathbf{Q}_{op} = \begin{bmatrix} 0 & 0 & \cdots & 0 & 1 \\ 0 & \cdots & 0 & 1 & -a_1 \\ \vdots & 0 & 1 & -a_1 & -a_2 + a_1^2 \\ 0 & \reflectbox{$\ddots$} & \reflectbox{$\ddots$} & \vdots & \vdots \\ 1 & -a_1 & \cdots & \cdots & \cdots \end{bmatrix} = \mathbf{Q}_o \mathbf{P}^{-1}$$

From this we have that

$$
\mathbf{P} = \mathbf{Q}_{op}^{-1} \mathbf{Q}_o =
\begin{bmatrix}
a_{n-1} & a_{n-2} & \cdots & a_2 & a_1 & 1 \\
a_{n-2} & a_{n-3} & \cdots & a_1 & 1 & 0 \\
a_{n-3} & a_{n-4} & \cdot^{\cdot} & 1 & 0 & 0 \\
\vdots & \vdots & \cdot^{\cdot} & \cdot^{\cdot} & \vdots & 0 \\
a_1 & 1 & 0 & \cdots & 0 & 0 \\
1 & 0 & 0 & \cdots & 0 & 0
\end{bmatrix}
\mathbf{Q}_o
\qquad (25.6)
$$

where the form of the inverse may be generated by structural arguments.

The above arguments yield the design algorithm as follows.

1. Determine the characteristic equation of the **A** matrix.

$$\lambda^n + a_1 \lambda^{n-1} + \cdots + a_n = 0$$

2. Determine the desired characteristic equation.

$$\lambda^n + d_1 \lambda^{n-1} + \cdots + d_n = 0$$

3. Form **P** as in (25.6) and find $\mathbf{P}^{-1}$.
4. Then for the observer

$$\mathbf{G} = \mathbf{P}^{-1} [\, d_n - a_n \quad \cdots \quad d_1 - a_1 \,]^{\mathrm{T}}$$

$$\mathbf{F} = \mathbf{A} - \mathbf{G}\mathbf{c}^{\mathrm{T}}$$

$$\mathbf{H} = \mathbf{b}$$

and the observer is

$$\hat{\mathbf{x}}(k+1) = \mathbf{F}\,\hat{\mathbf{x}}(k) + \mathbf{G}y(k) + \mathbf{H}u(k)$$

Example

For an example, we again take the system

$$\mathbf{x}(k+1) = \begin{bmatrix} 1 & 1 \\ 0 & 1 \end{bmatrix} \mathbf{x}(k) + \begin{bmatrix} 0 \\ 1 \end{bmatrix} u(k)$$

$$y(k) = x_1(k) = [\, 1 \quad 0 \,]\, \mathbf{x}(k)$$

It is easily computed that the characteristic equation is $\lambda^2 - 2\lambda + 1 = 0$ and the observability matrix is

$$\mathbf{Q}_o = \begin{bmatrix} 1 & 0 \\ 1 & 1 \end{bmatrix}$$

The transformed observability matrix must be

$$\mathbf{Q}_{op} = \begin{bmatrix} 0 & 1 \\ 1 & 2 \end{bmatrix}$$

and we again choose to have poles at ± 0.5, so that the desired characteristic equation is $\lambda^2 - 0.25 = 0$. Then

$$\mathbf{K} = [\, -0.25 - 1 \quad 0 - (-2) \,]^{\mathrm{T}} = [\, -1.25 \quad 2 \,]^{\mathrm{T}}$$

and

$$\mathbf{P} = \mathbf{Q}_{op}^{-1}\, \mathbf{Q}_o = \begin{bmatrix} -2 & 1 \\ 1 & 0 \end{bmatrix} \begin{bmatrix} 1 & 0 \\ 1 & 1 \end{bmatrix} = \begin{bmatrix} -1 & 1 \\ 1 & 0 \end{bmatrix}$$

Finally,

$$\mathbf{G} = \mathbf{P}^{-1}\mathbf{K} = \begin{bmatrix} 0 & 1 \\ 1 & 1 \end{bmatrix} \begin{bmatrix} -1.25 \\ 2 \end{bmatrix} = \begin{bmatrix} 2 \\ 0.75 \end{bmatrix}$$

This is seen to be the same gain as in the original example, and hence it leads to the same observer:

$$\hat{\mathbf{x}}(k+1) = \begin{bmatrix} -1 & 1 \\ -0.75 & 1 \end{bmatrix} \hat{\mathbf{x}}(k) + \begin{bmatrix} 2 \\ 0.75 \end{bmatrix} y(k) + \begin{bmatrix} 0 \\ 1 \end{bmatrix} u(k)$$

An alternative to this is careful use of Ackermann's formula (Chapter 23 and Appendix B), since the methods for observer pole placement are essentially the same as those for pole placement. In fact, one can use Ackermann's formula using transposes of $\mathbf{A}$, $\mathbf{C}$, and $\mathbf{G}$ so that

$$\mathbf{G}^{\mathrm{T}} = -[\,0 \quad 0 \quad \ldots \quad 0 \quad 1\,]\,\mathbf{Q}_{\mathrm{o}}^{-1\mathrm{T}} \left[\prod_{i=1}^{n} (\mathbf{A}^{\mathrm{T}} - \lambda_i \mathbf{I}) \right]$$

or

$$\mathbf{G} = -\left[\prod_{i=1}^{n} (\mathbf{A} - \lambda_i \mathbf{I}) \right] \mathbf{Q}_{\mathrm{o}}^{-1} \begin{bmatrix} 0 \\ \vdots \\ 0 \\ 1 \end{bmatrix}$$

25.2.3 Variation: the reduced-order observer

The identity observer is an n-dimensional system of equations for which the state $\hat{\mathbf{x}}$ is an estimate of the state $\mathbf{x}$ of the original system. It is arguably ridiculous to 'estimate' any of the state variables for which we have measurements, since the model has the measurements being noise-free. (In fact, the observer will have a smoothing property which may be of benefit with real, noisy measurements $\mathbf{y}$.)

To circumvent this philosophical problem, and also to reduce the computational load, Luenberger (1971) proposed a reduced-order observer, in which the dimension of the computational system is $(n - m)$ rather than n. His version, easily generalized, assumed that the m-vector $\mathbf{y}$ in (25.3) had the special $\mathbf{C}$ matrix

$$\mathbf{C} = [\,\mathbf{I}_m \,\vdots\, 0\,]$$

In this case the state can be written

$$\mathbf{x} = \begin{bmatrix} \mathbf{y} \\ \mathbf{w} \end{bmatrix}$$

and hence the dynamics equation may be written

$$\begin{bmatrix} \mathbf{y}(k+1) \\ \cdots \\ \mathbf{w}(k+1) \end{bmatrix} = \begin{bmatrix} \mathbf{A}_{11} & \vdots & \mathbf{A}_{12} \\ \cdots & & \cdots \\ \mathbf{A}_{21} & \vdots & \mathbf{A}_{22} \end{bmatrix} \begin{bmatrix} \mathbf{y}(k) \\ \cdots \\ \mathbf{w}(k) \end{bmatrix} + \begin{bmatrix} \mathbf{B}_1 \\ \cdots \\ \mathbf{B}_2 \end{bmatrix} \mathbf{u}(k)$$

As $\mathbf{y}$ is known from our measurements, this may be interpreted as a system with $(n-m)$-dimensional state vector $\mathbf{w}$ having dynamics

$$\mathbf{w}(k+1) = \mathbf{A}_{22}\mathbf{w}(k) + \{\mathbf{A}_{21}\mathbf{y}(k) + \mathbf{B}_2\mathbf{u}(k)\}$$

and measurement

$$\mathbf{z}(k) = \mathbf{C}'\,\mathbf{w}(k) = \mathbf{A}_{12}\,\mathbf{w}(k)$$

$$= \{\mathbf{y}(k+1) - \mathbf{A}_{11}\mathbf{y}(k) - \mathbf{B}_1\mathbf{u}(k)\} \tag{25.7}$$

where $\{\ \}$ denote quantities which are known numerically. Using this model and earlier results, (25.2) and (25.4), an identity observer may be designed for $\mathbf{w}(k)$ as

$$\hat{\mathbf{w}}(k+1) = (\mathbf{A}_{22} - \mathbf{L}\mathbf{A}_{12})\,\hat{\mathbf{w}}(k)$$

$$+\, \mathbf{L}\,\{\mathbf{y}(k+1) - \mathbf{A}_{11}\mathbf{y}(k) - \mathbf{B}_1\mathbf{u}(k)\}$$

$$+\, \{\mathbf{A}_{21}\mathbf{y}(k) + \mathbf{B}_2\mathbf{u}(k)\} \tag{25.8}$$

where $\mathbf{L}$ is the observer gain and can be chosen as desired provided that the system $(\mathbf{A}_{22}; \mathbf{A}_{12})$ is observable, i.e. that the matrix analogous to (25.5) with $\mathbf{A}_{22}$ interpreted as $\mathbf{A}$ and $\mathbf{A}_{12}$ interpreted as $\mathbf{C}$, has full rank $(n-m)$. The estimate of the original state $\mathbf{x}$ is now

$$\hat{\mathbf{x}}(k) = \begin{bmatrix} \mathbf{y}(k) \\ \hat{\mathbf{w}}(k) \end{bmatrix}$$

If the $\mathbf{y}(k+1)$ term in (25.7) is computationally inconvenient, a change of variable may be made by defining $\hat{\mathbf{z}}(k) = \hat{\mathbf{w}}(k) - \mathbf{L}\mathbf{y}(k)$. Then substitution in (25.8) yields

$$\hat{\mathbf{z}}(k+1) = (\mathbf{A}_{22} - \mathbf{L}\mathbf{A}_{12})\,\hat{\mathbf{z}}(k)$$

$$+ (\mathbf{A}_{22}\mathbf{L} - \mathbf{L}\mathbf{A}_{12}\mathbf{L} - \mathbf{L}\mathbf{A}_{11} + \mathbf{A}_{21})\,\mathbf{y}(k)$$

$$+ (\mathbf{B}_2 - \mathbf{L}\mathbf{B}_1)\,\mathbf{u}(k)$$

and the state estimate is

$$\hat{\mathbf{x}}(k) = \begin{bmatrix} \mathbf{I}_m & \mathbf{O} \\ -\mathbf{L} & \mathbf{I}_p \end{bmatrix}^{-1} \begin{bmatrix} \mathbf{y}(k) \\ \hat{\mathbf{z}}(k) \end{bmatrix}$$

where $p = (n-m)$.

Example

We reconsider the first example

$$\mathbf{x}(k+1) = \begin{bmatrix} 1 & 1 \\ 0 & 1 \end{bmatrix} \mathbf{x}(k) + \begin{bmatrix} 0 \\ 1 \end{bmatrix} u(k)$$

$$y(k) = x_1(k) = [\,1 \quad 0\,]\,\mathbf{x}(k)$$

and rewrite the state definition as

$$\mathbf{x}(k+1) = \begin{bmatrix} y(k+1) \\ w(k+1) \end{bmatrix} = \begin{bmatrix} a_{11}y(k) + a_{12}w(k) \\ a_{21}y(k) + a_{22}w(k) \end{bmatrix} + \begin{bmatrix} b_1 \\ b_2 \end{bmatrix} u(k)$$

$$= \begin{bmatrix} y(k) + w(k) \\ w(k) + u(k) \end{bmatrix}$$

The unknown state w has dynamics

$$w(k+1) = w(k) + u(k)$$

and a 'measurement' $y(k+1) - y(k)$. Its identity observer then is

$$\hat{w}(k+1) = [1 - g]\,\hat{w}(k) + g\,[y(k+1) - y(k)] + u(k)$$

where g is the arbitrary observer gain. Choosing the pole at -0.25 means that $g = 0.75$ and the observer is

$$\hat{w}(k+1) = -0.25\,\hat{w}(k) + 0.75\,[y(k+1) - y(k)] + u(k)$$

$$\hat{x}(k) = \begin{bmatrix} y(k) \\ \hat{w}(k) \end{bmatrix}$$

If basing $\hat{w}(k+1)$ on $y(k+1)$ is inconvenient for timing reasons (which is unlikely to be a problem for such a simple system), we can define

$$\hat{\hat{w}}(k+1) = \hat{w}(k) - g\,y(k) = \hat{w}(k) - 0.75\,y(k)$$

to obtain the alternative observer

$$\hat{\hat{w}}(k+1) = -0.25\,\hat{\hat{w}}(k) - 0.5625\,y(k) + 0.25\,u(k)$$

and

$$\hat{x}(k) = \begin{bmatrix} y(k) \\ \hat{\hat{w}}(k) + 0.75\,y(k) \end{bmatrix}$$

Simulation results are given in Fig. 25.2.

We remark that it appears that some users feel that reduced-order observers are not worth the design effort, particularly since the smoothing property on $y(k)$ is lost through their use.

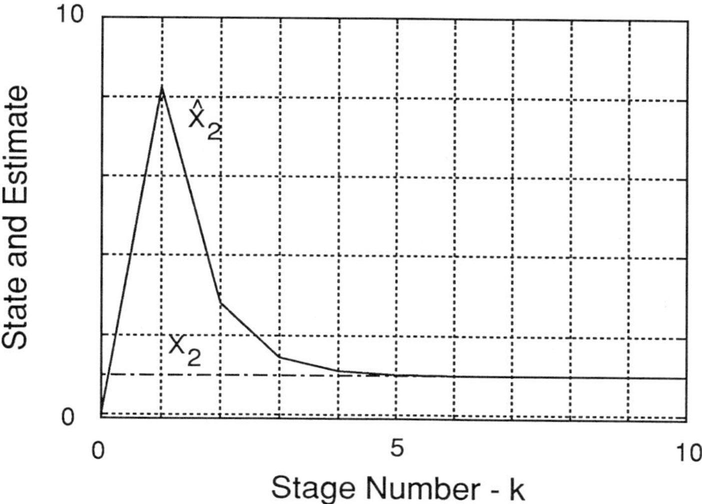

Figure 25.2 Actual (dashed line) and estimated (solid) value of unmeasured state variable for reduced-order observer example.

25.3 A GENERAL FORMULATION

The above sections have seen us consider the design of observers for the state $\mathbf{x}(k)$. Sometimes we may wish to estimate a function $\mathbf{L}\,\mathbf{x}(k)$. We see immediately that this is more general than the situation above; the direct estimation of $\mathbf{x}(k)$ is a special case in which $\mathbf{L} = \mathbf{I}$, and is our familiar identity observer.

As we are generalizing, we also allow the system to be time-varying. Thus let the system be described by

$$\mathbf{x}(k+1) = \mathbf{A}_k\mathbf{x}(k) + \mathbf{B}_k\mathbf{u}(k)$$

$$\mathbf{y}(k) = \mathbf{C}_k\mathbf{x}(k) \tag{25.9}$$

and we wish to design an observer/estimator with variable $\mathbf{z}(k)$ such that

$$\mathbf{z}(k+1) = \mathbf{F}_k\mathbf{z}(k) + \mathbf{G}_k\mathbf{y}(k) + \mathbf{H}_k\mathbf{u}(k)$$

and

$$\mathbf{w}(k) = [\ \mathbf{P}_k \quad \mathbf{V}_k\ ] \begin{bmatrix} \mathbf{z}(k) \\ \mathbf{y}(k) \end{bmatrix} \tag{25.10}$$

with $\mathbf{w}(k)$ estimating $\mathbf{L}_k\,\mathbf{x}(k)$ and $[\ \mathbf{z}^T(k) \quad \mathbf{y}^T(k)\ ]^T$ an n-vector or longer. Assume that $\mathbf{z}(k)$ estimates $\mathbf{T}_k\mathbf{x}(k)$ for some $\mathbf{T}_k$ to be determined. Then if

$$\mathbf{e}(k) = \mathbf{z}(k) - \mathbf{T}_k\mathbf{x}(k)$$

we may write, using this expression and (25.9) in (25.10), that

$$\mathbf{w}(k) = \mathbf{P}_k\,[\mathbf{e}(k) + \mathbf{T}_k\mathbf{x}(k)] + \mathbf{V}_k\,\mathbf{C}_k\mathbf{x}(k)$$
$$= \mathbf{P}_k\,\mathbf{e}(k) + [\mathbf{P}_k\,\mathbf{T}_k + \mathbf{V}_k\,\mathbf{C}_k\,]\,\mathbf{x}(k)$$

Then we see that if $\mathbf{e}(k) \to 0$ and $\mathbf{P}_k\,\mathbf{T}_k + \mathbf{V}_k\,\mathbf{C}_k = \mathbf{L}_k$, we have that $\mathbf{w}(k) \to \mathbf{L}_k\mathbf{x}(k)$. To make $\mathbf{e}(k) \to 0$, we need

$$\mathbf{e}(k+1) = \mathbf{z}(k+1) - \mathbf{T}_{k+1}\,\mathbf{x}(k+1)$$
$$= \mathbf{F}_k\,\mathbf{z}(k) + \mathbf{G}_k\,\mathbf{y}(k) + \mathbf{H}_k\,\mathbf{u}(k)$$
$$-\ \mathbf{T}_{k+1}\,\mathbf{A}_k\,\mathbf{x}(k) - \mathbf{T}_{k+1}\,\mathbf{B}_k\,\mathbf{u}(k)$$
$$= \mathbf{F}_k\ [\mathbf{z}(k) + \mathbf{T}_k\mathbf{x}(k)] + [\mathbf{F}_k\,\mathbf{T}_k + \mathbf{G}_k\,\mathbf{C}_k - \mathbf{T}_{k+1}\,\mathbf{A}_k]\,\mathbf{x}(k)$$
$$+\ [\mathbf{H}_k - \mathbf{T}_{k+1}\,\mathbf{B}_k]\,\mathbf{u}(k)$$
$$= \mathbf{F}_k\,\mathbf{e}(k) + [\mathbf{F}_k\,\mathbf{T}_k + \mathbf{G}_k\,\mathbf{C}_k - \mathbf{T}_{k+1}\,\mathbf{A}_k]\ \mathbf{x}(k)$$
$$+\ [\mathbf{H}_k - \mathbf{T}_{k+1}\,\mathbf{B}_k]\,\mathbf{u}(k)$$

to be a stable system unaffected by the actual state and inputs. Thus our design problem is solved by choosing:

1. $\mathbf{F}_k$ such that $\mathbf{e}(k)$ is always a stable sequence
2. $\mathbf{F}_k\,\mathbf{T}_k + \mathbf{G}_k\,\mathbf{C}_k - \mathbf{T}_{k+1}\,\mathbf{A}_k = 0$

3.　　$\mathbf{H}_k = \mathbf{T}_{k+1}\mathbf{B}_k$
4.　　$\mathbf{P}_k\mathbf{T}_k + \mathbf{V}_k\mathbf{C}_k = \mathbf{L}_k$

Provided $\dim(\mathbf{z}) \geq \dim(\mathbf{x}) - \dim(\mathbf{y}) = n - m$, a solution of this is usually possible (O'Reilly, 1983). We notice that if $\mathbf{L}_k = \mathbf{I}$, then the choices $\mathbf{F}_k = \mathbf{T}_{k+1}\mathbf{A}_k\mathbf{P}_k$ and $\mathbf{G}_k = \mathbf{T}_{k+1}\mathbf{A}_k\mathbf{V}_k$ are explicit values for $\mathbf{F}_k$ and $\mathbf{G}_k$.

We consider further the case $\mathbf{L}_k = \mathbf{I}$, and in particular the condition $\mathbf{z}(k) \rightarrow \mathbf{T}_k\mathbf{x}(k)$. The result is that provided $\mathbf{T}_k$ is $s \times n$, with $s \geq n-m$, and there exist matrices $\mathbf{P}_k$ and $\mathbf{V}_k$ such that $\mathbf{P}_k\mathbf{T}_k + \mathbf{V}_k\mathbf{C}_k = \mathbf{I}$, then

$$\mathbf{z}(k+1) = \mathbf{T}_{k+1}\mathbf{A}_k\mathbf{P}_k\mathbf{z}(k) + \mathbf{T}_{k+1}\mathbf{A}_k\mathbf{V}_k\mathbf{y}(k) + \mathbf{T}_{k+1}\mathbf{B}_k\mathbf{u}(k)$$

satisfies $\mathbf{z}(k) = \mathbf{T}_k\mathbf{x}(k)$, and $\mathbf{x}(k)$ is estimated by

$$\hat{\mathbf{x}}(k) = \mathbf{P}_k\mathbf{z}(k) + \mathbf{V}_k\mathbf{y}(k)$$

The above results may be interpreted and used in three ways:

1.　　to design a feedback law, with $\mathbf{L}_k$ as the desired feedback;
2.　　as a method to arrive at a reduced-order observer; and
3.　　to give 'optimal' observers for the case in which noise is explicitly considered (Chapter 28).

We start with an example demonstrating the first two interpretations.

Example

We again return to the simple example

$$\mathbf{x}(k+1) = \begin{bmatrix} 1 & 1 \\ 0 & 1 \end{bmatrix}\mathbf{x}(k) + \begin{bmatrix} 0 \\ 1 \end{bmatrix}u(k)$$

$$y(k) = x_1(k) = [\,1 \quad 0\,]\,\mathbf{x}(k)$$

We wish to estimate $\mathbf{L}_k\mathbf{x}(k) = x_2(k)$, so that

$$\mathbf{w}(k) = \mathbf{L}_k\mathbf{x}(k) = [0\ \ 1]\mathbf{x}(k)$$

We choose $F_k = 0.25$ and find that $\mathbf{T}_k = [\alpha \;\; \beta]$ must satisfy

$$0.25\,[\,\alpha \;\; \beta\,] + G_k[\,1 \;\; 0\,] - [\,\alpha \;\; \beta\,]\begin{bmatrix} 1 & 1 \\ 0 & 1 \end{bmatrix} = 0$$

yielding $\alpha = 4G_k/3$ and $\beta = -16G_k/9$. Examining the other two constraints yields

$$P_k \begin{bmatrix} \frac{4}{3} & \frac{-16}{9} \end{bmatrix} G_k + V_k\,[\,1 \;\; 0\,] = \mathbf{L}_k = [0 \;\; 1]$$

and

$$H_k = \beta = \frac{-16G_k}{9}$$

We find that $V_k = 0.75$ for any solution of this, with $P_k = 1/\beta$ related to G_k, with one or the other to be chosen. The choice $G_k = -9/16$ gives a comparison with our reduced-order observer above, yielding $H_k = 1$, and $P_k = 1$, and finally

$$z(k+1) = 0.25\,z(k) - 0.5625\,y(k) + u(k)$$

$$\hat{x}_2(k) = w(k) = P_k z(k) + V_k y(k) = z(k) + 0.75\,y(k)$$

The third interpretation requires more detail concerning random processes than we are prepared for at this time, so we settle for a demonstration that V_k may be considered the design parameter with no loss of generality. First, we elaborate upon the definitions above by taking

$$\begin{bmatrix} \mathbf{z}(k) \\ \mathbf{y}(k) \end{bmatrix} = \begin{bmatrix} \mathbf{T}_k \\ \mathbf{C}_k \end{bmatrix} \mathbf{x}(k)$$

and having this be invertible (a restriction on the generality of $\mathbf{T}_k$) so that

$$\mathbf{x}(k) = \begin{bmatrix} \mathbf{T}_k \\ \mathbf{C}_k \end{bmatrix}^{-1} \begin{bmatrix} \mathbf{z}(k) \\ \mathbf{y}(k) \end{bmatrix} = [\,\mathbf{P}_k \quad \mathbf{V}_k\,] \begin{bmatrix} \mathbf{z}(k) \\ \mathbf{y}(k) \end{bmatrix} \qquad (25.11)$$

We define the state error $\mathbf{e}(k)$ and compute its transition

$$\begin{aligned}
\mathbf{e}(k+1) &= \hat{\mathbf{x}}(k+1) - \mathbf{x}(k+1) \\[4pt]
&= \mathbf{P}_{k+1}\mathbf{z}(k+1) + \mathbf{V}_{k+1}\mathbf{y}(k+1) - \mathbf{x}(k+1) \\[4pt]
&= \mathbf{P}_{k+1}\mathbf{T}_{k+1}\left(\mathbf{A}_k\mathbf{P}_k\mathbf{z}(k) + \mathbf{B}_k\mathbf{u}(k) + \mathbf{A}_k\mathbf{V}_k\mathbf{y}(k)\right) \\
&\quad + \mathbf{V}_{k+1}\mathbf{C}_{k+1}\mathbf{x}(k+1) - \mathbf{x}(k+1) \\[4pt]
&= \mathbf{P}_{k+1}\mathbf{T}_{k+1}\left(\mathbf{A}_k\mathbf{P}_k\mathbf{z}(k) + \mathbf{B}_k\mathbf{u}(k) + \mathbf{A}_k\mathbf{V}_k\mathbf{y}(k)\right) \\
&\quad + \left(\mathbf{V}_{k+1}\mathbf{C}_{k+1} - \mathbf{I}\right)\left(\mathbf{A}_k\mathbf{x}(k) + \mathbf{B}_k\mathbf{u}(k)\right) \\[4pt]
&= \left(\mathbf{I} - \mathbf{V}_{k+1}\mathbf{C}_{k+1}\right)\mathbf{A}_k\left(\mathbf{P}_k\mathbf{z}(k) + \mathbf{V}_k\mathbf{y}(k) - \mathbf{x}(k)\right) \\[4pt]
&= \left(\mathbf{I} - \mathbf{V}_{k+1}\mathbf{C}_{k+1}\right)\mathbf{A}_k\mathbf{e}(k)
\end{aligned}$$

Hence we want eigenvalues of $[\mathbf{I} - \mathbf{V}_{k+1}\mathbf{C}_{k+1}]\mathbf{A}_k$ such that $\|\mathbf{e}(k)\|$ decreases. Thus $\mathbf{V}_k$ is the key to design, with $\mathbf{P}_k$ and $\mathbf{T}_k$ chosen to satisfy (25.11).

25.4 OBSERVERS IN CONTINUOUS TIME SYSTEMS

Observers for continuous time systems are essentially the same in concept as those for discrete time systems. Thus, for the system described by

$$\dot{\mathbf{x}} = \mathbf{A}\mathbf{x} + \mathbf{B}\mathbf{u}$$

$$\mathbf{y} = \mathbf{C}\mathbf{x}$$

we seek to find an estimate $\mathbf{z}$ of $\mathbf{T}\mathbf{x}$ using an algorithm of the form

$$\dot{z} = Fz + Gy + Hu$$

We consider the error $e = z - Tx$ and find that it evolves as

$$\dot{e} = Fz + Gy + Hu - TAx - TBu$$

$$= Fe + [GC + FT - TA]x + [H - TB]u$$

The choices

$$H = TB$$

$$GC + FT - TA = 0$$

ensure that the error is independent of the input u and the actual state x. Choosing F such that its eigenvalues are all in the left-half plane ensures $e \to 0$.

As usual, the simple case is the identity observer, for which $T = I$. Then we choose $H = B$ and G such that the eigenvalues of $(A - GC)$ (which are the eigenvalues of F) are stable. The latter is simply the pole-placement requirement and may be attacked as before or, by taking transposes, placed in the form $(A^T - C^T G^T)$ with unknown G^T which was met in Chapter 23.

Reduced-order observers and observers of functions are also analogous to the discrete-time case.

25.5 COMPUTER AIDS

As with pole placement, observer design is easy, as is the implementation of observers. Any algorithm which places poles can be adapted to observer design.

25.6 SUMMARY

With the goal of creating algorithms for estimating the state of dynamic systems, we have considered the deterministic Luenberger observer. We started with simple and then more general approaches to design of the most common form – the identity observer. This was followed by the reduced-order identity observer and then a fairly general observer formulation.

25.7 FURTHER READING

The original paper of Luenberger (1964) or his introduction paper (1971) are worth reading, as they give a view of the elements of the theory. Many texts, such as that of Franklin, Powell and Workman (1990), have at least some consideration of basic observer theory for computer systems. One elementary text with some discussion of continuous time systems is by Hostetter *et al.* (1989).

A book devoted entirely to observers is that of O'Reilly (1983).

26

Optimal control by multiplier-type methods

Optimal control theory is concerned with the mathematics of finding a function or set of parameters which cause a given functional (function of a function) to take on an extremal value – minimum or maximum – subject to constraints. Mathematically, the continuous time problem is often of the following form.

Find a function $\mathbf{u}(t)$ from a defined set of candidates U called the **admissible control** set such that a functional $J(\mathbf{u}(t))$, such as

$$J(\mathbf{u}(t)) = \int_{t_0}^{t_f} g(\mathbf{x}(\tau), \mathbf{u}(\tau)) \, d\tau$$

is minimized and for which the constraints

1. $\dot{\mathbf{x}}(t) = \mathbf{f}(\mathbf{x}(t), \mathbf{u}(t), t)$ for $t \in (t_0, t_f)$, and
2. $\mathbf{x}(t)$ is an admissible state, $\mathbf{x}(t) \in \mathbf{X}(t)$, $\forall \, t \in (t_0, t_f)$

are satisfied.

In this, t_0 is the initial time and is usually fixed, while t_f is the final or terminal time and may be fixed or flexible, finite or infinite, depending on the problem. The sets $\mathbf{X}(t)$ may specify only boundary conditions, or may represent ongoing constraints on the solutions, $\mathbf{x}(t)$, often called trajectories. Discrete time versions of such problems have also been studied. Part of the engineer's task is to use this mathematics to help in designing control laws for real systems. This is done for two reasons.

1. The optimal, or a near approximation, may be incorporated into the actual system control. This is true for space launcher trajectories, for example.

2. The optimal gives a standard against which the system can be judged, even if the result is too complicated for real-time implementation. For example, some optimal controllers in simple systems have been found to yield closed-loop responses which are rather like underdamped second-order systems for which the damping coefficient is around 0.6. Also, knowing the mathematical optimum payload of an airplane can help indicate the cost of legally-imposed constraints or engineering safety factors.

The standard results available are for linear systems and two criteria: minimum time to attain a desired state, and minimum average quadratic function of state and control for a fixed time of operation. Techniques are available for general problems, but then the results are seldom reducible to straightforward algorithms.

26.1 SYNOPSIS

One of the important methods of finding optimal trajectories is the **Pontryagin Maximum Principle**, which may be characterized as non-classical calculus of variations. Its basic formula is as follows.

The control $\mathbf{u}^{\#}(t)$ which causes a system

$$\dot{\mathbf{x}} = \mathbf{f}(\mathbf{x}(t), \mathbf{u}(t), t) \qquad \mathbf{x}(t_0) = \mathbf{x}_0$$

to follow an admissible trajectory $\mathbf{x}^{\#}(t)$ that minimizes the performance measure

$$J(\mathbf{u}) = h(\mathbf{x}(t_f), t_f) + \int_{t_0}^{t_f} g(\mathbf{x}(\tau), \mathbf{u}(\tau), \tau)\, d\tau$$

must necessarily satisfy the following conditions, expressed in terms of the Hamiltonian $\mathcal{H}$ defined as

$$\mathcal{H}(\mathbf{x}(t), \mathbf{u}(t), \mathbf{p}(t), t) \equiv -g(\mathbf{x}(t), \mathbf{u}(t), t) + \mathbf{p}^{\mathrm{T}}(t)\, \mathbf{f}(\mathbf{x}(t), \mathbf{u}(t), t)$$

such that

1. $$\dot{\mathbf{x}}^{\#}(t) = \left. \frac{\partial \mathscr{H}}{\partial \mathbf{p}} \right|_{(\mathbf{x}^{\#}(t),\,\mathbf{u}^{\#}(t),\,\mathbf{p}^{\#}(t),\,t)}$$

2. $$\dot{\mathbf{p}}^{\#}(t) = - \left. \frac{\partial \mathscr{H}}{\partial \mathbf{x}} \right|_{(\mathbf{x}^{\#}(t),\,\mathbf{u}^{\#}(t),\,\mathbf{p}^{\#}(t),\,t)}$$

3. $\mathscr{H}(\mathbf{x}^{\#}(t), \mathbf{u}^{\#}(t), \mathbf{p}^{\#}(t), t) \geq \mathscr{H}(\mathbf{x}^{\#}(t), \mathbf{u}(t), \mathbf{p}^{\#}(t), t)$ for all admissible $\mathbf{u}(t)$

4.   The final boundary conditions satisfy

$$\left[\left. \frac{\partial \mathbf{h}}{\partial \mathbf{x}} \right|_{(\mathbf{x}^{\#}(t_{\mathrm{f}}),\,t_{\mathrm{f}})} - \mathbf{p}^{\#}(t_{\mathrm{f}}) \right] \delta x_{\mathrm{f}} +$$

$$\left[\mathscr{H}(\mathbf{x}^{\#}(t), \mathbf{u}^{\#}(t), \mathbf{p}^{\#}(t), t) + \left. \frac{\partial \mathbf{h}}{\partial \mathbf{x}} \right|_{(\mathbf{x}^{\#}(t_{\mathrm{f}}),\,t_{\mathrm{f}})} \right] \delta t_{\mathrm{f}} = 0$$

A discretized version of this can be set out which is sometimes, but not always, helpful. In this case, for the system

$$\mathbf{x}(k+1) = \mathbf{f}(\mathbf{x}(k), \mathbf{u}(k), k) \qquad \mathbf{x}(0) = \mathbf{x}_0 \text{ given}$$

for which the function

$$J(\{\mathbf{u}(k), k=0, 1, \ldots, N-1\}, \mathbf{x}(0), N\,) \;=\; h\,(\,N\,, \mathbf{x}\,(\,N\,)) \;+\; \sum_{i=0}^{N-} $$

$1,)g(\mathbf{x}(i), \mathbf{u}(i), i)$

is to be minimized, define the Hamiltonian function

$$\mathscr{H}(\mathbf{x}(i), \mathbf{u}(i), \lambda(i+1), i) = -g(\mathbf{x}(i), \mathbf{u}(i), i) + \lambda^{\mathrm{T}}(i+1)\, \mathbf{f}(\mathbf{x}(i), \mathbf{u}(i), i)$$

where $\lambda(\cdot)$ is the sequence of Lagrange multipliers (sometimes called costates in this application) for this problem and must be determined.

Then the necessary conditions to be satisfied by the optimal state vector $\mathbf{x}^{\#}(i)$, control vector $\mathbf{u}^{\#}(i)$, and costate vector $\lambda^{\#}(i)$ are that

$$\mathbf{x}^{\#}(k+1) = \frac{\partial \mathcal{H}}{\partial \lambda(k+1)}\bigg|_{(\mathbf{x}^{\#}(k),\, \mathbf{u}^{\#}(k),\, \lambda^{\#}(k))} = \mathbf{f}(\mathbf{x}^{\#}(k), \mathbf{u}^{\#}(k), k)$$

$$\lambda^{\#}(k) = \frac{\partial \mathcal{H}}{\partial \mathbf{x}(k)}\bigg|_{(\mathbf{x}^{\#}(k),\, \mathbf{u}^{\#}(k),\, \lambda^{\#}(k))} \qquad k=0, 1, \ldots, N-1$$

$$\mathcal{H}(\mathbf{x}^{\#}(i), \mathbf{u}^{\#}(i), \lambda^{\#}(i), i) \geq \mathcal{H}(\mathbf{x}^{\#}(i), \mathbf{u}(i), \lambda^{\#}(i), i)$$

for all admissible $\mathbf{u}(i)$ with boundary conditions, depending upon the problem statement, such as

$$\mathbf{x}^{\#}(0) = \mathbf{x}_0$$

$$\lambda^{\#}(N) = \frac{\partial h(N, \mathbf{x})}{\partial \mathbf{x}}\bigg|_{\mathbf{x}^{\#}(N)}$$

A standard and useful result is that for the system described by

$$\mathbf{x}(k+1) = \mathbf{A}\mathbf{x}(k) + \mathbf{B}\mathbf{u}(k)$$

the control law which minimizes

$$\mathbf{J} = [\mathbf{x}(N) - \mathbf{x}_d(N)]^T \mathbf{S}\, [\mathbf{x}(N) - \mathbf{x}_d(N)]/2$$

$$+ \sum_{i=0}^{N-1} \big\{ [\mathbf{x}(i) - \mathbf{x}_d(i)]^T \mathbf{Q}\,[\mathbf{x}(i) - \mathbf{x}_d(i)] + \mathbf{u}^T(i)\, \mathbf{R}\mathbf{u}(i) \big\}/2$$

where $\mathbf{x}_d(i)$ is the desired state sequence, is given by

$$\mathbf{u}^{\#}(k) = -\,[\mathbf{R} + \mathbf{B}^T \mathbf{P}(k+1)\mathbf{B}]^{-1} \mathbf{B}^T \mathbf{b}(k+1) + \mathbf{K}(k)\mathbf{x}(k)$$

where

$$\mathbf{K}(k) = -\,[\mathbf{R} + \mathbf{B}^T \mathbf{P}(k+1)\mathbf{B}]^{-1} \mathbf{B}^T \mathbf{P}(k+1)\mathbf{A}$$

$$\mathbf{b}(k) = [\mathbf{A}^T + \mathbf{K}^T(k)\mathbf{B}^T]\,\mathbf{b}(k+1) - \mathbf{Q}\mathbf{x}_d(k)$$

$$\mathbf{P}(k) = \mathbf{Q} + \mathbf{A}^T \mathbf{P}(k+1)\mathbf{A} + \mathbf{A}^T \mathbf{P}(k+1)\mathbf{K}(k)$$

with boundary conditions

$$\mathbf{P}(N) = \mathbf{S} \quad \text{and} \quad \mathbf{b}(N) = -\mathbf{S}\mathbf{x}_\mathrm{d}(N)$$

This is the **linear quadratic tracking problem**; its special case in which $\mathbf{x}_\mathrm{d}(i) = \mathbf{0}$, and hence $\mathbf{b}(i) = \mathbf{0}$, for all i is the **linear quadratic regulator (LQR) problem**, although sometimes the latter term is restricted to the case for which $N \to \infty$ and the resulting gain $\mathbf{K}(k)$ is a constant.

We build up to this result and give examples, including linear and non-linear minimum time problems, in the following sections.

26.2 PERFORMANCE CRITERIA AND APPROACHES TO OPTIMIZATION

Optimization is usually relative to a scalar function of the system states and controls which ascribes a cost or payoff to the system operation. With a given initial condition (or equivalent initialization sequence) $\mathbf{x}(0)$, the history of the system, until we consider noise effects in later chapters, is entirely dependent upon the control sequence and the initial state. On the basis of the arguments the cost J is a function

$$J(\{\mathbf{u}(k), k=0, 1, ..., N-1\}, \mathbf{x}(0), N)$$

and is usually specified in a form such as

$$J(\{\mathbf{u}(k), k=0, 1, ..., N-1\}, \mathbf{x}(0), N) = h(N, \mathbf{x}(N))$$
$$+ \sum_{i=0}^{N-1} g(\mathbf{x}(i), \mathbf{u}(i), i)$$

subject to the system dynamics described by

$$\mathbf{x}(k+1) = \mathbf{f}(\mathbf{x}(k), \mathbf{u}(k), k) \qquad \mathbf{x}(0) = \mathbf{x}_0 \;\; \text{given} \qquad (26.1)$$

Additional constraints may include bounds on the control variables such as

$$\| \mathbf{u}(k) \| \leq U(k) \qquad (26.2.)$$

and constraints on the control and state such as

$$\mathbf{c}(\mathbf{u}(k), \mathbf{x}(k), k) \leq \mathbf{C}(k) \qquad (26.3)$$

The duration of the interval of interest, N, may be finite or infinite, fixed or variable; usually, however, N finite and variable (as in the minimum time case) must receive special treatment. N unbounded is characteristic of regulator problems in process control and is handled either using special methods or as a limiting case as $N \to \infty$; in this and the next section we shall assume N is fixed and finite. Equation (26.1) simply says the system must satisfy a known model, usually based on the problem's physics (or chemistry or biology, as the case may be). The constraints (26.2) and (26.3), defining admissible controls and admissible states, are often difficult to handle explicitly in the mathematics.

The functions $h(\cdot)$ and $g(\cdot)$ could be general, but in most of the literature they are constant or linear in the explicit time variable, may have time-varying weighting parameters, and may have additive terms in the absolute values or in quadratics in the control variable $\mathbf{u}(\cdot)$ and state variable $\mathbf{x}(\cdot)$. The most common state–space form in the literature has

$$J = \tfrac{1}{2}\mathbf{x}^{\mathrm{T}}(N)\,\mathbf{S}_N\,\mathbf{x}(N) + \tfrac{1}{2}\sum_{i=0}^{N-1} \{ \mathbf{x}^{\mathrm{T}}(i)\,\mathbf{Q}_i\,\mathbf{x}(i) + \mathbf{u}^{\mathrm{T}}(i)\,\mathbf{R}_i\,\mathbf{u}(i) \} \qquad (26.4)$$

with dynamics

$$\mathbf{x}(k+1) = \mathbf{A}_k\mathbf{x}(k) + \mathbf{B}_k\mathbf{u}(k) \qquad \mathbf{x}(0) = \mathbf{x}_0 \qquad (26.5)$$

This is called the **linear-quadratic (L-Q) problem** of automatic control: the dynamics are linear and the cost is quadratic in its elements. In the cost, the matrices $\mathbf{S}_N$ and $\mathbf{Q}_i$ are taken to be non-negative definite, and the matrices $\mathbf{R}_i$ are assumed positive definite. Whether the dynamics should be linear was argued in Chapter 9. We note that in much of what follows, the matrices $\mathbf{A}_k$, $\mathbf{B}_k$, $\mathbf{Q}_i$, and $\mathbf{R}_i$ are taken as constants $\mathbf{A}$, $\mathbf{B}$, $\mathbf{Q}$, $\mathbf{R}$, respectively. This is done for convenience in notation, as variation does not affect most of the general results.

The second major problem class is that of minimum time control. Formally, this is to find the minimum value of

$$J = \sum_{i=0}^{N-1} 1 = N \tag{26.6}$$

for which a given set of dynamics such as (26.1), or perhaps specialized to (26.5), is taken from a given initial condition x_0 to a final constraint $h(\mathbf{x}(N),N) = 0$. Usually there is a constraint such as (26.2).

To solve the above problems, there are three classes of methods plus numerical techniques. The first might be called a direct approach, involving manipulation of the solution matrices for the linear problem. The second is the **Principle of Optimality**, or **Bellman's Principle**, approach. The third uses methods based upon the calculus, the calculus of variations, and their extensions. These all have their place in the repertoire of the practising engineer; for the same problem, they all give the same result.

For difficult problems, the methods vary in the insight they provide and the direct utility of their results. Many users will find Bellman's method (see Chapter 27) interesting for the insights obtained, but burdened by the so-named 'curse of dimensionality', which is a need for excessive computer time and storage for problems of reasonable size (say 10 or so state variables). Most engineers find **Pontryagin's Maximum Principle** (non-classical calculus of variations) useful in a cookbook sort of way, but difficult to translate into closed-loop control laws.

Finding feedback control laws rather than optimal trajectories can be a serious problem. The mathematical methods often give the control as a function of time for a given set of initial conditions, $\mathbf{u}(t) = \mathbf{u}(\mathbf{x}_0,t)$, instead of a closed-loop law $\mathbf{u}(t) = \mathbf{u}(\mathbf{x}(t))$, and hence are inherently open-loop in nature. The engineers' fix is to recompute $\mathbf{u}(\mathbf{x}_0,t)$ every few time periods, so that the actual command at time τ is given by $\mathbf{u}(\tau) = \mathbf{u}(\mathbf{x}(t),\tau)$, $t \leq \tau < t+T$, $t=t_0, t_0+T, t_0+2T,\ldots,t_f$; such an approach is sometimes called open-loop feedback.

26.3 LAGRANGE THEORY AND THE PONTRYAGIN MAXIMUM PRINCIPLE

An important approach to optimization is the Lagrange theory of constrained optimization, leading in the case of continuous-time systems to the calculus of variations and its non-classical extension, the Pontryagin theory. The latter has a discrete-time version, but for our purposes it is mostly unnecessary, as the Lagrange theory will do quite nicely.

26.3.1 Parameter optimization and Lagrange multipliers

The problem of optimal control is one of constrained optimization over a function space, where the constraints are due to the dynamics relationships and to the various rules which determine admissible control commands. The basic ideas are developed in the ordinary calculus, however, where the optimization is over a set of numbers (rather than functions). The key idea is that of Lagrange multipliers to convert the constrained problem to an unconstrained one. We quickly review those ideas in this section.

If a function $f(x)$ is to be minimized by choice of a scalar x, it is well known that if the function is suitably 'nice' in that the derivatives exist, then the optimum choice of x for a 'local minimum' is $x^{\#}$ where

$$\left.\frac{\mathrm{d}f}{\mathrm{d}x}\right|_{x=x^{\#}} = 0$$

$$\left.\frac{\mathrm{d}^2f}{\mathrm{d}x^2}\right|_{x=x^{\#}} \geq 0 \tag{26.7}$$

In particular, the above are *necessary* conditions on the optimum $x^{\#}$, but may not be sufficient (because e.g. there may be more than one x which satisfies those conditions, or the second derivative may be 0).

If f is a scalar function of a vector $\mathbf{x}$, the necessary conditions are similar. In particular, it is necessary that the optimum vector $\mathbf{x}^{\#}$ satisfy

$$\left.\nabla_{\mathbf{x}} f(\mathbf{x})\right|_{\mathbf{x}=\mathbf{x}^{\#}} = \left.\left[\frac{\partial f}{\partial x_1} \quad \frac{\partial f}{\partial x_2} \quad \frac{\partial f}{\partial x_3} \quad \cdots \quad \frac{\partial f}{\partial x_n}\right]\right|_{\mathbf{x}=\mathbf{x}^{\#}} = 0$$

$$F_{\mathbf{xx}} = \left[\frac{\partial^2 f}{\partial x_i \, \partial x_j}\right]\Bigg|_{\mathbf{x}=\mathbf{x}^\#} \geq 0 \tag{26.8}$$

where the second condition means that the matrix is positive semi-definite.

Now consider the constrained problem of finding a vector $\mathbf{x}$ which minimizes a scalar function subject to a set of constraints. In particular, find a solution to $\min f(\mathbf{x})$, subject to

$$\mathbf{d}_i(\mathbf{x}) = 0 \qquad i = 1, 2, \ldots, m \tag{26.9}$$

(or $\mathbf{d}(\mathbf{x}) = 0$). To solve this, we look at the problem of finding

$$\min_{\mathbf{x}} \ \min_{\lambda} \ [f(\mathbf{x}) + \lambda^{\mathrm{T}} \mathbf{d}(\mathbf{x})] \tag{26.10}$$

without further constraints. It is a fundamental fact of mathematical programming that the $\mathbf{x}^\#$ which solves (26.9) also is part of the solution (along with $\lambda^\#$) of (26.10). The variable vector λ is called the **vector of Lagrange multipliers**, and using it relies upon a theorem (Luenberger, 1969) which states in essence that: if $\mathbf{x}_0$ is an extremum of the function f subject to the scalar constraints $\mathbf{d}_i(\mathbf{x}) = 0$, $i = 1, 2, \ldots, n$, and if $\nabla_{\mathbf{x}} \mathbf{d}_i(\mathbf{x}_0)$, $i = 1, 2, \ldots, n$ are linearly independent vectors, then there exist n scalars, λ_i, $i = 1, 2, \ldots, n$, such that the function

$$f(\mathbf{x}) + \sum_{i=1}^{n-1} \lambda_i \mathbf{d}_i(\mathbf{x})$$

is stationary at $\mathbf{x}_0$.

For inequality constraints, the **Kuhn–Tucker** theory applies; as seen in any text material on constrained optimization, such as Luenberger (1973).

Treating the problem (26.10) as a vector optimization problem in the extended vector $\mathbf{y}^{\mathrm{T}} = [\mathbf{x}^{\mathrm{T}} \ \lambda^{\mathrm{T}}]$ means we can apply the results for unconstrained optimization. The first-order necessary conditions are then

$$\nabla_x [f(\mathbf{x}) + \lambda^T \mathbf{d}(\mathbf{x})]\Big|_{(\mathbf{x}^\#, \lambda^\#)} = 0$$

$$\nabla_\lambda [f(\mathbf{x}) + \lambda^T \mathbf{d}(\mathbf{x})]\Big|_{(\mathbf{x}^\#, \lambda^\#)} = 0 = \mathbf{d}(\mathbf{x}^\#)$$

The second-order conditions are

$$\left[\frac{\partial^2 [f(\mathbf{x}) + \lambda^T \mathbf{d}(\mathbf{x})]}{\partial y_i\, \partial y_j} \right]_{(\mathbf{x}^\#, \lambda^\#)} \geq 0$$

from which, because λ appears linearly, we get a reduction to

$$\left[\frac{\partial^2 [f(\mathbf{x}) + \lambda^T \mathbf{d}(\mathbf{x})]}{\partial x_i\, \partial x_j} \right]_{(\mathbf{x}^\#, \lambda^\#)} \geq 0$$

Example

Consider the problem $\min (x_1^2 + x_2^2)$ subject to $x_1 + x_2 = 1$. Geometrically, this is seeking the smallest circle centred at the origin which touches the straight line that passes through $(1,0)$ and $(0,1)$; a simple graph shows clearly that this is minimized when $x_1 = x_2 = 1/2$ and the minimum value of $(x_1^2 + x_2^2)$ is $1/2$.

Proceeding formally, the augmented function is

$$x_1^2 + x_2^2 + \lambda(x_1 + x_2 - 1)$$

for which the first-order necessary conditions are

$$2x_1 + \lambda = 0$$

$$2x_2 + \lambda = 0$$

$$x_1 + x_2 - 1 = 0$$

Simultaneous solution yields $\lambda = 1$, $x_1 = x_2 = \lambda/2 = \tfrac{1}{2}$. The minimum value of the performance function is then found by direct calculation. This is the only extremal point to be found, and from the second-order condition

$$\left[\frac{\partial^2 \left(x_1^2 + x_2^2 + \lambda(x_1 + x_2 - 1) \right)}{\partial x_i \, \partial x_j} \right] = \begin{bmatrix} 2 & 0 \\ 0 & 2 \end{bmatrix}$$

which is positive definite, we know this is indeed a minimum. Clearly there is no other candidate for extremal; in fact it is obvious there is an unbounded maximum.

26.3.2 Treatment of discrete time optimum control as a problem in calculus

Let us consider our basic problem as formulated in the previous section. Recall that we wished to minimize

$$J(\{\mathbf{u}(k),\, k=0, 1, \ldots, N-1\}, \mathbf{x}(0), N) = h(N, \mathbf{x}(N)) + \sum_{i=0}^{N-1} g(\mathbf{x}(i), \mathbf{u}(i), i)$$

$$(26.11)$$

subject to

$$\mathbf{x}(k+1) = \mathbf{f}(\mathbf{x}(k), \mathbf{u}(k), k) \qquad \mathbf{x}(0) = \mathbf{x}_0 \text{ given} \qquad (26.12)$$

For a fixed-time problem, (26.12) specifies N constraints to be satisfied in finding the minimum and can be restated as

$$\mathbf{x}(k+1) - \mathbf{f}(\mathbf{x}(k), \mathbf{u}(k), k) = 0, \qquad k = 0, 1, \ldots, N-1$$

Heuristically, we define the augmented cost function, as

$$J_{\mathrm{a}}(\{\mathbf{x}\}, \{\mathbf{u}\}, \{\lambda\}) = -h(N, \mathbf{x}(N)) - \sum_{i=0}^{N-1} g(\mathbf{x}(i), \mathbf{u}(i), i)$$

$$+ \sum_{k=0}^{N-1} \lambda^{\mathrm{T}} \left[\mathbf{f}(\mathbf{x}(k), \mathbf{u}(k), k) - \mathbf{x}(k+1) \right] \qquad (26.13)$$

noting that the minus signs lead us to maximize this as an equivalent to minimizing the original cost, where the n-vectors $\lambda(k)$ are called Lagrange multipliers, or in the Pontryagin theory 'costates'. We may observe that the minimum of (26.11) when the constraints (26.12) are satisfied also gives a maximum of (26.13). The trick is to make the overall maximum of (26.13) lie at the minimum of the original problem. We use the Lagrange theory to find necessary conditions, and occasionally sufficient conditions, for this to occur.

We first rewrite (26.13) for convenience as

$$J_a(\{\mathbf{x}\},\{\mathbf{u}\},\{\lambda\}) = -h(N,\mathbf{x}(N))$$

$$+ \sum_{i=0}^{N-1} \{-g(\mathbf{x}(i),\mathbf{u}(i),i) + \lambda(i+1)^\mathrm{T} [\mathbf{f}(\mathbf{x}(i),\mathbf{u}(i),i) - \mathbf{x}(k+1)\} \qquad (26.14)$$

To find a maximum of J_a we look among the points for which the various first derivatives are 0, to find

$$\frac{\partial J_a}{\partial \mathbf{x}(N)} = -\frac{\partial h(N,\mathbf{x}(N))}{\partial \mathbf{x}(N)} - \lambda(N) = 0$$

$$\frac{\partial J_a}{\partial \mathbf{x}(i)} = -\frac{\partial g(\mathbf{x}(i)\,\mathbf{u}(i),i)}{\partial \mathbf{x}(i)} - \lambda(i) + \frac{\partial \mathbf{f}(\mathbf{x}(N)\,\mathbf{u}(i),i)}{\partial \mathbf{x}(i)} \lambda(i+1) = 0$$

$$i = 0, 1, ..., N-1$$

$$\frac{\partial J_a}{\partial \mathbf{u}(i)} = -\frac{\partial g(\mathbf{x}(i)\,\mathbf{u}(i),i)}{\partial \mathbf{u}(i)} + \frac{\partial \mathbf{f}(\mathbf{x}(N)\,\mathbf{u}(i),i)}{\partial \mathbf{u}(i)} \lambda(i+1) = 0$$

$$i = 0, 1, ..., N-1$$

$$\frac{\partial J_a}{\partial \lambda(i)} = - [\mathbf{x}(i) - \mathbf{f}(\mathbf{x}(i-1),\mathbf{u}(i-1),i-1)] \qquad i = 1, 2, ..., N$$

Satisfying these yields $N \times n$ equations in $\mathbf{x}$, i.e.

$$\mathbf{x}(k+1) = \mathbf{f}(\mathbf{x}(k),\mathbf{u}(k),k)$$

and $n \times (N-1)$ equations for λ, i.e.

$$\lambda(k) = -\frac{\partial g(\mathbf{x}(k), \mathbf{u}(k), k)}{\partial \mathbf{x}(k)} + \frac{\partial f(\mathbf{x}(k), \mathbf{u}(k), k)}{\partial \mathbf{x}(k)} \lambda(k+1)$$

$$k = 1, 2, \ldots, N-1$$

with the $2n$ boundary conditions

$$\mathbf{x}(0) = \mathbf{x}_0$$

$$\lambda(N) = \frac{\partial h((N), \mathbf{x}(N))}{\partial \mathbf{x}(N)}$$

The control is determined to satisfy the $m \times N$ constraints

$$-\frac{\partial g(\mathbf{x}(k), \mathbf{u}(k), k)}{\partial \mathbf{u}(k)} + \frac{\partial f(\mathbf{x}(k), \mathbf{u}(k), k)}{\partial \mathbf{u}(k)} \lambda(k+1) = 0$$

$$k = 0, 1, \ldots, N-1$$

which may or may not give an algebraic closed-form solution. To satisfy these we have $(n+1) \times N$ variables $\mathbf{x}(i)$, $i = 0, 1, \ldots, N$, along with $n \times N$ variables $\lambda(i+1)$, and $m \times N$ variables $\mathbf{u}(i)$, $i = 0, 1, \ldots, N-1$.

We note that although the above gives necessary conditions, they are not necessarily easy to solve. A very important reason for this, regardless of the fact that the functions may be non-linear, is that the N stages of dynamics equations have conditions specified at initial time 0, whereas the N stages of Lagrange multiplier equations have conditions specified on the final stage N. This situation is called a **two-point boundary-value problem (TPBVP)** and is one of the curses of the optimal control theory.

26.3.3 Special application of dynamic Lagrange multipliers: linear quadratic (LQ) problems

To see some of the implications of the above, we examine the LQ problem. The objective is to minimize

$$J = \frac{1}{2}\left(\mathbf{x}^T(N)\,\mathbf{S}\mathbf{x}(N) + \sum_{i=0}^{N-1}\{\mathbf{x}^T(i)\,\mathbf{Q}\mathbf{x}(i) + \mathbf{u}^T(i)\,\mathbf{R}\mathbf{u}(i)\}\right)$$

subject to dynamics

$$\mathbf{x}(k+1) = \mathbf{A}\mathbf{x}(k) + \mathbf{B}\mathbf{u}(k) \qquad \mathbf{x}(0) = \mathbf{x}_0$$

Making the identifications of the various functions, e.g.

$$g(\mathbf{x}(i), \mathbf{u}(i), i) = \frac{1}{2}\{\mathbf{x}^T(i)\,\mathbf{Q}\mathbf{x}(i) + \mathbf{u}^T(i)\,\mathbf{R}\mathbf{u}(i)\}$$

in equation (26.14) yields the set

$$\mathbf{x}(k+1) = \mathbf{A}\mathbf{x}(k) + \mathbf{B}\mathbf{u}(k)$$

$$\lambda(k) = \mathbf{A}^T\lambda(k+1) - \mathbf{Q}\mathbf{x}(k) \tag{26.15}$$

with boundary conditions

$$\mathbf{x}(0) = \mathbf{x}_0 \quad \text{and} \quad \lambda(N) = -\mathbf{S}\mathbf{x}(N) \tag{26.16}$$

and control law

$$-\mathbf{R}\mathbf{u}(k) + \mathbf{B}^T\lambda(k+1) = 0$$

or if $\mathbf{R}$ is invertible (as it will be if it is positive definite)

$$\mathbf{u}(k) = \mathbf{R}^{-1}\mathbf{B}^T\lambda(k+1) \tag{26.17}$$

Although this is a solution to the problem, it is not in a helpful form; to give easy computation from this solution, we need to know $\mathbf{x}(N)$, but we cannot know that until the optimal control is applied. To get a more helpful solution form, we must do some manipulation. Eliminating the control variable $\mathbf{u}(k)$ so that the equations can be solved, (26.15) becomes

$$\mathbf{x}(k+1) = \mathbf{A}\mathbf{x}(k) + \mathbf{B}\mathbf{R}^{-1}\mathbf{B}^{\mathrm{T}}\lambda(k+1)$$

$$\lambda(k) = \mathbf{A}^{\mathrm{T}}\lambda(k+1) - \mathbf{Q}\mathbf{x}(k) \tag{26.18}$$

In this special case, although the above does not look very promising, we use the properties of linear equations to generate a closed-loop feedback control.

We note that $\lambda(N)$ is a linear function of $\mathbf{x}(N)$ and conjecture that

$$\lambda(k) = -\mathbf{P}(k)\mathbf{x}(k)$$

for some suitable $n \times n$ matrix sequence $\{\mathbf{P}(k)\}$. We now substitute this into (26.18) to see if there exists a sequence consistent with this assumption, noting from (26.16) that it holds trivially for $k = N$ with $\mathbf{P}(N) = \mathbf{S}$. The substitution gives

$$\mathbf{x}(k+1) = \mathbf{A}\mathbf{x}(k) - \mathbf{B}\mathbf{R}^{-1}\mathbf{B}^{\mathrm{T}}\mathbf{P}(k+1)\mathbf{x}(k+1)$$

$$\mathbf{P}(k)\mathbf{x}(k) = \mathbf{A}^{\mathrm{T}}\mathbf{P}(k+1)\mathbf{x}(k+1) + \mathbf{Q}\mathbf{x}(k)$$

Solving the first for $\mathbf{x}(k+1)$, which needs the inverse matrix to exist, yields

$$\mathbf{x}(k+1) = [\mathbf{I} + \mathbf{B}\mathbf{R}^{-1}\mathbf{B}^{\mathrm{T}}\mathbf{P}(k+1)]^{-1}\mathbf{A}\mathbf{x}(k)$$

which is consistent with the second for all $\mathbf{x}(k)$ iff

$$\mathbf{P}(k) = \mathbf{A}^{\mathrm{T}}\mathbf{P}(k+1)\,[\mathbf{I} + \mathbf{B}\mathbf{R}^{-1}\mathbf{B}^{\mathrm{T}}\mathbf{P}(k+1)]^{-1}\mathbf{A} + \mathbf{Q} \tag{26.19}$$

Alternative forms are sometimes helpful. Using the matrix inversion lemma (Appendix B)

$$[\mathbf{E} + \mathbf{F}\mathbf{G}\mathbf{H}]^{-1} = \mathbf{E}^{-1} - \mathbf{E}^{-1}\mathbf{F}\,[\mathbf{G}^{-1} + \mathbf{H}\mathbf{E}^{-1}\mathbf{F}]^{-1}\mathbf{H}\mathbf{E}^{-1} \tag{26.20}$$

with $\mathbf{E} = \mathbf{I}$, $\mathbf{F} = \mathbf{B}$, $\mathbf{G} = \mathbf{R}^{-1}$, $\mathbf{H} = \mathbf{B}^{\mathrm{T}}\mathbf{P}(k+1)$, it can be written in the form called a Riccati equation.

$$P(k) = Q + A^T P(k+1) [I - B [R + B^T P(k+1) B]^{-1} B^T P(k+1)] A \quad (26.21)$$

Hence $P(\cdot)$ is computed backwards in its index from $P(N) = S$ (26.20).

Completing the derivation of the closed-loop control law, we find

$$\mathbf{u}^\#(k) = -R^{-1} B^T P(k+1) \mathbf{x}(k+1)$$
$$= -R^{-1} B^T P(k+1) [A\mathbf{x}(k) + B\mathbf{u}^\#(k)] \quad (26.22)$$

so that

$$\mathbf{u}^\#(k) = - [I + R^{-1} B^T P(k+1) B]^{-1} R^{-1} B^T P(k+1) A\mathbf{x}(k)$$
$$= - [R + B^T P(k+1) B]^{-1} B^T P(k+1) A\mathbf{x}(k) \quad (26.23)$$

Having $P(k)$ computed backwards in its index from $k = N$ can be a nuisance. If A^{-1} exists, however, we may use (26.19) to find

$$P(k+1)^{-1} = A [P(k) - Q]^{-1} A^T - B R^{-1} B^T$$

which rearranges to

$$P(k+1) = [A [P(k) - Q] A^T - B R^{-1} B^T]^{-1}$$

This can be convenient computationally once $P(0)$ is known, probably from backward computation from $k = N$.

We can also calculate the value of the 'cost to go from stage k', defined as

$$J_k = \frac{1}{2} \left(\mathbf{x}^T(N) S \mathbf{x}(N) + \sum_{i=k}^{N-1} \{ \mathbf{x}^T(i) Q \mathbf{x}(i) + \mathbf{u}^T(i) R \mathbf{u}(i) \} \right)$$

This is most easily done by induction. Assume

$$J_{k+1}^\# = \frac{1}{2} \mathbf{x}^T P(k+1) \mathbf{x}$$

with $\mathbf{P}(N) = \mathbf{S}$. Then we have

$$J_k = J_{k+1}^{\#} + \frac{1}{2}\left\{\mathbf{x}^{\mathrm{T}}(k)\,\mathbf{Q}\mathbf{x}(k) + \mathbf{u}^{\mathrm{T}}(k)\,\mathbf{R}\mathbf{u}(k)\right\}$$

Using the shorthand notations $\mathbf{u} = \mathbf{u}(k)$, $\mathbf{x} = \mathbf{x}(k)$, we have for any control $\mathbf{u}$

$$J_k = (\mathbf{u}^{\mathrm{T}}\mathbf{B}^{\mathrm{T}} + \mathbf{x}^{\mathrm{T}}\mathbf{A}^{\mathrm{T}})\,\mathbf{P}(k+1)\,(\mathbf{A}\mathbf{x} + \mathbf{B}\mathbf{u})/2 + \mathbf{x}^{\mathrm{T}}\mathbf{Q}\mathbf{x}/2 + \mathbf{u}^{\mathrm{T}}\mathbf{R}\mathbf{u}/2$$

$$\Rightarrow 2J_k = \mathbf{u}^{\mathrm{T}}(\mathbf{B}^{\mathrm{T}}\mathbf{P}(k+1)\mathbf{B} + \mathbf{R})\mathbf{u} + \mathbf{x}^{\mathrm{T}}(\mathbf{A}^{\mathrm{T}}\mathbf{P}(k+1)\mathbf{A} + \mathbf{Q})\mathbf{x}$$

$$+ \mathbf{u}^{\mathrm{T}}\mathbf{B}^{\mathrm{T}}\mathbf{P}(k+1)\mathbf{A}\mathbf{x} + \mathbf{x}^{\mathrm{T}}\mathbf{A}^{\mathrm{T}}\mathbf{P}(k+1)\mathbf{B}\mathbf{u}$$

Using the optimal control from (26.23)

$$\mathbf{u}^{\#} = -(\mathbf{R} + \mathbf{B}^{\mathrm{T}}\mathbf{P}(k+1)\mathbf{B})^{-1}\,\mathbf{B}^{\mathrm{T}}\mathbf{P}(k+1)\,\mathbf{A}\mathbf{x}(k)$$

then yields

$$2J_k^{\#} = \mathbf{x}^{\mathrm{T}}\mathbf{A}^{\mathrm{T}}\mathbf{P}(k+1)\mathbf{B}\,(\mathbf{R} + \mathbf{B}^{\mathrm{T}}\mathbf{P}(k+1)\mathbf{B})^{-1}\mathbf{B}^{\mathrm{T}}\mathbf{P}(k+1)\mathbf{A}\mathbf{x}$$

$$+ \mathbf{x}^{\mathrm{T}}(\mathbf{A}^{\mathrm{T}}\mathbf{P}(k+1)\mathbf{A} + \mathbf{Q})\mathbf{x}$$

$$- \mathbf{x}^{\mathrm{T}}\mathbf{A}^{\mathrm{T}}\mathbf{P}(k+1)\mathbf{B}\,(\mathbf{R} + \mathbf{B}^{\mathrm{T}}\mathbf{P}(k+1)\mathbf{B})^{-1}\mathbf{B}^{\mathrm{T}}\mathbf{P}(k+1)\mathbf{A}\mathbf{x}$$

$$- \mathbf{x}^{\mathrm{T}}\mathbf{A}^{\mathrm{T}}\mathbf{P}(k+1)\mathbf{B}\,(\mathbf{R} + \mathbf{B}^{\mathrm{T}}\mathbf{P}(k+1)\mathbf{B})^{-1}\mathbf{B}^{\mathrm{T}}\mathbf{P}(k+1)\mathbf{A}\mathbf{x}$$

$$= \mathbf{x}^{\mathrm{T}}[\mathbf{A}^{\mathrm{T}}\mathbf{P}(k+1)\mathbf{A} + \mathbf{Q}$$

$$- \mathbf{A}^{\mathrm{T}}\mathbf{P}(k+1)\mathbf{B}\,(\mathbf{R} + \mathbf{B}^{\mathrm{T}}\mathbf{P}(k+1)\mathbf{B})^{-1}\mathbf{B}^{\mathrm{T}}\mathbf{P}(k+1)\mathbf{A}]\mathbf{x}$$

Comparison with (26.22) yields $\mathbf{J}_k^{\#} = \mathbf{x}^{\mathrm{T}}\mathbf{P}(k)\mathbf{x}/2$ as was to be shown. To summarize, the optimal feedback control and cost are given by

1. $\quad \mathbf{u}^{\#}(k) = -[\mathbf{R} + \mathbf{B}^{\mathrm{T}}\mathbf{P}(k+1)\mathbf{B}]^{-1}\mathbf{B}^{\mathrm{T}}\mathbf{P}(k+1)\mathbf{A}\mathbf{x}(k)$

2. $\quad \mathbf{P}(k) = \mathbf{Q} + \mathbf{A}^{\mathrm{T}}\mathbf{P}(k+1)\mathbf{A}$

$\qquad - \mathbf{A}^{\mathrm{T}}\mathbf{P}(k+1)\mathbf{B}\,[\mathbf{R} + \mathbf{B}^{\mathrm{T}}\mathbf{P}(k+1)\mathbf{B}]^{-1}\mathbf{B}^{\mathrm{T}}\mathbf{P}(k+1)\mathbf{A}$

3. $\quad J_k^\# = \tfrac{1}{2}\mathbf{x}^T\mathbf{P}(k)\,\mathbf{x}$

with boundary condition $P(N) = S$.

We should note a number of features of the derivations and results in this section.

1. So far, we have satisfied necessary conditions; we do not yet know whether they are sufficient.
2. The raw solution gave us only a two-point boundary value to solve for $\mathbf{x}(\cdot)$ and $\lambda(\cdot)$. It took some cleverness to reach a feedback control law for this problem.
3. This has been a fixed duration problem. For most practical purposes, allowing N to vary puts us into a search mode, simply because N must be an integer and hence cannot have a useful time derivative (as in continuous time problems).

Example

Consider the control of the system

$$\mathbf{x}(k+1) = \begin{bmatrix} 0 & 1 \\ 1 & 1 \end{bmatrix}\mathbf{x}(k) + \begin{bmatrix} 0 \\ 1 \end{bmatrix}\mathbf{u}(k)$$

where the criterion to be minimized is

$$J = \mathbf{x}^T(20)\begin{bmatrix} 1 & 0 \\ 0 & 0 \end{bmatrix}\mathbf{x}(20) + \sum_{i=0}^{19} \mathbf{x}_1^2(i) + \mathbf{u}^2(i)$$

Then with

$$\mathbf{P}(20) = \begin{bmatrix} 2 & 0 \\ 0 & 0 \end{bmatrix}$$

and

$$\mathbf{P}(k) = \begin{bmatrix} 2 & 0 \\ 0 & 0 \end{bmatrix} + \mathbf{A}^T\left[\mathbf{P}(k+1) - \frac{\mathbf{P}(k+1)\begin{bmatrix} 0 & 0 \\ 0 & 1 \end{bmatrix}\mathbf{P}(k+1)}{2 + P_{22}(k+1)} \right]\mathbf{A}$$

we have a precomputable expression for $\{P(k)\}$. Hence, the gain $K(k)$ for $u(k) = K(k)\,x(k)$ is given (and precomputable) as

$$K(k) = -\frac{[\,0\ \ 1\,]\,P(k+1)\,A}{2 + P_{22}(k+1)}$$

The gain is plotted in Fig. 26.1(a); also shown are elements of P in 26.1(b) and a typical trajectory in 26.1(c). It is notable that for small k, that is, for a long time to go, $K(k)$ is nearly constant. This property may sometimes be used in applications. In this example the steady-state gain is $K = [-0.73\ \ -1]^T$. The dependence of this gain on $R = r$ is shown in Fig. 26.1(d).

The above example demonstrates that $K(k)$ may be nearly constant for $k \ll N$. For this to be true, $P(k)$ must also be constant, and hence equal a value P given by the solution of

$$P = Q + A^T P A - A^T P B\,[R + B^T P B]^{-1}\,B^T P A$$

The (relatively) easy way to solve this is numerically, by starting at $P(N)$ and working backwards until $P(k) \approx P(k+1)$. Matrix solution methods are used in CACSD (computer-aided control system design) packages.

(a)

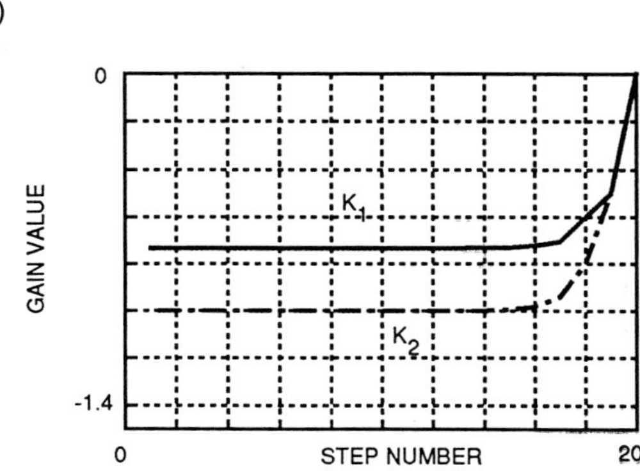

Figure 26.1(a) Optimal gains versus time for example of LQ optimal control.

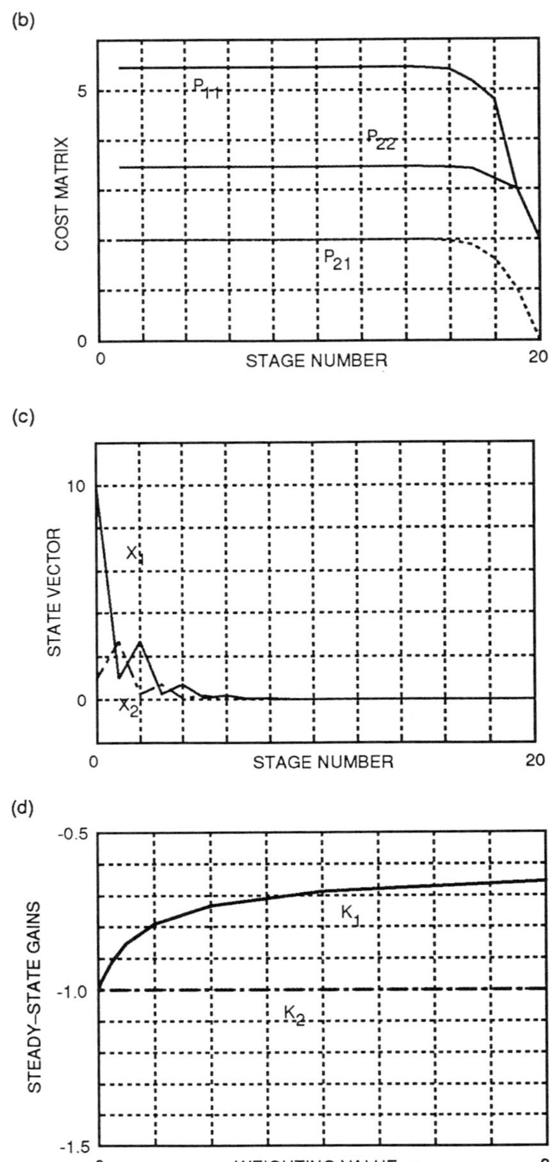

Figure 26.1(b)-(d) (b) Optimal weighting matrix elements for same problem. (c) Optimal evolution of state vector elements for given initial conditions. (d) Optimal steady-state gains (LQR) as function of control weighting *r*.

26.4 A PONTRYAGIN FORMULATION

The above has in fact used only a subset of the classical calculus of variations. In particular, the existence and utility of the first derivatives has been assumed. Pontryagin's contribution was to extend this to problems to which the assumption does not apply.

26.4.1 Basic approach

Recall that our basic problem is that we wish to minimize

$$J(\{\mathbf{u}(k), k=0, 1, \ldots, N-1\}, \mathbf{x}(0), N) = h(N, \mathbf{x}(N)) + \sum_{i=0}^{N-1} g(\mathbf{x}(i), \mathbf{u}(i), i)$$

subject to $\mathbf{x}(k+1) = \mathbf{f}(\mathbf{x}(k), \mathbf{u}(k), k)$ with $\mathbf{x}(0) = \mathbf{x}_0$ given. Then we define a function $\mathcal{H}$, called the Hamiltonian, as

$$\mathcal{H}(\mathbf{x}(i), \mathbf{u}(i), \lambda(i+1), i) = -g(\mathbf{x}(i), \mathbf{u}(i), i) + \lambda^{\mathrm{T}}(i+1)\, \mathbf{f}(\mathbf{x}(i), \mathbf{u}(i), i)$$

where $\{\lambda(\cdot)\}$ is the sequence of Lagrange multipliers, or costates, for this problem and are to be determined. The original problem, given in (26.1) is then equivalent to maximizing

$$J_a(\{\mathbf{x}\}, \{\mathbf{u}\}, \{\lambda\}) = -h(N, \mathbf{x}(N)) - \sum_{i=0}^{N-1} g(\mathbf{x}(i), \mathbf{u}(i), i)$$

$$- \sum_{k=0}^{N-1} \lambda^{\mathrm{T}}(k+1)\, [\mathbf{x}(k+1) - \mathbf{f}(\mathbf{x}(k), \mathbf{u}(k), k)]$$

which in terms of the Hamiltonian is

$$J_a = -h(N, \mathbf{x}(N)) + \sum_{i=0}^{N-1} [\mathcal{H}(\mathbf{x}(k), \mathbf{u}(i), \lambda(i+1), i) - \lambda^{\mathrm{T}}(i+1)\, \mathbf{x}(i+1)]$$

This function is to be maximized for each value of index i. To do so, a variation of $\delta x(i), \delta u(i), \delta\lambda(i)$ is made from the optimum $\mathbf{x}^\#(i)$, $\mathbf{u}^\#(i)$, $\lambda^\#(i)$ and a set of necessary conditions is found. In doing this carefully, in a manner partly demonstrated by Kuo (1980), we substitute $\mathbf{x} \leftarrow \mathbf{x}^\# + \varepsilon\,\delta\mathbf{x}$, $\mathbf{u} \leftarrow \mathbf{u}^\# + \gamma\,\delta\mathbf{u}$, etc. with scalars ε, γ, etc. and find that we are required to have

$$\mathbf{x}^\#(k+1) = \frac{\partial \mathscr{H}^\#(k)}{\partial \lambda^\#(k+1)} = \mathbf{f}(\mathbf{x}^\#(k), \mathbf{u}^\#(k), k)$$

$$\lambda^\#(k) = \frac{\partial \mathscr{H}^\#(k)}{\partial \mathbf{x}^\#(k)} \qquad k = 0, 1, \ldots, N-1 \tag{26.24}$$

with boundary conditions

$$\mathbf{x}^\#(0) = \mathbf{x}_0 \qquad \text{and} \qquad \lambda^\#(N) = \frac{\partial h(N, \mathbf{x}^\#(N))}{\partial \mathbf{x}(N)}$$

Furthermore, $\mathbf{u}^\#(i)$ must, for each i, minimize J_a and since it appears only in $\mathscr{H}$, must maximize $\mathscr{H}$. Hence

$$\mathscr{H}(\mathbf{x}^\#(i), \mathbf{u}^\#(i), \lambda^\#(i), i) \geq \mathscr{H}(\mathbf{x}^\#(i), \mathbf{u}(i), \lambda^\#(i), i) \tag{26.25}$$

and this expression determines $\mathbf{u}^\#(i)$, $i = 0, 1, 2, \ldots, N-1$. For many problems, $\mathbf{u}(\cdot)$ appears quadratically or higher in $\mathscr{H}$ and the maximum can be found by setting

$$\frac{\partial \mathscr{H}(\mathbf{x}^\#(i), \mathbf{u}, \lambda^\#(i), i)}{\partial \mathbf{u}} = 0 \tag{26.26}$$

In other cases, particularly those involving linear systems and time optimal control, (26.25) must be used.

26.4.2 Application to linear quadratic tracking problem

An important variation of the linear quadratic regulator problem of the previous section is the tracking problem, in which it is desired that the state follow a predetermined trajectory $\{\mathbf{x}_d(k)\}$. For this, the quadratic payoff (26.4) is modified to

$$J = [\mathbf{x}(N) - \mathbf{x}_d(N)]^T \mathbf{S}\, [\mathbf{x}(N) - \mathbf{x}_d(N)]/2$$

$$+ \sum_{i=0}^{N-1} \{ [\mathbf{x}(i) - \mathbf{x}_d(i)]^T \mathbf{Q}[\mathbf{x}(i) - \mathbf{x}_d(i)] + \mathbf{u}^T(i)\,\mathbf{R}\mathbf{u}(i)\}/2$$

We then find that

$$g(\mathbf{x}(i),\mathbf{u}(i),i) = \frac{1}{2} \{ [\mathbf{x}(i) - \mathbf{x}_d(i)]^T \mathbf{Q}\,[\mathbf{x}(i) - \mathbf{x}_d(i)] + \mathbf{u}^T(i)\,\mathbf{R}\mathbf{u}(i)\}$$

and (26.24) with its boundary conditions becomes

$$\mathbf{x}^{\#}(k+1) = \mathbf{A}\mathbf{x}^{\#}(k) + \mathbf{B}\mathbf{u}^{\#}(k)$$

$$\lambda^{\#}(k) = \frac{\partial \mathscr{H}^{\#}(k)}{\partial \mathbf{x}^{\#}(k)} \qquad k = 0,1,\dots$$

$$= \mathbf{A}^T \lambda^{\#}(k+1) - \mathbf{Q}\,[\mathbf{x}^{\#}(k) - \mathbf{x}_d(k)]$$

with revised boundary conditions

$$\mathbf{x}^{\#}(0) = \mathbf{x}_0$$

$$\lambda^{\#}(N) = -\,\mathbf{S}\,[\mathbf{x}^{\#}(N) - \mathbf{x}_d(N)] \tag{26.27}$$

and control law from (26.26) as

$$\mathbf{u}^{\#}(k) = \mathbf{R}^{-1}\mathbf{B}^T \lambda^{\#}(k+1)$$

By substitution,

$$\mathbf{x}^{\#}(k+1) = \mathbf{A}\mathbf{x}^{\#}(k) + \mathbf{B}\mathbf{R}^{-1}\mathbf{B}^T\lambda^{\#}(k+1)$$

$$\lambda^{\#}(k) = \mathbf{A}^T \lambda^{\#}(k+1) - \mathbf{Q}\,[\mathbf{x}^{\#}(k) - \mathbf{x}_d(k)] \tag{26.28}$$

We conjecture

$$\lambda^{\#}(k) = -\mathbf{P}(k)\,\mathbf{x}^{\#}(k) - \mathbf{b}(k) \qquad\qquad (26.29)$$

where $\mathbf{b}(k)$ is a bias term, and note that (26.27) implies

$$\mathbf{P}(N) = \mathbf{S}$$

$$\mathbf{b}(N) = -\mathbf{S}\mathbf{x}_{\mathrm{d}}(N)$$

Substitution into (26.28) then gives

$$\mathbf{x}^{\#}(k+1) = \mathbf{A}\mathbf{x}^{\#}(k) - \mathbf{B}\mathbf{R}^{-1}\mathbf{B}^{\mathrm{T}}[\mathbf{P}(k+1)\,\mathbf{x}^{\#}(k+1) + \mathbf{b}(k+1)]$$

$$\mathbf{P}(k)\,\mathbf{x}^{\#}(k) + \mathbf{b}(k) = \mathbf{A}^{\mathrm{T}}[\mathbf{P}(k+1)\,\mathbf{x}^{\#}(k+1) + \mathbf{b}(k+1)]$$
$$+ \mathbf{Q}\,[\mathbf{x}^{\#}(k) - \mathbf{x}_{\mathrm{d}}(k)]$$

Solving the first of these for $\mathbf{x}^{\#}(k+1)$ and substituting in the second yields

$$\mathbf{x}^{\#}(k+1) = [\mathbf{I} + \mathbf{B}\mathbf{R}^{-1}\mathbf{B}^{\mathrm{T}}[\mathbf{P}(k+1)]^{-1}\,[\mathbf{A}\mathbf{x}^{\#}(k) - \mathbf{B}\mathbf{R}^{-1}\mathbf{B}^{\mathrm{T}}\mathbf{b}(k+1)]$$

$$\mathbf{P}(k)\,\mathbf{x}^{\#}(k) + \mathbf{b}(k) = \mathbf{A}^{\mathrm{T}}\{\mathbf{P}(k+1)\,[\mathbf{I} + \mathbf{B}\mathbf{R}^{-1}\mathbf{B}^{\mathrm{T}}\mathbf{P}(k+1)]^{-1}$$
$$[\mathbf{A}\mathbf{x}^{\#}(k) - \mathbf{B}\mathbf{R}^{-1}\mathbf{B}^{\mathrm{T}}\mathbf{b}(k+1)]$$
$$+ \mathbf{b}(k+1)\} + \mathbf{Q}\,[\mathbf{x}^{\#}(k) - \mathbf{x}_{\mathrm{d}}(k)]$$

For this to hold for all $\mathbf{x}^{\#}(k)$ it is sufficient that

$$\mathbf{P}(k) = \mathbf{A}^{\mathrm{T}}\mathbf{P}(k+1)\,[\mathbf{I} + \mathbf{B}\mathbf{R}^{-1}\mathbf{B}^{\mathrm{T}}\mathbf{P}(k+1)]^{-1}\mathbf{A} + \mathbf{Q}$$

$$\mathbf{b}(k) = -\mathbf{A}^{\mathrm{T}}\mathbf{P}(k+1)\,[\mathbf{I} + \mathbf{B}\mathbf{R}^{-1}\mathbf{B}^{\mathrm{T}}\mathbf{P}(k+1)]^{-1}\mathbf{B}\mathbf{R}^{-1}\mathbf{B}^{\mathrm{T}}\mathbf{b}(k+1)$$
$$+ \mathbf{A}^{\mathrm{T}}\mathbf{b}(k+1) - \mathbf{Q}\mathbf{x}_{\mathrm{d}}(k) \qquad\qquad (26.30)$$

As before, the optimal control can be found from

$$\mathbf{u}^{\#}(k) = -\mathbf{R}^{-1}\mathbf{B}^{\mathrm{T}}[\mathbf{P}(k+1)\,\mathbf{x}^{\#}(k+1) + \mathbf{b}(k+1)]$$

$$= -\mathbf{R}^{-1}\mathbf{B}^{\mathrm{T}}\mathbf{P}(k+1)[\mathbf{A}\mathbf{x}^{\#}(k) + \mathbf{P}(k+1)\mathbf{B}\mathbf{u}^{\#}(k) + \mathbf{b}(k+1)]$$

to yield

$$\mathbf{u}^{\#}(k) = -[\mathbf{R} + \mathbf{B}^{\mathrm{T}}\mathbf{P}(k+1)\mathbf{B}]^{-1}\mathbf{B}^{\mathrm{T}}[\mathbf{P}(k+1)\mathbf{A}\mathbf{x}^{\#}(k) + \mathbf{b}(k+1)]$$

Defining a gain $\mathbf{K}(k)$ as

$$\mathbf{K}(k) = -[\mathbf{R} + \mathbf{B}^{\mathrm{T}}\mathbf{P}(k+1)\mathbf{B}]^{-1}\mathbf{B}^{\mathrm{T}}\mathbf{P}(k+1)\mathbf{A}$$

allows this to be written

$$\mathbf{u}^{\#}(k) = -[\mathbf{R} + \mathbf{B}^{\mathrm{T}}\mathbf{P}(k+1)\mathbf{B}]^{-1}\mathbf{B}^{\mathrm{T}}\mathbf{b}(k+1) + \mathbf{K}(k)\mathbf{x}^{\#}(k)$$

which is mainly helpful in that (26.30) becomes

$$\mathbf{P}(k) = \mathbf{Q} + \mathbf{A}^{\mathrm{T}}\mathbf{P}(k+1)\mathbf{A} + \mathbf{A}^{\mathrm{T}}\mathbf{P}(k+1)\mathbf{K}(k)$$

$$\mathbf{b}(k) = [\mathbf{A}^{\mathrm{T}} + \mathbf{K}^{\mathrm{T}}(k)\mathbf{B}^{\mathrm{T}}]\mathbf{b}(k+1) - \mathbf{Q}\mathbf{x}_{\mathrm{d}}(k)$$

In this the first is a rearrangement as was done to obtain (26.21).

In summary, since in the regulator $\mathbf{x}_{\mathrm{d}}(k) \equiv 0$ and hence $\mathbf{b}(k) \equiv 0$, both the regulator and tracking problems can be represented using the following algorithm:

$$\mathbf{P}(N) = \mathbf{S}$$

$$\mathbf{b}(N) = -\mathbf{S}\,\mathbf{x}_{\mathrm{d}}(N)$$

FOR $k = N-1, N-2, \ldots, 0$

$$\mathbf{K}(k) = -[\mathbf{R} + \mathbf{B}^{\mathrm{T}}\mathbf{P}(k+1)\mathbf{B}]^{-1}\mathbf{B}^{\mathrm{T}}\mathbf{P}(k+1)\mathbf{A}$$

$$\mathbf{P}(k) = \mathbf{Q} + \mathbf{A}^{\mathrm{T}}\mathbf{P}(k+1)\mathbf{A} + \mathbf{A}^{\mathrm{T}}\mathbf{P}(k+1)\mathbf{K}(k)$$

$$\mathbf{b}(k) = [\mathbf{A}^{\mathrm{T}} + \mathbf{K}^{\mathrm{T}}(k)\mathbf{B}^{\mathrm{T}}]\mathbf{b}(k+1) - \mathbf{Q}\,\mathbf{x}_{\mathrm{d}}(k)$$

FOR $j = 0, 1, 2, \ldots, N-1$

$$\mathbf{u}^{*}(j) = -[\mathbf{R} + \mathbf{B}^{\mathrm{T}}\mathbf{P}(j+1)\mathbf{B}]^{-1}\mathbf{B}^{\mathrm{T}}\mathbf{b}(j+1) + \mathbf{K}(j)\mathbf{x}(j)$$

26.4.3 Application to minimum time problems

It is not hard to demonstrate that minimum time problems with control magnitude constraints have 'bang-bang' solutions, in which the control elements usually switch back and forth from minimum to maximum admissible value. This means, incidentally, that the optimal controls may be relay-implemented and do not require throttling. To see the ideas, we consider a standard example.

Let $g(\mathbf{x},u,i) = 1$, so that N is to be minimized with

$$J = \sum_{i=0}^{N-1} 1 = N$$

Further, let $\mathbf{x}(k+1) = \mathbf{A}\mathbf{x}(k) + \mathbf{b}u(k)$ where $\mathbf{u}(k)$ is a scalar sequence (primarily for notational convenience). The boundary conditions are that $\mathbf{x}(0) = \mathbf{x}_0$ and $\mathbf{x}(N) = \mathbf{x}_f$, with N, the final time, unknown. Then

$$\mathscr{H} = -1 + \lambda^{\mathrm{T}}(k+1)[\mathbf{A}\mathbf{x}(k) + \mathbf{b}u(k)]$$

From the necessary conditions,

$$\mathbf{x}(k+1) = \mathbf{A}\mathbf{x}(k) + \mathbf{b}u(k)$$

$$\lambda(k) = \mathbf{A}^{\mathrm{T}}\lambda(k+1)$$

with $u(k)$ such that $\lambda^{\mathrm{T}}(k+1)\,\mathbf{b}u(k)$ is maximized.

Thus $u(k)$ takes on its maximum allowed value for $\lambda^{\mathrm{T}}(k+1)\,\mathbf{b} > 0$, takes on its minimum allowed value for $\lambda^{\mathrm{T}}(k+1)\,\mathbf{b} < 0$, and is undefined for $\lambda^{\mathrm{T}}(k+1)\,\mathbf{b} = 0$. If $u(k)$ is bounded, i.e. if $|u(k)| \leq 1$, then we have

$$u^{\#}(k) = \operatorname{sgn}\lambda^{\mathrm{T}}(k+1)\,\mathbf{b}$$

which is seen to be a 'bang-bang' control, implementable using relay actuators. Notice that because of the constraint on the magnitude of the control this is not the same problem as will be solved by placing poles at the origin (section 27.4); the latter does not restrict the control in this way.

It is arguable, and supportable for real systems by physical arguments and by the continuous time theory, that $u(k)$ should change signs between samples. This can be done in hardware by having the computer set a counter to signal when the switching should occur (see section 26.6).

The same ideas, in particular Lagrange multipliers, are also used for optimization of functionals, as we will see in the next section.

26.5 OPTIMIZATION OF CONTINUOUS TIME SYSTEMS

The mathematics of continuous time systems are sometimes easier and sometimes harder than for sampled data systems – harder because of the use of functionals (functions of functions) rather than functions, easier in that many of the derived expressions are less complex. The results are sometimes applicable to continuous time systems which are computer controlled, provided the engineer interprets them properly.

26.5.1 Basic Euler–Lagrange theory for continuous systems

The optimization theory for finding functions is similar in form to that for finding numbers, as we shall see. Here, instead of looking for a number x which minimizes a function J, we seek a function $x(t)$ which minimizes a functional $J(x(t))$. Thus for example we want a scalar function $x(t)$ for which the number J produced by

$$J(x(t)) = \int_{t_0}^{t_f} x(s)\, \mathrm{d}s$$

is minimized.

Example

Consider the performance functional

$$J(x(t)) = \int_0^3 (x(s) - s)^2\, \mathrm{d}s$$

If $x(s) = s^2$, for instance, then $J(x) = 1/12$. Obviously, the minimizing function is $x(s) = s$.

Let us consider a simple functional of a scalar function

$$J(x(t)) = \int_{t_0}^{t_f} F(x(s))\,\mathrm{d}s$$

and suppose that we consider a minimizing function $x^\#(t)$ and an increment of that function $h(t)$, i.e. $x(t) = x^\#(t) + h(t)$. Then provided t_0 and t_f are fixed

$$J(x^\# + h) = \int_{t_0}^{t_f} F(x^\#(s) + h(s))\,\mathrm{d}s$$

$$= \int_{t_0}^{t_f} \left[F(x^\#) + \frac{\mathrm{d}F}{\mathrm{d}x}\,h + \frac{\mathrm{d}^2 F}{\mathrm{d}^2 x}\,h^2 + \cdots \right]\mathrm{d}s$$

For h 'small'

$$J(x^\# + h) \approx \int_{t_0}^{t_f} \left[F(x^\#) + \frac{\mathrm{d}F}{\mathrm{d}x}\,h \right]\mathrm{d}s$$

$$\approx J(x^\#) + \int_{t_0}^{t_f} \left[\frac{\mathrm{d}F}{\mathrm{d}x}\,h \right]\mathrm{d}s$$

Since $h(s)$ is arbitrary and can be chosen, it is clear that for $J(x^\#)$ to be the minimum,

$$\left. \frac{\mathrm{d}F}{\mathrm{d}x} \right|_{x^\#(t)} = 0$$

is necessary.

Proceeding similarly, consider the functional

$$J(\mathbf{x}, u) = \int_{t_0}^{t_f} F(\mathbf{x}(s), \dot{\mathbf{x}}(s), u(s), s)\, ds \tag{26.31}$$

We again form $\mathbf{x}^{\#} + \delta\mathbf{x}$, $\dot{\mathbf{x}}^{\#} + \delta\dot{\mathbf{x}}$, $u^{\#} + \delta u$, all functions, and form the increment

$$J(\mathbf{x}^{\#} + \delta\mathbf{x}, u^{\#} + \delta u) = \int_{t_0}^{t_f} [F(\mathbf{x}^{\#} + \delta\mathbf{x}, \dot{\mathbf{x}}^{\#} + \delta\dot{\mathbf{x}}, u^{\#} + \delta u, s)]\, ds$$

$$\approx J(\mathbf{x}^{\#}, u^{\#}) + \int_{t_0}^{t_f} \left[\frac{\partial F}{\partial \mathbf{x}}\, \delta\mathbf{x} + \frac{\partial F}{\partial \dot{\mathbf{x}}}\, \delta\dot{\mathbf{x}} + \frac{\partial F}{\partial u}\, \delta u \right] ds$$

from which we find immediately that

$$\left. \frac{\partial F}{\partial x} \right|_{\mathbf{x}^{\#}, u^{\#}} = 0 \tag{26.32}$$

and

$$0 = \int_{t_0}^{t_f} \left[\frac{\partial F}{\partial \mathbf{x}}\, \delta\mathbf{x} + \frac{\partial F}{\partial \dot{\mathbf{x}}}\, \delta\dot{\mathbf{x}} \right] ds \tag{26.33}$$

Integrating this by parts gives

$$0 = \int_{t_0}^{t_f} \left[\frac{\partial F}{\partial \dot{\mathbf{x}}} - \int_{t_0}^{s} \frac{\partial F}{\partial \mathbf{x}}\, dt \right] \delta\dot{\mathbf{x}}\, ds \tag{26.34}$$

Since $\delta\dot{\mathbf{x}}$ can be arbitrary (and $\delta\mathbf{x}$ is its integral), we must have the term in [ ] identically equal to 0, so that

$$\frac{d}{dt} \frac{\partial F}{\partial \dot{\mathbf{x}}} - \frac{\partial F}{\partial \mathbf{x}} = 0 \tag{26.35}$$

This is the Euler–Lagrange equation for this problem. The boundary conditions, which come out of the integration by parts, are

$$\left[\nabla_{\dot{\mathbf{x}}} F \bigg|_{t_0} \right] \delta \mathbf{x}(t_0) = 0$$

$$\left[\nabla_{\dot{\mathbf{x}}} F \bigg|_{t_f} \right] \delta \mathbf{x}(t_f) = 0 \tag{26.36}$$

Since the problem often has a prescribed $\mathbf{x}(t_0)$, $\delta \mathbf{x}(t_0) = 0$ is required and $\nabla_{\dot{\mathbf{x}}} F$ can be arbitrary at this point. Similarly, if $\mathbf{x}(t_f)$ is arbitrary, then so is $\delta \mathbf{x}(t_f)$ and hence (26.35) requires $\nabla_{\dot{\mathbf{x}}} F = 0$ at $t = t_f$.

The next generalization we need is that of allowing for constraints. Thus we consider the problem of minimizing J where

$$J = \int_{t_0}^{t_f} g(\mathbf{x}, \dot{\mathbf{x}}, u, t)\, dt$$

and subject to constraints $\dot{\mathbf{x}} = \mathbf{f}(\mathbf{x}, u, t)$ and $\mathbf{x}(t_0) = \mathbf{x}_0$. It turns out we can again use Lagrange multipliers but that this time they are functions rather than numbers. Hence define

$$G(\mathbf{x}, \lambda, \dot{\mathbf{x}}, u, t) = g(\mathbf{x}, \dot{\mathbf{x}}, u, t) + \lambda^{\mathrm{T}} (\dot{\mathbf{x}} - \mathbf{f}(\mathbf{x}, u, t))$$

and find the extremum of

$$\hat{J} = \int_{t_0}^{t_f} G(\mathbf{x}, \lambda, \dot{\mathbf{x}}, u, t)\, dt$$

Considering the extended vector $\mathbf{y}$ defined by concatenating $\mathbf{x}$ and $\mathbf{y}$, i.e. $\mathbf{y} = [\mathbf{x}^{\mathrm{T}}\ \lambda^{\mathrm{T}}]^{\mathrm{T}}$, apply (26.31) to $(\mathbf{y}, \mathbf{u})$ instead of $(\mathbf{x}, \mathbf{u})$, use the results of (26.35) and (26.36), and then separate the components of **y** to find the necessary conditions

$$\frac{\partial[g - \lambda^{\mathrm{T}} \mathbf{f}]}{\partial u} = 0$$

$$\frac{\partial[g - \lambda^{\mathrm{T}} \mathbf{f}]}{\partial \mathbf{x}} - \frac{d\lambda}{dt} = 0$$

$$\frac{\partial[g + \lambda^{\mathrm{T}} (\dot{\mathbf{x}} - \mathbf{f})]}{\partial \lambda} = \dot{\mathbf{x}} - \mathbf{f} = 0 \tag{26.37}$$

with boundary conditions

$$\mathbf{x}(t_0) = \mathbf{x}_0$$

$$\lambda(t_{\mathrm{f}}) = 0 \tag{26.38}$$

26.5.2 Application to linear quadratic (LQ) problem

For the standard example, we consider

$$J = \frac{1}{2} \int_{t_0}^{t_{\mathrm{f}}} [\mathbf{x}^{\mathrm{T}}(t)\,\mathbf{Q}\mathbf{x}(t) + \mathbf{u}^{\mathrm{T}}(t)\,\mathbf{R}\mathbf{u}(t)]\,dt$$

subject to $\dot{\mathbf{x}} = \mathbf{A}\mathbf{x} + \mathbf{B}\mathbf{u}$ and $\mathbf{x}(t_0) = \mathbf{x}_0$.

This is the standard linear quadratic problem of optimal control and by assumption $\mathbf{Q} \geq 0$ and $\mathbf{R} > 0$. For $\mathbf{x}^{\#}$ and $\mathbf{u}^{\#}$ to be optimum, they must satisfy the necessary conditions (26.37) and (26.38). Substituting the obvious functions g and $\mathbf{f}$,

$$\mathbf{u}^{\#}(t) = \mathbf{R}^{-1}\,\mathbf{B}^{\mathrm{T}}\,\lambda^{\#}(t)$$

$$\dot{\lambda}^{\#}(t) = -\mathbf{A}^{\mathrm{T}} \lambda^{\#}(t) + \mathbf{Q}\mathbf{x}^{\#}(t)$$

$$\dot{\mathbf{x}}^{\#} = \mathbf{A}\mathbf{x}^{\#}(t) + \mathbf{B}\mathbf{u}^{\#}(t)$$

$$= \mathbf{A}\mathbf{x}^{\#}(t) + \mathbf{B}\mathbf{R}^{-1}\mathbf{B}^{\mathrm{T}}\lambda^{\#}(t)$$

with boundary conditions $\mathbf{x}^{\#}(t) = \mathbf{x}_0$ and $\lambda^{\#}(t_{\mathrm{f}}) = 0$.

The above is a standard result. It is notable that the system $[\mathbf{x}^{\mathrm{T}}\ \lambda^{\mathrm{T}}]^{\mathrm{T}}$ must satisfy a two-point boundary-value problem (TPBVP) and that $\mathbf{u}^{\#}$ is a function of time – i.e. is an open-loop control – rather than a feedback law. In this instance, as in the discrete time case, we are able to exploit the linearity of the differential equations to achieve an optimal feedback law.

We have

$$\frac{\mathrm{d}}{\mathrm{d}t} \begin{bmatrix} \mathbf{x} \\ \lambda \end{bmatrix} = \begin{bmatrix} \mathbf{A} & \mathbf{B}\mathbf{R}\mathbf{B}^{\mathrm{T}} \\ \mathbf{Q} & -\mathbf{A} \end{bmatrix} \begin{bmatrix} \mathbf{x} \\ \lambda \end{bmatrix}$$

Letting the transition matrix for this be $\phi(t, \tau)$ as in

$$\begin{bmatrix} \mathbf{x}(t_{\mathrm{f}}) \\ \lambda(t_{\mathrm{f}}) \end{bmatrix} = \begin{bmatrix} \phi_{11}(t_{\mathrm{f}}, t) & \phi_{12}(t_{\mathrm{f}}, t) \\ \phi_{21}(t_{\mathrm{f}}, t) & \phi_{22}(t_{\mathrm{f}}, t) \end{bmatrix} \begin{bmatrix} \mathbf{x}(t) \\ \lambda(t) \end{bmatrix}$$

we can find

$$\lambda(t) = -\phi_{22}(t, t_{\mathrm{f}})\,\phi_{21}(t_{\mathrm{f}}, t)\,\mathbf{x}(t)$$

so that the feedback law is

$$\mathbf{u}^{\#}(t) = -\mathbf{R}^{-1}\mathbf{B}^{\mathrm{T}}\phi_{22}(t, t_{\mathrm{f}})\,\phi_{21}(t_{\mathrm{f}}, t)\,\mathbf{x}(t)$$

$$= \mathbf{K}(t)\,\mathbf{x}(t)$$

Thus the optimal control law exhibits linear feedback with time-varying gain.

The above is the hard way to find the gain. An alternative is to try to find $\mathbf{H}(t)$ such that $\lambda(t) = -\mathbf{H}(t)\,\mathbf{x}(t)$ which in turn requires $\dot{\lambda} = -\dot{\mathbf{H}}\,\mathbf{x} - \mathbf{H}\,\dot{\mathbf{x}}$. From the differential equations

$$\dot{\lambda} = \mathbf{Q}\mathbf{x} - \mathbf{A}^{\mathrm{T}}\lambda = \mathbf{Q}\mathbf{x} + \mathbf{A}^{\mathrm{T}}\mathbf{H}\mathbf{x}$$

and

$$-\dot{\mathbf{H}}\mathbf{x} - \mathbf{H}\dot{\mathbf{x}} = -\dot{\mathbf{H}}\mathbf{x} - \mathbf{H}\left[\mathbf{A}\mathbf{x} - \mathbf{B}\mathbf{R}^{-1}\mathbf{B}^{\mathrm{T}}\mathbf{H}\mathbf{x}\right]$$

Since these hold and are equal for all $\mathbf{x}$, it must be that

$$\mathbf{Q} + \mathbf{A}^{\mathrm{T}}\mathbf{H} = -\dot{\mathbf{H}} - \mathbf{H}\mathbf{A} + \mathbf{H}\mathbf{B}\mathbf{R}^{-1}\mathbf{B}^{\mathrm{T}}\mathbf{H}$$

which is usually written as the Riccati equation

$$\dot{\mathbf{H}} = -\mathbf{Q} - \mathbf{A}^{\mathrm{T}}\mathbf{H} - \mathbf{H}\mathbf{A} + \mathbf{H}\mathbf{B}\mathbf{R}^{-1}\mathbf{B}^{\mathrm{T}}\mathbf{H}$$

Of course, from the definition of $\mathbf{H}$ we have

$$\mathbf{u}^{\#} = -\mathbf{R}^{-1}\mathbf{B}^{\mathrm{T}}\mathbf{H}(t)\,\mathbf{x}$$

We can show that

$$J^{\#}(\tau) = \frac{1}{2}\int_{\tau}^{t_f}\left[\mathbf{x}^{\#\tau}(t)\,\mathbf{Q}\mathbf{x}^{\#}(t) + \mathbf{u}^{\#\mathrm{T}}(t)\,\mathbf{R}\mathbf{u}^{\#}(t)\right]\,dt$$

$$= \frac{\mathbf{x}^{\mathrm{T}}(\tau)\,\mathbf{H}(\tau)\,\mathbf{x}(\tau)}{2}$$

in several ways. One verification is to show that both expressions have the same derivative with respect to τ when $\mathbf{H}$ satisfies the above differential equation and $\mathbf{u}^{\#}$ is as given. Other proofs involve, for example, other derivation techniques.

26.5.3 Pontryagin's maximum principle for continuous time models

The Pontryagin principle can be viewed as a variation of the classical theory in which the conditions

$$\frac{\partial G}{\partial \mathbf{u}}\bigg|_{\mathbf{x}^{\#}(t),\,\mathbf{u}^{\#}(t),\,\lambda^{\#}(t)} = [\nabla_{\mathbf{u}} G]\bigg|_{\mathbf{x}^{\#}(t),\,\mathbf{u}^{\#}(t),\,\lambda^{\#}(t)} = 0$$

are replaced by a local maximization of a function. Specifically, the statement is as follows. The control $\mathbf{u}^{\#}(t)$ which causes a system

$$\dot{\mathbf{x}} = \mathbf{f}(\mathbf{x}(t), \mathbf{u}(t), t) \qquad \mathbf{x}(t_0) = \mathbf{x}_0$$

to follow an admissible trajectory that minimizes the performance measure

$$J(\mathbf{u}) = h(\mathbf{x}(t_{\mathrm{f}}), t_{\mathrm{f}}) + \int_{t_0}^{t_{\mathrm{f}}} g(\mathbf{x}(\tau), \mathbf{u}(\tau), \tau)\, d\tau$$

must necessarily satisfy the following conditions, expressed in terms of the Hamiltonian $\mathscr{H}$ defined as

$$\mathscr{H}(\mathbf{x}(t), \mathbf{u}(t), \mathbf{p}(t), t) \equiv -g(\mathbf{x}(t), \mathbf{u}(t), t) + \mathbf{p}^{\mathrm{T}}(t)\, \mathbf{f}(\mathbf{x}(t), \mathbf{u}(t), t)$$

1. $$\dot{\mathbf{x}}^{\#}(t) = \frac{\partial \mathscr{H}}{\partial \mathbf{p}}\bigg|_{(\mathbf{x}^{\#}(t),\,\mathbf{u}^{\#}(t),\,\mathbf{p}^{\#}(t),\,t)}$$

2. $$\dot{\mathbf{p}}^{\#}(t) = \frac{\partial \mathscr{H}}{\partial \mathbf{x}}\bigg|_{(\mathbf{x}^{\#}(t),\,\mathbf{u}^{\#}(t),\,\mathbf{p}^{\#}(t),\,t)}$$

3. $$\mathscr{H}(\mathbf{x}^{\#}(t), \mathbf{u}^{\#}(t), \mathbf{p}^{\#}(t), t) \geq \mathscr{H}(\mathbf{x}^{\#}(t), \mathbf{u}(t), \mathbf{p}^{\#}(t), t)$$

4.   The boundary conditions satisfy

$$
\left[\frac{\partial h}{\partial \mathbf{x}} \bigg|_{(\mathbf{x}^{\#}(t_f),\, t_f)} - \mathbf{p}^{\#}(t_f) \right] \delta \mathbf{x}_f +
$$

$$
\left[\mathscr{H}(\mathbf{x}^{\#}(t), \mathbf{u}^{\#}(t), \mathbf{p}^{\#}(t), t) + \frac{\partial h}{\partial t} \bigg|_{(\mathbf{x}^{\#}(t_f),\, t_f)} \right] \delta t_f = 0
$$

In this, the modified condition is (3). It is tempting to view this as a recipe in a cookbook. We do so for some simple but instructive examples.

Example

Consider the trajectory shaping for an extra-atmospheric missile of constant mass and thrust. The state is described by vertical and horizontal positions y and x. The thrust attitude is ψ, so the dynamics are

$$
\ddot{x} = T\cos\psi \qquad \ddot{y} = T\sin\psi - g
$$

Defining the extended state vector

$$
\mathbf{z} = [\,x \quad \dot{x} \quad y \quad \dot{y}\,]^{\mathrm{T}}
$$

and taking thrust attitude ψ as the control variable u gives the state–space model

$$
\dot{z}_1 = z_2
$$

$$
\dot{z}_2 = T\cos u
$$

$$
\dot{z}_3 = z_4
$$

$$
\dot{z}_4 = T\sin u - g
$$

The object is to minimize time to achieve the state transition, so this is a time optimal problem. From this we have

$$\mathcal{H}(\mathbf{z}, u, \lambda, t) = -1 + \lambda^{\mathrm{T}} \mathbf{f}(\mathbf{z}, u, t) = -1 + \lambda_1 z_2 + \lambda_2 T \cos u$$

$$+ \lambda_3 z_4 + \lambda_4 T \sin u - \lambda_4 g$$

From the basic rules,

$$\dot{\mathbf{z}} = \mathbf{f}(\mathbf{z}, u)$$

$$\dot{\lambda} = \begin{bmatrix} 0 \\ -\lambda_1 \\ 0 \\ -\lambda_3 \end{bmatrix}$$

$$\lambda_2 T \sin u - \lambda_4 T \cos u = 0$$

The latter two of these relationships give

$$\lambda_1 = \lambda_1(t_0) \qquad \lambda_2 = \lambda_2(t_0) + \lambda_1(t_0)\,(t - t_0)$$

$$\lambda_3 = \lambda_3(t_0) \qquad \lambda_4 = \lambda_4(t_0) + \lambda_3(t_0)\,(t - t_0)$$

so that

$$\tan u = \frac{\lambda_4(t_0) + \lambda_3(t_0)\,(t - t_0)}{\lambda_2(t_0) + \lambda_1(t_0)\,(t - t_0)}$$

This is the basic 'bilinear tangent law' of rocket steering which has been used to motivate some trajectory designs in actual missile flights; the special case in which down range position is not fixed ($x(t_f)$ free) gives the result that $\lambda_1(t_f) = 0 = \lambda_1(t_0)$ and hence a 'linear tangent steering law'. Although the parameters could in principle be sought as solutions of the TPBVP, they will ordinarily be found using other methods.

Example

A simpler example than the above, and one in which we may discuss solution methods, is that for which

$$\ddot{x} = u \qquad \|u\| \le 1 \qquad x(0) = x_0 \qquad x(t_\mathrm{f}) = 0$$

The objective is minimum time control of the system. Using state–space models in which $z_1 = x$ and $z_2 = \dot{x}$ we have

$$\dot{\mathbf{z}} = \begin{bmatrix} 0 & 1 \\ 0 & 0 \end{bmatrix} \mathbf{z} + \begin{bmatrix} 0 \\ 1 \end{bmatrix} u = \begin{bmatrix} z_2 \\ u \end{bmatrix}$$

and hence $\mathcal{H}(\mathbf{z}, u, \mathbf{p}) = -1 + p_1 z_2 + p_2 u$. We find immediately that

$$\dot{\mathbf{z}} = \begin{bmatrix} z_2 \\ u \end{bmatrix} \quad \text{and} \quad \dot{\mathbf{p}} = \begin{bmatrix} 0 \\ p_1 \end{bmatrix}$$

For the control

$$-1 + p_1^\# z_2^\# + p_2^\# u^\# \ge -1 + p_1^\# z_2^\# + p_2^\# u$$

means that $u^\#$ must take its maximum positive value when $p_2^\#$ is positive and smallest value when $p_2^\#$ is negative. Since the constraint is that $-1 \le u \le 1$, this means

$$u^\#(t) = \mathrm{sgn}\,[\,p_2^\#(t)\,]$$

The solution of the system has $p_2^\#(t) = p_2^\#(0) + p_1^\#(0)t$ so that

$$u^\#(t) = \mathrm{sgn}\,[\,p_2^\#(0) + p_1^\#(0)t\,]$$

Notice that so far $u^\#$ is open-loop, i.e. a function of time. To learn more about it, we look at trajectories.

First we see that $u^{\#}$ may start at -1 and stay there, start at -1 and switch to $+1$, start at $+1$ and switch to -1, or start at $+1$ and stay at that value; these are the only possibilities admitted by the expression for $u^{\#}(t)$, except for the very special case when $p_2^{\#}(0) = p_1^{\#}(t) = 0$. We consider what trajectories look like in Fig. 26.2, which show $z_2 = \dot{z}_1$ versus z_1, a type of graph called the **phase plane**.

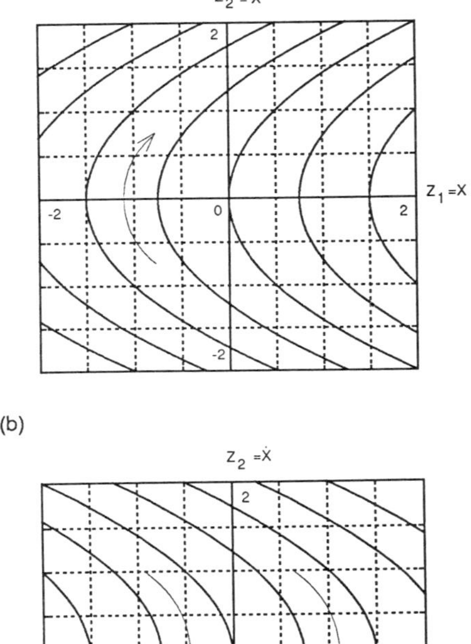

Figure 26.2(a)-(b) Development of optimal switching curves for minimum time transition to origin of example problem: (a) possible (upward moving) trajectories with $u = +1$; (b) possible (downward moving) trajectories with $u = -1$.

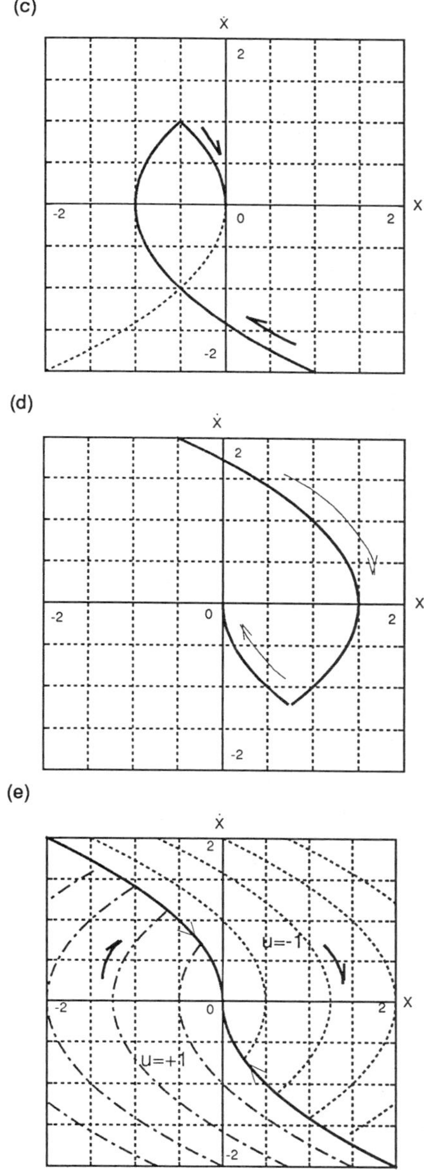

Figure 26.2(c)-(e) Development of optimal switching curves for minimum time transition to origin of example problem: (c) and (d) patching together curves from (a) and (b) to reach origin; (e) the switching and final approach curves (solid lines) and possible trajectories to reach switching curves (dotted).

One can show that

- when $u = +1$, then $z_1 = z_2^2/2 + C$ for a constant C; and
- when $u = -1$, then $z_1 = -z_2^2/2 + c$ for a constant c.

It is clear from the diagrams that to finish at $z = 0$, the trajectory must arrive from the second quadrant along a $u = -1$ curve, or it must arrive from the fourth quadrant along a $u = +1$ curve. Then from any position above those lines, a $u = -1$ command is needed to reach the fourth quadrant terminal line, while a $u = +1$ is needed from below these lines.

From the above arguments and some thinking and manipulation of the signs, one convinces oneself that

$$u = -1 \text{ when } z_1 + |z_2| z_2/2 > 0$$

$$u = +1 \text{ when } z_1 + |z_2| z_2/2 < 0$$

$$u = \operatorname{sgn} z_1 \text{ when } z_1 + |z_2| z_2/2 = 0$$

Hence one defines a switching line

$$z_1 = -|z_2| z_2/2$$

and uses $+1$ below this line and -1 above the line, and sgn (z_1) on the line. The control law is then

$$u^{\#}(z) = -\operatorname{sgn}(z_1 + |z_2| z_2/2)$$

with the implementation/interpretation that $u^{\#}(z) = \operatorname{sgn}(z_1)$ when $z_1 + |z_2| z_2/2 = 0$.

26.6 ENGINEERING OF SOLUTIONS

The optimal control theory so far has produced two-point boundary-value problems (TPBVPs) in the state and the Lagrange multipliers as the direct solutions. Producing feedback solutions from these has required further knowledge and manipulation, and implementation may require even more effort. Alternatively, even obtaining two-point boundary-value solutions may require numerical methods such as will be seen in Chapter 27. We comment below on various engineering aspects of the optimal control solutions.

26.6.1 Implementation of switching lines

The examples have already indicated trajectory shape considerations in one instance and switching lines in another. Either case may, with a bit of cleverness, be implemented with a digital computer. We show this with the switching line example.

Example

We saw in sections 26.4 and 26.5 examples in which the control is bang-bang in nature; sampling the data did not affect the nature of the control law. It is clear, however, that switching must take place precisely at the instant when $z_1 + |z_2| z_2/2 = 0$ for the origin to be reached. For the digital computer to do such a thing, it must predict when this instant will occur and either count down internally or set an external timer.

To predict time, we notice that since $\dot{z}_2 = u = \pm 1$, it must be that for any one arc segment, $\Delta T = \|\Delta z_2\|$. We consider the trajectories which start with $u = -1$ (i.e. above the switching lines) and observe they are described by

$$\frac{z_2^2(t)}{2} = -z_1(t) + \frac{z_2^2(0)}{2} + z_1(0)$$

while the $u = +1$ switching curve is

$$\frac{z_2^2(t)}{2} = z_1(t)$$

These intersect at the switching time t_g (time to go) for which

$$z_1(t_g) = \frac{1}{2}\left(\frac{z_2^2(0)}{2} + z_1(0)\right)$$

$$z_2(t_g) = \left(\frac{z_2^2(0)}{2} + z_1(0)\right)^{\frac{1}{2}}$$

From this we establish that

$$t_g = \|z_2(t_g) - z_2(0)\| = z_2(0) + \left(\frac{z_2^2(0)}{2} + z_1(0)\right)^{\frac{1}{2}}$$

The control command is to measure z and compute

$$t_g = z_2 + \left(\frac{z_2^2}{2} + z_1\right)^{\frac{1}{2}}$$

This number is 'time to go' from when the measurements were taken, and is used as the basis for loading the timer, which should count down its parameter and cause a switch when it has reached 0.

26.6.2 Choices of weighting matrices and solutions of linear quadratic problems

Setting of the matrices $\mathbf{S}, \mathbf{Q}, \mathbf{R}$ in the performance index is rarely an obvious matter, even for problems in which penalizing state errors or control commands in a quadratic manner appears natural. One way (Bryson and Ho, 1969) is to choose

$$\mathbf{S} = \mathbf{I}\,(1/(n \times \text{maximum allowable } \{\mathbf{x}^T(N)\,\mathbf{x}(N)\}))$$

$$\mathbf{Q} = \mathbf{I}\,(1/(nN \times \text{maximum allowable } \{\mathbf{x}^T(i)\,\mathbf{x}(i)\}))$$

$$\mathbf{R} = \mathbf{I}\,(1/(mN \times \text{maximum allowable } \{\mathbf{u}^T(i)\,\mathbf{u}(i)\}))$$

where $\mathbf{S}$ and $\mathbf{Q}$ are $n \times n$, $\mathbf{R}$ is $m \times m$, and the process has N stages; the changes for a tracking problem are obvious.

A variation on the above is that, if the *i*th element of $\mathbf{x}(k)$ has an allowable maximum of $\mathbf{x}_{i,\mathrm{max}}$, then let

$$\mathbf{Q} = \mathrm{diag}\,(1/x_{1,\mathrm{max}}^2,\, 1/x_{2,\mathrm{max}}^2,\,,\,...,\,1/x_{m,\mathrm{max}}^2)$$

and similarly

$$\mathbf{R} = \mathrm{diag}\,(1/u_{1,\mathrm{max}}^2,\, 1/u_{2,\mathrm{max}}^2,\,...,\,1/u_{n,\mathrm{max}}^2)$$

$$\mathbf{S} = \mathrm{diag}\,(1/x_{1,\mathrm{max}}^2(N),\, 1/x_{2,\mathrm{max}}^2(N),\,...,\,1/x_{n,\mathrm{max}}^2(N))$$

More detailed discussions are presented by Anderson and Moore (1989) and by Maciejowski (1989). They particularly emphasize the control magnitude/state error trade-off, noting that high control penalties relative to the state error penalties tend to allow an open-loop appearing system (if it is stable) while relatively low penalties force the appearance of poles near values determined by the steady-state gain (which depends upon the structure of $\mathbf{Q}$ and $\mathbf{R}$).

26.6.3 Properties of linear quadratic regulator (LQR) controllers

We have observed that the gain $\mathbf{K}(t)$ in LQ problems becomes constant when the situation is not time-varying and there is a long 'time to go'; the LQ optimization problem with infinite upper limit for the integral or summation cost function is called the linear quadratic regulator (LQR) problem. The solution to this problem is of the form

$$\mathbf{u} = \mathbf{K}\mathbf{x}$$

where, in the discrete-time case

$$\mathbf{K} = -[\mathbf{R} + \mathbf{B}^{\mathrm{T}}\mathbf{P}\mathbf{B}]^{-1}\mathbf{B}^{\mathrm{T}}\mathbf{P}\mathbf{A}$$

$$\mathbf{P} = \mathbf{Q} + \mathbf{A}^{\mathrm{T}}\mathbf{P}\mathbf{A} - \mathbf{A}^{\mathrm{T}}\mathbf{P}\mathbf{B}[\mathbf{R} + \mathbf{B}^{\mathrm{T}}\mathbf{P}\mathbf{B}]^{-1}\mathbf{B}^{\mathrm{T}}\mathbf{P}\mathbf{A}$$

and in the continuous-time case

$$\mathbf{K} = -\mathbf{R}^{-1}\mathbf{B}^T\mathbf{H}$$

$$\mathbf{H}\mathbf{B}\mathbf{R}^{-1}\mathbf{B}^T\mathbf{H} - \mathbf{A}^T\mathbf{H} - \mathbf{H}\mathbf{A} - \mathbf{Q} = \mathbf{O}$$

The expressions for $\mathbf{P}$ and $\mathbf{H}$ are called algebraic Riccati equations. The resulting transfer function of the system is

$$\mathbf{G}(q) = \mathbf{C}\,(q\mathbf{I} - \mathbf{A} - \mathbf{B}\mathbf{K})^{-1}\,\mathbf{B}$$

where q is either the Laplace transform operator s or the z-transform variable z. $\mathbf{G}(q)$ can also be written in terms of the open-loop transfer function $\mathbf{K}(q\mathbf{I} - \mathbf{A})^{-1}\,\mathbf{B}$ or the return difference $\mathbf{I} - \mathbf{K}\,(q\mathbf{I} - \mathbf{A})^{-1}\,\mathbf{B}$.

Some of the many interesting facts about the approach and the control law are as follows.

1. For any $\mathbf{K}$, there exist weighting matrices $\mathbf{Q}$ and $\mathbf{R}$ such that $\mathbf{K}$ is an optimal controller. (Strictly, there exist $\mathbf{Q}$ and $\mathbf{R}$ such that $\mathbf{K}(q\mathbf{I} - \mathbf{A})^{-1}\mathbf{B}$ is obtained.)
2. For the continuous-time case, when $\mathbf{R} = \rho\mathbf{I}$, there is at least $60°$ of phase margin in each input channel. This is somewhat less (down to very small) in the discrete-time case.
3. The system is robust against some types of model errors. For example, a single-input system remains stable for non-linear gain variations $\phi(e)$ which are sector bounded by

$$0.5\,e < \phi(e) < \varepsilon^{-1}\,e \qquad 0 < \varepsilon < 2$$

26.7 COMPUTER AIDS

Computer packages can usually solve the LQ problem and variations of it. The LQR problem is solved in the Control Systems Toolbox of MATLAB® and by Ctrl-C, for examples.

26.8 SUMMARY

Optimal control is the theory associated with finding the best control according to a mathematical criterion, and may be compared, not always favourably, to the control laws generated by the traditional techniques such as root locus, frequency domain methods, and even manual tuning. The latter have tended to be the laws of choice in process control, while optimal control is more associated with the aerospace industry.

In this chapter we have presented an introduction to the Pontryagin and related theories for designing optimal control laws, and two important types of results of applying those theories to linear systems. The two results are for the time optimal (or minimum time to traverse from one state to another state) problem, for which the result is that the control will usually be 'bang-bang' in nature, and the quadratic performance index, for which the result is that the control law is linear in the state; all problems are occasionally subject to intervals called **singular arcs** in which the control is not defined by the first level arguments here.

26.9 FURTHER READING

Further reading on Pontryagin methods perhaps best starts with Pontryagin *et al.* (1962). Bryson and Ho (1969) and Lee and Markus (1967) are old standbys. Very accessible introductions appear in many standard textbooks at second course level and above, including those by Kuo (1980) and Shinners (1992). One very readable and more specialized text is the relatively early book by Kirk (1970) while a compendium of results on linear quadratic problems is given by Anderson and Moore (1989). Somewhat broader coverage, including industrial examples, is given by Grimble and Johnson (1988).

27

Other optimal control methods

Chapter 26 discussed Lagrange multiplier methods of converting constrained optimization problems to unconstrained problems, with the particular application being control system problems in which the state is to traverse an optimum path subject to the constraint that the equations of motion (usually the laws of physics) are not to be violated. There are other methods, and we look briefly at some of them in this section.

27.1 SYNOPSIS

There are many more methods than Pontryagin and calculus of variations approaches to designing optimal control laws. The first method we look at has its origins in transfer function theory. Given a model

$$A(z)\,Y(z) = z^{-k} B(z)\,U(z) + \lambda C(z)\,V(z)$$

one seeks a control such that the variance of $y(mT)$ is minimized for any m if $\{v(mT)\}$ is a noise sequence. The solution has

$$U(z) = - \left\{ \frac{G(z)}{B(z)\,F(z)} \right\} Y(z)$$

where $G(z)$ and $F(z)$ are related by

$$C(z) = A(z)\,F(z) + z^{-k} G(z)$$

A second approach is to choose parameters in a structured situation so that a criterion is optimized. This may involve algebra or numerical techniques such as mathematical programming or simulation.

A third approach uses **dynamic programming**, based upon the **(Bellman's) Principle of Optimality**, which states roughly that an optimal control law is optimum for any state reached, and is independent of how that state was reached. It is applied to the problem of minimizing

$$J(\{u(k), k=0, 1, \ldots, N-1\}, x_0, N) = h(N, x(N)) + \sum_{i=0}^{N-1} g(x(i), u(i), i)]$$

subject to the dynamics

$$x(k+1) = f(x(k), u(k), k) \quad x(0) = x_0$$

The result is that the optimum cost from state x at stage k, $J^\#(k, x)$ satisfies

$$J^\#(N, x) = h(N, x)$$

$$J^\#(k, x) = \min_u \left[J^\#(k+1, f(x, u, k)) + g(x, u, k) \right]$$

for $k = N-1, N-2, \ldots, 0$. This is applied to the linear-quadratic (LQ) problem as an illustration, and of course it gives the same result as the Pontryagin approach.

Finally, we look briefly at the essence of some numerical approaches used alongside the optimal theory when the theory does not give closed-form solutions. These trajectory determination methods include the methods of neighbouring extremals and steepest ascent.

27.2 MINIMUM VARIANCE CONTROL

One attractive performance indicator for optimization is the variance of the output, where the output (or its measurement) can be viewed as being random. One way to treat this is based on classical and transfer function ideas, rather than state-space.

27.2.1 Basic result

One form of SISO control that is intuitively appealing is **minimum variance control**, similar to ISE control, in which the designer seeks a control law which minimizes the integral of the squared error between an input and an output. We consider one such method, following Åström (1970).

Consider the system modelled by

$$A(z)\, Y(z) = z^{-k} B(z)\, U(z) + \lambda C(z)\, V(z) \tag{27.1}$$

where $\{y(n)\}$ is the output variable sequence, $\{u(n)\}$ is the input variable sequence, and $\{v(n)\}$ is a sequence of random variables, i.e. disturbances and noises input to the system. $A(z)$, $B(z)$, and $C(z)$ are all taken to be polynomials which, with no loss of generality, all have the same order. Thus

$$A(z) = 1 + \sum_{i=1}^{n} a_i z^{-i}$$

$$B(z) = \sum_{i=0}^{n} b_i z^{-i}$$

$$C(z) = \sum_{i=0}^{n} c_i z^{-i}$$

The object of the control law is to minimize the variance of $y(mT)$ for any m, i.e. to find a control law which yields

$$\min \mathscr{E}[y(mT) - \mathscr{E}[y(mT)]^2 \tag{27.2}$$

The solution of the problem is to set

$$U(z) = - \left\{ \frac{G(z)}{B(z)\, F(z)} \right\} Y(z) \tag{27.3}$$

where $F(z)$ and $G(z)$ are defined as

$$F(z) = \sum_{i=0}^{k-1} f_i z^{-i}$$

$$G(z) = \sum_{i=0}^{n} g_i z^{-i} \tag{27.4}$$

and have the property

$$C(z) = A(z) F(z) + z^{-k} G(z) \tag{27.5}$$

In implementation, (27.3) becomes

$$u(mT) = -\frac{1}{f_0 b_0} \sum_{i=0}^{n-1} g_i y(mT - iT) - \frac{1}{f_0 b_0} \sum_{i=1}^{n+k-1} \beta_i u(mT - iT)$$

where

$$\beta_1 = f_1 + b_1,$$

$$\beta_2 = f_2 + f_1 b_1 + b_2, \text{ etc.}$$

Finding $F(z)$ and $G(z)$ is reasonably straightforward. One way is to expand (27.5) and match coefficients. Another is to let $F(z)$ be $C(z)/A(z)$ with $G(z)$ the appropriate remainder.

It is noteworthy that if $k=1$ and $C(z) = 1 = c_0$, then

$$G(z) = (1 - A(z))z = -\sum_{i=1}^{n} a_i z^{-i+1}$$

Thus the control simply cancels the effect of the dynamics in propagating past values – presumably measured – of the output and

$$Y(z) = \lambda V(z)$$

which is random.

We remark that if the system is non-minimum phase, i.e. has zeros outside the unit circle, then the control law will be unstable because $B(z)$ is in the denominator of (27.3).

27.2.2 Justification

Rearranging (27.1) gives

$$\lambda V(z) = \frac{A(z)}{C(z)} Y(z) - z^{-k} \frac{B(z)}{C(z)} U(z) \tag{27.6}$$

Then, if we substitute for $C(z)$ in (27.1) using a form

$$C(z) = A(z) F(z) + z^{-k} G(z) \tag{27.7}$$

and solve for $Y(z)$ we find

$$Y(z) = z^{-k} \frac{B(z)}{A(z)} U(z) + \lambda F(z) V(z) + \lambda z^{-k} \frac{G(z)}{A(z)} V(z)$$

Replacing the last term on the right-hand side using (27.6) and rearranging gives

$$Y(z) = \lambda F(z) V(z) + z^{-k} \frac{G(z)}{C(z)} Y(z)$$

$$+ z^{-k} \left\{ \frac{B(z)}{A(z)} - \frac{B(z)}{A(z)} \frac{G(z)}{C(z)} z^{-k} \right\} U(z) \tag{27.8}$$

Now, using (27.7), it can be shown that

$$\frac{B(z)}{A(z)} - \frac{B(z)}{A(z)} \frac{G(z)}{C(z)} z^{-k} = \frac{B(z)}{C(z)} \left\{ \frac{C(z)}{A(z)} - \frac{G(z)}{A(z)} z^{-k} \right\}$$

$$= \frac{B(z) F(z)}{A(z)}$$

This simplifies (27.8) giving

$$z^k Y(z) = \lambda z^k F(z)V(z) + \frac{G(z)}{C(z)} Y(z) + \frac{B(z)\,F(z)}{C(z)} U(z)$$

In the time domain, this becomes

$$y(mT + kT) = \lambda \sum_{i=0}^{k-1} f_i\, v(mT + kT - iT)$$

$$+\, \mathscr{Z}^{-1}\left[\frac{B(z)\,F(z)\,U(z) + G(z)\,Y(z)}{C(z)}\right]$$

where $\mathscr{Z}^{-1}$ is the inverse z-transform. For the system to be causal, the highest indices within the inverse transform must be $y(mT)$ and $u(mT)$, whereas the lowest in the first sum on the right is $v(mT+T)$. Hence the expected values of the cross-terms in squaring the right-hand side are 0, and we have

$$\mathscr{E}\{y^2(mT + kT)\} = \lambda^2 \mathscr{E}\left[\sum_{i=0}^{k-1} f_i^2 v^2(mT + kT - iT)\right] \qquad (27.9)$$

$$+\, \mathscr{E}\left\{\mathscr{Z}^{-1}\left[\frac{B(z)\,F(z)\,U(z) + G(z)\,Y(z)}{C(z)}\right]^2\right\}$$

The two terms are independent and non-negative. Only the second term depends on the control $\{u(mT)\}$ sequence. Hence, the only way to minimize the mean-square value of $y(\cdot)$ is to choose the control to make the second term on the right zero, i.e. to choose

$$U(z) = -\left\{\frac{G(z)}{B(z)\,F(z)}\right\} Y(z)$$

This is an interesting proof (an alternative demonstration is given by Tzafestas, 1985) in that we chose a pair of functions $F(z)$ and $G(z)$ satisfying certain conditions and then found the optimal control in terms of those functions. It is a legitimate thing to do, but one always wonders how anyone would have dreamed up the functions to use. Clearly (in hindsight) part of the key was to isolate the disturbances from the input and output (in (27.9)).

27.2.3 Application

We may well know from physical arguments that in the absence of disturbances the system is modelled by

$$Y(z) = z^{-k} \frac{B_1(z)}{A_1(z)} U(z)$$

If, in the absence of input commands, the output has spectral density $\Phi(\omega)$ (see Appendix C), this can be factored into

$$\Phi(\omega) = \lambda^2 \frac{C_1(e^{j\omega}) C_1(e^{-j\omega})}{A_2(e^{j\omega}) A_2(e^{-j\omega})}$$

where $A_2(z)$ and $C_1(z)$ have all of their zeros inside the unit circle. Our complete model is then

$$Y(z) = z^{-k} \frac{B_1(z)}{A_1(z)} U(z) + \lambda \frac{C_1(z)}{A_2(z)} V(z)$$

where $\{v(kT)\}$ is white, zero-mean, unit-variance gaussian noise. If we define

$$A(z) = A_1(z) A_2(z)$$

$$B(z) = B_1(z) A_2(z)$$

$$C(z) = C_1(z) A_1(z)$$

we end up with the form of (27.1)

$$A(z) Y(z) = z^{-k} B(z) U(z) + \lambda C(z) V(z)$$

Example

Let a system have dynamics given by

$$A(z) = 1 - z^{-1} - z^{-2}$$

$$B(z) = 1$$

$$C(z) = \tfrac{1}{2}(1 + z^{-1})$$

$$k = 2$$

which implies a basic system driven by its input plus a moving average noise process. Then (27.4) and (27.5) imply we need

$$\tfrac{1}{2}(1 + z^{-1}) = (1 - z^{-1} - z^{-2})\,(f_0 + f_1 z^{-1}) + z^{-2}(g_0 + g_1 z^{-1} + g_2 z^{-2})$$

This may be solved to yield $f_0 = \tfrac{1}{2}$, $f_1 = 1$, $g_0 = \tfrac{3}{2}$, $g_1 = 1$, $g_2 = 0$, and hence from (27.3) the minimum variance control law is

$$u(kT) = -2u(kT - T) - 3y(kT) - 2y(kT - T)$$

27.3 OPTIMUM PARAMETER SETTING

An alternative which is not really optimal control in the modern sense, but rather is parameter optimization with a prescribed control law, has already been mentioned in another context (Chapter 19 for prototype pole placement to ISE, etc. criteria.) Here one chooses a form such as

$$H(z) = \frac{b_0 + b_1 z^{-1} + \cdots + b_p z^{-p}}{1 - z^{-1}}$$

for $U(z)/E(z)$ where $E(z)$ may be the transform of the error involved $(e(k) = y(k) - y_d(k))$. Then for a fixed number of steps and a particular reference signal, one chooses (probably numerically) the parameters, here $b_0, b_1, \ldots, b_p$, which minimize a criterion such as

$$J = \sum_{k=0}^{N} \{ e^2(k) + (u(k) - u^{\#})^2 \}$$

in which $u^{\#}$ is the steady-state value of $u(k)$ for the given reference signal. The computations are in the realm of mathematical programming.

Example

Consider a motor with transfer function

$$G(s) = \frac{a}{s(s + a)}$$

which is to be controlled in a unity feedback structure, using gain only and a zero-order hold.

The criterion is to be

$$J = \sum_{i=0}^{\infty} e^2 (iT)$$

for a step applied at time 0. By simulation of the discrete-time equations with transfer function

$$H(z) = K \left[\frac{T}{z - 1} - \frac{1 - e^{-aT}}{a(z - e^{-aT})} \right]$$

a curve of J versus K can be developed as in Fig. 27.1. Also shown there are step responses resulting from using optimal and non-optimal gains and mis-modelled systems.

The simulations were done for $a = 10$, $T = 1$, and summation over 100 steps. The optimum step response illustrates a damping coefficient (in continuous-time terms) of about $\zeta \approx 0.61$.

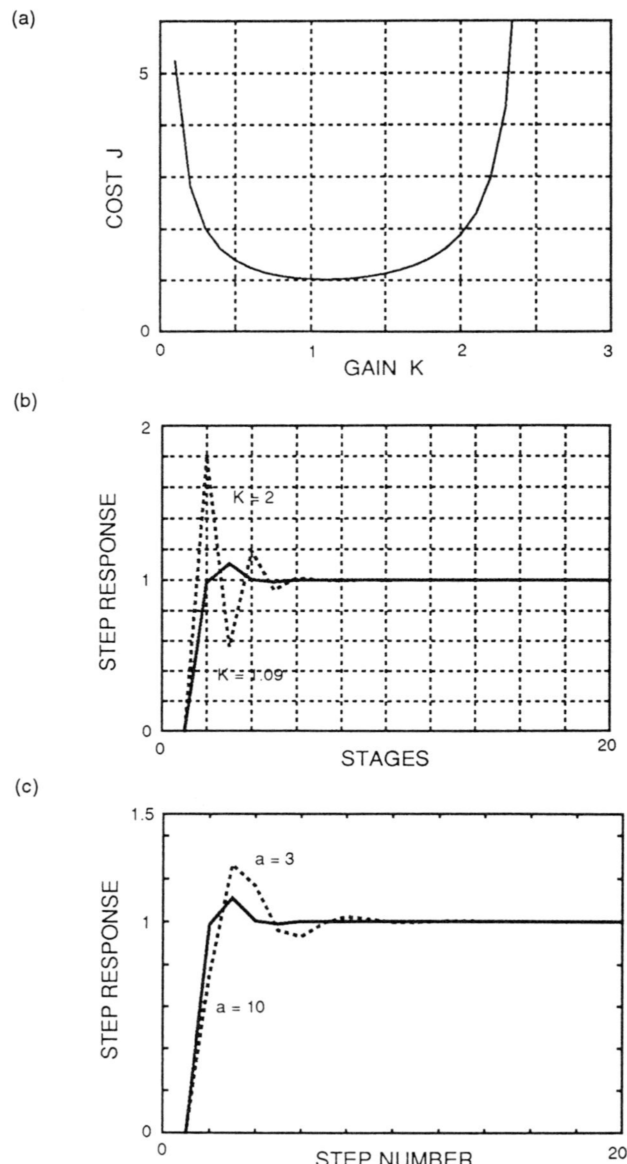

Figure 27.1 Optimal parameter setting for motor control example.
(a) Cost J as a function of gain K. (b) Step responses for the optimal gain $K = 1.09$ and another gain $K = 2$. (c) Robustness of optimal control gain $K = 1.09$ shown by step responses for designed system ($a = 10$) and erroneous system ($a = 3$).

27.4 MINIMUM TIME PROBLEMS USING LINEAR ALGEBRA

The deadbeat control problem (Chapter 21) can also be considered a minimum time problem, and we here look at it in this way with the goal of illustrating the use of linear algebraic methods to derive a control law.

The problem here is most easily stated as one of finding a control sequence $\{\mathbf{u}(k)\}$ such that an arbitrary initial state $\mathbf{x}(0)$ is driven to another arbitrary state $\mathbf{x}_f$ (without loss of generality this final state can be zero, but we choose not to impose that restriction at this point) in the minimum number of stages N, i.e. such that $\mathbf{x}(N) = \mathbf{x}_f$, when the system is described by

$$\mathbf{x}(k+1) = \mathbf{A}\mathbf{x}(k) + \mathbf{B}\mathbf{u}(k) \tag{27.10}$$

In fact we would much prefer a feedback law $\mathbf{u}(k) = f(\mathbf{x}(k))$ or preferably $\mathbf{u}(k) = f(\mathbf{y}(k))$ where

$$\mathbf{y}(k) = \mathbf{C}\mathbf{x}(k)$$

but we shall ignore this at first in looking for insight. We saw in Chapter 26 that seeking an initial condition dependent time sequence and then a control law is not uncommon in optimal control theory,

Consider the solution to (27.10), given as usual by

$$\mathbf{x}(k) = \mathbf{A}^k\mathbf{x}(0) + \sum_{i=1}^{k} \mathbf{A}^{i-1}\mathbf{B}\mathbf{u}(k-i)$$

For our desired N value,

$$\Delta\mathbf{x}(N) = \mathbf{x}(N) - \mathbf{A}^N\mathbf{x}(0) = \sum_{i=1}^{N} \mathbf{A}^{i-1}\mathbf{B}\mathbf{u}(N-i)$$

will be arbitrary, because $\mathbf{x}(0)$ is arbitrary. Hence, the right-hand side sum must be such that, by proper choice of $\mathbf{u}(k)$, $k=0,1,...,N-1$, it can take on any value. For this to happen, there must be among the

Nm columns of the *N* matrices $\mathbf{A}^{i-1}\mathbf{B}, i=1,2,...,N$, *n* columns which are linearly independent, i.e. whose sum can form any possible *n*-vector. It is known (from the Cayley–Hamilton theorem) that $\mathbf{A}^n$ depends upon $\{\mathbf{A}^i, i=0,1,...,n-1\}$, so it follows that $N < n$. In fact, if we look back at the controllability theory (Chapter 22) *N* is the controllability index and $(\mathbf{A},\mathbf{B})$ must be controllable for us to have the time-optimal control required.

We consider first the case $\mathbf{x}_f = \mathbf{0}$ and let the control variable $u(k)$ be a scalar quantity at each time instant. For convenience we denote

$$\mathbf{K} = [\, 1 \quad 0 \quad 0 \quad \cdots \quad 0 \,]\,[\, \mathbf{A}^{n-1}\mathbf{B} \,\vdots\, \mathbf{A}^{n-2}\mathbf{B} \,\vdots\, \cdots \,\vdots\, \mathbf{AB} \,\vdots\, \mathbf{B} \,]^{-1} \quad (27.11)$$

i.e. $\mathbf{K}$ is the first row of the indicated inverse $n \times n$ matrix. Notice that the inverse exists if the system is controllable (Chapter 22). Then we contend that if

$$u(k) = -\mathbf{K}\mathbf{A}^n\mathbf{x}(k) \qquad k = 0,1,...,n-1 \qquad (27.12)$$

then $\mathbf{x}(n) = \mathbf{0}$. The demonstration of this is instructive, so we shall pursue it briefly. We argue that the *n* column matrices $\mathbf{A}^{n-1}\mathbf{B}$, $\mathbf{A}^{n-2}\mathbf{B}, ..., \mathbf{AB}, \mathbf{B}$ must span the *n*-dimensional space. Hence any *n*-vector, and in particular $\mathbf{A}^n\mathbf{x}(k)$, must have a representation in terms of these vectors, i.e. there must exist real numbers $a_i, i=1,2,...,n$, such that

$$\mathbf{A}^n\mathbf{x}(k) = \mathbf{A}^{n-1}\mathbf{B}a_1(k) + \mathbf{A}^{n-2}\mathbf{B}a_2(k) + \cdots + \mathbf{AB}a_{n-1}(k) + \mathbf{B}a_n(k) \quad (27.13)$$

Choose $u(k) = -a_1(k)$. Then

$$\mathbf{x}(k+1) = \mathbf{A}\mathbf{x}(k) - \mathbf{B}a_1(k)$$

implies

$$\mathbf{A}^{n-1}\mathbf{x}(k+1) = \mathbf{A}^n\mathbf{x}(k) - \mathbf{A}^{n-1}\mathbf{B}a_1(k)$$

$$= \mathbf{A}^{n-1}\mathbf{B}a_1(k) + \mathbf{A}^{n-2}\mathbf{B}a_2(k) + \cdots + \mathbf{AB}a_{n-1}(k)$$

$$+ \mathbf{B}a_n(k) - \mathbf{A}^{n-1}\mathbf{B}a_1(k)$$

Thus

$$\mathbf{A}^n \mathbf{x}(k+1) = \mathbf{A}^{n-1}\mathbf{B}a_2(k) + \mathbf{A}^{n-2}\mathbf{B}a_3(k) + \cdots + \mathbf{A}\mathbf{B}\,a_n(k)$$

But, from (27.13)

$$\mathbf{A}^n\mathbf{x}(k+1) = \mathbf{A}^{n-1}\mathbf{B}a_1(k+1) + \mathbf{A}^{n-2}\mathbf{B}a_2(k+1) + \cdots$$

$$+ \mathbf{A}\mathbf{B}\,a_{n-1}(k+1) + \mathbf{B}\,a_n(k+1)$$

so we notice that $a_n(k+1)$ will be zero. From this observation it follows that if we start at $k = 0$, then

$$a_n(1) = a_n(2) = \cdots = a_n(n) = a_{n-1}(2) = a_{n-1}(3) = \cdots = a_{n-1}(n) =$$

$$a_{n-2}(3) = a_{n-2}(4) = \cdots = a_{n-2}(n) = \cdots = a_1(n) = 0$$

Thus, provided that we keep choosing $u(k) = -a_1(k)$, we will obtain

$$\mathbf{A}^n\mathbf{x}(n) = \mathbf{0} \tag{27.14}$$

The calculation of $a_1(k)$ is easily done from the matrix representation of (27.13), i.e. from

$$\mathbf{A}^n\mathbf{x}(k) = [\ \mathbf{A}^{n-1}\mathbf{B}\ \vdots\ \mathbf{A}^{n-2}\mathbf{B}\ \vdots\ \cdots\ \vdots\ \mathbf{A}\mathbf{B}\ \vdots\ \mathbf{B}\ ] \begin{bmatrix} a_1(k) \\ a_2(k) \\ \vdots \\ a_n(k) \end{bmatrix} \tag{27.15}$$

Thus provided the inverse exists, $a_1(k)$ is given by the first element of

$$[\ \mathbf{A}^{n-1}\mathbf{B}\ \vdots\ \mathbf{A}^{n-2}\mathbf{B}\ \vdots\ \cdots\ \vdots\ \mathbf{A}\mathbf{B}\ \vdots\ \mathbf{B}\ ]^{-1}\, \mathbf{A}^n\mathbf{x}(k)$$

which, with the minus sign, is the same as the control given by (27.12) when the gain $\mathbf{K}$ is given by (27.11).

The above arguments have given us a feedback law for driving $x(0)$ to 0 in n steps, and we note that generally we cannot expect to require fewer steps than this. Rather than a feedback law, a sequence $\{u(k)\}$ can be derived from noting

$$\mathbf{x}(n) = \mathbf{A}^n \mathbf{x}(0) + \sum_{i=0}^{n-1} \mathbf{A}^i \mathbf{B} u(n - 1 - i)$$

yielding for $\mathbf{x}(n) = \mathbf{0}$ the sequence

$$\begin{bmatrix} u(0) \\ u(1) \\ \vdots \\ u(n-1) \end{bmatrix} = -[\,\mathbf{A}^{n-1}\mathbf{B} \vdots \mathbf{A}^{n-2}\mathbf{B} \vdots \cdots \vdots \mathbf{B}\,]^{-1} \mathbf{A}^n \mathbf{x}(0) \qquad (27.16)$$

We might argue that our feedback control law amounts to resetting the time origin and recalculating (27.16) each cycle and transmitting only $u(0)$ from the computation; this approach is called open-loop feedback. Alternatively, one computation of the above yields the entire sequence $\{u(k)\}$ to accomplish the task, which is pure open-loop control.

In the above, the control law was a linear gain independent of time, and when applied the law would always lead to the origin in n or fewer stages (or time steps). Applying the same techniques to a situation with $\mathbf{x}_f \neq \mathbf{0}$ changes this. In the batch or precomputation case, the fact that $\mathbf{x}(n) = \mathbf{x}_f$ is desired leads to (following (27.14) and (27.15))

$$\begin{bmatrix} u(0) \\ u(1) \\ \vdots \\ u(n-1) \end{bmatrix} = -[\,\mathbf{A}^{n-1}\mathbf{B} \vdots \mathbf{A}^{n-2}\mathbf{B} \vdots \cdots \vdots \mathbf{B}\,]^{-1} [\mathbf{x}_f - \mathbf{A}^n \mathbf{x}(0)]$$

This is fairly straightforward. In the feedback case, however, it may be shown by the same methods as used earlier that if

$$u(k) = - \mathbf{K} \left[\mathbf{A}^n \mathbf{x}(k) - \mathbf{A}^k \mathbf{x}_f \right]$$

then $\mathbf{x}(n) = \mathbf{x}_f$, where we have assumed that the system starts at $\mathbf{x}(0)$ at time 0. The varying nature of the feedback is of course due to the dynamics, which 'move' the 'target' $\mathbf{x}_f$ depending upon the number of stages remaining.

The demonstration of the above notes, as in the previous derivation, that if

$$\mathbf{A}^n \mathbf{x}(k) - \mathbf{A}^k \mathbf{x}_f = \sum_{i=1}^{n} a_i(k) \, \mathbf{A}^{n-i} \mathbf{B}$$

then

$$\mathbf{A}^{n-1} \mathbf{x}(k+1) = \mathbf{A}^n \mathbf{x}(k) - \mathbf{A}^{n-1} \mathbf{B} \, a_1(k)$$

$$= \mathbf{A}^k \mathbf{x}_f + \sum_{i=2}^{n} a_i(k) \, \mathbf{A}^{n-i} \mathbf{B}$$

Hence

$$\mathbf{A}^n \mathbf{x}(k+1) - \mathbf{A}^{k+1} \mathbf{x}_f = \sum_{i=2}^{n} a_i(k) \, \mathbf{A}^{n-i+1} \mathbf{B}$$

$$= \sum_{i=1}^{n} a_i(k+1) \, \mathbf{A}^{n-i} \mathbf{B}$$

with $a_n(k+1) = 0$ and after n stages

$$\mathbf{A}^n \left[\mathbf{x}(n) - \mathbf{x}_f \right] = 0$$

which is the desired result.

When we have more than one input variable, the situation is straightforward but notationally messy. In fact, we will need in general q stages, where q is the controllability index.

27.5 DYNAMIC PROGRAMMING FOR OPTIMAL CONTROL

Dynamic programming was made famous in the 1950s by Richard Bellman, but the underlying ideas were known well before that. Bellman stated them in his **Principle of Optimality** and generated numerous examples at a time when there was a need for the methods.

> An optimal policy has the property that, whatever the initial state and initial decision are, the remaining decisions must constitute an optimal policy with regard to the state resulting from the first decision (Bellman and Kalaba, 1965).

We might remark that the principle is concerned with the policy (control law) rather than the optimal path (trajectory). Thus the policy tells the optimal thing to do when one arrives in state $\mathscr{A}$ at time T, regardless of how that point was reached.

27.5.1 Basic approach

The usage of the principle is straightforward: one finds the optimum from all points (say $\mathscr{A}, \mathscr{B}, \mathscr{C}, ...$) at stage K to the end (stage N). One considers all points $(a, \ell, c, ...)$ at stage $K-1$ and determines for each one the cost of going to $(\mathscr{A}, \mathscr{B}, \mathscr{C}, ...)$ in one step. Then the cost from each of $(a, \ell, c, ...)$ to the end is given by

$$\text{optimum cost } (a \to \text{END}) =$$

$$\min_{(\mathscr{I}=\mathscr{A}, \mathscr{B}, \mathscr{C}, ...)} \left(\text{Cost } (a \to \mathscr{I}) + \text{optimum cost } (\mathscr{I} \to \text{END}) \right)$$

This is repeated for each $i = a, \ell, c, ...$. From this we now know the optimum cost (and policy) from $(a, \ell, c, ...)$ to the end starting from stage $K-1$.

Let us make the above specific to our needs by applying it to the generic control systems problem given by Chapter 26 and repeated here for convenience.

$$J(\{\mathbf{u}(k), k=j, ..., N-1\}, \mathbf{x}(j), N) = h(N, \mathbf{x}(N))$$

$$+ \sum_{i=j}^{N-1} g(\mathbf{x}(i), \mathbf{u}(i), i)] \qquad (27.17)$$

subject to

$$\mathbf{x}(k+1) = f(\mathbf{x}(k), \mathbf{u}(k), k) \tag{27.18}$$

Define $J^\#(j, \mathbf{x})$ as cost in (27.17) when the control is optimal. Clearly for any final state $\mathbf{x} = \mathbf{x}(N)$,

$$J^\#(N, \mathbf{x}) = h(N, \mathbf{x})$$

Suppose we have arrived at state $\mathbf{x}$ at time $N-1$. No matter how we arrived there, the overall cost will be minimized by minimizing the last stage using our only remaining control, $\mathbf{u}(N-1)$. Thus

$$J^\#(N-1, \mathbf{x}) = \min_{\mathbf{u}(N-1)} [h(N, \mathbf{x}(N)) + g(\mathbf{x}, \mathbf{u}(N-1), N-1)]$$

and the seeming dependence on $\mathbf{x}(N)$ is removed by using the dynamics equation (27.18) to get

$$J^\#(N-1, \mathbf{x}) = \min_u [h(N, f(\mathbf{x}, \mathbf{u}, N-1)) + g(\mathbf{x}, \mathbf{u}, N-1)] \tag{27.19}$$

By the principle of optimality, the optimum from $(k, \mathbf{x}(k))$ to the end is independent of how it was reached. In particular, this is true for $(N, \mathbf{x}(N))$ and thus (27.19) becomes

$$J^\#(N-1, \mathbf{x}) = \min_u [J^\#(N, f(\mathbf{x}, \mathbf{u}, N-1)) + g(\mathbf{x}, \mathbf{u}, N-1)]$$

If the functions are differentiable, then the optimum $\mathbf{u}^\#$ is found where

$$\frac{\partial \left([J^\#(N, f(\mathbf{x}, \mathbf{u}, N-1)) + g(\mathbf{x}, \mathbf{u}, N-1)] \right)}{\partial \mathbf{u}} = 0$$

Frequently this can be solved explicitly for $\mathbf{u}^\#$, which will be then a function of $\mathbf{x}$. We denote this as $\mathbf{u}^\#(N-1, \mathbf{x})$ and see that regardless of how it was obtained, $J^\#$ can be written

$$J^{\#}(N-1,\mathbf{x}) = J^{\#}(N,f(\mathbf{x},\mathbf{u}^{\#}(N-1,\mathbf{x}),N-1)) + g(\mathbf{x},\mathbf{u}^{\#}(N-1,\mathbf{x}),N-1)$$

We can continue in this same vein to find $J^{\#}(N-2,\mathbf{x})$, then $J^{\#}(N-3,\mathbf{x})$, ..., $J^{\#}(0,\mathbf{x})$. The generic expression, often called Bellman's equation, is

$$J^{\#}(N-k,\mathbf{x}) = \min_{u} [J^{\#}(N-k+1,f(\mathbf{x},\mathbf{u},N-k)) + g(\mathbf{x},\mathbf{u},N-k)]$$

$$= J^{\#}(N-k+1,f(\mathbf{x},\mathbf{u}^{\#}(N-k),\mathbf{x}),N-k) + g(\mathbf{x},\mathbf{u}^{\#}(N-k,\mathbf{x}),N-k)$$

$$k=1,2,...,N$$

where the boundary condition is

$$J^{\#}(N,\mathbf{x}) = h(N,\mathbf{x})$$

The original problem has the optimum cost $J^{\#}(0,\mathbf{x}(0))$ and the control sequence is

$$\mathbf{u}(0) = \mathbf{u}^{\#}(0,\mathbf{x}(0))$$

$$\mathbf{u}(1) = \mathbf{u}^{\#}(1,f(\mathbf{x}(0),\mathbf{u}^{\#}(0,\mathbf{x}(0)),0))$$

$$\mathbf{u}(2) = \mathbf{u}^{\#}(2,f(\mathbf{x}(1),\mathbf{u}^{\#}(1,\mathbf{x}(1)),1))$$

$$\vdots$$

In fact we have a closed-loop control law, because with k stages remaining, if we know the state $\mathbf{x}$, then the control to command is $\mathbf{u}^{\#}(N-k,\mathbf{x})$. The cost remaining is $J^{\#}(N-k,\mathbf{x})$.

27.5.2 Application to a special case: the LQ problem

Let us specialize further to the LQ problem, i.e. the problem of minimizing

$$J = \tfrac{1}{2}\,\mathbf{x}^{\mathrm{T}}(N)\,\mathbf{S}\mathbf{x}(N) + \tfrac{1}{2}\sum_{i=0}^{N-1} [\mathbf{x}^{\mathrm{T}}(i)\,\mathbf{Q}\mathbf{x}(i) + \mathbf{u}^{\mathrm{T}}(i)\,\mathbf{R}\mathbf{u}(i)]$$

with dynamics

$$\mathbf{x}(k+1) = \mathbf{A}\mathbf{x}(k) + \mathbf{B}\mathbf{u}(k) \qquad \mathbf{x}(0) = \mathbf{x}_0$$

Defining $J^{\#}$ as before, we find

$$J^{\#}(N, \mathbf{x}) = \tfrac{1}{2}\mathbf{x}^{\mathrm{T}}\mathbf{S}\mathbf{x}$$

Continuing with stage $N=1$, we derive

$$J^{\#}(N-1, \mathbf{x}) = \min_{u} \tfrac{1}{2}\,[\mathbf{x}^{\mathrm{T}}(N)\,\mathbf{S}\mathbf{x}(N) + \mathbf{x}^{\mathrm{T}}\mathbf{Q}\mathbf{x} + \mathbf{u}^{\mathrm{T}}\mathbf{R}\mathbf{u}]$$

$$= \min_{u} \tfrac{1}{2}\,[(\mathbf{A}\mathbf{x} + \mathbf{B}\mathbf{u})^{\mathrm{T}}\mathbf{S}\,(\mathbf{A}\mathbf{x} + \mathbf{B}\mathbf{u}) + \mathbf{x}^{\mathrm{T}}\mathbf{Q}\mathbf{x} + \mathbf{u}^{\mathrm{T}}\mathbf{R}\mathbf{u}] \tag{27.20}$$

Taking the vector derivative

$$0 = \frac{\partial[(\mathbf{A}\mathbf{x} + \mathbf{B}\mathbf{u})^{\mathrm{T}}\mathbf{S}\,(\mathbf{A}\mathbf{x} + \mathbf{B}\mathbf{u}) + \mathbf{x}^{\mathrm{T}}\mathbf{Q}\mathbf{x} + \mathbf{u}^{\mathrm{T}}\mathbf{R}\mathbf{u}]}{\partial\mathbf{u}}$$

$$= \mathbf{B}^{\mathrm{T}}\mathbf{S}\,(\mathbf{A}\mathbf{x} + \mathbf{B}\mathbf{u}) + [(\mathbf{A}\mathbf{x} + \mathbf{B}\mathbf{u})^{\mathrm{T}}\mathbf{S}\mathbf{B}]^{\mathrm{T}} + \mathbf{R}\mathbf{u} + [\mathbf{u}^{\mathrm{T}}\mathbf{R}]^{\mathrm{T}}$$

Provided $\mathbf{S}$ and $\mathbf{R}$ are symmetric, which is the usual assumption for matrices used in quadratic forms and is one made without loss of generality, and the inverse exists, this requirement becomes

$$(\mathbf{B}^{\mathrm{T}}\mathbf{S}\mathbf{B} + \mathbf{R})\mathbf{u} + \mathbf{B}^{\mathrm{T}}\mathbf{S}\mathbf{A}\mathbf{x} = 0$$

and then

$$\mathbf{u}^{\#}(N-1, \mathbf{x}) = -\,(\mathbf{B}^{\mathrm{T}}\mathbf{S}\mathbf{B} + \mathbf{R})^{-1}\,\mathbf{B}^{\mathrm{T}}\mathbf{S}\mathbf{A}\mathbf{x}$$

Substituting this in (27.20) yields

$$J^{\#}(N-1, \mathbf{x}) = \tfrac{1}{2}\,\big\{\,[\mathbf{A}\mathbf{x} - \mathbf{B}(\mathbf{B}^{\mathrm{T}}\mathbf{S}\mathbf{B} + \mathbf{R})^{-1}\mathbf{B}^{\mathrm{T}}\mathbf{S}\mathbf{A}\mathbf{x}]^{\mathrm{T}}$$

$$\times \mathbf{S}\,[\mathbf{A}\mathbf{x} - \mathbf{B}(\mathbf{B}^{\mathrm{T}}\mathbf{S}\mathbf{B} + \mathbf{R})^{-1}\mathbf{B}^{\mathrm{T}}\mathbf{S}\mathbf{A}\mathbf{x}] + \mathbf{x}^{\mathrm{T}}\mathbf{Q}\mathbf{x}$$

$$+ [(\mathbf{B}^{\mathrm{T}}\mathbf{S}\mathbf{B} + \mathbf{R})^{-1}\,\mathbf{B}^{\mathrm{T}}\mathbf{S}\mathbf{A}\mathbf{x}]^{\mathrm{T}}\mathbf{R}(\mathbf{B}^{\mathrm{T}}\mathbf{S}\mathbf{B} + \mathbf{R})^{-1}\mathbf{B}^{\mathrm{T}}\mathbf{S}\mathbf{A}\mathbf{x}\big\}$$

After considerable algebra this becomes

$$J^\#(N-1,\mathbf{x}) = \tfrac{1}{2}\mathbf{x}^\mathsf{T}\mathbf{P}\mathbf{x}$$

where

$$\mathbf{P} = \mathbf{Q} + \mathbf{A}^\mathsf{T}\mathbf{S}\mathbf{A}$$
$$+ [(\mathbf{B}^\mathsf{T}\mathbf{S}\mathbf{B} + \mathbf{R})^{-1}\mathbf{B}^\mathsf{T}\mathbf{S}\mathbf{A}]^\mathsf{T}\mathbf{B}^\mathsf{T}\mathbf{S}\mathbf{B}[(\mathbf{B}^\mathsf{T}\mathbf{S}\mathbf{B} + \mathbf{R})^{-1}\mathbf{B}^\mathsf{T}\mathbf{S}\mathbf{A}]$$
$$+ [(\mathbf{B}^\mathsf{T}\mathbf{S}\mathbf{B} + \mathbf{R})^{-1}\mathbf{B}^\mathsf{T}\mathbf{S}\mathbf{A}]^\mathsf{T}\mathbf{R}[(\mathbf{B}^\mathsf{T}\mathbf{S}\mathbf{B} + \mathbf{R})^{-1}\mathbf{B}^\mathsf{T}\mathbf{S}\mathbf{A}]$$
$$- [(\mathbf{B}^\mathsf{T}\mathbf{S}\mathbf{B} + \mathbf{R})^{-1}\mathbf{B}^\mathsf{T}\mathbf{S}\mathbf{A})^\mathsf{T}\mathbf{B}^\mathsf{T}\mathbf{S}\mathbf{A}$$
$$- \mathbf{A}^\mathsf{T}\mathbf{S}\mathbf{B}[(\mathbf{B}^\mathsf{T}\mathbf{S}\mathbf{B} + \mathbf{R})^{-1}\mathbf{B}^\mathsf{T}\mathbf{S}\mathbf{A}]$$

Using symmetry of the $\mathbf{R}, \mathbf{S}$, and inverse matrices then gives

$$\mathbf{P} = \mathbf{Q} + \mathbf{A}^\mathsf{T}\mathbf{S}\mathbf{A} - \mathbf{A}^\mathsf{T}\mathbf{S}\mathbf{B}(\mathbf{B}^\mathsf{T}\mathbf{S}\mathbf{B} + \mathbf{R})^{-1}\mathbf{B}^\mathsf{T}\mathbf{S}\mathbf{A}$$

Calling this $\mathbf{P}(N-1)$ and letting $\mathbf{P}(N) = \mathbf{S}$, we can find by exactly the same steps that

$$\mathbf{u}^\#(k,\mathbf{x}) = -(\mathbf{B}^\mathsf{T}\mathbf{P}(k+1)\mathbf{B} + \mathbf{R})^{-1}\mathbf{B}^\mathsf{T}\mathbf{P}(k+1)\mathbf{A}\mathbf{x} \qquad (27.21)$$

and

$$\mathbf{P}(k) = \mathbf{Q} + \mathbf{A}^\mathsf{T}\mathbf{P}(k+1)\mathbf{A}$$
$$- \mathbf{A}^\mathsf{T}\mathbf{P}(k+1)\mathbf{B}(\mathbf{B}^\mathsf{T}\mathbf{P}(k+1)\mathbf{B} + \mathbf{R})^{-1}\mathbf{B}^\mathsf{T}\mathbf{P}(k+1)\mathbf{A} \qquad (27.22)$$

with

$$J^\#(k,\mathbf{x}) = \tfrac{1}{2}\mathbf{x}^\mathsf{T}\mathbf{P}(k)\mathbf{x}$$

We note that the control law (27.21) is linear in $\mathbf{x}$ but with time-varying gain. That the result is the same as found in Chapter 26 is not surprising; we note that this approach gave the feedback form of the control without intermediate steps.

The discussion of the previous chapter about letting $N \to \infty$ and achieving a constant regulator gain also applies, of course. We elaborate upon that briefly.

If we have the problem of the linear digital regulator, given as the LQ case with $N \to \infty$,

$$J = \tfrac{1}{2} \sum_{i=0}^{\infty} \{ \mathbf{x}^{\mathrm{T}}(i)\,\mathbf{Q}\mathbf{x}(i) + \mathbf{u}^{\mathrm{T}}(i)\,\mathbf{R}\mathbf{u}(i) \} \tag{27.23}$$

with dynamics

$$\mathbf{x}(k+1) = \mathbf{A}\mathbf{x}(k) + \mathbf{B}\mathbf{u}(k) \quad \text{and} \quad \mathbf{x}(0) = \mathbf{x}_0$$

then heuristically, if the pair $(\mathbf{A};\mathbf{B})$ is controllable we can take $\mathbf{x}(k) \to 0$ in a few steps and stop controlling. Hence, $J \to$ constant. However, in this case the Riccati equation (27.22) must for sufficiently small k (remember the computation 'starts' with large $k=N$ and works toward smaller k) take on a steady-state solution P, given by the solution of the matrix quadratic equation, or algebraic Riccati equation,

$$\mathbf{P} = \mathbf{Q} + \mathbf{A}^{\mathrm{T}}\mathbf{P}\mathbf{A}$$

$$-\mathbf{A}^{\mathrm{T}}\mathbf{P}\mathbf{B}(\mathbf{B}^{\mathrm{T}}\mathbf{P}\mathbf{B} + \mathbf{R})^{-1}\mathbf{B}^{\mathrm{T}}\mathbf{P}\mathbf{A}$$

The optimal performance value J and the control law are given by

$$J = \tfrac{1}{2}\mathbf{x}_0^{\mathrm{T}}\,\mathbf{P}\mathbf{x}_0 \tag{27.24}$$

$$\mathbf{u}^*(k,\mathbf{x}) = -(\mathbf{B}^{\mathrm{T}}\mathbf{P}\mathbf{B} + \mathbf{R})^{-1}\mathbf{B}^{\mathrm{T}}\mathbf{P}\mathbf{A}\mathbf{x}(k)$$

$$= -\mathbf{G}\mathbf{x}(k)$$

where $\mathbf{G}$ is a constant.

Example

Let $x(k+1) = ax(k) + u(k)$ with

$$J = \tfrac{1}{2} \sum_{i=0}^{\infty} \{ x^2(i) + ru^2(i) \}$$

From (27.24), we find the optimal control is

$$u^* = - \frac{pax(k)}{p + r}$$

where p is the positive root which solves the expression given by the Riccati equation

$$p = 1 + a^2 p - \frac{a^2 p^2}{p + r}$$

or its more convenient form

$$p^2 + (r - 1 - a^2 r)\, p - r = 0$$

We see that for very small r, $u^* \approx -a\, x(\mathrm{k})$, the solution $\rightarrow 0$ in one step, and $p = 1$. For large r (u penalized), $u^* \approx 0$ and

$$p \approx \frac{1}{1 - a^2} = 1 + a^2 + a^4 + \cdots + a^{2n} + \cdots$$

reflecting that control is not applied and the state simply tends to 0 as a^n.

27.5.3 Numerical dynamic programming

Dynamic programming, like most optimization methods, is seldom able to produce closed-form solutions. The method lends itself well to numerical methods, however, except for the 'curse of dimensionality', which we will meet below.

The essence of the numerical implementation is straightforward.

1. We form a grid $\mathscr{G}$ in state space by discretizing the state variables, so that $\mathscr{G}$ is the array of size m^n if each state variable can take on m values and the system has dimension n. Then for each point $i \in \mathscr{G}$ on the array, we associate a state $^i\mathbf{x}$.
2. For each $i \in \mathscr{G}$, initialize $J^\#(N, {}^i\mathbf{x}) = h(N, {}^i\mathbf{x})$.
3. For each remaining stage $k = 1, 2, \ldots, N$, calculate for each $i \in \mathscr{G}$

$$J^\#(N-k, {}^i\mathbf{x}) = \min_u \; [J^\#(N-k+1, f({}^i\mathbf{x}, \mathbf{u}, N-k)) + g({}^i\mathbf{x}, \mathbf{u}, N-k)]$$

Since table lookup is involved, $J^\#(N-k+1, f({}^i\mathbf{x}, \mathbf{u}, N-k))$ may require interpolation or other approximation. In any case, store for future use the minimizing control $\mathbf{u}^\#(N-k, {}^i\mathbf{x})$. The optimal cost array $J^\#(N-k, {}^i\mathbf{x})$ is needed for the next stage, but $J^\#(N-k+1, {}^i\mathbf{x})$ could if necessary be discarded or stored.

The problem with the technique becomes readily apparent here and has two aspects. First and most obvious is the size of the grid, which for a three-dimensional system with 1% discretization (101 numbers) requires more than 10^6 elements in each array of $J^\#$ and $\mathbf{u}^\#$. Second, the minimization in step 3 requires that the new grid be computed for m^n starting states, and since the search may cover most of the target grid, needs a numerical check of up to m^n points for each. These two problems quickly turn a simple approach into a huge computational problem, and they constitute the **curse of dimensionality**.

Many useful results have been obtained with the principle of optimality, but ordinarily much engineering insight and many approximations are needed to make the computational burden bearable.

27.6 NUMERICAL METHODS AND APPROACHES

Few optimal control problems have a useful explicit solution, so we are often in the position of needing to find numerical solutions. In section 27.5.3 we looked briefly at numerical dynamic programming. In this section, we look at two methods: direct optimization using what is called steepest descent; and solutions of the two-point boundary-value problems from the Pontryagin approach using neighbouring extremals and quasi-linearization. We first introduce the basic ideas of these in the next section.

27.6.1 Basics

One of the recurring problems with Pontryagin and related optimal control solutions is that we must solve two-point boundary-value problems. In the simplest cases, for example, the initial state vector $\mathbf{x}(0)$ and the final Lagrange multipliers/costates $\lambda(t_f)$ must take on certain values. To solve these, we have a choice of two iterative numerical methods:

1. find, among the state and costate histories which satisfy the boundary conditions, the functions which satisfy the system differential equations; and
2. find, among all functions which satisfy the differential equations, those which satisfy the boundary conditions.

To understand a method called **quasi-linearization** which applies to the first of these, we consider the problem of finding a function $\mathbf{x}(t)$ which satisfies

$$\dot{\mathbf{x}}(t) = f(\mathbf{x}(t), t) \tag{27.25}$$

and boundary conditions

$$\mathbf{B}_0 \mathbf{x}(0) = \mathbf{c}_0 \quad \text{and} \quad \mathbf{B}_f \mathbf{x}(t_t) = \mathbf{c}_f \tag{27.26}$$

It is assumed that $\dim(\mathbf{c}_0) + \dim(\mathbf{c}_f) = n = \dim(\mathbf{x})$. We will develop an iteration for the solution by creating a sequence $\mathbf{x}(i, t)$ of functions such that we hope $\mathbf{x}(k, t) \to \mathbf{x}(t)$ as k increases. Let $\mathbf{x}(i, t)$ satisfy

$$\mathbf{B}_0 \mathbf{x}(i, 0) = \mathbf{c}_0 \quad \text{and} \quad \mathbf{B}_f \mathbf{x}(i, t_t) = \mathbf{c}_f$$

If $\mathbf{x}(i, t)$ also satisfies (27.25) then we are finished. If not, then we observe that

$$\dot{\mathbf{x}}(t) = f(\mathbf{x}(t), t) = f(\mathbf{x}(i, t), t)$$

$$+ \left. \frac{\partial f(\mathbf{x} \, t)}{\partial \mathbf{x}} \right|_{\mathbf{x}(i, t)} (\mathbf{x}(t) - (\mathbf{x}(i, t)))$$

$$+ \text{ higher-order terms}$$

This may be rearranged to appear as

$$\dot{\mathbf{x}}(t) \approx \left.\frac{\partial f(\mathbf{x},t)}{\partial \mathbf{x}}\right|_{\mathbf{x}(i,t)}\mathbf{x}(t)$$

$$+ \left[f(\mathbf{x}(i,t)) - \left.\frac{\partial f(\mathbf{x},t)}{\partial \mathbf{x}}\right|_{\mathbf{x}(i,t)}\mathbf{x}(i,t)\right]$$

Thus it is reasonable to take the next guess of $\mathbf{x}(t)$ as

$$\dot{\mathbf{x}}(i+1,t) = \left.\frac{\partial f(\mathbf{x},t)}{\partial \mathbf{x}}\right|_{\mathbf{x}(i,t)}\mathbf{x}(i+1,t))$$

$$+ \left[f(\mathbf{x}(i,t)) - \left.\frac{\partial f(\mathbf{x},t)}{\partial \mathbf{x}}\right|_{\mathbf{x}(i,t)}\mathbf{x}(i,t)\right] \tag{27.27}$$

Since when the time function $\mathbf{x}(i,t)$ is inserted the partial derivatives also become time functions, the non-linearity of the differential equation has been removed and the above is a forced time-varying linear differential equation. We may find its solution by computing the transition matrix $\phi(i+1,t)$ using

$$\dot{\phi}(i+1,t) = \left.\frac{\partial f(\mathbf{x},t)}{\partial \mathbf{x}}\right|_{\mathbf{x}(i,t)}\phi(i+1,t) \quad\text{and}\quad \phi(i+1,0) = \mathbf{I} \tag{27.28}$$

and the computation is done by either perturbation or by n integrations with the successive initial conditions

$$[\,1 \ \ 0 \ \ 0 \ \cdots \ 0\,]^{\mathrm{T}}, [\,0 \ \ 1 \ \ 0 \ \cdots \ 0\,]^{\mathrm{T}}, ..., [\,0 \ \ 0 \ \ 0 \ \cdots \ 1\,]^{\mathrm{T}}$$

Another numerical integration is done using the forcing function

$$\left[f(\mathbf{x}(i,t)) - \left.\frac{\partial f(\mathbf{x},t)}{\partial \mathbf{x}}\right|_{\mathbf{x}(i,t)}\mathbf{x}(i,t)\right]$$

and zero initial conditions to yield the particular solution $\mathbf{x}_{\mathrm{p}}(i+1,t)$.

Then the initial conditions for $\mathbf{x}(i+1,t)$ are chosen to solve

$$\mathbf{B}_0\,\mathbf{x}(i+1,0) = c_0$$

$$\mathbf{B}_f\left(\phi(i+1,t_f)\,\mathbf{x}(i+1,0) + \mathbf{x}_p(i+1,t_f)\right) = c_f$$

The iteration continues until $\mathbf{x}(i+1,\,t) \approx \mathbf{x}(i,\,t)$.

An alternative approach is to guess at initial conditions $\mathbf{x}(i,\,t)$, ensuring that they satisfy $\mathbf{B}_0\,\mathbf{x}(i,\,t) = c_0$, and then integrate (27.25) to find $\mathbf{x}(i,\,t_f)$. If this satisfies (27.26), we are done, but if not, then another initial condition set must be chosen. We again linearize in the vicinity of $\mathbf{x}(i,\,t)$ using the transition matrix (27.28). We know that

$$\frac{\partial \mathbf{B}_f\,\mathbf{x}(i+1,\,t_f)}{\partial \mathbf{x}(i+1,0)} = \mathbf{B}_f\,\phi(i+1,\,t_f)$$

The change $\Delta\mathbf{x}(i+1,0)$ required is then clearly given by the n equations in n unknowns

$$\mathbf{B}_0\,\Delta\mathbf{x}(i+1,0) = 0$$

$$\mathbf{B}_f\,\phi(i+1,t_f)\,\Delta\mathbf{x}(i+1,0) = c_f - \mathbf{B}_f\,\mathbf{x}(i,t_f)$$

We are finished when $\Delta\mathbf{x}(i+1,0) = 0$, or nearly so.

The above are in many respects similar. The difference is that in the first, a solution $\mathbf{x}(i,t)$ satisfies a linear differential equation (27.27) and the boundary conditions (27.26), while in the second case, the solution satisfies the nonlinear differential equation (27.25).

Steepest descent (or ascent) is, in contrast to the above, an optimization method in which we attempt to find a vector $\mathbf{x}^{\#}$ which minimizes a function $g(\mathbf{x})$. Using Taylor series, we write for a guess $\mathbf{x}(i)$, with $\nabla_{\mathbf{x}}g(\mathbf{x}) = \left[\dfrac{\partial g}{\partial \mathbf{x}_i}\right]^{\mathrm{T}}$,

$$g(\mathbf{x}^{\#}) = g(\mathbf{x}(i)) + \nabla_{\mathbf{x}}g\Big|_{\mathbf{x}(i)}(\mathbf{x}^{\#} - \mathbf{x}(i))$$

$$+ (\mathbf{x}^{\#} - \mathbf{x}(i))^{\mathrm{T}}\left\{\frac{\partial^2 g}{\partial \mathbf{x}_i \partial \mathbf{x}_j}\right\}\Big|_{\mathbf{x}(i)}(\mathbf{x}^{\#} - \mathbf{x}(i)) + \cdots$$

If $(\mathbf{x}^{\#} - \mathbf{x}(i))$ is 'small', then

$$g(\mathbf{x}^{\#}) \approx g(\mathbf{x}(i)) + \nabla_{\mathbf{x}}g\Big|_{\mathbf{x}(i)} (\mathbf{x}^{\#} - \mathbf{x}(i))$$

If we choose

$$(\mathbf{x}^{\#} - \mathbf{x}(i)) = -\alpha \, \nabla_{\mathbf{x}}g\Big|^{\mathrm{T}}_{\mathbf{x}(i)} \tag{27.29}$$

then we have, for α small enough for the higher-order terms to be ignored, that

$$g(\mathbf{x}^{\#}) \approx g(\mathbf{x}(i)) - \alpha(\nabla_{\mathbf{x}}g)(\nabla_{\mathbf{x}}g)^{\mathrm{T}}$$

$$\leq g(\mathbf{x}(i))$$

Rearranging (27.29), it seems reasonable to choose

$$\mathbf{x}(i+1) = \mathbf{x}(i) - \alpha \, \nabla_{\mathbf{x}}g\Big|^{\mathrm{T}}_{\mathbf{x}(i)}$$

as a better approximation to $\mathbf{x}^{\#}$. We repeat that it is required that the parameter α be small enough that the linear approximation holds, but the scheme is readily modified to incorporate constraints on the step size $(\mathbf{x}(i+1) - \mathbf{x}(i))$ and side constraints such as $h(\mathbf{x}) = 0$.

27.6.2 Quasi-linearization

The basic conditions resulting from the maximum principle are of the form

$$\mathbf{x}(k+1) = f(\mathbf{x}(k), \mathbf{u}(k), k)$$

$$\lambda(k) = \frac{\partial f(\mathbf{x}, \mathbf{u}(k), k)}{\partial \mathbf{x}} \, \lambda(k+1)$$

$$\mathbf{u}(k) = \mathbf{u}(\mathbf{x}(k), \lambda(k))$$

In differential equation form, which is often more useful to us and which has easier notation,

$$\dot{\mathbf{x}} = f(\mathbf{x}, \mathbf{u}, t)$$

$$\dot{\lambda} = -\nabla_{\mathbf{x}}H^{\mathrm{T}} = -\frac{\partial f}{\partial \mathbf{x}}\lambda$$

$$\mathbf{u} = \mathbf{u}(\mathbf{x}, \lambda) \tag{27.30}$$

with boundary conditions (for free end point $\mathbf{x}(t_{\mathrm{f}})$)

$$\mathbf{x}(0) = \mathbf{x}_0 \qquad \lambda(t_{\mathrm{f}}) = -\nabla_{\mathbf{x}}\mathbf{h}^{\mathrm{T}}$$

Defining the extended vector $\mathbf{y}$

$$\mathbf{y} = \begin{bmatrix} \mathbf{x} \\ \lambda \end{bmatrix}$$

gives, because $\mathbf{u}$ is a function of $\mathbf{x}$ and λ,

$$\dot{\mathbf{y}} = \begin{bmatrix} \hat{f}(\mathbf{y}, t) \\ -\frac{\partial f}{\partial \mathbf{x}}\lambda \end{bmatrix} = g(\mathbf{y}, t)$$

with boundary conditions

$$\mathbf{y}(0) = \begin{bmatrix} \mathbf{x}_0 \\ ? \end{bmatrix} \qquad \mathbf{y}(t_{\mathrm{f}}) = \begin{bmatrix} ? \\ -\nabla_{\mathbf{x}}\mathbf{h}^{\mathrm{T}} \end{bmatrix} \tag{27.31}$$

where ? denotes unknown and unspecified values. Now choose functions $\mathbf{x}(t)$ and $\lambda(t)$ which satisfy $\mathbf{x}(0)$ and $\lambda(t_{\mathrm{f}})$ respectively; they could be constants, for example. This constitutes the initial guess $\mathbf{y}(0, t)$. We keep adjusting these functions until the differential equations are satisfied. The iteration follows from

$$\dot{\mathbf{y}}(i+1,t) \approx g(\mathbf{y}(i,t),t) + \{\nabla_{\mathbf{y}}g\}^{\mathrm{T}} (\mathbf{y}(i+1,t) - \mathbf{y}(i,t))$$

for which a particular solution solves

$$\dot{\mathbf{y}}_{\mathrm{p}} = g(\mathbf{y}(i,\,t),t) - \{\nabla_{\mathbf{y}}g\}\,\mathbf{y}(i,\,t) \qquad \mathbf{y}_{\mathrm{p}}(0) = \mathbf{0}$$

and the homogeneous solution follows from

$$\mathbf{y}_{\mathrm{h}}(t) = \Phi(t,0)\,\mathbf{y}_{\mathrm{h}}(0)$$

$$\dot{\Phi}(t,0) = \{\nabla_{\mathbf{y}}g\}\,\Phi(t,0) \qquad \Phi(0,\,0) = \mathbf{I}$$

Then

$$\mathbf{y}(i+1,t) = \Phi(t,0)\,\mathbf{y}_{\mathrm{h}}(0) + \mathbf{y}_{\mathrm{p}}(t)$$

and the initial conditions $\mathbf{y}_{\mathrm{h}}(0)$ are chosen so that the boundary conditions (27.31) are satisfied.

27.6.3 Variation of extremals – the shooting method

The above may seem a difficult method to find a solution to the differential equations, and it can be so. An alternative is to choose initial conditions, run out the solution to (27.30), and then check the final conditions. We modify the initial conditions available (here $\lambda(0)$) systematically until the final conditions are satisfied. Each solution in the iteration is then optimal for the set of boundary conditions it satisfies (even if they are the wrong ones) and, since the iteration tends to work in a region of the state space, the method is called that of variation of extremals.

The simplified method of doing this is to choose $\mathbf{x}(0) = \mathbf{x}_0$ and some $\lambda(0)$, run out trajectories to find $\mathbf{x}(t)$ and $\lambda(t)$ and particularly $\mathbf{x}(t_{\mathrm{f}})$ and $\lambda(t_{\mathrm{f}})$, find (perhaps by perturbation) the values $d\mathbf{x}(t_{\mathrm{f}})/d\lambda(0)$ and $d\lambda(t_{\mathrm{f}})/d\lambda(0)$, and make appropriate adjustments in $\lambda(0)$ along the lines of

$$\lambda(i+1,0) = \lambda(i,0) + \left[\frac{d\lambda(t_f)}{d\lambda(0)}\right]^{-1} (\lambda_{des}(i,t_f) - \lambda(i,t_f))$$

where

$$\lambda_{des}(i,t_f) = -\nabla_x^T \mathbf{h}(\mathbf{x}(i,t_f),\lambda(i,t_f))$$

The problem with this is that it can be very sensitive to the values of $\lambda(0)$, which may range over several orders of magnitude.

Linearization methods have been proposed for this problem, also, but are beyond the scope of this book; see Kirk (1970).

27.6.4 Steepest descent in the calculus of variations

One early and successful method of determining the optimal trajectories (but not necessarily the closed-loop feedback law) was the method of steepest descent applied to the calculus of variations (Bryson and Denham, 1962).

Consider again the basic problem: find a function $\mathbf{u}(t)$ (actually we would prefer the function $\mathbf{u}(\mathbf{x})$) which minimizes

$$J = \mathbf{h}(\mathbf{x}(t_f)) + \int_{t_0}^{t_f} g(\mathbf{x}(t),\mathbf{u}(t),t)\, dt$$

subject to

$$\dot{\mathbf{x}} = f(\mathbf{x},\mathbf{u},t)$$

with $\mathbf{x}(t_0)$ specified and t_f specified.

From the calculus of variations, the first variation of this (essentially the first derivative) is

$$\Delta J = [\nabla_x \mathbf{h} - \lambda]^T \Big|_{t_f} \Delta \mathbf{x}(t_f)$$

$$+ \int_{t_0}^{t_f} \left\{ \left[\nabla_x(g + \lambda^T f) + \frac{d\lambda}{dt} \right]^T \Delta \mathbf{x} + [\nabla_u(g + \lambda^T f)]^T \Delta \mathbf{u} \right\} dt$$

Suppose **u** is arbitrary but that we force

$$\dot{\mathbf{x}} = f(\mathbf{x}, \mathbf{u}, t) \qquad\qquad \mathbf{x}(0) = \mathbf{x}_0$$

$$\dot{\lambda} = -\nabla_{\mathbf{x}}\{g + \lambda^{\mathrm{T}} f\} = -\nabla_{\mathbf{x}}\mathbf{H} \qquad\qquad \lambda(t_{\mathrm{f}}) = \nabla_{\mathbf{x}}\mathbf{h}\Big|_{t_{\mathrm{f}}}$$

to hold. Then

$$\Delta J = \int_{t_0}^{t_{\mathrm{f}}} [\nabla_{\mathbf{u}}^{\mathrm{T}} \mathbf{H} \Delta \mathbf{u}]\, \mathrm{d}t$$

If $\nabla_{\mathbf{u}} \mathbf{H} = \mathbf{0}$, we are finished in that the first variation is zero and hence we have an extremal of the problem. Otherwise, we choose

$$\Delta \mathbf{u} = \mathbf{u}(i+1, t) - \mathbf{u}(i, t) = -\alpha \, [\nabla_{\mathbf{u}} \mathbf{H}]$$

If α is 'small enough', this will lead to $\Delta J < 0$, i.e. a decrease in the performance functional J. The iteration process continues until the result is 'good enough'.

27.7 COMPUTER AIDS

Computer aids for the techniques presented here are not standard, as each problem seems to have its peculiarities. Some simulation packages, such as PSI, have the capability to find good parameter values and hence implement section 27.3 for special cases.

27.8 SUMMARY

We have only touched on some important techniques and results: minimum variance control as a transfer function design problem, an alternative approach to minimum time (deadbeat) control, dynamic programming, and numerical methods for trajectory design problems. All except occasionally dynamic programming are outside the realm of the usual introductory courses in control theory; they are presented here to indicate some results and philosophies that the control engineer may find useful even if not a theory specialist.

27.9 FURTHER READING

Parameter selection is among the topics covered by Iserman (1981). Textbooks on optimal control theory usually have at least some of these techniques in more detail. A quite readable old book is Kirk (1970), while an ageing classic with a very helpful treatment of numerical methods is Bryson and Ho (1969). More advanced texts and monographs are many and varied: a quite different point of view related to their industrial experiences is presented by Grimble and Johnson (1988).

Old papers in the research literature, such as Bryson and Denham (1962) using steepest descent and the many articles co-authored by Bellman using his principle are also interesting both for the results and style of presentation.

28

State estimation in noise

There are a number of signal processing approaches called filtering:

1. signal frequency content shapers, including notch filters for removing power line noise and band-pass filters for extracting desirable signals such as AM radio broadcasts from the environment;
2. detection filters such as matched filters for indicating the presence or absence of certain signal types, such as radar pulses and discrete symbols in communication networks; and
3. state estimation filters for inferring estimates of signal states from available measurements and noise.

Of these, the first two are predominantly the realm of communications, except for the analogue guard filters sometimes used with sampled data. The third, however, is of critical interest to control systems for two applications: for plant monitoring and event reconstruction, and with state feedback controllers. In both cases, state estimates are used because the entire state vector of a physical system is seldom directly measured.

State estimation filters are philosophically different from signal extraction filters, even though similarities can be shown. The object is to process measurements $\{y(t)\}$ made of a system with state vector $x(t)$ in order to derive a 'good' estimate $\hat{x}(t)$ of the state. We have already met deterministic observers, which constitute one approach to this task. In this section, we consider the famous Kalman–Bucy filters for linear systems and meet some of the extensions of the ideas to non-linear systems. We also present observers for noisy situations.

28.1 OVERVIEW

The basic theory is associated with the problem of deriving a state estimate $\hat{x}(kT)$ of the state vector $x(kT)$ of a system with dynamics

$$\mathbf{x}(kT+T) = \mathbf{A}\mathbf{x}(kT) + \mathbf{B}\mathbf{u}(kT) + \mathbf{v}(kT)$$

from the measurements

$$\mathbf{y}(kT) = \mathbf{C}\mathbf{x}(kT) + \mathbf{w}(kT)$$

or, more precisely, from a set of measurements such as $\mathbf{Y}(k) = \{\mathbf{y}(0), \mathbf{y}(T), \ldots, \mathbf{y}(kT)\}$. The quantities $\mathbf{v}(\cdot)$ and $\mathbf{w}(\cdot)$ are taken as white noises (see Appendix C) with means $\mathbf{0}$ and autocovariances

$$\mathbf{Q}_k = \mathscr{E}[\mathbf{v}(kT)\mathbf{v}^{\mathrm{T}}(mT)] = \mathbf{C}_{\mathbf{v}\mathbf{v}}(kT, mT)\,\delta_{km}$$

$$\mathbf{R}_k = \mathscr{E}[\mathbf{w}(kT)\mathbf{w}^{\mathrm{T}}(kT)] = \mathbf{C}_{\mathbf{w}\mathbf{w}}(kT, mT)\,\delta_{km}$$

The noises are assumed uncorrelated with each other, and with the (random) initial condition on $\mathbf{x}$, so that

$$\mathscr{E}[\mathbf{v}(kT)\mathbf{w}^{\mathrm{T}}(mT)] = \mathbf{0}$$

$$\mathscr{E}[\mathbf{x}(0)] = \mathbf{x}_0$$

$$\Sigma_0 = \mathscr{E}[\mathbf{x}(0) - \mathbf{x}_0][\mathbf{x}(0) - \mathbf{x}_0]^{\mathrm{T}}$$

$$\mathbf{0} = \mathscr{E}[\mathbf{x}(0)\mathbf{w}^{\mathrm{T}}(kT)] \qquad \mathbf{0} = \mathscr{E}[\mathbf{x}(0)\mathbf{v}^{\mathrm{T}}(kT)]$$

Under these circumstances, the Kalman–Bucy filter given by the calculations

$$\hat{\mathbf{x}}(k+1|k) = \mathbf{A}\hat{\mathbf{x}}(k) + \mathbf{B}\mathbf{u}(k)$$

$$\Sigma(k+1|k) = \mathbf{A}\Sigma(k)\mathbf{A}^{\mathrm{T}} + \mathbf{Q}_k$$

$$\mathbf{K}(k+1) = \Sigma(k+1|k)\,\mathbf{C}^{\mathrm{T}}\,[\mathbf{C}\,\Sigma(k+1|k)\,\mathbf{C}^{\mathrm{T}} + \mathbf{R}_{k+1}]^{-1}$$

$$\hat{\mathbf{x}}(k+1) = \hat{\mathbf{x}}(k+1|k) + \mathbf{K}(k+1)(\mathbf{y}(k+1) - \mathbf{C}\hat{\mathbf{x}}(k+1|k))$$

$$\Sigma(k+1) = [\mathbf{I} - \mathbf{K}(k+1)\mathbf{C}]\,\Sigma(k+1|k) \tag{28.1}$$

is optimal in several senses, including the best (minimum variance) linear unbiased estimate, the maximum likelihood estimate, and the minimum variance Bayes' estimate. The equations are computed for each k as the measurements $y(k)$ arrive (although in fact $\Sigma(k+1|k)$, $K(k+1)$, $\Sigma(k)$ could be computed in advance). The algorithm is clearly a straightforward one.

A similar result is obtained by modifying observers to consider noise explicitly, and the optimal such observers are under certain circumstances the same as Kalman filters.

28.2 A LINEAR UNBIASED MINIMUM VARIANCE ESTIMATE

The Kalman filter may be derived in several ways and justified in several more. We choose here to do a justification as a best linear unbiased estimate (BLUE) of the state, given the received data. It is appropriate to linear systems driven by uncorrelated noises and with measurements corrupted by uncorrelated noises. We will not pursue the many extensions or alternative derivations.

28.2.1 Linear combinations of estimates

The basic idea of the Kalman filter can be expressed from many points of view. Here we consider the following situation as motivation: let $\mathbf{x}_1$ and $\mathbf{x}_2$ both be unbiased estimates of an unknown vector $\mathbf{x}$, with

$$\mathscr{E}[\mathbf{x}_1] = \mathscr{E}[\mathbf{x}_2] = \mathbf{x}$$

$$\mathrm{cov}(\mathbf{x}_1) = \Sigma_1 \qquad \mathrm{cov}(\mathbf{x}_2) = \Sigma_2$$

We attempt to find a linear combination of $\mathbf{x}_1$ and $\mathbf{x}_2$ such that the new estimate, call it $\mathbf{x}_3$, is also unbiased and so its expected square error $\mathscr{E}[\mathbf{x}_3-\mathbf{x}]^\mathrm{T}(\mathbf{x}_3-\mathbf{x})$ is minimized. Since this latter quantity is $\mathrm{tr}[\mathscr{E}(\mathbf{x}_3-\mathbf{x})(\mathbf{x}_3-\mathbf{x})^\mathrm{T}]$ = trace of covariance of $\mathbf{x}_3$, we will seek to optimize the latter. Thus we seek matrices $\mathbf{A}$ and $\mathbf{B}$ such that

$$\mathbf{x}_3 = \mathbf{A}\mathbf{x}_1 + \mathbf{B}\mathbf{x}_2$$

$$\mathscr{E}[\mathbf{x}_3] = \mathbf{x}$$

and tr(cov($\mathbf{x}_3$)) is minimized. Imposing the mean value condition first,

$$\mathbf{x} = \mathscr{E}[\mathbf{x}_3] = \mathbf{A}\mathscr{E}[\mathbf{x}_1] + \mathbf{B}\mathscr{E}[\mathbf{x}_2] = \mathbf{A}\mathbf{x} + \mathbf{B}\mathbf{x}$$

For this to hold regardless of $\mathbf{x}$ requires $\mathbf{A} = \mathbf{I} - \mathbf{B}$. With this constraint, cov($\mathbf{x}_3$) is given by direct computation as

$$\text{cov}(\mathbf{x}_3) = \mathscr{E}[(\mathbf{I}-\mathbf{B})\mathbf{x}_1 + \mathbf{B}\mathbf{x}_2 - \mathbf{x}][(\mathbf{I}-\mathbf{B})\mathbf{x}_1 + \mathbf{B}\mathbf{x}_2 - \mathbf{x}]^{\mathrm{T}}$$

$$= \mathscr{E}[(\mathbf{I}-\mathbf{B})(\mathbf{x}_1-\mathbf{x}) + \mathbf{B}(\mathbf{x}_2-\mathbf{x})][(\mathbf{I}-\mathbf{B})(\mathbf{x}_1-\mathbf{x}) + \mathbf{B}(\mathbf{x}_2-\mathbf{x})]^{\mathrm{T}}$$

$$= (\mathbf{I}-\mathbf{B})\,\text{cov}(\mathbf{x}_1)\,(\mathbf{I}-\mathbf{B})^{\mathrm{T}} + \mathbf{B}\,\text{cov}(\mathbf{x}_2)\,\mathbf{B}^{\mathrm{T}}$$

$$+ \mathbf{B}\mathscr{E}([\mathbf{x}_2 - \mathbf{x}][\mathbf{x}_1 - \mathbf{x}]^{\mathrm{T}})\,(\mathbf{I}-\mathbf{B})^{\mathrm{T}}$$

If the errors in $\mathbf{x}_2$ and in $\mathbf{x}_1$ are uncorrelated, the last term is zero and this becomes

$$\text{cov}(\mathbf{x}_3) = (\mathbf{I}-\mathbf{B})\,\Sigma_1\,(\mathbf{I}-\mathbf{B})^{\mathrm{T}} + \mathbf{B}\,\Sigma_2\,\mathbf{B}^{\mathrm{T}}$$

$$= \Sigma_1 - \mathbf{B}\,\Sigma_1 - \Sigma_1\,\mathbf{B}^{\mathrm{T}} + \mathbf{B}\,[\Sigma_1 + \Sigma_2]\,\mathbf{B}^{\mathrm{T}}$$

By any of several methods (see Appendix B), minimizing the trace of this with respect to the matrix $\mathbf{B}$ (and it is a minimum because Σ_1 and Σ_2 are, like all covariance matrices, at least positive semi-definite) yields that the optimum must satisfy

$$\Sigma_1 - \mathbf{B}^{\#}[\Sigma_1 + \Sigma_2] = \mathbf{0}$$

Using a matrix pseudo-inverse if necessary, or the inverse if possible (we assume the latter)

$$\mathbf{B}^{\#} = \Sigma_1\,[\Sigma_1 + \Sigma_2]^{-1}$$

from which it follows that

$$\mathbf{A}^{\#} = \Sigma_2\,[\Sigma_1 + \Sigma_2]^{-1}$$

and that $\mathbf{x}_3$ may be written several ways, including

$$\mathbf{x}_3 = \mathbf{x}_1 + \Sigma_1[\Sigma_1 + \Sigma_2]^{-1}\ (\mathbf{x}_2 - \mathbf{x}_1)$$

$$= \Sigma_2[\Sigma_1 + \Sigma_2]^{-1}\ \mathbf{x}_1 + \Sigma_1[\Sigma_1 + \Sigma_2]^{-1}\ \mathbf{x}_1$$

With some manipulation

$$\mathrm{cov}(\mathbf{x}_3) = \Sigma_2\,[\Sigma_1 + \Sigma_2]^{-1}\,\Sigma_1\,[\Sigma_1 + \Sigma_2]^{-1}\,\Sigma_2$$

$$+\ \Sigma_1\,[\Sigma_1 + \Sigma_2]^{-1}\,\Sigma_2\,[\Sigma_1 + \Sigma_2]^{-1}\,\Sigma_1$$

$$=\ \Sigma_2\,[\Sigma_1 + \Sigma_2]^{-1}\,\Sigma_1 = \Sigma_1\,[\Sigma_1 + \Sigma_2]^{-1}\,\Sigma_2$$

where extensive use has been made of the symmetry property of covariance matrices.

The combining of two estimates linearly to give a better one (the covariance of the sum is less than the covariance of either individual term) is fundamental to us. The filter just becomes a matter of organizing the two estimates to combine. For this we will in effect use an estimate based upon past data combined with some noisy new data, in effect a second estimate.

28.2.2 Application to state estimation

Consider a linear system with state vector $\mathbf{x}$, input vector $\mathbf{u}$, and measurement vector $\mathbf{y}$, described by the dynamics equation

$$\mathbf{x}(k+1) = \mathbf{A}_k\mathbf{x}(k) + \mathbf{B}_k\mathbf{u}(k) + \mathbf{G}_k\mathbf{v}(k) \tag{28.2}$$

with the measurements taken according to

$$\mathbf{y}(k) = \mathbf{C}_k\mathbf{x}(k) + \mathbf{w}(k)$$

where $\{\mathbf{w}(k)\}$ and $\{\mathbf{v}(k)\}$ are noise sequences with the properties

$$\mathscr{E}[\mathbf{w}(k)] = \mathbf{0} \qquad\qquad \mathscr{E}[\mathbf{v}(k)] = \mathbf{0}$$

$$\mathrm{cov}(\mathbf{v}(k)) = \mathbf{Q}_k \qquad\qquad \mathrm{cov}(\mathbf{w}(k)) = \mathbf{R}_k$$

$$\mathscr{E}[\mathbf{v}(i)\mathbf{v}^{\mathrm{T}}(j)] = \mathbf{Q}_i\,\delta_{ij} \qquad\qquad \mathscr{E}[\mathbf{w}(i)\mathbf{w}^{\mathrm{T}}(j)] = \mathbf{R}_i\,\delta_{ij}$$

$$\mathscr{E}[\mathbf{w}(k)\mathbf{v}^{\mathrm{T}}(j)] = \mathbf{0}$$

$$\mathscr{E}[\mathbf{x}(0)] = \mathbf{x}_0 \qquad\qquad \mathrm{cov}(\mathbf{x}(0)) = \Sigma_0$$

where δ_{ij} is the Kronecker delta (which equals 0 for $i \neq j$ and 1 for $i = j$). The problem is to generate a 'good' estimate $\hat{\mathbf{x}}(k)$ of $\mathbf{x}(k)$; good is taken to mean

1. $\mathscr{E}[\hat{\mathbf{x}}(k)] = \mathscr{E}[\mathbf{x}(k)]$;
2. $\Sigma(k) = \mathrm{cov}(\hat{\mathbf{x}}(k))$ is as small as possible, in that its trace is minimized; and
3. the estimate is generated using linear operations.

Assume that at some step k we have a 'good' estimate $\hat{\mathbf{x}}(k)$ and that we progress to step $k+1$ and take a new measurement $\mathbf{y}(k+1)$. Then from step 3, we attempt to generate a new estimate $\hat{\mathbf{x}}(k+1)$ as

$$\hat{\mathbf{x}}(k+1) = \mathbf{F}\hat{\mathbf{x}}(k) + \mathbf{H}\mathbf{y}(k+1) + \mathbf{K}\mathbf{u}(k)$$

We first apply condition (1) and the effect of $\mathscr{E}[\ ]$ on (28.2) to find

$$\mathscr{E}[\hat{\mathbf{x}}(k+1)] = \mathscr{E}[\mathbf{x}(k+1)]$$

$$= \mathbf{A}_k\mathscr{E}[\mathbf{x}(k)] + \mathbf{B}_k\mathbf{u}(k)$$

$$= \mathbf{F}\mathscr{E}[\hat{\mathbf{x}}(k)] + \mathbf{H}\mathbf{C}_{k+1}\mathscr{E}[\mathbf{x}(k+1)] + \mathbf{K}\mathbf{u}(k)$$

$$= \mathbf{F}\mathscr{E}[\mathbf{x}(k)] + \mathbf{H}\mathbf{C}_{k+1}\{\mathbf{A}_k\mathscr{E}[\mathbf{x}(k)] + \mathbf{B}_k\mathbf{u}(k)\} + \mathbf{K}\mathbf{u}(k)$$

Rearranging gives

$$[\mathbf{F} - (\mathbf{I} - \mathbf{H}\mathbf{C}_{k+1})\mathbf{A}_k]\mathscr{E}(\mathbf{x}(k)) - (\mathbf{I} - \mathbf{H}\mathbf{C}_{k+1})\mathbf{B}_k\mathbf{u}(k) + \mathbf{K}\mathbf{u}(k) = \mathbf{0}$$

The choices $\mathbf{F} = (\mathbf{I} - \mathbf{H}\mathbf{C}_{k+1})\mathbf{A}_k$ and $\mathbf{K} = (\mathbf{I} - \mathbf{H}\mathbf{C}_{k+1})\mathbf{B}_k$ ensure that the new estimate is unbiased if the old one was. Using these, the estimator becomes

$$\hat{\mathbf{x}}(k+1) = \mathbf{A}_k \hat{\mathbf{x}}(k) + \mathbf{B}_k \mathbf{u}(k) + \mathbf{H}(\mathbf{y}(k+1) - \mathbf{C}_{k+1}(\mathbf{A}_k \hat{\mathbf{x}}(k) + \mathbf{B}_k \mathbf{u}(k)))$$

$$(28.3)$$

For convenience we define

$$\hat{\mathbf{x}}(k+1|k) = \mathbf{A}_k \hat{\mathbf{x}}(k) + \mathbf{B}_k \mathbf{u}(k)$$

and rewrite (28.3) as

$$\hat{\mathbf{x}}(k+1) = \hat{\mathbf{x}}(k+1|k) + \mathbf{H}(\mathbf{y}(k+1) - \mathbf{C}_{k+1}\hat{\mathbf{x}}(k+1|k))$$

The quantity $\hat{\mathbf{x}}(k+1|k)$ may be interpreted as the prediction of $\mathbf{x}(k+1)$ based upon $\hat{\mathbf{x}}(k)$. Our new estimate is now to be a prediction based upon old data, updated by a linear weighting of the difference between the new measurement and the predicted new measurement. The goal is to find a good weighting $\mathbf{H}$.

We assume that $\mathrm{cov}(\hat{\mathbf{x}}(k) - \mathbf{x}(k)) = \Sigma(k)$ is known. Defining

$$\Sigma(k+1|k) = \mathrm{cov}(\hat{\mathbf{x}}(k+1|k) - \mathbf{x}(k+1))$$

we compute

$$\Sigma(k+1|k) = \mathrm{cov}(\mathbf{A}_k \hat{\mathbf{x}}(k) + \mathbf{B}_k \mathbf{u}(k) - \mathbf{A}_k \mathbf{x}(k) - \mathbf{B}_k \mathbf{u}(k) - \mathbf{G}_k \mathbf{v}(k))$$

$$= \mathrm{cov}(\mathbf{A}_k(\hat{\mathbf{x}}(k) - \mathbf{x}(k)) - \mathbf{G}_k \mathbf{v}(k))$$

$$= \mathbf{A}_k \Sigma(k) \mathbf{A}_k^{\mathrm{T}} + \mathbf{G}_k \mathbf{Q}_k \mathbf{G}_k^{\mathrm{T}}$$

We are able to compute $\Sigma(k+1)$ from

$$\Sigma(k+1) = \mathrm{cov}(\hat{\mathbf{x}}(k+1) - \mathbf{x}(k+1))$$

$$= \mathrm{cov}(\hat{\mathbf{x}}(k+1|k) + \mathbf{H}(\mathbf{y}(k+1) - \mathbf{C}_{k+1}\hat{\mathbf{x}}(k+1|k)) - \mathbf{x}(k+1))$$

$$= \mathrm{cov}(\hat{\mathbf{x}}(k+1|k) - \mathbf{x}(k+1) + \mathbf{H}(\mathbf{C}_{k+1}\hat{\mathbf{x}}(k+1) + \mathbf{w}(k+1)$$

$$- \mathbf{C}_{k-1}\hat{\mathbf{x}}(k+1|k))$$

Evaluating and using the lack of cross correlation between the measurement noise $\{v(k)\}$ and the dynamics noise $\{w(k)\}$

$$\Sigma(k+1) = (\mathbf{I} - \mathbf{H}C_{k+1})\,\Sigma(k+1|k)\,(\mathbf{I} - \mathbf{H}C_{k+1})^{\mathrm{T}} + \mathbf{H}R_{k+1}\,\mathbf{H}^{\mathrm{T}}$$

Notice that this holds for any gain $\mathbf{H}$. Minimizing the trace with respect to $\mathbf{H}$ shows that the optimum $\mathbf{H}$ is given by

$$\mathbf{H}^{\#} = \Sigma(k+1|k)\,C_{k+1}^{\mathrm{T}}\,[C_{k+1}\,\Sigma(k+1|k)\,C_{k+1} + R_{k+1}]^{-1}$$

This is usually called the Kalman gain and written

$$\mathbf{K}(k+1) = \Sigma(k+1|k)\,C_{k+1}^{\mathrm{T}}\,[C_{k+1}\,\Sigma(k+1|k)\,C_{k+1} + R_{k+1}]^{-1} \qquad (28.4)$$

Using this, one of the many possible expressions for $\Sigma(k+1)$ is

$$\Sigma(k+1) = [\mathbf{I} - \mathbf{K}(k+1)\,C_{k+1}]\,\Sigma(k+1|k)$$

In summary, the algorithm is

Step 1. Initialize $\hat{\mathbf{x}}(0) = \mathbf{x}_0$, $\Sigma(0) = \Sigma_0$
Step 2. Cycle as follows, for $k = 0, 1, 2, \ldots$

    (a)   Take measurement $\mathbf{y}(k+1)$

    (b)   Predict $\mathbf{x}(k+1)$ as $\hat{\mathbf{x}}(k+1|k)$

$$\hat{\mathbf{x}}(k+1|k) = \mathbf{A}_k\,\hat{\mathbf{x}}(k) + \mathbf{B}_k\,\mathbf{u}(k)$$

    (c)   Compute $\Sigma(k+1|k)$ and gain $\mathbf{K}(k+1)$

$$\Sigma(k+1|k) = \mathbf{A}_k\,\Sigma(k)\,\mathbf{A}_k^{\mathrm{T}} + \mathbf{G}_k\,\mathbf{Q}_k\,\mathbf{G}_k^{\mathrm{T}}$$

$$\mathbf{K}(k+1) = \Sigma(k+1|k)\,C_{k+1}^{\mathrm{T}}\,[C_{k+1}\,\Sigma(k+1|k)\,C_{k+1}^{\mathrm{T}} + R_{k+1}]^{-1}$$

    (d)   Compute $\hat{\mathbf{x}}(k+1)$ and update $\Sigma(k+1)$

$$\hat{\mathbf{x}}(k+1) = \hat{\mathbf{x}}(k+1|k) + \mathbf{K}(k+1)\,(\mathbf{y}(k+1) - C_{k+1}\,\hat{\mathbf{x}}(k+1|k))$$

$$\Sigma(k+1) = [\mathbf{I} - \mathbf{K}(k+1)\,C_{k+1}]\,\Sigma(k+1|k)$$

The algorithm is clearly straightforward. We now consider some properties and examples, since perhaps this is the best way to illustrate Kalman filtering.

Example

We choose to consider a simple system

$$\mathbf{x}(k+1) = \begin{bmatrix} 0 & 1 \\ -0.63 & 1.6 \end{bmatrix} \mathbf{x}(k) + \begin{bmatrix} 1 & 0 \\ 0 & 1 \end{bmatrix} \mathbf{v}(k)$$

$$\mathbf{y}(k) = [\, 1 \quad 0\,]\, \mathbf{x}(k) + \mathbf{w}(k)$$

This may be interpreted as a system with stable poles (at 0.9 and 0.7) driven by noise affecting both 'position' $\mathbf{x}_1$ and 'velocity' $\mathbf{x}_2$. Measurements of the position are taken, but they are noisy (due to $\mathbf{w}(k)$). We are assumed to have no information concerning initial condition mean $\mathbf{x}_0$ and covariance matrix $\Sigma(0)$. Various noise covariances $\mathbf{Q}$ and $\mathbf{R}$ are to be studied, with $\mathbf{R} \neq \mathbf{0}$.

Some simulation runs showing component errors and bounds resulting from the standard deviations $P_{ii}^{\frac{1}{2}}$ are shown in Fig. 28.1 for $i = 1, 2$.

We remark that in the example the gains $\mathbf{K}(k)$ and covariance matrices $\Sigma(k)$ appear to become near constant, in a way that is similar to, although reversed in time from, the optimal control situation (Chapter 26). This useful property can lead to a quite useful sub-optimal but justifiable and easily computed state estimation filter. One of the many formulations of this property assumes

$$\Sigma(k+1|k) \to \Sigma = \text{constant} \qquad \text{and} \qquad \mathbf{K}(k) \to \mathbf{K} = \text{constant}$$

in which case they must satisfy

$$\Sigma = \mathbf{A}\,\Sigma\,\mathbf{A}^{\mathrm{T}} - \mathbf{A}\,\Sigma\,\mathbf{C}^{\mathrm{T}}(\mathbf{C}\,\Sigma\,\mathbf{C}^{\mathrm{T}} + \mathbf{R})^{-1}\mathbf{C}\,\Sigma\,\mathbf{A}^{\mathrm{T}} + \mathbf{G}\mathbf{Q}\mathbf{G}^{\mathrm{T}}$$

$$\mathbf{K} = \Sigma\,\mathbf{C}^{\mathrm{T}}(\mathbf{C}\,\Sigma\,\mathbf{C}^{\mathrm{T}} + \mathbf{R})^{-1}$$

Matrix manipulations allow other representations.

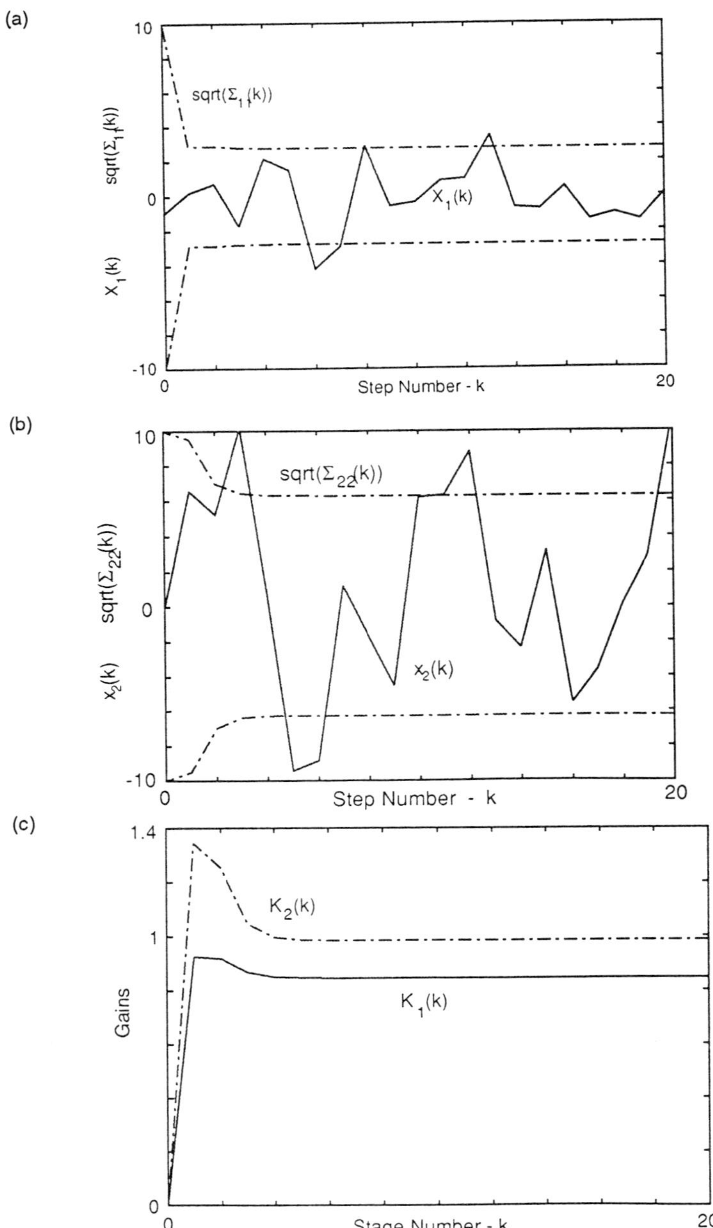

Figure 28.1 Operation of the Kalman filter. (a) and (b) Error in estimates of states x_1 and x_2 (solid lines) and bounds determined from error covariance matrix (dash-dot lines). (c) Kalman gain elements $\mathbf{K}_1$ and $\mathbf{K}_2$.

28.3 EXTENDED KALMAN FILTERING FOR NON-LINEAR SYSTEMS

The Kalman–Bucy filter applies only to linear systems, albeit time-varying systems are allowed. The general theory for non-linear systems simply does not exist. The popular engineering alternative is to linearize the Kalman filter where necessary and retain the non-linearities where possible, and even this has several variants. We review the most popular, called the extended Kalman filter, and then indicate how the others work.

Let the system be described by dynamics

$$\mathbf{x}(k+1) = f(\mathbf{x}(k), \mathbf{u}(k), \mathbf{v}(k), k)$$

and measurement equations

$$\mathbf{y}(k) = g(\mathbf{x}(k), \mathbf{w}(k), k)$$

where $\mathbf{x}$ is an n-dimensional state, $\mathbf{y}$ is an m-dimensional measurement/output, and $\mathbf{v}$ and $\mathbf{w}$ are noises, presumably white with covariances $\mathbf{Q}$ and $\mathbf{R}$ respectively. Let $\hat{\mathbf{x}}(k)$ represent a 'good' estimate of $\mathbf{x}(k)$.

Recall that the structure of the Kalman filter is

$$\hat{\mathbf{x}}(k+1) = \hat{\mathbf{x}}(k+1|k) + \mathbf{K}(k)(\mathbf{y}(k+1) - \mathbf{C}\hat{\mathbf{x}}(k+1|k))$$

where $\hat{\mathbf{x}}(k+1|k) = \mathbf{A}\hat{\mathbf{x}}(k) + \mathbf{B}\mathbf{u}(k)$ and $\mathbf{A}, \mathbf{B}, \mathbf{C}$, have the usual meanings. $\hat{\mathbf{x}}(k+1|k)$ is, in words, the best estimate of $\mathbf{x}(k+1)$ given data up to time k; $\hat{\mathbf{x}}(k+1)$ is the best estimate of the same quantity given data to time $k+1$.

It seems reasonable to use a similar structure for the non-linear situation. Doing this, a good guess appears to be

$$\hat{\mathbf{x}}(k+1|k) = f(\hat{\mathbf{x}}(k), \mathbf{u}(k), 0, k)$$

and

$$\hat{\mathbf{x}}(k+1) = \hat{\mathbf{x}}(k+1|k) + \mathbf{K}(k+1)(\mathbf{y}(k+1) - g(\hat{\mathbf{x}}(k+1|k), 0, k+1))$$

with gain $\mathbf{K}(k+1)$ to be determined.

In the linear case, $\mathbf{K}(k)$ results in essence from a ratio of the covariances of $\mathbf{y}(k)$ – essentially depending on the covariances of $\hat{\mathbf{x}}(k)$ and $\mathbf{v}(k)$ – and of $\hat{\mathbf{x}}(k|k-1)$ – depending on $\hat{\mathbf{x}}(k-1)$ and $\mathbf{w}(k-1)$. Our problem is to find those covariances for the non-linear case; the approximate solution is to use linearization, and the particular character of the extended Kalman filter is bound up in the fact that it linearizes about its estimate rather than about some nominal.

Hence we define

$$\hat{\mathbf{A}}(k) = \left. \frac{\partial f(\mathbf{x}, \mathbf{u}, \mathbf{v}, k)}{\partial \mathbf{x}} \right|_{(\hat{\mathbf{x}}(k),\mathbf{u}(k),0,k)}$$

$$\hat{\mathbf{B}}(k) = \left. \frac{\partial f(\mathbf{x}, \mathbf{u}, \mathbf{v}, k)}{\partial \mathbf{u}} \right|_{(\hat{\mathbf{x}}(k),\mathbf{u}(k),0,k)}$$

$$\hat{\mathbf{G}}(k) = \left. \frac{\partial f(\mathbf{x}, \mathbf{u}, \mathbf{v}, k)}{\partial \mathbf{v}} \right|_{(\hat{\mathbf{x}}(k),\mathbf{u}(k),0,k)}$$

$$\hat{\mathbf{C}}(k) = \left. \frac{\partial g(\mathbf{x}, \mathbf{w}, k)}{\partial \mathbf{x}} \right|_{(\hat{\mathbf{x}}(k),0,k)}$$

We use the definitions

$$\Sigma(k|k) = \mathrm{cov}(\hat{\mathbf{x}}(k) \text{ given } k \text{ measurements})$$

$$\Sigma(k+1|k) = \mathrm{cov}(\hat{\mathbf{x}}(k+1|k) \text{ given } k \text{ measurements})$$

To approximate the second of these, we see that

$$\mathbf{x}(k+1) = f(\mathbf{x}(k), \mathbf{u}(k), \mathbf{v}(k), k)$$

$$\approx f(\hat{\mathbf{x}}(k), \mathbf{u}(k), 0, k)$$

$$+ \left. \frac{\partial f(\mathbf{x}, \mathbf{u}, \mathbf{v}, k)}{\partial \mathbf{x}} \right|_{(\hat{\mathbf{x}}(k),\mathbf{u}(k),0,k)} (\mathbf{x}(k) - \hat{\mathbf{x}}(k))$$

$$+ \left. \frac{\partial f(\mathbf{x}, \mathbf{u}, \mathbf{v}, k)}{\partial \mathbf{v}} \right|_{(\hat{\mathbf{x}}(k),\mathbf{u}(k),0,k)} (\mathbf{v}(k) - 0)$$

$$\approx \hat{\mathbf{x}}(k+1|k) + \hat{\mathbf{A}}(k)(\mathbf{x}(k) - \hat{\mathbf{x}}(k)) + \hat{\mathbf{G}}(k)\mathbf{v}(k)$$

Provided the estimator is working well, we have

$$\mathscr{E}(\hat{\mathbf{x}}(k+1|k)) = \mathbf{x}(k+1) = f(\mathbf{x}(k), \mathbf{u}(k), 0, k)$$

and hence

$$\Sigma(k+1|k)) = \mathscr{E}\left[\hat{\mathbf{x}}(k+1|k) - \mathscr{E}(\hat{\mathbf{x}}(k+1|k))\right]\left[\hat{\mathbf{x}}(k+1|k) - \mathscr{E}(\hat{\mathbf{x}}(k+1|k))\right]^{\mathrm{T}}$$

is given approximately by

$$\mathscr{E}\left[\hat{\mathbf{A}}(k)(\hat{\mathbf{x}}(k) - \mathbf{x}(k)) - \hat{\mathbf{G}}(k)\mathbf{v}(k)\right]\left[\hat{\mathbf{A}}(k)(\hat{\mathbf{x}}(k) - \mathbf{x}(k)) - \hat{\mathbf{G}}(k)\mathbf{v}(k)\right]^{\mathrm{T}}$$

Hence an approximation is

$$\Sigma(k+1|k) = \hat{\mathbf{A}}(k)\,\Sigma(k|k)\,\hat{\mathbf{A}}^{\mathrm{T}}(k) + \hat{\mathbf{G}}(k)\,\mathbf{Q}(k)\,\hat{\mathbf{G}}^{\mathrm{T}}(k)$$

from which

$$\mathbf{K}(k+1) = \Sigma(k+1|k)\,\hat{\mathbf{C}}^{\mathrm{T}}(k)\,[\hat{\mathbf{C}}(k+1)\,\Sigma(k+1|k)\,\hat{\mathbf{C}}^{\mathrm{T}}(k+1) + \mathbf{R}(k+1)]^{-1}$$

Proceeding similarly

$$\Sigma(k+1|k+1) = [\mathbf{I} - \mathbf{K}(k+1)\,\hat{\mathbf{C}}(k+1)]\,\Sigma(k+1|k)$$

The filter thus linearizes about its current estimates, and has a clear potential to wander away from the 'actual' trajectory, although then the differences $\{\mathbf{y}(k) - g(\hat{\mathbf{x}}, 0, k)\}$, called the innovations, would be expected to become large.

Extended Kalman filtering via linearization, either about the nominal (which gives an ordinary Kalman filter) or about the computed actual trajectory (which gives the extended filter) is a common and useful method. In it lie the seeds of problems, however. Extended Kalman filters may in operation diverge badly from the actual trajectory: residuals are monitored and the filter restarted in one type of fix; gains are kept artificially high in another type. Once again the engineer is taken from using nice formulae into the real world of applying these, with both judgement and experience among his guides.

28.4 OPTIMAL OBSERVERS FOR NOISE SUPPRESSION

We have already considered observers for situations modelled as deterministic (Chapter 25). Now we will consider noise explicitly in the observer approach to estimation. For a system, we now allow time variation, i.e.

$$\mathbf{x}(k+1) = \mathbf{A}_k \mathbf{x}(k) + \mathbf{B}_k \mathbf{u}(k) + \mathbf{v}(k)$$

$$\mathbf{y}(k) = \mathbf{C}_k \mathbf{x}(k) + \mathbf{w}(k)$$

where the initial condition $\mathbf{x}(0)$ is random, as are the 'noises' $\mathbf{v}(k)$ and $\mathbf{w}(k)$. They are all assumed normally distributed and are independent of each other. Our notation is as follows:

$$\mathbf{x}(0) \sim N(\mathbf{x}_0, \Sigma_0) \qquad \Sigma_0 > \mathbf{0}$$

$$\mathbf{v}(k) \sim N(\mathbf{0}, \mathbf{Q}_k) \qquad \mathbf{Q}_k \geq \mathbf{0}$$

$$\mathbf{w}(k) \sim N(\mathbf{0}, \mathbf{R}_k) \qquad \mathbf{R}_k > \mathbf{0}$$

As we saw in Chapter 25, the system

$$\mathbf{z}(k+1) = \mathbf{T}_{k+1} \mathbf{A}_k \mathbf{P}_k \mathbf{z}(k) + \mathbf{T}_{k+1} \mathbf{A}_k \mathbf{V}_k \mathbf{y}(k) + \mathbf{T}_{k+1} \mathbf{B}_k \mathbf{u}(k)$$

$$\hat{\mathbf{x}}(k) = \mathbf{P}_k \mathbf{z}(k) + \mathbf{V}_k \mathbf{y}(k)$$

is an observer for the original system provided that

$$\mathbf{P}_k \mathbf{T}_k + \mathbf{V}_k \mathbf{C}_k = \mathbf{I}_n$$

Defining the error

$$\mathbf{e}(k) \approx \hat{\mathbf{x}}(k) - \mathbf{x}(k)$$

we can show that

$$\mathbf{e}(k+1) = [\mathbf{I} - \mathbf{V}_{k+1}\mathbf{C}_{k+1}]\,\mathbf{A}_k\,\mathbf{e}(k) + \mathbf{V}_{k+1}\mathbf{w}(k+1)$$

$$+ [\mathbf{V}_{k+1}\mathbf{C}_{k+1} - \mathbf{I}]\,\mathbf{v}(k)$$

$$\mathbf{e}(0) = [\mathbf{I} - \mathbf{V}_0\mathbf{C}_0]\,[\hat{\mathbf{x}}(0) - \mathbf{x}_0] + \mathbf{V}_0\mathbf{v}(0) \tag{28.5}$$

Clearly, if $\hat{\mathbf{x}}(0) = \mathbf{x}_0$ then $\mathscr{E}\{\mathbf{e}(0)\} = \mathbf{0}$ and hence, because of (28.5), $\mathscr{E}\{\mathbf{e}(k)\} = \mathbf{0}$. We assume the initial $\hat{\mathbf{x}}(0)$ is so chosen. Then the covariance of $\mathbf{e}(k)$, Σ_k, satisfies

$$\Sigma_{k+1} = [\mathbf{I} - \mathbf{V}_{k+1}\mathbf{C}_{k+1}]\,[\mathbf{A}_k\Sigma_k\mathbf{A}_k^{\mathrm{T}} + \mathbf{Q}_k]\,[\mathbf{I} - \mathbf{V}_{k+1}\mathbf{C}_{k+1}]^{\mathrm{T}}$$

$$+ \mathbf{V}_{k+1}\mathbf{R}_{k+1}\mathbf{V}_{k+1}^{\mathrm{T}}$$

$$\Sigma_0 = [\mathbf{I} - \mathbf{V}_0\mathbf{C}_0]\,\Sigma_0\,[\mathbf{I} - \mathbf{V}_0\mathbf{C}_0]^{\mathrm{T}} + \mathbf{V}_0\mathbf{R}_0\mathbf{V}_0^{\mathrm{T}} \tag{28.6}$$

We wish to minimize the covariance of the error by proper choice of $\mathbf{V}$. Hence we take (not rigorously, but easily made rigorous)

$$\frac{1}{2}\frac{\mathrm{d}\Sigma_{k+1}}{\mathrm{d}\mathbf{V}_{k+1}} = \mathbf{V}_{k+1}\,[\mathbf{C}_{k+1}\{\mathbf{A}_k\Sigma_k\mathbf{A}_k^{\mathrm{T}} + \mathbf{Q}_k\}\mathbf{C}_{k+1}^{\mathrm{T}} + \mathbf{R}_{k+1}]$$

$$- [\mathbf{A}_k\Sigma_k\mathbf{A}_k^{\mathrm{T}} + \mathbf{Q}_k]\,\mathbf{C}_{k+1}^{\mathrm{T}}$$

$$= 0 \tag{28.7}$$

Any $\mathbf{V}_{k+1}$ for which this holds is optimal and may be denoted $\mathbf{V}_{k+1}^{\#}$. If either

$$\mathbf{C}_{k+1}\{\mathbf{A}_k\Sigma_k\mathbf{A}_k^{\mathrm{T}} + \mathbf{Q}_k\}\mathbf{C}_{k+1}^{\mathrm{T}} > \mathbf{0}$$

or

$$\mathbf{R}_{k+1} > \mathbf{0} \tag{28.8}$$

or if their sum happens to be positive definite, then $\mathbf{V}_{k+1}$ is unique and is given by

$$\mathbf{V}_{k+1} = [\mathbf{A}_k \, \Sigma_k \, \mathbf{A}_k^{\mathrm{T}} + \mathbf{Q}_k] \, \mathbf{C}_{k+1}^{\mathrm{T}} [\mathbf{C}_{k+1}^{\mathrm{T}} \{\mathbf{A}_k \Sigma_k \mathbf{A}_k^{\mathrm{T}} + \mathbf{Q}_k\} \, \mathbf{C}_{k+1}^{\mathrm{T}}$$
$$+ \mathbf{R}_{k+1}]^{-1} \tag{28.9}$$

The resulting Σ_{k+1} is given by (28.6) and (28.9) as

$$\Sigma_{k+1} = [\mathbf{A}_k \Sigma_k \mathbf{A}_k^{\mathrm{T}} + \mathbf{Q}_k] \left[\mathbf{I} - \mathbf{C}_{k+1}^{\mathrm{T}} [\mathbf{C}_{k+1} \{\mathbf{A}_k \Sigma_k \mathbf{A}_k^{\mathrm{T}} + \mathbf{Q}_k\} \, \mathbf{C}_{k+1}^{\mathrm{T}} \right.$$
$$\left. + \mathbf{R}_{k+1}]^{-1} \mathbf{C}_{k+1} [\mathbf{A}_k \Sigma_k \, \mathbf{A}_k^{\mathrm{T}} + \mathbf{Q}_k] \, \right] \tag{28.10}$$

The gain turns out to be the Kalman filter gain expression ((28.1) and (28.4)). This may be seen in part by looking at the expression and in part by noticing

$$\hat{\mathbf{x}}(k+1) = \mathbf{P}_{k+1} \mathbf{z}(k+1) + \mathbf{V}_{k+1} \mathbf{y}(k+1)$$
$$= \mathbf{P}_{k+1} \mathbf{T}_{k+1} \mathbf{A}_k \mathbf{P}_k \mathbf{z}(k) + \mathbf{P}_{k+1} \mathbf{T}_{k+1} \mathbf{A}_k \mathbf{V}_k \mathbf{y}(k)$$
$$+ \mathbf{P}_{k+1} \mathbf{T}_{k+1} \mathbf{B}_k \mathbf{u}(k) + \mathbf{V}_{k+1} \mathbf{y}(k+1)$$
$$= \mathbf{A}_k \, \hat{\mathbf{x}}(k) + \mathbf{B}_k \mathbf{u}(k)$$
$$+ \mathbf{V}_{k+1} [\mathbf{y}(k+1) - \mathbf{C}_{k+1} \{\mathbf{A}_k \hat{\mathbf{x}}(k) + \mathbf{B}_k \mathbf{u}(k)\}]$$

is generally true for the observer, regardless of $\mathbf{V}_{k+1}$.

If the relationships (28.8) do not hold, then the choice of optimal $\mathbf{V}$ is not unique. One possible design is to obtain the minimal order observer (smallest dimension of $\mathbf{z}(k)$), a problem addressed by Yoshikawa and Kobayashi (1972). For their results, suppose we want an observer of rank $p = n - m_2$, where $m_2 = \dim(\mathbf{y}(k))$, and rank$(\mathbf{Q}(k)) = m_1 \leq m_2$. We assume $\mathbf{R}$ is partitioned as

$$\mathbf{R} = \begin{bmatrix} \mathbf{R}_1 & 0 \\ 0 & 0 \end{bmatrix}$$

where $\mathbf{R}_1 > \mathbf{0}$ has rank m_1, and that

$$\mathbf{y}(k) = \begin{bmatrix} \mathbf{C}_{11}(k) & \mathbf{C}_{12}(k) \\ \mathbf{C}_{21}(k) & \mathbf{C}_{22}(k) \end{bmatrix} \mathbf{x}(k) + \begin{bmatrix} \mathbf{w}(k) \\ \mathbf{0} \end{bmatrix}$$

with $\mathbf{C}_{22}$, which is $m_2 \times m_2$, having full rank and thus being invertible. Then for any $\mathbf{V}_k^{\#}$, i.e. any $\mathbf{V}$ satisfying (28.7), we partition it as

$$\mathbf{V}_k^{\#} = \begin{bmatrix} \mathbf{V}_{ak}^{\#} \\ \mathbf{V}_{bk}^{\#} \end{bmatrix}$$

and can show that if

$$\hat{\mathbf{P}}_k^{\#} = \begin{bmatrix} \mathbf{I}_{n-m_2} \\ -\mathbf{C}_{22}^{-1}(k)\,\mathbf{C}_{21}(k) \end{bmatrix}$$

$$\hat{\mathbf{T}}_k^{\#} = \begin{bmatrix} \mathbf{I}_{n-m_2} - \mathbf{V}_{ak}^{\#}\mathbf{C}_{11}(k) & \vdots & \mathbf{V}_{bk}^{\#}\mathbf{C}_{12}(k) \\ \mathbf{C}_{21}(k) & & \mathbf{C}_{22}(k) \end{bmatrix}$$

then the observer is given by $\mathbf{V}_k^{\#}$ and, for any invertible $\mathbf{L}_k$, $\mathbf{P}^{\#}(k) = \hat{\mathbf{P}}_k^{\#}\mathbf{L}_k$ and $\mathbf{T}_k^{\#} = \mathbf{L}_k^{-1}\hat{\mathbf{T}}_k^{\#}$.

Example

Once more we consider the system of the first example of Chapter 25, but this time with noisy measurements.

$$\mathbf{x}(k+1) = \begin{bmatrix} 1 & 1 \\ 0 & 1 \end{bmatrix} \mathbf{x}(k) + \begin{bmatrix} 0 \\ 1 \end{bmatrix} \mathbf{u}(k)$$

$$\mathbf{y}(k) = \mathbf{x}_1(k) + \mathbf{w}(k) = [\,1 \quad 0\,]\,\mathbf{x}(k) + [1]\,\mathbf{w}(k)$$

We assume that $\mathbf{w}(k) \sim \mathrm{N}(0,4)$ and $\mathbf{x}(0) \sim \mathrm{N}(\mathbf{x}_0, 25)$. Then the optimal filter has gain (28.9)

$$\mathbf{V}_{k+1} = \frac{\begin{bmatrix} \sigma_{11} + \sigma_{12} + \sigma_{21} + \sigma_{22} \\ \sigma_{21} + \sigma_{22} \end{bmatrix}}{\sigma_{11} + \sigma_{12} + \sigma_{21} + \sigma_{22} + 1}$$

where $\Sigma_k = \{ \sigma_{ij} \}$ and from (28.10)

$$\Sigma_{k+1} = \begin{bmatrix} \sigma_{11} + \sigma_{12} + \sigma_{21} + \sigma_{22} & \sigma_{22} + \sigma_{12} \\ \sigma_{21} + \sigma_{22} & \sigma_{22} \end{bmatrix}$$

$$\begin{bmatrix} 1 - \frac{\begin{bmatrix} \sigma_{11} + \sigma_{12} + \sigma_{21} + \sigma_{22} & \sigma_{22} + \sigma_{12} \\ 0 & 0 \end{bmatrix}}{\sigma_{11} + \sigma_{12} + \sigma_{21} + \sigma_{22} + 4} \end{bmatrix}$$

and $\Sigma_0 = \mathrm{diag}\,[25, 25]$. This can be arranged into another form if desired. It is noteworthy that the gain is time-varying even though the system is time invariant.

28.5 CONTINUOUS-TIME KALMAN FILTERS

Although derivable by limiting arguments, the Kalman filter for continuous-time systems requires more mathematical sophistication to derive than does that for discrete-time systems. We simply summarize the results here.

For the continuous-time system described by

$$\dot{\mathbf{x}}(t) = \mathbf{A}\mathbf{x}(t) + \mathbf{B}\mathbf{u}(t) + \mathbf{G}\mathbf{v}(t)$$

$$\mathbf{y}(t) = \mathbf{C}\mathbf{x}(t) + \mathbf{w}(t)$$

where the noises have the properties

$$\mathscr{E}[\mathbf{v}(t)] = \mathbf{0} \qquad\qquad \mathscr{E}[\mathbf{w}(t)] = \mathbf{0}$$

$$\mathrm{cov}(\mathbf{v}(t)) = \mathbf{Q}\delta(t) \qquad\qquad \mathrm{cov}(\mathbf{w}(t)) = \mathbf{R}\delta(t)$$

$$\mathscr{E}[\mathbf{x}(0)] = \mathbf{x}_0 \qquad\qquad \mathrm{cov}(\mathbf{x}(0)) = \Sigma_0$$

the optimal state estimate $\hat{\mathbf{x}}(t)$ is given by the solution of

$$\dot{\hat{\mathbf{x}}}(t) = \mathbf{A}\hat{\mathbf{x}}(t) + \mathbf{B}\mathbf{u}(t) + \mathbf{K}(t)[\mathbf{y}(t) - \mathbf{C}\hat{\mathbf{x}}(t)]$$

where

$$\hat{\mathbf{x}}(0) = \mathbf{x}_0$$

$$\mathbf{K}(t) = \Sigma(t)\mathbf{C}^{\mathrm{T}}\mathbf{R}^{-1}$$

$$\dot{\Sigma}(t) = \mathbf{A}\Sigma(t) + \Sigma(t)\mathbf{A}^{\mathrm{T}} + \mathbf{G}\mathbf{Q}\mathbf{G}^{\mathrm{T}} - \Sigma(t)\mathbf{C}^{\mathrm{T}}\mathbf{R}^{-1}\mathbf{C}\Sigma(t)$$

$$\Sigma(0) = \Sigma_0$$

If the error is defined as $\mathbf{e}(t) = \mathbf{x}(t) - \hat{\mathbf{x}}(t)$, it can be shown that $\mathscr{E}[\mathbf{e}(t)] = \mathbf{0}$ and $\mathrm{cov}(\mathbf{e}(t)) = \Sigma(t)$. As with the discrete time filter, all of the matrices $\mathbf{A}$, $\mathbf{B}$, $\mathbf{G}$, $\mathbf{C}$, $\mathbf{Q}$, and $\mathbf{R}$ may be time-varying in the above. It is necessary that $\mathbf{R}$ be non-singular, which has the implication that there are no noise-free measurements; the singular case is treated in the literature, however. In practice, Σ_0 may be made 'large' and $\mathbf{x}_0$ may be taken to be $\mathbf{0}$ if no other information is available. Constant gains and covariances are possible after an initial transient with time invariant plants. Then the derivative of $\Sigma(t)$ becomes zero and we find the constant value of $\Sigma(t) = \Sigma$ is the solution of

$$\mathbf{0} = \mathbf{A}\Sigma + \Sigma\mathbf{A}^{\mathrm{T}} + \mathbf{G}\mathbf{Q}\mathbf{G}^{\mathrm{T}} - \Sigma\mathbf{C}^{\mathrm{T}}\mathbf{R}^{-1}\mathbf{C}\Sigma$$

and the resulting constant Kalman gain is $\mathbf{K} = \Sigma\mathbf{C}^{\mathrm{T}}\mathbf{R}^{-1}$.

28.6 COMPUTER AIDS

Kalman filters are very easy to program for the basic algorithm; engineering is involved in choosing the $\mathbf{Q}$ and $\mathbf{R}$ matrices (particularly $\mathbf{Q}$, as it represents 'noise' in disturbances, modelling errors, etc., and in any case is often selected to keep the Kalman gain from vanishing and hence ruining the filter's ability to track the measurements.) CAD packages should have little difficulty with the computations; the steady-state gains are computed with standard toolbox outputs of such progams as MATLAB® and Ctrl-C.

28.7 FURTHER READING

We have only introduced a few basic ideas of Kalman filtering. Books such as Meditch (1969) treat the linear case very thoroughly. Applications are explored by Gelb (1974), while both basic and advanced ideas are presented by Jazwinski (1970) and Anderson and Moore (1979).

Examples of extended Kalman filtering are given by Gelb (1974). Variations of the above schemes include linearization about a nominal state history, thus yielding predetermined gains, and iterations of $\hat{x}$ to attempt to reduce the measurement residuals; see Jazwinski (1970).

We should remark that the justification above is only one possible approach to derivation. Other proofs yielding similar results may be derived as maximum likelihood estimates, best estimates with gaussian noises. The ideas may also be applied for optimal smoothing, in which state $\mathbf{x}(kT)$ is estimated when data from $\mathbf{y}(kT + mT)$ is available for $m > 0$, and optimal prediction, when the state $\mathbf{x}(kT + mT)$ is predicted for times far ahead of available data $\{\mathbf{y}(0), \mathbf{y}(T), ..., \mathbf{y}(kT)\}$, $m > 1$. Another variation is the 'moving window' estimate, which always uses only a fixed interval of data around the moving time of interest.

A recent discussion of the state estimation problem for linear systems is presented by Grimble and Johnson (1988).

29

State feedback using state estimates

Modern optimal control laws, whether state feedback or open-loop feedback in nature, almost invariably have the control commands based ultimately upon the state variables. In addition, state-space design methods such as pole placement are most easily done using full state feedback. The full state is rarely measured, however, so it is necessary for the engineer to modify the design. A very useful approach to doing this modification is to use state estimates in place of the state variables, with the estimates coming from an observer or Kalman filter. In this chapter we consider some implications of this approach.

29.1 SYNOPSIS

State feedback controllers are sometimes relatively easy to design. In pole placement with linear systems

$$\mathbf{u}(k) = \mathbf{K}\mathbf{x}(k)$$

is the relationship of feedback control to state, while in the linear quadratic problem

$$\mathbf{u}(k) = \mathbf{K}(k)\mathbf{x}(k)$$

i.e. the gain is time-varying. More generally, we may have

$$\mathbf{u}(k) = g(\mathbf{x}(k); k)$$

as the solution to a control problem. When we have only an estimate $\hat{\mathbf{x}}(k)$ of the state available, the usual engineering solution is to substitute this, i.e. to make

$$\mathbf{u}(k) = g(\hat{\mathbf{x}}(k); k)$$

the operational law. This is 'correct' in two cases.

1. In linear systems with no noise, pole placement controllers, and an observer as the state estimator, the control algorithm for

$$\mathbf{x}(k+1) = \mathbf{A}\mathbf{x}(k) + \mathbf{B}\mathbf{u}(k)$$

$$\mathbf{y}(k) = \mathbf{C}\mathbf{x}(k)$$

is

$$\hat{\mathbf{x}}(k+1) = (\mathbf{A} - \mathbf{G}\mathbf{C})\hat{\mathbf{x}}(k) + \mathbf{G}\mathbf{y}(k) + \mathbf{B}\mathbf{u}(k)$$

$$\mathbf{u}(k) = \mathbf{K}\hat{\mathbf{x}}(k)$$

where the eigenvalues of $(\mathbf{A} - \mathbf{G}\mathbf{C})$ and $(\mathbf{A} + \mathbf{B}\mathbf{K})$ are chosen to meet design specifications.

2. In linear systems, influenced by gaussian white noise, in which the control is optimal according to a quadratic criterion, the system

$$\mathbf{x}(k+1) = \mathbf{A}\mathbf{x}(k) + \mathbf{B}\mathbf{u}(k) + \mathbf{G}\mathbf{v}(k)$$

$$\mathbf{y}(k) = \mathbf{C}\mathbf{x}(k) + \mathbf{w}(k)$$

has control law of the form

$$\hat{\mathbf{x}}(k+1|k) = \mathbf{A}\hat{\mathbf{x}}(k) + \mathbf{B}\mathbf{u}^{\#}(k)$$

$$\hat{\mathbf{x}}(k+1) = \hat{\mathbf{x}}(k+1|k) + \mathbf{G}(k+1)(\mathbf{y}(k+1) - \mathbf{C}\hat{\mathbf{x}}(k+1|k))$$

$$\mathbf{u}^{\#}(k) = \mathbf{K}(k)\hat{\mathbf{x}}(k)$$

where $\mathbf{K}(k)$ and $\mathbf{G}(k)$ are respectively the optimal gains for the deterministic optimal control problem and the Kalman filter problem.

This chapter demonstrates these claims and introduces some of the terminology of the issue.

29.2 INTERACTION WITH STATE ESTIMATORS – SEPARATION

If the state is needed for the feedback control law, there are a number of possibilities in structuring the control algorithm.

If the state is in fact available, we have a **deterministic** situation. The system is noise-free and its parameters are known. Using optimal or other control theory, the control law is found to be

$$\mathbf{u}^{\#}(t) = \Phi(\mathbf{x}(t), t) \qquad\qquad (29.1)$$

When the state is not available, then we must either use a measurement feedback law in which $\mathbf{u}(t) = \Theta(\mathbf{y}(t), t)$, a possibly difficult design task, or attempt to use a state estimator. This latter has three possibilities: certainty equivalent, separation and dual control. Suppose first that we have a case, possibly stochastic, i.e. noisy, in which it turns out that the best control law has the form

$$\mathbf{u}^{\#}(t) = \Phi(\hat{\mathbf{x}}(t), t)$$

$$\hat{\mathbf{x}}(t) = \mathscr{E}(\mathbf{x}(t) \mid \mathbf{Y}(t), \mathbf{U}(t))$$

$$\mathbf{Y}(t) = \text{history of measurement} = \{\mathbf{y}(s); s \le t\}$$

$$\mathbf{U}(t) = \text{history of commands} = \{\mathbf{u}(s); s < t\}$$

where Φ is the same function as in (29.1). Then it is said that the **certainty equivalence (CE) property** holds. Now, consider the optimal control law which has the form

$$\mathbf{u}^{\#}(t) = \Psi(\hat{\mathbf{x}}(t), t)$$

where $\hat{\mathbf{x}}$ is defined as above, then the **separation property** is said to hold. Clearly separation is weaker than certainty equivalence, because Ψ may be different from the control law of the deterministic case.

Finally, when

$$\mathbf{u}^{\#}(t) = \Gamma(\hat{\mathbf{x}}(t), \hat{\mathbf{p}}(t), t)$$

where $\hat{\mathbf{p}}$ are elements of a hyperstate, and represent indicators of the uncertainty in $\hat{\mathbf{x}}(t)$, we have **dual control**.

Certainty equivalence has considerable appeal to engineers, for they have reasonably workable methods for finding the deterministic optimal control law (e.g. Pontryagin principle, Chapter 26) and the conditional mean of the state (e.g. extended Kalman filters, Chapter 28). If CE holds, the controller may be built as a filter and optimal control law in cascade. It is known to hold for the *l*inear optimal control problem with *q*uadratic payoff and *g*aussian noise (LQG) problem (as is argued below) and certain extensions to non-linear measurements and non-additive noise in which the control does not affect the covariance of the conditional mean estimate (i.e. there is no dual effect on second order statistics), but in general it cannot be expected that even separation, let alone CE, will be optimal, and in fact the true optimum in most cases involving both noise and optimization of a performance index is not known.

We will show the optimality of CE for the LQG problem in section 29.4. The simpler problem of determining the effects of using a deterministic observer (Chapter 25) with a pole placement controller (Chapter 23) is discussed in the next section.

29.3 COMBINED USE OF OBSERVERS AND POLE PLACEMENT CONTROLLERS

We have seen that state feedback allows all of the poles of a controllable linear system to be placed at the designer's discretion.

Although the obvious thing to do is use an estimate of the state instead of the full state in the feedback, as in Fig. 29.1, the question is, what does this do to system stability and to all of those carefully chosen poles? We will show that the closed system will now have the set of poles consisting of the designed poles from the state feedback design done independently and the designed poles from the observer. To show this we take the simple but useful case of an identity state observer.

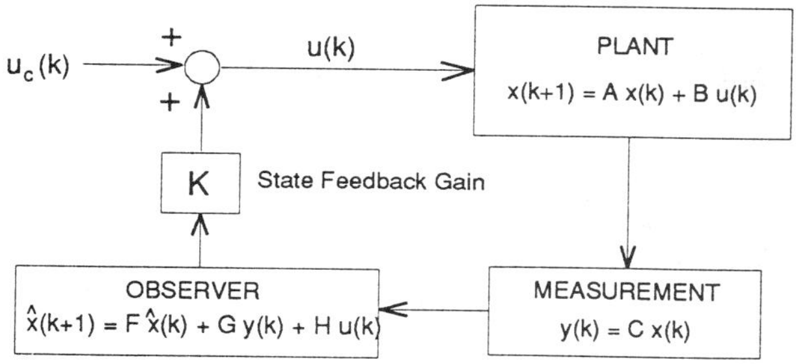

Figure 29.1 Structure of controller using state feedback with observer estimate of state.

Suppose a state feedback gain $\mathbf{K}$ has been selected such that the eigenvalues of

$$\mathbf{x}(k+1) = \mathbf{A}\mathbf{x}(k) + \mathbf{B}\mathbf{u}(k) \tag{29.2}$$

$$\mathbf{u}(k) = -\mathbf{K}\mathbf{x}(k) + \mathbf{u}_c(k)$$

are located at $\lambda_1, \lambda_2, \ldots, \lambda_n$. Furthermore, assume the gain $\mathbf{G}$ in the identity observer

$$\hat{\mathbf{x}}(k+1) = (\mathbf{A} - \mathbf{G}\mathbf{C})\hat{\mathbf{x}}(k) + \mathbf{B}\mathbf{u}(k) + \mathbf{G}\mathbf{y}(k)$$

is chosen so that the observer's eigenvalues are $\mu_1, \mu_2, \ldots, \mu_n$.

When the observer state estimate $\hat{\mathbf{x}}(k)$ is used instead of the state $\mathbf{x}(k)$ in (29.2), the resulting system has a state of dimension $2n$ modelled by

$$\begin{bmatrix} \mathbf{x}(k+1) \\ \hline \hat{\mathbf{x}}(k+1) \end{bmatrix} = \begin{bmatrix} \mathbf{A} & -\mathbf{B}\mathbf{K} \\ \hline \mathbf{G}\mathbf{C} & \mathbf{A} - \mathbf{G}\mathbf{C} - \mathbf{B}\mathbf{K} \end{bmatrix} \begin{bmatrix} \mathbf{x}(k) \\ \hline \hat{\mathbf{x}}(k) \end{bmatrix} + \begin{bmatrix} \mathbf{B} \\ \hline \mathbf{B} \end{bmatrix} \mathbf{u}_c(k)$$

A similarity transform

$$\begin{bmatrix} \mathbf{x}(k) \\ \mathbf{e}(k) \end{bmatrix} = \mathbf{P} \begin{bmatrix} \mathbf{x}(k) \\ \hat{\mathbf{x}}(k) \end{bmatrix}$$

where

$$\mathbf{P} = \begin{bmatrix} \mathbf{I} & \mathbf{0} \\ \mathbf{I} & -\mathbf{I} \end{bmatrix}$$

corresponding to a change of variables to $\mathbf{x}(k)$ and $\mathbf{e}(k) = \mathbf{x}(k) - \hat{\mathbf{x}}(k)$ converts this to

$$\begin{bmatrix} \mathbf{x}(k+1) \\ \hdashline \mathbf{e}(k+1) \end{bmatrix} = \left[\begin{array}{c:c} \mathbf{A} - \mathbf{BK} & \mathbf{BK} \\ \hdashline \mathbf{0} & \mathbf{A} - \mathbf{GC} \end{array} \right] \begin{bmatrix} \mathbf{x}(k) \\ \hdashline \mathbf{e}(k) \end{bmatrix} + \begin{bmatrix} \mathbf{B} \\ \hdashline \mathbf{0} \end{bmatrix} \mathbf{u}_c(k)$$

This system, which has the same eigenvalues as the original because of the nature of similarity transforms, clearly has eigenvalues which are the solution of

$$\det\left((\lambda\mathbf{I} - (\mathbf{A} - \mathbf{BK}))(\lambda\mathbf{I} - (\mathbf{A} - \mathbf{GC}))\right) = 0$$

This requires that one of

$$\det(\lambda\mathbf{I} - (\mathbf{A} - \mathbf{BK})) = 0$$

$$\det(\lambda\mathbf{I} - (\mathbf{A} - \mathbf{GC})) = 0$$

must hold. These of course are the eigenvalues of the design for the pole placement feedback state controller, $\lambda_1, \lambda_2, ..., \lambda_n$, and of the design for the state observer, $\mu_1, \mu_2, ..., \mu_n$ respectively.

The messages of the above are that:

1. the closed-loop system which uses the identity observer has $2n$ eigenvalues (n for the plant and n for the observer); and

2. the use of the observer does not move the designed poles from the pole placement algorithm.

Example

As an example of the above we consider the very simple system

$$\mathbf{x}(k+1) = \begin{bmatrix} 1 & \frac{1}{2} \\ 0 & 1 \end{bmatrix} \mathbf{x}(k) + \begin{bmatrix} 0 \\ 1 \end{bmatrix} \mathbf{u}(k)$$

$$\mathbf{y}(k) = \begin{bmatrix} 1 & 0 \end{bmatrix} \mathbf{x}(k)$$

This is observable, controllable, and unstable. We first design a state feedback controller

$$\mathbf{u}(k) = -\mathbf{K}\mathbf{x}(k)$$

to place poles at ± 0.5. This is easily found to require

$$\mathbf{K} = \begin{bmatrix} \frac{3}{2} & 2 \end{bmatrix}$$

Since the state is not available, we next design an identity observer with poles at ± 0.25. This turns out to require a gain of

$$\mathbf{G} = \begin{bmatrix} 2 \\ \frac{15}{8} \end{bmatrix}$$

The entire control system is now described, for external commands $\mathbf{u}_c(k)$, by

$$\mathbf{x}(k+1) = \begin{bmatrix} 1 & \frac{1}{2} \\ 0 & 1 \end{bmatrix} \mathbf{x}(k) + \begin{bmatrix} 0 \\ 1 \end{bmatrix} \mathbf{u}(k)$$

$$\mathbf{y}(k) = \begin{bmatrix} 1 & 0 \end{bmatrix} \mathbf{x}(k)$$

$$\mathbf{u}(k) = [\ \tfrac{-3}{2}\ \ -2\ ]\,\hat{\mathbf{x}}(k) + \mathbf{u}_c(k)$$

$$\hat{\mathbf{x}}(k+1) = \begin{bmatrix} -1 & \tfrac{1}{2} \\ \tfrac{-15}{8} & 1 \end{bmatrix}\hat{\mathbf{x}}(k) + \begin{bmatrix} 2 \\ \tfrac{15}{8} \end{bmatrix}\mathbf{y}(k) + \begin{bmatrix} 0 \\ 1 \end{bmatrix}\mathbf{u}(k)$$

Substituting for $\mathbf{u}(k)$ and $\mathbf{y}(k)$ in the system and observer equations yields the closed-loop system description

$$\begin{bmatrix} \mathbf{x}(k+1) \\ \hat{\mathbf{x}}(k+1) \end{bmatrix} = \begin{bmatrix} 1 & \tfrac{1}{2} & 0 & 0 \\ 0 & 1 & \tfrac{-3}{2} & -2 \\ 2 & 0 & -1 & \tfrac{1}{2} \\ \tfrac{15}{8} & 0 & \tfrac{-27}{8} & -1 \end{bmatrix}\begin{bmatrix} \mathbf{x}(k) \\ \hat{\mathbf{x}}(k) \end{bmatrix} + \begin{bmatrix} 0 \\ 1 \\ 0 \\ 1 \end{bmatrix}\mathbf{u}_c(k)$$

$$\mathbf{y}(k) = [\ 1\ \ 0\ \ 0\ \ 0\ ]\begin{bmatrix} \mathbf{x}(k) \\ \hat{\mathbf{x}}(k) \end{bmatrix}$$

Direct computation of the eigenvalues of the 4×4 matrix shows that they are at ± 0.5, ± 0.25 as expected.

The result is readily extended to using reduced-order observers. In fact it may be stated that: if a system is completely observable and controllable with m linearly independent outputs, a composite system may be designed in which $2n - m$ eigenvalues are chosen arbitrarily.

29.4 THE LINEAR-QUADRATIC-GAUSSIAN CONTROL PROBLEM

We have in previous chapters considered the problems of optimal control of a deterministic linear system with a quadratic performance index and of determination of a state estimate of a linear system subject to random noise added to both the dynamics and the measurements. The results were a linear state feedback control law and a Kalman filter, respectively. We now consider the problem of optimal control of a linear system subject to noise in which only a linear function of the state is measured and the performance index is quadratic.

The system is taken to be described by

$$\mathbf{x}(k+1) = \mathbf{A}\mathbf{x}(k) + \mathbf{B}\mathbf{u}(k) + \mathbf{v}(k)$$

$$\mathbf{y}(k) = \mathbf{C}\mathbf{x}(k) + \mathbf{w}(k) \tag{29.3}$$

where as usual the state at sample k (or time kT) is represented by the n-vector $\mathbf{x}(k)$, $\mathbf{u}(k)$ is the m-vector of commands, and $\mathbf{y}(k)$ is the p-vector of measurements. The noises $\mathbf{w}(k)$ and $\mathbf{v}(k)$ are taken to be gaussian with mean value 0 and covariance matrices $\mathrm{cov}(\mathbf{v}(k)) = \mathbf{Q}_n$ and $\mathrm{cov}(\mathbf{w}(k)) = \mathbf{R}_n$. The initial state $\mathbf{x}(0)$ is assumed to have gaussian distribution with mean $\bar{\mathbf{x}}$ and $\mathrm{cov}(\mathbf{x}(0)) = \Sigma_0$. The random variables $\mathbf{w}(k)$, $\mathbf{w}(j)$, $\mathbf{v}(k)$, $\mathbf{v}(j)$, $\mathbf{x}(0)$ are assumed to be uncorrelated for all k and j, $k \neq j$.

The object is to find an optimal control law $\mathbf{u}^*(\cdot)$ for which the performance index J is minimized, where

$$J = \tfrac{1}{2}\mathscr{E}\left\{\mathbf{x}^{\mathrm{T}}(N)\mathbf{S}_{\mathrm{PI}}\mathbf{x}(N) + \sum_{i=0}^{N-1}[\mathbf{x}^{\mathrm{T}}(i)\mathbf{Q}_{\mathrm{PI}}\mathbf{x}(i)\,\mathbf{u}^{\mathrm{T}}(i)\,\mathbf{R}_{\mathrm{PI}}\mathbf{u}(i)]\right\} \tag{29.4}$$

The mean value is taken because the state $\mathbf{x}(k)$ is a random variable; if it were not random, the problem would be that of Chapter 26.

To attack this problem, we use two fundamental facts from probability and matrix theory.

1.   If $\bar{\mathbf{x}} = \mathscr{E}\{\mathbf{x}\}$, then

$$\mathscr{E}\{\mathbf{x}^{\mathrm{T}}\mathbf{M}\mathbf{x}\} = \mathscr{E}\left\{(\mathbf{x}-\bar{\mathbf{x}})^{\mathrm{T}}\mathbf{M}(\mathbf{x}-\bar{\mathbf{x}})\right\} + \bar{\mathbf{x}}^{\mathrm{T}}\mathbf{M}\bar{\mathbf{x}}$$

$$= \bar{\mathbf{x}}^{\mathrm{T}}\mathbf{M}\bar{\mathbf{x}} + \mathscr{E}\,\mathrm{tr}\,\{\mathbf{M}(\mathbf{x}-\bar{\mathbf{x}})(\mathbf{x}-\bar{\mathbf{x}})^{\mathrm{T}}\}$$

$$= \bar{\mathbf{x}}^{\mathrm{T}}\mathbf{M}\bar{\mathbf{x}} + \mathrm{tr}\,\{\mathbf{M}\,\mathrm{cov}\,(\mathbf{x})\} \tag{29.5}$$

2. $\mathscr{E}\{\mathbf{x}\} = \mathscr{E}\{\mathscr{E}\{\mathbf{x}|\mathbf{y}\}\}$ (29.6)

Our control law is required to be causal, so that $\mathbf{u}^*(k) = f(\mathbf{Y}(k), k)$, where $\mathbf{Y}(k)$ denotes the entire set of measured data to time k, e.g. $\mathbf{Y}(k) = \{\mathbf{y}(k), \mathbf{y}(k-1), \ldots, \mathbf{y}(0)\}$.

We first apply (29.6) to (29.4) to obtain

$$J = \tfrac{1}{2}\,\mathscr{E}\{\mathscr{E}\{\mathbf{x}^{\mathrm{T}}(N)\mathbf{S}_{\mathrm{PI}}(N)\mid\mathbf{Y}(N)\}$$

$$+ \sum_{i=0}^{N-1}\left[\mathscr{E}\{\mathbf{x}^{\mathrm{T}}(i)\,\mathbf{Q}_{\mathrm{PI}}\,\mathbf{x}(i)\mid\mathbf{Y}(i)\} + \mathscr{E}\{\mathbf{u}^{\mathrm{T}}(i)\,\mathbf{R}_{\mathrm{PI}}\,\mathbf{u}(i)\mid\mathbf{Y}(i)\}\right]\}$$

We now apply (29.5) to this on a term-by-term basis, use the notation $\hat{\mathbf{x}}(k)$ for the conditional mean $\mathscr{E}\{\mathbf{x}(k)\mid\mathbf{Y}(k)\}$, and note that $\mathbf{u}(k)$ is not random as far as our selection is concerned.

$$2J = \mathscr{E}\left\{\hat{\mathbf{x}}^{\mathrm{T}}(N)\mathbf{S}_{\mathrm{PI}}\hat{\mathbf{x}}(N) + \sum_{i=0}^{N-1}[\hat{\mathbf{x}}^{\mathrm{T}}(i)\,\mathbf{Q}_{\mathrm{PI}}\,\hat{\mathbf{x}}(i) + \mathbf{u}^{\mathrm{T}}(i)\,\mathbf{R}_{\mathrm{PI}}\,\mathbf{u}(i)]\right.$$

$$\left. + \operatorname{tr}\left(\mathbf{S}\operatorname{cov}\left(\mathbf{x}(N)\mid\mathbf{Y}(N)\right)\right) + \sum_{i=0}^{N-1}\operatorname{tr}\left(\mathbf{Q}_{\mathrm{PI}}\operatorname{cov}\left(\mathbf{x}(i)\mid\mathbf{Y}(i)\right)\right)\right\}$$

$$(29.7)$$

We write this as $2J = 2\hat{J} + 2J_{\mathrm{cov}}$ where

$$\hat{J} = \mathscr{E}\left\{\hat{\mathbf{x}}^{\mathrm{T}}(N)\mathbf{S}_{\mathrm{PI}}\hat{\mathbf{x}}(N) + \sum_{i=0}^{N-1}\left[\hat{\mathbf{x}}^{\mathrm{T}}(i)\,\mathbf{Q}_{\mathrm{PI}}\hat{\mathbf{x}}(i) + \mathbf{u}^{\mathrm{T}}(i)\,\mathbf{R}_{\mathrm{PI}}\,\mathbf{u}(i)\right]\right\}$$

$$J_{\mathrm{cov}} = \mathscr{E}\left\{\operatorname{tr}\left(\mathbf{S}\operatorname{cov}\left(\mathbf{x}(N)\mid\mathbf{Y}(N)\right)\right) + \sum_{i=0}^{N-1}\operatorname{tr}\left(\mathbf{Q}_{\mathrm{PI}}\operatorname{cov}\left(\mathbf{x}(i)\mid\mathbf{Y}(i)\right)\right)\right\}$$

so that we may emphasize the following two observations, distilled from our studies of the Kalman filter.

1. The expansion (29.7) is general. Hence, $\mathbf{u}^*$ may be expected to depend upon conditional covariances.
2. In the case of the linear system (29.3), the covariances of $\mathbf{x}(k)$ are not dependent upon the control or upon the measurements,

but only upon the covariances $\mathbf{Q}_n$ and $\mathbf{R}_n$ of the system disturbances and the measurement noise. Hence minimizing J is equivalent to minimizing $\hat{J}$ in this special case.

On the basis of the latter claim, we then consider the problem of minimizing $\hat{J}$ and immediately see that, because of the zero mean property of the noises, we have the problem of finding

$$\min\left[\hat{\mathbf{x}}^{\mathrm{T}}(N)\,\mathbf{S}_{\mathrm{PI}}\hat{\mathbf{x}}(N) + \sum_{i=0}^{N-1}\left[\hat{\mathbf{x}}^{\mathrm{T}}(i)\,\mathbf{Q}_{\mathrm{PI}}\hat{\mathbf{x}}(i) + \mathbf{u}^{\mathrm{T}}(i)\,\mathbf{R}_{\mathrm{PI}}\mathbf{u}(i)\right]\right]$$

subject to $\hat{\mathbf{x}}(k+1) = \mathbf{A}\,\hat{\mathbf{x}}(k) + \mathbf{B}\mathbf{u}(k)$. We know the solution to this (from section 26.3.3, for example) is

$$\mathbf{u}^*(k) = -\left[\mathbf{R}_{\mathrm{PI}} + \mathbf{B}^{\mathrm{T}}\mathbf{P}(k+1)\mathbf{B}\right]^{-1}\mathbf{B}^{\mathrm{T}}\mathbf{P}(k+1)\,\hat{\mathbf{x}}(k)$$

$$\mathbf{P}(k) = \mathbf{Q}_{\mathrm{PI}} + \mathbf{A}^{\mathrm{T}}\left[\mathbf{P}(k+1)\right.$$

$$\left. - \mathbf{P}(k+1)\mathbf{B}\left[\mathbf{R}_{\mathrm{PI}} + \mathbf{B}^{\mathrm{T}}\mathbf{P}(k+1)\mathbf{B}\right]^{-1}\mathbf{B}^{\mathrm{T}}\mathbf{P}(k+1)\right]\mathbf{A}$$

$$\mathbf{P}(N) = \mathbf{S}_{\mathrm{PI}}$$

The conditional mean is generated from the Kalman filter (Chapter 28) as

$$\hat{\mathbf{x}}(k+1|k) = \mathbf{A}\hat{\mathbf{x}}(k) + \mathbf{B}\mathbf{u}^*(k)$$

$$\Sigma_{k+1|k} = \mathbf{A}\,\Sigma_{k|k}\,\mathbf{A}^{\mathrm{T}} + \mathbf{Q}_n$$

$$\mathbf{K}(k+1) = \Sigma_{k+1|k}\,\mathbf{C}^{\mathrm{T}}\left[\mathbf{C}\,\Sigma_{k+1|k}\,\mathbf{C}^{\mathrm{T}} + \mathbf{R}_n\right]^{-1}$$

$$\hat{\mathbf{x}}(k+1) = \hat{\mathbf{x}}(k+1|k) + \mathbf{K}(k+1)\left[\mathbf{y}(k+1) - \mathbf{C}\,\hat{\mathbf{x}}(k+1|k)\right]$$

$$\Sigma_{k+1|k+1} = \left[\mathbf{I} - \mathbf{K}(k+1)\mathbf{C}\right]\Sigma_{k+1|k}$$

The control is precisely that of the deterministic case except that the state estimate is substituted for the state in the computation. Also, the state estimator is not affected by the fact that a feedback control is in use. Hence, we have a case of certainty equivalence.

If the control somehow affects the covariances in J_{cov}, either directly or via the state, then the optimum control must account for this and at best we would have a possibility of separation and more likely we would need to account for a dual effect; see Chapter 31.

Finally, we note that the noise levels indeed affect the performance index even though they do not affect the control law in this case. This is true because the noise levels affect J_{cov}.

The structure of the controller is shown in Fig. 29.2.

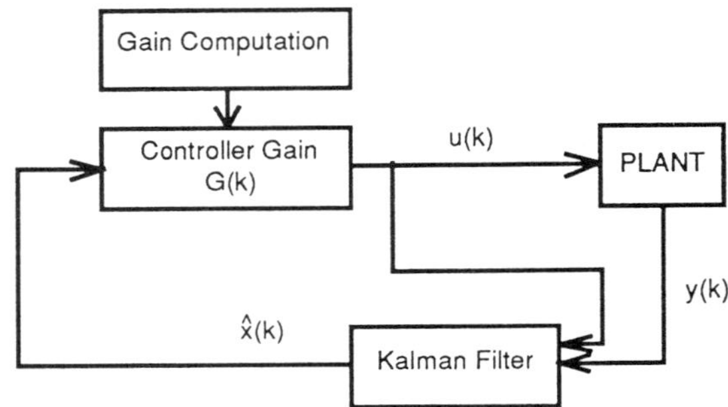

Figure 29.2 Typical structure of certainty equivalent control system.

Unfortunately, controllers of this type are not necessarily robust to model errors; see Anderson and Moore (1989) and Chapter 33.

29.5 COMPUTER AIDS

The algorithms for pole placement/observer design are so well known for the above problems that they are part of a number of CACSD packages. Steady-state Kalman gains and optimal control gains are also commonly available in such programs. Code for Kalman filters and LQ controllers is easy to write, which is part of the reason CE controllers are so attractive.

29.6 SUMMARY AND REFERENCES

The engineer's inclination is to split design into two parts: control law and filters. It is certainly arguable that this is suboptimal, as one could expect that the filter response might interact with the control law. However, in some cases this approach has been shown to cause no loss in performance:

1. in noise free cases of linear systems, state feedback pole placement is not affected by use of an observer's state estimate in place of the actual state; and
2. with linear systems in gaussian noise using a quadratic cost criterion, an optimal controller is still optimal if its input is from an optimal (Kalman) filter rather than from exact state measurements.

In this section, we demonstrated the former and justified the latter. The latter is proven in textbooks such as Meditch (1969); the issues of separation and certainty equivalence were addressed by Tse and Bar-Shalom (1975). The matter is pursued more generally in the adaptive control literature; see Chapter 31.

LQG controllers are discussed further in Grimble and Johnson (1988) and the controllers, the CE principle, etc., are addressed in Anderson and Moore (1989).

30
System identification

In previous chapters it was assumed that a model of the process exists and that the designer's task is to use that model in the creation of a feedback controlled system. In fact, although a model structure may exist, it is frequently the case that its parameters are unknown, or at least are not known to the required precision. Hence a motor model may be derivable from basic electromagnetics and physics, but the moment of inertia of a particular motor may be only approximately known; a missile's mass may decrease linearly with burn time of the rocket motor, but the rate and initial mass may be guaranteed to only a few per cent. This may or may not be a problem – after all, the reason for using feedback control in the first place is to reduce errors due to such factors as imprecisely known parameters – but accurate estimation of the parameters is sometimes important.

A more serious problem arises when the process is complex to model, as with a sequence of rollers in a steel rolling mill or the crushers in an ore crusher, or is not even very well understood, as in chemical manufacture. Here it may be necessary to guess at the system's structural relationships as well as the parameters.

In these cases, system identification methods are brought to bear. That these have received a great deal of attention is evidenced from the size of the literature, consisting of many papers and books dating back a number of years. The techniques of this chapter exemplify some of the approaches to system identification, with most being implicitly or explicitly parameter estimation techniques, including batch methods, incremental methods, special input (sinusoidal, step, or noise responses) and opportunistic methods. Most are suitable for linear shift-invariant (time invariant) systems, but will not of themselves yield even the system order, let alone the existence or form of a non-linearity.

30.1 SYNOPSIS

There are many methods (and books) about system identification. Most reduce to parameter estimation. One traditional class uses frequency response data, in which one infers a system model from examining frequency response data such as in Fig. 30.1, using the knowledge of Bode plot construction such as in Chapter 20.

Many time domain approaches develop from least-squares fits of input–output data. Here a system model

$$\mathbf{y}(k) = -\sum_{i=1}^{n} a_i \mathbf{y}(k-i) + \sum_{j=0}^{m} b_j \mathbf{u}(k-j)$$

is assumed. This is represented in a vector form as

$$\mathbf{y}(k+1) = \phi(k)\,\theta$$

where

$$\theta = \begin{bmatrix} a_1 \\ a_2 \\ \vdots \\ a_n \\ b_0 \\ \vdots \\ b_m \end{bmatrix}$$

$$\phi(k) = [-\mathbf{y}(k-1) \;\; -\mathbf{y}(k-2) \;\; ... \;\; -\mathbf{y}(k-n) \;\; \mathbf{u}(k) \;\; \mathbf{u}(k-1) \;\; ... \;\; \mathbf{u}(k-m)]$$

Then if we define

$$\mathbf{Y}_N = \begin{bmatrix} \mathbf{y}(1) \\ \mathbf{y}(2) \\ \mathbf{y}(3) \\ \vdots \\ \mathbf{y}(N) \end{bmatrix} \qquad \Phi_N = \begin{bmatrix} \phi(1) \\ \phi(2) \\ \phi(3) \\ \vdots \\ \phi(N) \end{bmatrix}$$

we can show that a best least-squares fit $\hat{\theta}$ of the data is

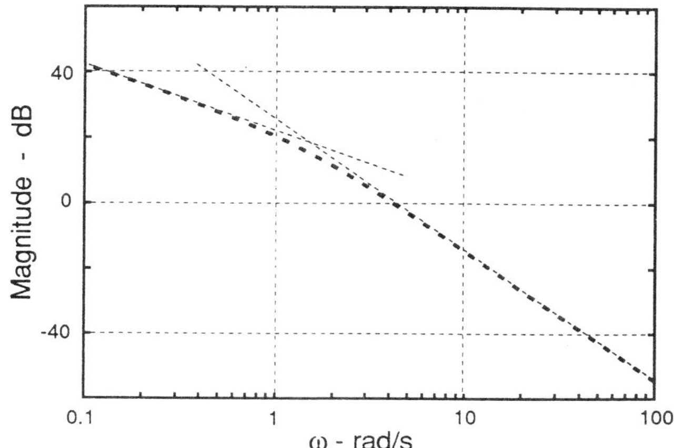

Figure 30.1 Identification from frequency response. 'Experimental data' is heavily dotted, while asymptotes are lighter.

$$\hat{\theta} = (\Phi_N^T \, \Phi_N)^{-1} \Phi_N^T \, Y$$

and hence this is a 'good' estimate of the parameters θ. This basic result can be extended in many ways, including making it recursive, modifying the details of the recursion, and using a Kalman filter interpretation to achieve the same result.

30.2 IDENTIFICATION OF TRANSFER FUNCTION PARAMETERS

In the following short subsections we indicate and exemplify some common and effective methods for identification. The first two are particularly suited to 'black box' identification and the extraction of essential features of the input–output response characteristics. The third is a common and useful approach for finding numerical values of parameters which appear linearly in a model of known size.

For some purposes a sophisticated mathematical model of a system may be required; it may even be the ultimate goal of the identification process to produce such a model. For many applications, however, it is arguable that only input–output information is needed and that the internal operation of the system is essentially irrelevant. This latter end is particularly served by various simple schemes of relating output to input. All of these typically use one or more simple inputs and assume linearity and superposition so that the results are applicable to other inputs. The simple inputs are typically pulses, steps, or sets of sinusoids; the outputs are then pulse responses, step responses, or frequency responses. These may or may not be further associated with mathematical models.

Thus a system may have a pulse response as in Fig. 30.2(a), a unit step response as in Fig. 30.2(b) and a frequency response as in Fig. 30.2(c). Notice that all of the graphs in Fig. 30.2 can in principle be easily established experimentally, although some inputs may not be allowed with some plants for operational reasons. In this section we look briefly at several simple identification schemes. These are useful, but could hardly be called profound. Their characteristic is that they involve little manipulation of the data, and the response data is virtually the model.

30.2.1 Frequency response – a batch special input modelless approach

The ideas of frequency response batch identification methods are straightforward: one obtains the Bode plots of the system under study and fits linear constant-coefficient transfer functions to the data. The fit may be done manually or by computer (using e.g. least-squares); the plots may be obtained by using sinusoidal input signals and finding the magnitude ratio and phase shift of the output, or by using a long sequence of noise-like inputs and computing the discrete Fourier transform of the output signal. The method can be done by any undergraduate with a signal generator and oscilloscope, but has at least two very important faults.

1. Because only steady-state operation is assessed, transients and in particular delays are likely to go undiscovered unless special care is taken. Non-linearities may manifest themselves with obvious clipping or harmonics in the outputs.
2. With real equipment, and particularly large processes, one may not be able to do the experiments.

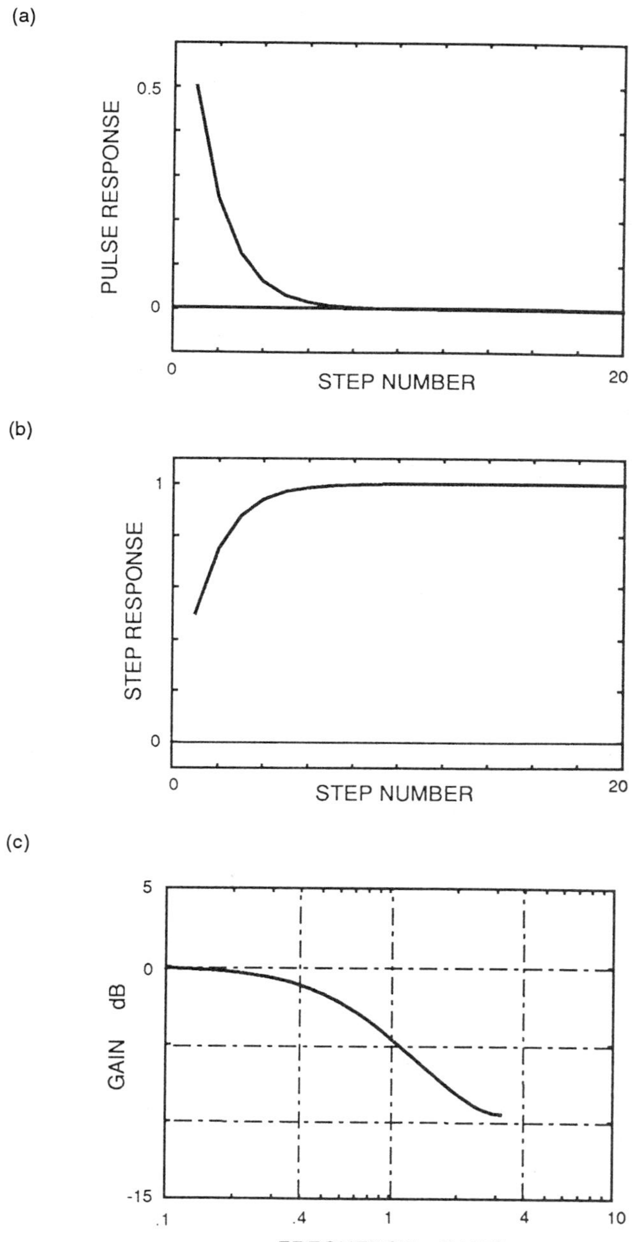

Figure 30.2 Possible types of data for identification purposes: (a) pulse response; (b) step response; and (c) frequency response.

Example

From Fig. 30.1, the rolloff of 20 dB/decade at lower frequencies and 40 dB/decade at higher ones indicates a pole excess of one in the lower region and two in the upper. In the absence of further information, this indicates a model of the form

$$G(s) = \frac{K}{s(\tau s + 1)}$$

Then the intersection of the asymptotes at frequency ω_i implies that $\tau = 1/\omega_i$, while the value of K is given by evaluation of the above at some frequency ω. The value of K is given, for low frequency $0.1\,\text{rad}\,\text{s}^{-1}$, say, by the fact that

$$20\log\left[\frac{K}{(0.1)}\right] \approx |G(0.1\text{j})|$$

as read from the graph; in this case the results are that $K \approx 13$ and $\tau \approx 0.7$.

A variation of the explicit frequency response method is to excite the system with a random noise signal and get frequency response using Fourier transform methods. Thus let

$$Y(z) = H(z)\,U(z)$$

be an input–output relation in which $H(z)$ is unknown, and in which we have implicitly assumed the system is linear. The power spectral density (Appendix C) of the output is

$$\Phi_y(\omega) = \mathscr{E}[Y(\text{e}^{\text{j}\omega T})\,Y(\text{e}^{-\text{j}\omega T})]$$

$$= H(\text{e}^{\text{j}\omega T})\,H(\text{e}^{-\text{j}\omega T})\,\mathscr{E}[U(\text{e}^{\text{j}\omega T})\,U(\text{e}^{-\text{j}\omega T})]$$

$$= H(\text{e}^{\text{j}\omega T})\,H(\text{e}^{-\text{j}\omega T})\,\Phi_u(\omega)$$

$$= |H(\text{e}^{\text{j}\omega T})|^2\,\Phi_u(\omega)$$

If $\{u(m)\}$ is a white noise sequence, then $\Phi_u(\omega) = \text{constant} = N$ and we have

$$|H(\mathrm{e}^{\mathrm{j}\omega T})| = \left(\frac{\Phi_y(\omega)}{N}\right)^{\frac{1}{2}}$$

where the spectral density of y, $\Phi_y(\omega)$, can be estimated using the discrete Fourier transform (DFT) of $\{y(m)\}$. The details of this are sometimes subtle and are found in the digital signal processing literature. Notice that the phase of the transfer function is not given from these data and must be otherwise established.

30.2.2 Step and impulse responses – batch or on-line special input methods

This method is similar in ease of concept to the frequency response methods. One makes a step or impulsive change in the input signal of the system under study and observes the output. For machinery, the step change might simply be a change in the operating set point; in medicine, a (near) impulsive change can be provided by an injection of a drug. A single experiment can be sufficient for the method, but therein lies its problem: lack of data. A step response as in Fig. 30.2(b) has perhaps half-a-dozen parameters readily visible (rise time, delay, overshoot, settling time, oscillation frequency, damping (Chapter 13)) so one can at best extract a similar number of unknown parameters. Even more than with frequency response, the method relies upon matching known responses with the unknown one. This is done by proposing a system model structure and adjusting its parameters until the model's step response approximates the observed data.

Example

The data of Fig. 30.3 were obtained as an impulse response and could for example represent a gradual using up and also excretion of a drug after injection. We examine it by a method sometimes called 'peeling off'. In this, we first consider only the tail and, assuming this is due to only a single slow effect (such as excretion, say), we fit a curve $\alpha \exp(-\beta t)$ to a few points; to do this it is convenient to take the logarithm of the output and fit a straight line to the tail, by which from Fig. 30.3(b) we establish that $\alpha \approx 0.33$ and $\beta \approx 0.16$. We then

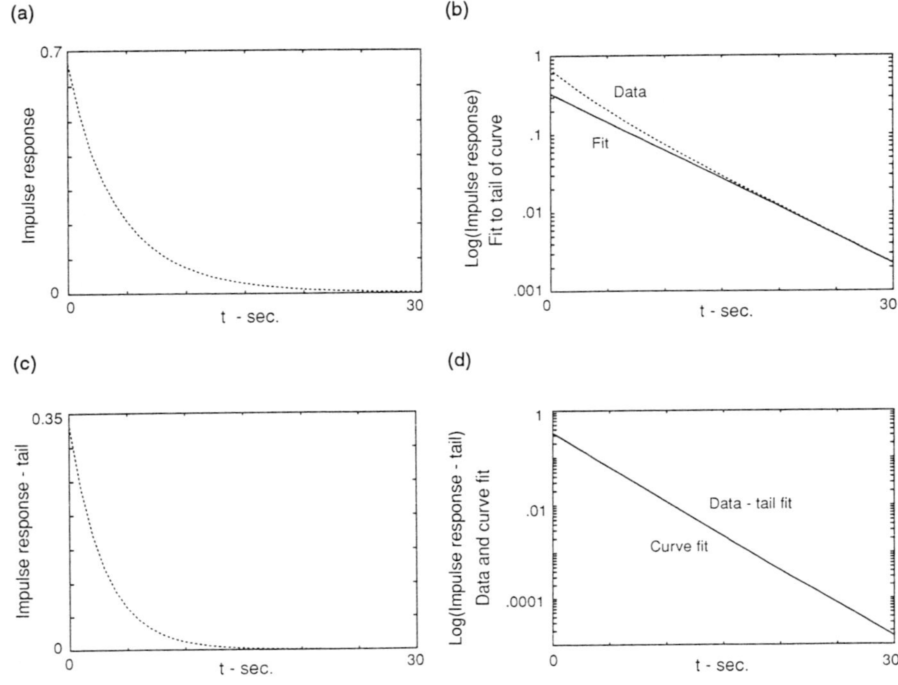

Figure 30.3 The 'peeling off' of tails of the impulse response:
(a) original curve; (b) logarithm of original data, to allow line fit of tail
(shown as solid line); (c) data after subtraction of term fitted in (b); and
(d) logarithm of points in part (c), with heavy line fit.

subtract this, yielding Fig. 30.3(c), and repeat the process on this
curve for the curve $\gamma \exp(-\delta t)$, obtaining $\gamma \approx 0.33$ and $\delta \approx 0.33$. This
could be repeated if necessary, but subtracting this function leaves a
very small remainder. Thus we obtain the approximate transfer
function

$$G(s) = \frac{\alpha}{s + \beta} + \frac{\gamma}{s + \delta} = \frac{2}{6s + 1} + \frac{1}{3s + 1}$$

and proceed, if desired, to make interpretations as to physical effects
(e.g. drug usage and excretion in the medical case).

30.2.3 Auslander's method

One of the simplest approaches, but an intuitively appealing one for which the presenters (Auslander *et al.*, 1978) claim many extensions and refinements have been worked out, results from careful examination of the step response of the plant in question. Since the step response at time nT is the sum of the pulse responses $p(iT)$, $i = 0, 1, \ldots, n$, one can find the pulse response (and therefore a complete characterization of a time-invariant linear system) by, in effect, differentiating the step response. More particularly, consider Fig. 30.4.

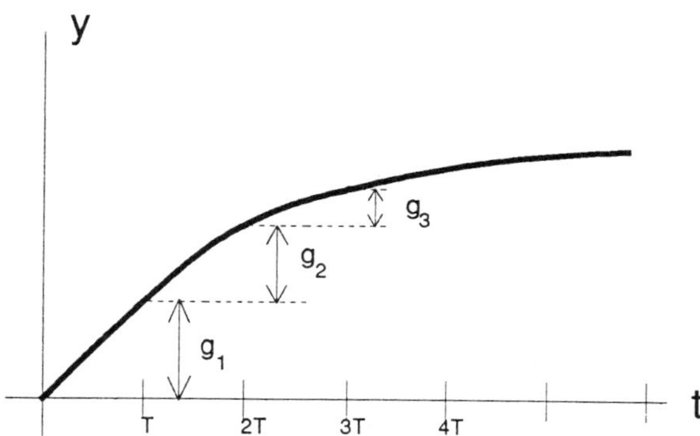

Figure 30.4 Unit step response and definition of parameters for Auslander's method.

The pulse transfer function, from the above arguments, is simply given by

$$G(z) = g_0 + g_1 z^{-1} + g_2 z^{-2} + \cdots + g_m z^{-m} + \cdots$$

After a useful number of samples, this can be ended with a term of the form

$$\frac{g_n z^{-n}}{1 - pz^{-1}}$$

which has an asymptotically zero pulse response. The alternative form of $G(z)$ is, with g_0 taken as $g_0 = 0$ for causality,

$$G(z) = \frac{b_1 z^{-1} + b_2 z^{-2} + \cdots + b_n z^{-n}}{1 - pz^{-1}}$$

which is a system with $n-1$ zeros and one pole, where

$$b_1 = g_1$$

$$b_i = g_i - pg_{i-1} \qquad i = 2, 3, \ldots, n$$

The pole p is determined from the plant gain. Thus for a unit step, if the steady-state plant output is k, then

$$k = \sum_{i=1}^{n-1} g_i + \frac{g_n}{1 - p}$$

or

$$p = 1 - \frac{g_n}{k - \sum_{i=1}^{n-1} g_i}$$

The above equations yield an input–output model of the plant. A state-space form may be written directly as

$$\mathbf{x}(k+1) = \mathbf{P}\mathbf{x}(k) + \mathbf{q}u(k)$$

$$\mathbf{y}(k) = \mathbf{c}^T \mathbf{x}(k)$$

using the definitions

$$\mathbf{P} = \begin{bmatrix} 0 & 1 & 0 & 0 & \cdots & \cdots & 0 \\ 0 & 0 & 1 & 0 & \cdots & \cdots & 0 \\ 0 & 0 & 0 & 1 & 0 & \cdots & 0 \\ \vdots & \vdots & \vdots & \vdots & & & \vdots \\ 0 & 0 & 0 & 0 & \cdots & 0 & 1 \\ 0 & 0 & 0 & 0 & \cdots & 0 & p \end{bmatrix} \qquad \mathbf{q} = \begin{bmatrix} g_1 \\ g_2 \\ g_3 \\ \vdots \\ g_{n-1} \\ g_n \end{bmatrix}$$

$$\mathbf{c}^T = [\,1 \quad 0 \quad 0 \dots 0\,]$$

Using these models of the system, control laws and controllers may be developed in standard ways (Auslander *et al.*, 1978). The examples therein indicate a certain robustness of the approach.

30.2.4 Cauer form identification – on-line with successive models and sequences of steps and impulses

More sophisticated versions of the above have been proposed, including the method of Shieh *et al.* (1971). This method is rather of an experimental type, in that it involves a sequence of tests of a compensated system, with the compensation leading to knowledge of the original system model.

We consider a standard sampled-data transfer function, as in

$$H(z) = \frac{b_0 z^m + b_1 z^{m-1} + \cdots + b_m}{z^n + a_1 z^{n-1} + \cdots + a_n}$$

we must first rewrite it as

$$H(z) = \frac{c_m + c_{m-1}(z-1) + c_{m-2}(z-1)^2 + \cdots + c_0(z-1)^m}{d_n + d_{n-1}(z-1) + d_{n-2}(z-1)^2 + \cdots + (z-1)^n}$$

where the easy coefficients are

$$d_1 = 1 + a_1 + a_2 + \cdots + a_n$$

$$c_m = b_0 + b_1 + b_2 + \cdots + b_m$$

and the others are somewhat more difficult to write in general form. The Cauer second canonical form for this in terms of $(z-1)$ is

$$H(z) = \cfrac{1}{h_1 + \cfrac{1}{\cfrac{h_2}{(z-1)} + \cfrac{1}{h_3 + \cfrac{1}{\cfrac{h_4}{(z-1)} + \cfrac{1}{\cfrac{h_{2n}}{(z-1)}}}}}}$$

Since a unit step response in steady-state is given by

$$y_{ss} = \lim_{z \to 1} H(z)$$

and a unit pulse response by

$$y_{ss} = \lim_{z \to 1} (1 - z^{-1}) H(z)$$

it is obvious from the above that we can obtain h_1 from the unit step response of the original system, which is $1/h_1$. If in our experiment we connect the now known gain h_1 with the (unknown) system $H(z)$ as in Fig. 30.5, we find that the result has transfer function $H_1(z)$ given by

$$H_1(z) = \frac{h_2}{(z-1)} + \cfrac{1}{h_3 + \cfrac{1}{\cfrac{h_4}{(z-1)} + \cfrac{1}{\vdots}}}$$

The unit pulse response of this clearly, from y_{ss} defined above, has steady-state value h_2. With this known, we now experimentally configure the system as in Fig. 30.6, which can be shown to have transfer function $H_2(z)$ given by

$$H_2(z) = \cfrac{1}{h_3 + \cfrac{1}{\cfrac{h_4}{(z-1)} + \cfrac{1}{\vdots}}}$$

This is of the same form, so we can find h_3 from its steady-state step response, feed back h_3 as we did h_1 earlier, find h_4 from its unit pulse response, etc.

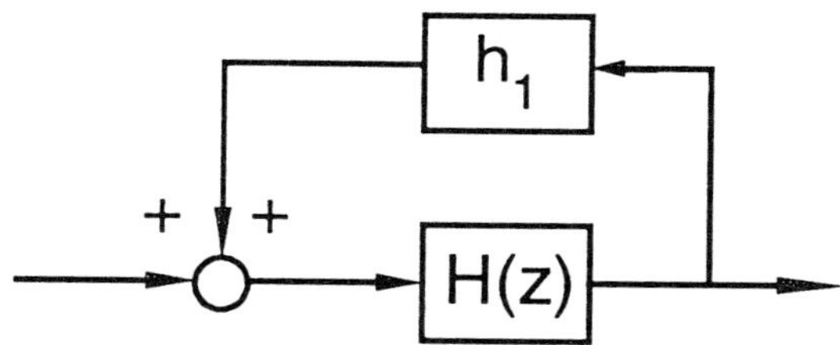

Figure 30.5 Feedback of first element in Cauchy expansion identification.

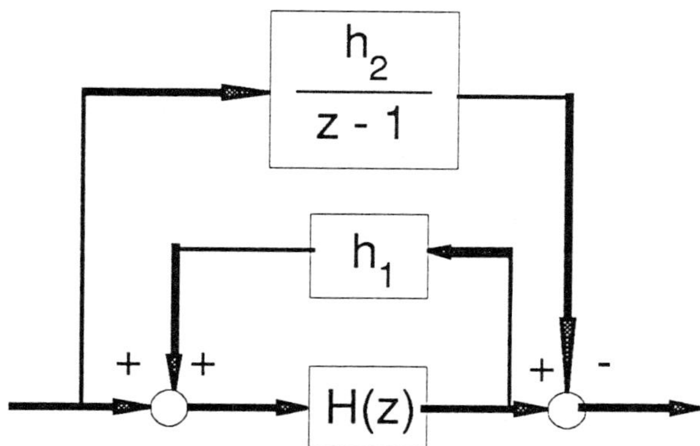

Figure 30.6 Feedforward of second element in Cauchy expansion identification.

The unravelling of the coefficients in particular cases is of course much easier than trying to come up with a general representation of the coefficients. It is clear that this is an experimental, as opposed to data processing, form of 'peeling off' of coefficients.

30.3 DIRECT PARAMETER ESTIMATION FOR POSTULATED MODELS

The above techniqes tended not to require *a priori* knowledge of the model size, but only of its general nature (as a model of a linear system). The more common alternative, especially for on-line recursive techniques, is to postulate a model which is linear in its parameters and then estimate the values of those parameters.

30.3.1 Least-squares data fits for transfer function models: batch form

In cases in which either there is no system model available or parameters of a linear system are unknown, it may be possible to use an input sequence $\{\mathbf{u}(k)\}$ to generate a set of measurements $\{\mathbf{y}(k)\}$ and from these to compute a set of parameters which are consistent with this information. To do so, we conjecture that the data fit a model

$$\mathbf{y}(k) = -\sum_{i=1}^{n} a_i \mathbf{y}(k-i) + \sum_{j=0}^{m} b_j \mathbf{u}(k-j)$$

and that the coefficients $\{a_i\}$ and $\{b_j\}$ are to be determined. To do this it is convenient to define the vectors

$$\theta = \begin{bmatrix} a_1 \\ a_2 \\ \vdots \\ a_n \\ b_0 \\ \vdots \\ b_m \end{bmatrix}$$

$$\phi(k) = [\, -y(k-1) \ \ -y(k-2) \ \ \dots \ \ -y(k-n) \ \ u(k) \ \ u(k-1) \ \ \dots \ \ u(k-m)\,]$$

for which we observe that $\mathbf{y}(k) = \phi(k)\,\theta$.

If we only have an estimate $\hat{\theta}$ of θ, where

$$\hat{\theta} = \begin{bmatrix} \hat{a}_1 \\ \hat{a}_2 \\ \vdots \\ \hat{a}_n \\ \hat{b}_0 \\ \vdots \\ \hat{b}_m \end{bmatrix}$$

then at any time kT we will have

$$\mathbf{y}_{\text{pred}}(k) = \phi(k)\,\hat{\theta}$$

and with proper choice of $\hat{\theta}$ the error

$$\mathbf{e}(k) = \mathbf{y}(k) - \mathbf{y}_{\text{pred}}(k) = \phi(k)\,[\,\theta - \hat{\theta}\,]$$

will be small. We choose to select $\hat{\theta}$ such that the sum square error over a number of samples is minimized, i.e. minimize $\mathbf{J}_N$ where

$$\mathbf{J}_N = \sum_{i=1}^{N} \mathbf{e}^2(k)$$

To do this, it is most convenient to use matrix notation, remembering that $\{\mathbf{y}(k)\}$ and $\{\mathbf{u}(k)\}$ are known to us. Define

$$\mathbf{Y}_N = \begin{bmatrix} \mathbf{y}(1) \\ \mathbf{y}(2) \\ \mathbf{y}(3) \\ \vdots \\ \mathbf{y}(N) \end{bmatrix} \qquad \Phi_N = \begin{bmatrix} \phi(1) \\ \phi(2) \\ \phi(3) \\ \vdots \\ \phi(N) \end{bmatrix}$$

where $\mathbf{Y}_N$ is $N \times 1$ and Φ_N is $N \times (n+m+1)$. Then clearly $\mathbf{J}_N$ can be written as

$$J_N = [\, \mathbf{Y}_N - \Phi_N \hat{\theta} \,]^T [\, \mathbf{Y}_N - \Phi_N \hat{\theta} \,] \tag{30.1}$$

To minimize this with respect to the unknown $\hat{\theta}$, we take the first derivative, or gradient, with respect to $\hat{\theta}$ and set it to 0. Then

$$\nabla_\theta J_N = 0 = -2\Phi_N^T [\, \mathbf{Y}_N - \Phi_N \hat{\theta} \,]$$

which has the solution, when the inverse exists (otherwise use a pseudo-inverse)

$$\hat{\theta} = (\Phi_N^T \Phi_N)^{-1} \Phi_N^T \mathbf{Y} \tag{30.2}$$

and in any case gives the normal equations of least squares. We notice, incidentally, that

$$\frac{\partial^2 J_N}{\partial \hat{\theta}^2} = 2\Phi_N^T \Phi_N \geq 0$$

so that we indeed have a minimum.

Example

Suppose we have the data in the following table:

k	1	2	3	4	5
$\mathbf{y}(k)$	1	1.25	0.75	0	1.5
$\mathbf{u}(k)$	0	1	0.1667	−0.5	2

We conjecture that there is one lagged value of $\mathbf{y}$ and the present value of $\mathbf{u}$ contributing to the output, i.e. that coefficients $a(1)$ and $b(0)$ are to be found. Using the data, we can set up that

$$
\begin{bmatrix} 1.25 \\ 0.75 \\ 0 \\ 1.5 \end{bmatrix} - \begin{bmatrix} -1 & 1 \\ -1.25 & 0.1667 \\ -0.75 & -0.5 \\ -0 & 2 \end{bmatrix} \begin{bmatrix} a(1) \\ b(0) \end{bmatrix} = \mathbf{Y}_N - \Phi_N \hat{\theta}
$$

is the error to be minimized. Then

$$
\Phi_N^\mathrm{T} \Phi_N = \begin{bmatrix} 3.125 & 0.833 \\ 0.833 & 5.278 \end{bmatrix} \qquad \Phi_N^\mathrm{T} \mathbf{Y}_N = \begin{bmatrix} -2.1875 \\ 4.375 \end{bmatrix}
$$

and the result is that

$$
\hat{\theta} = \begin{bmatrix} -0.5 \\ 0.75 \end{bmatrix}
$$

so that $\hat{a}(1) = -0.5$ and $\hat{b}(0) = 0.75$. It is easy to calculate that the resulting $\mathbf{J}$ is approximately $\mathbf{0}$, indicating a good data fit.

30.3.2 Least-squares data fits for transfer function models: recursive form

The above can be done in recursive form, which is particularly handy in real time. This form is derived below using matrix manipulation.

When more data are taken after an estimate $\hat{\theta}$ is computed, one can simply recompute the estimate using (30.2) with the new value of N. An alternative is to combine the old estimate with the new data in a manner which can decrease the amount of computation needed. To see this, we define $\hat{\theta}_N$ as the estimate after N data points $\{\mathbf{y}(k), \mathbf{u}(k)\}$ are taken:

$$
\hat{\theta}_N = (\Phi_N^\mathrm{T} \Phi_N)^{-1} \Phi_N^\mathrm{T} \mathbf{Y}_N \tag{30.3}
$$

Now consider taking one more data point. We have

$$\hat{\theta}_{N+1} = (\Phi_{N+1}^{T}\Phi_{N+1})^{-1}\Phi_{N+1}^{T}\,\mathbf{Y}_{N+1} \tag{30.4}$$

We note, however, that

$$\Phi_{N+1} = \begin{bmatrix} \Phi_N \\ \cdots \\ \phi_{N+1} \end{bmatrix}$$

$$\mathbf{Y}_{N+1} = \begin{bmatrix} \mathbf{Y}_N \\ \cdots \\ \mathbf{Y}_{N+1} \end{bmatrix}$$

Hence by substitution in (30.4) we find

$$\hat{\theta}_{N+1} = (\Phi_N^{T}\Phi_N + \phi^{T}(N+1)\,\phi(N+1))^{-1}(\Phi_N^{T}\mathbf{Y}_N + \phi(N+1)\,\mathbf{y}(N+1))$$

Using a standard matrix inversion identity (Appendix B.1) which states

$$(\mathbf{A} + \mathbf{B}\mathbf{B}^{T})^{-1} = \mathbf{A}^{-1} - \mathbf{A}^{-1}\mathbf{B}(\mathbf{I} + \mathbf{B}^{T}\mathbf{A}^{-1}\mathbf{B})^{-1}\mathbf{B}^{T}\mathbf{A}^{-1} \tag{30.5}$$

and identifying $\mathbf{A} = (\Phi_N^{T}\Phi_N)$ and $\mathbf{B} = \phi^{T}(N+1) = \phi^{T}$ gives

$$\hat{\theta}_{N+1} = (\Phi_N^{T}\Phi_N)^{-1}\,\Phi_N^{T}\mathbf{Y}_N + (\Phi_N^{T}\Phi_N)^{-1}\phi^{T}\,\mathbf{y}(N+1)$$

$$- (\Phi_N^{T}\Phi_N)^{-1}\phi^{T}(1 + \phi(\Phi_N^{T}\Phi_N)^{-1}\phi^{T})^{-1}\phi(\Phi_N^{T}\Phi_N)^{-1}\phi^{T}\,\mathbf{y}(N+1)$$

$$- (\Phi_N^{T}\Phi_N)^{-1}\phi^{T}(1 + \phi(\Phi_N^{T}\Phi_N)^{-1}\phi^{T})^{-1}\phi(\Phi_N^{T}\Phi_N)^{-1}\Phi_N\,\mathbf{Y}_N$$

Noticing (30.3) and using a small amount of manipulation yield

$$\hat{\theta}_{N+1} = \hat{\theta}_N + \frac{(\Phi_N^{T}\Phi_N)^{-1}\phi^{T}}{1 + \phi(\Phi_N^{T}\Phi_N)^{-1}\phi^{T}}\,[\mathbf{y}(N+1) - \phi\hat{\theta}_N]$$

It is convenient to make the definition

$$\mathbf{P}_N = (\Phi_N^T \Phi_N)^{-1}$$

and then to notice that from the identity (30.5) it follows easily that

$$\hat{\theta}_{N+1} = \hat{\theta}_N + \frac{\mathbf{P}_N \phi^T}{1 + \phi \mathbf{P}_N \phi^T} [y(N+1) - \phi\hat{\theta}_N]$$

$$\mathbf{P}_{N+1} = \mathbf{P}_N - \mathbf{P}_N \phi^T (1 + \phi\mathbf{P}_N\phi^T)^{-1} \phi\mathbf{P}_N$$

$$= \mathbf{P}_N - \frac{\mathbf{P}_N \phi^T \phi \mathbf{P}_N}{1 + \phi\mathbf{P}_N\phi^T}$$

where

$$\phi = \phi(N+1)$$

$$= [\, -y(N) \;\; -y(N-1) \;\; ... \;\; -y(N-n) \;\; u(N+1) \;\; u(N) \;\; ... \;\; u(N+1-m) \,]$$

This constitutes a recursive algorithm for $\hat{\theta}$. Initial conditions may be derived by taking a small batch of data at the start and using the estimate (30.3) and its associated $\mathbf{P}_N$ matrix, or alternatively by selecting a guessed $\hat{\theta}$ along with an initial $\mathbf{P}$ matrix; e.g. the latter may be a diagonal matrix with large elements.

30.3.3 Variations – weighted least squares

Of the possible variations to least-squares fits, we mention the weighted least squares method. Here, the cost function J allows the errors in various predictions to have differing weights, so that

$$\mathbf{J}_N = \sum_{i=1}^{N} w(i)\, \mathbf{e}^2(i) \tag{30.6}$$

Defining the matrix

$$\mathbf{W} = \text{diag}\,(w(1), w(2), \ldots, w(N))$$

we find (30.1) becomes

$$\mathbf{J}_N = [\mathbf{Y}_N - \mathbf{\Phi}_N \hat{\theta}]^{\mathrm{T}}\,\mathbf{W}\,[\mathbf{Y}_N - \mathbf{\Phi}_N \hat{\theta}]$$

and the solution (30.2) becomes

$$\hat{\theta} = (\mathbf{\Phi}_N^{\mathrm{T}} \mathbf{W} \mathbf{\Phi}_N)^{-1} \mathbf{\Phi}_N^{\mathrm{T}} \mathbf{W} \mathbf{Y}_N$$

One particularly attractive version of this, philosophically, is the case in which $w(i) = \lambda^{N-i}$ with $0 < \lambda \le 1$. Here λ is a 'forgetting factor', as it is clear from (30.6) that early errors have less influence on $\mathbf{J}$ than those with larger indices. This allows the technique to incorporate slowly varying parameters and, in the recursive case, to 'forget' initial condition errors, particularly in $\mathbf{P}_0$. It is easy to incorporate this type of weighting into the recursive form, which can be shown using the methods of the previous section to become

$$\hat{\theta}_{N+1} = \hat{\theta}_N + \frac{\mathbf{P}_N \phi^{\mathrm{T}}}{\lambda + \phi \mathbf{P}_N \phi^{\mathrm{T}}}\,[y(N+1) - \phi \hat{\theta}_N]$$

$$\mathbf{P}_{N+1} = \left[\mathbf{P}_N - \frac{\mathbf{P}_N \phi^{\mathrm{T}} \phi \mathbf{P}_N}{\lambda + \phi \mathbf{P}_N \phi^{\mathrm{T}}}\right]\frac{1}{\lambda}$$

where now $\mathbf{P}_N$ denotes $(\mathbf{\Phi}_N^{\mathrm{T}} \mathbf{W} \mathbf{\Phi}_N)^{-1}$.

The above forgetting factor uses a time-varying weighting matrix $\mathbf{W}$, as can be seen by the nature of the weighting factor $\lambda^{N-i} = w(i)$. If we use the matrix

$$\mathbf{W}_N = \text{diag}\,(w(1), w(2), \ldots, w(N))$$

then the recursive form can be shown to be

$$\hat{\theta}_{N+1} = \hat{\theta}_N + \frac{\mathbf{P}_N \phi^{\mathrm{T}}}{\frac{1}{w(N+1)} + \phi \mathbf{P}_N \phi^{\mathrm{T}}} [\mathbf{y}(N+1) - \phi \hat{\theta}_N]$$

$$\mathbf{P}_{N+1} = \left[\mathbf{P}_N - \frac{\mathbf{P}_N \phi \phi^{\mathrm{T}} \mathbf{P}_N}{\frac{1}{w(N+1)} + \phi \mathbf{P}_N \phi^{\mathrm{T}}} \right]$$

where $\mathbf{P}_N$ denotes $(\Phi_N^{\mathrm{T}} \mathbf{W}_N \Phi_N)^{-1}$. We remark that both weighting schemes have the intuitively satisfying characteristic that 'disbelieving' recent data in the sense of a small $w(N+1)$ or large λ yields a 'small' gain and $\hat{\theta}_{N+1} \approx \hat{\theta}_N$.

30.3.4 Kalman filtering for parameters when state measured

Suppose we have a system with input sequence $\{\mathbf{u}(k)\}$ and output sequence $\{\mathbf{y}(k)\}$ and hypothesize that they are related by

$$\mathbf{y}(k) = -\sum_{i=1}^{n} a_i \mathbf{y}(k-i) + \sum_{j=0}^{m} b_j \mathbf{u}(k-j)$$

with n and m known. Define a state vector

$$\Theta(k) = [\, a_1 \quad a_2 \quad \dots \quad a_n \quad b_0 \quad \dots \quad b_m \,]^{\mathrm{T}}$$

and an estimated state $\hat{\Theta}(k)$

$$\hat{\Theta}(k) = [\hat{a}_1(k) \quad \hat{a}_2(k) \quad \dots \quad \hat{a}_n(k) \quad \hat{b}_0(k) \quad \dots \quad \hat{b}_m(k)\,]^{\mathrm{T}}$$

where $\hat{a}_1(k)$ is the estimate of the unknown parameter a_i at step k, etc. Further, define a measurement matrix $\mathbf{C}(k)$ by

$$\mathbf{C}(k) = [\, -\mathbf{y}(k-1) \quad \dots \quad -\mathbf{y}(k-n) \quad \mathbf{u}(k) \quad \dots \quad \mathbf{u}(k-m)\,]^{\mathrm{T}}$$

Then in Kalman filter terms we have dynamics

$$\Theta(k) = \mathbf{I}\Theta(k-1)$$

with scalar measurements

$$\mathbf{y}(k) = \mathbf{C}^{\mathrm{T}}(k)\,\Theta(k)$$

Then a recursive filter for θ is, if we assume some noise on the measurement model (not unreasonable) of variance r and on the state model (less reasonable) of covariance matrix $\mathbf{Q}$,

$$\hat{\Theta}(k) = \hat{\Theta}(k-1) + \mathbf{K}(k)\,(\mathbf{y}(k) - \mathbf{C}^{\mathrm{T}}(k)\,\hat{\Theta}(k-1))$$

$$\mathbf{K}(k) = \Sigma(k)\,\mathbf{C}(k)\,r^{-1}$$

$$\Sigma(k) = [(\Sigma(k-1) + \mathbf{Q})^{-1} + \mathbf{C}(k)\,\mathbf{C}^{\mathrm{T}}(k)/r]^{-1}$$

Presented this way, it is arguable that choosing $\mathbf{Q}$ is an art form, but in fact we may let $\mathbf{Q} \to \mathbf{0}$ without too much trouble provided that the initial estimate $\Sigma(0)$ of $\mathrm{cov}(\hat{\Theta}(0))$ is reasonably large. This technique amounts to a recursive least-squares estimate of Θ in this context.

30.3.5 An adaptive observer for on-line parameter estimation with state measurements

Another approach to parameter identification of multiple-input–multiple-output (MIMO) systems is based upon observer ideas and illustrates also the use of Lyapunov functions. We hypothesize a state-variable form for the system

$$\mathbf{x}(k+1) = \mathbf{A}\mathbf{x}(k) + \mathbf{B}\mathbf{u}(k)$$

$$\mathbf{y}(k) = \mathbf{I}\mathbf{x}(k)$$

and attempt to determine $\mathbf{A}$, and $\mathbf{B}$ from input–output data, i.e. the sequences $\{\mathbf{u}(k)\}$ and $\{\mathbf{x}(k)\}$. Thus define (Kudva and Narendra, 1974)

$$\hat{\mathbf{A}}(k) = \text{estimate of } \mathbf{A} \text{ at time } k$$

$$\hat{\mathbf{B}}(k) = \text{estimate of } \mathbf{B} \text{ at time } k$$

$$\theta = [\, \mathbf{A} \,\vdots\, \mathbf{B} \,]$$

$$\hat{\theta}(k) = [\, \hat{\mathbf{A}}(k) \,\vdots\, \hat{\mathbf{B}}(k) \,]$$

$$\mathbf{z}(k) = \begin{bmatrix} \mathbf{x}(k) \\ \cdots\cdots \\ \mathbf{u}(k) \end{bmatrix}$$

$$\hat{\mathbf{x}}(k+1) = \hat{\mathbf{A}}(k+1)\mathbf{x}(k) + \hat{\mathbf{B}}(k+1)\mathbf{u}(k)$$

$$\mathbf{e}(k) = \hat{\mathbf{x}}(k) - \mathbf{x}(k)$$

$$\Theta(k) = \hat{\theta}(k) - \theta$$

Using this, and noting that $\mathbf{z}(k)$ is a known quantity, we find

$$\mathbf{e}(k+1) = \Theta(k+1)\,\mathbf{z}(k)$$

For application of the Lyapunov theory, we define the functions

$$V(k) = \text{tr}\,[\Theta(k)\,\Theta^{\mathrm{T}}(k)]$$

$$= \sum_{i,j} (\hat{a}_{i,j} - a_{i,j})^2 + \sum_{i,j} (\hat{b}_{i,j} - b_{i,j})^2$$

$$\Delta V(k) = V(k+1) - V(k)$$

We wish to find a scheme by which the quadratic $V(k) \rightarrow 0$ as k becomes large. We propose the form

$$\hat{\theta}(k+1) = \hat{\theta}(k) - \alpha\, \frac{\mathbf{P}\mathbf{e}(k)\,\mathbf{z}^{\mathrm{T}}(k-1)}{\mathbf{z}^{\mathrm{T}}(k-1)\,\mathbf{z}(k-1)} \tag{30.7}$$

and attempt to find a scalar α and matrix $\mathbf{P}$ such that $\Delta V(k) < 0$ for all k. With considerable algebra, it can be shown that

$$\Delta V(k) = \alpha \, \frac{\mathbf{e}^{\mathrm{T}}(k)\,[\alpha\mathbf{P}-2\mathbf{I}]\,\mathbf{P}\mathbf{e}(k)}{\mathbf{z}^{\mathrm{T}}(k-1)\,\mathbf{z}(k-1)}$$

This is less than or equal to zero provided $\alpha\,[\alpha\mathbf{P}-2\mathbf{I}]\,\mathbf{P}$ is negative semidefinite, or equivalently

1. $\quad 0 < \alpha < 2$

2. $\quad \mathbf{P} > \mathbf{0}$ (positive definite)

3. $\quad \|\mathbf{P}\| < 1$ $\hfill$ (30.8)

Thus under conditions (30.8) the algorithm (30.7) will find $\mathbf{A}$ and $\mathbf{B}$ in the sense that $\hat{\mathbf{A}}(k) \to \mathbf{A}$ and $\hat{\mathbf{B}}(k) \to \mathbf{B}$.

A potential application of this was presented by Woo and Rootenberg (1975), but use of real data for their problem presented difficulties for Bates (1976).

30.3.6 The general form for recursive parameter estimation

For the many forms of parameter estimation for linear systems, with various approaches, it has been argued (Ljung and Söderström, 1983) that a general form is possible. The basic structure is

$$\hat{\theta}(k) = \hat{\theta}(k-1) + \gamma(k)\,\mathbf{P}(k)\,\eta(k)\,[\mathbf{y}(k) - \hat{\mathbf{y}}(k)]$$

in which $\hat{\theta}(k)$ is the parameter estimate at time k, $\mathbf{y}(k)$ is the measured data, and $\hat{\mathbf{y}}(k)$ is the predicted data based upon the measurements up to time $k-1$. The technique-dependent features are the scalar sequence $\{\gamma(k)\}$, which should go to zero or a small value as k increases, the matrix $\mathbf{P}(k)$ derived from measurements up to time k, and the vector (if $\mathbf{y}(k)$ is a scalar) or matrix $\eta(k)$ which is derived from the measurements and typically is related to the gradient of $\hat{\mathbf{y}}$ with respect to $\hat{\theta}$. We may readily identify in the recursive least-squares approach, for example, that

$$\hat{\mathbf{y}}(k) = \phi(k)\,\hat{\theta}_{k-1}$$

$$\gamma(k) = 1$$

$$\eta(k) = \phi^{\mathrm{T}}(k)$$

$$\mathbf{P}(k) = \frac{(\Phi_{k-1}^{\mathrm{T}}\Phi_{k-1})^{-1}}{1 + \phi(k)\,(\Phi_{k-1}^{\mathrm{T}}\Phi_{k-1})^{-1}\phi^{\mathrm{T}}(k)}$$

The basic recursive algorithm in the reference is

$$\varepsilon(k) = \mathbf{y}(k) - \hat{\mathbf{y}}(k)$$

$$\hat{\Lambda}(k) = \hat{\Lambda}(k-1) + \gamma(k)\,[\varepsilon(k)\,\varepsilon^{\mathrm{T}}(k) - \hat{\Lambda}(k-1)]$$

$$\mathbf{R}(k) = \mathbf{R}(k-1) + \gamma(k)\,[\psi(k)\,\hat{\Lambda}^{-1}(k)\,\psi^{\mathrm{T}}(k) - \mathbf{R}(k-1)]$$

$$\hat{\theta}(k) = \hat{\theta}(k-1) + \gamma(k)\mathbf{R}^{-1}(k)\,\psi(k)\,\hat{\Lambda}^{-1}(k)\,\varepsilon(k)$$

$$\xi(k+1) = \mathbf{A}(\hat{\theta}(k))\,\xi(k) + \mathbf{B}(\hat{\theta}(k))\,\mathbf{z}(k)$$

$$\begin{bmatrix} \hat{\mathbf{y}}(k+1) \\ \mathrm{col}\,\psi(k+1) \end{bmatrix} = \mathbf{C}(\hat{\theta}(k))\,\xi(k+1)$$

A rough interpretation of this – helpful here is Ljung (1981) – with details depending upon the version of the algorithm, is that

$$\mathbf{y}(k) = \text{measurement vector}$$

$$\hat{\mathbf{y}}(k) = \text{predicted measurement vector}$$

$$\hat{\theta}(k) = \text{parameter estimate vector; dimension } n_p$$

$$\mathbf{z}(k) = \begin{bmatrix} \mathbf{y}(k) \\ \mathbf{u}(k) \end{bmatrix}$$

$$\mathbf{u}(k) = \text{system input (or command) vector}$$

$$\psi(k) = \text{matrix of } \frac{\mathrm{d}\hat{\mathbf{y}}^{\mathrm{T}}(k)}{\mathrm{d}\theta_i}\,,\ i = 1, 2, \dots , n_p$$

$$\xi(k) = \text{filter for } \hat{y}(k) \text{ and } \psi(k)$$

$$\mathbf{R}(k) = \text{search direction adjustment matrix}$$

$$\Lambda(k) = \text{relative error weighting matrix}$$

$\mathbf{A}(k), \mathbf{B}(k) = $ generalized dynamics equation matrices, derived from system dynamics, but extended to generalized state $\xi(k)$.

A considerable amount is known about the convergence of such algorithms and their properties. It is presented here only to guide the reader into this branch of the literature.

30.3.7 Partitioned adaptive filters

An interesting variation which keeps arising, partly because computer power is becoming so inexpensive, is the partitioned adaptive filter approach. In such systems, the philosophy is straightforward even if some of the theory is not.

Assume the system is modelled by a form such as

$$\mathbf{x}(k+1) = \mathbf{A}_\theta \mathbf{x}(k) + \mathbf{B}_\theta \mathbf{u}(k) + \mathbf{G}_\theta \mathbf{v}(k)$$

$$\mathbf{y}(k) = \mathbf{C}_\theta \mathbf{x}(k) + \mathbf{w}(k)$$

where the defintions are the usual Kalman filter ones (Chapter 28), with state vector $\mathbf{x}$, input $\mathbf{u}$, measurements $\mathbf{y}$, and noises $\mathbf{v}$ and $\mathbf{w}$. If θ were known precisely, there would be no problem in forming a state estimate $\hat{\mathbf{x}}$ of $\mathbf{x}$ using a Kalman filter, and it would be expected that the errors $\mathbf{y}(k) - \mathbf{C}_\theta \hat{\mathbf{x}}(k|k-1)$ would be 'small' and white noise-like. If θ is not known, then a Kalman filter based upon a guessed value $\hat{\theta}$ would have such errors be large or small depending upon how good the guess is. The technique may be summarized as follows: set up several Kalman filters, each designed based upon a different parameter vector $\theta_i \in \Theta$, a set of candidates, and select as the correct estimate of θ that vector whose filter gives the smallest, whitest errors.

The algorithm is basically, for N vectors $\theta_i \in \Theta$, as follows.

1. Initialize: for $r = 1, 2, \ldots, N$

$$\mathbf{P}_{\theta_r}(0|-1) = \text{cov}(\mathbf{x}(0))$$

$$\mathbf{p}(\theta_r | \ldots) = \frac{1}{N}$$

$$\hat{\mathbf{x}}_{\theta_r}(0) = \mathbf{0}$$

2. Kalman filter: for $r = 1, 2, \ldots, N$

$$\mathbf{K}_{\theta_r}(k) = \mathbf{P}_{\theta_r}(k|k-1)\,\mathbf{C}_{\theta_r}^{\mathrm{T}}[\mathbf{C}_{\theta_r}\mathbf{P}_{\theta_r}(k|k-1)\mathbf{C}_{\theta_r}^{\mathrm{T}} + \mathbf{R}]^{-1}$$

$$\mathbf{P}_{\theta_r}(k|k) = \mathbf{P}_{\theta_r}(k|k-1) - \mathbf{K}_{\theta_r}(k)\,\mathbf{C}_{\theta_r}\mathbf{P}_{\theta_r}(k|k-1)$$

$$\mathbf{P}_{\theta_r}(k+1|k) = \mathbf{A}_{\theta_r}\mathbf{P}_{\theta_r}(k|k)\,\mathbf{A}_{\theta_r}^{\mathrm{T}} + \mathbf{G}_{\theta_r}\mathbf{Q}_{\theta_r}\mathbf{G}_{\theta_r}^{\mathrm{T}}$$

$$\hat{\mathbf{x}}_{\theta_r}(k|k-1) = \mathbf{A}_{\theta_r}\mathbf{x}_{\theta_r}(k-1) + \mathbf{B}_{\theta_r}\mathbf{u}(k-1)$$

$$\hat{\mathbf{x}}_{\theta_r}(k) = \hat{\mathbf{x}}_{\theta_r}(k|k-1) + \mathbf{K}_{\theta_r}(\mathbf{y}(k) - \mathbf{C}_{\theta_r}\hat{\mathbf{x}}_{\theta_r}(k|k-1))$$

3. Non-linear decision mechanism
   (a) Calculate likelihood ratios, for $r = 1, 2, \ldots, N$

$$\mathbf{e}_r(k) = \mathbf{y}(k) - \mathbf{C}_{\theta_r}\hat{\mathbf{x}}_{\theta_r}(k)$$

$$\mathbf{P}_{\mathbf{e}_r} = \mathbf{C}_{\theta_r}\mathbf{P}_{\theta_r}(k|k-1)\mathbf{C}_{\theta_r}^{\mathrm{T}} + \mathbf{R}$$

$$\mathbf{L}_{\theta_r} = |\mathbf{P}_{\theta_r}(k|k-1)|^{-\frac{1}{2}} \exp\left[-\tfrac{1}{2}\| \mathbf{e}_r(k) \|^2_{\mathbf{P}_{\mathbf{e}_r}}\right]$$

(b) Calculate *a posteriori* probability densities, for $r = 1, 2, \ldots, N$

$$p(\theta_r | \mathbf{Y}(k)) = \frac{\mathbf{L}_{\theta_r}}{\displaystyle\sum_{i=1}^{N} \mathbf{L}_{\theta_r} p(\theta_i | \mathbf{Y}(k-1))}\, p(\theta_r | \mathbf{Y}(k-1))$$

If desired, a state estimate may be calculated as

$$\hat{\mathbf{x}}(k) = \sum_{i=1}^{N} p(\theta_i \mid \mathbf{Y}(k)) \hat{\mathbf{x}}_{\theta_r}(k)$$

Given this simple idea, which is in some ways like the matched filter methods of communications, it is possible to show that $p(\theta_i \mid \mathbf{Y}(k)) \to 1$ for the 'closest' θ_i to the actual θ. Furthermore it is straightforward to manipulate the set Θ by adding and deleting members until a good estimate of θ is found; this is done by replacing θ_i for which $p(\theta_i \mid \mathbf{Y}(k)) < \varepsilon$ with a different and hopefully better value and restarting the process.

The partitioned adaptive filter is traceable to Lainiotis (1976). A discussion under the name multimodel identification is presented by Anderson and Moore (1979). Search algorithms were added by Lamb and Westphal (1979).

30.4 COMPUTER AIDS

There are system identification packages with MATLAB® and other commercial software. These tend to contain algorithms of the recursive least-squares family.

30.5 SUMMARY AND FURTHER READING

We have only touched the surface of a small part of the systems identification topic. There are a number of books on the topic, dating back to the 1970s; among the more recent works are those of Ljung, including Ljung (1981) and Ljung and Söderström (1983). Quite readable is the book by Graupe (1972). In addition, texts such as that by Franklin, Powell and Workman (1990) have chapters on identification. The issue is also a portion of one approach to the problem of adaptive control, discussed in Chapter 31, and is being attacked by the methods of artificial neural networks (Chapter 32).

31

Adaptive and self-tuning control

Feedback control was introduced for the purpose of compensating for variability in components and raw materials, so that the quality of product output could be improved. A properly designed feedback control system is expected to be rather insensitive to plant errors, component drift, and input disturbances and variability. Furthermore, the system should be stable, robust, and cost effective. Traditional techniques have usually been satisfactory for these purposes.

In some special circumstances, there is a desire for accuracy or productivity that is difficult to provide with traditional fixed controller methods. In a space launch, for example, the variability in rocket motors, even if only a per cent or so of thrust acceleration, may lead to a several $m\,s^{-1}$ error in orbital injection velocity which would take considerable satellite fuel to correct the resulting orbit. In a ship autopilot, the variability in operating conditions of the ship may mean that a loss of several per cent in either speed or fuel will result unless the autopilot is carefully tuned to exact water depth, wave conditions, wind conditions, and desired speed. In a number of processes, raw material variability may make the output product vary by an undesirable amount if a fixed controller setting is used. Even the time taken to tune controllers on plant start-up can induce an expense which operators would like to eliminate if possible by making the loops self-tuning.

For reasons such as the above, there has been interest for a number of years in so-called adaptive control, in which the controller appears to recognize the actual plant parameters and adjust itself accordingly. The increasing availability of relatively inexpensive computing power has made such schemes more and more attractive financially, and control engineers find the concepts stimulating and entertaining, with a flavour of artificial intelligence and expert systems drifting into the field from other specialties. We should be aware of two facts, however.

1. Performance improvements have been shown to be obtainable, but sometimes the amount of improvement is of the order of a few per cent; larger improvements may be due more to inadequacies or oversimplifications of the original fixed laws than to the introduction of adaptive laws.
2. The schemes ultimately reduce to a rather complicated non-linear control law for which stability and robustness may not be guaranteed.

With these comments always at the back of our mind, we introduce adaptive control in this chapter. Because adaptive control theory is the meeting point of many of the other aspects of control theory, particularly control law design, non-linear stability, and state and parameter estimation, there exist many schemes for 'adapting' the control to the situation. We look at only a few here, to give the flavour of the specialty.

31.1 SYNOPSIS

There are many different approaches to adaptive control, and in fact there are many different definitions of adaptive control. In addition, there are several ways of classifying adaptive control approaches. One useful distinction is between direct adaptive controllers, in which control law parameters are determined directly from the sensed data, and indirect laws, in which the system is identified and the control law parameters are determined in terms of the identified mode. An alternative used here is to group and discuss several approaches on the basis of the system structure.

Gain scheduling is a possibility when a single set of parameter values is not sufficient for the entire range of operating conditions. The concept is straightforward and arguably open-loop.

More advanced are **optimalizing schemes**, which assume that for each set of operating conditions, there is an optimal gain set. The controller then searches for this optimum. Thus an index of performance (IP) is chosen and assumed to vary with operating conditions. The controller searches systematically for the optimum.

We have already studied various control law design methods, both classical and modern, both *ad hoc* and optimal, and found that they are in principle straightforward to find if the controlled plant is well modelled. We have also looked at system identification techniques. It seems obvious that if we identify our system in its operating regime, we can then use that information to choose controller parameters. That is the principle of the adaptive approach often called **self-tuning**. Self-tuning amounts to a separation of the adaptive problem into two obvious parts: identification and control. The problem is to be certain these do not interact strangely.

Model reference automatic control (MRAC) can be implemented in a number of ways. The idea is to have a model of the plant and to attempt to make the actual plant behave in this way. Hence an aircraft controller may have a desired 'feel' to it, in that pilots like a certain stick deflection to produce a certain attitude rate; the gains are adjusted so that the aircraft has this feel, regardless of the actual dynamics. There are many forms of model reference controllers, of which Fig. 31.6 shows only a basic one. The heart of the system, of course, is the 'controller adjuster'.

The final approach we mention is the optimal control with uncertain parameters, or **stochastic optimal control**, approach. This is in some ways the purest adaptive controller, but results are few. Its block diagram is the simplest because the structure is at its core a *gestalt* – a unified one.

In the following sections, all of the above ideas are pursued further.

31.2 DEFINITIONS AND CONCEPTS

'Adaptive' is not a rigorously defined term, although many authors have made attempts. When all is said, done, and analysed, most adaptive controllers either make internal changes to their parameters (such as gains) or change plant parameters (such as damping) in response to changes in operating conditions (such as aircraft flight regime – subsonic high altitude, landing, etc.). As such, they can be considered just another kind of control law, albeit perhaps non-linear and complicated.

The usual definition assumes certain controller parameters are to be selected by the 'adaptation' process. This adaptation is associated with information processing, and an adaptive controller is one which 'learns'. The conditions are non-*a priori* quantities in probability densities associated with the plant or measurement subsystems. With this in mind we consider a few of the many proposed definitions, including a formal mathematical definition (item 6).

1. Adaptive in common usage means 'tending to fit [itself] to new or different conditions' (Davis, 1970).

2. An adaptive system is one which measures its performance relative to a given index of performance and modifies its parameters to approach an optimum set of values (Eveleigh, 1967).
3. An adaptive system is any physical system which has been designed with an adaptive viewpoint (Truxal, 1961).
4. Adaptation is the ability of self-adjustment or self-modification in accordance with changing conditions of environment or structure (Aseltine *et al.*, 1958).
5. Adaptation is a process of continual optimization in the face of complex and time-varying environmental effects (Jarvis, 1975).
6. A system A is adaptive with respect to a family $\{S_g\}$, $g \in G$, of input sets $S_g = \{U(t)\}$, where each $U(t)$ is a vector of inputs to A, if the performance function $P(g)$ obtained when S_g is applied to A is acceptable, i.e. belongs to the set W of acceptable performance function values. More briefly, A is adaptive with respect to G and W if it maps G into W.

    Thus a process/controller system is called adaptive with respect to a specified set of environmental changes, external disturbances, etc., if its performance function is acceptable for all such changes.

Without defining adaptive directly, Tsypkin (1971) described some characteristic features of adaptation.

> The most characteristic feature of adaptation is an accumulation and a slow usage of the current information to eliminate the uncertainty due to insufficient *a priori* information and for the purpose of optimizing a certain selected performance index.

We see from the range of definitions and properties – from an adaptive system is 'one that performs satisfactorily' through 'one designed from an adaptive point of view' to 'one which measures and tries to optimize its index of performance over a set of parameters' – that the issue is undecided. We tend to like a definition which includes 'learning' about the system, but for our purposes perhaps Truxal's definition is as good as any and better than some.

An input to a partly known plant will often demonstrate the **dual effect** first cited by Fel'dbaum (1965):

1. it will cause the output to move toward a desired value, and
2. it will cause the output to change in a manner which, if measured,

will allow us to determine something about the unknown parameters of the system.

In many systems, this requires a control law trade-off, as explicit dual control is almost impossible to do optimally. It will frequently require the occasional application of a probing signal to aid the adaptation.

In the following sections, we will elaborate upon several adaptive control schemes. They are presented as if the boundaries between them are distinct, but of course they are not. In addition, some are more or less adaptive depending upon the definition of 'adaptive' which is used.

31.3 GAIN SCHEDULING

All engineers should be aware of gain scheduling as a possibility when a single set of parameter values is not sufficient for the entire range of operating conditions. The concept is straightforward, as shown in Fig. 31.1.

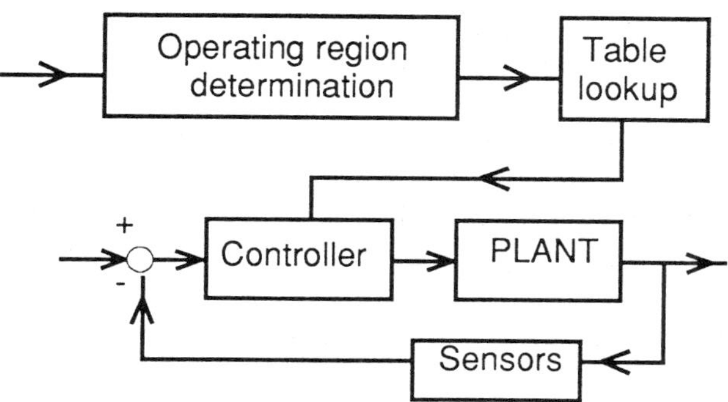

Figure 31.1 Structure of control law for adaptation using gain scheduling.

Gain scheduling can be seen to be one step removed from having gains selected from tables by human operators; the switching is done on the basis of sensed data. Conceptually, the operating regimes are based upon a partitioning of the state space (or more properly measurement space), as in Fig. 31.2.

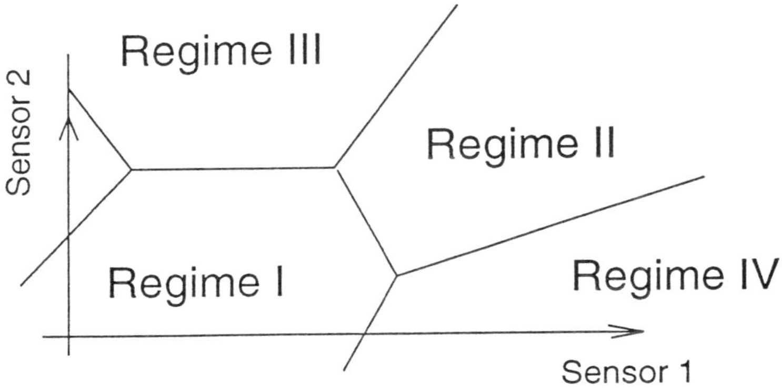

Figure 31.2 Operating regimes using different gains with gain scheduling.

We try to clarify and motivate the idea of gain schedules using aircraft autopilots as an example. It is no surprise to find that the autopilot and other gains are very different, perhaps crucially so, between a landing configuration (gear and flaps down, altitude and speed low, etc.) and a cruise configuration (gear and flaps up, altitude high, speed fairly high, etc.). In the latter configuration, for example, the airframe will respond much more rapidly to a given elevator deflection, so to keep pilot 'stick feel' constant, the stick-to-elevator gain must be decreased in going from landing configuration to cruise configuration. A computer can be programmed to check altitude, attitude, airspeed, throttle, etc. select the proper gains from a table stored in memory for that configuration, and implement those gains in the autopilot loops.

We remark that the scheme is little different in principle from having human operators change the gains according to tables of operating conditions, and it has been argued that gain scheduling is not truly adaptive.

31.4 OPTIMALIZING CONTROLS

Optimalizing controls, also called extremum adaptive controls, had their principles first presented by Draper and Li (1951) and Li (1952) and have been extensively examined. The underlying concept is that the available controller parameters are adjusted to find an extremum (optimum) of an index of performance (IP).

Optimalizing schemes assume that for each set of operating conditions, there is an optimal gain set. The controller then searches for this optimum. Thus an IP is chosen and assumed to have the form such as Fig. 31.3 for each possible operating condition.

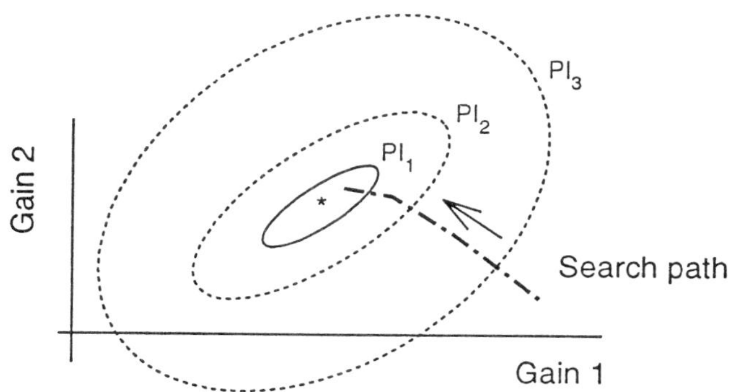

Figure 31.3 Contours of an Index of Performance PI, $PI_1 < PI_2 < PI_3$, showing a search path for the miniminum. The contours may not be known explicitly, in which case the controller must 'bindly' seek decreases in the cost.

In the simplest versions, in which adjustment uses gradients, the partial derivatives $\partial IP/\partial \theta_i$ are found, perhaps numerically by perturbation and measurement, and then, for a gain K,

$$\theta(m+1) = \theta(m) + K\nabla_{\theta_i} IP$$

For example, the IP might be

$$IP = \sum_m (r(m) - y(m))^2$$

and the derivatives found by perturbation of the parameters successively, with the sum taken over some number M of samples. In fact, the early implementations were more subtle and used more basic engineering knowledge of the underlying system than this. One famous example was the Draper and Li (1951) carburettor adjuster, with structure as shown in Fig. 31.4.

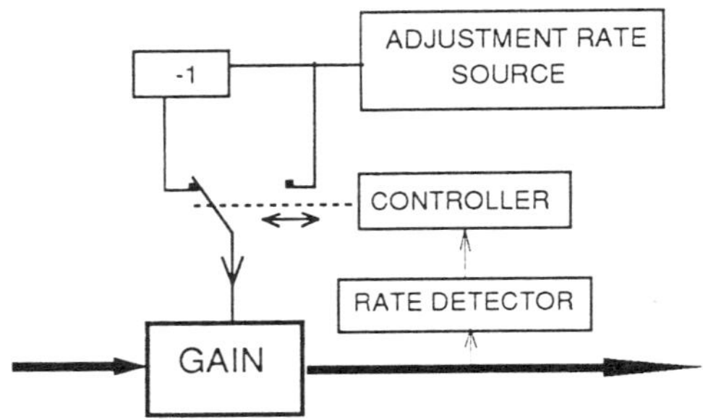

Figure 31.4 The Draper and Li optimalizing controller. The gain is increased or decreased at a constant rate, depending upon whether the IP computed or inferred from the data is increasing or decreasing.

Here the system parameter to be optimized was steady-speed torque of an internal combustion engine, and the control parameter was carburettor adjustment. The carburettor adjustment was varied at constant rate, either up or down, and the rate of change of engine torque was measured. The adjustment was changed in sign whenever the rate of change of torque changed sign, so that change continued while torque increased, but changed sign when torque started to decrease, the latter indicating that peak (optimum) torque had been passed. In experiments, the system tended to 'hunt' around the optimum.

Alternative applications and details are summarized by Aseltine *et al.* (1958). One scheme, for example, injected a small sinusoidal command which would, at optimum parameter setting, be filtered out by the system dynamics and not appear in the output; hence the amount of sinusoid in the output served as an IP.

A survey of the issues involved with the search for the optimum is given by Jarvis (1975).

31.5 SELF-TUNING REGULATORS

The principle of the adaptive approach often called 'self-tuning' is that the system is identified (using a method as in Chapter 30) and this information is used to design the parameters of a controller by a method such as in Chapter 8, Chapter 23, or Chapter 26. The resulting structure is shown in Fig. 31.5.

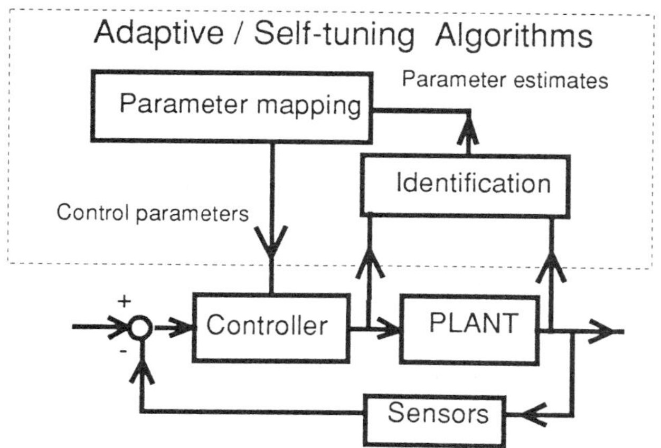

Figure 31.5 Structure of the adaptive/self-tuning algorithm. This is an indirect adaptation, because it relies on system identification.

Self-tuning amounts to a separation of the adaptive problem into two obvious parts: identification and control. The problem is to be certain these do not interact strangely.

This alternative to computerized tables, which in principle can always yield a good choice of gains, was pursued by Åström in the late 1960s and has been studied by many researchers since then, with commercial implementations now available. Self-tuning regulator laws entail a system identification (often of the Kalman filter type, Chapters 28 and 30) for the system parameters $s_1, s_2, ..., s_L$ and then the mapping of these parameters into the control parameters $p_1, p_2, ..., p_k$ by some known function, e.g. $p_i = f_i(s_j; j = 1, 2, ..., L)$.

An example should help clarify the concepts involved.

Example

Suppose a plant is known to be described by the transfer function

$$G(z) = \frac{1}{(z - a)(z - b)} = \frac{Y(z)}{U(z)}$$

where $0 < a < b < 0.8$. The feedback controller is chosen to be a simple gain K, so that $u(k) = K(y_{des}(k) - y(k))$. The closed-loop poles are to have as fast an oscillation as possible within the constraints that $K > 0$ and pole magnitude ≤ 0.8.

A quick root locus sketch shows that the closed-loop poles will be at the intersection of the vertical line bisecting the $a-b$ interval on the positive real axis with the circle of radius 0.8. Thus

$$z_p = 0.8\, e^{j\omega T}$$

where

$$\mathrm{Re}\,(z_p) = 0.8 \cos(\omega T) = \frac{a+b}{2}$$

The closed-loop poles are at the solution of

$$z^2 - (a+b)\,z + ab + K = 0$$

for which the real parts yield

$$|z_p|^2 \cos(2\omega T) - (a+b)\,|z_p|\cos(\omega T) + ab + K = 0$$

Substituting the requirements $|z_p|=0.8$ and $\cos(\omega T)=(a+b)/1.6$ gives

$$ab - 0.64 + K = 0 \Rightarrow K = 0.64 - ab$$

Thus a reasonable thing to do in a self-tuning regulator is to have an identifier (such as in Chapter 30) with output $\hat{a}$ and $\hat{b}$ as estimates of a and b respectively and to set the controller gain at

$$K = 0.64 - \hat{a}\hat{b}$$

The above example demonstrates the ideas involved with the self-tuning regulator: identification of parameters followed by a mapping of the estimates into controller parameters. The mapping is an equation (or set thereof) rather than a table look-up procedure. If the estimator is wrong about the plant parameters, of course, the whole scheme could go terribly wrong. For example, if ever the identifier should give numbers such that $\hat{a}\hat{b} = 0.64$, then $K = 0$ and no inputs will be applied to the plant; without input, the identifier may be unexcited and stall at the improper values. Thus logic checks, minimum and maximum gains, etc., may be needed in the implementation, and significant simulation efforts may be required to verify proper operation.

Industrial self-tuners often are based upon a PID regulator formulation. Reasonable rules of thumb exist for PID controllers, such as Ziegler–Nichols, Cohen–Coon, and variations (see Chapter 8). In one variation, the regulator finds three parameters, the system gain K, the delay before response t_d, and the time constant τ of the plant by having the controller open the feedback loop and apply a small change in input to the plant. The measured K, t_d, and τ are used along with an appropriate rule to compute gains for a PI or PID controller.

The self-tuning philosophy can be applied to many different combinations of estimators and control laws and can be done on start-up only, periodically, or on command from a supervisor. Control laws studied have included pole placement controllers, minimum variance controllers, LQG controllers, and phase and gain margin controllers. Estimators have included least squares and some of its extensions, instrumental variables, extended Kalman filters, and maximum likelihood methods. A variation which can save some calculations in the real-time situation is the implicit self-tuner, in which the estimated parameters feed directly without mappings into the controller computations.

31.6 MODEL REFERENCE ADAPTIVE CONTROL (MRAC)

The MRAC scheme dates back to the late 1950s and the work of Whittaker at MIT (Whittaker *et al.* 1958; Whittaker, 1959a; 1959b). The structure of the approach is shown in the block diagram of Fig. 31.6.

There are many forms of model reference controllers, of which the figure shows only a basic one. The heart of the system, of course, is the 'controller adjuster'. The idea is that the controller parameters are adjusted so that the controlled process 'behaves like' a model system.

The regulator in Fig. 31.6 is characterized by a number of parameters in a vector θ. The adjustment rule, a supervisory

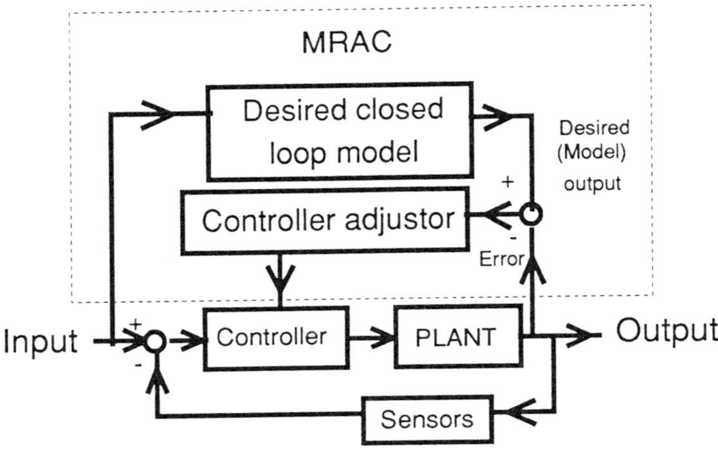

Figure 31.6 The Model Reference Adaptive Controller (MRAC) structure.

computer program, is allowed access to much information, including current parameter values, error between actual and desired output, and perhaps actual input, output, and command values.

Given the structure of the model reference adaptive system (MRAS) as above, the adjustment rule becomes the important part of the design problem. Whittaker proposed the 'MIT rule' in which

$$\frac{d\theta}{dt} = -\alpha e \frac{de}{d\theta}$$

where

$$e = \text{error} = y_{\text{actual}} - y_{\text{model}}$$

α is an algorithm gain parameter, and the gradient on the right can be generated either from the equations for y_{actual} and y_{model} or as solutions of an appropriate linear system. The algorithm is digitized as

$$\theta(m+1) = \theta(m) - \alpha' e(m)\frac{de}{d\theta} \tag{31.1}$$

$$e(m) = y_{\text{actual}}(m) - y_{\text{model}}(m)$$

Example

We explore some of the issues with an example. As done similarly by Åström and Wittenmark (1989), let the system model be

$$y(m+1) = ay(m) + bu(m)$$

and let the desired model be

$$y_d(m+1) = a_m y_d(m) + b_m r(m)$$

The goal of matching the responses can be met by using feedback

$$u(m) = k_r r(m) - k_y y(m)$$

provided

$$k_r = \frac{b_m}{b} \qquad \text{and} \qquad k_y = \frac{a - a_m}{b}$$

The problem with doing this is that a and b are unknown. Now using the nominated control law form

$$y(m+1) = ay(m) + bk_r r(m) - bk_y y(m)$$

and since

$$e(m) = y(m) - y_m(m)$$

using z-transforms gives

$$Y(z) = \frac{bk_r}{z - a + bk_y} R(z)$$

$$Y_m(z) = \frac{b_m}{z - a_m} R(z)$$

$$E(z) = \left[\frac{bk_r}{z - a + bk_y} - \frac{b_m}{z - a_m}\right] R(z)$$

We then establish the z-transforms of the gradients of e with respect to the parameters k_y and k_r as

$$\left[\frac{\partial e}{\partial k_r}\right](z) = \frac{b}{z - a + bk_y} R(z)$$

$$\left[\frac{\partial e}{\partial k_y}\right](z) = -\frac{b^2 k_r}{(z - a + bk_y)^2} R(z) = -\frac{b}{z - a + bk_y} Y(z)$$

These cannot be computed directly because a and b are unknown, so approximations are necessary before they can be used in the updating equations (31.1) Typically two assumptions are made.

1. The system is close enough to zero error that the approximations

$$k_r \approx \frac{b_m}{b} \qquad k_y \approx \frac{a - a_m}{b}$$

hold and hence $a - bk_y \approx a_m$.
2. When used in (31.1) the gradients may be erroneous in magnitude by some amount with no deterioration because the gain α is essentially arbitrary. Hence there should be no problem (provided b and b_m have the same sign) in assigning $b = b_m$ in the gradient.

Given the above approximations, the MRAC parameter setting becomes

$$\left[\frac{\widehat{\partial e}}{\partial k_r}\right](m+1) = a_m \left[\frac{\widehat{\partial e}}{\partial k_r}\right](m) + b_m r(m)$$

$$\left[\frac{\widehat{\partial e}}{\partial k_y}\right](m+1) = a_m \left[\frac{\widehat{\partial e}}{\partial k_y}\right](m) - b_m y(m)$$

$$e(m) = y(m) - y_m(m)$$

$$y(m+1) = a_m y_m(m) + b_m r(m)$$

$$k_r(m+1) = k_r(m) - \alpha \left[\frac{\widehat{\partial e}}{\partial k_r}\right](m) e(m)$$

$$k_y(m+1) = k_y(m) - \alpha \left[\frac{\widehat{\partial e}}{\partial k_y}\right](m) e(m)$$

and the control command becomes

$$u(m) = k_r(m) r(m) - k_y(m) y(m)$$

The results of a simulation of this scheme are shown in Fig. 31.7 for the case in which $a = 0.9$, $b = 0.1$, $a_m = 0.5$, $b_m = 0.5$. The command signal $r(m)$ was a square wave of period 20 steps, and the initial conditions had $y(0) = y_m(0) = 0$, $k_r(0) = 1$, and $k_y(0) = 0$. The 'proper' values are $k_r = 5$ and $k_y = 4$, and convergence occurred in this case with $\alpha = 1$. Smaller α gave slower response, while large α led to oscillations.

The above example is illustrative of several aspects of the model reference approach. Three different aspects should be noted: the actual computations and approximations, the underlying unmentioned problems of stability, and the philosophical base.

In the above example there was nothing special about either the use of the z-transforms or of the linear model. The partial derivatives

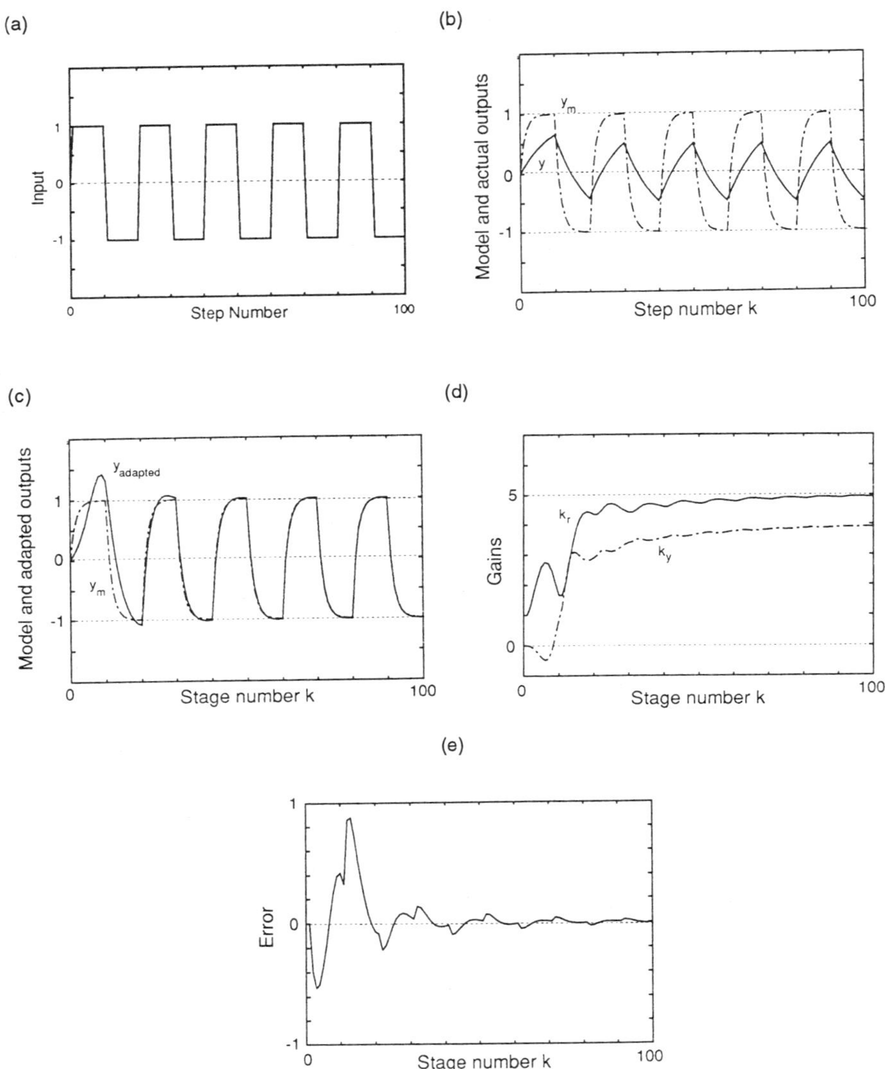

Figure 31.7 Example of MRAC: (a) the input signal; (b) actual response y (solid) and model response y_m (dashed) to input with no MRAC; (c) adapted (solid) and model (dashed) responses to input; (d) evolution of the adaptive controller gains; and (e) the system output error in response, showing decrease as adaptation improves.

could have been evaluated in the time domain, a simple exercise for the reader. In the worst case, the sensitivity of errors to parameter changes can be explored numerically, but in any case it does not require linearity of the mathematical models used.

The MRAC approach often works, especially if α is small relative to the size of the input reference signal, implying a quite slow change of parameters and requiring that the initial parameter choice be close. The issue, since the control commands are clearly non-linearly dependent on the input reference $\{r(m)\}$, is that stability is very hard to establish.

Philosophically, the MRAC attempts to make a real system behave like some desired model system. It overlaps heavily into what we have called self-tuning schemes if the parameter adjustment block actually identifies the real system.

31.7 STOCHASTIC OPTIMAL CONTROL THEORY

All of the above schemes have had structure imposed upon them by the designers using heuristic arguments. For reasons varying from mathematical curiosity to hopes of yielding improved performance, researchers have sought a unified theoretical framework for such problems. The generic field name is stochastic optimal control, and the results are regrettably few. The structure is general, as in Fig. 31.8.

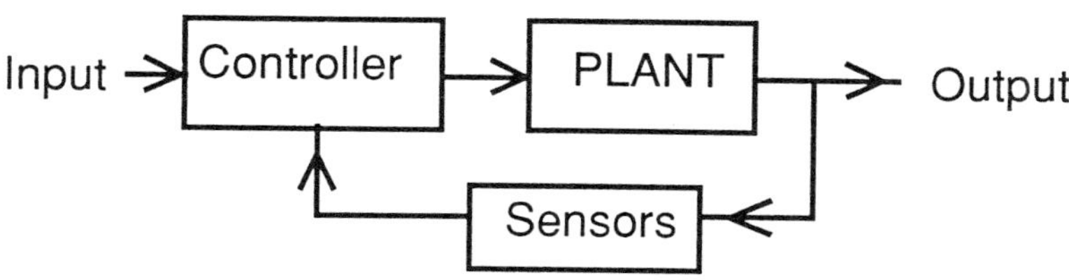

Figure 31.8 General system structure for stochastic optimal control studies.

The problem is taken as one of finding a control, and preferably a feedback control law, which minimizes a loss function. This is so far similar to deterministic optimal control theory. The extra difficulty arises when randomness is introduced; this stochastic effect may be due not only to noise in measurements, in initial conditions, and in the system dynamics, as in the Kalman filtering problem, but in the parameters of the system, both those of the physical equipment (time constants, etc.) and those of the probability distributions (covariances, biases, etc.).

It is not even clear that in general a solution to this problem exists, i.e. conditions for existence of a solution are not known. If it does exist, however, a functional equation based upon Bellman's dynamic programming can be used to study it. All of the unknowns, both state and parameters of various types, are combined into a 'hyperstate' which is a probability distribution of the state from the measurements. This hyperstate is estimated in one section of the controller, and it is mapped into the control by a second section. The issues are illustrated in an example from Åström (1983).

Example

Consider a scalar system described by

$$y(n+1) = y(n) + bu(n) + e(n)$$

in which y is the output, u is the control, e is a white noise with mean 0 and standard deviation σ_e, and b is a random variable (constant) chosen from a normal distribution. The index of performance IP is taken as minimizing the mean square deviation of the output, i.e.

$$\text{IP} = \sum_k y^2(k)$$

is to be minimized.

It turns out that b can be estimated using a Kalman-type filter as in the system identification sections. The model is

$$b(n+1) = b(n)$$

$$y(n+1) - y(n) = \Delta y(n) = u(n)b + e(n)$$

and the Kalman filter yields an estimate $\hat{b}(n)$ with standard deviation $\sigma(n)$.

Introduce the N-step look-ahead loss function

$$V_N = \min \mathscr{E}\left\{ \sum_{k=i+1}^{i+N} y^2(k) \mid Y_i \right\}$$

where Y_i is the data available at time i, i.e.

$$Y_i \equiv \{y(i), y(i-1), \ldots\}$$

The hyperstate for this problem, because $\hat{b}$ is gaussian with mean $\hat{b}(n)$ and standard deviation $\sigma(n)$, is $\{y(n), \hat{b}(n), \sigma(n)\}$. Åström uses the normalized variables

$$\eta = \frac{y}{\Sigma_e} \qquad \beta = \frac{\hat{b}(n)}{\sigma(n)} \qquad \mu = -\frac{u\hat{b}}{y}$$

and argues that it can be shown V_N depends only on η and β. In particular, the Bellman equation for increasing N is

$$V_N(\eta, \beta) = \min_{\mu}\left\{ 1 + \eta^2 \left[(1-\mu)^2 + \left(\frac{\mu}{\beta}\right)^2 \right] \right.$$
$$\left. + \int_{-\infty}^{\infty} V_{N-1}\left(\eta(1+\mu) + e\, \frac{\sqrt{(\beta^2 + \mu^2\eta^2)}}{\beta}, \sqrt{(\beta^2 + \mu^2\eta^2)} - \frac{e\mu\eta}{\beta} \right) \phi(e)\, de \right\}$$

where $\phi \sim N(0,1)$ probability density. Minimization yields $\mu_N(\eta, \beta)$ and $V_N(\eta, \beta)$. The control is then, for fixed look-ahead interval length N, given by

$$u(k) = -\frac{y(k)}{\hat{b}(n)} \mu_N(\eta(k), \beta(k)) \qquad k = i, i+1, \ldots, N+i-1$$

If N becomes large (here $N > 20$ or so is 'large'), μ_N becomes almost independent of N in that increasing N does not numerically affect μ. The answer found by Åström (1983) was determined numerically to be a complicated surface defined over the (η, β) plane.

Åström remarks that the certainty equivalence control law is

$$u(n) = -\frac{y(k)}{\hat{b}(n)}$$

i.e. the known parameter optimal control is used with a substitution of the estimate $\hat{b}$ for the unknown parameter b. This amounts to $\mu = 1$.

Another approximation (see also Aoki, 1968) chooses

$$\mu = \frac{\beta^2}{1 + \beta^2}$$

or

$$u(n) = -\frac{1}{\hat{b}(n)} \frac{(\hat{b}(n))^2}{\hat{b}(n)^2 + \sigma^2(n)} y(n)$$

This is called 'cautious control' because the gain is small when b is not well known, i.e. $\sigma(n)$ is large.

Åström's solution found that when b is poorly known, the control μ is large. This may be interpreted as a 'probe' of the system to find b, and is an example of the exploitation of the dual control property.

There are two problems (at least) with this approach. First, the formulation and setup of the Bellman dynamic programming problem may be difficult. Second, the numerical solution for the control may require excessive computer time; Åström states that 130 hours of CPU time on a VAX 11/780 were needed for the above example.

31.8 COMPUTER ASSISTANCE

It would seem that there are few standard algorithms for adaptive control design, and hence there are few general CAD programs. The best technique is perhaps simulation, and even these quickly become intricate.

31.9 COMMENTS AND FURTHER READING

We have looked at a few approaches to adaptive control in the above sections. Issues hardly mentioned include the following.

- *Stability* When all is said and done, the adaptive control laws are non-linear controllers. One would like to be assured the resulting systems are stable, but it is very difficult to demonstrate this.
- *Expert systems and artificial intelligence (AI)* It would seem that the controller should be 'smart enough' to 'recognize' what the situation is and act accordingly, as perhaps a human operator might. This is a new research area. One particular area of research is neural networks for control, discussed briefly in Chapter 32.
- *Dual control and probing signals* Given that the dual effect exists but is too difficult to handle optimally, it is useful sometimes to have probing signals to aid identification. This is sometimes done in practice, but we have barely mentioned it.
- *Adaptive filtering* This is a specialty of its own, and it is particularly pursued by the signal processing specialists in communications. Many of the ideas overlap with those of adaptive control and system identification.

To pursue the ideas further, a starting point is perhaps the survey article of Åström (1987). Several books concerning algorithms are now appearing, including those by Åström and Wittenmark (1989) and Goodwin and Sin (1984).

We remark that adaptive controllers are real: they are being implemented in ship autopilots, where gain changes made according to water depth, wind, etc., can lead to increased operating efficiency worth thousands of dollars, and as parts of off-the-shelf control systems such as process controllers where typically the capability is for Ziegler–Nichols type automatic tuning of PID regulators. A number of applications are listed by Åström (1987). Commercial self-tuning regulators are considered by Åström and Hägglund (1988).

32

Learning control

Adaptive control and system identification of the standard types usually reduce to parameter estimation. Since the control law is ultimately a mapping from the measurement history to commands to the plant, it is intriguing to consider attempting to establish an appropriate such mapping without using the specialized structure that the standard methods use. Several techniques have been suggested which, to a greater or lesser extent, self-learn or are taught the proper control. Among these are:

- artificial neural networks,
- functional learning,
- expert systems, and
- fuzzy systems.

In this chapter we try to look at a the essence of each of these.

32.1 SYNOPSIS

There are several approaches to control law design in which the action of the controller might be said to show 'intelligence'. In artificial neural networks, the controller is configured as many similar elements, called 'neurons', which have fixed input–output relationships

$$\mathbf{u} = f\left(\sum_i \alpha_i \mathbf{y}_i\right)$$

and typical functions for f are such as

$$f(z) = \frac{1}{1 + e^{-z}}$$

$$f(z) = \text{sat}(z) = \begin{cases} 1 & 1 \le z \\ z & -1 \le z \le 1 \\ -1 & z \le -1 \end{cases}$$

By connecting a great many such elements in parallel and series a very complex relationship between measurement vectors $\mathbf{y}$ and command vectors $\mathbf{u}$ can be implemented. If the elementary parameters α_i are not known *a priori*, a supervisory routine somewhat like a mathematical programming algorithm can be used, along with a set of desired input–output pairs, to 'teach' the network. In 'learning' its parameters, the network might be said to show intelligence; in addition, the interconnection of the identically formed simple elements is rather like an animal's nervous system. For these reasons, the field is called artificial neural networks (ANN) or simply neural networks.

A less tightly structured scheme in which the good control law is assumed representable by an unknown function $U(\mathbf{y}(t); t)$ where $\mathbf{y}$ are measurements may be formulated. If we use an algorithm such that an estimate of the function $\hat{U}(\mathbf{y}; t)$ is formed and such that $\hat{U} \rightarrow U$, the system can be said to learn to do good control, and hence to show intelligence.

A completely different approach arises when humans 'teach' a controller by informing the controller, in a rough way suitable for teaching human operators, of their own operational rules. Since the rough way usually involves a lack of precision, at least in terms of usual computer algorithms, this is called fuzzy control and is based upon fuzzy set theory. The latter is a set manipulation algebra in which elements have a degree of membership of a set which is other than the binary degree (it belongs or it does not belong) in conventional set theory. The set theory allows the controller to be taught in terms such as

> IF {the speed is *TOO HIGH*}
> THEN {decrease the input current *SLIGHTLY*}

A large set of such rules, when processed using the fuzzy set logical operations, yields a 'taught' control function $U(\mathbf{y}; t)$.

Related philosophically to fuzzy set applications because of its reliance upon experts, but usually utilizing standard logical processing of a set of rules, is expert systems theory. Here again the controller is taught by experts to evaluate a situation and act accordingly. A typical rule is:

> IF {motor is on} AND {excess temperature is indicated}
> THEN {turn motor off} AND {sound alarm signal}

We briefly look at each of the above, the two taught system approaches and the two self-learning approaches, in this chapter. We also indicate a couple of other schemes which show promise.

32.2 ARTIFICIAL NEURAL NETWORKS (ANN)

Artificial neural networks represent a confluence of ideas from biology and computing; users of the latter have a desire to do rapid computing using computers in parallel, while the former field suggests that brains operate by using highly interconnected brain cells called neurons, each performing relatively simple processes. The resulting methods all work by processing inputs through a simple set of quite structured computations to give an output; learning appears in the use of the inputs and desired outputs to derive the parameters of the structures.

32.2.1 Some basic notions

More specifically, and to take a simple example, let $\{e(k)\}$ be an error signal input to the control law and let $\{u(k)\}$ be the output command, as in Fig. 32.1.

In this scheme, the function $f(\cdot)$ is prechosen and is of a generally sigmoid character. For pattern recognition, for example, and also to correspond to notions of brain operation, the choice $f(y) = \text{sgn}(y)$, or a scaled and biased version of it, is often made. Partly for reasons of ease of using some of the algorithms, the smooth function

$$f(y) = \frac{1}{1 + e^{-y}} \tag{32.1}$$

is frequently used. The control law is determined by choosing a structure such as this and then selecting appropriate parameters $\alpha_i, \beta_i, \gamma_i, \delta$ for $i = 1, 2, \ldots, N$ and N a chosen number of elementary

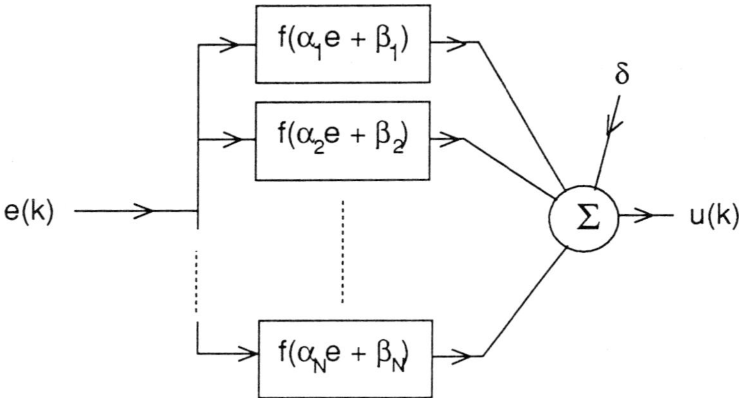

Figure 32.1 A simple SISO ANN structure with one layer.

blocks. Because the blocks are entirely internal to the operation, they constitute a 'hidden layer' in the network; the inputs *e* are the input layer, and the outputs *u* are from the output layer of the computation. In the figure, it is clear that

$$u(x(k)) = \delta + \sum_{i=1}^{N} \gamma_i f(\alpha_i e(k) + \beta_i) \qquad (32.2)$$

It is arguable that in the usual case there exists an ideal, but unknown, function $u(x)$ which gives 'good' control and that the goal is to 'learn' this function. Such is the point of view of ANN models.

Example

An example used in this chapter is

$$x(k+1) = (1 - x(k))^2 - u(k)$$

with the goal being to find a control law which would give one-step deadbeat control.

The claim often seen is that ANN methods can be used. We consider a one-hidden layer method for the above problem and generate an approximate solution using heuristics. The point is not to illustrate the learning, but the ANN structure and parameters which might give a control law.

In this example, the input is $x(k)$ at each stage. It is fed (with weight = 1) to a number N of neural elements with varying biases; these have outputs denoted by $y_j, j = 1, 2, ..., N$. The control output $u(k)$ is then given by

$$u(k) = \sum_{i=1}^{N} \alpha_i y_i$$

Since the solution to the example problem is obviously

$$u^*(x(k)) = (1 - x(k))^2$$

the object is to find weights such that $u(k) \approx u^*(k)$.
We observe that $f(y)$ in (32.1) has

$$f(0) = 0.5 \qquad f(-5) = 0.0007 \approx 0 \qquad f(5) = 0.993 \approx 1$$

and decide to define

$$g(x,b,\Delta) = \frac{1}{1 + e^{-5(x-b)/\Delta}}$$

We then choose

$$y_1 = 1 - g(x, 0, 0.5)$$

$$y_2 = g(x, 2, 0.5)$$

$$y_3 = 1 - g(x, -1, 0.5)$$

$$y_4 = g(x, 3, 0.5)$$

$$y_5 = 1 - g(x, -2, 0.5)$$

$$y_6 = g(x, 4, 0.5)$$

$$y_7 = 1 - g(x, -3, 0.5)$$

$$y_8 = g(x, 5, 0.5)$$

Ordinarily the b components would be chosen by the learning algorithm.

With the above, the control

$$u(k) = 2y_1 + 2y_2 + 4y_3 + 4y_4 + 6y_5 + 6y_6 + 8y_7 + 8y_8$$

is a rough but reasonable approximation to $u^*(k)$. As shown in Fig. 32.2, it shows errors up to about 0.5 over the range $-3 \le x \le 5$, and is almost exact for the integer values in this range.

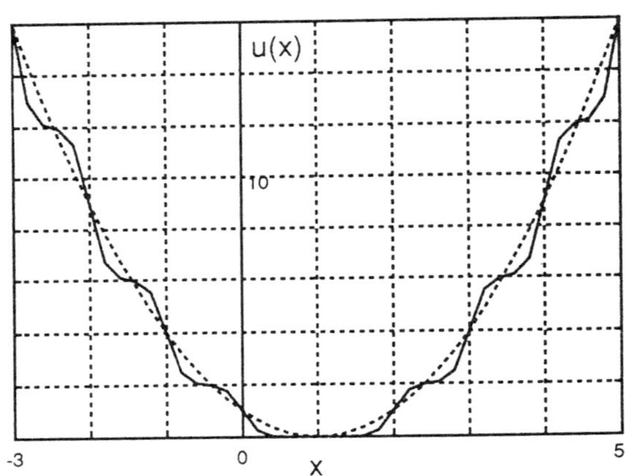

Figure 32.2 Example of a possible ANN curve fit with only a few neurons.

For the learning situation, a typical approach is to have a cost function $\mathcal{F}(\alpha, \beta, \gamma, \delta)$ which is to be optimized by choosing $\alpha = (\alpha_1, \alpha_2, ..., \alpha_N)$, $\beta = (\beta_1, \beta_2, ..., \beta_N)$, $\gamma = (\gamma_1, \gamma_2, ..., \gamma_N)$, δ. A common such function is a quadratic motivated by

$$\mathcal{J}(\alpha, \beta, \gamma, \delta) = \sum_{k=k_1}^{k_2} (u^*(x(k)) - u(x(k)))^2$$

which in this instance is

$$\mathcal{J}(\alpha, \beta, \gamma, \delta) = \sum_{k=k_1}^{k_2} \left(u^*(x(k)) - \left(\delta + \sum_{i=1}^{N} \gamma_i f(\alpha_i x(k) + \beta_i)\right)\right)^2$$

Since the very reason for using this method is that $u^*(x)$ is unknown, the model must be developed further. One case in which the simplest calculations can work is when the system dynamics are

$$x(k+1) = g(x(k)) + u(k) \tag{32.3}$$

The desired value $x_{\text{des}}(k+1; x(k))$ is presumably known; for example, if we seek deadbeat control for any $x(k)$, then we have

$$u(x(k)) = x_{\text{des}}(k+1; x(k)) - g(x(k))$$

With such a definition, we may operate the system, for we can measure $x(k+1)$ and derive

$$\mathcal{J}(\alpha, \beta, \gamma, \delta) = \sum_{k=k_1}^{k_2} (x_{\text{des}}(k+1; x(k)) - x(k+1))^2$$

With this as a measure of the efficacy of the proposed controller, we may proceed to determine improved parameters. The straight-forward approach is a gradient descent method. With $\theta = (\alpha \vdots \beta \vdots \gamma \vdots \delta)^{\text{T}}$, find $\nabla_\theta \mathcal{J}$. Then, choose

$$\Delta\theta = -\frac{\mu}{2} \nabla_\theta \mathcal{J} = \mu \sum_{k=k_1}^{k_2} (x_{\text{des}}(k+1; x(k)) - x(k+1))(\nabla_\theta x(k+1))$$

where scalar step size μ is a parameter.

Continuing with the gradient descent, $\nabla_\theta x(k+1)$ can be found because of the special structures (32.3), (32.1) and (32.2). In particular, since

$$x(k+1) = g(x(k)) + \delta + \sum_{i=1}^{N} \gamma_i\, f(\alpha_i x(k) + \beta_i)$$

with $f(\cdot)$ given by (32.1), we can find

$$\frac{\partial x(k+1)}{\partial \delta} = 1$$

$$\frac{\partial x(k+1)}{\partial \gamma_i} = f(\alpha_i\, x(k) + \beta_i)$$

$$\frac{\partial x(k+1)}{\partial \alpha_i} = \gamma_i\, x(k)\, f(\alpha_i x(k) + \beta_i)\, (1 - f(\alpha_i x(k) + \beta_i))$$

$$\frac{\partial x(k+1)}{\delta \beta_i} = \gamma_i\, f(\alpha_1 x(k) + \beta_1)\, (1 - f(\alpha_1 x(k) + \beta_1))$$

Rather than batch processing the above, it is frequent that we choose $k_1 = k_2$. In this way, the parameters are readjusted after each cycle of the process, with equations such as

$$\delta(k+1) = \delta(k) - \mu\,(x_{\text{des}}(k+1) - x(k+1))$$

$$\gamma_i(k+1) = \gamma_i(k) - \mu\,(x_{\text{des}}(k+1) - x(k+1))\, f(\alpha_1 x(k) + \beta_i)$$

$$\alpha_i(k+1) = \alpha_i(k) - \mu\,(x_{\text{des}}(k+1) - x(k+1))$$

$$\times\, \gamma_i x(k)\, f(\alpha_i x(k) + \beta_i)(1 - f(\alpha_i x(k) + \beta_i))$$

$$\beta_i(k+1) = \beta_i(k) - \mu\,(x_{\text{des}}(k+1) - x(k+1))$$

$$\times\, \gamma_i f(\alpha_i x(k) + \beta_i)\, (1 - f(\alpha_i x(k) + \beta_i))$$

32.2.2 Multilayers

There are many variations of the above simple layout. The most obvious are to have multiple inputs, so that (32.2) becomes

$$u(x(k)) = \delta + \sum_{i=1}^{N} \gamma_i f\left(\sum_{j=1}^{M} \alpha_{ij} x_j(k) + \beta_i \right)$$

or multiple outputs, so that (32.2) becomes

$$u_j(x(k)) = \delta_j + \sum_{i=1}^{N} \gamma_{ji} f(\alpha_{ji} x(k) + \beta_j)$$

Another variation, and perhaps the most interesting, is to have multiple intermediate, or hidden, layers. In this case, with multiple inputs and outputs, one possible form is

$$u_j(k) = {}^{N-1}\gamma_{j,\,N_N+1} + \sum_{i=1}^{N_{N-1}} {}^{N-1}\gamma_{j,i} \; {}^{N-1}y_i(n) \qquad j=1,2,...,N$$

$$^iy_j(n) = f\left(\sum_{k=1}^{N_{i-1}} {}^i\gamma_{j,\,k}{}^{i-1} y_k(n) + {}^i\gamma_{j,\,N_{i-1}+1} \right) = f({}^ix_j) \; i=2,3,...,N-1$$

$$j=1,2,...,N_i$$

$$^1y_i(n) = x_i \qquad\qquad i=1,2,...,N_1$$

where $N-2$ = number of hidden layers, N_i = number of nodes in layer $i-1$; N_1 inputs and N_N outputs are assumed. Also

$$u_j(n) = \text{output } j \text{ of network at step } n$$

$$^iy_j(n) = \text{output of the } j\text{th node of layer } i \qquad\qquad (32.4)$$

$$^ix_j(n) = \text{net input to the } j\text{th node of layer } i$$

$$^i\gamma_{j,\,k} = \text{weighting of input } k \text{ to node } j \text{ of layer } i$$

Although there is no necessity that the function $f(\cdot)$ be the same for all nodes or that the sigmoid logistic function (32.1) be used, it is convenient because it has the property

$$\left.\frac{\partial f(x(\alpha))}{\partial \alpha}\right|_{\alpha*} = f(x(\alpha^*)) \ \{1 - f(x(\alpha^*))\} \ \left.\frac{\partial x(\alpha)}{\partial \alpha}\right|_{\alpha*} \qquad (32.5)$$

Using this, it is straightforward in principle, but tedious and notationally obscure, to find the various partial derivatives for the gradient descent or similar technique. We show this by the following.

$$\mathcal{F}(\theta) = \frac{1}{2} \sum_n \sum_{i=1}^{N_N} (u_{i,\,\text{des}}(n) - u_i(n;\theta))^2$$

$$\frac{\partial \mathcal{F}(\theta)}{\partial \theta} = -\sum_n \sum_{i=1}^{N_N} (u_{i,\,\text{des}}(n) - u_i(n;\theta)) \frac{\partial u_i(n;\theta)}{\partial \theta} \qquad (32.6)$$

Then we start enumerating the coefficients. For convenience, we choose $n=1$ and suppress it in the notation.

$$\frac{\partial u_i(\theta)}{\partial (^N\gamma_{j,\,N_j+1})} = \begin{cases} 1 & j=1 \\ 0 & j \neq i \end{cases}$$

$$\frac{\partial u_i(\theta)}{\partial (^N\gamma_{j,k})} = \begin{cases} {}^{N-1}y_k & j=1 \\ 0 & j \neq i \end{cases}$$

In terms of the inputs to the previous layer, we have

$$\frac{\partial u_i(\theta)}{\partial (^{N-1}x_j)} = {}^N\gamma_{i,\,j} \ {}^{N-1}y_j \ (1 - {}^{N-1}y_j)$$

where we have used (32.5). Because of the definition of $^j x_k$ implied by (32.4), we have

$$\frac{\partial(^{N-1}x_j)}{\partial(^{N-1}\gamma_{i,k})} = \begin{cases} 1 & i=j \text{ and } k=N_{N-1}+1 \\ 0 & i\neq j \text{ or } k>N_{N-1}+1 \\ {^{N-1}y_k} & i=j \text{ and } 1\leq k\leq N_{N-1} \end{cases}$$

$$\frac{\partial(^{N-1}x_j)}{\partial(^{N-2}x_k)} = {^{N-1}\gamma_{i,k}} \; {^{N-2}y_k} \; (1 - {^{N-2}y_k})$$

We may continue backwards through the network, from output back to input, using analogous relationships. Now we apply this to the problem at hand by reconsidering $\mathcal{F}(\theta)$ from (32.6), specialized to this application.

$$\frac{\partial\mathcal{F}(\theta)}{\partial\theta} = \sum_{i=1}^{N_N} (u_{i,\,\text{des}} - u_i(\theta)) \frac{\partial u_i(\theta)}{\partial\theta}$$

We see that

$$\frac{\partial\mathcal{F}(\theta)}{\partial(^{N}\gamma_{j,k})} = \begin{cases} -(u_{j,\,\text{des}}(n) - u_j(n;\theta)) & k = N_{N-1} + 1 \\ -(u_{j,\,\text{des}}(n) - u_j(n;\theta))^{N-1}y_k & 1 \leq k \leq N_{N-1} \end{cases}$$

$$\frac{\partial\mathcal{F}(\theta)}{\partial(^{N-1}\gamma_{j,k})} = \sum_{i=1}^{N_N} (u_{i,\,\text{des}} - u_i(\theta)) \frac{\partial u_i(\theta)}{\partial(^{N-1}\gamma_{j,k})}$$

$$= -\sum_{i=1}^{N_N} (u_{i,\,\text{des}} - u_i(\theta)) \sum_{m=1}^{N_{N-1}} \frac{\partial u_i(\theta)}{\partial(^{N-1}x_m)} \frac{\partial(^{N-1}x_m)}{\partial(^{N-1}\gamma_{j,k})}$$

$$-\sum_{i=1}^{N_N}(u_{i,\,\text{des}} - u_i(\theta)) \sum_{m=1}^{N_{N-1}} {^{N}\gamma_{i,m}}{^{N-1}y_m}(1 - {^{N-1}y_m}){^{N-2}y_k}\,\delta(m,k)$$

$$= - {^{N-1}y_k}(1 - {^{N-1}y_k}){^{N-2}y_k} \sum_{i=1}^{N_N} {^{N}\gamma_{i,k}}(u_{i,\,\text{des}} - u_i(\theta))$$

The next stage can be computed to be

$$\frac{\partial \mathcal{F}(\theta)}{\partial (^{N-2}\gamma_{j,k})} = \sum_{i=1}^{N_N} (u_{i,\,\mathrm{des}} - u_i(\theta)) \frac{\partial u_i(\theta)}{\partial (^{N-2}\gamma_{j,k})}$$

$$= -\sum_{i=1}^{N_N} (u_{i,\,\mathrm{des}} - u_i(\theta)) \sum_{m=1}^{N_{N-1}} \frac{\partial u_i(\theta)}{\partial (^{N-1}x_m)} \sum_{p=1}^{N_{N-2}} \frac{\partial (^{N-1}x_m)}{\partial (^{N-2}x_p)} \frac{\partial (^{N-2}x_p)}{\partial (^{N-2}\gamma_{j,k})}$$

$$= -\sum_{i=1}^{N_N} (u_{i,\,\mathrm{des}} - u_i(\theta)) \sum_{m=1}^{N_{N-1}} {}^N\gamma_{i,m}\, {}^{N-1}y_m\, (1 - {}^{N-1}y_m)$$

$$\times \sum_{p=1}^{N_{N-2}} {}^{N-1}\gamma_{m,p}\, {}^{N-2}y_p\, (1 - {}^{N-2}y_p)\, {}^{N-3}y_j\, \delta(p,j)$$

$$= - {}^{N-2}y_j\, (1 - {}^{N-2}y_j)\, {}^{N-3}y_j \sum_{m=1}^{N_{N-1}} {}^{N-1}\gamma_{m,j}\, {}^{N-1}y_m\, (1 - {}^{N-1}y_m)$$

$$\sum_{i=1}^{N_N} {}^N\gamma_{i,m}(u_{i,\,\mathrm{des}} - u_i(\theta))$$

We now define the 'effective error' at node j of level i as ${}^i\mu_j$, with

$${}^N\mu_i = (u_{i,\,\mathrm{des}} - u_i(\theta))$$

$${}^{N-1}\mu_j = {}^{N-1}y_j\, (1 - {}^{N-1}y_j) \sum_{i=1}^{N_N} {}^N\gamma_{i,j}\, (u_{i,\,\mathrm{des}} - u_i(\theta))$$

$$= {}^{N-1}y_j\, (1 - {}^{N-1}y_j) \sum_{i=1}^{N_N} {}^N\gamma_{i,j}\, {}^N\mu_i$$

$${}^{N-2}\mu_j = {}^{N-2}y_j\, (1 - {}^{N-2}y_j) \sum_{m=1}^{N_{N-1}} {}^{N-1}\gamma_{m,j}\, {}^{N-1}y_m(1 - {}^{N-1}y_m) \times$$

$$\sum_{i=1}^{N_N} {}^N\gamma_{i,m}\, (u_{i,\,\mathrm{des}} - u_i(\theta))$$

$$= {}^{N-2}y_j \, (1 - {}^{N-2}y_j) \sum_{m=1}^{N_{N-1}} {}^{N-1}\gamma_{m,j} \, {}^{N-1}y_m \, (1 - {}^{N-1}y_m) \sum_{i=1}^{N_N} {}^{N}\gamma_{i,m} \, {}^{N}\mu_i$$

$$= {}^{N-2}y_j \, (1 - {}^{N-2}y_j) \sum_{m=1}^{N_{N-1}} {}^{N-1}\gamma_{m,j} \, {}^{N-1}\mu_m$$

The general term is then

$$^{k-1}\mu_j = {}^{k-1}y_j \, (1 - {}^{k-1}y_j) \sum_{i=1}^{N_k} {}^{k}\gamma_{i,j} \, {}^{k}\mu_i$$

It is seen that

$$\frac{\partial \mathscr{F}(\theta)}{\partial({}^{k}\gamma_{j,k})} = - {}^{k}\mu_j \, {}^{k-1}y_j$$

A gradient descent on $\mathscr{F}$ will have in general

$$^{k}\gamma_{j,k} \leftarrow {}^{k}\gamma_{j,k} - \eta \, \frac{\partial \mathscr{F}(\theta)}{\partial({}^{k}\gamma_{j,k})}$$

with a constant η chosen such that the linearization approximation holds. Specifically, we have

$$^{m}\gamma_{j,k} \leftarrow {}^{m}\gamma_{j,k} + \eta \, {}^{m}\mu_j \, {}^{m-1}y_k \qquad\qquad 1 \le k \le N_{m-1}$$

$$^{m}\gamma_{j,k} \leftarrow {}^{m}\gamma_{j,k} + \eta \, {}^{m}\mu_j \qquad\qquad k = N_{m-1} + 1$$

with

$$^{N}\mu_i = (u_{i,\,\text{des}} - u_i(\theta))$$

$$^{m-1}\mu_j = {}^{m-1}y_j \, (1 - {}^{m-1}y_j) \sum_{i=1}^{N_k} {}^{m}\gamma_{i,j} \, {}^{m}\mu_i$$

and where the indices range over $k = 1, 2, ..., N_{m-1}$, $j = 1, 2, ..., N_m$ and $m = 2, 3, ..., N$, as the adjustment algorithm for the parameters. This, because of the way in which the error sensitivities $^k\mu_j$ are generated, is called a **backward propagation** algorithm. It turns out to work fairly well, which can be considered a surprise for a technique which is basically a gradient descent algorithm of mathematical programming.

Example

The system described in the example of section 32.2.1 (page 764) may be arranged for back propagation, with an input layer (of the scalar $x(n)$), a hidden layer with 21 nodes, and an output layer of the scalar $u_{\mathrm{des}}(n)$, which was taken as $x(n+1)$. The set-up had

$$u(k) = \sum_{i=1}^{N} {}^2\gamma_i \, {}^2y_i + {}^2\gamma_{N+1}$$

$${}^2y_i = f({}^1\gamma_{i,1} x(k) + {}^1\gamma_{i,2})$$

Then the update equations were

$$e = x(k+1) - u(k)$$

$${}^2\gamma_i = {}^2\gamma_i + \eta e \, {}^2y_i \qquad\qquad i = 1, 2, ..., N$$

$${}^2\gamma_{N+1} = {}^2\gamma_{N+1} + \eta e$$

$${}^1\gamma_{i,1} = {}^1\gamma_{i,1} + \eta' e \, {}^2y_i \, (1 - {}^2y_i) \, x(k) \qquad\qquad i = 1, 2, ..., N$$

$${}^1\gamma_{i,N+1} = {}^1\gamma_{i,N+1} + \eta' e \, {}^2y_i \, (1 - {}^2y_i)$$

with η' adjusted at each step so that the maximum value

$$\max_i \left\{ |\eta' e \, {}^2y_i \, (1 - {}^2y_i) \, x(k)|, \, |\eta' e \, {}^2y_i \, (1 - {}^2y_i)| \right\}$$

was limited (to keep the adjustments within the range such that the linearization held) to about 0.05. This was run with inputs from the range $[-5, 5]$, applied either systematically or randomly, and the output error computed. Typically, the learning gain was about 0.05, and a nearly stable solution was reached after 5000–10000 trials, with the system incapable of achieving a fit closer than ±0.5–0.7 of ideal.

32.2.3 Comments

ANNs can be used for system identification by having their inputs the same as the system to be identified and their outputs attempting to match those of the actual system. The issue of system dynamics (memory) is handled by using state vector inputs or by putting delays into the network. An alternative puts the ANN in 'backwards', with system outputs as ANN input and attempts to have the ANN output match the actual system inputs (Fig. 32.3); then, of course, it is easy to find the system input which will give a desired system output.

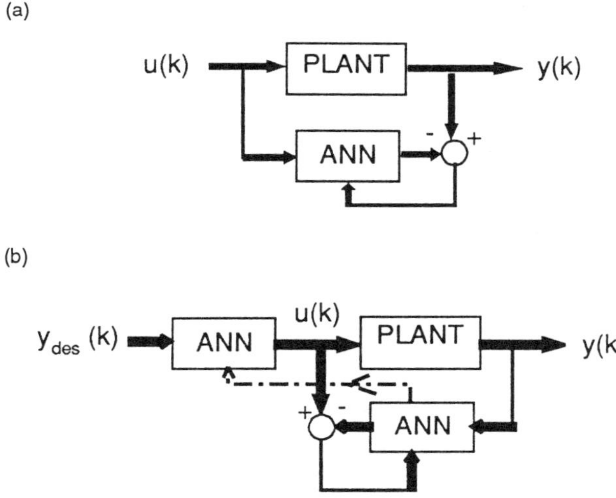

Figure 32.3 Structures for using ANNs: (a) for identification; and (b) for adaptive control.

32.2.4 Further reading

Much of the neural network activity of interest to engineers has been in pattern recognition, as in recognition of printed characters. One starting point for studies of control systems which use artificial neural networks is the book by Narendra and Parthasarathy (1990). A now-classical demonstration is of the inverted pendulum, or broom-stick balancing, control system, as given by, among others, Anderson (1989). Another demonstration is of truck-backing, described by Nguyen and Widrow (1990).

Somewhat more general, with emphasis on basic terminology and a survey of methods and applications, is the introductory book of Simpson (1990). Engineering students may find the special issues of the IEEE Proceedings (Lau and Widrow, 1990) helpful.

ANNs are the subject of intense research at this time, with some interesting applications being commercialized. The literature is thus rapidly expanding, with both research articles and textbooks proliferating.

32.3 ADAPTATION OF NON-PARAMETRIC CONTROL LAWS

When the control law for a system is treated as an unparameterized function, several advantages in implementation accrue. In this section the problem of determining such a law for an unknown system is considered and a learning scheme is proposed. An example demonstrates the operation of the scheme, and Lyapunov theory is presented which demonstrates the convergence of the law.

32.3.1 Introduction

The control theory literature contains many articles which in the opening stages define the feedback law for a system with state $\mathbf{x} \in \mathbf{X} \subseteq \mathbf{R}^n$ = set of n-dimensional real vectors, and measurements $\mathbf{y} \in \mathbf{Y} \subseteq \mathbf{R}^p$ as a mapping such as $\mathbf{U} : \mathbf{X}$ (or $\mathbf{Y}$) $\times \mathbf{T} \rightarrow \mathbf{R}^m$, where $\mathbf{T}$ is the time interval, either $[0, t_f]$ or $\{t_1, t_2, ..., t_N\}$. A more common notation is $u(\mathbf{x}, t)$ (or $u(\mathbf{y}, t)$), with all of the sets, etc. suppressed. Few papers follow up this approach in its generality; the function quickly becomes a linear law such as $u(\mathbf{x}, t) = K\,\mathbf{x}$, or some other parameterized law, perhaps a dynamic one.

Suppose we decide to pursue the implications of the general formulatio 1. We would appear to accrue some advantages: implementation becomes table look-up rather than computation; the control function is 'learned' for an unknown system; the function is not constrained to be of any particular form, and hence the technique should be applicable for non-linear systems.

32.3.2   The learning algorithm

Suppose we are presented with a plant described by

$$\mathbf{x}(k+1) = f_1(\mathbf{x}(k), u(k), k)$$

$$\mathbf{y}(k) = \mathbf{C}\mathbf{x}(k)$$

for which f_1 is only partially known, its state $\mathbf{x}$ is an n-vector and measurements $\mathbf{y}$ are p-vectors. The control $u(k)$ is taken for convenience to be scalar, and we seek a function

$$u(k) = u^*(\mathbf{x}(k), \mathbf{y}(k), \mathbf{x}_{\text{des}}(k+1), \mathbf{x}_{\text{des}}(k), k)$$

$$= u^*(s(k))$$

where $s(k)$ is a notational convenience indicating the information available and needed by the control law. We assume that u^* has a representation

$$u^*(s) = \int_\Gamma K(\gamma, s) g(\gamma) \, d\gamma$$

for a known scalar kernel $K(\gamma, s)$, some unknown function $g(\gamma)$, and a known region Γ which contains s. Then we estimate $u^*(s)$ at stage k by $\hat{u}(k; s)$ and in particular send the command $\hat{u}(k; s(k))$ to the plant. The error at stage $k+1$ is then

$$\mathbf{e}(k+1) = \mathbf{x}_{\text{des}}(k+1) - \mathbf{x}(k+1)$$

and the control function is improved by taking

$$\hat{g}(k+1;\gamma) = \hat{g}(k;\gamma) + \alpha\, K(\gamma, s(k))\,\mathbf{R}\mathbf{e}(k+1)$$

$$\hat{u}(k+1;s) = \int_{\Gamma} K(\gamma, s)\,\hat{g}(k+1;\gamma)\,d\gamma$$

where α is a scalar, $\mathbf{R}$ is a $1 \times n$ matrix, and $K(\gamma, s)$ is a chosen 'spread' function.

The motivation for the algorithm is direct: from a certain position $\mathbf{x}(k)$ with required position $\mathbf{x}_{des}(k+1)$ (encoded as $s(k)$) a command $u(k)$ is selected by using a function $u(k; s)$ evaluated at $s = s(k)$. The plant then outputs $\mathbf{x}(k+1)$ which is erroneous by the amount $\mathbf{e}(k+1)$. To improve the function $u(k; s)$ at the point $s(k)$, a correction proportional to $\mathbf{e}(k+1)$ is made; in particular $\alpha\,\mathbf{R}\mathbf{e}(k+1)$ is added to the old function at point $s(k)$. To 'spread' the knowledge so that better controls are also used in the vicinity of $s(k)$, a 'spread function' $G(s, s(k))$ derived from $K(s, s(k))$ is used so that $\alpha\,\mathbf{R}\mathbf{e}(k+1)$ also affects points s 'near' to $s(k)$.

32.3.3 Convergence and stability

For analysis of convergence of the estimate $\hat{u}(k; s)$ to $u^*(s)$ and also for study of the stability of the system which has such an adaptive control, there appear to be a number of alternatives, mostly based on non-linear stability. We outline a Lyapunov demonstration of the convergence.

We consider that the desired trajectory $\{\mathbf{x}_{des}(k)\}$ is known to be attainable for some unknown control function $u^*(\mathbf{x})$ and that this function has a representation

$$u^*(s) = \int_{\Gamma} K(\gamma, s)\,g(\gamma)\,d\gamma$$

where $K(\gamma, s)$ is a known scalar kernel, $g(\gamma)$ is the unknown core function, and Γ is the finite region upon which the superstate $s(k)$ is defined. We assume further that the original system is stable, and that $\mathbf{e}(k) = \mathbf{x}_{des}(k) - \mathbf{x}(k)$ satisfies

$$\mathbf{e}(k+1) = f(\mathbf{e}(k), \mathbf{x}_{\mathrm{des}}(k)) + h(\mathbf{e}(k), \mathbf{x}_{\mathrm{des}}(k))\,\tilde{u}(k)$$

where $\tilde{u}(k)$ is the scalar control error, defined by

$$\tilde{u}(k) = u^*(s(k)) - \hat{u}(k; s(k))$$

where, with the definitions

$$\hat{u}(k; s) = \int_\Gamma K(\gamma, s)\,\hat{g}(k; \gamma)\,\mathrm{d}\gamma$$

$$\tilde{g}(k; s) = g(s) - \hat{g}(k; s)$$

we have

$$\tilde{u}(k) = \int_\Gamma K(\gamma, s(k))\,\tilde{g}(k; \gamma)\,\mathrm{d}\gamma$$

We choose the algorithm

$$\hat{g}(k+1; \gamma) = \hat{g}(k; \gamma) + \alpha K(\gamma, s(k))\,\mathbf{e}(k+1) \qquad \forall\, \gamma \in \Gamma$$

from which it follows that

$$\tilde{g}(k+1; \gamma) = \tilde{g}(k; \gamma) - \alpha K(\gamma, s(k))\,\mathbf{e}(k+1)$$

We now consider that we would like $\{\hat{u}(k; s)\} \to u^*(s)$ as a function, and would also like the sequence $\{\mathbf{e}(k)\} \to \mathbf{0}$. We choose the functional

$$V(k) = \int_\Gamma \tilde{g}^2(k; \gamma)\,\mathrm{d}\gamma$$

and examine its progression with k. For notational convenience, we in most places will drop the index when it is k. Then

$$V(k+1) = \int_\Gamma \tilde{g}^2 \, d\gamma - 2\alpha \int_\Gamma K(\gamma, s(k)) \, \tilde{g} \, d\gamma \, \mathbf{R}(\mathbf{f} + \mathbf{h}\tilde{u})$$

$$+ \alpha^2 \int_\Gamma K^2(\gamma, s(k)) \, d\gamma \, (\mathbf{f} + \mathbf{h}\tilde{u})^{\mathrm{T}} \mathbf{R}^{\mathrm{T}} \mathbf{R} \, (\mathbf{f} + \mathbf{h}\tilde{u})$$

Using the definition of $\tilde{u}(k)$ and defining for convenience

$$\mathscr{K}^2 = \int_\Gamma K^2(\gamma, s(k)) \, d\gamma$$

puts this in the form

$$V(k+1) = \int_\Gamma \tilde{g}^2 \, d\gamma - 2\,\alpha\tilde{u}\,\mathbf{R}(\mathbf{f} + \mathbf{h}\tilde{u})$$

$$+ \alpha^2 \mathscr{K}^2 (\mathbf{f} + \mathbf{h}\tilde{u})^{\mathrm{T}} \mathbf{R}^{\mathrm{T}} \mathbf{R} (\mathbf{f} + \mathbf{h}\tilde{u})$$

which rearranges to

$$\Delta V = V(k+1) - V(k) = \mathbf{f}^{\mathrm{T}}(\alpha^2 \mathscr{K}^2 \mathbf{R}^{\mathrm{T}} \mathbf{R})\mathbf{f}$$

$$+ \tilde{u}^2(-2\alpha \mathbf{R}\mathbf{h} + \alpha^2 \mathscr{K}^2 \mathbf{h}^{\mathrm{T}} \mathbf{R}^{\mathrm{T}} \mathbf{R}\mathbf{h})$$

$$+ 2\tilde{u}(\alpha^2 \mathscr{K}^2 \mathbf{h}^{\mathrm{T}} \mathbf{R}^{\mathrm{T}} \mathbf{R}\mathbf{f} - \alpha \mathbf{R}\mathbf{f})$$

We would like to find conditions which ensure that $\Delta V \leq 0$. Our degrees of freedom appear to be α $\mathbf{R}$ and the kernel K. We can look at three terms:

$$T_1 = \mathbf{f}^{\mathrm{T}}(\alpha^2 \mathscr{K}^2 \mathbf{R}^{\mathrm{T}} \mathbf{R})\mathbf{f}$$

$$T_2 = -2\alpha \mathbf{R}\mathbf{h} + \alpha^2 \mathscr{K}^2 \mathbf{h}^{\mathrm{T}} \mathbf{R}^{\mathrm{T}} \mathbf{R}\mathbf{h}$$

$$T_3 = \alpha^2 \mathscr{K}^2 \mathbf{h}^{\mathrm{T}} \mathbf{R}^{\mathrm{T}} \mathbf{R}\mathbf{f} - \alpha \mathbf{R}\mathbf{f} \tag{32.7}$$

so that

$$\Delta V = T_1 + T_2 \tilde{u}^2 + 2T_3 \tilde{u}$$

Clearly a minimal requirement for convergence is that $T_2 \leq 0$. This is obtained for

$$0 \leq \alpha \mathbf{R} \mathbf{h} \leq \frac{2}{\mathcal{K}^2}$$

with the minimum of T_2 at

$$\alpha \mathbf{R} \mathbf{h} = \frac{1}{\mathcal{K}^2} \tag{32.8}$$

where the resulting minimum is

$$T_2 = \frac{1}{\mathcal{K}^2}$$

Since $\mathbf{Re}(k+1)$ can be interpreted as a weighting of the available state errors (with $r_i = 0$ for x_i unavailable), these constraints help determine the magnitude of the weightings. $\alpha \mathbf{R}$ can then be specified more or less well, depending upon the state of knowledge of $\mathbf{h}(s(k))$.

Having minimized T_2, we survey the implications. In this case, this means (from (32.8) and (32.7)) that

$$T_3 \tilde{u} = \left(\frac{\mathcal{K}^2}{\mathcal{K}^2} - 1 \right)(\alpha \mathbf{R} \mathbf{f})\tilde{u} = 0$$

The final consideration is that T_1 be small, as clearly $T_1 \geq 0$. This is essentially a need for the state error to have only a small dynamics contribution regardless of the control error, i.e. for $f(\mathbf{e}(k), \mathbf{x}_{\text{des}}(k))$ to be small. To assure this with unknown dynamics is difficult. A partial expedient may be to have the original system controlled in a feedback manner by a PID control, say, involving the error. Thus we propose that the system model

$$\mathbf{x}(k+1) = f_s(\mathbf{x}(k)) + h_s(\mathbf{x}(k))\,u(k)$$

be replaced by

$$\mathbf{x}(k+1) = f_s(\mathbf{x}(k)) + h_s(\mathbf{x}(k))\,\{K_p\,\mathbf{e}(k) + K_D\,\dot{\mathbf{e}}(k) + u_r(k)\}$$

so that the error equation is strongly stable even in the absence of ideal control.

Example

In this section we repeat the example of section 32.2.1 (page 764) involving a non-linear system.

Recall that the scalar system has dynamics equation

$$\mathbf{x}(k+1) = (1 - \mathbf{x}(k))^2 - u(k)$$

and the desire is that, for any $\mathbf{x}(k)$, $\mathbf{x}_{\text{des}}(k+1) = 0$. The required control is obviously

$$u^*(k) = - (1 - x(k))^2$$

In the simulation results, we have chosen

$$K(\gamma, s) = \exp\left(- \frac{(\gamma - s)^2}{2\,(0.2)^2}\right)$$

$$\hat{g}(0; s) = 5\,|\,s\,|$$

$$\alpha R = 0.5$$

and x is stored on a grid of size 41 elements: $\Gamma = [-4, 4]$.

Simulations were run from initial conditions $x(0)$ chosen randomly and uniformly on $[-3, 3]$ and were run until the first of the conditions $k = 5$ and $|x(k)| < 0.01$ was reached. Unstable cases were limited to avoid overflow (the initial $\hat{g}$ was not sufficient to assure stability over Γ).

Simulation results are shown in Fig. 32.4, which shows the curves $\hat{u}(j, x)$ for $j=1$, 10, and 50.

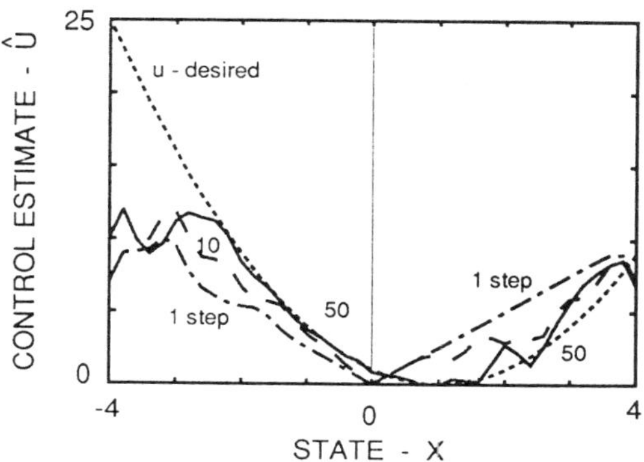

Figure 32.4 Evolution of learning of control law for simple example of non-parametric learning. Shown are actual (but unknown) ideal law and the law learned after 1, 10, and 50 trials.

32.3.4 Issues and discussion

A number of issues become apparent when the technique is applied. It is clear that the method works for simulations of simple systems, but the theoretical underpinning at this time is weak. Little is known about stability and convergence, about algorithm parameters, control law information requirements in terms of the minimal vector $s(k)$, or the rate of learning.

The algorithm here has been reported by Westphal (1990). A related continuous time algorithm has been investigated by Messner *et al.* (1989, 1991), for learning the actual joint torques needed for attaining a desired robot motion sequence; in that application, convergence results and successful implementation were shown.

Other methods of learning would seem to be somewhat related in philosophy but not detail to this one. Distantly related are the linearization methods of neighbouring optimal control (Chapter 27) or even optimalizing control (Chapter 31), with its efforts to optimize an unknown function by probing. A further possibility is the genetic algorithm approach, discussed for a robotic application by Davidor (1991).

32.4 FUZZY CONTROL

The intent in using so-called fuzzy control is to attempt to encode the knowledge of expert plant operators, expressed in terms such as

IF the temperature is *high*, THEN decrease the heat input *a bit*

into algorithms suitable for the binary world of computers, so that the computer becomes a tireless expert operator. The key ideas are as follows.

- Variables, rather than being processed in terms of their numerical values, are evaluated in terms of their degrees of membership of sets.
- Logic describing the relationships of variables has degree of membership as an output rather than a binary TRUE/FALSE output.
- The actual numerical output applied as a control to the plant is determined from the degrees of membership.

32.4.1 Fuzzy sets and logic

The basic notion of fuzzy logic is the concept of fuzzy set membership. Thus, for example, to say 'John is a tall person' or 'John belongs to the set of tall persons' in conventional set operations has a value TRUE (or 1) if HEIGHT(John) > 1.83 m (say) and FALSE (or 0) otherwise. It may be observed that the choice of the number 1.83 is perhaps arbitrary and in any case we might not be inclined to say 'John is not tall' if HEIGHT(John) = 1.82 m. The problems carry over in an obvious way to plant operations, where 'temperature is high' or 'speed is slow' have similar problems.

Because of the above problems, the set membership evaluation is changed to allow a degree of membership (DoM) function to be defined. This function is selected to have a value 1 when the criterion is definitely met and a value 0 if it is clearly not met. Thus we define for any set a DoM function, as

$$\mu_A(x) = \begin{cases} 0 & \text{when } x \text{ is not in the set } A \\ \text{intermediate value} & \text{when } x \text{ is arguably almost in the set } A \\ 1 & \text{when } x \text{ is in the set } A \end{cases}$$

Some membership functions are shown in Fig. 32.5.

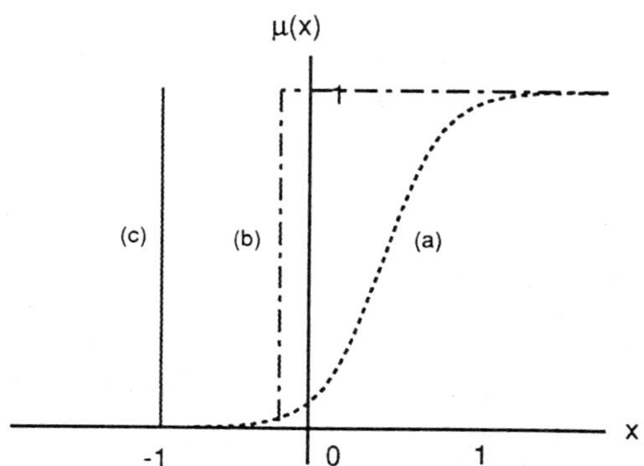

Figure 32.5 The degree of membership function: (a) typical fuzzy membership; (b) hard membership, with DoM = 1 for $x \geq -0.2$; and (c) hard membership for the single value $x = -1$.

Having defined the DoM of sets and auxiliary operators such as VERY (by interviewing operators, for example), we then are in a position to perform logical operations on the fuzzy sets. The main operations are OR, AND, NOT when using a particular variable vector, and $\Rightarrow$ (or linguistic implication, i.e. IF ... THEN ...) to relate different variables.

- The OR (or + or $\cup$) of two fuzzy sets A and B yields a fuzzy set C, $C = A + B$, with membership function defined by

$$\mu_{A+B}(x) = \max(\mu_A(x); \mu_B(x))$$

- The AND (or $\cdot$ or $\cap$) of two fuzzy sets A and B yields a fuzzy set C, $C = A \cdot B$, with membership function defined by

$$\mu_{A \cdot B}(x) = \min(\mu_A(x); \mu_B(x))$$

- The NOT (or complement or $\neg$) of a fuzzy set A is a fuzzy set C with membership function defined by

$$\mu_{\neg A}(x) = 1 - \mu_A(x)$$

- If two sets A and B are defined over different variables x and y (as in x for an input measurement and y for an output command) then the fuzzy conditional statement IF A THEN B generates a relation R with membership function defined on these two variables. This relation can be denoted $R = A \times B$, and its membership function is defined as

$$\mu_R(x, y) = \mu_{A \times B}(x, y) = \min(\mu_A(x); \mu_B(y))$$

Finally, a composition rule of inference is needed, so that the logic can be used. (The composition rule of inference is the rule by which we establish inferred information. For example, if we have a rule

IF a person is short THEN that person is a jockey

then the rule of inference tells us that the particularization 'John is short' allows us to claim 'John is a jockey'. Note that the rule need not be correct for the argument to be valid.)

Formally the composition rule is: if the fuzzy set A takes on value A' with membership function $\mu_{A'}(x)$, then the corresponding value of B, call it B', is defined by a membership function $\mu_{B'}(y)$ computed as

$$\mu_{B'}(y) = \max_{x} \min \left[\mu_{A'}(x); \mu_{A \cdot B}(x, y)\right]$$

In our application, the measurement of a variable will amount to defining a DoM which has value 0 except for $x = $ measurement; the output will still be a function, however.

32.4.2 Application to control

With these definitions, plus the agreement that a rule such as

> IF temperature is high AND flow rate is low
> THEN pump power is increased

is encoded as

> IF temperature is high THEN
> IF flow rate is low THEN pump power is increased

we are in a position to develop a fuzzy control law. In particular, for each output command variable, we develop rules in the format such as

> IF fluid temperature is high THEN
> IF flow rate is low THEN pump power is increased
>   OR
> IF fluid temperature is low
> THEN pump power is NOT increased
>   OR
> IF pump is overheating THEN pump power is decreased

etc. The next step is to take numerical values of the input quantities (by reading instruments) and evaluate the above using the composition rule of inference.

When this is done, we will have a DoM for the output variable, such as perhaps in Fig. 32.6.

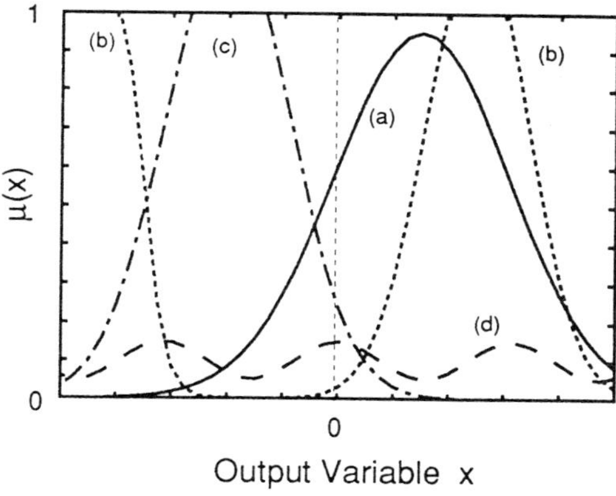

Figure 32.6 Inferring a value of an output variable is relatively easy if its DoM is single modal, as in (a). (b) shows two modes and is contradictory, (c) has a flat function and is ambiguous, while (d) is so low as to be indeterminate.

With the output variable's DoM established for the particular inputs supplied, the final step is to choose the command to be given. Advice on this varies, although provided there is a dominant value the result may not be too different for the candidates. The typical rule is to choose either the maximizing value or the centroid value

$$y_{max} = \arg \left\{ \max_{y} \left[\mu_{command}(y) \right] \right\}$$

$$y_{centroid} = \frac{\int \sigma \mu_{command}(\sigma)\, d\sigma}{\int \mu_{command}(\sigma)\, d\sigma}$$

The latter's particular appeal lies in its averaging of controls in which the maximizing value is not unique. Usually one would examine the rules further if contradictory or indeterminate results were indicated.

Example

We demonstrate with a trivial example. Measurements T of temperature are taken, and a change in heater command u is given. The logic relations of the operators are reduced to the following:

> IF $\{T$ is very high$\}$ THEN $\{$decrease u a lot$\}$
>   OR
> IF $\{T$ is low$\}$ THEN $\{$increase u somewhat$\}$
>   OR
> IF $\{T$ is about right$\}$ THEN $\{u$ is unchanged$\}$

The first step is to encode the T and u phrases above.

For T we choose the forms shown in Fig. 32.7. Although direct numerical work with the graphs is possible, these are encoded as the functions

A: $\mu_A(T) = \{T$ is very high$\} = \dfrac{1}{1 + \exp(-5(T-33))}$

B: $\mu_B(T) = \mu\{T$ is low$\} = \dfrac{1}{1 + \exp(-6.25(T-22))}$

C: $\mu_C(T) = \mu\{T$ is about right$\} = \exp(-2.5(T-26)^2)$

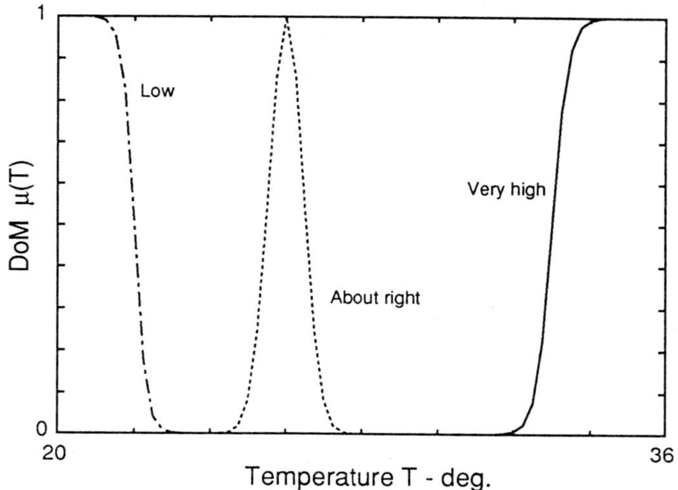

Figure 32.7 DoM describing the temperature T for text example.

and it is assumed $20 \le T \le 36$. For u, we have

$$1: \qquad \mu_1(\Delta u) = \mu(\text{decrease } u \text{ a lot}\} = \frac{1}{1 + \exp(-16.33(\Delta u + 0.6))}$$

$$2: \qquad \mu_2(\Delta u) = \mu(\text{increase } u \text{ somewhat}\} = \exp(-50(\Delta u + 0.4)^2$$

$$3: \qquad \mu_3(\Delta u) = \mu(u \text{ unchanged}) = \begin{cases} 1 & \Delta u = 0 \\ 0 & \Delta u \ne 0 \end{cases}$$

with Δu defined over the range $-1 \le \Delta u \le 1$, as shown in Fig. 32.8. The rules are then encoded as

> IF A THEN 1
>   OR
> IF B THEN 2
>   OR
> IF C THEN 3

The DoM of these are

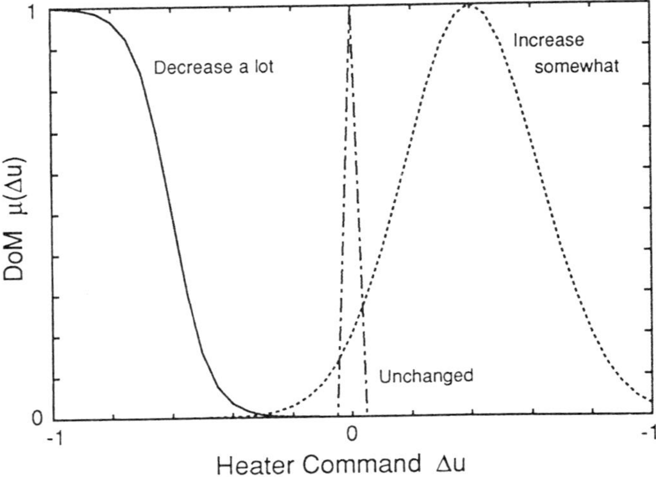

Figure 32.8 DoM implied by the three rules for change Δu in heater setting for text example.

$$\mu_{A1}(T, \Delta u) = \min \{ \mu_A(T), \mu_1(\Delta u) \}$$

$$\mu_{B2}(T, \Delta u) = \min \{ \mu_B(T), \mu_2(\Delta u) \}$$

$$\mu_{C3}(T, \Delta u) = \min \{ \mu_C(T), \mu_3(\Delta u) \}$$

and are noteworthy for being functions of two variables. Finally, we develop the function of two variables giving the contingencies

$$\mu_{OR}(T, \Delta u) = \max \{\mu_{A1}(T, \Delta u), \mu_{B2}(T, \Delta u), \mu_{C3}(T, \Delta u)\}$$

as shown in Fig. 32.9 which encapsulates the possibilities and for each measurement.

For a measurement T_{meas}, we define the DoM function corresponding to this instantiation.

$$\mu_{\mathrm{meas}}(T) = \begin{cases} 1 & T = T_{\mathrm{meas}} \\ 0 & T \neq T_{\mathrm{meas}} \end{cases}$$

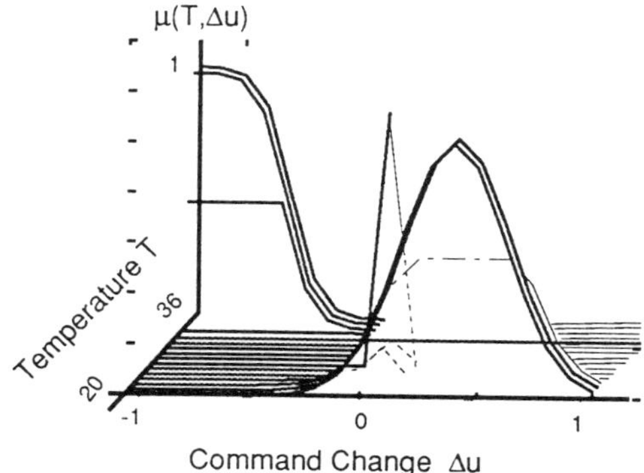

Figure 32.9 DoM as a function of temperature and heater change resulting from the three rules.

We can then develop the DoM for the heater command at this value as

$$\mu_{controller}(\Delta u \; @ \; T_{meas})$$

$$= \max_{T} \min \left\{ \mu_{meas}(T), \; \mu_{OR}(T, \Delta u) \right\}$$

Choosing Δu as the argument of the maximization of this (rather than the centroid) gives

$$\Delta u(T) = \arg \max_{\Delta u} \mu_{controller}(\Delta u \; @ \; T)$$

Such a function is shown in Fig. 32.10. It will be seen to be poorly defined in some regions, simply because in fact $\mu_{controller}(\Delta u \; @ \; T)$ is small at values such as $T = 24.5$. This indicates an incomplete set of rules.

32.4.3 Comments and further reading

A control law is ultimately a rule or mapping from a set of measurements and required-variable values to a command to a plant. Fuzzy control laws ultimately yield just that, although the motivations

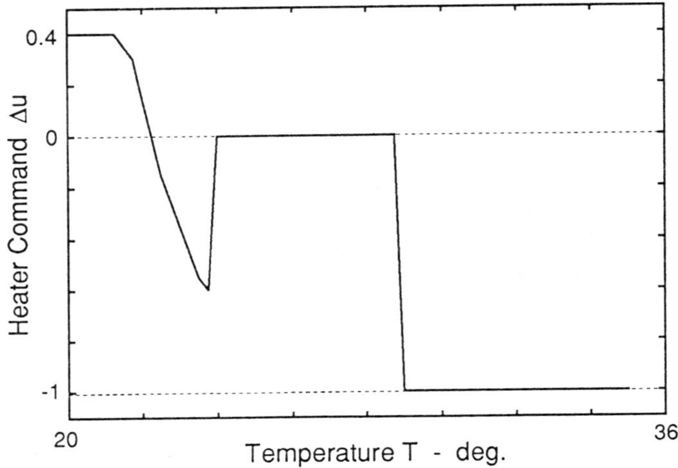

Figure 32.10 The command change as a function of temperature when peak value of DoM is the selection rule. Note that the DoM is in some regions of $(T, \Delta u)$ too small for a reliable decision and more rules are needed.

are not mathematical and the implementation may (unnecessarily) utilize fuzzy logic computations in real-time. The rules are usually tuned (by changing the DoMs for inputs or outputs or by adding more rules), just as PIDs, LQGs, and other rules are tuned so that the actual (rather than theoretical) system gives 'good' performance. As a way of generating and adjusting non-linear laws for non-linear systems, fuzzy control theory shows promise and, we are told, is actually used in some systems. Nevertheless, in spite of the rhetoric, ultimately this is another method for generating commands as functions of measurements, and the function can be precomputed and stored in computer memory along with interpolation rules.

Since this intriguing alternative to conventional approaches to control law design is based upon fuzzy sets, one starting point for serious study is the book by Zadeh (1965); early application thoughts were pursued by Zadeh (1973).

Examples are in the literature. Pilot plant-size boiler–steam engine and stirred tank systems have been studied, and comparisons to direct digital control of PI type have been made which are favourable to fuzzy control (King and Mamdani, 1977). A small nuclear reactor was controlled by a system described by Bernard (1988); another recent study and description is by Li and Lau (1989).

32.5 EXPERT SYSTEMS

The branch of AI research called expert systems is fairly recent, and its attempted application to control systems even more recent. It seems to be applicable to the supervisory level of control and to PLC decision-making rather than to algorithmic device commanding. We look briefly at the concept and possiblilities.

32.5.1 The notion of the artificial expert

An expert system is a piece of software which attempts to perform like a human expert within a very constrained field of knowledge but at considerable depth. Some of the most famous examples of expert systems are in medicine, where a list of symptoms and laboratory results is presented to a computer which then suggests diagnoses and treatments; a commercial example is the widely reported DEC® system used to configure computer systems to meet customer requirements.

The reasons for using a software expert system vary, although the goal ultimately is to emulate a human expert. The motivation may be to replace the human in a situation where

- the human cannot always be available (doctors in remote areas);
- the situation is a routine one in which the human expert would be under-utilized (routine examination of X-ray films or electrocardiograms in medicine, plant operating data in engineering);
- augmentation is needed, i.e. when the knowledge base is too broad for a single human (diagnosis of obscure diseases, selection of configuration for complex systems); or
- consistency of application of expertise is desirable (human operators make mistakes when they tire or have information overloads), etc.

To be useful, the expert system should have a number of properties. It must be based on a large body of knowledge contained in rules of an `IF ... THEN ... ELSE ...` type. It should have a traceable line of reasoning, i.e. it must indicate why a certain decision has been made, and do so in an understandable way. It must be flexible enough to accept new and altered rules.

To build an expert system, the knowledge must exist in an empirical manner to which heuristic rules are applied. Also it is fundamental that the system is likely to be less than 100% successful, just as are the humans it emulates.

32.5.2 Industrial expert systems

Expert systems in industrial applications are generally of two types: diagnostic systems and design systems. Diagnostic systems are used to suggest reasons for a failure or malfunction, and their primary economic benefit is the reduction of down-time by warning of problems or assisting in isolating causes for rapid repair. In design systems, the benefit seems to derive from suggesting alternatives to human designers for tasks such as product configuration or management of production flow.

A new type of system is being investigated for process control on-line. It may be observed, and perhaps should have been emphasized, that control is not a simple matter of a few, possibly complicated algorithms and their associated input–output relationships. In fact, even simple algorithms have a number of associated heuristically based tests. For example, we note the following.

- Sensor data must be tested. Not only must it be verified that the sensor has not failed absolutely, but the data received must be at least roughly consistent with other data.
- Commands should be reasonable, which usually means at least that they are magnitude-limited.
- In the event of a malfunction, the controller should recognize that there may be a problem and do something 'reasonable.' No control law should make a problem worse.
- The control law should react properly to predictable transients. For example, special startup sequences should be performed under appropriate circumstances.

In addition, there are certain tasks performed by expert human operators which might arguably be done by a software system. Among these are decisions as to when to retune a PID controller (whether a self-tuner or a manually tuned controller), when to consider changeover to a backup system, etc.

One suggested structure for viewing the problem was presented by Åström *et al.* (1986). Their study, using OPS4 as the expert system shell, used rules such as

```
RULE 5
    ( (State is upstart) (Goal is PID-control)
      (PID parameters available)
→
      (<DELETE> (State is upstart)
      (<ADD> (State is PID-control) (Start PID))))
RULE 12
    ( (State is =X)
      (Alarm has occurred)
→
      (<DELETE> (State is =X)
      (<ADD> (State is alarm)(Old state is =X))))
```

The system used 70 such rules. The plant studied had a choice of three algorithms (PID, relay, kc–tc) with a choice of two sets of PID parameters, depending on set point, or self-tuning of a Ziegler–Nichols type.

32.5.3 Comments and further reading

Expert systems of this type (rather than the fuzzy set encoding of the previous section) are still a field with much potential but few reported results. The paper by Åström *et al.* (1986) gives an outline of the potential and discusses some of the problems. Otherwise, expert systems are a standard topic of computer science/AI. A brief summary is given by Tanimoto (1987).

32.6 FINAL REMARKS

Learning control is, rather like adaptive control, a philosophy more than a method. We have looked at four different approaches in this chapter. For each, we have given leads to more reading. Overall, the reader might be advised to follow the progress of the field in the literature. The robotics conferences are among those presenting interesting results, as are the journals such as *IEEE Transactions on Robotics and Automation*.

33

Robust control

Feedback control is necessary to compensate for unpredictable disturbances to a plant or process and for inaccuracies in predicting plant response because of errors or approximations in plant models. There have always been several techniques for designing compensators: feedforward control based upon measurements of disturbances and placing margins in the design specifications are among the basic ones, while adaptive and self-tuning controllers are among the more complicated ways of improving plant models. An alternative, philosophically related to gain and phase margins, is the rapidly developing field of robust control theory, which entails mathematical design of control laws to meet defined uncertainty levels. In this chapter we briefly and superficially introduce this subject.

33.1 SYNOPSIS

Consider the system portrayed in Fig. 33.1. From this it is easy to show that

$$\mathbf{Y}(s) = (\mathbf{I} + \mathbf{G}(s)\,\mathbf{C}(s))^{-1}\,\mathbf{G}(s)\,\mathbf{C}(s)\,\mathbf{R}(s) + (\mathbf{I} + \mathbf{G}(s)\,\mathbf{C}(s))^{-1}\,\mathbf{D}(s)$$

$$- (\mathbf{I} + \mathbf{G}(s)\,\mathbf{C}(s))^{-1}\mathbf{G}(s)\,\mathbf{C}(s)\,\mathbf{N}(s)$$

There are several goals to be met through choice of $\mathbf{C}(s)$:

1. the system should be stable, even in the presence of errors in the implementation of $\mathbf{G}(s)$ and $\mathbf{C}(s)$;
2. the response $\mathbf{Y}(s)$ should be satisfactory, even when the system is disturbed (by $\mathbf{D}(s)$) or the measurements are noisy (through $\mathbf{N}(s)$); and
3. preferably, both 1 and 2 should hold simultaneously.

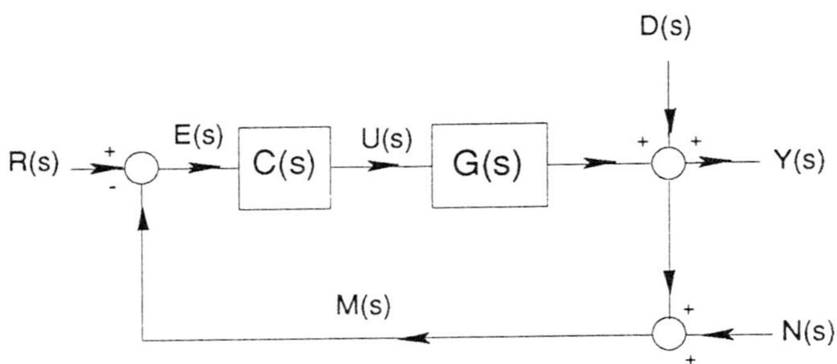

Figure 33.1 System structure for transfer function domain robust controller studies.

If the above are to be done explicitly, then the size of model errors and exogenous disturbances and noise must be defined in some way. This means that measures of these factors must be developed, and then the extent of the errors defined for each situation. For these purposes, the theory works with mathematical norms, which are generalizations of absolute values of scalars and are denoted $\|\cdot\|$ (Appendix B), and the problem definition includes restrictions such as

$$\|\mathbf{D}(j\omega)\| \leq \Delta(\omega)$$

and

$$\mathbf{G}(s) = \mathbf{G}_{\text{nom}}(s)\,(\mathbf{I} + \mathbf{L}(s)) \qquad \text{where } \|\mathbf{L}(j\omega)\| < l(\omega)$$

Under these circumstances, typical results are small gain principles of the following type

- For robust stability, it is required that

$$\|\mathbf{G}(j\omega)\,\mathbf{C}(j\omega)\| < l^{-1}(\omega)$$

- For optimal performance, it is required that we form

$$\min_{\mathbf{C}} \|(\mathbf{I} + \mathbf{G}(s)\,\mathbf{C}(s))^{-1}\|$$

- For robust performance, it is required that conditions such as

$$1 > \| w(\omega) \, (\mathbf{I} + \mathbf{G}(s) \, \mathbf{C}(s))^{-1} \|$$

$$+ \| l_\mathrm{m}(\omega)(\mathbf{I} + \mathbf{G}(s) \, \mathbf{C}(s))^{-1} \mathbf{G}(s) \, \mathbf{C}(s) \|$$

hold.

Another important approach to robust performance while maintaining stability uses the Youla parameterization of stabilizing controllers.

The state–space method using loop transfer recovery (LTR) to improve the parameter robustness of LQR controllers (Chapter 28) by systematically detuning the filter's model is relatively easy conceptually and is computable in a straightforward manner.

33.2 BASIC NOTIONS AND KEY RESULTS

The approaches used in robust control theory require more advanced mathematical methods than we have explicitly used in previous chapters, although many of those chapters could be expressed in terms of the notations used here. Thus we define the distance measures called norms and present the most important ones in this section.

It is arguable that there have been two particularly key notions used in robust control theory: small gain theorems and parameterization of the set of stabilizing gains. Those are also met in this section.

33.2.1 Notation and error models

When discussing errors, we need some sort of measure of them. When the quantities involved are scalars this is relatively easy, but when they are (or can be considered as) vectors, it is necessary to use some more advanced mathematics. The basic notion of size is the **norm** of a vector. For a vector $\mathbf{v}$, the norm is a scalar function, denoted $\|\mathbf{v}\|$, with the following mathematical properties.

1. $\| \mathbf{v} \| > 0$ and $\| \mathbf{v} \| = 0$ if $\mathbf{v} = \mathbf{0}$.
2. $\| \alpha \, \mathbf{v} \| = | \alpha | \, \| \mathbf{v} \|$ for any scalar α.
3. $\| \mathbf{v} + \mathbf{y} \| \leq \| \mathbf{v} \| + \| \mathbf{y} \|$.

There are many different norms, of which the most common are defined as

$$\|\mathbf{v}\|_p = \left(\sum_{i=1}^{n} |v_i|^p \right)^{\frac{1}{p}}$$

A very common norm has $p = 2$ and is called the euclidean norm. Another useful norm uses $p = 1$. The limit $p \to \infty$ turns out to yield the $\max_i |v_i|$ and is called the ∞-norm,

$$\|\mathbf{v}\|_\infty = \max_i |v_i|$$

Matrices can be viewed in either of two ways: as vectors having their own norms or as operators on vectors $\mathbf{v}$ which yield other vectors $\mathbf{y}$ and hence having their norms defined by the relationship of the $\mathbf{v}$-norm and the $\mathbf{y}$-norm. This latter interpretation gives an **induced norm**, and the usual definition has

$$\|\mathbf{A}\|_p = \sup_{\mathbf{v} \neq \mathbf{0}} \frac{\|\mathbf{A}\mathbf{v}\|_p}{\|\mathbf{v}\|_p}$$

where 'sup' denotes the supremum, or least upper bound, and is effectively 'almost' the max, or maximum. (The difference is subtle and mostly does not concern us. We see, however, that $f(x) = 1 - e^{-x}$ has a largest possible value, or sup, of 1, whereas $\max(f(x))$ is undefined – there is no value of x for which it takes on a value which cannot be exceeded.) We notice that

$$\|\mathbf{A}\mathbf{v}\|_p \leq \|\mathbf{A}\|_p \|\mathbf{v}\|_p$$

The matrix $\mathbf{A}$ will also have a singular value decomposition (SVD) given by

$$\mathbf{A} = \mathbf{U}\Sigma\mathbf{V}^*$$

with

$$UU^* = I \qquad VV^* = I \qquad \Sigma = \begin{bmatrix} \Sigma_1 & 0 \\ 0 & 0 \end{bmatrix}$$

$$\Sigma_1 = \mathrm{diag}\,(\sigma_1, \sigma_2, \ldots, \sigma_r) \qquad \sigma_1 \geq \sigma_2 \geq \ldots \geq \sigma_r > 0$$

The largest singular value is σ_1 and is denoted $\bar{\sigma}(A)$; the minimum singular value is denoted $\underline{\sigma}(A)$; the number r of singular values is the rank of the matrix. The connection with norms is that it turns out that

$$\bar{\sigma}(A) = \|\,A\,\|_2$$

In the above, the vectors and matrices could be functions of parameters, in which cases the norms are also functions. We will most commonly meet this with transfer function matrices $G(s)$, and we will be interested in frequency responses, and in particular $\|G(j\omega)\|_2$ and its limits $\bar{\sigma}(G(j\omega))$ and $\underline{\sigma}(G(j\omega))$. These latter bound the frequency response and allow us to evaluate frequency domain compensators.

The ideas above extend to signals, which are functions of time, and 'averaging' of those functions, a notion particularly related to noises and disturbances. Of the extensions, the common one is the quadratic extension, which for a signal $y(t)$ is

$$\|\,y\,\|_2 = \left(\int_0^\infty y^T(t)\, y(t)\, dt \right)^{\frac{1}{2}}$$

and which because of Parseval's theorem becomes

$$\|\,y\,\|_2 = \left(\frac{1}{2\pi} \int_{-\infty}^\infty Y^T(-j\omega)\, Y(j\omega)\, d\omega \right)^{\frac{1}{2}}$$

The important result of all this is that for an input signal $u(s)$ into a system $G(s)$, the norm induced in G by the signal norm $\|\cdot\|_2$ is

$$\| G \|_2 = \sup_{\omega} \bar{\sigma}(G(j\omega)) = \| G \|_{\infty}$$

which is the ∞-norm. We emphasize that the latter is an induced norm.

This all becomes important when model errors are to be defined in explicit terms by specifying the limits of accuracy of matrices and transfer functions. For a state–space example, consider a model

$$\dot{x} = Ax + Bu$$

$$y = Cx \tag{33.1}$$

in which A is imprecisely known. Then a possible model is that

$$A = A_0 + \delta A \tag{33.2}$$

where δA may be one of several possible types. If its structure is unspecified, but its size is known to be limited, then the model is said to have unstructured variations and the error is assumed to satisfy a constraint such as

$$\| \delta A \| \leq a \tag{33.3}$$

for a specified scalar a. An alternative has structured variations, such as

$$\delta A = \sum_i \alpha_i A_i \tag{33.4}$$

where $-1 \leq \alpha_i \leq 1$ and of course other possibilities, such as randomly varying α_i, can be considered.

With multivariable transfer functions, similar possibilities exist. Thus the plant model $G(s)$ may be

$$G(s) = G_0(s) + \delta G(s)$$

with possibilities including

$$\delta G(s) = L(s)\,G_o(s)$$

$$\delta G(s) = G_o(s)\,L(s)$$

$$\delta G(s) = \Delta(s)$$

where the uncertainties are multiplicative on the output, multiplicative on the input, and additive, respectively. The error matrices may be structured, in that elements are known to within an unknown parameter as in (33.4), or unstructured but bounded. In the latter case, for example, we may have

$$\|L(j\omega)\| \le l_m(\omega) \tag{33.5}$$

or

$$\|\Delta(j\omega)\| \le l_a(\omega) \tag{33.6}$$

where l_m and l_a are specified scalar functions, but nothing further is known about the individual elements of the matrices L or Δ.

In the above, the matrix norm used will vary with the theory being used, although in much of the theory both the 2-norm and the ∞-norm appear. In all cases, the ultimate use is that the plant $G(s)$ is said to be a realization from a set $\mathscr{G}$ defined as, for example,

$$\mathscr{G} = \{\, G(s)\,|\,G(s) = (I + L(s))\,G_o(s)\,;\,\|L(s)\|_2 \le l_m(\omega)\,\} \tag{33.7}$$

33.2.2 Small gain theory

One of the fundamental notions in the study of robustness is small gain theory. In some cases, this can seem obvious and almost incidental, but several key results in robust stability theory are built upon it. An over-simplified characterization says that if the loop gain $G(s)\,C(s)$ is stable, then the closed-loop system is stable if, for all ω

$$\| \mathbf{G}(j\omega)\,\mathbf{C}(j\omega) \| < 1 \tag{33.8}$$

This is arguable using a Nyquist approach concerning origin encirclements by $\det(\mathbf{I} + \mathbf{G}(s)\,\mathbf{C}(s))$, and superficially would seem to be sufficient since there would be no encirclements under the stated condition. In any case, the result tells us that if the loop gain is 'small', then closing the loop on a stable open-loop system will yield a stable closed-loop system; more precisely and more relevant to robustness, a feedback loop composed of stable subsystems will remain stable in the presence of uncertainties provided that the transfer function gain is 'small enough'.

33.2.3 Gain parameterization

One of the difficulties with design is that the compensator which meets specifications for any given plant is rarely unique; we hesitate to design compensators for all possible plants in the set $\mathscr{G}$ and then choose from the intersection of these to obtain a robust design. One way to avoid this exhaustive approach is to parametrize the stabilizing controllers for one of the elements of $\mathscr{G}$. Such parameterization can be approached several ways, of which we show two. The first, due partly to Youla (Youla *et al.*, 1976) and bearing his name, proceeds along the following lines.

The idea here is that all compensators $\mathbf{C}(s)$ which stabilize a plant $\mathbf{G}(s)$, i.e. for which $(\mathbf{I} + \mathbf{G}(s)\,\mathbf{C}(s))^{-1}$ has only left-half plane zeros, can be written in terms of transfer functions $\mathbf{K}(s)$ which are stable. In fact, if we can find a stabilizing compensator $\mathbf{C}_0(s)$, and if

$$\mathbf{C}_0(s) = \mathbf{U}_0(s)\,\mathbf{V}_0^{-1}(s) = \tilde{\mathbf{V}}_0^{-1}(s)\,\tilde{\mathbf{U}}_0(s)$$

$$\mathbf{G}(s) = \mathbf{N}(s)\,\mathbf{D}^{-1}(s) = \tilde{\mathbf{D}}^{-1}(s)\,\tilde{\mathbf{N}}(s)$$

where $\mathbf{U}_0, \mathbf{V}_0, \mathbf{N}, \mathbf{D}$ are appropriately sized matrices of polynomials, then any stable matrix $\mathbf{K}(s)$ of appropriate size will generate a stabilizing compensator $\mathbf{C}(s)$ through the calculation

$$\mathbf{C}(s) = (\mathbf{U}_0(s) + \mathbf{D}(s)\,\mathbf{K}(s))\,(\mathbf{V}_0(s) - \mathbf{N}(s)\,\mathbf{K}(s))^{-1}$$

This can also be written

$$\mathbf{C}(s) = \mathbf{C}_0(s) + \tilde{\mathbf{V}}_0^{-1}(s)\,\mathbf{K}(s)\,(\mathbf{I} - \mathbf{V}_0^{-1}(s)\,\mathbf{N}(s)\,\mathbf{K}(s))^{-1}\,\mathbf{V}_0^{-1}$$

and several other ways.

To find the right-hand matrix fraction description of matrices such as $\mathbf{G}(s)$, we use the Smith–McMillan form $\mathbf{M}(s)$, with $\mathbf{L}(s)$ and $\mathbf{R}(s)$ the appropriate unimodular transformation matrices (see sections 10.2.2 and 11.3.2). This follows from the basic form

$$\mathbf{G}(s) = \mathbf{L}(s)\,\mathbf{M}(s)\,\mathbf{R}(s)$$

where

$$\mathbf{M}(s) = \mathrm{diag}\,\{e_1(s), ..., e_r(s), ...\}\,(\mathrm{diag}\,\{f_1(s), ..., f_r(s), ...\})^{-1}$$
$$= \mathbf{N}'(s)\,\mathbf{D}'^{-1}(s)^{-1}$$

Using this, we may define

$$\mathbf{N}(s) = \mathbf{L}(s)\,\mathbf{N}'(s) \qquad \mathbf{D}(s) = \mathbf{R}^{-1}(s)\,\mathbf{D}'(s)$$

so that

$$\mathbf{G}(s) = \mathbf{N}(s)\,\mathbf{D}^{-1}(s)$$

A similar representation may be obtained for $\mathbf{C}_0(s)$.

Another approach, related to the parameterization above but in a different form and having the incidental feature of alternatively characterizing the compensator, defines, using $\mathbf{G}_0(s)$, the matrix $\mathbf{Q}(s)$ which implies the desired controller $\mathbf{C}(s)$. Thus let

$$\mathbf{Q}(s) = \mathbf{C}(s)\,(\mathbf{I} + \mathbf{G}_0(s)\,\mathbf{C}(s))^{-1}$$

$$\mathbf{C}(s) = \mathbf{Q}(s)\,(\mathbf{I} + \mathbf{G}_0(s)\,\mathbf{Q}(s))^{-1}$$

and note that the closed-loop transfer function is given by $G_0(s) Q(s)$; if the closed-loop system is stable, we may say $Q(s)$ stabilizes $G_0(s)$. Then, under certain restrictions on the set $\mathscr{G}$ of possible plants, i.e.

* $\mathscr{G} = \{G(s) \mid G(s) = G_0(s) + \delta G(s); \delta G(s)$ suitably bounded,
* all RHP poles of G_0 are of order 1, and
* all poles of $G(s)$ on the real axis are at the origin$\}$,

we have the following:

> if $Q_0(s)$ stabilizes the system with plant $G_0(s)$, then all $Q(s)$ which stabilize $G_0(s)$ are given by
>
> $$Q(s) = Q_0(s) + Q_1(s)$$
>
> where $Q_1(s)$ is stable and $G_0(s) Q_1(s) G_0(s)$ is stable.

In either case above, if we can find a stabilizing controller, we know a parameterization of all stabilizing controllers and can seek among these for a controller with further properties, such as robust stability or robust performance. A sub-optimal design method to attempt is to find a simple representation with a variable parameter, such as a gain, and vary that gain to meet the specifications; thus if $G_0(s)$ is stable, the choices $Q_0(s) = 0$ and $Q_1(s)$ equals anything stable are adequate for the theorem if not in practice. The latter is because the implications of $Q_0(s) = 0$ are that $C(s) = G_0^{-1}(s)$, which may be impossible to implement and be unstable if the plant is not minimum phase.

33.3 TRANSFER FUNCTION APPROACHES TO ROBUSTNESS

Much recent robustness theory utilizes transfer function models for the mathematical arguments, although it has been remarked that the actual calculations frequently use state–space operations. In this section, we consider robust stability and robust performance, with emphasis on the notions involved. We remark that design methods seem largely of the cut and try or iteration of analysis variety.

33.3.1 Robust stability

Robust stability results depend in detail, but not in essence, on the uncertainty type – whether input or output multiplicative, for example. To develop the techniques used and illustrate a gain limit yielding robust stability, we consider a set $\mathscr{G}$ of plants characterized by a nominal plant $\mathbf{G}_0(s)$ with output multiplicative unstructured uncertainty

$$\mathbf{G}(s) = (\mathbf{I} + \mathbf{L}(s))\,\mathbf{G}_0(s)$$

Then the system will remain stable with compensator $\mathbf{C}(s)$ provided that

$$\bar{\sigma}(\mathbf{G}_0(j\omega)\,\mathbf{C}(j\omega)\,[\mathbf{I} + \mathbf{G}_0(j\omega)\,\mathbf{C}(j\omega)]^{-1}) < \frac{1}{\bar{\sigma}(\mathbf{L}(j\omega))} \qquad (33.9)$$

A short version of the argument used for this is that the closed-loop transfer function frequency response is

$$\mathbf{T}(j\omega) = \mathbf{G}(j\omega)\,\mathbf{C}(j\omega)\,[\,\mathbf{I} + \mathbf{G}(j\omega)\,\mathbf{C}(j\omega)\,]^{-1}$$

and hence (dropping the arguments $j\omega$) we need that

$$\det\,[\,\mathbf{I} + \mathbf{GC}\,] \neq 0$$

This requires that the singular values all be positive, so that

$$\underline{\sigma}\,[\,\mathbf{I} + \mathbf{GC}\,] = \underline{\sigma}\,[\,\mathbf{I} + \mathbf{G}_0\mathbf{C} + \mathbf{LG}_0\mathbf{C}\,] > 0$$

Thus we need

$$\underline{\sigma}\,[\,[(\mathbf{LG}_0\mathbf{C})^{-1} + \mathbf{L}^{-1} + \mathbf{I}\,]\,\mathbf{LG}_0\mathbf{C}\,] > 0$$

which holds, since $\mathbf{L}\,\mathbf{G}_0\,\mathbf{C}$ is stable by assumption, if

$$\underline{\sigma}\,[\,[(\mathbf{G}_0\mathbf{C})^{-1} + \mathbf{I}\,]\,\mathbf{L}^{-1} + \mathbf{I}\,] > 0$$

This holds provided that

$$\underline{\sigma} \, [\, [(G_0 C)^{-1} + I \,] \, L^{-1} \,] > 1$$

for which it is sufficient that

$$\underline{\sigma} \, [(G_0 C)^{-1} + I \,] \, \underline{\sigma} \, [L^{-1}] > 1$$

This is the same as

$$\bar{\sigma} \, [\, [(G_0 C)^{-1} + I \,]^{-1}] \, \bar{\sigma} \, [L] < 1 \qquad (33.10)$$

from which (33.9) follows.

An alternative argument can be developed using the small gain theorem. For variety, we consider the case of input multiplicative errors as in

$$G(s) = G_0(s) \, [I + L(s)] \qquad \| L(j\omega) \| \le l(\omega) \qquad (33.11)$$

Our objective is to evaluate the robustness of stability associated with a controller $C(s)$. The system is diagrammed in Fig. 33.2(a) with an alternative form in 33.2(b).

In the alternative form, we make the definitions

$$M(s) = - \, G_0(s) \, C(s) \, (I + G_0(s) \, C(s))^{-1} \, l_0$$

$$\Delta(s) = \frac{L(s)}{l_0}$$

where

$$l_0 = \sup_{\omega} \, (\bar{\sigma}(L(j\omega)))$$

and hence $\| \Delta(s) \| \le 1$.

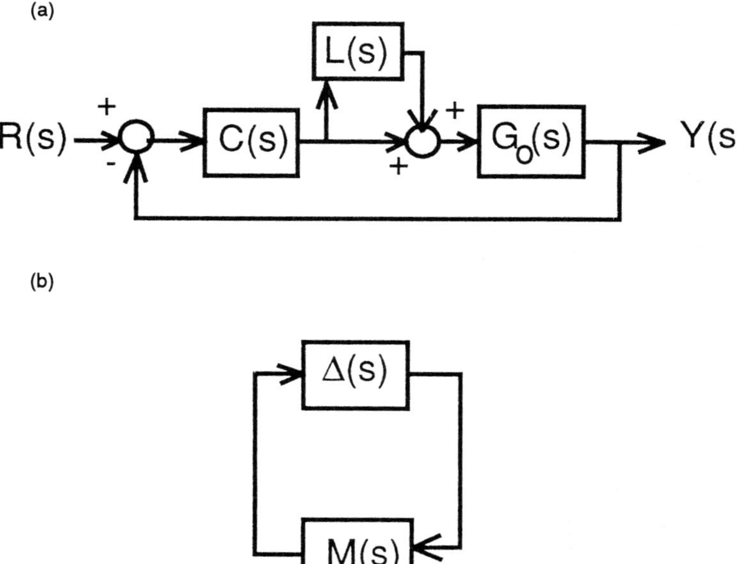

Figure 33.2 Uncertainty description models (after Morari and Zafiriou, 1989): (a) multiplicative uncertainty block diagram; and (b) general model.

We use the small gain result and consider a system as in Fig. 33.2(b). If $\mathbf{M}(s)$ is stable, the implication of the small gain theorem is that the indicated loop is stable provided $\|\mathbf{M}(s)\,\Delta(s)\| < 1$ for all ω; an alternative statement of this for the induced ∞-norm is that, for all ω

$$\bar{\sigma}(\mathbf{M}(j\omega)\,\Delta(j\omega)) < 1$$

$$\sup_{\omega} \, (\bar{\sigma}(\mathbf{M}(j\omega)\,\Delta(j\omega))) < 1$$

or, because of the restriction on $\|\Delta(s)\|$, that

$$\sup_{\omega} \, (\bar{\sigma}(\mathbf{M}(j\omega))) < 1$$

or

$$\|\mathbf{M}(s)\|_{\infty} < 1$$

Immediate application of this to the system of Fig. 33.2(a) yields that, for all ω

$$\bar{\sigma}(\mathbf{G}_0(j\omega)\,\mathbf{C}(j\omega)\,(\mathbf{I} + \mathbf{G}_0(j\omega)\,\mathbf{C}(j\omega))^{-1})\,l_0 < 1 \qquad (33.12)$$

or

$$\|\,\mathbf{G}_0(s)\,\mathbf{C}(s)\,(\mathbf{I} + \mathbf{G}_0(s)\,\mathbf{C}(s))^{-1}\,\|_\infty < \frac{1}{l_0}$$

is required for robust stability. Furthermore, since at high frequencies $\mathbf{G}_0(j\omega)\,\mathbf{C}(j\omega)$ is usually 'small', we can approximate (33.12) to yield the requirement that, for ω large

$$\bar{\sigma}(\mathbf{G}_0(j\omega)\,\mathbf{C}(j\omega)) < \frac{1}{l_0}$$

It is worth noting that any of these is easily computed (using SVD algorithms) for a given $\mathbf{C}(s)$.

The problem with the above is not only finding $\mathbf{C}(s)$ which stabilizes the nominal plant $\mathbf{G}_0(s)$, but finding one with sufficient stability margin to stabilize all the possible plants meeting the requirements (33.5–8).

33.3.2 Robust performance

Discussion of performance requires a definition of performance goals. For the robust theory, this is defined in terms of the error. Thus we take the system description

$$\mathbf{Y}(s) = (\mathbf{I} + \mathbf{G}(s)\,\mathbf{C}(s))^{-1}\,\mathbf{G}(s)\,\mathbf{C}(s)\,\mathbf{R}(s) + (\mathbf{I} + \mathbf{G}(s)\,\mathbf{C}(s))^{-1}\,\mathbf{D}(s)$$

$$- (\mathbf{I} + \mathbf{G}(s)\,\mathbf{C}(s))^{-1}\,\mathbf{G}(s)\,\mathbf{C}(s)\,\mathbf{N}(s)$$

and define the error $\mathbf{e}(t)$ and its transfer function $\mathbf{E}(s)$ through

$$\mathbf{E}(s) = \mathbf{Y}(s) - \mathbf{R}(s)$$

$$= (\mathbf{I} + \mathbf{G}(s)\,\mathbf{C}(s))^{-1}\,(\mathbf{D}(s) - \mathbf{R}(s))$$

$$- (\mathbf{I} + \mathbf{G}(s)\,\mathbf{C}(s))^{-1}\,\mathbf{G}(s)\,\mathbf{C}(s)\,\mathbf{N}(s) \tag{33.13}$$

Then we choose to evaluate the error in the square sense

$$\| \mathbf{e} \|_2 = \left(\int_0^\infty (\mathbf{e}^{\mathrm{T}}(t)\,\mathbf{e}(t)\,\mathrm{d}t\,) \right)^{\frac{1}{2}}$$

and which because of Parseval's theorem becomes

$$\| \mathbf{e} \|_2 = \left(\frac{1}{2\pi} \int_{-\infty}^{\infty} \mathbf{E}^{\mathrm{T}}(-\mathrm{j}\omega)\,\mathbf{E}(\mathrm{j}\omega)\,\mathrm{d}\omega \right)^{\frac{1}{2}} \tag{33.14}$$

We can extend this by

1. frequency weighting the error, with lower frequencies being emphasized compared to higher frequencies;
2. selecting some components of $\mathbf{e}(t)$ as being more important than others; and
3. emphasizing certain input signals.

The first two of these can be done with a matrix $\mathbf{W}_{\mathrm{output}}(\omega)$ and the third with a matrix $\mathbf{W}_{\mathrm{input}}(\omega)$ so that instead of (33.14) we work with

$$\| e' \|_2 = \left(\frac{1}{2\pi} \int_{-\infty}^{\infty} (\mathbf{W}_{\mathrm{output}}\,\mathbf{E}(\mathrm{j}\omega)\,\mathbf{W}_{\mathrm{input}})^{\mathrm{H}}\,(\mathbf{W}_{\mathrm{output}}\,\mathbf{E}(\mathrm{j}\omega)\,\mathbf{W}_{\mathrm{input}})\,\mathrm{d}\omega \right)^{\frac{1}{2}}$$

where $(\ )^{\mathrm{H}}$ denotes the complex conjugate transpose. In these terms, we seek the controller $\mathbf{C}(s)$ which minimizes this error, or the solution of

$$\min_{\mathbf{C}(s)} \| \mathbf{W}_{\text{output}}(s)\, \mathbf{E}(s)\, \mathbf{W}_{\text{input}}(s) \|_2^2$$

$$= \min_{\mathbf{C}(s)} \frac{1}{2\pi} \int_{-\infty}^{\infty} (\mathbf{W}_{\text{output}}\, \mathbf{E}(j\omega)\, \mathbf{W}_{\text{input}})^{\text{H}}\, (\mathbf{W}_{\text{output}}\, \mathbf{E}(j\omega)\, \mathbf{W}_{\text{input}})\, d\omega$$

For computation, the above needs the inputs $\mathbf{D}(s)$, $\mathbf{R}(s)$, $\mathbf{N}(s)$ defined. It is standard to consider the noise as a separate problem, so that $\mathbf{N}(s) = \mathbf{0}$, and to note that the error in (33.13) is the same whether the input is a disturbance $\mathbf{D}(s)$ or a reference $\mathbf{R}(s)$; hence it is unnecessary to separate the issues of tracking command inputs from those of suppressing errors. In fact, then

$$\mathbf{E}(s) = \varepsilon(s)\, \mathbf{V}(s)$$

where

$$\varepsilon(s) = (\mathbf{I} + \mathbf{G}(s)\, \mathbf{C}(s))^{-1}$$

is called the **sensitivity function**. Now we consider the problem of minimizing error when the inputs are norm-bounded. In particular, we design for inputs $\mathbf{v}(t)$ for which the frequency weighted 2-norm is bounded, as in

$$\mathscr{V} = \{\, \mathbf{v} \mid \| \mathbf{W}_{\text{input}}^{-1}\, \mathbf{v} \|_2^2 \le 1 \,\}$$

or its equivalent, $\mathbf{v}' = \mathbf{W}_{\text{input}}^{-1}\, \mathbf{v}$, $\mathscr{V}' = \{\, \mathbf{v}' \mid \| \mathbf{v}' \|_2^2 \le 1 \}$. We then seek the maximum error due to such inputs, i.e.

$$\max_{\mathbf{v} \in \mathscr{V}} \| \mathbf{W}_{\text{output}} \mathbf{E} \|_2 = \max_{\mathbf{v} \in \mathscr{V}} \| \mathbf{W}_{\text{output}}\, \varepsilon\, \mathbf{W}_{\text{input}}\, \mathbf{v}' \|_2$$

By careful choice of $\mathbf{v}'$ and examination of the definition of the 2-norm, we can argue

$$\max_{\mathbf{v'} \in \mathscr{V'}} \| \mathbf{W}_{\text{output}}\, \varepsilon\, \mathbf{W}_{\text{input}}\, \mathbf{v'} \|_2$$

$$= \sup_{\omega} \bar{\sigma}(\mathbf{W}_{\text{output}}(j\omega)\, \varepsilon(j\omega)\, \mathbf{W}_{\text{input}}(j\omega))$$

$$= \| \mathbf{W}_{\text{output}}\, \varepsilon\, \mathbf{W}_{\text{input}} \|_\infty$$

Thus the optimal design problem becomes

$$\min_{\mathbf{C}(s)} \| \mathbf{W}_{\text{output}}\, (\mathbf{I} + \mathbf{GC})^{-1}\, \mathbf{W}_{\text{input}} \|_\infty$$

$$= \min_{\mathbf{C}(s)} \sup_{\omega} \bar{\sigma}(\mathbf{W}_{\text{output}}(j\omega)(\mathbf{I} + \mathbf{G}(j\omega)\, \mathbf{C}(j\omega))^{-1}\, \mathbf{W}_{\text{input}}(j\omega))$$

The optimum value will be some (unknown) value J^*. Hence, we may state that, for all w,

$$\min_{\mathbf{C}(s)} \| \mathbf{W}_{\text{output}}\, (\mathbf{I} + \mathbf{GC})^{-1}\, \mathbf{W}_{\text{input}} \|_\infty \leq J^*$$

or, with 'reasonable' scaling on the weighting functions $\mathbf{W}_{\text{input}}$ and $\mathbf{W}_{\text{output}}$, that

$$\min_{\mathbf{C}(s)} \| \mathbf{W}_{\text{output}}\, (\mathbf{I} + \mathbf{GC})^{-1}\, \mathbf{W}_{\text{input}} \|_\infty < 1$$

Hence we have met the performance requirement satisfactorily if we find a compensator $\mathbf{C}(s)$ such that

$$\| \mathbf{W}_{\text{output}}\, (\mathbf{I} + \mathbf{GC})^{-1}\, \mathbf{W}_{\text{input}} \|_\infty < 1$$

The robust performance problem is then of finding $\mathbf{C}(s)$ which meets this requirement for any $\mathbf{G}(s) \in \mathscr{G}$. To simplify the algebra, we combine the two weighting functions into a single scalar function $w(s)$, so that the requirement is that

$$\| w\, (\mathbf{I} + \mathbf{GC})^{-1} \|_\infty < 1$$

We use the model

$$G(s) = (I + lL(s))\, G_0(s) \qquad \|L(j\omega)\| \le 1$$

and note that we require, for all ω and for all $G \in \mathcal{G}$

$$\bar{\sigma}(w(j\omega)\,(I + G(j\omega)\,C(j\omega))^{-1}) < 1$$

Defining

$$\varepsilon_0(s) = (I + G_0(s)\,C(s))^{-1}$$

$$\eta_0(s) = (I + G_0(s)\,C(s))^{-1}\,G_0(s)\,C(s)$$

let us compute

$$\varepsilon(s) = (I + G(s)\,C(s))^{-1} = \varepsilon_0(s)\,(I + lL(s)\,\eta_0(s))^{-1}$$

Then we compute that the robust performance requirement is

$$\bar{\sigma}(w(j\omega)\,\varepsilon(j\omega)) = \bar{\sigma}(w(j\omega)\,\varepsilon_0(j\omega)\,(I + lL(j\omega)\,\eta_0(j\omega))^{-1}) < 1$$

and this is implied by

$$\bar{\sigma}(w(j\omega)\,\varepsilon_0(j\omega))\,\bar{\sigma}((I + lL(j\omega)\,\eta_0(j\omega))^{-1}) < 1 \tag{33.15}$$

Using the equivalent relationship

$$\bar{\sigma}((I + lL(j\omega)\,\eta_0(j\omega))^{-1}) = \underline{\sigma}(I + lL(j\omega)\,\eta_0(j\omega))$$

$$= 1 - \bar{\sigma}(lL(j\omega)\,\eta_0(j\omega))$$

$$= 1 - \bar{\sigma}(l\eta_0(j\omega))$$

with (33.15) leads to

$$\bar{\sigma}(w(j\omega)\,\varepsilon_0(j\omega)) + \bar{\sigma}(l\eta_0(j\omega)) < 1$$

or explicitly

$$\bar{\sigma}(w(j\omega)\,(\mathbf{I} + \mathbf{G}_0(j\omega)\,\mathbf{C}(j\omega))^{-1})$$

$$+ \bar{\sigma}(l(\mathbf{I} + \mathbf{G}_0(j\omega)\,\mathbf{C}(j\omega))^{-1}\mathbf{G}_0(j\omega)\,\mathbf{C}(j\omega)) < 1 \qquad (33.16)$$

It is easy to consider this expression as having a robust stability term (the second term, since l is associated with the model bounds) and a nominal performance term (the first term, with its weighted sensitivity emphasis). Thus we get robust performance by leaving a performance margin (first term $< \alpha < 1$) and also being 'stable enough' (second term $< 1 - \alpha$). This is a classic engineering trade-off made mathematically explicit.

33.3.3 A design approach

The problem in the above is to find a compensator $C(s)$ so that (33.16) is satisfied. One approximate method is to find the compensator which yields optimal nominal performance, and then to detune this until the robustness constraint is satisfied. We might note that this is essentially what loop transfer recovery (section 33.4) does. The search can be aided by using the controller parameterization to obtain

$$\bar{\sigma}(w(j\omega)\,(\mathbf{I} - \mathbf{G}_0(j\omega)\,\mathbf{Q}(j\omega))) + \bar{\sigma}(l\mathbf{G}_0(j\omega)\,\mathbf{Q}(j\omega)) < 1$$

One special case for which there is an explicit solution for C has G_{nom} stable and of the form

$$G_{\text{nom}}(s) = G_{\text{A}}(s)\,G_{\text{M}}(s)$$

and the input similarly

$$R(s) = R_A(s) R_M(s)$$

where the subscript A denotes all-pass, i.e. of the form for example

$$G_A(s) = e^{-s\theta} \prod_i \frac{-s + a_i}{s + a_i^*}$$

where $\mathrm{Re}\,(a_i) > 0$ and $\theta \geq 0$. The subscript M indicates that the factor is minimum phase, i.e. that the zeros have negative real parts; the poles have negative real parts because of the assumption that the function is stable. Using the above factorizations, it has been shown by Morari and Zafiriou (1989, Thm. 4.1–1) that the optimal ISE control is given by

$$Q_0(s) = (G_M(s) R_M(s))^{-1} (G_A^{-1}(s) R_M(s))_*$$

In this, the notation $(\ )_*$ indicates that after partial fraction expansion, all terms involving poles of $G_A^{-1}(s)$ are omitted. As an example, if

$$G_{\mathrm{nom}}(s) = G_A(s) = \frac{-s+2}{s+2}$$

and

$$R_M(s) = \frac{s+1}{s}$$

then

$$Q_0(s) = \frac{s}{s+1} \left\{ \frac{s+2}{-s+2} \frac{s+1}{s} \right\}_*$$

$$= \frac{s}{s+1} \left\{ -1 + \frac{1}{s} + \frac{6}{-s+2} \right\}_* = \frac{-s+1}{s+1}$$

With Q_0 determined, we now detune using a filter $F(s)$, so that finally

$$Q(s) = F(s)\,Q_0(s)$$

One straightforward technique is to choose a filter form and adjust one or more parameters of the filter to meet the robustness conditions. A candidate which retains the error constants for a Type 1 system is

$$F(s) = \frac{1}{(\lambda s + 1)^n}$$

while a Type 2 system can be retained using the form

$$F(s) = \frac{n\lambda s + 1}{(\lambda s + 1)^n}$$

in which cases n and λ are parameters. The order n is chosen so that $Q(s)$ is proper, i.e. so that the order of the denominator is higher than that of the numerator. Then for a chosen λ, check the following:

- robust stability – ensure that, for $\omega = 0$

$$|G_{\text{nom}}(j\omega)\,Q_0(j\omega)\,F(j\omega)\,lL(\omega)| < 1$$

- robust performance – increase λ until, for all ω

$$|G_{\text{nom}}(j\omega)\,Q_0(j\omega)\,F(j\omega)\,lL(\omega)|$$
$$+\,|(1 - G_{\text{nom}}(j\omega)Q_0(j\omega)\,F(j\omega))\,w(\omega)| < 1$$

Example

We are given a process model

$$G_0(s) = ke^{-s\theta} \prod_{i=1}^{m} \frac{1}{\tau_i s + 1}$$

in which the delay θ, which is nominally equal to 1, is uncertain up to 15%. The performance is to be optimized for a step input, to show a Type 1 characteristic, and to have a maximum weighting of $w_{max}(\omega) = 0.4$.

The set $\mathscr{G}$ of potential actual plants is defined by

$$\mathscr{G} = \left\{ G(s): \left| \frac{G - G_0}{G_0} \right| = |e^{-j\delta} - 1| < lL(\omega) \right\}$$

from which it follows from the specifications that

$$l = 2 \qquad L(s) = \begin{cases} \dfrac{1 - e^{-0.15s}}{2} & 0.15\omega \le \pi \\ 1 & 0.15\omega > \pi \end{cases}$$

The optimum non-robust and possibly unrealizable controller can be found as

$$Q_0(s) = (G_M(s) R_M(s))^{-1} (G_A^{-1}(s) R_M(s))_*$$

$$= \left(\frac{k}{s} \prod_{i=1}^{m} \frac{1}{\tau_i s + 1} \right)^{-1} \left(\frac{1}{e^{-s\theta}} \frac{1}{s} \right)$$

$$= \frac{1}{ke^{-s}} \prod_{i=1}^{m} (\tau_i s + 1)$$

and we choose to use the filter

$$F(s) = \frac{1}{(\lambda s + 1)^m}$$

This yields the controller

$$C(s) = \frac{1}{k((\lambda s + 1)^m - 1)} \prod_{i=1}^{m} (\tau_i s + 1)$$

We now choose λ so as to achieve robust stability. To do this, we search for λ such that, for $\omega = 0$

$$|G_0(j\omega)\, Q_0(j\omega)\, F(j\omega)\, lL(\omega)| < 1$$

which, since $G_0(j\omega)\, Q_0(j\omega) = 1$, is easy to compute. Using any calculator (such as MathCad) one can compute this for various m as

m	λ
1	0.1
2	0.062
3	0.05

The robust performance requirement simplifies, for all ω to

$$|F(j\omega)| + |(1 - e^{-j\omega} F(j\omega))\, W(\omega)| < 1$$

With $m = 1$, the robust stability value $\lambda = 0.1$ from the table above yields the maximum of the left-hand side ≈ 1.6; checking other λ reveals that the robust performance inequality requires $\lambda > 0.33$.

33.4 STATE–SPACE APPROACHES

One of the most used robust control law development methods is the loop transfer recovery (LTR) method, which is basically a linear quadratic regulator (LQR) with output from a suitably detuned Kalman filter. We review this before looking briefly at one of the early and alternative state–space approaches to robust performance, that of guaranteeing a maximum cost in a LQ problem with state feedback.

33.4.1 A special technique: loop transfer recovery (LTR)

In Chapter 29, it was argued that a system modelled as

$$\dot{\mathbf{x}} = \mathbf{A}\mathbf{x} + \mathbf{B}\mathbf{u} + \mathbf{\Gamma}\mathbf{v}$$

$$\mathbf{y} = \mathbf{C}\mathbf{x} + \mathbf{w}$$

with $\operatorname{cov}(\mathbf{v}(t)) = \mathbf{Q}_n \delta(t)$, $\operatorname{cov}(\mathbf{w}(t)) = \mathbf{R}_n \delta(t)$, $\mathbf{x}(0) \sim \mathrm{N}(\mathbf{x}_0, \Sigma_0)$ is optimally controlled according to the cost index

$$J = \mathscr{E}\left\{ \int_0^\infty [\mathbf{x}^{\mathrm{T}}(\tau)\,\mathbf{Q}_{\mathrm{PI}}\,\mathbf{x}(\tau) + \mathbf{u}^{\mathrm{T}}(\tau)\,\mathbf{R}_{\mathrm{PI}}\,\mathbf{u}(\tau)]\,\mathrm{d}\tau \right\}$$

by the control law

$$\mathbf{u}(t) = -\mathbf{K}\hat{\mathbf{x}}(t) = -\mathbf{R}_{\mathrm{PI}}^{-1}\,\mathbf{B}^{\mathrm{T}}\,\mathbf{P}\,\hat{\mathbf{x}}(t)$$

in which **P** solves

$$\mathbf{0} = \mathbf{Q}_{\mathrm{PI}} + \mathbf{A}^{\mathrm{T}}\mathbf{P} + \mathbf{P}\mathbf{A} - \mathbf{P}\mathbf{B}\mathbf{R}_{\mathrm{PI}}^{-1}\mathbf{B}^{\mathrm{T}}\mathbf{P}$$

and $\dot{\hat{x}}$ is the output of the steady-state Kalman filter

$$\dot{\hat{x}} = A\hat{x} + Bu + G\,[y - C\,[A\hat{x} + Bu]\,]$$

where

$$G = \Sigma\,C^T R_n^{-1}$$

$$0 = A\Sigma + \Sigma A^T - \Sigma C^T R_n^{-1} C\Sigma + \Gamma Q_n \Gamma^T$$

Now if noise-free state feedback is possible, the use of the filter estimate $\hat{x}$ is unnecessary, the compensator is $C(s) = K$ in Fig. 33.1, and the loop transfer function is

$$G_1(s) = K\,(sI - A)^{-1} B \tag{33.17}$$

It is known that this transfer function exhibits gain margin of at least $\frac{1}{2}$ and phase margin of $60°$; it is robust in that it can tolerate a gain error of $\frac{1}{2}$ or more.

If the Kalman filter must be used, however, the comparable loop transfer function becomes

$$G_2(s) = K\,(sI - A - BK - GC)^{-1} GC\,(sI - A)^{-1} B \tag{33.18}$$

It is known (Anderson and Moore, 1989) that in this case it is possible that for certain filters the margins can become very small, so that the method cannot be used without care. Usually, however, (33.18) approaches (33.17) if $G \rightarrow \infty$ and thereby approaches the original state feedback design properties. G can be made large by making Q_n large, which corresponds to implying that either disturbances can be large or the model errors are large enough that disturbance effects are apparent. Either way, the method is to design a succession of filters, using a parameterized Q_n as in

$$Q_n = Q_{n,0} + q B^T Q_\delta B \qquad Q_\delta > 0$$

with q a scalar parameter.

Example

To see the effect of the LTR methods, we consider a simple problem using a motor model. In particular, assume

$$\dot{x} = \begin{bmatrix} 0 & 1 \\ 0 & -1 \end{bmatrix} x + \begin{bmatrix} 0 \\ 1 \end{bmatrix} u = Ax + bu$$

$$y = [\, 1 \quad 0\,]\, x = Cx$$

If the optimal control criterion is

$$J = \int_0^\infty \left(x^T \begin{bmatrix} 100 & 0 \\ 0 & 1 \end{bmatrix} x + [\, 1\ ] u^2 \right) d\tau = \int_0^\infty (x^T Q x + R u^2)\, d\tau$$

then the optimal control law (Chapter 26) for state feedback is

$$u = - K_c x$$

where

$$K_c = [\, 10 \quad 3.69\,]$$

The loop transfer return difference $K_c(j\omega I - A)^{-1} b$ is shown (as dotted lines) in Fig. 33.3.

Since only a scalar measurement is involved, a filter or observer is required. We use a Kalman filter, assuming that noise terms

$$\begin{bmatrix} 0 \\ 1 \end{bmatrix} v \text{ and } [1]\, w$$

with variances q_n and r_n are respectively added to the dynamics and measurement equation models. Choosing

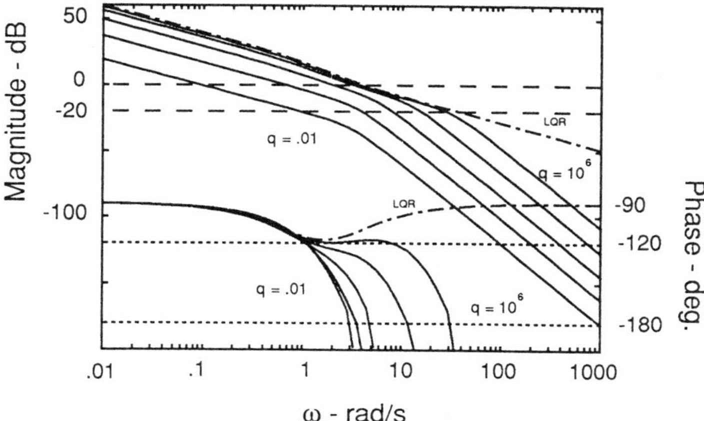

Figure 33.3 Loop transfer function recovery by using LQG controller with various system noise parameters q in the Kalman filter.

$$r_n = 1$$

$$q_n = q$$

and letting q be a parameter, we may find the steady-state Kalman filter gains $\mathbf{K}_f(q)$ and from this the return difference

$$\mathbf{K}_c(j\omega\mathbf{I} - a + b\mathbf{K}_c + \mathbf{K}_f(q)\,\mathbf{C})^{-1}\,\mathbf{K}_f(q)\,\mathbf{C}\,(j\omega\mathbf{I} - \mathbf{A})^{-1}b$$

is computed. These are shown in Fig. 33.1 for $q = 0.01, 1, 100, 10^4, 10^6$. The gains are

q \	0.01	1	100	10^4	10^6
$\mathbf{K}_f$ =	0.095	0.73	3.58	13.18	43.73
	0.0046	0.27	6.47	86.82	956.3

From the figure, it is clear that as q increases, the filter plus controller system becomes closer in loop transfer characteristic to the state feedback controller case. In this instance, the gain cross-over frequency increases with q, implying an input robustness; the infinite gain margin and 60° + phase margin of the original system become 20 dB or so and about 60° respectively in all cases using the filter, implying a gain and time constant robustness in all cases.

The above demonstates only the influence of tuning the Kalman filter design to achieve closer to the desired transfer function. In MIMO cases, the principal gains are used. In a more complicated or special case, the gain margins when the filter is used may be almost zero (Anderson and Moore, 1989), clarifying that (a) such a structure may exhibit no parameter robustness and hence (b) the Kalman filter must be carefully designed if the loop recovery is to be effective.

33.4.2 Robust performance

Robust performance has also been approached in state–space. One such problem was defined as in (33.1–4) with the structured errors

$$\dot{\mathbf{x}} = \left(\mathbf{A} + \sum_{i=1}^{p} \theta_i \mathbf{A}_i\right)\mathbf{x} + \theta_b \mathbf{B}\mathbf{u}$$

where $\mathbf{A}$ and $\mathbf{A}_i$ ($i = 1, 2, ..., p$) are $n \times n$ matrices, and $\mathbf{B}$ is an $n \times 1$ vector, while the θ_i ($i = 1, 2, ..., p$ with $-1 \le \theta_i \le 1$ and $1 \le \theta_b \le b$) and b are scalars. Then for the cost criterion

$$J = \tfrac{1}{2} \mathbf{x}^{\mathrm{T}} \mathbf{H} \mathbf{x} + \tfrac{1}{2} \int_0^T (\mathbf{x}^{\mathrm{T}} \mathbf{Q} \mathbf{x} + \mathbf{u}^{\mathrm{T}} \mathbf{R} \mathbf{u})\, \mathrm{d}t$$

the cost J will be a function of the parameters $\theta = [\theta_1\ \theta_2\ ...\ \theta_p\ \theta_b]$, and we wish to choose a feedback control law $\mathbf{u} = -\mathbf{K}(t)\mathbf{x}$ such that the maximum cost, regardless of the instantiation of θ, is minimized. Thus we need

$$\min_{\mathbf{K}(t)} \max_{\theta} J$$

Alternatively, let $V(K)$ be the least upper bound of J as θ varies; we wish to minimize V by choice of $\mathbf{K}$. Then it can be shown (Dorato, 1987) that

$$\mathbf{u} = -\mathbf{R}^{-1}\mathbf{B}^{\mathrm{T}}\mathbf{S}(t)\,\mathbf{x}$$

with cost guarantee

$$V = \tfrac{1}{2}\mathbf{x}^{\mathrm{T}}\mathbf{S}(t)\,\mathbf{x}$$

provided that $\mathbf{S}(t)$ solves the modified Ricatti equation

$$-\dot{\mathbf{S}} = \mathbf{Q} - \mathbf{S}\mathbf{B}\mathbf{R}^{-1}\mathbf{B}^{\mathrm{T}}\mathbf{S} + \mathbf{S}\mathbf{A} + \mathbf{A}^{\mathrm{T}}\mathbf{S} + \sum_{i=1}^{p} \mathbf{P}_i \, \mathrm{abs}(\Lambda_i)\, \mathbf{P}_i^{\mathrm{T}}$$

$$\mathbf{S(T)} = \mathbf{H}$$

where the orthonormal matrix $\mathbf{P}_i$ diagonalizes $[\mathbf{SA} + \mathbf{A}_i^{\mathrm{T}}\mathbf{S}]$ to yield Λ_i, as in

$$\mathbf{P}_i^{\mathrm{T}}[\mathbf{SA}_i + \mathbf{A}_i^{\mathrm{T}}\mathbf{S}]\,\mathbf{P}_i = \Lambda_i$$

and $\mathrm{abs}\,(\Lambda)$ has elements equal to the absolute values of elements of Λ. Steady-state solutions can be found from backward integration.

One notes that in the absence of uncertainty, this is simply the LQ problem of optimal control.

33.5 COMPUTER AIDS

Computer aids are both necessary and available for these problems. Finding the singular values such as $\bar{\sigma}(\mathbf{G}(j\omega))$, for example, is easy numerically using SVD algorithms. The LTR iterations (on the parameter) are also straightforward using the LQR and LQE algorithms. Packages such as MATLAB® and Ctrl-C also have special robust control packages. Manufacturers literature and worked examples can be more helpful here than some of the textbooks.

33.6 SUMMARY

The above has only touched the surface of this intensively researched new topic in an attempt to convey a feel for the nature of the approaches and a couple of important results. One might attempt to be aware of the modelling methods used for the uncertainties, the key steps of small gain theorems and stabilizing controller parameterizations, and the nature and ubiquity of the LTR approach. One might also be aware of the fact that the theory is mostly associated with linear systems and, to some extent, the way in which the linear model is good enough for non-linearities. Many engineers will perhaps check any design with simulations, including a set of runs with parameters randomly chosen from within the specification envelope (a relative of so-called Monte Carlo techniques.)

33.7 FURTHER READING

Selected research literature is readily available in reprint books such as those by Dorato (1987) and Dorato and Vedavali (1990).

An advanced textbook devoted to the subject, particularly the multivariable transfer function approach, is that by Morari and Zafiriou (1989), while Maciejowski (1989) includes helpful discussions of robust control and Anderson and Moore (1989) have studies of loop transfer recovery (LTR).

34

Structures of multivariable controllers

Multi-input–multi-output (MIMO) control is in principle handled using the theory presented in earlier chapters. The classical theory has been extended to systems described by plant models $\mathbf{G}(s)$ where $\mathbf{G}$ is a matrix, and much state–space theory, such as optimal control and pole-placement feedback, is intrinsically multivariable in capability. The result is that, in principle, control can be done as in Fig. 34.1(a).

For a number of reasons, however, the engineering applications may not use the full theory. These range from theoretical reasons, such as models too large to handle well, through reasons of need to manipulate the models into forms to which the theory and engineers' judgements apply, to practical reasons of controllers too complicated to be readily implemented.

Engineering judgement has led to real systems such as the following.

- In industrial robots such as the PUMA series by Unimate/Kawasaki, a hierarchical arrangement is used. The individual joints are each controlled as SISO systems using a servo-motor under microprocessor control. The joints are co-ordinated by a supervisory minicomputer.
- In some process control, the input and output sets of variables are decomposed into pairs. Then a particular input (called a manipulated) variable is generated based primarily upon a selected output (called a controlled) variable. The fact that variables necessarily interact – e.g. the output temperature of a fluid depends upon both the heat input and the flow rate – is accounted for in the design of the controllers.

In essence, the reality is that large MIMO systems are likely to be controlled as in Figs 34.1(b) or 34.1(c). Reasons include engineering convenience, judgement, and understanding, plus the ready availability of computers of many sizes and prices, some built into measurement devices, actuators, or low level process controllers, and others used for tasks such as plant control and electric grid supervision and control.

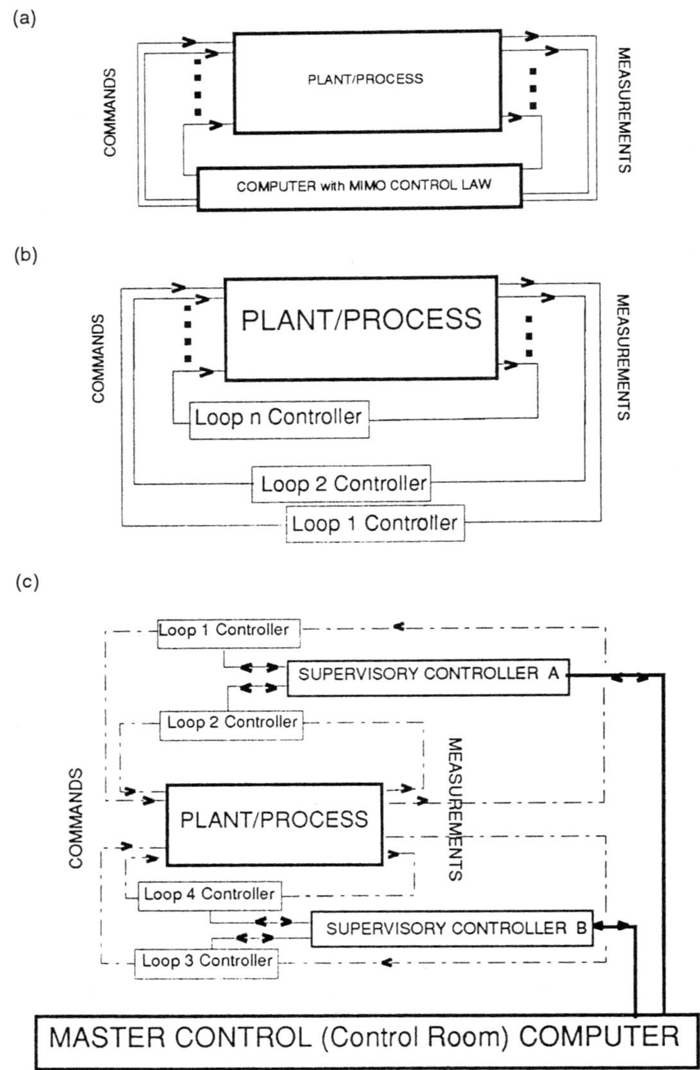

Figure 34.1 Structuring of control of multivariable systems: (a) MIMO controller; (b) decoupled into set of SISO loops; and (c) typical hierarchical/distributed control.

Although many of these systems have been designed primarily based on engineering judgement, as in Chapter 2, there exists a patchwork of theory which is helpful. Such theory is met in this chapter. The various techniques will do one or more of the following:

- allow approximation of the plant so that designs are more practically studied;
- allow preliminary design of controllers so that laws can be designed using extensions of SISO methods; and
- allow simplification and approximation of controllers designed using and reflecting the full power of the design methods of earlier chapters.

34.1 SUMMARY

Standard in the engineers' repertoire, if not in teaching, are special structures for compensation of SISO systems. Especially common are the notions of feedforward control, of either inputs or disturbances, and cascade controls to handle multiple loops in a special case of MIMO control.

More generally, decomposition of the MIMO problem into smaller problems depends ultimately on exploiting special structural relationships of the system. These potentially include different time-scales of the subsystems' responses and (partial) decoupling of the internal variables from each other.

In matrix terms for linear system models, we will be concerned with the basic plant/process model

$$\dot{x} = Ax + Bu$$

$$y = Cx \tag{34.1}$$

in which the A, B, and C matrices have special character, such as is represented in the partitionings

$$B = \begin{bmatrix} B_1 & 0 \\ 0 & B_2 \end{bmatrix} \qquad C = \begin{bmatrix} C_1 & 0 \\ 0 & C_2 \end{bmatrix}$$

For convenience, we use partitioning and the notations

$$\mathbf{x} = \begin{bmatrix} \mathbf{x}_1 \\ \mathbf{x}_2 \end{bmatrix} \qquad \mathbf{u} = \begin{bmatrix} \mathbf{u}_1 \\ \mathbf{u}_2 \end{bmatrix} \qquad \mathbf{y} = \begin{bmatrix} \mathbf{y}_1 \\ \mathbf{y}_2 \end{bmatrix} \qquad \mathbf{A} = \begin{bmatrix} \mathbf{A}_{11} & \mathbf{A}_{12} \\ \mathbf{A}_{21} & \mathbf{A}_{22} \end{bmatrix}$$

where $\mathbf{x}_1$ is an n_1-vector, $\mathbf{y}_1$ is a p_1-vector, etc. In some such cases, we would like to be able to achieve adequate control with

$$\mathbf{u}_1 = \mathbf{f}_1(\mathbf{y}_1) \qquad \mathbf{u}_2 = \mathbf{f}_2(\mathbf{y}_2)$$

instead of the general MIMO

$$\mathbf{u} = \mathbf{f}(\mathbf{y})$$

This will of course require special assumptions or properties of the coupling matrices $\mathbf{A}_{12}$ and $\mathbf{A}_{21}$, or alternatively special functions $\mathbf{f}_1$ and $\mathbf{f}_2$ to account for the coupling.

In transfer function terms we need the matrix transfer functions $\mathbf{G}_{ij}(s)$ and vector transforms $\mathbf{Y}_i(s)$ and $\mathbf{U}_j(s)$ in

$$\mathbf{Y}(s) = \begin{bmatrix} \mathbf{Y}_1(s) \\ \mathbf{Y}_2(s) \end{bmatrix} = \begin{bmatrix} \mathbf{G}_{11}(s) & \mathbf{G}_{12}(s) \\ \mathbf{G}_{21}(s) & \mathbf{G}_{22}(s) \end{bmatrix} \begin{bmatrix} \mathbf{U}_1(s) \\ \mathbf{U}_2(s) \end{bmatrix}$$

$$= \mathbf{G}(s)\,\mathbf{U}(s) \tag{34.2}$$

to be such that a pairing of inputs and outputs, such as in specializing the general case

$$\mathbf{U}(s) = \mathbf{G}_c(s)\,\mathbf{E}(s) = \mathbf{G}_c(s)\,[\mathbf{R}(s) - \mathbf{Y}(s)]$$

$$= \begin{bmatrix} \mathbf{G}_{c11}(s) & \mathbf{G}_{c12}(s) \\ \mathbf{G}_{c21}(s) & \mathbf{G}_{c22}(s) \end{bmatrix} [\mathbf{R}(s) - \mathbf{Y}(s)]$$

to the special case represented by

$$U(s) = \begin{bmatrix} G_{c11} & 0 \\ 0 & G_{c22} \end{bmatrix} [R(s) - Y(s)]$$

$$= \begin{bmatrix} G_{c11}(s) & (R_1(s) - Y_1(s)) \\ G_{c22}(s) & (R_2(s) - Y_2(s)) \end{bmatrix}$$

gives adequate control.

With either of the above, and with variations of them, there has arisen a body of knowledge, containing both practical and theoretical elements, which is helpful to the engineer. It is unfortunate that this body is seldom definitive as to what should be done.

We remark that in many cases, because the state–space representation (34.1) has transfer function representation

$$Y(s) = G(s)\,U(s) = C(sI - A)^{-1}BU(s)$$

and a transform relationship (34.2) has many state–space representations, the techniques below apply to both types of problems even though motivated by only one of them.

34.2 CASCADE AND FEEDFORWARD STRUCTURES

Two common and straightforward structures for basic systems are feedforward and cascade controllers. Both have the aim of improving performance, the former by anticipating and precompensating for errors, the latter by using extra measurements in a natural way.

The concept of feedforward control is straightforward, even though the implementations are sometimes not; it can require extra measurements and careful tuning. There are two situations in which it is useful:

1. reduction of disturbance effects by measuring the disturbance and applying a compensatory command before the error grows to the point at which the feedback controller will cancel it, e.g. one might measure the temperature of fluids input to a boiler and adjust the boiler flame to maintain the output fluid at constant temperature; and

2. open-loop anticipation of commands needed to attain a specified output s, e.g. generate robot limb commands based upon the input

of desired position – feedback is used to compensate for position errors resulting from inexactly known robot parameters.

Both are, in effect, the generation of a control based directly upon the measured quantity, whether disturbance or input, in lieu of waiting until a feedback error arises and then compensating. The controller structures appear in Figs 34.2 and 34.3.

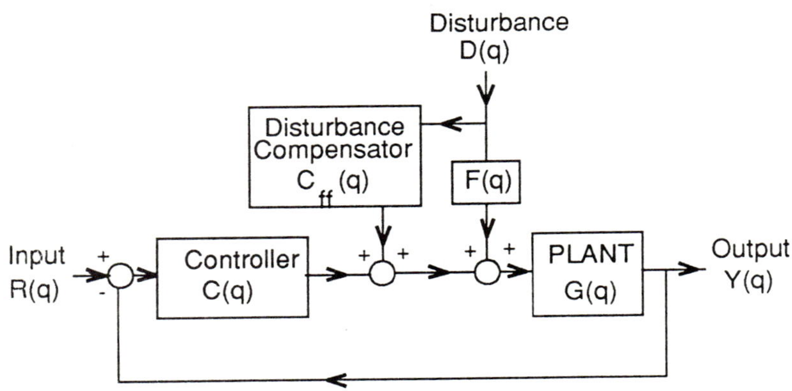

Figure 34.2 Structure of feedforward disturbance compensation, when the disturbance is measured.

The case of feedforward control of disturbances can be analysed in the usual manner. Using Fig. 34.2, we have

$$\mathbf{Y}(q) = (\mathbf{I} + \mathbf{G}(q)\,\mathbf{C}(q))^{-1}$$

$$(\mathbf{G}(q)\,(\mathbf{F}(q) + \mathbf{C}_{ff}(q))\,\mathbf{D}(q) + \mathbf{G}(q)\,\mathbf{C}(q)\,\mathbf{R}(q))$$

and it is clear that if we can measure the disturbance $d(t)$, and hence by implication treat its transform $\mathbf{D}(s)$ as a potentially known input, then designing $\mathbf{C}_{ff}(q)$ such that

$$(\mathbf{I} + \mathbf{G}(q)\,\mathbf{C}(q))^{-1}\,(\mathbf{G}(q)\,(\mathbf{F}(q) + \mathbf{C}_{ff}(q))\,\mathbf{D}(q))$$

is small, and using the controller with transfer function

$$\mathbf{U}(q) = \mathbf{C}(q)\,\mathbf{E}(s) + \mathbf{C}_{ff}(q)\,\mathbf{D}(q)$$

will reduce the effects of disturbances.

Command feedforward is also intended to reduce the error transients. It is inherently an open-loop command, as can be seen by considering Fig. 34.3.

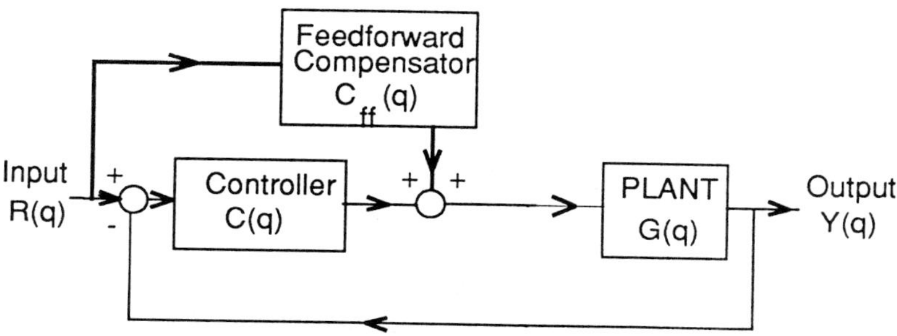

Figure 34.3 Structure of feedforward input compensation.

Here we have

$$\mathbf{Y}(q) = (\,\mathbf{I} + \mathbf{G}(q)\,\mathbf{C}(q))^{-1}\,(\mathbf{G}(q)\,\mathbf{C}(q) + \mathbf{G}(q)\,\mathbf{C}_{ff}(q))\,\mathbf{R}(q)$$

and formally (or for scalars)

$$\mathbf{Y}(q) = (\,\mathbf{I} + \mathbf{G}(q)\,\mathbf{C}(q))^{-1}\mathbf{G}(q)\,\mathbf{C}(q)\,(\mathbf{I} + \mathbf{C}^{-1}(q)\,\mathbf{C}_{ff}(q))\,\mathbf{R}(q)$$

$$= (\,\mathbf{I} + \mathbf{G}(q)\,\mathbf{C}(q))^{-1}\mathbf{G}(q)\,\mathbf{C}(q)\,\mathbf{R}'(q)$$

We see that by the choice $\mathbf{C}_{ff}(q) = \mathbf{G}^{-1}(q)$, we have $\mathbf{Y}(q) = \mathbf{R}(q)$, a provocative result which is often impractical because of the large commands needed for step responses. It is possible to use lower performance feedback loops, i.e. simpler $\mathbf{C}(q)$, with feedforward because the errors to be removed are smaller due to the use of $\mathbf{C}_{ff}(q)$. It is noteworthy that implementation can be done directly as in

$$\mathbf{U}(q) = \mathbf{C}(q)\,\mathbf{E}(q) + \mathbf{C}_{ff}(q)\,\mathbf{R}(q)$$

or indirectly by modifying the reference input, as in

$$\mathbf{U}(q) = \mathbf{C}(q)\left[(\mathbf{I} + \mathbf{C}'_{ff}(q))\,\mathbf{R}(q) - \mathbf{Y}(q)\right]$$

$$= \mathbf{C}(q)\,(\mathbf{R}'(q) - \mathbf{Y}(q))$$

The second common configuration uses cascaded controllers to implement multiple loops. Ordinarily this comes about when process variables other than the main controlled outputs are available; a standard example is in a servo-motor, in which the shaft angle is to be controlled, but where using a tachometer allows shaft speed to be fed back in a control loop. A simple configuration is shown in Fig. 34.4.

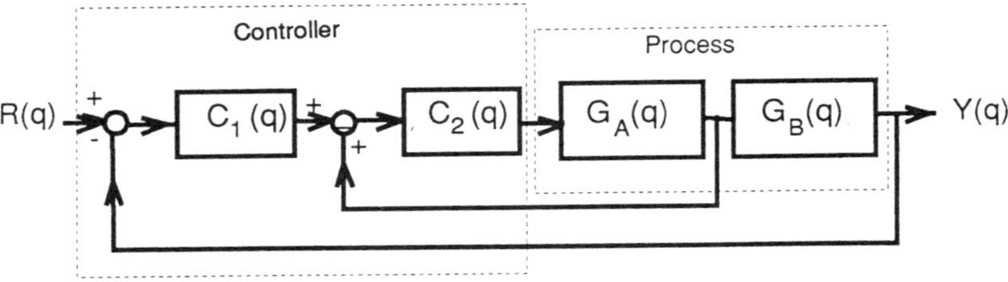

Figure 34.4 Structure of cascade compensation.

It is easy to show in the scalar case that

$$Y(q) = \frac{G_B(q)\,G_A(q)\,C_2(q)\,C_1(q)}{1 - C_2(q)\,G_A(q) + G_B(q)\,G_A(q)\,C_2(q)\,C_1(q)}\,R(q)$$

and hence clearly the use of the inner loop gives extra freedom in designing the system. We remark that one of the costs of doing this is that we must ensure that inaccuracies or failures in either the inner or outer loop are not catastrophic.

34.3 TRANSFER FUNCTION APPROACHES TO MULTIVARIABLE SYSTEMS I: INPUT–OUTPUT PAIRING

A process control operation can easily have hundreds of manipulated (input, control) variables and a similar number of output (measured) variables. It is not practical to do MIMO control on this entire system, nor would the results of doing so necessarily be reasonably implementable. Problems with implementation would range from the obvious need for considerable centralized computer power, and extensive communications to and from it, to the less obvious problems of assuring fail-safe operation of subsystems, if for example a communication line were to fail. An additional and important factor is that operating personnel may have extensive experience with, say, PID control of certain valves which would be lost to no good advantage if an elaborate centralized controller were used.

To investigate the problems and what can be done here, we consider only 2-in–2-out (2I–2O) processes and their linear time-invariant system models. We look first at the problem, then at the relative gain away method of considering pairing, and finally the issue of decoupling compensators.

34.3.1 The issue

A process modelled by (34.2) and specialized to scalar use in the form

$$Y(s) = \begin{bmatrix} Y_1(s) \\ Y_2(s) \end{bmatrix} = \begin{bmatrix} G_{11}(s) & G_{12}(s) \\ G_{21}(s) & G_{22}(s) \end{bmatrix} \begin{bmatrix} U_1(s) \\ U_2(s) \end{bmatrix}$$

$$= G(s)\,U(s) \tag{34.3}$$

can be depicted as in Fig. 34.5

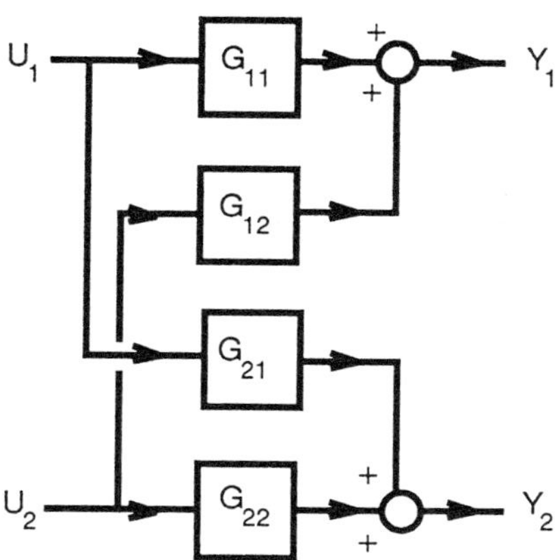

Figure 34.5 A 2I–2O process/plant.

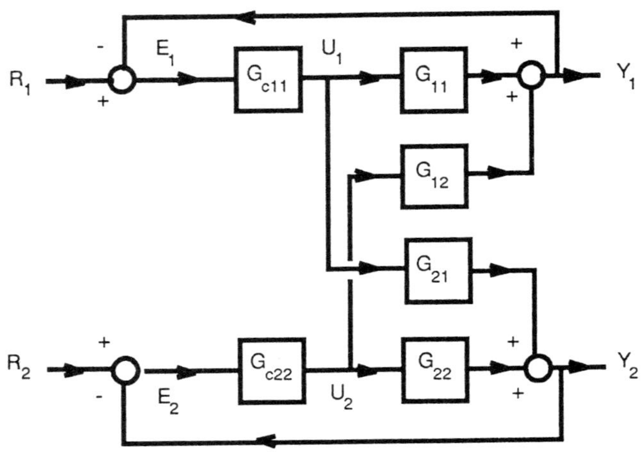

Figure 34.6 Two-loop control of the 2I–2O plant using input/output pairing.

Figure 34.6 depicts the closed-loop situation when the plant is controlled through pairings of input 1 with output 1, and input 2 with output 2. Figure 34.7 shows the alternative pairing.

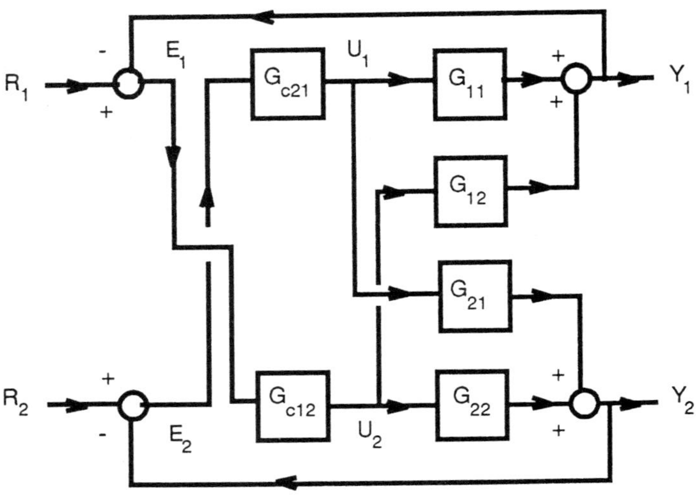

Figure 34.7 Alternative pairing of two-loop control of 2I–2O plant.

It is straightforward to show that in the first case, since

$$\mathbf{U}(s) = \begin{bmatrix} G_{c11}(s) & 0 \\ 0 & G_{c22}(s) \end{bmatrix} [R(s) - Y(s)]$$

$$= \begin{bmatrix} G_{c11}(s) & (R_1(s) - Y(s)) \\ G_{c22}(s) & (R_2(s) - Y(s)) \end{bmatrix}$$

the transfer functions for the closed-loop become the following:

$$Y_1(s) = \frac{G_{11}G_{c11}(1 + G_{c22}G_{22}) - G_{12}G_{21}G_{c11}G_{c22}}{(1 + G_{11}G_{c11})(1 + G_{22}G_{c22}) - G_{21}G_{12}G_{c11}G_{c22}} R_1$$

$$+ \frac{G_{12}G_{c22}}{(1 + G_{11}G_{c11})(1 + G_{22}G_{c22}) - G_{21}G_{12}G_{c11}G_{c22}} R_2$$

$$Y_2(s) = \frac{G_{21}G_{c11}}{(1 + G_{11}G_{c11})(1 + G_{22}G_{c22}) - G_{21}G_{12}G_{c11}G_{c22}} R_1$$

$$+ \frac{G_{22}G_{c22}(1 + G_{c11}G_{11}) - G_{12}G_{21}G_{c11}G_{c22}}{(1 + G_{11}G_{c11})(1 + G_{22}G_{c22}) - G_{21}G_{12}G_{c11}G_{c22}} R_2$$

$$(34.4)$$

It may be observed that to the extent that the coupling terms $G_{12}(s)$ and $G_{21}(s)$ are 'small', the control will be dominated by the 1–1 and 2–2 feedback loops. A number of observations can be made, however, when these terms cannot be ignored.

1. If controllers are designed or tuned as if no coupling occurs, i.e. as if control design only considers selecting $G_{cii}(s)$, $i=1,2$, to give good closed-loop transfer functions

$$\frac{G_{ii}(s)G_{cii}(s)}{1 + G_{ii}(s)G_{cii}(s)} \qquad i = 1,2$$

and hence good poles at the zeros of

$$1 + G_{ii}(s)G_{cii}(s)$$

we may be very disappointed with performance, and even have an unstable system because the actual system poles will be at

$$(1 + G_{11}(s)\,G_{c11}(s))\,(1 + G_{22}(s)\,G_{c22}(s))$$

$$- G_{21}(s)\,G_{12}(s)\,G_{c11}(s)\,G_{c22}(s) = 0$$

2.   We would like to have a scientific way of selecting the proper, or 'best', pairings of inputs with outputs. It would also be helpful to have a way of knowing when pairing is inappropriate for some subsystems.
3.   Compensation for the presence of coupling would be a useful design option.

34.3.2   Pairing selection using the relative gain array

The relative gain array provides a methodology for selecting pairs of input and output variables in order to minimize interactions among the loops. Although straightforward and experimentally implementable, it has the disadvantage of considering only the steady-state character of the interactions.

To see how it works, we consider again the 2I–2O system

$$Y(s) = \begin{bmatrix} Y_1(s) \\ Y_2(s) \end{bmatrix} = \begin{bmatrix} G_{11}(s) & G_{12}(s) \\ G_{21}(s) & G_{22}(s) \end{bmatrix} \begin{bmatrix} U_1(s) \\ U_2(s) \end{bmatrix}$$

$$= G(s)\,U(s)$$

For a small step input with one variable, say u_1, we can hold either u_2 or y_2 constant and find the effect on y_1. Specifically, we find

$$\left[\frac{\Delta y_1}{\Delta u_1}\right]_{u_2} = \lim_{s \to 0} sY_1(s) = \lim_{s \to 0} G_{11}(s)$$

when $u_2 = 0$ and $U_1(s) = 1/s$. Also, with the same input but the constraint $y_2 = 0$, we find

$$0 = Y_2(s) = G_{21}(s)\,\frac{1}{s} + G_{22}(s)\,U_2(s)$$

so that

$$U_2(s) = -\frac{G_{21}(s)}{sG_{22}(s)}$$

and hence

$$\left(\frac{\Delta y_1}{\Delta u_1}\right)_{y_2} = \lim_{s\to 0} sY_1(s) = \lim_{s\to 0}\left\{G_{11}(s) - \frac{G_{12}(s)\,G_{21}(s)}{G_{22}(s)}\right\}$$

We then define the relative gain λ_{ij} between y_i and u_j as

$$\lambda_{ij} = \left(\frac{\Delta y_i}{\Delta u_j}\right)_{y_{\bar i}}^{-1}\left(\frac{\Delta y_i}{\Delta u_j}\right)_{u_{\bar j}}$$

The bar notation represents the 'other value' so when $i=2$, $\bar i=1$. Thus for $i=1,2$, $j=1,2$

$$\lambda_{11} = \left(\frac{\Delta y_1}{\Delta u_1}\right)_{y_2}^{-1}\left(\frac{\Delta y_1}{\Delta u_1}\right)_{u_2}$$

$$\lambda_{12} = \left(\frac{\Delta y_1}{\Delta u_2}\right)_{y_2}^{-1}\left(\frac{\Delta y_1}{\Delta u_2}\right)_{u_1}$$

$$\lambda_{21} = \left(\frac{\Delta y_2}{\Delta u_1}\right)_{y_1}^{-1}\left(\frac{\Delta y_2}{\Delta u_1}\right)_{u_2}$$

$$\lambda_{22} = \left(\frac{\Delta y_2}{\Delta u_2}\right)_{y_1}^{-1}\left(\frac{\Delta y_2}{\Delta u_2}\right)_{u_1}$$

Looking at these, we see that:

1. if $\lambda_{11} = 0$, then output 1 does not respond to input 1, so input 1

should not be used to control output 1;

2. if $\lambda_{11} = 1$, then the numerator and denominator of the relative gain expression are equal and input 2 does not couple to output 1;
3. if $0 < \lambda_{11} < 1$, then an interaction exists and u_2 will affect y_1 – the closer λ_{11} is to zero, the stronger the interaction; and
4. if $\lambda_{11} < 0$, then u_2 strongly affects y_1 and in the opposite direction from u_1. (This is a potentially difficult interaction to deal with.)

In addition, by working from the definitions, we may show

5. $\quad\lambda_{11} + \lambda_{12} = 1 \qquad \lambda_{11} + \lambda_{21} = 1$
$\quad\lambda_{21} + \lambda_{22} = 1 \qquad \lambda_{12} + \lambda_{22} = 1$

The above may be extended to larger systems (with N outputs and N controls) by defining

$$\lambda_{ij} = \frac{\left(\dfrac{\Delta y_i}{\Delta u_j}\right)_{\text{all } u_m \text{ constant, } m \neq j}}{\left(\dfrac{\Delta y_i}{\Delta u_j}\right)_{\text{all } y_k \text{ constant, } k \neq i}}$$

It then turns out that the relative gain array matrix $\Lambda = \{\lambda_{ij}\}$ has the property that its row sums and its column sums are all equal to one:

$$\sum_{i=1}^{N} \lambda_{ij} = 1 \qquad j = 1, 2, \ldots, N$$

$$\sum_{j=1}^{N} \lambda_{ij} = 1 \qquad i = 1, 2, \ldots, N$$

In selecting pairings, then, we

1. attempt to pair outputs i with inputs j for which λ_{ij} is relatively small and positive; and
2. use caution when any output i has input j such that λ_{ij} is negative.

34.3.3 Sequential loop closing

With the output–input pairings decided, there is still the issue of designing the feedback laws for each pair. If we assume (without loss of generality) it has been decided that $G_{cii}(s)$ are to be the only non-zero controllers, one approach is to design the compensators successively. Thus from (34.4) we first design $G_{c11}(s)$ in

$$Y_1(s) = \frac{G_{11}\,G_{c11}\,(1 + G_{22})}{1 + G_{11}G_{c11}}\,R_1$$

to give good response. Then $G_{c22}(s)$ is designed so that

$$Y_2(s) = \frac{G_{22}\,G_{c22}\,(1 + G_{c11}G_{11}) - G_{12}G_{21}\,G_{c11}\,G_{c22}}{(1 + G_{11}G_{c11})\,(1 + G_{22}G_{c22}) - G_{21}G_{12}\,G_{c11}\,G_{c22}}\,R_2$$

also responds 'well'. The procedure obviously can continue for higher numbers of inputs and outputs.

The basic advantage of the approach is that each stage is a SISO design problem for which methods are known and stability can be evaluated. There are several disadvantages:

- a good pairing is important;
- early design choices, such as the choice of $G_{c11}(s)$ above, will affect the later design problems in ways that can be hard to predict and may be deleterious; and
- the disturbance effects of inputs other than the one current at each stage are difficult to account for.

Notice that this is essentially the same as the cascade control of section 34.2.

34.3.4 Decoupling compensators

If interactions are significant, we may be forced to compensate for them by using decoupling compensators, which are in some ways similar to feedforward compensators. We might consider this a design method which avoids true MIMO control by using sets of SISO controllers with cross-compensation built in.

From (34.3)

$$\mathbf{Y}(s) = \begin{bmatrix} Y_1(s) \\ Y_2(s) \end{bmatrix} = \begin{bmatrix} G_{11}(s) & G_{12}(s) \\ G_{21}(s) & G_{22}(s) \end{bmatrix} \begin{bmatrix} U_1(s) \\ U_2(s) \end{bmatrix}$$

it is clear that if we choose

$$U_1(s) = U_{1,r}(s) - \frac{G_{12}(s)}{G_{11}(s)} U_{2,r}(s)$$

(where the r subscript indicates the command reference value) then

$$Y_1(s) = G_{11}(s) U_{1,r}(s) + G_{12}(s) [U_2(s) - U_{2,r}(s)]$$

Similarly, if we also choose

$$U_2(s) = U_{2,r}(s) - \frac{G_{21}(s)}{G_{22}(s)} U_{1,r}(s)$$

we obtain finally

$$Y_1(s) = \left(G_{11}(s) - G_{12}(s) \frac{G_{21}(s)}{G_{22}(s)} \right) U_{1,r}(s)$$

$$Y_2(s) = \left(G_{22}(s) - G_{21}(s) \frac{G_{12}(s)}{G_{11}(s)} \right) U_{2,r}(s)$$

Defining

$$U_{i,r}(s) = G_{cii}(s) \left(R_i(s) - Y_i(s) \right) \qquad i = 1, 2$$

then gives the closed-loop transfer functions

$$Y_1(s) = \frac{G_{c11}(s) [G_{11}(s) - G_{12}(s) G_{21}(s)/G_{22}(s)]}{1 + G_{c11}(s) [G_{11}(s) - G_{12}(s) G_{21}(s)/G_{22}(s)]} R_1(s)$$

$$Y_2(s) = \frac{G_{c22}(s)\,[G_{22}(s) - G_{12}(s)\,G_{21}(s)/G_{11}(s)]}{1 + G_{c22}(s)\,[G_{22}(s) - G_{12}(s)\,G_{21}(s)/G_{11}(s)]}\,R_2(s)$$

Thus the structural change has made the control loop appear independent. The structure is shown in Fig. 34.8.

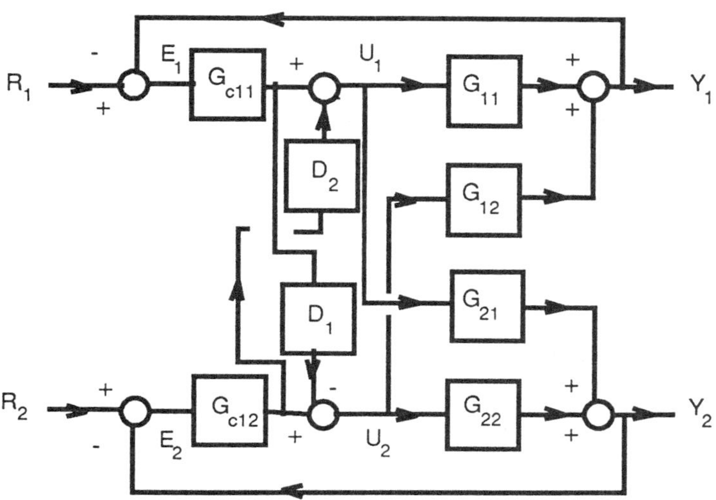

Figure 34.8 Pairing of input and output plus explicit decouplers.

In the above theory, the decoupling controls are given by

$$U_{1,\,\text{dec}}(s) = -D_2(s)\,U_{2,r}(s) = -\frac{G_{12}(s)}{G_{11}(s)}\,U_{2,r}(s)$$

$$U_{2,\,\text{dec}}(s) = -D_1(s)\,U_{1,r}(s) = -\frac{G_{21}(s)}{G_{22}(s)}\,U_{1,r}(s)$$

We note that if cancellation is not exact, which is likely for various reasons including the imprecision with which the transfer functions are known, then

$$Y_1(s) = (G_{11}(s) - G_{12}(s) D_1(s)) U_{1,r}(s)$$

$$+ (G_{12}(s) - G_{11}(s) D_2(s)) U_{2,r}(s)$$

$$Y_2(s) = (G_{21}(s) - G_{22}(s) D_1(s)) U_{1,r}(s)$$

$$+ (G_{22}(s) - G_{21}(s) D_2(s)) U_{2,r}(s)$$

Hence the attempt at decoupling has simply created a different, albeit perhaps more favourable, cross-coupling of the variables. Decouplers are in essence feedforward control terms, effectively cancelling the effects of the 'disturbances' to a loop caused by the cross-coupling imposed by the physical process.

It is always possible to compensate for only some of the cross-couplers, e.g. by choosing to use only $D_1(s)$ above; this situation is called **partial decoupling**.

34.4 TRANSFER FUNCTION APPROACHES II: DIAGONALIZATION OF TRANSFER MATRICES

An important notion in the compensation of plants described by transfer function matrices is that the compensator transfer function matrix $\mathbf{K}(s)$ can be considered structured.

$$\mathbf{K}(s) = \mathbf{K}_a \mathbf{K}_b(s) \mathbf{K}_c(s) = \mathbf{K}'_c(s) \mathbf{K}'_b(s) \mathbf{K}'_a$$

where $\mathbf{K}_a$ and $\mathbf{K}'_a$ are permutation matrices (which reorder the inputs and outputs by permuting rows and columns), $\mathbf{K}_c(s)$ and $\mathbf{K}'_c(s)$ are diagonal matrices of rational functions of s,

$$\mathbf{K}_c(s) = \operatorname{diag} \{k_i(s)\}$$

and $\mathbf{K}_b(s)$ and $\mathbf{K}'_b(s)$ are products of elementary matrices each having the form

$$\mathbf{I} + \alpha_{ij}(s) \mathbf{e}_i \mathbf{e}_j^T \qquad j \neq i$$

The design algorithm first has $\mathbf{K}_a$ and $\mathbf{K}_b(s)$ selected to make the plant transfer function $\mathbf{G}(s)$ diagonally dominant. Then $\mathbf{K}_c(s)$ is chosen in a set of SISO designs to achieve performance goals. More precisely, choose $\mathbf{K}_a$ and $\mathbf{K}_b(s)$ so that

$$\mathbf{G}'(s) = \mathbf{G}(s)\,\mathbf{K}_a\,\mathbf{K}_b(s)$$

is diagonally dominant (see Chapter 15). Next, choose $\mathbf{K}_c(s)$, which is by definition diagonal, so that $\mathbf{G}'(s)\,\mathbf{K}_c(s)$ has the desired properties (using Nyquist methods with Gershgorin bands as necessary to ensure stability margins). Finally, the controller becomes

$$\mathbf{U}(s) = \mathbf{K}_a\,\mathbf{K}_b(s)\,\mathbf{K}_c(s)\,\mathbf{E}(s)$$

The first problem here is to choose $\mathbf{K}_a$ and $\mathbf{K}_b(s)$. Occasionally cut and try methods are possible. The next stage is to try

$$\mathbf{K}_a\,\mathbf{K}_b(j\omega) = \mathbf{G}^{-1}(j\omega)$$

for a chosen frequency for which the inverse exists.

Unfortunately, $\mathbf{G}(j\omega)$ will usually be complex except at $\omega = 0$, and the inverse will not exist at 0 if the system has integrators (poles at the origin); an approximation may work, however. Otherwise, certain other theory can by tried; see Maciejowski (1989).

34.5 MIMO STRUCTURES: STATE–SPACE APPROACHES

For this section, we use continuous time models because it is primarily plant models which are being decomposed. Necessarily, because of the nature of the modern theory's search for optimal controllers, some continuous time controllers are also presented.

Essentially, the state–space model of a plant is given by (34.1), repeated here.

$$\dot{\mathbf{x}} = \mathbf{A}\mathbf{x} + \mathbf{B}\mathbf{u}$$

$$\mathbf{y} = \mathbf{C}\mathbf{x}$$

The goal is to decompose this in some useful and sensible way into subsystems for which the controllers can be derived individually. There are two situations of interest: decomposition into (nearly) uncoupled subsystems, and decomposition into 'fast' and 'slow' subsystems. These are modelled as

$$\dot{\mathbf{x}} = \begin{bmatrix} \dot{\mathbf{x}}_1 \\ \dot{\mathbf{x}}_2 \end{bmatrix} = \begin{bmatrix} A_{11} & \varepsilon A_{12} \\ \varepsilon A_{21} & A_{22} \end{bmatrix} \begin{bmatrix} \mathbf{x}_1 \\ \mathbf{x}_2 \end{bmatrix} + \begin{bmatrix} B_1 & 0 \\ 0 & B_2 \end{bmatrix} \begin{bmatrix} \mathbf{u}_1 \\ \mathbf{u}_2 \end{bmatrix}$$

$$\mathbf{y} = C_1 \mathbf{x}_1 + C_2 \mathbf{x}_2$$

and

$$\begin{bmatrix} \dot{\mathbf{x}}_1 \\ \varepsilon \dot{\mathbf{x}}_2 \end{bmatrix} = \begin{bmatrix} A_{11} & A_{12} \\ A_{21} & A_{22} \end{bmatrix} \begin{bmatrix} \mathbf{x}_1 \\ \mathbf{x}_2 \end{bmatrix} + \begin{bmatrix} B_1 \\ B_2 \end{bmatrix} \mathbf{u}$$

$$\mathbf{y} = C_1 \mathbf{x}_1 + C_2 \mathbf{x}_2$$

respectively. In these $\varepsilon > 0$ is 'small' and $\mathbf{x}_i$, $\mathbf{u}_i$ are vectors of lengths n_i and m_i, respectively, for $i = 1,2$, where we have limited the decomposition to two levels only for notational convenience. In either case, the theory then considers a hierarchy and suggests whether the decomposed system is near to optimal according to the quadratic criterion

$$\mathbf{J} = \int_{t_0}^{t_f} \mathbf{x}^{\mathrm{T}}(\tau)\,\mathbf{Q}\mathbf{x}(\tau) + \mathbf{u}^{\mathrm{T}}(\tau)\,\mathbf{R}\mathbf{u}(\tau)\,\mathrm{d}\tau$$

34.5.1 Loosely coupled systems

The state–space handling of systems which are somewhat independent is different in philosophy from the classical methods. One result which demonstrates this is that for the optimal LQ control of a system modelled as

$$\dot{\mathbf{x}} = \begin{bmatrix} \mathbf{x}_1 \\ \mathbf{x}_2 \end{bmatrix} = \begin{bmatrix} A_{11} & \varepsilon A_{12} \\ \varepsilon A_{21} & A_{22} \end{bmatrix} \begin{bmatrix} \mathbf{x}_1 \\ \mathbf{x}_2 \end{bmatrix} + \begin{bmatrix} B_{11} & \varepsilon B_{12} \\ \varepsilon B_{21} & B_{22} \end{bmatrix} \begin{bmatrix} \mathbf{u}_1 \\ \mathbf{u}_2 \end{bmatrix}$$

$$\mathbf{y} = C_1 \mathbf{x}_1 + C_2 \mathbf{x}_2 \tag{34.5}$$

The object is to minimize

$$J = \tfrac{1}{2} \int_{t_0}^{t_f} \left\{ \mathbf{x}^T \begin{bmatrix} Q_{11} & \varepsilon Q_{12} \\ \varepsilon Q_{21} & Q_{22} \end{bmatrix} \mathbf{x} + \mathbf{u}^T \begin{bmatrix} R_{11} & 0 \\ 0 & R_{22} \end{bmatrix} \mathbf{u} \right\} dt \qquad (34.6)$$

The general solution is that (Chapter 26)

$$\mathbf{u}(t) = - \mathbf{R}^{-1} \mathbf{B}^T \mathbf{K}(t) \mathbf{x}(t)$$

where the gain $\mathbf{K}(t)$ is the solution to the matrix Riccati equation

$$\dot{\mathbf{K}} = - \mathbf{K}\mathbf{A} - \mathbf{A}^T \mathbf{K} + \mathbf{K}\mathbf{B}\mathbf{R}^{-1} \mathbf{B}^T \mathbf{K} - \mathbf{Q} \qquad \mathbf{K}(t_f) = 0 \qquad (34.7)$$

If $t_f \to \infty$, then $\mathbf{K}(t) \to \mathbf{K}$ is the positive definite root of the algebraic equation

$$0 = \mathbf{K}\mathbf{A} + \mathbf{A}^T \mathbf{K} - \mathbf{K}\mathbf{B}\mathbf{R}^{-1}\mathbf{B}^T \mathbf{K} + \mathbf{Q}$$

It is not unusual for the dimensions of the state to be too large for this to be reasonably solved, requiring as it does the solution of $n(n+1)/2$ coupled quadratic equations. An alternative is to design a control for each of the two systems defined by $\varepsilon = 0$ and then modify the controller for the presence of the interactions. In particular, one assumes that the gain $\mathbf{K}$ is a function of both t and ε, and then defines the approximate gain $\mathbf{G}$ as

$$\mathbf{G} = \mathbf{K}(t,0) + \varepsilon \frac{\partial \mathbf{K}(t,0)}{\partial \varepsilon} + \frac{\varepsilon^2}{2} \frac{\partial^2 \mathbf{K}(t,0)}{\partial \varepsilon^2} + \cdots + \frac{\varepsilon^p}{p!} \frac{\partial^p \mathbf{K}(t,0)}{\partial \varepsilon^p}$$

for an appropriate truncation parameter p. It is obvious that the steady-state values as $t_f \to \infty$ must be

$$\mathbf{K} = \begin{bmatrix} \mathbf{K}_1 & 0 \\ 0 & \mathbf{K}_2 \end{bmatrix}$$

where

$$0 = K_1 A_{11} + A_{11}^T K_1 - K_1 B_{11} R_1^{-1} B_{11}^T K_1 + Q_1$$

$$0 = K_2 A_{22} + A_{22}^T K_2 - K_2 B_{22} R_2^{-1} B_{22}^T K_2 + Q_2$$

Not so obvious, but proven by Kokotovic *et al.* (1969) is that

$$\frac{\partial K(t, 0)}{\partial \varepsilon} = \begin{bmatrix} 0 & K_{12}^1 \\ K_{12}^{1\,T} & 0 \end{bmatrix}$$

where K_{12}^1 solves the linear equations

$$0 = -K_{12}^1 (A_{22} - B_{22} R_2^{-1} B_{22}^T K_2) - (A_{11} - B_{11} R_1^{-1} B_{11}^T K_1)^T K_{12}$$

$$- K_1 A_{12} - A_{12}^T K_2 + K_1 (B_{11} R_{11}^{-1} B_{21}^T + B_{12} R_{22}^{-1} B_{22}^T) K_2 - Q_{12}$$

The method of proof is by direct expansion of (34.7) using (34.5) and (34.6). The use of the expansion helps in three aspects.

1. It allows a reduction in computation. Computing the constant terms, for subsystems of equal size $n/2 \times n/2$ requires solution of only $2\,(n/2)(n/2 + 1)$ simultaneous quadratic equations. The next term then needs computation of n linear equations.
2. The cost criterion in using a pth order approximation in ε to K is a $(2p + 1)$ order approximation (in ε) to J. At the extreme, the separation approximation is roughly linear in ε for small ε; we might have only a 10% cost penalty if $\varepsilon \approx 0.1$.
3. We might use only K_1 and K_2, so that $u_i = K_i x_i$, $i = 1, 2$, is selected as a reasonable control law.

34.5.2 Systems with slow and fast modes

Many systems which are in principle of high dimension are decomposable into subsystems which operate at vastly different time scales. For one example, an aircraft's roll attitude can be changed much more rapidly than its altitude; for another, motor shaft response is much faster than the liquid tank level for which the shaft operates an outlet valve. In such instances, the fast mode may be considered virtually instantaneous by the designer of the controller for the slow mode, while the designer of the fast mode controller may treat the state of the slow mode as a constant parameter.

A matrix model is given by

$$\begin{bmatrix} \dot{\mathbf{x}}_1 \\ \varepsilon\dot{\mathbf{x}}_2 \end{bmatrix} = \begin{bmatrix} \mathbf{A}_{11} & \mathbf{A}_{12} \\ \mathbf{A}_{21} & \mathbf{A}_{22} \end{bmatrix} \begin{bmatrix} \mathbf{x}_1 \\ \mathbf{x}_2 \end{bmatrix} + \begin{bmatrix} \mathbf{B}_1 \\ \mathbf{B}_2 \end{bmatrix} \mathbf{u}$$

$$\mathbf{y} = \mathbf{C}_1\mathbf{x}_1 + \mathbf{C}_2\mathbf{x}_2 \tag{34.8}$$

In this, the submatrices $\mathbf{A}_{ij}$ are taken to have elements of 'similar' size, and hence the dynamics of $\mathbf{x}_2$ are fast relative to those of $\mathbf{x}_1$.

Example

Consider the system with transfer function

$$G(s) = \frac{1}{(0.01s + 1)(s + 1)}$$

A state–space model of this is

$$\begin{bmatrix} \dot{x}_1 \\ \dot{x}_2 \end{bmatrix} = \begin{bmatrix} -1 & 1 \\ 0 & 100 \end{bmatrix} \begin{bmatrix} x_1 \\ x_2 \end{bmatrix} + \begin{bmatrix} 0 \\ 100 \end{bmatrix} \mathbf{u}$$

$$\mathbf{y} = \begin{bmatrix} 1 & 0 \end{bmatrix} \mathbf{x}$$

An alternative form, similar to (34.8) is

$$\begin{bmatrix} \dot{x}_1 \\ 0.01\dot{x}_2 \end{bmatrix} = \begin{bmatrix} -1 & 1 \\ 0 & 1 \end{bmatrix} \begin{bmatrix} x_1 \\ x_2 \end{bmatrix} + \begin{bmatrix} 0 \\ 1 \end{bmatrix} \mathbf{u}$$

$$\mathbf{y} = \begin{bmatrix} 1 & 0 \end{bmatrix} \mathbf{x}$$

A special control law for x_1 might simply use x_2 as a commandable quantity (because $x_2 \approx -u$). Also, a control law for x_2 could be designed assuming that x_1 is known and constant.

In the more general terms of (34.8), it seems reasonable to assume that $\varepsilon \approx 0$, so that if A_{22} is non-singular, then

$$x_2 = - A_{22}^{-1}(A_{21}x_1 + B_2 u)$$

Using this, the slow system description becomes

$$\dot{x}_1 = (A_{11} - A_{12}A_{22}^{-1}A_{21})x_1 + (B_1 - A_{12}A_{22}^{-1}B_2)u$$

$$y = (C_1 - C_2 A_{22}^{-1}A_{21})x_1 - C_2 A_{22}^{-1}B_2 u \tag{34.9}$$

We rewrite this slow mode model as

$$\dot{x}_s = A_s x_s + B_s u_s$$

$$y_s = C_s x_s + D_s u_s$$

On the other hand, the slow system state becomes essentially a constant input value $x_{1,\text{ref}}$ to the fast system. Under these circumstances, we treat $x_{2,\text{ref}}$ as the resulting steady-state value (if it were allowed to converge to it) and u_{ref} as the associated command. Then it must be that

$$\dot{x}_{2,\text{ref}} = 0$$

$$x_{2,\text{ref}} = -A_{22}^{-1}(A_{21}x_{1,\text{ref}} + B_2 u_{\text{ref}}) \tag{34.10}$$

Then if we define the excursion from reference values as

$$\delta x_f = x_2 - x_{2,\text{ref}} \qquad \delta u = u - u_{\text{ref}}$$

we find that because of (34.10)

$$\delta \dot{\mathbf{x}}_f = \mathbf{A}_{22}\,\delta \mathbf{x}_f + \mathbf{B}_2\,\mathbf{u}_{ref}$$

$$\delta \mathbf{y}_f = \mathbf{C}_2\,\delta \mathbf{x}_f \tag{34.11}$$

The design problem then can be considered one of finding good state feedback controllers for the separate slow and fast systems and combining those to give the appropriate controller for the original system; the analysis problem is to determine whether this 'engineering' solution is 'good'.

A partial answer to this is given by considering any state feedback controller. Let

$$\mathbf{u}_s = \mathbf{G}_s\mathbf{x}_s$$

$$\delta \mathbf{u}_f = \mathbf{G}_f\,\delta \mathbf{x}_f$$

be state feedback controllers for the slow and fast modes separately, i.e. for (34.9) and (34.11), where it is assumed the state feedback gives stability, and in particular that $(\mathbf{A}_{22} + \mathbf{B}_2\mathbf{G}_f)$ is stable. Then the control to the original system defined by

$$\mathbf{u} = \left[(\mathbf{I} + \mathbf{G}_f\mathbf{A}_{22}^{-1}\mathbf{B}_2)\mathbf{G}_s + \mathbf{G}_f\mathbf{A}_{22}^{-1}\mathbf{A}_{21}\right]\mathbf{x}_1 + \mathbf{G}_f\mathbf{x}_2$$

will yield a state vector

$$\mathbf{x}_1(t) = \mathbf{x}_s(t) + O(\varepsilon)$$

$$\mathbf{x}_2(t) = -\mathbf{A}_{22}^{-1}(\mathbf{A}_{21} + \mathbf{B}_2\mathbf{G}_s)\mathbf{x}_s(t) + \delta \mathbf{x}_f(t) + O(\varepsilon)$$

where $O(\varepsilon) \to 0$ faster than ε. It is also true that

$$\mathbf{u}(t) = \mathbf{u}_s(t) + \delta \mathbf{u}_f(t) + O(\varepsilon)$$

$$\mathbf{y}(t) = \mathbf{y}_s(t) + \delta \mathbf{y}_f(t) + O(\varepsilon)$$

34.5.3 Model aggregation

The aggregation approach involves the replacement of a true model

$$\dot{\mathbf{x}} = \mathbf{A}\mathbf{x} + \mathbf{B}\mathbf{u} \tag{34.12}$$

with n-dimensional state vector $\mathbf{x}$ with a model with state $\mathbf{z}$ defined, using a $p \times n$ matrix $\mathbf{D}$ with $p < n$, by

$$\mathbf{z} = \mathbf{D}\mathbf{x}$$

and hence a dynamics model

$$\dot{\mathbf{z}} = \mathbf{F}\mathbf{z} + \mathbf{G}\mathbf{u} \tag{34.13}$$

provided

$$\mathbf{F}\mathbf{D} = \mathbf{D}\mathbf{A} \qquad \mathbf{G} = \mathbf{D}\mathbf{B}$$

If $\mathbf{A}$ and $\mathbf{D}$ satisfy

$$\mathbf{D}\mathbf{A} = \mathbf{D}\mathbf{A}\mathbf{D}^{\mathrm{T}}(\mathbf{D}\mathbf{D}^{\mathrm{T}})^{-1}\mathbf{D}$$

then an appropriate $\mathbf{F}$ is

$$\mathbf{F} = \mathbf{D}\mathbf{A}\mathbf{D}^{\mathrm{T}}(\mathbf{D}\mathbf{D}^{\mathrm{T}})^{-1}$$

$\mathbf{D}$ is called the **aggregation matrix**, and $\mathbf{z}$ is the **aggregated state vector**. The aggregation matrix is the primary design factor in studies of the system. Its use allows inferences to be made about smaller dimensional systems than the original, and it is chosen so that the error made by evaluating (34.13) instead of (34.12) is 'small'.

Aggregation of weakly coupled fast and slow systems

If the matrix $\mathbf{A}$ has eigenvalues $\lambda_1, \lambda_2, ..., \lambda_n$ and corresponding eigenvectors $\mathbf{e}_1, \mathbf{e}_2, ..., \mathbf{e}_n$, then

$$\mathbf{DAe}_i = \mathbf{FDe}_i = \lambda_i \mathbf{De}_i \qquad (34.14)$$

Hence if the vector $\mathbf{De}_i \neq \mathbf{0}$, then it is an eigenvector of $\mathbf{F}$ with eigenvalue λ_i. This may be used to select the aggregation system as one with fast or slow modes. In particular, let $\lambda_1, \lambda_2, ..., \lambda_p$ be the desired aggregated eigenvalues and choose $\mathbf{D}$ such that

$$\mathbf{De}_i \neq \mathbf{0} \qquad i = 1, 2, ..., p$$

$$\mathbf{De}_i = \mathbf{0} \qquad i = p+1, ..., n$$

In general, this task is not easy for large systems, partly because of the need to find the eigenvalues and eigenvectors of the matrix $\mathbf{A}$.

The task is somewhat simpler, at least in approximations, if the system has 'weakly coupled' subsystems. Consider two subsystems coupled into a system described by

$$\mathbf{A} = \begin{bmatrix} \Lambda & \mathbf{A}_{12} \\ \mathbf{A}_{21} & \mathbf{M} \end{bmatrix}$$

in which $\Lambda = \mathrm{diag}(\lambda_1, \lambda_2, ..., \lambda_{n_1})$, $\mathbf{M} = \mathrm{diag}(\mu_1, \mu_2, ..., \mu_{n_2})$, with $n_1 + n_2 = n$. Defining the bounds of the eigenvectors by

$$r = \max_i |\lambda_i| \qquad R = \min_i |\mu_i|$$

then the subsystems are weakly coupled, and the $\mathbf{x}_2$ subsystem is fast relative to the $\mathbf{x}_1$ subsystem, if

$$\frac{r}{R} \ll 1 \qquad \frac{n_1 \max_{ij} |\{\mathbf{A}_{21}\}_{ij}| \max_{ij} |\{\mathbf{A}_{12}\}_{ij}|}{R^2} \ll 1$$

Then it has been shown that the eigenvalues of **A** are approximately given by the solution of

$$\det (v_j \mathbf{I} - \Lambda + \mathbf{A}_{12} \mathbf{M}^{-1} \mathbf{A}_{21}) = 0 \qquad j = 1, \ldots, n_1$$

$$v_j = \mu_{j-n_1} \qquad j = n_1 + 1, \ldots, n_2$$

The eigenvectors of **A** are approximately

$$\mathbf{e}_j = \begin{pmatrix} \xi_j \\ -\Lambda^{-1} \mathbf{A}_{21}\, \xi_j \end{pmatrix} \qquad j = 1, \ldots, n_1$$

$$\mathbf{e}_{j+n_1} = \begin{pmatrix} \dfrac{\mathbf{A}_{12}}{\mu_j}\, \eta_j \\ \eta_j \end{pmatrix} \qquad j = 1, 2, \ldots, n_2$$

where

$$(\Lambda - \mathbf{A}_{12} \mathbf{M}^{-1} \mathbf{A}_{21})\, \xi_j = v_j\, \xi_j \qquad j = 1, 2, \ldots, n_1$$

$$\mathbf{M} \eta_j = \mu_j \eta_j \qquad j = 1, 2, \ldots, n_2$$

These approximations may make the computation of an appropriate aggregation matrix as in (34.14) easier.

Suboptimal control through aggregation

The object of the control may be to minimize

$$J = \int_0^\infty \left(\mathbf{x}^T(\tau)\, \mathbf{Q} \mathbf{x}(\tau) + \mathbf{u}^T(\tau)\, \mathbf{R} \mathbf{u}(\tau) \right) d\tau$$

for the system

$$\dot{\mathbf{x}} = \mathbf{A}\mathbf{x} + \mathbf{B}\mathbf{u}$$

We consider instead, for computational reasons, the optimal control of

$$\dot{z} = Fz + Gu$$

with criterion

$$J = \int_0^\infty \left[z^T(\tau)\, Q_F\, z(\tau) + u^T(\tau)\, Ru(\tau) \right] d\tau$$

We know that the solution to the former is

$$u = -R^{-1} B^T P x$$

where P satisfies

$$A^T P + PA - PBR^{-1}B^T P + Q = 0$$

while the solution to the latter is

$$u = -R^{-1} G^T P_F z$$

where P satisfies

$$F^T P_F + P_F F - P_F G R^{-1} G^T P_F + Q_F = 0$$

Since $G = DB$ and $FD = DA$, if we multiply the final equation above by D^T and D, we find

$$A^T D^T P_F + P_F DA - D^T P_F DBR^{-1}B^T D^T P_F + D^T Q_F D = 0$$

Comparing this with (34.12), we see that $D^T P_F D$ corresponds to P if $D^T Q_F D$ corresponds to Q, although they cannot generally be equal because of their probable difference in rank. One possible choice is

$$Q_F = (DD^T)^{-1}\, DQD^T\, (DD^T)^{-1}$$

The final suboptimal control is given by

$$\mathbf{u} = -\mathbf{R}^{-1}\mathbf{G}^T\mathbf{P}_F\mathbf{D}\mathbf{x}$$

and it may be shown that the resulting cost of

$$\mathbf{J} = \mathbf{x}^T(0)\mathbf{P}_A\mathbf{x}(0) \tag{34.15}$$

compares to the system optimal

$$\mathbf{x}^T(0)\mathbf{P}\mathbf{x}(0)$$

and the subsystem optimal

$$\mathbf{x}^T(0)\mathbf{D}^T\mathbf{P}_F\mathbf{D}\mathbf{x}(0)$$

through

$$0 \le \mathbf{x}^T(0)\mathbf{D}^T\mathbf{P}_F\mathbf{D}\mathbf{x}(0) \le \mathbf{x}^T(0)\mathbf{P}\mathbf{x}(0) \le \mathbf{x}^T(0)\mathbf{P}_A\mathbf{x}(0)$$

In this, $\mathbf{P}_A$ is the solution of

$$0 = (\mathbf{A} - \mathbf{B}\mathbf{R}^{-1}\mathbf{G}\mathbf{P}_F\mathbf{D})^T\,\mathbf{P}_A\,(\mathbf{A} - \mathbf{B}\mathbf{R}^{-1}\,\mathbf{G}\mathbf{P}_F\mathbf{D}) + \mathbf{Q}$$
$$+ (\mathbf{G}\mathbf{P}_F\mathbf{D})^T\mathbf{R}^{-1}\mathbf{G}\mathbf{P}_F\mathbf{D}$$

In principle, $\mathbf{D}$ may be manipulated to minimize the actual cost as in (34.15) and hence to get near the lower bound.

34.5.4 Periodic co-ordination

The notion of periodic co-ordination is that of a simple hierarchy in which low levels of control, designed only for their subsystems and seeing interactions as disturbances, are periodically commanded by a coordinator. Thus, for example, the system properly described by

$$
\begin{bmatrix} \mathbf{x}_1(k+1) \\ \mathbf{x}_2(k+1) \end{bmatrix} = \begin{bmatrix} \mathbf{A}_{11} & \mathbf{A}_{12} \\ \mathbf{A}_{21} & \mathbf{A}_{22} \end{bmatrix} \begin{bmatrix} \mathbf{x}_1(k) \\ \mathbf{x}_2(k) \end{bmatrix} + \begin{bmatrix} \mathbf{B}_1 & 0 \\ 0 & \mathbf{B}_2 \end{bmatrix} \begin{bmatrix} \mathbf{u}_1(k) \\ \mathbf{u}_2(k) \end{bmatrix}
$$

$$
\begin{bmatrix} \mathbf{y}_1(k) \\ \mathbf{y}_2(k) \end{bmatrix} = \begin{bmatrix} \mathbf{C}_1 & 0 \\ 0 & \mathbf{C}_2 \end{bmatrix} \begin{bmatrix} \mathbf{x}_1(k) \\ \mathbf{x}_2(k) \end{bmatrix} \tag{34.16}
$$

will have two subsystems modelled as

$$
\mathbf{x}_i(k+1) = \mathbf{A}_{ii}\mathbf{x}_i(k) + \mathbf{B}_i\mathbf{u}_i + \text{disturbances}
$$

A control law of the form

$$
\mathbf{u}_i(k) = f(\mathbf{Y}_i(k), k) + \mathbf{u}_{i,\,\text{ext}}(k) \qquad i=1,2
$$

is designed for each, where by definition $\mathbf{Y}_i(k) = \{\mathbf{y}_i(k-j),\ j = 0, 1, ..., k\}$, $i = 1, 2$. Periodically, such as whenever $k = pN$ for some integer p and integer period N, the supervisor interacts with the subsystems by considering their measurements and commands and the overall system model (34.16). The supervisor then generates commands based upon its knowledge of the model and its knowledge of the measurements (whether the entire history or the history at supervisory instants). These are then given to the subsystems as

$$
\mathbf{u}_{i,\,\text{ext}}(pN) = \mathbf{F}_i\,(\mathbf{Y}_1\,(pN),\ \mathbf{Y}_2\,(pN),\ \mathbf{U}_1\,(pN),\ \mathbf{U}_2\,(pN)) \qquad i=1,2
$$

or

$$
\mathbf{u}_{i,\,\text{ext}}(pN) = \mathbf{F}'_i\,(\mathbf{Y}'_1\,(pN),\ \mathbf{Y}'_2\,(pN),\ \mathbf{U}'_1\,(pN),\ \mathbf{U}'_2\,(pN)) \qquad i=1,2
$$

where

$$
\mathbf{Y}'_i\,(pN) = \{\mathbf{y}_i((p-j)N,\ j=0,1,...,p)\}
$$

$$
\mathbf{U}_i\,(pN) = \{\mathbf{u}_i((pN-j),\ j=0,1,...,pN)\}
$$

$$
\mathbf{U}'_i\,(pN) = \{\mathbf{u}_i((p-j)N,\ j=0,1,...,p)\} \qquad i=1,2
$$

Notice that the subsystems see the effects of other subsystems, i.e. the terms $\mathbf{A}_{ij}\mathbf{x}_j(k)$ for $j \neq i$, as environmental disturbances. The co-ordinator, however, which has complete information, can plan for this and send appropriate compensatory commands.

34.6 COMPUTER AIDS

Among the decomposition algorithms in commonly available CACSD programs are those for model reduction based upon fast and slow modes, as in section 34.5.3.

34.7 FURTHER READING

The standard teaching with regard to introduction to decomposition of systems through pairing is given in texts such as Seborg, Edgar and Mellichamp (1989 Chs 19 & 28) and Stephanopoulos (1984 Chs 24 & 25). Multivariable transfer function methods are in Maciejowski (1989).

The fast/slow mode discussion here follows principally Chow and Kokotovic (1976), while aggregation and ε-coupling follow Aoki (1968) and Kokotovic *et al.* (1969), respectively. These and other interesting research articles are reprinted in Larson (1979).

Appendix A
z-Transform

A.1 DEFINITION AND IMPORTANT PROPERTIES

One way of representing discrete time systems and their inputs uses the z-transform. This is defined for a sequence $\{x(k),\ k=\ldots,-2,-1, 0,1,2,\ldots\}$ by the double-sided series

$$X(z) = \mathcal{Z}(\{x(k)\}) = \sum_{i=-\infty}^{\infty} x(i)\, z^{-i}$$

or, on the presumption we shall use that all signals are causal and hence have $x(i) = 0$, $i < 0$, by the single-sided series

$$X(z) = \mathcal{Z}(\{x(k)\}) = \sum_{i=0}^{\infty} x(i)\, z^{-i}$$

where the $X(z)$ and $\mathcal{Z}\{\cdot\}$ notations are both used to represent the transform, depending upon the application. Here we say that $X(z)$ forms a transform pair with $\{x(k)\}$, for which we often use the notation $\{x(k)\} \Leftrightarrow X(z)$. For a great many common signals and systems, a rational (i.e. ratio of polynomials) representation is available for the transform. Thus, for example, if

$$x(k) = \begin{cases} 1 & k = 0 \\ 0 & k \neq 0 \end{cases}$$

then

$$X(z) = 1\,z^0 + 0\,z^{-1} + 0\,z^{-2} + \cdots = 1$$

If

$$x(k) = \begin{cases} 0 & k < 0 \\ 1 & k \geq 0 \end{cases}$$

then

$$X(z) = \sum_{i=-\infty}^{\infty} z^{-i} = \frac{1}{1 - z^{-1}} = \frac{z}{z - 1}$$

These are usually derived by direct evaluation for the simpler expressions and given in tables such as Table A.2 (page 870).

Table A.1 Properties of *z*-transform

Linearity	$\mathscr{Z}(ax(k) + by(k)) = a\,\mathscr{Z}(x(k)) + b\,\mathscr{Z}(y(k))$
Time shift	$\mathscr{Z}(x(k-n)) = z^{-n}\,\mathscr{Z}(x(k))$
Final value	$\lim_{k \to \infty} x(k) = \lim_{z \to 1} (1 - z^{-1})\mathscr{Z}(x(k))$
Initial value	$x(0) = \lim_{z \to 0} \mathscr{Z}(x(k))$
Convolution	$\mathscr{Z}\left(\sum_{n=0}^{k} x(n)\,y(k - n)\right) = \mathscr{Z}(x(k))\,\mathscr{Z}(y(k))$

The *z*-transform is a linear operation and hence its properties are similar to those for any linear transformation, and in particular to the Laplace transform. In fact, the *z*-transform is often conveniently thought of as the discrete analog of the Laplace transform. The commonly quoted properties are in Table A.1, but there are three properties which we use, implicitly or explicitly, very frequently. Some of these are demonstrated below.

Property 1: Linearity The sum of *z*-transforms is the *z*-transform of the sum of the sequences. Consider the two sequences $\{x(k)\}$ and $\{y(k)\}$ with *z*-transforms $X(z)$ and $Y(z)$, respectively. Then if

$w(k) = ax(k) + by(k)$, for any constants a and b, it is true that

$$W(z) = aX(z) + bY(z)$$

and vice versa. It is this property which will allow us to do superposition of inputs, partial fraction expansions, etc.

Property 2: Backward shift If for sequences $\{x(k)\}$ and $\{y(k)\}$ the relationship holds that

$$y(k) = x(k-1)$$

for all k (assuming $x(-1)$ is defined somehow) then

$$Y(z) = \sum_{k=0}^{\infty} y(k)\, z^{-k} = \sum_{k=0}^{\infty} x(k-1)\, z^{-k} = x(-1) + z^{-1} \sum_{i=0}^{\infty} x(k)\, z^{-k}$$

$$= z^{-1}X(z) + x(-1)$$

and in fact if $y(k) = x(k-p)$ then

$$Y(z) = z^{-p} X(z) + x(-p) + z^{-1}x(-p+1) + z^{-2}x(-p+2)$$

$$+ \cdots + z^{-p+1}x(-1)$$

Thus, if $y(k) = -ay(k-1) + u(k-1)$ and if $\{u(k)\} \Leftrightarrow U(z)$ and $\{y(k)\} \Leftrightarrow Y(z)$, then

$$Y(z) = -az^{-1} Y(z) - ay(-1) + z^{-1} U(z) + u(-1)$$

or

$$Y(z) = \left(\frac{z^{-1}}{1 + az^{-1}} \right) U(z) + \frac{-ay(-1) + u(-1)}{1 + az^{-1}}$$

This shows an explicit relationship between the transforms of the $u(k)$ and $y(k)$ sequences. Notice the use of the linearity assumption on the right-hand side. This method generalizes immediately to the general case of a difference equation such as

$$y(k) = -\sum_{i=1}^{n} a_i y(k-i) + \sum_{i=0}^{m} b_i u(k-i)$$

to yield

$$Y(z) = \frac{b_0 + b_1 z^{-1} + b_2 z^{-2} + \cdots + b_m z^{-m}}{1 + a_1 z^{-1} + a_2 z^{-2} + \cdots + a_n z^{-n}} U(z)$$

$$+ \frac{(\text{terms in initial conditions on } y() \text{ and } u())}{1 + a_1 z^{-1} + a_2 z^{-2} + \cdots + a_n z^{-n}}$$

$$= H(z) U(z) + \frac{IC}{D(z)}$$

The expression $H(z)$ is the discrete transfer function of the system of equations, or simply the transfer function. The denominator polynomial is denoted $D(z)$, and the initial condition terms are denoted IC.

We should remark here that we have been a bit crude in not spelling out the causality assumption about the system – no outputs prior to the arrival of the requisite inputs.

Property 2a: Forward shift The above backward shift also has a forward shift version. Again working directly from the definition of the one-sided transform, if $y(k) = x(k+1)$, then

$$Y(z) = \sum_{i=0}^{\infty} y(i) z^{-i}$$

$$= z \sum_{i=1}^{\infty} x(i) z^{-i}$$

$$= zX(z) - zx(0)$$

Similarly,

$$\mathcal{Z}(\{x(k+p)\}) = z^p X(p) - z^p x(0) - z^{p-1}x(1) - z^{p-2}x(2) - \cdots - zx(p-1)$$

Property 3: The convolution-product relationship If two sequences $\{x(k)\}$ and $\{y(k)\}$ are convolved, the z-transformation of the convolution is the product of the z-transforms of the sequences. Thus if

$$w(k) = \sum_{i=0}^{\infty} x(k-i)\,y(i)$$

then

$$W(z) = \sum_{k=0}^{\infty} \left[\sum_{i=0}^{\infty} x(k-i)\,y(i) \right] z^{-k}$$

$$= \sum_{i=0}^{\infty} y(i) \sum_{k=0}^{\infty} x(k-i)\,z^{-k}$$

$$= \sum_{i=0}^{\infty} y(i)[z^{-i} X(z) + x(-i) + z^{-1}x(-i+1) + z^{-2}x(-i+2)$$

$$+ \cdots + z^{-i+1}x(-1)]$$

$$= Y(z)\,X(z) \tag{A.1}$$

provided $x(i)$ is also causal, i.e. $x(i) = 0$ for $i < 0$. More terms appear if this is not true; for example if only $x(-1)$ and $x(-2)$ are non-zero for $i < 0$, (A.1) becomes

$$= Y(z)\,X(z) + x(-1)\,[zY(z) - y(0)]$$

$$+ x(-2)[z^2\,Y(z) - zy(0) + y(1)]$$

The pattern extends, but if it becomes of infinite extent, then two-sided z-transforms, i.e. those which index from $-\infty$ to $+\infty$, are used, and the basic result of (A.1) follows easily.

Notationally, if ** denotes convolution, then the rule is

$$\{y(k)\} ** \{x(k)\} \Leftrightarrow Y(z)X(z)$$

$$\{y(k)\}\{x(k)\} \Leftrightarrow Y(z) ** X(z)$$

where the left-hand side of the second relation denotes the sequence $\{z(k)\} = \{y(k)x(k)\}$.

The application of this arises from our observation that the output of our simple linear system is the convolution of the input with the pulse response. Thus, for a system with input $\{u(k)\}$ and pulse response $\{h(k)\}$, the transform of the output $\{y(k)\}$ is given by $H(z)U(z)$.

A.2   THE INVERSE Z-TRANSFORM

The inverse z-transform is defined by the contour integral

$$y(k) = \frac{1}{2\pi j} \oint_C Y(z) z^{k-1} \, dz$$

where the contour C is a circle in the region of convergence of $Y(z)$. For our purposes this becomes, because $Y(z)$ is usually a rational function of z,

$$y(k) = \sum_i \text{Residues}(Y(z) z^{k-1} \text{ at singularities } z_i \text{ of } Y(z))$$

In fact, for our purposes it is seldom necessary to perform an inversion in this way. Again because $Y(z)$ is rational in z, we can make good progress with either of two methods: long division, and partial fraction expansion plus tables.

Using long division, we find

$$Y(z) = \frac{N(z)}{D(z)} = \sum_{i=0}^{\infty} y(i) z^{-i}$$

where $N(z)$ and $D(z)$ are numerator and denominator polynomials in z^{-1}, which by long division yield a sum with constant coefficients $f(i)$. The coefficients are then the same as $y(i)$ because our assumptions are sufficient to ensure uniqueness of the defining z-transform.

Example

Let

$$Y(z) = \frac{z}{z+a} = \frac{1}{1+az^{-1}}$$

Then

$$1 + az^{-1} \,\overline{\big)\, 1} \quad \begin{array}{l} 1 - az^{-1} + a^2 z^{-2} - \cdots + (-a)^k z^{-k} \cdots \\[4pt] \end{array}$$

$$
\begin{aligned}
&\underline{1 + az^{-1}} \\
&\;\; -az^{-1} \\
&\;\; \underline{-az^{-1} - a^2 z^{-2}} \\
&\qquad\quad a^2 z^{-2} \\
&\qquad\quad \underline{a^2 z^{-2} + a^3 z^{-3}} \\
&\qquad\qquad\quad a^3 z^{-3} \\
&\qquad\qquad\qquad \vdots
\end{aligned}
$$

which immediately gives $y(i) = (-a)^i$.

The alternative to the above is to use partial fraction expansion of $Y(z)$ and interpret the resulting terms using tables. That this works follows from the Fundamental Theorem of Algebra, which tells us that the denominator can be factored into a product of factors linear in z (or quadratic in z if the coefficients are to be real), the linearity of the z-transform, and the uniqueness of the transform under our assumptions. The method works by finding suitable rational expressions $Y_1(z), Y_2(z), \ldots, Y_n(z)$, where suitable means we have already computed their inverses $\{y_1(k)\}, \{y_2(k)\}, ..., \{y_n(k)\}$ in tables, such that

$$Y(z) = Y_1(z) + Y_2(z) + Y_3(z) + \cdots + Y_n(z)$$

Then we immediately have

$$y(k) = y_1(k) + y_2(k) + \cdots + y_n(k)$$

Example

Suppose

$$Y(z) = \frac{z^2 + (a + b)z/2}{z^2 + (a + b)z + ab}$$

Then we immediately find

$$Y(z) = \frac{z^2 + (a + b)z/2}{(z + a)(z + b)} = \frac{z}{2(z + a)} + \frac{z}{2(z + b)}$$

from which it follows that

$$y(k) = \frac{(-a)^k + (-b)^k}{2}$$

Notice that, if a few examples are tried, it becomes obvious that the partial fraction method gives expressions for $y(k)$, whereas the long-division method gives numerical values.

An alternative method to get numerical values is to use simulation techniques. In these, the rational fraction is converted to a difference equation which is solved using a computer. Here, then, if $Y(z)$ is to be found, we observe that it is a pulse response of a linear system. Let

$$X(z) = Y(z)\,U(z) = \frac{N(z)}{D(z)}\,U(z)$$

Then $D(z)X(z) = N(z)U(z)$ from which

$$(z^n + a_1 z^{n-1} + a_2 z^{n-2} + \cdots + a_n)X(z)$$

$$= (b_0 z^n + b_1 z^{n-1} + \cdots + b_m z^{n-m})U(z)$$

which agrees with the difference equation

$$x(k+n) + a_1 x(k+n-1) + a_2 x(k+n-2) + \cdots + a_n x(k)$$

$$= b_0 u(k+n) + b_1 u(k+n-1) + \cdots b_m u(k+n-m)$$

or alternatively

$$x(k) = -a_1 x(k-1) - a_2 x(k-2) - \cdots - a_n x(k-n)$$

$$+ b_0 u(k) + b_1 u(k-1) + \cdots + b_m u(k-m)$$

The latter is easily computed for $k \geq 0$ taking $u(0)=1$, $u(k)=0$, $k \neq 0$; $x(k) = 0$, $k < 0$. The numbers $x(k)$, $k \geq 0$, then are the unit pulse response of the system, and hence are the same as the numbers associated with the definition of $Y(z)$.

A.3   THE 'EQUIVALENT' TRANSFORM $\mathscr{Z}_{eq}$

Consider a system with a differential equation model

$$u^{(n)} + \alpha_1 u^{(n-1)} + \cdots + \alpha_{n-1}\dot{u} + \alpha_n u$$

$$= \beta_m e + \beta_{m-1}\dot{e} + \cdots + \beta_0 e^{(m)}$$

and hence the Laplace transform transfer function

$$G_a(s) = \frac{U(s)}{E(s)} = \frac{\beta_m + \beta_{m-1}s + \cdots + \beta_0 s^m}{s^n + \alpha_1 s^{n-1} + \cdots \alpha_n}$$

This system will have a unit impulse response $h(t)$ and therefore a set of samples $\{h(0), h(T), h(2T), ...\}$. Hence we have

$$h(t) = \mathcal{L}^{-1}[G_a(s)]$$

$$h(nT) = \mathcal{L}^{-1}[G_a(s)]\Big|_{t=nT}$$

By definition, the z-transform of this sequence is

$$G_d(z) = \sum_{i=0}^{\infty} h(iT)\, z^{-i} = \mathcal{Z}\left\{\mathcal{L}^{-1}[G_a(s)]\Big|_{t=nT}\right\}$$

To make the notation more convenient, we write the right-hand side as

$$\mathcal{Z}_{eq}[G(s)] = \mathcal{Z}\left\{\mathcal{L}^{-1}[G_a(s)]\Big|_{t=nT}\right\}$$

or as

$$G^*(z) = \mathcal{Z}\left\{\mathcal{L}^{-1}[G_a(s)]\Big|_{t=nT}\right\}$$

This may be taken as defining the 'equivalent' z-transform for the Laplace transform.

By careful work with particular systems, we are able to work out the equivalents as shown in these two examples.

1. Let $G_c(s) = 1/s$. Then $h(t) = 1$, $h(iT) = 1$, and

$$G_d(s) = \sum_{i=0}^{\infty} z^{-i} = \frac{1}{1 - z^{-1}} = \frac{z}{z - 1}$$

$$\mathcal{Z}_{eq}\left(\frac{1}{s}\right) = \frac{z}{z - 1}$$

2. Let $G_c(s) = 1/(s+a)$. Then $h(t) = \mathrm{e}^{-at}$, $h(iT) = \mathrm{e}^{-aiT}$ and

$$G_d(s) = \sum_{i=0}^{\infty} (\mathrm{e}^{-aT}z)^{-i} = \frac{z}{z - \mathrm{e}^{-aT}}$$

$$\mathscr{Z}_{\mathrm{eq}}\left(\frac{1}{s+a}\right) = \frac{z}{z - \mathrm{e}^{-aT}}$$

Continuing in a like mode, we find Table A.2. This table also shows the z-transform 'equivalent' of $(1 - \mathrm{e}^{-sT})G_c(s)/s$, which is both the step-invariant response and arguably the 'correct' representation for a plant model (Chapter 12).

Particular properties to note are that

$$\mathscr{Z}_{\mathrm{eq}}[(1 - \mathrm{e}^{-Ts})\, G(s)] = (1 - z^{-1})\mathscr{Z}_{\mathrm{eq}}[G(s)]$$

$$\mathscr{Z}_{\mathrm{eq}}[G(s) + H(s)] = \mathscr{Z}_{\mathrm{eq}}[G(s)] + \mathscr{Z}_{\mathrm{eq}}[H(s)]$$

$$\mathscr{Z}_{\mathrm{eq}}[G(s)\, H(s)] \neq \mathscr{Z}_{\mathrm{eq}}[G(s)]\,\mathscr{Z}_{\mathrm{eq}}[H(s)]$$

The first two are easily shown using time domain arguments (essentially from the definition of $\mathscr{Z}_{\mathrm{eq}}$) and the last can be demonstrated with counter-examples. All also make sense intuitively, as the student is invited to consider.

A.4 FURTHER READING

Several textbooks have elaborate explanations of the z-transform; students may like Ogata (1987), for example. Other books have more or less extensive tables containing information similar to Table A.2. Among the latter are Kuo (1980), Ogata (1987), and Franklin, Powell, and Workman (1990). We remark, however, that by using partial fraction expansions it is possible to generate most transforms from a table going only up to second-order terms.

The modified z-transform is discussed and tables are presented by Kuo (1980).

Table A.2 z and Laplace transforms

$G(s)$	$g(t)$ $t \geq 0$	$g(kT)$ $k \geq 0$	*Impulse equivalent* $G_1(z) = \mathscr{Z}_{eq}(G(s))$
$\dfrac{1}{s}$	1	1	$\dfrac{1}{1-z^{-1}}$
$\dfrac{1}{s+a}$	e^{-at}	e^{-akT}	$\dfrac{1}{1-e^{-aT}z^{-1}}$
$\dfrac{1}{s^2}$	t	kT	$\dfrac{Tz^{-1}}{\left(1-z^{-1}\right)^2}$
$\dfrac{a}{s(s+a)}$	$1-e^{-at}$	$1-e^{-akT}$	$\dfrac{\left(1-e^{-aT}\right)z^{-1}}{\left(1-z^{-1}\right)\left(1-e^{-aT}z^{-1}\right)}$
$\dfrac{\omega}{s^2+\omega^2}$	$\sin \omega t$	$\sin k\omega T$	$\dfrac{z^{-1}\sin \omega T}{1-2z^{-1}\cos \omega T + z^{-2}}$
$\dfrac{s}{s^2+\omega^2}$	$\cos \omega t$	$\cos k\omega T$	$\dfrac{1-z^{-1}\cos \omega T}{1-2z^{-1}\cos \omega T + z^{-2}}$
$\dfrac{\omega}{(s+a)^2+\omega^2}$	$e^{-at}\sin \omega t$	$e^{-k\omega T}\sin k\omega T$	$\dfrac{e^{-aT}z^{-1}\sin \omega T}{1-2e^{-aT}z^{-1}\cos \omega T + e^{-2aT}z^{-2}}$
$\dfrac{s+a}{(s+a)^2+\omega^2}$	$e^{-at}\cos \omega t$	$e^{-k\omega T}\cos k\omega T$	$\dfrac{1-e^{-aT}z^{-1}\cos \omega T}{1-2e^{-aT}z^{-1}\cos \omega T + e^{-2aT}z^{-2}}$

ZOH-equivalent

$$G_2(z) = (1 - z^{-1})\,\mathscr{Z}_{\text{eq}}\left(\frac{G(s)}{s}\right)$$

$$\frac{Tz^{-1}}{1-z^{-1}}$$

$$\frac{1}{a}\,\frac{\left(1-e^{-aT}\right)z^{-1}}{1-e^{-aT}z^{-1}}$$

$$\frac{T^2}{2}\,\frac{\left(1+z^{-1}\right)z^{-1}}{\left(1-z^{-1}\right)^2}$$

$$\frac{\left(aT-1+e^{-aT}\right)z^{-1}+\left(1-e^{-aT}-aTe^{-aT}\right)z^{-2}}{a\left(1-z^{-1}\right)\left(1-e^{-aT}z^{-1}\right)}$$

$$\frac{(1-\cos\omega T)}{\omega}\,\frac{z^{-1}+z^{-2}}{1-2z^{-1}\cos\omega T+z^{-2}}$$

$$\frac{\sin\omega T}{\omega}\,\frac{z^{-1}-z^{-2}}{1-2z^{-1}\cos\omega T+z^{-2}}$$

$$\frac{\left(1-\frac{a}{\omega}e^{-aT}\sin\omega T-e^{-aT}\cos\omega T\right)z^{-1}+\left(e^{-2aT}+\frac{a}{\omega}e^{-aT}\sin\omega T-e^{-aT}\cos\omega T\right)z^{-2}}{1-2e^{-aT}z^{-1}\cos\omega T+e^{-2aT}z^{-2}}\left(\frac{\omega}{a^2+\omega^2}\right)$$

$$\frac{\left(1+\frac{\omega}{a}e^{-aT}\sin\omega T-e^{-aT}\cos\omega T\right)z^{-1}+\left(e^{-2aT}-\frac{\omega}{a}e^{-aT}\sin\omega T-e^{-aT}\cos\omega T\right)z^{-2}}{1-2e^{-aT}z^{-1}\cos\omega T+e^{-2aT}z^{-2}}\left(\frac{a}{a^2+\omega^2}\right)$$

Appendix B
Review of matrices

In the use of state–space representations and in multi-dimensional transfer functions, matrices are used regularly and with great effect. In this appendix we review a few definitions, properties, and common facts (theorems) about them. We should try to remember, however, that matrices are a notational device rather than an advanced method of analysis. Hence they can be built up and dissected at will for the purposes of aiding our understanding.

B.1 DEFINITIONS AND NOTATIONS

A **matrix** is a rectangular array of numbers, with n rows and m columns. We often denote such an $n \times m$ array by a bold capital letter or by a generic element. An element from the ith row and jth column is denoted with subscripts ij or i,j. Thus we have

$$
\mathbf{A} = [a_{ij}] = \begin{bmatrix}
a_{11} & a_{12} & a_{13} & \cdots & a_{1m} \\
a_{21} & a_{22} & a_{23} & \cdots & a_{2m} \\
a_{31} & a_{32} & a_{33} & \cdots & a_{3m} \\
\vdots & \vdots & \vdots & \ddots & \vdots \\
a_{n1} & a_{n2} & a_{n3} & \cdots & a_{nm}
\end{bmatrix}
$$

The special case of a 1-column matrix, i.e. an $n \times 1$ matrix, we shall often call an n-vector and denote it by a lower-case bold letter, with elements given only one subscript. Thus the n-vector $\mathbf{x}$ is

$$
\mathbf{x} = [x_i] = \begin{bmatrix}
x_1 \\
x_2 \\
x_3 \\
\vdots \\
x_n
\end{bmatrix}
$$

Many matrices of interest to us will be square, i.e. have $m = n$. The elements a_{11}, a_{22}, a_{33}, ..., a_{nn} of a square matrix are called the main diagonal of the matrix. One very special matrix is the **identity matrix**, which is square, usually denoted $\mathbf{I}$ or, if its dimension must be made clear, by $\mathbf{I}_n$ for an $n \times n$ identity, and in which all elements are zero except the diagonal elements, which are 1. Thus

$$\mathbf{I} = \begin{bmatrix} 1 & 0 & 0 & \cdots & 0 & 0 \\ 0 & 1 & 0 & & . & 0 \\ 0 & 0 & 1 & & . & 0 \\ \vdots & & & & & \vdots \\ 0 & . & . & & 1 & 0 \\ 0 & . & . & \cdots & 0 & 1 \end{bmatrix}$$

A matrix may always be multiplied by a scalar by multiplying it element by element times the scalar, so that

$$\alpha \mathbf{A} = [\alpha\, a_{ij}] = \{\alpha \mathbf{A}\}_{ij}$$

and the complex conjugate $\mathbf{A}^*$ of the matrix $\mathbf{A}$ is the matrix of the complex conjugates of the elements of $\mathbf{A}$.

Two matrices $\mathbf{A}$ and $\mathbf{B}$ may be multiplied to yield a matrix $\mathbf{C}$, provided $\mathbf{A}$ is $n \times m$ and $\mathbf{B}$ is $m \times p$, i.e. their sizes are compatible. The result is a matrix of dimension $n \times p$. Notationally,

$$\mathbf{C} = \mathbf{AB} = [c_{ij}] = \left[c_{ij} = \sum_{k=1}^{m} a_{ik} b_{kj} \right]$$

In general, $\mathbf{BA}$ may not even be defined (because the dimensions are not compatible) and if it is defined, usually $\mathbf{BA} \neq \mathbf{AB}$.

The **transpose** of an $n \times m$ matrix $\mathbf{A}$, denoted by $\mathbf{A}^{\mathrm{T}}$, is the $m \times n$ matrix which is a mirror image of $\mathbf{A}$ around its diagonal terms. Thus,

$$\mathbf{A}^{\mathrm{T}} = [\, a_{ij}\,]^{\mathrm{T}} = [\, a_{ji}\,]$$

and a numerical example is

$$\begin{bmatrix} 1 & 2 & 3 \\ 4 & 5 & 6 \end{bmatrix}^{\mathrm{T}} = \begin{bmatrix} 1 & 4 \\ 2 & 5 \\ 3 & 6 \end{bmatrix}$$

The **Hermitian conjugate** of the matrix $\mathbf{A}$ is the transpose of its complex conjugate, and is denoted $\mathbf{A}^{\mathrm{H}}$.

The **determinant** is a scalar function of a square matrix which combines its elements in a particular way: it adds all possible products of n elements of the matrix where one element is taken from each row and column. It can be evaluated recursively using the ideas of cofactors, which we develop here. For a matrix $\mathbf{A}$, the matrix of dimension $n-1 \times n-1$ defined by deleting row i and column j is denoted M_{ij}. Its determinant is called the ijth minor of $\mathbf{A}$. The principal minors are those for which $i = j$. The cofactor corresponding to an element a_{ij} is defined as $(-1)^{i+j} \det(M_{ij})$. Using these ideas, the determinant of $\mathbf{A}$ may be written using Laplace's expansion as

$$\det(\mathbf{A}) = \sum_{j=1}^{n} a_{ij} (-1)^{i+j} \det(\mathbf{M}_{ij})$$

for any $i = 1, 2, \ldots, n$. Thus the determinant of order n is computed in terms of determinants of order $n - 1$. This can continue until a low-enough level is reached:

- if $n = 1$, $\det(\mathbf{A}) = a_{11}$;
- if $n = 2$, $\det(\mathbf{A}) = a_{11}a_{22} - a_{12}a_{21}$;
- if $n = 3$, $\det(\mathbf{A}) = a_{11}a_{22}a_{33} + a_{12}a_{23}a_{31} + a_{13}a_{21}a_{32} - a_{12}a_{21}a_{33}$
$$- a_{13}a_{22}a_{31} - a_{11}a_{23}a_{32}$$

and so on. A few manipulation rules may be derived for determinants, of which the most important are

1. $\det(\mathbf{AB}) = \det(\mathbf{A})\det(\mathbf{B})$ provided $\mathbf{A}$ and $\mathbf{B}$ are square;
2. $\det(c\mathbf{A}) = c^n \det(\mathbf{A})$ where c is a scalar and $\mathbf{A}$ is $n \times n$;
3. $\det(\mathbf{B}) = c \det(\mathbf{A})$ if $\mathbf{B}$ is obtained by multiplying one row or column of $\mathbf{A}$ by a scalar c;

4. $\det(\mathbf{A}^T) = \det(\mathbf{A})$;
5. $\det(\mathbf{A}\sim) = -\det(\mathbf{A})$ if $\mathbf{A}\sim$ is $\mathbf{A}$ with two rows or columns interchanged; and
6. $\det(\mathbf{A}\sim) = \det(\mathbf{A})$ if $\mathbf{A}\sim$ is $\mathbf{A}$ with a multiple of one row or column added to another.

One function of a square matrix which yields a matrix is the **adjoint matrix**, defined using the co-factors as

$$\mathrm{adj}\,(\mathbf{A}) = [\,(-1)^{i+j}\,\det(\mathbf{M}_{ij})\,]^T$$

where the transpose should be noted. A second square matrix function is the **inverse matrix**, which is that matrix $\mathbf{B}$ which, if it exists for a given $\mathbf{A}$, results in

$$\mathbf{AB} = \mathbf{BA} = \mathbf{I}$$

This is usually denoted $\mathbf{A}^{-1}$. The inverse may in principle be computed from Cramer's Rule

$$\mathbf{A}^{-1} = \frac{\mathrm{adj}\,(\mathbf{A})}{\det(\mathbf{A})}$$

although for large matrices it is more efficient to use other methods. Note that $\det(\mathbf{A})$ must be non-zero for this to hold. We remark that it is possible to define special inverses for non-square matrices and those for which the determinant is zero (see below). A matrix $\mathbf{A}$ for which $\mathbf{A}\,\mathbf{A}^H = \mathbf{I}$, i.e. for which the Hermitian conjugate is also the inverse, is called **unitary**.

A square matrix for which $\mathbf{A}^T = \mathbf{A}$ is said to be **symmetric**, while one for which $\mathbf{A}^T = -\mathbf{A}$ is **skew-symmetric**. A matrix for which a_{ij} is a function of $(i - j)$ is said to be **Toeplitz**; thus in a Toeplitz matrix we will have for instance $a_{13} = a_{25} = a_{36} \ldots$. A square matrix for which $a_{ij} = 0$ for $i > j$ (or $i < j$) is said to be **upper triangular (lower triangular)**.

We now continue with some notes on more advanced concepts in matrix manipulation and discussion.

B.2   RANK

Let $a_1, a_2, ..., a_p$ be a set of p n-vectors. These are said to be **linearly independent** if there are no scalar coefficients c_i, not all zero, for which

$$c_1 a_1 + c_2 a_2 + \cdots + c_p a_p = 0$$

It is a result of linear algebra that there can be at most $p = n$ linearly independent vectors in the set (and there may be fewer). The maximum value of p for which the set $\{a_i\}$ is linearly independent is called the **dimension** of the set. It turns out that any other vector in the set can be written as a linear combination of the basic p vectors, so they are called the **basis** of the set and are said to **span the set**.

Now consider an $n \times m$ matrix A. Consider the m columns of the matrix as a set of n-vectors. Then the dimension of this set is called the **rank** of A. If the rows are considered as n different m-vectors (by transposing), it turns out that the dimension of this set is the same as the rank of A. The rank is always less than or equal to $\min(m, n)$, as a look at the definitions will disclose. An $m \times n$ matrix A for which rank$(A) = \min(m, n)$ is said to have **full rank**.

Important rules about rank as a mapping from the matrix to the integers are

1. rank$(AB) \le \min(\text{rank}(A), \text{rank}(B))$;
2. if A is $n \times m$, $n > m$, and rank$(A) = m$, then $A^T A$ has an inverse, i.e. it is non-singular;
3. if A is $n \times m$ and B is $m \times p$, then (Sylvester's Inequality)

$$\text{rank}(A) + \text{rank}(B) - m \le \text{rank}(AB) \le \min(\text{rank}(A), \text{rank}(B))$$

4. if $A = 0$, then rank$(A) = 0$;
5. if $A = xy^T$ where x is an n-vector and y is an m-vector and neither is zero, then rank$(A) = 1$; and
6. if A is $n \times m$ and rank$(A) = q$, then there exist n-vectors x_i and m-vectors y_j, $n \ge q$ and $m \ge q$, such that

$$A = \sum_{i=1}^{q} x_i y_i^T$$

B.3 MATRIX INVERSES AND DECOMPOSITIONS

Matrices may be factored, and the most basic factorization is of a square matrix into a product of two (nearly) triangular matrices, where 'nearly' allows a permutation of the rows. The representation in terms of a permutation of a lower triangular matrix multiplied by an upper triangular matrix is called an LU (or occasionally LR) decomposition. The reason for doing this is that it makes other operations easy.

Lower–upper factorization

Suppose that we have found matrices $\mathbf{L}$ and $\mathbf{U}$ such that

$$\mathbf{A} = \mathbf{L}\mathbf{U}$$

where $\mathbf{L}$ is a permuted lower diagonal matrix with ones on the permuted diagonal, and $\mathbf{U}$ is upper diagonal. For example

$$\begin{bmatrix} 1 & 2 & 3 \\ 4 & 5 & 6 \\ 7 & 8 & 0 \end{bmatrix} = \begin{bmatrix} \frac{1}{7} & 1 & 0 \\ \frac{4}{7} & \frac{1}{2} & 1 \\ 1 & 0 & 0 \end{bmatrix} \begin{bmatrix} 7 & 8 & 0 \\ 0 & \frac{6}{7} & 3 \\ 0 & 0 & \frac{9}{2} \end{bmatrix}$$

It is easy to show that if the inverses exist, then

$$(\mathbf{A}\mathbf{B})^{-1} = \mathbf{B}^{-1}\mathbf{A}^{-1}$$

For the decomposition,

$$\mathbf{A}^{-1} = \mathbf{U}^{-1}\mathbf{L}^{-1}$$

which are relatively easy to compute. Even easier is the determinant

$$\det(\mathbf{A}) = \det(\mathbf{L})\det(\mathbf{U})$$

which is easily shown to be

$$\det(\mathbf{A}) = \prod_{i=1}^{n} u_{ii} \prod_{i=1}^{n} l_{ii}$$

This above is called an LU-factorization.

Orthonormal factorization

Another useful factorization is the orthonormal factorization, in which a rectangular matrix is factored into the product of an orthonormal matrix, i.e. a square matrix $\mathbf{Q}$ for which $\mathbf{QQ}^{\mathrm{T}} = \mathbf{I}$, and an upper triangular matrix. For an $n \times m$ matrix $\mathbf{A}$,

$$\mathbf{A} = \mathbf{QR}$$

where $\mathbf{Q}$ is $n \times n$ and $\mathbf{R}$ is $n \times m$. It is useful for computing the best least squares solution to an overdetermined set of linear equations, i.e. a set for which there are more equations than unknowns.

Singular value decomposition

A very powerful decomposition is the singular value decomposition (SVD). If $\mathbf{A}$ is an $n \times m$, $n \geq m$, complex matrix, then there exist unitary matrices $\mathbf{Y}$ and $\mathbf{U}$, of dimension $n \times m$ and $m \times m$, respectively, such that

$$\mathbf{A} = \mathbf{YSU}^{\mathrm{H}} \tag{B.1}$$

where $\mathbf{S} = \mathrm{diag}\,\{\sigma_1, \sigma_2, ..., \sigma_m\}$ is the matrix of eigenvalues of $\mathbf{AA}^{\mathrm{H}}$ and it is assumed $\sigma_1 \geq \sigma_2 \geq ... \geq \sigma_m$. The unitary property means

$$\mathbf{Y}^{\mathrm{H}}\mathbf{Y} = \mathbf{I}_m$$

$$\mathbf{U}^{\mathrm{H}}\mathbf{U} = \mathbf{UU}^{\mathrm{H}} = \mathbf{I}_m$$

The matrices are all complex, and (B.1) constitutes a singular value decomposition of **A**. This amounts to a representation of **A** in terms of m rank one matrices

$$\mathbf{A} = \sum_{i=1}^{m} \sigma_i \mathbf{Y}_i \mathbf{U}_i^H$$

where $\mathbf{Y}_i$ and $\mathbf{U}_i$ are the ith columns of **Y** and **U**, respectively.

It is direct to show that

$$\mathbf{A}^H \mathbf{A} = \mathbf{U}\mathbf{S}^2 \mathbf{U}^H$$

Because $\mathbf{U}^H = \mathbf{U}^{-1}$, clearly **U** are the eigenvectors of $\mathbf{A}^H \mathbf{A}$ and $\{\sigma_i^2\}$ are its eigenvalues.

Similar properties involving **Y** hold if $n \leq m$.

When **A** is a function of frequency, then the transformation is also such a function. One name which has been suggested for the singular values under those circumstances is **principal gains**. The notations $\bar{\sigma}$ and $\underline{\sigma}$ are frequently used for the largest and smallest singular values, σ_1 and σ_m respectively.

Modal decomposition

Finally, there is the modal decomposition

$$\mathbf{A} = \mathbf{M}\mathbf{L}\mathbf{M}^{-1} \tag{B.2}$$

where **L** is the matrix with the eigenvalues of **A** on its main diagonal and (possibly) ones on the diagonal above the main, i.e. **L** is the Jordan form for **A**. **M** is the matrix of (generalized) eigenvectors of **A**. We elaborate slightly because this representation will be of considerable use to us. For a matrix **A**, a vector **e** and scalar λ such that

$$\mathbf{A}\mathbf{e} = \lambda\mathbf{e} = \mathbf{e}\lambda \tag{B.3}$$

are called an eigenvector and eigenvalue of **A**, respectively. A square matrix **A** of dimension $n \times n$ has n eigenvectors and up to n distinct eigenvalues. Supposing the eigenvalues are distinct and calling them $\lambda_1, \lambda_2, ..., \lambda_n$ and that the corresponding eigenvectors are called $e_1, e_2, ..., e_n$, we may write them all in the form

$$\mathbf{A} \, [\, \mathbf{e}_1 \vdots \mathbf{e}_2 \vdots ... \vdots \mathbf{e}_n] = [\mathbf{e}_1 \vdots \mathbf{e}_2 \vdots ... \vdots \mathbf{e}_n \,] \begin{bmatrix} \lambda_1 & 0 & 0 \\ 0 & \lambda_2 & 0 \\ . & . & . \\ 0 & 0 & \lambda_n \end{bmatrix}$$

Calling the matrix of eigenvectors $\mathbf{M}$ and the matrix of eigenvalues Λ gives

$$\mathbf{AM} = \mathbf{M}\Lambda$$

from which (B.2) follows because the eigenvectors are always linearly independent and hence $\mathbf{M}$ is invertible. The eigenvalues are computed from noting that the definition (B.3) requires that

$$(\mathbf{A} - \lambda \mathbf{I})\, \mathbf{e} = \mathbf{0} \tag{B.4}$$

which for $\mathbf{e}$ non-zero (and hence non-trivial) means $(\mathbf{A} - \lambda\mathbf{I})^{-1}$ cannot exist and therefore

$$\det(\mathbf{A} - \lambda\mathbf{I}) = 0$$

This is a scalar equation in the unknown scalar λ. In general, it yields an nth order polynomial in λ and hence gives n values, possibly distinct, for the eigenvalues. For each eigenvalue, we can then find a solution of (B.4) for a corresponding eigenvector $\mathbf{e}$.

B.4 SIMILARITY TRANSFORMATIONS

Suppose that we have an n-vector $\mathbf{x}$ which is the state vector in describing a certain linear system, so that the system is modelled in the form

$$x(k+1) = \mathbf{A}x(k) + \mathbf{B}u(k)$$

$$\mathbf{y}(k) = \mathbf{C}\mathbf{x}(k) \tag{B.5}$$

Suppose further that we decide to change the coordinate system in the representation, and to do this linearly. Then the new coordinates are a state vector $\mathbf{w}$, where

$$\mathbf{w} = \mathbf{T}\mathbf{x} \tag{B.6}$$

where presumably the matrix $\mathbf{T}$ is invertible, so that we can change back to the old coordinates if desired. To compute the dynamics for $\mathbf{w}$, we substitute (B.5) into (B.6) to find

$$\mathbf{w}(k+1) = \mathbf{T}\mathbf{x}(k+1) = \mathbf{T}\mathbf{A}\mathbf{x}(k) + \mathbf{T}\mathbf{B}u(k)$$

$$= \mathbf{T}\mathbf{A}\mathbf{T}^{-1}\mathbf{w}(k) + \mathbf{T}\mathbf{B}u(k)$$

and

$$\mathbf{y}(k) = \mathbf{C}\mathbf{T}^{-1}\mathbf{w}(k)$$

The matrix $\mathbf{T}\mathbf{A}\mathbf{T}^{-1}$ is said to be similar to $\mathbf{A}$, and the change is called a similarity transformation.

We first verify a couple of facts about the effects of the transformation.

1. The similarity transformation does not change the eigenvalues of the system. This follows from considering the characteristic equation

$$0 = \det(\mathbf{T}\mathbf{A}\mathbf{T}^{-1} - z\mathbf{I}) = \det(\mathbf{T})\det(\mathbf{A} - z\mathbf{I})\det(\mathbf{T}^{-1})$$

so that the eigenvalues of the transformed system are the same as those of the original system.

e input–output relation is unchanged. Consider the z-transform he output of the **w**-system.

$$\mathbf{Y}(z) = \mathbf{C}\mathbf{T}^{-1}(z\mathbf{I} - \mathbf{T}\mathbf{A}\mathbf{T}^{-1})^{-1}\mathbf{B}\mathbf{U}(z)$$

$$= \mathbf{C}\,(z\mathbf{I} - \mathbf{A})^{-1}\mathbf{B}\mathbf{U}(z)$$

as in the original.

It is particularly useful to do a tranformation into certain canonical forms. This is done in section B.12.

B.5   QUADRATIC FORMS, POSITIVE DEFINITE MATRICES, ETC.

An expression of the form

$$\sum_{i=1}^{n}\sum_{j=1}^{n} a_{ij}x_i y_j$$

where the a_{ij} are scalars, is called a **bilinear form** in the $2n$ variables $x_1,\ldots,x_n$ and $y_1,\ldots,y_n$, where we note that the expression is linear in the elements separately, i.e. with other elements fixed. It may be written with n-vectors $\mathbf{x}$ and $\mathbf{y}$ and square coefficient matrix $\mathbf{A}$ as

$$\mathbf{x}^{\mathrm{T}}\mathbf{A}\mathbf{y} \tag{B.7}$$

If $\mathbf{A} = \mathbf{I}$, then this is the inner product of $\mathbf{x}$ and $\mathbf{y}$. If $\mathbf{x} = \mathbf{y}$, then (B.7) is called a **quadratic form** in the n variables $x_1, x_2, \ldots, x_n$ and may be written

$$Q = \mathbf{x}^{\mathrm{T}}\mathbf{A}\mathbf{x} \tag{B.8}$$

Writing out the terms, we can find

$$Q = a_{11}x_1^2 + (a_{12} + a_{21})x_1 x_2 + (a_{13} + a_{31})x_1 x_3$$

$$+ \cdots + (a_{1n} + a_{n1})x_1 x_n + a_{22}x_2^2 + (a_{23} + a_{32})x_2 x_3$$

$$+ (a_{2n} + a_{n2})x_2 x_n + \cdots + a_{nn}x_n^2$$

If we denote $c_{jk} = (a_{jk} + a_{kj})/2$, then we see this can be written as

$$Q = c_{11} x_1^2 + (c_{12} + c_{21}) x_1 x_2 + \cdots + (c_{1n} + c_{n1}) x_1 x_n$$

$$+ c_{22} x_2^2 + (c_{23} + c_{32}) x_2 x_3 + \cdots + c_{nn} x_n^2$$

or in matrix form

$$Q = \mathbf{x}^T \mathbf{C} \mathbf{x}$$

where $c_{jk} = c_{kj}$, i.e. $\mathbf{C}$ is symmetric. Thus, any real quadratic form may be written in the form (B.8) where $\mathbf{A}$ is symmetric. With matrices, the above can be derived with

$$Q = \mathbf{x}^T \mathbf{A} \mathbf{x} = \mathbf{x}^T \frac{(\mathbf{A} + \mathbf{A}^T)}{2} \mathbf{x} = \mathbf{x}^T \mathbf{C} \mathbf{x}$$

We will usually assume that a quadratic form uses a symmetric matrix in its representation.

Any scalar function $f(x)$ is called **positive definite** if it satisfies the conditions

$$f(x) > 0 \qquad x \neq 0$$

$$f(0) = 0$$

It is called **positive semidefinite** if the first inequality is $f(x) \geq 0$. It is **negative definite** if $-f(x)$ is positive definite, and **negative semidefinite** if $-f(x)$ is positive semidefinite. If the function is not one of the above, it is said to be **indefinite**.

For the special function represented by a quadratic form,

$$f(x) = \mathbf{x}^T \mathbf{C} \mathbf{x}$$

the definiteness is determined by the properties of the matrix $\mathbf{C}$. If the quadratic form is positive definite, for example, then we often will say that $\mathbf{C}$ is a positive definite matrix. This is slightly abusive of the terminology, but is very common.

It can be shown that the quadratic form is positive definite (the most usual case of interest) only if the symmetric matrix **C** is such that:

1. **C** has full rank, **C** is invertible, **C** has a non-zero determinant, and other equivalent statements;
2. **C** has positive eigenvalues; and
3. the principal minor determinants, i.e. those whose diagonals are from the main diagonal of **C**, are all positive. In fact, only the n leading principal minors need be checked. The leading principal minors are

$$[c_{11}] \quad \begin{bmatrix} c_{11} & c_{12} \\ c_{21} & c_{22} \end{bmatrix} \quad \begin{bmatrix} c_{11} & c_{12} & c_{13} \\ c_{21} & c_{22} & c_{23} \\ c_{31} & c_{32} & c_{33} \end{bmatrix} \cdots \mathbf{C}$$

i.e. those $k \times k$ minors which have the first k elements of the main diagonal as their diagonals.

Similar statements for positive semidefinite are:

1. **C** has non-negative eigenvalues; and
2. the principal minors (not just the leading ones) are all non-negative.

For negative definite quadratic forms, the matrix should exhibit:

1. full rank, etc.;
2. negative eigenvalues; and
3. positive leading principal minors of even dimension and negative leading principal minors of odd dimension.

For negative semidefiniteness, the matrix **C** should exhibit:

1. non-positive eigenvalues;
2. non-positive odd-dimensioned principal minors and non-negative even-dimensioned principal minors.

When we choose positive definite matrices, we often simply select diagonal matrices with positive diagonal elements. This is particularly common in linear-quadratic optimal control problems.

The other place where quadratic forms appear is in estimation problems. Here, we are concerned with expected values of functions of random variables. It is easily shown that for a random n-vector $\mathbf{y}$, the matrix given by

$$\mathbf{P} = \mathscr{E}\{\mathbf{y}\mathbf{y}^T\}$$

is positive semidefinite. To show this, we consider a quadratic form, for arbitrary n-vector $\mathbf{x}$

$$\mathbf{x}^T\mathbf{P}\mathbf{x} = \mathbf{x}^T\mathscr{E}\{\mathbf{y}\mathbf{y}^T\}\mathbf{x} = \mathscr{E}\{\mathbf{x}^T\mathbf{y}\mathbf{y}^T\mathbf{x}\}$$

$$= \mathscr{E}\{(\mathbf{x}^T\mathbf{y})^2\} \geq 0$$

which follows because $\mathbf{x}^T\mathbf{y}$ is a scalar and the expected value of a scalar squared is necessarily non-negative.

B.6   PROJECTION MATRICES

Let $\mathscr{X}_1$ be a set of vectors which are a linear combination of columns of a matrix M, i.e. $\mathscr{X}_1$ is in the range space of $\mathbf{M}$. This means there exists a vector $\mathbf{q}$ such that, if $\mathbf{x}$ is in $\mathscr{X}_1$, then $\mathbf{x} = \mathbf{M}\mathbf{q}$. Let $\mathscr{X}_2$ be all other vectors of the same dimension as $\mathbf{x}$; these vectors have no such representation in terms of $\mathbf{M}$. We assume that $\mathbf{M}$ is $n \times m$ dimensional, and without loss of generality take $m \leq n$ and $\mathbf{M}$ with full rank $(= m)$.

For any vector $\mathbf{x}$, it must be a sum of two vectors, one in $\mathscr{X}_1$ and one in $\mathscr{X}_2$, i.e.

$$\mathbf{x} = \mathbf{x}_M + \mathbf{x}_{NM}$$

where for some $\mathbf{q}$, $\mathbf{x}_M = \mathbf{M}\mathbf{q}$. To find $\mathbf{q}$ and hence $\mathbf{x}_M$, we compute

$$\mathbf{x} = \mathbf{M}\mathbf{q} + \mathbf{x}_{NM}$$

$$\mathbf{M}^T\mathbf{x} = \mathbf{M}^T\mathbf{M}\mathbf{q} + \mathbf{M}^T\mathbf{x}_{NM}$$

Now x_{NM} must be orthogonal to the columns of M by contradiction with its definition. (If for some column j, $x_{NM}^T M_j \neq 0$, then choosing q with $q_j \neq 0$, $q_i = 0$, gives a portion of x_{NM} in $\mathscr{X}_1$; a contradiction.) From this it follows that $M^T x_{NM} = 0$. Thus

$$q = (M^T M)^{-1} M^T x \quad \text{and} \quad x = Px + (I - P)x$$

where $P = M(M^T M)^{-1} M^T$ is the projection matrix of x onto $\mathscr{X}_1$.
We also see that

$$x = Px + (I - P)x = x_M + x_{NM}$$

A common notation denotes the projection of x orthogonal to M by

$$P_\perp = (I - M(M^T M)^{-1} M^T)$$

and $P_\perp P = P P_\perp = 0$.
Using this, if we have $\mathscr{X}_3 = \{x : Mx = 0\}$ and $\mathscr{X}_4$ as its complement, then since any vector has a representation

$$x = x_3 + x_4 = x_3 + Mq$$

where $x_3 \in \mathscr{X}_3$ and $x_4 \in \mathscr{X}_4$, we find

$$x_3 = P_\perp x$$

B.7 MATRIX IDENTITIES

Among the useful matrix identities are the **Householder matrix inversion identities**

$$(A + BCB^T)^{-1} = A^{-1} - A^{-1}BC(C + CB^T A^{-1} BC)^{-1} CB^T A^{-1}$$

$$= A^{-1} - A^{-1}B(C^{-1} + B^T A^{-1} B)^{-1} B^T A^{-1} \qquad (B.9)$$

which are frequently used in obtaining computationally convenient forms for optimal filters and controllers. (It is assumed that all of the indicated inverses exist.)

Another useful expression is the form for the inverse of a matrix in terms of submatrices. In particular,

$$\begin{bmatrix} \mathbf{Q} & \mathbf{S} \\ \mathbf{T} & \mathbf{R} \end{bmatrix}^{-1} = \begin{bmatrix} \mathbf{A} & \mathbf{B} \\ \mathbf{C} & \mathbf{D} \end{bmatrix}$$

satisfies

$$\mathbf{A} = (\mathbf{Q} - \mathbf{S}\mathbf{R}^{-1}\mathbf{T})^{-1}$$

$$\mathbf{B} = -\mathbf{A}\mathbf{S}\mathbf{R}^{-1}$$

$$\mathbf{C} = -\mathbf{R}^{-1}\mathbf{T}\mathbf{A}$$

$$\mathbf{D} = \mathbf{R}^{-1} + \mathbf{R}^{-1}\mathbf{T}\mathbf{A}\mathbf{S}\mathbf{R}^{-1}$$

The above can all be verified by direct evaluation. For example, multiplying the second line of (B.9) by $(\mathbf{A} + \mathbf{B}\mathbf{C}\mathbf{B}^{\mathrm{T}})$ confirms

$$\mathbf{I} = \mathbf{I} + \mathbf{A}^{-1}\mathbf{B}\mathbf{C}\mathbf{B}^{\mathrm{T}} - \mathbf{A}^{-1}\mathbf{B}\,(\mathbf{C}^{-1} + \mathbf{B}^{\mathrm{T}}\mathbf{A}^{-1}\mathbf{B})^{-1}(\mathbf{B}^{\mathrm{T}} + \mathbf{B}^{\mathrm{T}}\mathbf{A}^{-1}\mathbf{B}\mathbf{C}\mathbf{B}^{\mathrm{T}}) = \mathbf{I}$$

B.8 THE CAYLEY–HAMILTON THEOREM

The version of the Cayley–Hamilton theorem which we will use frequently is one which allows us to express power series in the matrix $\mathbf{A}$ as polynomials of no higher degree than $n-1$.

Theorem: (Cayley–Hamilton)
A square matrix satisfies its characteristic equation. Thus if

$$\lambda^n + a_1\lambda^{n-1} + a_2\lambda^{n-2} + \cdots + a_{n-1}\lambda + a_n = 0$$

is the characteristic equation of an $n \times n$ matrix $\mathbf{A}$, then

$$\mathbf{A}^n + a_1\mathbf{A}^{n-1} + a_2\mathbf{A}^{n-2} + \cdots + a_{n-1}\mathbf{A} + a_n\mathbf{I} = 0$$

Corollary

A square matrix $\mathbf{A}$ has the property that any power of $\mathbf{A}$, say $\mathbf{A}^k$, can be written as a linear combination of the matrices $\mathbf{A}^j$, $j = 0, 1, \ldots, n-1$.

Corollary

The matrix $[\mathbf{B} \vdots \mathbf{A}\,\mathbf{B} \vdots \mathbf{A}^2\mathbf{B} \vdots \ldots \vdots \mathbf{A}^k\mathbf{B}]$ has no more linearly independent rows (has no greater rank) than the matrix $[\mathbf{B} \vdots \mathbf{A}\,\mathbf{B} \vdots \mathbf{A}^2\mathbf{B} \vdots \ldots \vdots \mathbf{A}^{n-1}\mathbf{B}]$.

B.9 FUNCTIONS OF MATRICES

Let a function of a scalar have a power-series expansion

$$f(x) = \sum a_i x^i$$

Then, subject to convergence, we define for a square matrix $\mathbf{A}$ that

$$f(\mathbf{A}) = \sum a_i \mathbf{A}^i$$

is the meaning of the notation $f(\mathbf{A})$.

It is straightforward to demonstrate that, if the eigenvalues of $\mathbf{A}$ are λ_i, $i = 1, 2, \ldots, n$ and are distinct, and if the modal transformation is $\mathbf{M}$, so that

$$\mathbf{A} = \mathbf{M}\Lambda\mathbf{M}^{-1}$$

then

$$f(\mathbf{A}) = \mathbf{M} \,\mathrm{diag}\,[\,f(\lambda_1), \ldots, f(\lambda_n)]\,\mathbf{M}^{-1}$$

Example

The most useful example for us is the matrix exponential. Since

$$e^\alpha = 1 + \frac{\alpha}{1!} + \frac{\alpha^2}{2!} + \cdots + \frac{\alpha^k}{k!} + \cdots$$

for scalars, we define the matrix exponential to mean

$$e^{\mathbf{A}} = 1 + \frac{\mathbf{A}}{1!} + \frac{\mathbf{A}^2}{2!} + \cdots + \frac{\mathbf{A}^k}{k!} + \cdots$$

If $\mathbf{A}$ has distinct eigenvalues $\lambda_1, \lambda_2, \ldots, \lambda_n$, then we may compute the matrix exponential as

$$e^{\mathbf{A}} = \mathbf{M}\ \mathrm{diag}(\exp(\lambda_1), \exp(\lambda_2), \ldots, \exp(\lambda_n))\ \mathbf{M}^{-1}$$

with suitable modifications using the Jordan form if the eigenvalues are repeated.

B.10 MINIMIZATION

There are several ways to justify an extremum of a matrix form. We consider the form

$$F(\mathbf{U}) = \mathbf{U}^{\mathrm{T}}\mathbf{S}\mathbf{U} + \mathbf{U}^{\mathrm{T}}\mathbf{T}\mathbf{X} + \mathbf{X}^{\mathrm{T}}\mathbf{T}^{\mathrm{T}}\mathbf{U} + \mathbf{A}$$

and argue that the extremum of this over $\mathbf{U}$ as a function of $\mathbf{X}$ is given by

$$\mathbf{U}^{\#} = -\mathbf{S}^{-1}\mathbf{T}\mathbf{X} \tag{B.10}$$

provided $\mathbf{S}^{-1}$ exists. In this, $\mathbf{X}$ is $n \times p$, $\mathbf{S}$ is $m \times m$, $\mathbf{T}$ is $m \times n$, $\mathbf{U}$ is $m \times p$, and $\mathbf{A}$ is $p \times p$. One way of justifying (B.10) is to complete the square in $F(\mathbf{U})$. Doing this, we find that

$$F(\mathbf{U}) = (\mathbf{U} + \mathbf{S}^{-1}\mathbf{T}\mathbf{X})^{\mathrm{T}}\mathbf{S}\,(\mathbf{U} + \mathbf{S}^{-1}\mathbf{T}\mathbf{X}) - \mathbf{X}^{\mathrm{T}}\mathbf{T}^{\mathrm{T}}\mathbf{S}^{-1}\mathbf{T}\mathbf{X} + \mathbf{A}$$

Since only the first of the right-hand side terms depends upon $\mathbf{U}$, it is clear that if $\mathbf{S}$ is positive definite, $F(\mathbf{U})$ is minimized in some sense by choosing $(\mathbf{U} + \mathbf{S}^{-1}\mathbf{T}\mathbf{X}) = \mathbf{0}$. That sense is of course that the first term is zero and hence its trace, eigenvalues, and other scalar measures are zero. If $p = 1$, this is an ordinary quadratic form which is minimized by the given choice.

Alternative demonstrations result from formal differentiation and from expansion with a scalar parameter which is zero at the optimum. For the first, we take (formally)

$$\frac{\partial F(\mathbf{U})}{\partial \mathbf{U}} = 2\mathbf{SU} + 2\mathbf{TX} = 0 \Rightarrow \mathbf{U}^\# = -\mathbf{S}^{-1}\mathbf{TX}$$

For the second, we assume $\mathbf{U}^\#$ is the optimum and let

$$\mathbf{U} = \mathbf{U}^\# + \alpha\,\Delta \tag{B.11}$$

where Δ is an arbitrary matrix and α is a scalar. Using this, it must be that the optimum occurs for $\alpha = 0$. Thus we have

$$F(\mathbf{U}) = (\mathbf{U}^\# + \alpha\Delta)^{\mathrm{T}}\,\mathbf{S}\,(\mathbf{U}^\# + \alpha\Delta) + (\mathbf{U}^\# + \alpha\Delta)^{\mathrm{T}}\mathbf{TX}$$

$$+\ \mathbf{X}^{\mathrm{T}}\mathbf{T}^{\mathrm{T}}(\mathbf{U}^\# + \alpha\Delta) + \mathbf{A}$$

$$= F(\mathbf{U}^\#) + \alpha\Delta^{\mathrm{T}}\mathbf{SU}^\# + \alpha\mathbf{U}^{\#\mathrm{T}}\mathbf{S}\Delta + \alpha^2\Delta^{\mathrm{T}}\mathbf{S}\Delta + \alpha\Delta^{\mathrm{T}}\mathbf{TX}$$

$$+\ \alpha\mathbf{X}^{\mathrm{T}}\mathbf{T}^{\mathrm{T}}\Delta$$

From this,

$$\frac{\partial F(\mathbf{U})}{\partial \alpha} = \Delta^{\mathrm{T}}\mathbf{SU}^\# + \mathbf{U}^{\#\mathrm{T}}\mathbf{S}\Delta + 2\alpha\Delta^{\mathrm{T}}\mathbf{S}\Delta + \Delta^{\mathrm{T}}\mathbf{TX} + \mathbf{X}^{\mathrm{T}}\mathbf{T}^{\mathrm{T}}\Delta$$

The minimizing value is known to be at $\alpha = 0$. Hence we have

$$\frac{\partial F(\mathbf{U})}{\partial \alpha} = 0 = \Delta^{\mathrm{T}}\mathbf{SU}^\# + \mathbf{U}^{\#\mathrm{T}}\mathbf{S}\Delta + \Delta^{\mathrm{T}}\mathbf{TX} + \mathbf{X}^{\mathrm{T}}\mathbf{T}^{\mathrm{T}}\Delta$$

$$= \Delta^{\mathrm{T}}\,(\mathbf{SU}^\# + \mathbf{TX}) + (\mathbf{SU}^\# + \mathbf{TX})^{\mathrm{T}}\Delta$$

For this to be true for arbitrary Δ, it must be that $(\mathbf{SU}^\# + \mathbf{TX}) = \mathbf{0}$. This gives (B.10) as before when $\mathbf{S}^{-1}$ exists.

The last argument above can also be used to minimize a trace of a matrix. Thus, if we wish to find

$$\min_{\mathbf{U}} \mathrm{tr}\left((\mathbf{U}+\mathbf{S}^{-1}\mathbf{T}\mathbf{X})^T\mathbf{S}(\mathbf{U}+\mathbf{S}^{-1}\mathbf{T}\mathbf{X})-\mathbf{X}^T\mathbf{T}^T\mathbf{S}^{-1}\mathbf{T}\mathbf{X}+\mathbf{A}\right) \qquad (\mathrm{B.12})$$

we may substitute as in (B.11) and look for an extremum with respect to the scalar α. Because of the symmetry of the matrix expression (B.12), the trace is non-negative and clearly has a minimum when its matrix function has a minimum. The latter has already been shown to happen when (B.10) holds, but could also be done directly by taking the partial derivative of (B.12) with respect to α when (B.11) is substituted, etc.

B.11 ACKERMANN'S FORMULA

Given the $n \times n$ matrix $\mathbf{A}$ with characteristic equation

$$\lambda^n + a_1\lambda^{n-1} + a_2\lambda^{n-2} + \cdots + a_{n-1}\lambda + a_n = 0$$

and the $n \times 1$ matrix $\mathbf{B}$, we wish to find a $1 \times n$ matrix $\mathbf{K}$ such that the matrix $\Phi = \mathbf{A} - \mathbf{B}\mathbf{K}$ has characteristic equation

$$\lambda^n + \alpha_1\lambda^{n-1} + \alpha_2\lambda^{n-2} + \cdots + \alpha_{n-1}\lambda + \alpha_n = 0$$

To do this, we first place $\mathbf{A},\mathbf{B}$ into a controllable canonical form using a similarity transform. In particular, let $\mathbf{T}$ be such that

$$\mathbf{A}_T = \mathbf{T}^{-1}\mathbf{A}\mathbf{T} = \begin{bmatrix} -a_1 & -a_2 & -a_3 & \cdots & -a_{n-1} & -a_n \\ 1 & 0 & 0 & \cdots & 0 & 0 \\ 0 & 1 & 0 & \cdots & 0 & 0 \\ \vdots & \vdots & \vdots & & \vdots & \vdots \\ 0 & 0 & 0 & \cdots & 0 & 0 \\ 0 & 0 & 0 & \cdots & 1 & 0 \end{bmatrix}$$

$$\mathbf{B}_T = \mathbf{T}^{-1}\mathbf{B} = \begin{bmatrix} 1 \\ 0 \\ \vdots \\ 0 \end{bmatrix} \qquad (\mathrm{B.13})$$

or alternatively the form

$$\mathbf{A'_T} = \mathbf{T^{-1}AT} = \begin{bmatrix} 0 & 1 & 0 & \cdots & 0 & 0 \\ 0 & 0 & 1 & \cdots & 0 & 0 \\ 0 & 0 & 0 & \cdots & 0 & 0 \\ \vdots & \vdots & \vdots & & \vdots & \vdots \\ 0 & 0 & 0 & \cdots & 0 & 1 \\ -a_n & -a_{n-1} & -a_{n-2} & \cdots & -a_2 & -a_1 \end{bmatrix}$$

$$\mathbf{B'_T} = \mathbf{T^{-1}B} = \begin{bmatrix} 0 \\ \vdots \\ 0 \\ 1 \end{bmatrix}$$

We define the controllability matrices for each representation, so that

$$\mathbf{Q_c} = [\ \mathbf{B}\ \vdots\ \mathbf{AB}\ \vdots\ ...\ \vdots\ \mathbf{A}^{n-1}\mathbf{B}\ ]$$

$$\mathbf{Q_{cT}} = [\ \mathbf{B_T}\ \vdots\ \mathbf{A_T B_T}\ \vdots\ ...\ \vdots\ \mathbf{A_T^{n-1} B_T}\ ]$$

$$= \mathbf{T^{-1}}\ [\ \mathbf{B}\ \vdots\ \mathbf{AB}\ \vdots\ \cdots\ \vdots\ \mathbf{A}^{n-1}\mathbf{B}\ ] = \mathbf{T^{-1} Q_c}$$

Clearly, we have

$$\mathbf{T} = \mathbf{Q_c Q_{cT}^{-1}}$$

and we remark the transformation exists if and only if $\mathbf{Q_c}$ is non-singular, which means the pair $(\mathbf{A}, \mathbf{B})$ is controllable.

We notice that for any $\mathbf{K_T}$, in the first model we have

$$\mathbf{A_T} - \mathbf{B_T K_T} = \begin{bmatrix} -a_1 - K_{T1} & -a_2 - K_{T2} & \cdots & -a_{n-1} - K_{Tn-1} & -a_n - K_{Tn} \\ 1 & 0 & \cdots & 0 & 0 \\ 0 & 1 & \cdots & 0 & 0 \\ \vdots & \vdots & & \vdots & \vdots \\ 0 & 0 & \cdots & 0 & 0 \\ 0 & 0 & \cdots & 1 & 0 \end{bmatrix}$$

so that the characteristic equation is

$$\lambda^n + (a_1 + K_{T1})\lambda^{n-1} + (a_2 + K_{T2})\lambda^{n-2} + \cdots + (a_n + K_{Tn}) = 0$$

Matching the desired closed-loop characteristic equation involves simply making

$$a_1 + K_{T1} = \alpha_1 \quad a_2 + K_{T2} = \alpha_2 \quad a_3 + K_{T3} = \alpha_3 \quad \ldots \quad a_n + K_{Tn} = \alpha_n$$

and then adjusting for the transformation $\mathbf{T}$ by taking

$$\mathbf{K} = \mathbf{K_T}\,\mathbf{T}^{-1}$$

Ackermann's formula simplifies some of the calculations, and in particular removes the need to compute the open-loop characteristic equation. To do this, it exploits the special structure of the $\mathbf{A_T}$ matrix.

To derive the formula, we first note that, because a matrix satisfies its characteristic equation, $\mathbf{A_T}$ satisfies

$$\mathbf{A_T}^n + a_1\mathbf{A_T}^{n-1} + a_2\mathbf{A_T}^{n-2} + \ldots + a_{n-1}\mathbf{A_T} + a_n\mathbf{I} = 0 \tag{B.14}$$

The same matrix substituted in the desired characteristic equation is not zero, but is a matrix

$$\mathbf{D}(\mathbf{A_T}) = \mathbf{A_T}^n + \alpha_1\mathbf{A_T}^{n-1} + \alpha_2\mathbf{A_T}^{n-2} + \ldots + \alpha_{n-1}\mathbf{A_T} + \alpha_n\mathbf{I}$$

Substituting for $\mathbf{A_T}^n$ from (B.14), we find

$$\mathbf{D}(\mathbf{A_T}) = (\alpha_1 - a_1)\mathbf{A_T}^{n-1} + (\alpha_2 - a_2)\mathbf{A_T}^{n-2} + \cdots + (\alpha_n - a_n)\mathbf{I}$$

Let $\mathbf{e}_k = [\,0 \quad 0 \ldots 0 \quad 1 \quad 0 \ldots 0\,]^T$ be the unit vector with only the kth element non-zero. We observe that

$$\mathbf{e}_n^T\mathbf{A_T} = \mathbf{e}_{n-1}^T \quad \mathbf{e}_{n-1}^T\mathbf{A_T} = \mathbf{e}_{n-2}^T \quad \cdots \quad \mathbf{e}_2^T\mathbf{A_T} = \mathbf{e}_1^T$$

and hence that

$$\mathbf{e}_n^{\mathrm{T}} \mathbf{A}_{\mathrm{T}}^{k} = \mathbf{e}_{n-k}^{\mathrm{T}}$$

From this it follows that

$$\mathbf{e}_n^{\mathrm{T}} \mathbf{D}(\mathbf{A}_{\mathrm{T}}) = (\alpha_1 - a_1)\,\mathbf{e}_1^{\mathrm{T}} + (\alpha_2 - a_2)\,\mathbf{e}_2^{\mathrm{T}} + \cdots + (\alpha_n - a_n)\,\mathbf{e}_n^{\mathrm{T}}$$

$$= \mathbf{K}_{\mathrm{T}}$$

Using all of the above, we finally have that

$$\mathbf{K} = \mathbf{K}_{\mathrm{T}}\mathbf{T}^{-1} = \mathbf{e}_n^{\mathrm{T}}\mathbf{D}(\mathbf{A}_{\mathrm{T}})\,\mathbf{T}^{-1} = \mathbf{e}_n^{\mathrm{T}}\mathbf{T}^{-1}\mathbf{D}\,(\mathbf{T}\mathbf{A}_{\mathrm{T}}\mathbf{T}^{-1})\,\mathbf{T}\mathbf{T}^{-1}$$

$$= \mathbf{e}_n^{\mathrm{T}}\mathbf{T}^{-1}\mathbf{D}(\mathbf{A}) = \mathbf{e}_n^{\mathrm{T}}(\mathbf{Q}_{c\mathrm{T}}\,\mathbf{Q}_c^{-1})\,\mathbf{D}(\mathbf{A})$$

$$\mathbf{K} = \mathbf{e}_n^{\mathrm{T}}\mathbf{Q}_c^{-1}\mathbf{D}(\mathbf{A})$$

where the last follows from the easily shown fact that $\mathbf{e}_n^{\mathrm{T}}\mathbf{Q}_{c\mathrm{T}} = \mathbf{e}_n^{\mathrm{T}}$. We notice that this can be significantly easier to compute than actually performing all of the transformations.

In the case of the second model, it is easy to show that

$$\mathbf{e}_1^{\mathrm{T}}\mathbf{A}_{\mathrm{T}}' = \mathbf{e}_2^{\mathrm{T}} \qquad \mathbf{e}_2^{\mathrm{T}}\mathbf{A}_{\mathrm{T}}' = \mathbf{e}_3^{\mathrm{T}} \qquad \cdots \qquad \mathbf{e}_{n-1}^{\mathrm{T}}\mathbf{A}_{\mathrm{T}}' = \mathbf{e}_n^{\mathrm{T}}$$

Hence

$$\mathbf{e}_1^{\mathrm{T}}\mathbf{D}(\mathbf{A}_{\mathrm{T}}') = (\alpha_1 - a_1)\,\mathbf{e}_n^{\mathrm{T}} + (\alpha_2 - a_2)\,\mathbf{e}_{n-1}^{\mathrm{T}} + \cdots + (\alpha_n - a_n)\,\mathbf{e}_1^{\mathrm{T}}$$

$$= \mathbf{K}_{\mathrm{T}}'$$

where we note that $\mathbf{K}_{\mathrm{T}}'$ has its components in reverse order from those above. Finally

$$\mathbf{K} = \mathbf{K}_{\mathrm{T}}'\mathbf{T}^{-1} = \mathbf{e}_1^{\mathrm{T}}\mathbf{D}(\mathbf{A}_{\mathrm{T}}')\,\mathbf{T}^{-1} = \mathbf{e}_n^{\mathrm{T}}\mathbf{T}^{-1}\mathbf{D}\,(\mathbf{T}\mathbf{A}_{\mathrm{T}}'\mathbf{T}^{-1})\,\mathbf{T}\mathbf{T}^{-1}$$

$$= \mathbf{e}_1^{\mathrm{T}}\mathbf{T}^{-1}\mathbf{D}(\mathbf{A}) = \mathbf{e}_1^{\mathrm{T}}(\mathbf{Q}_{c\mathrm{T}}'\,\mathbf{Q}_c^{-1})\,\mathbf{D}(\mathbf{A})$$

which, because

$$Q'_{cT} = [\; B'_T \;\vdots\; A'_T B'_T \;\vdots\; ... \;\vdots\; A_T'^{n-1} B'_T \;]$$

gives

$$e_1^T Q'_{cT} = e_n^T$$

and hence yields again the formula

$$K = e_n^T Q_c^{-1} D(A)$$

Thus the derivation is seen to be independent of the intermediate canonical representation.

B.12 SPECIAL SIMILARITY TRANSFORMATIONS INTO THE STANDARD CANONICAL FORMS

We may use canonical forms for the system representation when this is convenient, as by doing so we are not affecting the system behaviour. The important canonical forms are the controllable or phase variable form, the observable form, and the Jordan form, which often becomes a diagonal form.

In the following subsections, we examine how to obtain these canonical forms. All reduce to having a similarity transformation, of course; it is only a matter of specifying which similarity transformation.

For these transformations, it will be assumed that the system is described by

$$\dot{x} = Ax + Bu$$

$$y = Cx$$

where the characteristic equation of the **A** matrix is given by

$$\det(\lambda \mathbf{I} - \mathbf{A}) = \lambda^n + a_1 \lambda^{n-1} + a_2 \lambda^{n-2} + \dots + a_{n-1}\lambda + a_n$$

$$= 0$$

and u and y are scalars. Below, the $\mathbf{W}$ matrix is defined by

$$\mathbf{W} = \begin{bmatrix} a_{n-1} & a_{n-2} & \cdots & a_2 & a_1 & 1 \\ a_{n-2} & a_{n-3} & & a_1 & 1 & 0 \\ \vdots & \vdots & \vdots & & \vdots & \vdots \\ a_1 & 1 & 0 & & 0 & 0 \\ 1 & 0 & \cdots & 0 & 0 & 0 \end{bmatrix}$$

The controllable canonical form

To obtain the controllable, or phase variable, canonical form, we define the $n \times n$ matrix

$$\mathbf{Q_c} = [\, \mathbf{B} \;\vdots\; \mathbf{AB} \;\vdots\; \mathbf{A}^2\mathbf{B} \;\vdots\; \mathbf{A}^3\mathbf{B} \;\vdots\; \dots \;\vdots\; \mathbf{A}^{n-1}\mathbf{B} \,]$$

(We will see this in Chapter 22 as the controllability matrix.) Assume that $\mathbf{Q_c}$ is invertible. Then the transformation $\mathbf{x} = \mathbf{Tz}$, with $\mathbf{T}$ defined as $\mathbf{T} = \mathbf{Q_c}\mathbf{W}$ gives the state equations

$$\dot{\mathbf{z}} = \mathbf{A_c}\mathbf{z} + \mathbf{B_c}u$$

$$y = \mathbf{CTz}$$

where

$$\mathbf{A_c} = \mathbf{T}^{-1}\mathbf{AT} = \begin{bmatrix} 0 & 1 & 0 & & \cdots & 0 \\ 0 & 0 & 1 & 0 & \cdots & 0 \\ \vdots & \vdots & & \vdots & & \vdots \\ -a_n & -a_{n-1} & \cdots & & \cdots & -a_1 \end{bmatrix}$$

$$\mathbf{B_c} = \mathbf{T}^{-1}\mathbf{B} = \begin{bmatrix} 0 \\ 0 \\ \vdots \\ 0 \\ 1 \end{bmatrix} \tag{B.15}$$

and **CT** is not of fixed form, but is given by direct calculation. It might be remarked that $\mathbf{A_c}$ is a **companion form** of matrix **A**; companion forms are matrices with either first or last row or first or last column consisting of the negative coefficients of the characteristic equation and the remainder of the matrix zero except for an identity matrix of size $(n-1) \times (n-1)$ placed off the diagonal. (Other examples are in (B.16) and (B.18).)

The proof is possible in several ways; we choose to do a simple verification as in Ogata (1987). We first notice that the matrix **A** satisfies

$$\mathbf{A}\,\mathbf{Q_c} = [\,\mathbf{AB} \mid \mathbf{A^2B} \mid ... \mid \mathbf{A^nB}\,]$$

$$= [\,\mathbf{AB} \mid \mathbf{A^2B} \mid ... \mid \mathbf{A^{n-1}B} \mid -a_n\mathbf{B} - a_{n-1}\mathbf{AB}$$

$$- \cdots - a_1\mathbf{A^{n-1}B}\,]$$

where the last follows directly from the Cayley–Hamilton Theorem. Rewriting the last in matrix notation gives

$$\mathbf{AQ_c} = [\mathbf{B} \mid \mathbf{AB} \mid ... \mid \mathbf{A^{n-1}B}]\begin{bmatrix} 0 & 0 & 0 & ... & 0 & -a_n \\ 1 & 0 & 0 & ... & 0 & -a_{n-1} \\ 0 & 1 & 0 & ... & 0 & -a_{n-2} \\ \vdots & & & & & \vdots \\ 0 & & & \ddots & 0 & -a_2 \\ 0 & 0 & 0 & ... & 1 & -a_1 \end{bmatrix}$$

$$= \mathbf{Q_c}\mathbf{A_c^T} \tag{B.16}$$

Now we have from the definition of $\mathbf{A_c}$ that

$$\mathbf{A}_c = \mathbf{T}^{-1}\mathbf{A}\mathbf{T} = \mathbf{W}^{-1}\mathbf{Q}_c^{-1}\mathbf{A}\mathbf{Q}_c\mathbf{W}$$

but from (B.16) we can substitute for $\mathbf{A}\,\mathbf{Q}_c$ to get

$$\mathbf{A}_c = \mathbf{W}^{-1}\mathbf{A}_c^{\mathrm{T}}\mathbf{W}$$

which must be verified. It is easy but tedious to show that

$$\mathbf{W}\mathbf{A}_c = \mathbf{A}_c^{\mathrm{T}}\mathbf{W}$$

as required, so that $\mathbf{A}_c$ is correctly given by the transformation. To verify $\mathbf{B}_c$, we need to show that

$$\mathbf{B}_c = \mathbf{T}^{-1}\mathbf{B}$$

or

$$\mathbf{B} = \mathbf{T}\mathbf{B}_c$$

Computing the right-hand side of this we have

$$\mathbf{T}\mathbf{B}_c = \mathbf{Q}_c\mathbf{W}\mathbf{B}_c$$

$$= \mathbf{Q}_c \begin{bmatrix} 1 \\ 0 \\ 0 \\ \vdots \\ 0 \end{bmatrix}$$

$$= \mathbf{B}$$

as required. The demonstration is very straightforward, using only the definitions of the quantities involved.

An alternative controllable canonical form is given by (B.13), and its verification is similar.

The observable canonical form

The observable canonical form is obtained much as the controllable form was. We define the observability matrix, an $n \times n$ matrix (in this case) given by

$$\mathbf{Q}_o = \begin{bmatrix} \mathbf{C} \\ \mathbf{CA}^2 \\ \mathbf{CA} \\ \vdots \\ \mathbf{CA}^{n-1} \end{bmatrix}$$

Then our transformation is given by $\mathbf{x} = \mathbf{Tz}$ where

$$\mathbf{T} = (\mathbf{WQ}_o)^{-1}$$

With this, we have

$$\dot{\mathbf{z}} = \mathbf{A}_o\,\mathbf{z} + \mathbf{B}_o\,u$$

$$y = \mathbf{C}_o\,\mathbf{z}$$

where

$$\mathbf{A}_o = \mathbf{T}^{-1}\mathbf{A}\mathbf{T}$$

$$= \begin{bmatrix} 0 & 0 & 0 & \dots & 0 & -a_n \\ 1 & 0 & 0 & \dots & 0 & -a_{n-1} \\ 0 & 1 & 0 & \dots & 0 & -a_{n-2} \\ \vdots & \vdots & & & \vdots & \vdots \\ 0 & 0 & 0 & \dots & 1 & -a_1 \end{bmatrix} \tag{B.17}$$

$$\mathbf{C}_o = \mathbf{CT}$$

$$= [\,0\ 0\ 0\ 0\ 0\ \dots\ 0\ 1\,]$$

The proof is almost identical to that given for the controllable form. $\mathbf{B_o}$ is in this case not of fixed form; the controllable and observable canonical forms are designed to fix the structure of the $\mathbf{A}$ and one of the $\mathbf{B}$ and $\mathbf{C}$ matrices.

An alternative form has

$$\mathbf{A_o'} = \mathbf{T'^{-1}AT'}$$

$$= \begin{bmatrix} -a_1 & 1 & 0 & \dots & 0 & 0 \\ -a_2 & 0 & 1 & \dots & 0 & 0 \\ -a_3 & 0 & 0 & \dots & 0 & 0 \\ \vdots & \vdots & \vdots & \dots & & \vdots \\ -a_{n-1} & 0 & 0 & & 0 & 1 \\ -a_n & 0 & 0 & \dots & 0 & 0 \end{bmatrix} \tag{B.18}$$

$$\mathbf{C_o'} = \mathbf{CT'}$$

$$= [\, 1 \quad 0 \quad 0 \quad \dots \quad 0 \quad 0 \,] \tag{B.19}$$

This is developed in the same manner as above.

The Jordan canonical form

Most of the similarity transformation work in the text assumes that the eigenvalues of a square $n \times n$ matrix $\mathbf{A}$ are distinct, so that a modal matrix $\mathbf{M}$, whose columns are eigenvectors of $\mathbf{A}$, is suitable for the transformation

$$\mathbf{M^{-1}AM} = \Lambda = \text{diag}\,\{\lambda_1, \lambda_2, \dots, \lambda_n\}$$

From this follows that

$$\mathbf{e^{AT}} = \mathbf{Me^{\Lambda T}M^{-1}} = \mathbf{M}\,\text{diag}\,\{\exp(\lambda_1 T), \exp(\lambda_2 T), \dots, \exp(\lambda_n T)\}\mathbf{M^{-1}}$$

$$\mathbf{A^k} = \mathbf{M\Lambda^K M^{-1}} = \mathbf{M}\,\text{diag}\,\{\lambda_1^k, \lambda_2^k, \dots, \lambda_n^k\}\mathbf{M^{-1}}$$

More generally, some eigenvalues will be repeated. In this case, the similarity transformation is to a Jordan canonical form, given by a diagonal matrix consisting of Jordan blocks and exemplified by

$$J = \begin{bmatrix} \mathbf{J}_{n_1}(\lambda_1) & 0 & 0 & 0 \\ 0 & \mathbf{J}_{n_2}(\lambda_1) & 0 & 0 \\ 0 & 0 & \mathbf{J}_{n_3}(\lambda_2) & 0 \\ & \cdot & \cdot & \cdot \\ 0 & 0 & 0 & \mathbf{J}_{n_p}(\lambda_q) \end{bmatrix}$$

where each Jordan block $\mathbf{J}_k(\lambda)$ is given by the $k \times k$ matrix

$$\mathbf{J}_k(\lambda) = \begin{bmatrix} \lambda & 1 & 0 & 0 & \cdots & 0 \\ 0 & \lambda & 1 & 0 & \cdots & \vdots \\ 0 & 0 & \lambda & 1 & \cdots & 1 \\ \vdots & \vdots & \vdots & \vdots & \ddots & \vdots \\ 0 & 0 & 0 & 0 & \cdots & \lambda \end{bmatrix}$$

Thus $n_1 + n_2 + \cdots + n_p = n$, p is the number of Jordan blocks, and the number of distinct eigenvalues is $q \leq p$. The number of Jordan blocks associated with each eigenvalue is at least 1 and is up to the number n_r, the multiplicity of the eigenvalue; it is given by the number of independent eigenvectors associated with the eigenvalue. The size of a given Jordan block depends upon the number of generalized eigenvectors generated by a particular ordinary eigenvector.

Let us expand the above by elaborating the definitions. A matrix $\mathbf{A}$ which has characteristic equation

$$\det(\lambda\mathbf{I} - \mathbf{A}) = 0$$

will have n eigenvalues if the matrix is $n \times n$. Some of these eigenvalues may be repeated, so that in fact the characteristic equation is

$$(\lambda - \lambda_1)^{n_{r1}} (\lambda - \lambda_2)^{n_{r2}} \cdots (\lambda - \lambda_q)^{n_{rq}} = 0$$

where

$$n_{r1} + n_{r2} + \cdots + n_{rq} = n$$

There are q distinct eigenvalues and the eigenvalue λ_i has multiplicity n_{r_i}. A non-trivial vector $\mathbf{e}$ is a generalized eigenvector of the matrix if

$$(\mathbf{A} - \lambda_1 \mathbf{I})^{n_{r1}} (\mathbf{A} - \lambda_2 \mathbf{I})^{n_{r2}} \cdots (\mathbf{A} - \lambda_q \mathbf{I})^{n_{rq}} \mathbf{e} = 0$$

This means that for some i, and some $k \leq i$, we have

$$(\mathbf{A} - \lambda \mathbf{I})^k \mathbf{e} = 0$$

If $k = 1$, then $\mathbf{e}$ is an (ordinary) eigenvector of the matrix $\mathbf{A}$ corresponding to the eigenvalue λ_i.

Let $\mathbf{e}_1$ be an ordinary eigenvector of $\mathbf{A}$ corresponding to an eigenvalue λ_e. Then

$$(\mathbf{A} - \lambda_e \mathbf{I}) \mathbf{e}_1 = 0$$

Let $\mathbf{e}_2, \mathbf{e}_3, \ldots, \mathbf{e}_k$ be the generalized eigenvectors of $\mathbf{A}$ for this eigenvalue. Then

$$(\mathbf{A} - \lambda_e \mathbf{I})^2 \mathbf{e}_2 = 0$$
$$(\mathbf{A} - \lambda_e \mathbf{I})^3 \mathbf{e}_3 = 0$$
$$\vdots$$
$$(\mathbf{A} - \lambda_e \mathbf{I})^k \mathbf{e}_k = 0$$

Subtraction of one of these from the one below it and factoring out $(\mathbf{A} - \lambda_e \mathbf{I}) \neq 0$ means

$$(\mathbf{A} - \lambda_e \mathbf{I}) \mathbf{e}_2 = \mathbf{e}_1$$
$$(\mathbf{A} - \lambda_e \mathbf{I}) \mathbf{e}_3 = \mathbf{e}_2$$
$$\vdots$$
$$(\mathbf{A} - \lambda_e \mathbf{I}) \mathbf{e}_k = \mathbf{e}_{k-1} \tag{B.20}$$

The above gives a method for generating the successive generalized eigenvectors of $\mathbf{A}$: first find (ordinary) eigenvectors for each eigenvalue λ_e, then generate the generalized eigenvectors up to the number given by the multiplicity n_r of the eigenvalue. Continue until n eigenvectors have been found for the matrix.

It is straightforward to convince oneself that the relationships (B.20) will lead to a Jordan block of dimension $k \times k$ for the eigenvalue λ_e.

Notice that if the Jordan form has all Jordan blocks of dimension 1, then the Jordan matrix is diagonal.

For the Jordan block, it may be shown that

$$e^{\mathbf{J}T} = \mathbf{I} + \mathbf{J}T + \mathbf{J}^2 T^2/2 + \cdots$$

$$= \begin{bmatrix} e^{\lambda T} & Te^{\lambda T} & T^2 e^{\lambda T}/2 & \cdots & T^{k-1} e^{\lambda T}/k! \\ 0 & e^{\lambda T} & Te^{\lambda T} & \cdots & \vdots \\ \vdots & \vdots & & & \\ 0 & 0 & 0 & \cdots & e^{\lambda T} \end{bmatrix}$$

and that

$$\mathbf{J}^m = \begin{bmatrix} \lambda^m & m\lambda^{m-1} & \binom{m}{2}\lambda^{m-2} & \binom{m}{3}\lambda^{m-3} & \binom{m}{k-1}\lambda^{m-2} \\ 0 & \lambda^m & m\lambda^{m-1} & \binom{m}{2}\lambda^{m-2} & \vdots \\ \vdots & \vdots & \vdots & \vdots & m\lambda^{m-1} \\ 0 & 0 & 0 & \cdots & \lambda^m \end{bmatrix}$$

where by convention the combinations $\binom{m}{j} = 0$ for $j > m$. The matrix with Jordan blocks will also be of block form when raised to a power or used in the exponential.

Change of variable ordering

Sometimes the companion matrix forms in the observable and controllable canonical forms look slightly different: the a_i row (or column) is the first rather than the last. This is a result of an ordering of variables and can be easily changed. The similarity transformation is

$$
\mathbf{P} = \begin{bmatrix}
0 & 0 & 0 & 0 & \cdots & 0 & 0 & 1 \\
0 & 0 & 0 & 0 & & 0 & 1 & 0 \\
0 & 0 & 0 & 0 & \cdots & 1 & 0 & 0 \\
 & & \vdots & & & \vdots & & \vdots \\
0 & 0 & 0 & 1 & \cdots & 0 & 0 & 0 \\
0 & 0 & 1 & 0 & & 0 & 0 & 0 \\
0 & 1 & 0 & 0 & & 0 & 0 & 0 \\
1 & 0 & 0 & 0 & \cdots & 0 & 0 & 0
\end{bmatrix}
$$

for which we see that $\mathbf{P} = \mathbf{P}^{-1}$. Applying this, we see for example that

$$
\mathbf{P} \begin{bmatrix}
0 & 0 & 0 & \cdots & 0 & -a_n \\
1 & 0 & 0 & \cdots & 0 & -a_{n-1} \\
0 & 1 & 0 & \cdots & 0 & -a_{n-2} \\
\vdots & \vdots & \vdots & & & \vdots \\
0 & 0 & 0 & \cdots & 1 & -a_1
\end{bmatrix} \mathbf{P}
$$

$$
= \begin{bmatrix}
-a_1 & 1 & 0 & 0 & \cdots & 0 \\
-a_2 & 0 & 1 & 0 & \cdots & 0 \\
-a_3 & 0 & 0 & 1 & \cdots & 0 \\
\vdots & \vdots & & & \ddots & \vdots \\
-a_n & 0 & 0 & 0 & \cdots & 0
\end{bmatrix}
$$

when applied to (B.17). Similar results hold for (B.15).

B.13 METRICS FOR MATRICES

The **norm** of an element of a vector space is a scalar function, usually denoted $\lVert \mathbf{x} \rVert$ for an element $\mathbf{x}$ of the space, which has the following properties:

1. $\|\mathbf{x}\| \geq 0$ and $\|\mathbf{x}\| = 0$ if $\mathbf{x} = \mathbf{0}$
2. $\|-\mathbf{x}\| = \|\mathbf{x}\|$
3. $\|\mathbf{x} + \mathbf{y}\| \leq \|\mathbf{x}\| + \|\mathbf{y}\|$

It can be seen that this is a generalization of the notion of absolute value for scalars.

If a vector norm has been defined, an induced matrix norm is defined by

$$\|\mathbf{A}\| = \sup_{\mathbf{x} \neq 0} \frac{\|\mathbf{A}\mathbf{x}\|}{\|\mathbf{x}\|}$$

The commonly used Euclidean norm for vectors is given by

$$\|\mathbf{x}\| = \sqrt{\mathbf{x}^T \mathbf{x}}$$

and can be shown to induce the spectral norm

$$\|\mathbf{A}\|_s = \bar{\sigma}$$

where $\bar{\sigma}$ denotes the positive root of the largest eigenvalue of $\mathbf{A}^T\mathbf{A}$. Another norm is defined directly as

$$\|\mathbf{A}\| = \sum_{i,j} |a_{ij}|$$

Other norms can be defined. Two are of particular interest when $\mathbf{A}$ is a function. With minor restrictions (no poles on the imaginary axis), we define $\|\mathbf{A}(s)\|_2$ as

$$\|\mathbf{A}(s)\|_2 = \left(\frac{1}{2\pi} \int_{-\infty}^{\infty} \mathrm{tr}\,[\mathbf{A}(j\omega)\,\mathbf{A}^T(j\omega)]\,d\omega \right)^{\frac{1}{2}}$$

and $\|\mathbf{A}(s)\|_\infty$ as

$$\|\mathbf{A}(s)\|_\infty = \sup_{\omega} \bar{\sigma}(\mathbf{A}(j\omega))$$

where $\bar{\sigma}$ is the positive square root of the largest eigenvalue of $\mathbf{A}(j\omega)\mathbf{A}^H(j\omega)$, and the superscript H denotes the complex conjugate transpose. We remark that

$$\bar{\sigma}(\mathbf{A}(j\omega))) = \|\mathbf{A}(j\omega)\|_s$$

The set of all matrix functions $\mathbf{A}(s)$ which are proper, in that each element $\mathbf{a}_{ij}(s)$ has no more finite zeros than poles, and for which $\|\mathbf{A}(s)\|_\infty < \infty$ constitutes a Lebesgue space (denoted L_∞). If in addition the functions are exponentially stable, i.e. have no poles with real parts non-negative, they constitute a Hardy space, denoted H_∞.

The use of norms is common in advanced control theory and has been for many years. Currently it is part of the foundation of robust control theory using transfer functions. Accordingly, more information has been presented in section 33.3.

B.14 FURTHER READING

Matrix methods appear as needed in most texts, with many of them having reviews. One starts with linear algebra texts, but control theory specific results are often in textbooks such as Anderson and Moore (1989).

Appendix C
Description of random processes

The modelling of systems sometimes requires the explicit modelling of randomness in the system or its signals. Such models are based in the notions of probability theory, but for the purposes of this book and indeed for much of control theory only the means and variances of the signals are ordinarily needed.

C.1 EVENTS, PROBABILITIES, AND PROBABILITY DENSITY FUNCTIONS

For describing random situations, we use probability theory. This is built up from three axioms concerning **events:**

1. $\Pr(\text{any event}) \geq 0$;
2. $\Pr(\Omega) = 1$, where Ω denotes the union of all possible events; and
3. $\Pr(A \text{ and } B) = \Pr(A) + \Pr(B)$, when $A \cap B = \phi$, i.e. the intersection of A and B is the null set ϕ.

The above notions are consistent with a relative frequency interpretation of probability and also with our intuitive notions.

If an event is associated with a number, that number is called a **random variable.** For example, the event that a certain face of a die shows when the die is rolled may be associated with the number of spots on that face. For the random variable x, we may describe its event probabilities using the cumulative probability function P, as in

$$P(x \leq X) = \Pr(\text{event that the random number is } \leq X)$$

$$= F(X)$$

It follows that

$$0 \leq P(x \leq X) \leq 1$$

$$0 \leq F(X) \leq 1$$

It is often true that there exists a function $p(x)$, called the **probability density function** (p.d.f.), such that

$$F(X) = P(x \leq X) = \int_{-\infty}^{X} p(\xi)\, d\xi$$

If this holds, then it follows that

$$p(x) = \frac{dF(x)}{dx}$$

$$\int_{-\infty}^{\infty} p(\xi)\, d\xi = 1$$

$$P(X_1 < x \leq X_2) = F(X_2) - F(X_1) = \int_{x_1}^{x_2} p(x)\, dx$$

If an event structure can provide us with two (or more) random variables x and y, then we may define a **joint cumulative distribution function**

$$P(y \leq Y, x \leq X) = F(Y,X)$$

and

$$F(Y,X) = \int_{-\infty}^{Y} \int_{-\infty}^{X} p(\upsilon,\xi)\, d\xi\, d\upsilon$$

$$p(y,x) = \frac{d^2 F(y,x)}{dy\, dx}$$

If the random variables are **independent**, then

$$F(Y,X) = F(Y)F(X)$$

Whether or not they are independent, a **conditional** joint cumulative distribution function may be defined as

$$F(X \mid Y) \equiv \frac{F(Y, X)}{F(Y)} = P(x \leq X \mid y \leq Y)$$

This is the probability that $x \leq X$ when it is known that the random variable y has value $\leq Y$. Using this, a **conditional joint probability density function** for $y = Y$ may be derived, assuming the relevant quantities exist and are non-zero, as

$$p(x \mid y) = \frac{p(y, x)}{p(y)}$$

These, we note, are functions of Y and not distributions over y. It turns out that

$$\int_{-\infty}^{\infty} p(x \mid Y) \, dx = 1$$

$$p(x \mid Y) \geq 0$$

$$p_x(x) = \int_{-\infty}^{\infty} p(x, \xi) \, d\xi = \int_{-\infty}^{\infty} p(x \mid \xi) p(\zeta) \, d\zeta$$

where the last is called the **marginal density** (or marginal p.d.f.) of x, as it is in fact a probability density function.

C.2   AVERAGES AND MOMENTS: MEANS AND VARIANCES

If $g(\cdot)$ is any function, we define the **expected value**, or expectation, of the random variable $g(x)$ as

$$\mathscr{E}[g(x)] = \mathscr{E}g(x) \equiv \int_{-\infty}^{\infty} g(\xi) p(\xi) \, d\xi$$

$$= \overline{g(x)}$$

and similarly

$$\mathscr{E}g(x,y) = \int\limits_{-\infty}^{\infty} \int\limits_{-\infty}^{\infty} g(\xi, \upsilon) p(\xi, \upsilon) \, d\xi \, d\upsilon$$

Notice that although x (and y) may be random variables, $\mathscr{E}g(x)$ is a number (constant). Furthermore, since

$$\mathscr{E}[\alpha g(x) + \beta f(x)] = \alpha \mathscr{E}g(x) + \beta \mathscr{E}f(x)$$

we see that $\mathscr{E}[\,\cdot\,]$ is a linear operator.

The expectation operator is particularly useful in manipulating **moments**. The most common are given here.

- Mean

$$\mathscr{E}[x] = \int\limits_{-\infty}^{\infty} x p(x) \, dx = \bar{x}$$

- Mean square

$$\mathscr{E}[x^2] = \int\limits_{-\infty}^{\infty} x^2 p(x) \, dx$$

- nth moment

$$\mathscr{E}[x^n] = \int\limits_{-\infty}^{\infty} x^n p(x) \, dx$$

Central moments, defined relative to the mean, are particularly useful:

- nth central moment

$$\mathscr{E}[(x - \bar{x})^n] = \int\limits_{-\infty}^{\infty} (x - \bar{x})^n p(x) \, dx$$

- First central moment

$$\mathscr{E}(x - \bar{x}) = 0$$

• Variance or second central moment

$$\sigma^2 = \mathscr{E}[(x - \bar{x})^2] = \int_{-\infty}^{\infty} (x - \bar{x})^2 p(x)\, \mathrm{d}x$$

$$= \mathscr{E}x^2 - \bar{x}^2$$

$$= (\text{standard deviation})^2$$

C.3 RANDOM PROCESSES

If an event generates or is associated (in principle) with a function of time (or some other variable), then we have a random process. If we take that random process x at some point in time t, the result $x(t)$ is a random variable. To completely describe a random process, we need in principle to know the joint p.d.f.

$$p(x(t_1), x(t_2), \ldots, x(t_n))$$

for all $n = 1, 2, \ldots$, and all times t_i, $i = 1, 2, \ldots, n$. This task is reduced somewhat if the process is of special type: gaussian, Poisson, stationary, ergodic, etc., as discussed below.

Moments

The commonly used moments of a random process are the first and second moments.

• Mean value function

$$m(t) = \int_{-\infty}^{\infty} x(t)\, p(x(t))\, \mathrm{d}x(t)$$

• Correlation function

$$\mathscr{R}_{xy}(t_1, t_2) = \mathscr{E}[x(t_1)y(t_2)]$$

$$= \int_{-\infty}^{\infty} x(t_1)\, y(t_2)\, p(x(t_1), y(t_2))\, \mathrm{d}x(t_1)\, \mathrm{d}y(t_2)$$

- Autocorrelation function

$$\mathscr{R}_x(t_1,t_2) = \mathscr{R}_{xx}(t_1, t_2)$$

In the special case called a **stationary random process**, the p.d.f. is not dependent upon the value of time, but only on the time between samples. In equation form

$$p(x(t_1+\tau), x(t_2+\tau), \ ..., \ x(t_n+\tau)) = p(x(t_1), x(t_2), ..., x(t_n))$$

for all τ, n, and t_i. If stationarity cannot be shown, it may be sufficient to show or assume the process is wide sense stationary. This needs

$$m(t) = \overline{x(t)} = m = \text{constant}$$

$$\mathscr{R}_x(t_1+\tau, t_2+\tau) = \mathscr{R}_x(t_1, t_2) = \mathscr{R}_x(t_1-t_2,0) = \mathscr{R}_x(t_1-t_2)$$

where the last is a notational convenience.

The second central moment of a stationary random process is given by

$$\mathscr{C}_x(t_1, t_2) = \mathscr{E}[(x(t_1) - m)(x(t_2) - m)]$$

$$= \mathscr{R}_x(t_1, t_2) - m^2$$

$$= \mathscr{C}_x(t_1, t_2)$$

Vector cases

The above generalizes nicely when a random variable or function is in fact a vector: we simply use matrix notations to keep track of the elements, which after all are jointly distributed random variables. Thus let $x(t)$ be a vector function of time. We look at its moments and find that the mean value function is easily found:

$$\overline{x(t)} = \mathscr{E}[x(t)] = n\text{-vector with elements } \mathscr{E}x_i(t)$$

For the second moment, we write

$$\mathscr{R}_x(t_1, t_2) = \left\{ r_{ij} = \mathscr{E} x_i(t_1) x_j(t_2) \right\}$$

$$= \mathscr{E}[\mathbf{x}(t_1) \mathbf{x}^{\mathrm{T}}(t_2)]$$

which is clearly an $n \times n$ matrix; it is termed the **(auto)-correlation matrix**; the corresponding central moment

$$\mathscr{C}_x(t_1, t_2) = \left\{ c_{ij} = \mathscr{E}(x_i(t_1) - m_i(t_1)) \, (x_j(t_2) - m_j(t_2)) \right\}$$

$$= \mathscr{E}[\mathbf{x}(t_1) \mathbf{x}^{\mathrm{T}}(t_2)] - \mathbf{m}(t_1) \mathbf{m}^{\mathrm{T}}(t_2)$$

is called the **(auto)-covariance matrix**.

If the process is stationary, then the mean value function is constant and

$$\mathscr{R}_x(t_1, t_2) = \mathscr{R}_x(t_1 - t_2)$$

$$\mathscr{C}_x(t_1, t_2) = \mathscr{C}_x(t_1 - t_2)$$

are symmetric.

We notice that $\mathscr{C}_x(0)$ has terms $c_{ij} = \mathscr{E}(x_i - \bar{x}_i)(x_j - \bar{x}_j)$. Hence

$$c_{ii} = \sigma_{x_i}^2 = \operatorname{var}(x_i), \text{ while } c_{ij} = \rho_{ij} \, \sigma_{x_i} \, \sigma_{x_j}$$

where ρ_{ij} has the property $-1 \leq \rho_{ij} \leq 1$ and is called the correlation coefficient. The matrix is called the **covariance matrix**.

In the important case in which the noise is white (see below), or at least uncorrelated, we have

$$\mathscr{R}_x(\tau) = \mathbf{R} \qquad \tau = 0$$

$$= 0 \qquad \tau \neq 0$$

with the notation

$$\mathscr{R}_x(\tau) = \mathbf{R}\delta(\tau)$$

where δ is the Dirac delta function in continuous time or the Kronecker delta in discrete time (in which case the notation is usually $\delta(iT, jT)$ or δ_{ij}). We notice that it follows from the definitions that the matrices $\mathcal{R}_x$ are positive semi-definite.

In the above, no mention has been made of the p.d.f. of the random processes. In particular, a process may be white or uncorrelated without being gaussian, in spite of a common presumption to the contrary.

C.4 SPECTRA

For a stationary random process $x(t)$, we can compute the Fourier transform of the auto-correlation function to yield the power spectrum $\mathcal{S}(\omega)$

$$\mathcal{S}_x(\omega) = \int_{-\infty}^{\infty} \mathcal{R}_x(\tau)\, e^{-j\omega\tau}\, d\tau$$

from which it follows that

$$\mathcal{R}_x(\tau) = \frac{1}{2\pi} \int_{-\infty}^{\infty} \mathcal{S}_x(\omega)\, e^{-j\omega\tau}\, d\omega$$

and in particular

$$\mathcal{R}_x(0) = \frac{1}{2\pi} \int_{-\infty}^{\infty} \mathcal{S}_x(\omega)\, d\omega$$

We note from the definition that if $\mathcal{R}_x(\tau) = \mathbf{R}\,\delta(\tau)$, then $\mathcal{S}_x(\omega) = \mathbf{R} = $ constant. The process is then called a white noise process, because it contains all frequencies equally as white light is modelled, and we see that it is uncorrelated in time.

C.5 EFFECT OF LINEAR SYSTEMS

If the stationary scalar random process $x(t)$ is passed through a linear system with impulse response $h(t) \Leftrightarrow H(j\omega)$, then it follows from the definitions of the quantities that the scalar output process $y(t)$, given by

$$y(t) = \int h(\alpha)\, x(t - \alpha)\, d\alpha = x(t) \,^{**}\, h(t)$$

has

$$R_y(\tau) = h(\tau) \,^{**}\, h(\tau) \,^{**}\, R_x(\tau)$$

$$\mathscr{S}_y(\omega) = H(j\omega)\, H(-j\omega)\, \mathscr{S}_x(\omega) \tag{C.1}$$

These representations are vital in the study of random processes. Particularly noteworthy in (C.1) is that, given a spectral density $\mathscr{S}_y(\omega)$, we can in principle produce noise with this spectrum by passing white noise (with $\mathscr{S}_x(\omega) = 1$) through a linear system with transfer function $H(s)$, where the transfer function is chosen so that $\mathscr{S}_y(\omega) = |H(j\omega)|^2$.

C.6 GAUSSIAN PROCESSES

An n-vector $\mathbf{x}$ is a gaussian (or normal) random variable if its probability density is of the form

$$p(x) = \frac{1}{(2\pi)^{n/2}} \frac{1}{|\mathbf{S}|^{1/2}} \exp\left[-\tfrac{1}{2}(x - m)^{\mathrm{T}}\, \mathbf{S}^{-1/2}(x - m)\right]$$

for some n-vector m and positive semidefinite $n \times n$ matrix $\mathbf{S}$. One can show that

$$\mathscr{E}(\mathbf{x}) = \mathbf{m}$$

$$\mathscr{E}((\mathbf{x}) - \mathbf{m})(\mathbf{x} - \mathbf{m})^{\mathrm{T}}) = \mathbf{S}$$

A common notation for this is $N(\mathbf{m}, \mathbf{S})$.

A random process $\mathbf{x}(t)$ is gaussian if vectors $\mathbf{x}(t_i)$, $i = 1, 2, \ldots$ are jointly gaussian for any times t_i. This latter means that the vectors formed by, e.g.

$$\mathbf{y} = [\; \mathbf{x}^{\mathrm{T}}(t_1) \quad \mathbf{x}^{\mathrm{T}}(t_2) \quad \mathbf{x}^{\mathrm{T}}(t_3) \quad \ldots \quad \mathbf{x}^{\mathrm{T}}(t_k) \;]^{\mathrm{T}}$$

are $N(\mathbf{m}_y(t), \mathbf{S}_y(t))$, where $t = [t_1\; t_2 \ldots t_k]$, i.e. the mean and covariance matrices may depend upon the times involved. We frequently restrict the discussion to only a single time t, so that $N(\mathbf{m}_x(t), \mathbf{S}_x(t))$ are of primary interest.

C.7 TIME AVERAGES AND ERGODICITY

We frequently estimate properties, especially the first and second moments, of random processes. In a single experiment, we may attempt to find $\mathscr{E}(x(t))$ from measurements $y(t_i)$ of $x(t_i)$ and do processing such as

$$\frac{1}{N} \sum_{i=1}^{N} y(t_i)$$

A process for which all such time averages can be expected to yield good estimates of the process moments is said to be **ergodic**. Clearly an ergodic process must be stationary, but the reverse need not hold.

C.8 FURTHER READING

Random processes and probability theory have entire books devoted to them. Among the classics are those by Doob (1953) and Loève (1963). More recent and readable for beginners is Papoulis (1977, see Part 3, especially Ch. 9).

In the same type of context as we use in this book, a review of random processes is given by Anderson and Moore (1979, Appendix A).

References

Ahson, S.I. (1986) Real-time programming languages for process control, in *Microprocessor-Based Control Systems*, (ed. N.K. Sinha), D. Reidel, Holland.

Allworth, S.T. (1981) *Introduction to Real-time Software Design*, Macmillan, Hampshire.

Anderson, B.D.O. and Moore, J.B. (1979) *Optimal Filtering*, Prentice-Hall, New Jersey.

Anderson, B.D.O. and Moore, J.B (1989) *Optimal Control: Linear Quadratic Methods*, Prentice-Hall, New Jersey.

Anderson, C.W. (1989) Learning to control an inverted pendulum using neural networks. *IEEE Control Systems Magazine*, **9**(3), April, 31–9.

Aoki, M. (1968) Control of large-scale dynamic systems by aggregation. *IEEE Transactions on Automatic Control*, **13**, 246–53.

Aseltine, J.A., Mancini, A.R. and Sarture, C.W. (1958) A survey of adaptive control systems. *IRE Transactions on Automatic Control*, **6**, December, 102–8.

Åström, K.J. (1970) *Introduction to Stochastic Control Theory*, Academic Press, New York.

Åström, K.J. (1983) Theory and application of adaptive control – a survey. *Automatica*, **19**(5), 471–86.

Åström, K.J. (1987) Adaptive feedback control. *IEEE Proceedings*, **75**(2), Feb, 185–217.

Åström, K.J. and Hägglund, T. (1988) *Automatic Tuning of PID Controllers*, Instrument Society of America, North Carolina.

Åström, K.J. and Wittenmark, B. (1989) *Adaptive Control*, Addison-Wesley, Massachusetts.

Åström, K.J. and Wittenmark, B. (1990) *Computer Controlled Systems: Theory and Design*, 2nd edn, Prentice-Hall, New York.

Åström, K.J., Anton, J.J. and Arzen, K.E. (1986) Expert control. *Automatica*, **22**(3), 277–86.

Åström, K.J., Borisson, U., Ljung, L. and Wittenmark, B. (1977) Theory and applications of self-tuning regulators. *Automatica*, **13**(5), 457–76.

Auslander, D.M., Takahashi, Y. and Tomizuka, M. (1978) Direct digital process control: practice and algorithms for microprocessor application. *IEEE Proceedings*, **66**(2), 199–208.

Barney, G.C. (1988) *Intelligent Instrumentation*, 2nd edn, Prentice-Hall International, London.

Bates, B.D. (1976) Design of equipment to allow on-line measurement of respiratory impedance (both resistance and compliance), BE (Hons) thesis, University of Queensland, Elec. Eng. Dept.

Bellman, R. (1957) *Dynamic Programming*, Princeton University Press, Princeton, New Jersey.

Bellman, R. and Kalaba, R. (1965) *Dynamic Programming and Modern Control Theory*, Academic Press, New York.

Bennett, S. (1979) *A History of Control Engineering, 1800–1930*, Pergamon, Stevenage.

Bennett, S. (1988) *Real-Time Computer Control: An Introduction*, Prentice-Hall International, New Jersey.

Bernard, J.A. (1988) Use of a rule-based system for process control. *IEEE Control Systems Magazine*, **8**(3), 3–13.

Black, H.S. (1934) Stabilized feedback amplifiers. *Bell System Technical Journal*, **13**, 1–18.

Bode, H.W. (1940) Relations between attenuation and phase in feedback amplifier design. *Bell System Technical Journal*, **19**, 427–54; see also Bode, 1945.

Bode, H.W. (1945) *Network Analysis and Feedback Amplifier Design*, Van Nostrand, Princeton, New Jersey.

Bollinger, J.G. and Duffie, N.A. (1988) *Computer Control of Machines and Processes*, Addison-Wesley, Massachusetts.

Borer, J.R. (1985) *Instrumentation and Control for the Process Industries*, Elsevier, London.

Bryson, A.E. Jnr and Denham, W.F. (1962) A steepest-ascent method for solving optimum programming problems. *Transactions of ASME: Journal of Applied Mechanics*, June, 247–57.

Bryson, A.E. Jnr and Ho, Y.-C. (1969) *Applied Optimal Control: Optimization, Estimation and Control*, Blaisdell, Massachusetts.

Butler, J.L. (1970) Comparative criteria for minicomputers. *Instrum. Tech.*, **17**, 67–82.

Chen, C.-T. (1975) *Analysis and Synthesis of Linear Control Systems*, Holt, Rinehart and Winston, New York.

Chow, J.H. and Kokotovic, P.V. (1976) A decomposition of near-optimum regulators for systems with slow and fast modes. *IEEE Transactions on Automatic Control*, **21**(5), 701–5.

Clarke, D. (1986) Automatic tuning of PID regulators, in *Expert Systems and Optimization in Process Control* (eds A. Mamdani and J. Efstathiou), Gower Technical Press, England, 85–104.

Davidor, Y. (1991) *Genetic Algorithms and Robotics*, World Scientific, London.

Davis, P. (ed.) (1970) *American Heritage Dictionary of English Language*, Dell, New York.

Davison, E.J. (1970) On pole assignment in linear systems with incomplete state feedback. *IEEE Transactions on Automatic Control*, **15**(3), 348–51.

D'Azzo, J.J. and Houpis, C.H. (1968) *Feedback Control System Analysis and Synthesis*, 2nd edn, International Student Edition, McGraw-Hill International, Sydney.

D'Azzo, J.J. and Houpis, C.H. (1988) *Linear Control System Analysis and Design*, 3rd edn, McGraw-Hill, New York.

Derenzo, S.E. (1990) *Interfacing: A Laboratory Approach Using the Microcomputer for Instrumentation, Data Analysis and Control*, Prentice-Hall International, New Jersey.

DiStefano, J.J. Jnr, Stubberud, A.R. and Williams, I.J. (1976) *Feedback and Control Systems*, SI edn, McGraw-Hill Schaum's Outline, New York .

Doebelin, E.O. (1990) *Measurement Systems: Application and Design*, 4th edn, McGraw-Hill, New York.

Doob, J.L. (1953) *Stochastic Processes*, Wiley, New York.

Dorato, P. (ed.) (1987) *Robust Control*, IEEE Press, New York.

Dorato, P. and Yedavalli, R.K. (eds) (1990) *Advances in Robust Control*, IEEE Press, New York.

Dorf, R.C. (1989) *Modern Control Systems*, 5th edn, Addison-Wesley, Massachusetts.

Doyle, J.C., Bruce, A.F. and Tannenbaum, A.R. (1992) *Feedback Control Theory*, Maxwell Macmillan International, New York.

Draper, C.S. and Li, Y.T. (1951) *Principles of Optimalizing Control Systems and an Application to the Internal Combustion Engine*, ASME publication, September.

Eveleigh, V.W. (1967) *Adaptive Control and Optimization Techniques*, McGraw-Hill, New York.

Faradzhev, R.C., Ngoc, P.V. and Shapiro, A.V. (1986) Controllability theory of discrete dynamic systems. *Automation and Remote Control*, **47**(1), 1–20.

Fel'dbaum, A.A. (1965) *Optimal Control Systems*, Academic Press, New York.

Flint, D.C. (1983) *The Data Ring Main: An Introduction to Local Area Networks*, Wiley, Chichester.

Flynn, W.R. (1990) 7th annual programmable controller update. *Control Engineering*, **32**(2), 65–75.

Francis, G. (1990) Software quality management, in *Software Reliability Handbook* (ed. P. Rook), Elsevier Science, Essex.

Franklin, G.F. (1985) Direct digital control design using pole placement techniques, in *Applied Digital Control*, (ed. S.G. Tzafestas), North-Holland, Amsterdam.

Franklin, G.F. and Powell, J.D. (1980) *Digital Control of Dynamic Systems*, Addison-Wesley, Reading, Massachusetts.

Franklin, G.F., Powell, J.D. and Workman, M.L. (1990) *Digital Control of Dynamic Systems*, 2nd edn, Addison-Wesley, New York.

Furuta, K. and Kim, S. B. (1987) Pole assignment in a specified disk. *IEEE Transactions on Automatic Control*, **32**(5), May, 423–27.

Gawthrop, P.J. (1987) *Continuous-Time Self-Tuning Control*, Wiley, New York.

Gee, K.C.E. (1983) *Introduction to Local Area Networks*, Macmillan, London.

Gelb, A. (1974) *Applied Optimal Estimation*, MIT Press, Massachusetts.

Gershwin, S.B. and Jacobson, D.H. (1971) A controllability theory for nonlinear systems. *IEEE Transactions on Automatic Control*, **6**(1), 37–46.

Gibson, J.E. (1963) *Nonlinear Automatic Control*, McGraw-Hill, New York.

Gibson, J.R. (1987) *Electronic Processor Systems*, Edward Arnold, London.

Goodwin, G.C. and Sin, K.S. (1984) *Adaptive Filtering, Prediction and Control*, Prentice-Hall, New Jersey.

Goodwin, G.C., Middleton, R.H. and Poor, H.V. (1992) High speed digital signal processing and control. *IEEE Proceedings*, **80**(2), 240–59.

Graham, D. and Lathrop, R. C. (1953) The synthesis of optimum response: criteria and standard forms. *Transactions of AIEE*, **72**(II), 273–88.

Graupe, D. (1972) *Identification of Systems*, Van Nostrand, New York.

Gressang, R.V. (1977) Discussion. *IEEE Transactions on Automatic Control*, **22**(5), 805.

Grimble, M.J. and Johnson, M.A. (1988) *Optimal Control and Stochastic Estimation* (2 volumes), Wiley, Chichester.

Groover, M.P. (1987) *Automation, Production Systems and Computer Integrated Manufacturing*, Prentice-Hall International, New Jersey.

Gupta, M.M. (ed.) (1986) *Adaptive Methods for Control Systems Design,* IEEE Press, New York.

Haddad, W.M. and Bernstein, D.S. (1992) Controller design with regional pole constraints. *IEEE Transactions on Automatic Control*, **37**(1), 54–69.

Hall, E.L. and Hall B.C. (1985) *Robotics: A User Friendly Introduction*, Holt, Rinehart and Winston, New York.

Heizer, J. and Render, B. (1988) *Production and Operations Management*, Allyn and Bacon, Boston.

Holland, R.C. (1983) *Microcomputers for Process Control*, Pergamon, Oxford.

Horch, J.W. (1987) *Software Engineering Standards*, IEEE.

924 *References*

Hostetter, G.H., Savant, C.J. Jnr and Stefani, R.T. (1989) *Design of Feedback Control Systems*, 2nd edn, Holt, Rinehart and Winston, New York.

Hunter, R.P. (1987) *Automated Process Control Systems: Concepts and Hardware*, Prentice-Hall International, New Jersey.

IBM (1983–4) *IBM System 9000 Operating System Reference Manual, Part 2: Logical Input/Output and System Services*, Release 1.13, IBM.

IBM (1984) *IBM System 9000 BASIC Reference Manual*, Release 1.2, 3rd edn, IBM.

IBM Instruments Inc. (1983) *Computer System Operation System Reference Manual*, IBM Instruments.

Iserman, R. (1981) *Digital Control Systems*, Springer-Verlag, New York.

Jacob, J.M. (1988) *Industrial Control Electronics*, Prentice-Hall, New Jersey.

Jarvis, R.A. (1975) Optimization strategies in adaptive control: a selective survey, *IEEE Transactions on Systems, Man and Cybernetics*, **5**(1), 83–94.

Jazwinski, A.H. (1970) *Stochastic Processes and Filtering Theory*, Academic Press, New York.

Joseph, B. (1989) *Real-Time Personal Computing*, Prentice-Hall International, New Jersey.

Jury, E.I. (1958) *Sampled Data Control Systems*, Wiley, New York.

Jury, E.I. (1987) On the history and progress of sampled-data systems. *IEEE Control Systems Magazine*, **7**(1), 16–21.

Kalman, R.E. (1960) A new approach to linear filtering and prediction problems. *Transactions of ASME: Journal of Basic Engineeering*, Series D, **82**, 35–45.

Kalman, R.E. (1961) On the general theory of control systems, in *Proceedings, First IFAC Congress, Moscow*, Butterworths, 481–92.

Kalman, R.E. and Bucy, R.S. (1961) New results in linear filtering and prediction theory. *Transactions of ASME: Journal of Basic Engineering*, Series D, **83**, 95–108.

Kernighan, B.W. and Ritchie, D.M. (1978) *The C Programming Language*, Prentice-Hall, New Jersey.

King, P.J. and Mamdani, E.H. (1977) The application of fuzzy control systems to industrial processes. *Automatica*, **13**(3), 235–42.

Kirk, D.E. (1970) *Optimal Control Theory*, Prentice-Hall, New Jersey.

Kissell, T.E. (1986) *Understanding and Using Programmable Controllers*, Prentice-Hall International, New Jersey.

Klamka, J. (1991) *Controllability of Dynamical Systems*, Kluwer, The Netherlands.

Knowles, J.B. (1978) Comprehensive, yet computationally simple, direct digital control-system design technique, *Proceedings of IEE*, **125**(12), 1383–95.

Kokotovic, P.V., Perkins, W.R., Cruz, J.B. and D'Ans, G. (1969) ε-coupling method for near-optimum design of large-scale linear systems, *Proceedings of IEE*, May, **116**(5), 889–92.

Kowalczuk, Z. (1989) Finite register length issue in the digital implementation of discrete PID algorithms. *Automatica*, **25**(3), 393–405.

Kudva, P. and Narendra, K.S. (1974) An identification procedure for discrete multivariable systems. *IEEE Transactions on Automatic Control*, **19**(5), 549–52.

Kuo, B.C. (1980) *Digital Control Systems*, Holt, Rinehart and Winston, New York. (A second edition (1992) has been advertised by Saunders College Publishing, Philadelphia.)

Kuo, B.C. (1987) Automatic Control Systems, 5th edn, Prentice-Hall.

Lainiotis, D.G. (1976) Partitioning: a unifying framework for adaptive systems, I: estimation. *IEEE Proceedings*, **64**(8), 1126–43.

Lamb, P. R. and Westphal, L. C. (1979) Simplex-directed partitioned adaptive filters. *International Journal of Control*, **30**(4), 617–27.

Larson, R.E. (1979) *Tutorial: Distributed Control*, IEEE Computer Society, New Jersey.

Lau, C.G.Y. and Widrow, B. (1990) Special issues on neural networks, I: Theory and modeling, II: Analysis, techniques, and applications. *IEEE Proceedings*, **78**(9 and 10).

Lee, E.B. and Markus, L. (1967) *Foundations of Optimal Control Theory*, Wiley, New York.

LeFebvre, K. (1987) Communications hierarchy for the factory. *Electronics & Power*, March, 192–4.

Leigh, J.R. (1988) *Applied Digital Control*, Prentice-Hall, London.

Li, Y.F. and Lau, C.C. (1989) Development of fuzzy algorithms for servo systems. *IEEE Control Systems Magazine*, **9**(3), 65–71.

Li, Y.T. (1952) Optimalizing system for process control. *Instruments*, **25**, 72–77, 190, 324.

Lister, A.M. and Eager, R.D. (1988) *Fundamentals of Operating Systems*, 4th edn, Macmillan, London.

Ljung, L. (1981) Analysis of a general recursive prediction error identification algorithm. *Automatica*, **17**(1), 89–100.

Ljung, L. and Söderström, T. (1983) *Theory and Practice of Recursive Identification*, MIT Press, Massachusetts.

Lodzinsky, A.J. (1990) PLCs develop new hardware and software personalities. *Control Engineering*, **37**(2), 53–4.

Loève, M.M. (1963) *Probability Theory*, Van Nostrand, Princeton, New Jersey.

Luenberger, D.G. (1964) Observing the state of a linear system, *IEEE Trans. Mil. Electron.*, **8**, April, 74–80.

Luenberger, D.G. (1969) *Optimization by Vector Space Methods*, Wiley, New York.

Luenberger, D.G. (1971) An introduction to observers, *IEEE Transactions on Automatic Control*, **16**(6), 596–602.

Luenberger, D.G. (1973) *Introduction to Linear and Nonlinear Programming*, Addison-Wesley, Massachusetts.

MacFarlane, A.G.J. (ed.) (1979) *Frequency-Response Methods in Control Systems*, IEEE Press, New Jersey.

Maciejowski, J.M. (1989) *Multivariable Feedback Design*, Addison-Wesley, Wokingham, England.

Mansfield, P.H. (1973) *Electrical Transducers for Industrial Measurement*, Butterworths, London.

McClamroch, N.H. (1980) *State Models of Dynamic Systems: A Case Study Approach*, Springer-Verlag, New York.

McCloy, D. and Martin, H.R. (1980) *Control of Fluid Power: Analysis and Design*, 2nd edn, Ellis Horwood, Chichester.

McCormick, E.J. and Sanders, M.S. (1983) *Human Factors in Engineering and Design,* 5th edn, McGraw-Hill International, Sydney.

Meditch, J.S. (1969) *Stochastic Optimal Linear Estimation and Control*, McGraw-Hill, New York.

Messner, W., Horowitz, R., Kao W-W. and Boals, M. (1989) *A New Adaptive Learning Rule*, University of California RAMP Report 89–10/ESRC 89–12. Also published (1991) in *IEEE Transactions on Automatic Control*, **36**(2), 188–97.

Moore, B.C. (1981) Principal component analysis in linear systems: controllability, observability, and model reduction. *IEEE Transactions on Automatic Control*, **26**(1), 17–31.

Morari, M. and Zafiriou, E. (1989) *Robust Process Control*, Prentice-Hall, New Jersey.

Murthy, D.N.P., Page, N.W. and Rodin E.Y. (1990) *Mathematical Modeling: A Tool for Problem Solving in Engineering, Physical, Biological and Social Sciences*, Pergamon, Oxford.

Narendra, K.S. and Parthasarathy, K. (1990) Identification and control of dynamical systems using neural networks. *IEEE Transactions on Neural Networks*, **1**(1), 4–27.

Nguyen, D. and Widrow, B. (1990) Neural networks for self-learning control systems. *IEEE Control Systems Magazine*, **10**(3), 18–23.

Nyquist, H. (1932) Regeneration theory. *Bell System Technical Journal*, **11**, 126–47. Reprinted in MacFarlane (1979).

Ogata, K. (1987) *Discrete-Time Control Systems*, Prentice-Hall International, New Jersey.

O'Reilly, J. (1983) *Observers for Linear Systems*, Academic Press, London.

Papoulis, A. (1977) *Signal Analysis*, McGraw-Hill, New York.

Paul, R.P. (1981) *Robot Manipulators: Mathematics, Programming, and Control*, MIT Press, Massachusetts.

Pessen, D.W. (1989) Ladder-diagram design for programmable controllers. *Automatica*, **25**(3), May, 407–12.

Phillips, C.L. and Harbor, R.D. (1991) *Feedback Control Systems,* 2nd edn, Prentice-Hall, New Jersey.

Pontryagin, L.S., Boltyanski, V.G., Gamkrelidze, R.V. and Mischenko, E.F. (1962) *The Mathematical Theory of Mathematical Processes,* Wiley, New York.

Pressman, R.S. and Williams, J.E. (1977) *Numerical Control and Computer-Aided Manufacturing,* Wiley, New York.

Quirk, W.J., (ed.) (1985) *Verification and Validation of Real-Time Software,* Springer-Verlag, Berlin.

Rabiner, L.R. and Gold, B. (1975) *Theory and Application of Digital Signal Processing,* Prentice-Hall, New Jersey.

Ragazzini J.R. and Franklin G.F. (1958) *Sampled Data Control Systems,* McGraw-Hill, New York.

Ragazzini, J.R. and Zadeh, L.A. (1952) The analysis of sampled-data systems. *AIEE Transactions,* **71**(II), 225–34.

Refai, M.K. (1986) Microprocessor-based digital controller for DC motor speed control. *Microprocessor and Microsystems,* **10**(10), December, 543–52.

Roberts, R.A. and Mullis, C.T. (1987) *Digital Signal Processing,* Addison-Wesley, Massachusetts.

Rosenbrock, H.H. (1970) *State–space and Multivariate Theory,* Nelson, London.

Sage, A. (ed.) (1977) *Systems Engineering: Methodology and Applications,* IEEE Press, New York.

Samson, D. (ed.) (1989) *Management for Engineers,* Longman Cheshire, Melbourne, Australia.

Saucedo, R.S. and Schiring, E.E. (1968) *Introduction to Continuous and Digital Control Systems,* Macmillan.

Schildt, H. (1988) *Using Turbo C,* Borland-Osborne/McGraw-Hill, California.

Seborg, D.E., Edgar, T.F. and Mellichamp, D.A. (1989) *Process Dynamics and Control,* Wiley, New York.

Shieh, L.S., Chen, C.F. and Huang, C.J. (1971) Simple methods for identifying linear systems from frequency or time response data. *International Journal of Control,* **13**(6), 1027–39.

Shimano, B. (1979) VAL: A versatile robot programming and control system, in *Proceedings of COMPSAC 79*, IEEE, New York 878–83.

Shinners, S.M. (1992) *Modern Control System Theory and Design*, Wiley, New York.

Simpson, P.K. (1990) *Artificial Neural Systems*, Pergamon, New York.

Singh, M.G. (1980) *Dynamical Hierarchical Control*, North-Holland, Amsterdam.

Singh, M.G., Elloy, J.-P., Mezencev, R. and Munro, N. (1980) *Applied Industrial Control*, Pergamon, Oxford.

Sinha, N.K. (ed.) (1986) *Microprocessor Based Control Systems*, D. Reidel, Holland.

Sinha, P.K. (1987) *Microprocessors for Engineers: Interfacing for Real Time Applications*, Ellis Horwood (Wiley), Chichester.

Skelton, R.E. (1988) *Dynamic Systems Control*, Wiley, New York.

Sloman, M.S. (1982) Communications for Computer Control, in *Computer Control of Industrial Processes* (eds S. Bennett and D. A. Linkens), Pregrinus, Stevenage, 112–30.

Snyder, W.E. (1985) *Industrial Robots*, Prentice-Hall International, New York.

Sommerville, I. (1989) *Software Engineering*, 3rd edn, Addison-Wesley, Wokingham.

Sorenson, H.W. (ed.) (1985) *Kalman Filtering: Theory and Application*, IEEE Press, New York.

St. Clair, D.W. (1991) Improving control loop performance – without the math. *Control Engineering*, October, 141–3.

Stankovic, J.A. (1988) Misconceptions about real-time computing. *Computer*, **21**(10), 10–19.

Stephanopoulos, G. (1984) *Chemical Process Control*, Prentice-Hall International, New Jersey.

Tanimoto, S.L. (1987) *The Elements of Artificial Intelligence*, Computer Science Press, Maryland.

Tannenbaum, A.S. (1988) *Computer Networks*, 2nd edn, Prentice-Hall, New Jersey.

Thayer, R.H. (1990) *Tutorial: Software Engineering Project Management*, IEEE Computer Society Press, Los Alamitos, California.

Truxal, J. (1961) The concept of adaptive control, in *Adaptive Control Systems* (eds E. Mishkin and L. Braun, Jnr), McGraw-Hill, New York, 1–19.

Tse, E. and Bar-Shalom, Y. (1975) Adaptive dual control for stochastic nonlinear systems with free end–time. *IEEE Transactions on Automatic Control*, **20**(5), 670–5.

Tsypkin, Y.Z. (1971) *Adaptation and Learning in Adaptive Systems*, Academic Press, New York.

Tzafestas, S.G. (1985) Digital PID and self-tuning control, in *Applied Digital Control* (ed. S.G. Tzafestas), North–Holland, Amsterdam, 1–49.

Unimation Inc. (1979) *User's Guide to VAL*, Unimation Inc., Connecticut.

Webb, J. and Greshock, K. (1990) *Industrial Control Electronics*, Merrill, Colombus.

Westphal, L.C. (1990) Adaptation of nonparametric control laws, *Preprints of Control 90*, The Fourth Conference on Control Engineering, Institution of Engineers, Australia, 1–3 August, 210–14.

Whittaker, H.P., Yamron, J. and Kezer, A. (1958) Design of model reference adaptive control systems for aircraft, *MIT Instrumentation Laboratory Report R–164*, Massachusetts Institute of Technology, Cambridge, Massachusetts.

Whittaker, H.P. (1959a) The MIT adaptive autopilot, in *Proceedings of the Self Adaptive Flight Control Systems Symposium* (ed. P. C. Gregory), *WADC Tech. Rept. 59–44*, Wright Air Development Center, Wright–Patterson Air Force Base, Ohio.

Whittaker, H.P. (1959b) An adaptive system for control of the dynamics performance of aircraft and spacecraft. *Inst. Aeronautical Sci.*, Paper 59–100.

Wiberg, D.M. (1971) *State Space and Linear Systems*, McGraw-Hill Schaum's, New York.

Wiener, N. (1949) *Extrapolation, Interpolation and Smoothing of Stationary Time Series*, MIT Press, Cambridge, Massachusetts.

Willems, J.C. and Mitter, S.K. (1971) Controllability, observability, pole allocation and state reconstruction. *IEEE Transactions on Automatic Control*, **16**(12), 582–95.

Williamson, D. (1991) *Digital Control and Implementation*, Prentice-Hall, Sydney, Australia.

Williamson, D. and Kadiman, K. (1989) Optimal finite wordlength linear quadratic regulation, *IEEE Transactions on Automatic Control*, **34**(12), 1218–28.

Wonham, M. (1967) *IEEE Transactions on Automatic Control*, **12**(6), 660–65.

Woo, J.L. and Rootenberg, J. (1975) Analysis and simulation of an adaptive system for forced ventilation of the lungs. *IEEE Transactions on Biomedical Engineering*, **22**(5), 400–11.

Yeung, K.S. and Chaid, K.Q. (1986) Bode design formulas for discrete compensators. *Electronics Letters*, **20**(24), 1304–5.

Yoshikawa, T. and Kobayashi, H. (1972) Comments on optimal minimal-order observer-estimators for discrete linear time-varying systems, *IEEE Transactions on Automatic Control*, **17**(2), 272–3.

Youla, D.C., Jabr, H.A. and Bongiorno, J.H. Jnr (1976) Modern Weiner–Hopf design of optimal controllers – Part II: The multivariable case. *IEEE Transactions on Automatic Control*, **21**(3), 319–38.

Zadeh, L. (1965) Fuzzy sets. *Information and Control*, **8**, 338–53.

Zadeh, L. (1973) Outline of a new approach to the analysis of complex systems and decision processes. *IEEE Transactions on Systems, Man and Cybernetics*, **3**(1), 26–44.

Ziegler, J.G. and Nichols, N.B. (1942) Optimum settings for automatic controllers. *Transactions of ASME*, **64**, 759–68.

Index